Reactive Intermediates

Volume 2

Reactive Intermediates

Volume 2

Edited by

R.A. Abramovitch

Clemson University
Clemson, South Carolina

PLENUM PRESS · NEW YORK AND LONDON

Library of Congress Cataloging in Publication Data

Main entry under title:

Reactive intermediates, 2

Includes bibliographies and index.
1. Chemical reaction, Conditions, and laws of—Addresses, essays, lectures. I.
Abramovitch, R. A., 1930–
QD501.R345   541'39   79-344
ISBN-13: 978-1-4613-3194-0   e-ISBN-13: 978-1-4613-3192-6
DOI: 10.1007/ 978-1-4613-3192-6

© 1982 Plenum Press, New York
Softcover reprint of the hardcover 1st edition 1982
A Division of Plenum Publishing Corporation
233 Spring Street, New York, N.Y. 10013

Contributors

André Baretta, Université d'Aix-Marseille, Centre de Saint-Jérôme, Laboratoire de Stéréochimie, rue H. Poincaré, 13397 Marseille Cédex 4, France.

Per H. J. Carlsen, Department of Chemistry, Emory University, Atlanta, Georgia 30322.

Albert Padwa, Department of Chemistry, Emory University, Atlanta, Georgia 30322.

Manfred G. Reinecke, Department of Chemistry, Texas Christian University, Fort Worth, Texas 76129.

E. F. V. Scriven, Department of Chemistry and Applied Chemistry, University of Salford, Salford M5 4WT, United Kingdom. Present address: Reilly Tar and Chemical Corporation, 1510 Market Square Center, 151 N. Delaware Street, Indianapolis, Indiana 46204.

J-M. Surzur, U.D.E.S.A.M., Faculté des Sciences et Technique (St. Jérôme), 13 397 Marseille Cédex 13, France.

Yi-Noo Tang, Department of Chemistry, Texas A&M University, College Station, Texas 77840.

Bernard Waegell, Université d'Aix-Marseille, Centre de Saint-Jérôme, Laboratoire de Stéréochimie, rue H. Poincaré, 13397 Marseille Cédex 4, France.

Preface

The field of reactive intermediates has been blossoming at a rapid rate in recent years and its impact on chemistry, both "pure" and "applied," as well as on biology, astronomy, and other areas of science, is enormous. Several books have been published which cover the area; one, edited by McManus,* surveys the subject in general at the senior undergraduate or beginning graduate level. In addition, a number of monographs have appeared which deal with individual topics such as carbenes, nitrenes, free radicals, carbanions, carbenium ions, and so on, in great depth.

Our objective is somewhat different. We hope that these *Advances in ...* type of volumes will appear at irregular intervals of a year to 18 months each. We intend to publish up-to-date reviews in relatively new areas of the chemistry of reactive intermediates. These will be written by world authorities in the field, each one of whom will give the reader a current in-depth review of all aspects of the chemistry of each of these species. It is our plan that the subjects to be reviewed will cover not only organic chemistry but also inorganic, physical, bio-, industrial, and atmospheric chemistry. The volumes themselves, we hope, will end up being reasonably interdisciplinary, though this need not and probably will not be the case for the individual reviews. It is our intention to give readers ideas about the importance of reactive intermediates in fields other than their own, as well as to bring them up to date in their own individual areas.

I welcome suggestions of topics (and authors) that should be discussed in future volumes.

Clemson, S.C.   R. A. Abramovitch

---

* Samuel P. McManus, Ed., *Organic Reactive Intermediates*, Vol. 26 in *Organic Chemistry*, Academic Press, New York (1973).

Contents

2. Nitrile Ylides and Nitrenes From 2*H*-Azirines
Albert Padwa and Per H. J. Carlsen

3. Radical Cyclizations by Intramolecular Additions
J-M. Surzur

4. Reactions of Silicon Atoms and Silylenes
Yi-Noo Tang

5. Five-Membered Hetarynes

Manfred G. Reinecke

6. A Survey of Favorskii Rearrangement Mechanisms. Influence of
    the Nature and Strain of the Skeleton
    André Baretta and Bernard Waegell

Current Aspects of the Solution Chemistry of Arylnitrenes

E. F. V. Scriven

I. INTRODUCTION

1. Definition and Historical

Nitrenes (RN) are electroneutral electron-deficient monovalent nitrogen species. The parent of this series of intermediates is nitrene (imidogen according to Chemical Abstracts nomenclature) (NH) itself, derivatives of which are formulated by replacing hydrogen by various functionalities. Thus replacement of hydrogen by aryl- gives arylnitrenes. The nitrogen atom in nitrenes contains six electrons in its valence shell, and these species may exist either in the singlet state (**1**) (an electrophile) or in the triplet state (**2**) (a diradical). This review seeks to highlight recent important developments and current trends in the solution chemistry of aryl- and hetarylnitrenes from both a mechanistic and synthetic viewpoint. Some closely related transformations that do not necessarily involve nitrenes will be discussed for comparative purposes.

Ar N̈:   Ar N̈·
1   **2**

E. F. V. Scriven ● Department of Chemistry and Applied Chemistry, University of Salford, Salford M5 4WT, United Kingdom. Present address: Reilly Tar & Chemical Corporation, 1510 Market Square Center, 151 N. Delaware Street, Indianapolis, Indiana 46204.

Nitrenes were first proposed as reactive intermediates in the Lossen rearrangement by Tiemann[1] in 1891, and subsequently by Stieglitz[2] (1896) to account for the mechanisms of the related Hofmann, Curtius, and Beckmann rearrangements. Apart from an extensive amount of work by Curtius[3] on the thermal decomposition of aryl and sulphonyl azides, interest in nitrenes declined until the reemergence of carbene chemistry in the 1950s. Reviews on nitrene chemistry by Kirmse (1959),[4] Horner (1963),[5] and Abramovitch (1964)[6] reflected this renewed interest in nitrenes and provided a stimulus for the immense research effort undertaken in the 1960s, which has been surveyed in many reviews[7-16] and two books.[17,18] Of particular note are two chapters by Abramovitch[11,14] describing all facets of nitrenes, and an excellent chapter devoted specifically to aryl- and hetarylnitrenes by P. A. S. Smith.[19] The present chapter is intended to update these reviews and it is hoped that it will stimulate further work, particularly on the synthetic aspects of intermolecular arylnitrene reactions.

2. Nomenclature

Nitrenes were originally called "residues" by early workers and no accepted name existed when the renewal of effort took place in this field in the 1950s. Consequently, many names appeared in the literature, viz. imene,[20] phenylstickstoff,[21] imin,[4] imine radical,[22] azene,[23] azacarbene,[24] azylen,[25] imidogen,[6] and nitrene[26]; finally the name *nitrene* won the day and is now in almost universal use. *Nitrene* seems a good name as it is closely analogous to *carbene*, a species with which it is isoelectronic, but criticisms can be leveled against it.[18] Protonated and alkylated nitrenes are referred to as *nitrenium ions*.

3. Methods of Generation and Evidence for Arylnitrenes

The ready availability of aryl azides and aromatic nitro (and nitroso) compounds has led to the decomposition of the former and deoxygenation of the latter becoming by far the most used methods of generation of arylnitrenes.

A. Thermolysis of Aryl Azides

When aryl azides are heated above $\sim 130°C$ in "inert" aromatic or aliphatic solvents nitrogen is evolved and an arylnitrene generated that reacts with the solvent, "dimerizes" or, if suitably substituted, cyclizes. Thermal decomposition of aryl azides generates arylnitrenes in the singlet state which usually collapse rapidly to the triplet ground state before reaction unless certain structural features are present in the molecule.

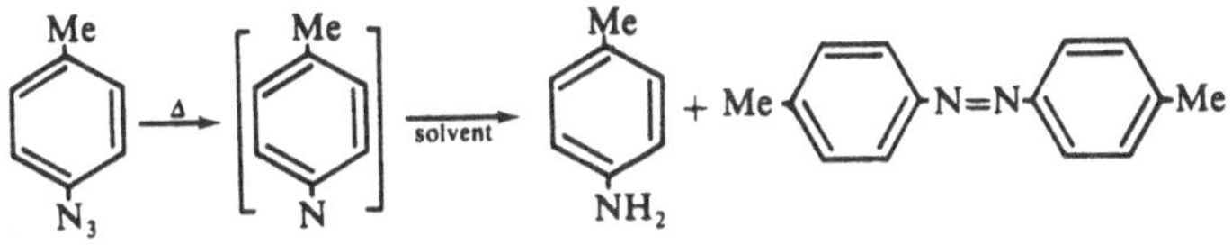

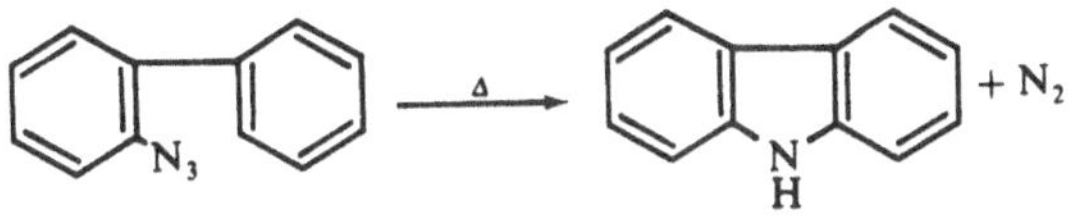

Evidence for nitrene intermediacy comes from kinetic analyses that show reactions of this type to be strictly first order in aryl azide.[27,28] Thermolysis of aryl azides in more reactive solvents, which contain an olefinic double bond, does not involve the intermediacy of a nitrene.[27] A kinetic study[27] of the thermolysis of *p*-anisyl azide in indene indicated a concerted loss of nitrogen and no evidence could be found for the formation of a triazoline. This is surprising as azides are well known to add to olefins to give triazolines in other systems.[29] Loss of nitrogen without nitrene involvement is also observed[30] in the decomposition of aryl azides bearing certain *o*-substituents (e.g., nitro, phenylazo, and carbonyl or thiocarbonyl[31]); for instance, thermolysis of *o*-nitrophenyl azide gives benzofuroxan:

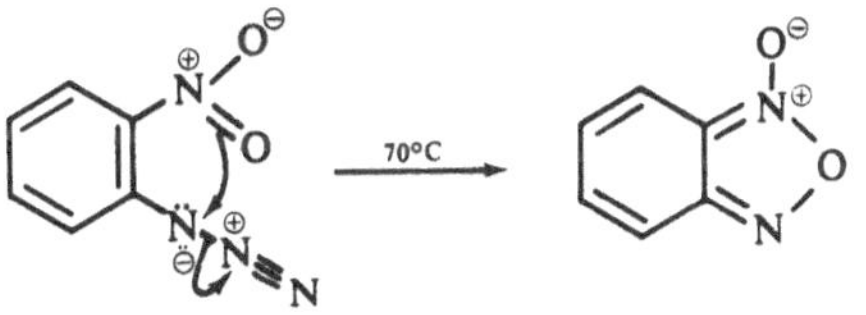

Activation parameters point to an intramolecular 1,3-dipolar addition mechanism for the conversion of *o*-azidobenzophenone to 3-phenylanthranil:

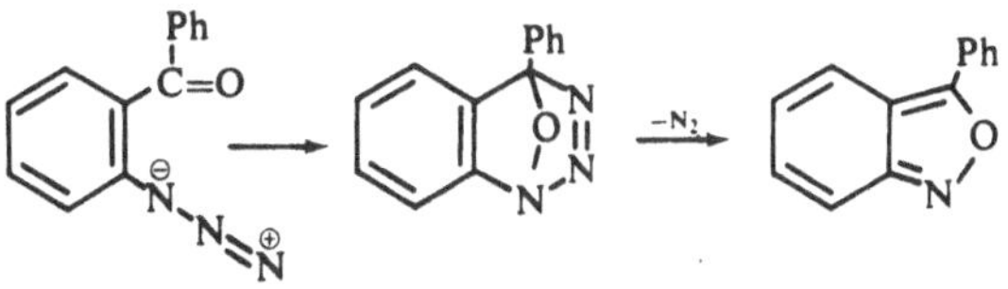

This conclusion has been questioned, however, and a mechanism involving neighboring group participation proposed[30c]:

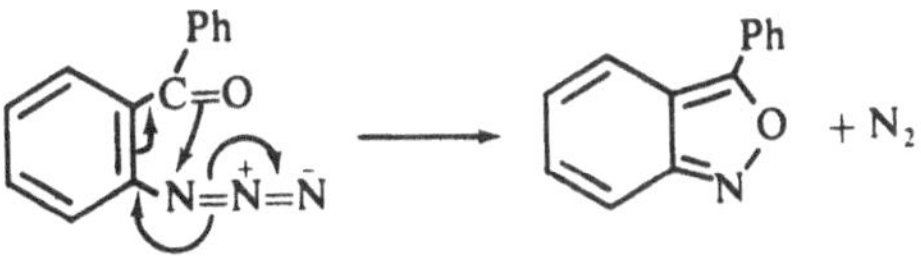

B. *Photolysis of Aryl Azides*

Many thermal reactions of aryl azides also take place photochemically and often photolysis is preferable to thermolysis as milder conditions may be employed, and the use of sensitizers allows the generation of the nitrene specifically in the singlet or triplet state:

$$
ArN_3 \xrightarrow{h\nu} {}^1ArN_3 \longrightarrow \begin{array}{l} {}^1ArN + {}^1N_2 \\ {}^3ArN + {}^3N_2 \end{array}
$$

$$
\downarrow \text{ISC}
$$

$$
{}^3ArN_3 \longrightarrow \begin{array}{l} {}^3ArN + {}^1N_2 \\ {}^1ArN + {}^3N_2 \end{array}
$$

Arylnitrene triplet states have been observed by uv[33] and esr[34] when aryl azides are irradiated at low temperature in solid matrices. Flash photolysis of 1-azidoanthracene in ethanol at room temperature gives a short-lived intermediate (with a half-life of 3–10 μsec) that has the same uv spectrum as the nitrene generated by the matrix photolysis of the same azide at 77 K.[35] When these solid matrices are allowed to warm up to room temperature the nitrene spectrum collapses and "typical nitrene products" (*vide supra*) are obtained. The nature of the products formed by analysis of reaction mixtures is the main way used to adduce nitrene intermediacy on photolysis.

C. *Reduction of Nitro and Nitrosoarenes*

Most commonly, deoxygenation is effected with tervalent phosphorus reagents such as $(EtO)_3P$, Ph_3P, $(EtO)_2PMe$, and others. Whether **3** or a discrete nitrene is the reactive species still remains in doubt. Evidence in favor[36] of a nitrene comes from the very similar yields of products obtained from the deoxygenation of nitroaromatics and thermolysis of corresponding aryl azides. On the other hand, a kinetic analysis[37] and a labeling study[38] in other systems come

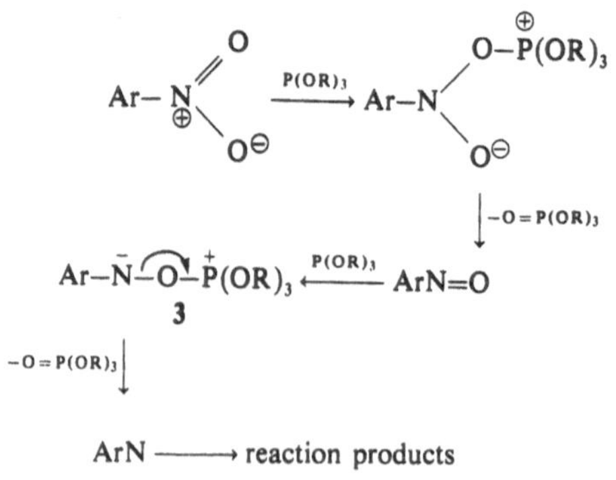

$$
ArN \longrightarrow \text{reaction products}
$$

out against the involvement of a discrete nitrene. The formation of **4** in a triethylphosphite deoxygenation also is hard to explain by a nitrene mechanism,[39] but ethylation following a nitrene-induced cyclization was not precluded.

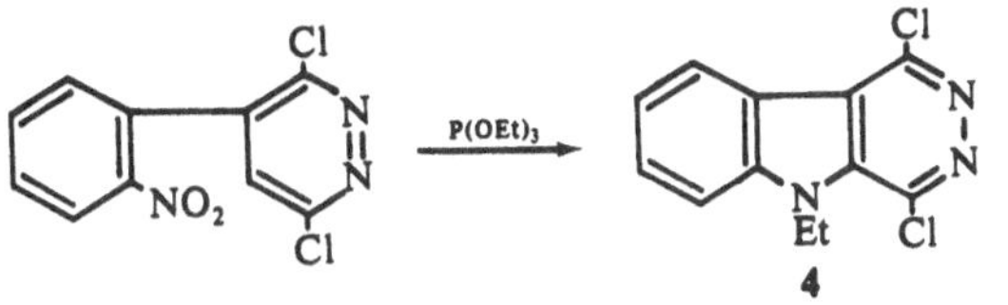

Thus, it is apparent that generalizations about nitrene involvement in deoxygenations, or any other reaction discussed in this section, cannot be made from studies of any one system, and ultimately each case must be treated on its merit. Where in this review nitrene intermediacy is postulated for reactions by analogy to other processes this point should be borne in mind.

Recently,[40] tributylphosphine has been found to be a most efficient deoxygenating agent permitting the conversion of nitrobenzene in alcoholic solution to 2-alkoxy-3*H*-azepines in 60% yield:

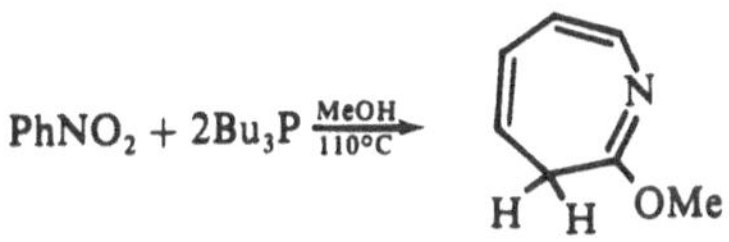

Deoxygenations have been carried out with other than phosphorus reagents. Phenazine is formed on cyclization of 2-nitrodiphenylamine in the presence of ferrous oxalate in the Waterman–Vivian synthesis[41] but a metal–nitrene complex (a nitrenoid) is thought to be involved rather than a true nitrene[42]:

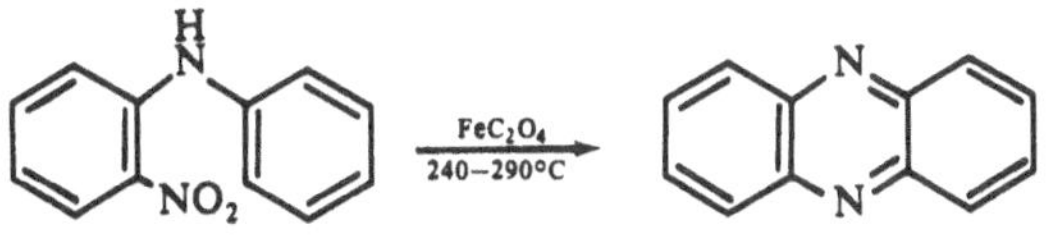

Treatment of 2-nitrobiphenyl with hexamethyldisilane at 240°C gives carbazole, providing a new method of generating nitrenes from nitro compounds.[43]

D. Fragmentation of Heterocycles

Heterocyclic fragmentations which formally lead to arylnitrenes are far less important synthetically than the methods described above. Photolysis of *N*-aryloxaziridines in a quartz apparatus is thought to give a nitrene[44]:

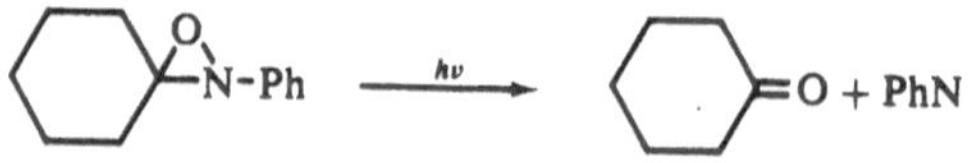

Nitrenes have also been postulated as intermediates in the ring opening of 3-substituted anthranils, before they initiate a series of interesting heterocyclic transformations. Thermolysis of 3-phenylanthranils (prepared from *o*-azido-benzophenone[45]) gives acridones,[46] whereas photolysis in methanol affords methoxyazepines via a benzazirine (5) intermediate.[47] When the 3-phenyl group is replaced by a methylimidazolo group some even more complex rearrangements take place on thermolysis of the anthranil (6).[48]

Photolysis of 2-alkylindazoles (aza analogs of anthranils) in aqueous media gives either azepinones (7) or 2-amino-5-hydroxyacetophenones (8) depending upon the pH — a nitrene $\rightleftharpoons$ nitrenium ion equilibrium is thought to be involved.[49]

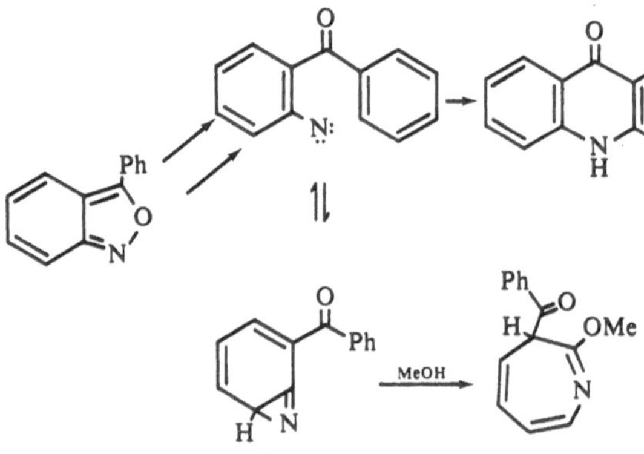

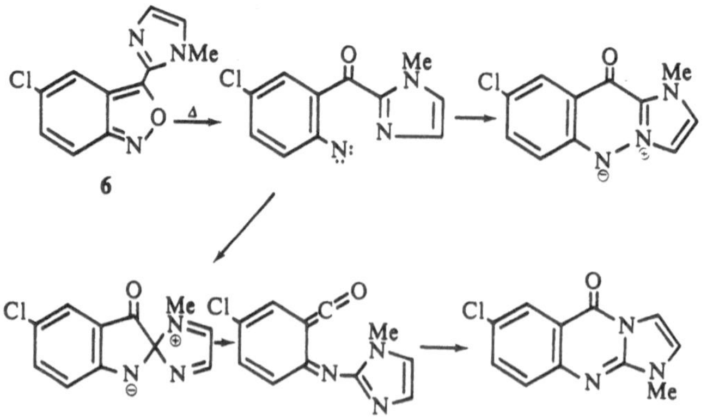

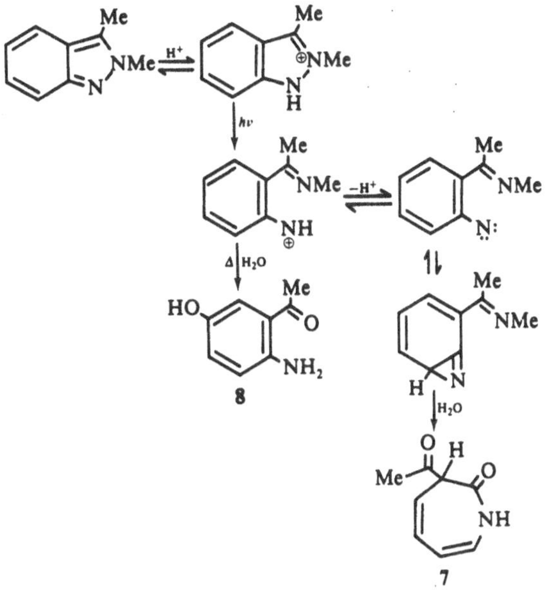

E. Oxidation of Amines

Formation of phenazine by oxidation of *o*-aminodiphenylamine with litharge is a very old reaction[50]:

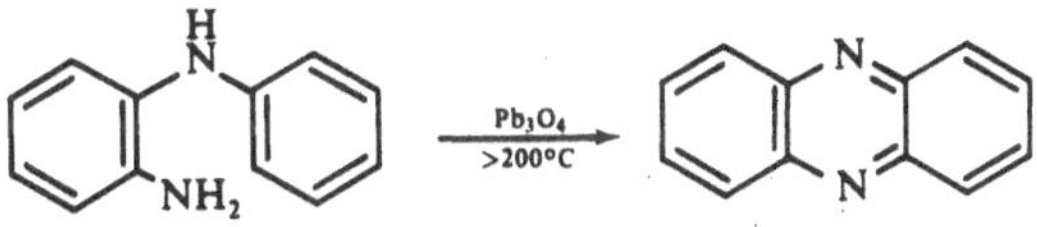

Many oxidants have been used since, but not all the oxidations observed involve a nitrene, particularly in cases where the nature of the oxidizing agent influences the products of the reaction. A free nitrene is thought to be involved in the oxidation of 4-amino-3,5-diphenyltriazole as the products are the same when different oxidants [Pb(OAc)$_4$, Pb(OBz)$_4$, PhI(OAc)$_2$, and PhI(O$_2$CCHCl$_2$)$_2$] are used[51]:

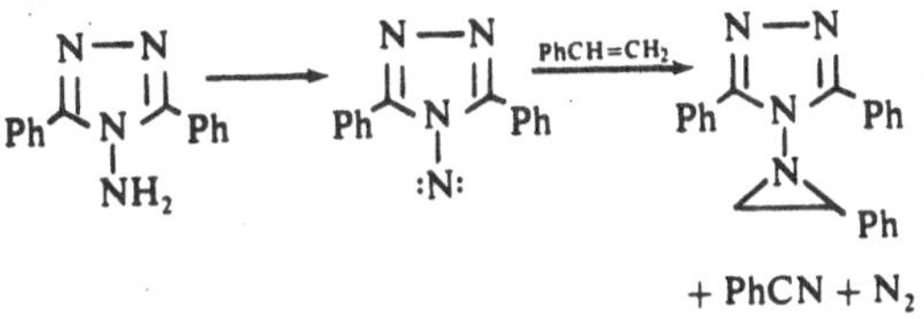

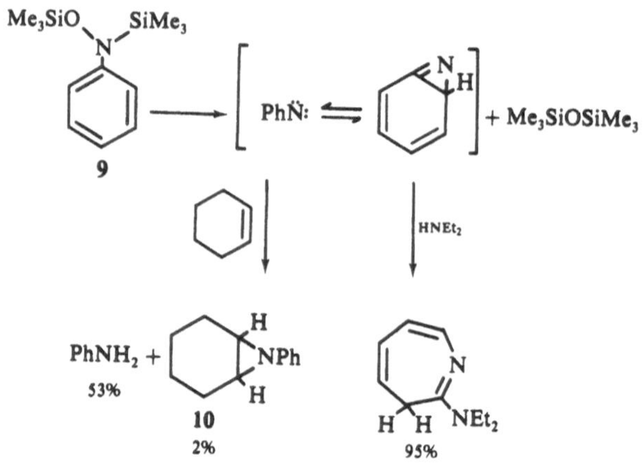

F. Newer Methods of Generation

One of the most promising new methods of generation to appear recently is the thermal deoxysilylation of *N,O*-bis(trimethylsilyl)hydroxylamine[52] (9). Deoxysilylation in diethylamine gives the highest reported yield of ring expansion of a nitrene to an azepine. When deoxysilylation is carried out in cyclohexene a 2% yield of the aziridine (10) is obtained, providing the first instance of the addition of phenylnitrene itself to an olefinic double bond. (Several instances of addition of highly electrophilic nitrenes to double bonds are known; see Section II.1.B.) The failure of phenylnitrene generated by deoxygenation of nitrosobenzene to add to alkenes raises the question as to whether discrete nitrenes are indeed involved in the above reactions. The generality of this method at present suffers from lack of availability of many hydroxylamines.

N-Chloroanilines have sometimes been used to prepare nitrenes (or nitrenoids)[53]; recently a nitrenoid (11), which is produced by treating *N*-chloroaniline with butyllithium at −100°C, has been trapped with trichlorosilane.[54]

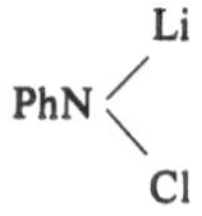

4. Spectral, Structural, and Theoretical Considerations

As the major spectral, structural, and theoretical features of arylnitrenes have been dealt with at some length in previous reviews only salient points of older work will be stated here.

Several aryl azides have been irradiated in rigid matrices and give uv spectra that are similar to those of aromatic radicals. Irradiation of phenyl azide at 77 K produces bands at 314 and 402 nm, attributed to triplet phenylnitrene, which correspond to the bands of the anilino radical at 309 and 400 nm.[55] Flash photolysis of 1-azidoanthracene in ethanol generates a short-lived intermediate ($t_{1/2} = 3$–$10 \mu sec$) which has the same uv spectrum as that of the same nitrene in a glass at 77 K.[56] Reiser[57] has found that the lifetimes of triplet arylnitrenes determined from rates of hydrogen abstraction in a polystyrene matrix decrease with increase in net negative charge on nitrogen (calculated by the Hückel MO method). This observation is consistent with the observed reluctance of arylnitrenes to add to C=C bonds[58] and to abstract hydrogen atoms compared with carbonyl- and sulphonylnitrenes.[14] Again, unlike highly electrophilic azides, phenyl azide only inserts slowly and highly selectively into primary, secondary, and tertiary C–H bonds in alkanes.[58]

Many years ago Wolff[59] isolated a compound that he called "dibenzamil" from the thermolysis of phenyl azide in aniline. This compound was finally characterized independently by Huisgen[60] and confirmed by von E. Doering[61] as 2-anilino-3H-azepine; many other azepines have since been prepared both thermally and photolytically by this method. Almost without exception, the mechanism of these ring expansions has been considered to involve initial attack by a nucleophile (usually an amine) on a benzazirine (**12**) intermediate that was considered to be in equilibrium with a singlet phenylnitrene. Such an intermediate should be amenable to spectroscopic observation but this has been singularly unforthcoming. Now, Chapman[62] has carried out an ir study of the photolysis of phenyl azide matrix isolated in argon at 80 K and found no species corresponding to benzazirine which might be expected to have an absorption in the range $\nu_{C=N}$ 1786–1724 cm^{-1} by analogy to aliphatic azirines[63]:

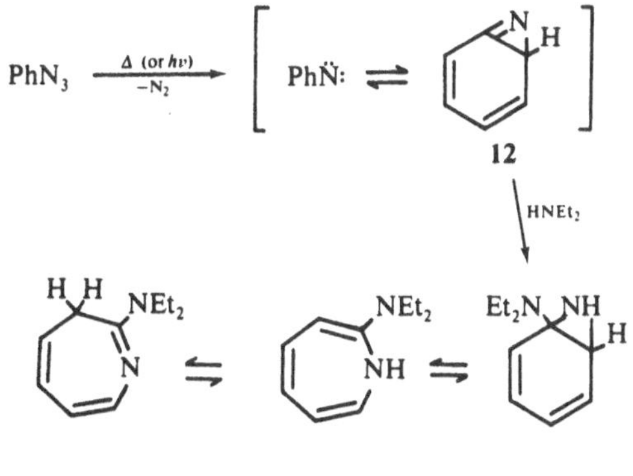

Instead, an intense absorption at 1895 cm^{-1} was observed and attributed to 1-aza-1,2,4,6-cycloheptatetraene (**13**). Irradiation of α-azidostyrene (**14**) under the above conditions with light of wavelength above 336 nm gave the azirine (**15**)

(*ν*$_{CN}$ 1755 cm^{-1}) which, itself, on further irradiation (>216 nm) gave a photostationary state containing the tautomer (**16**) which has a heterocumulene stretching band at 1930 cm^{-1}. This compares well with the main absorption at 1895 cm^{-1} in the intermediate (**13**).

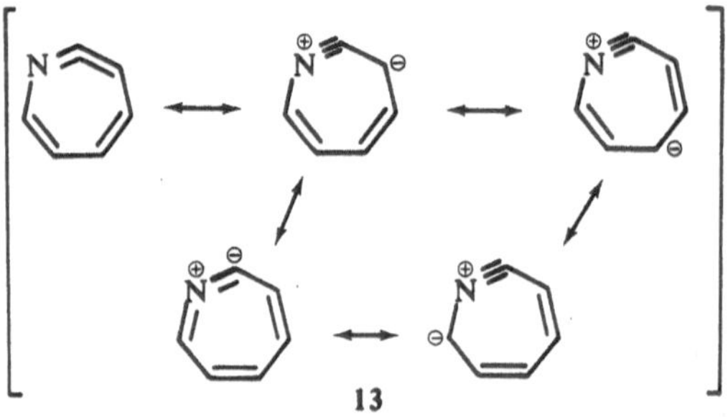

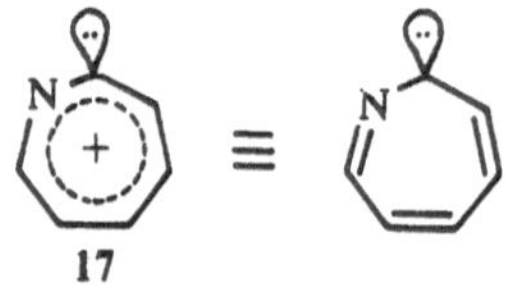

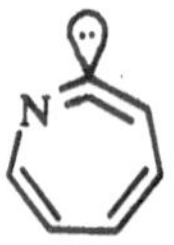

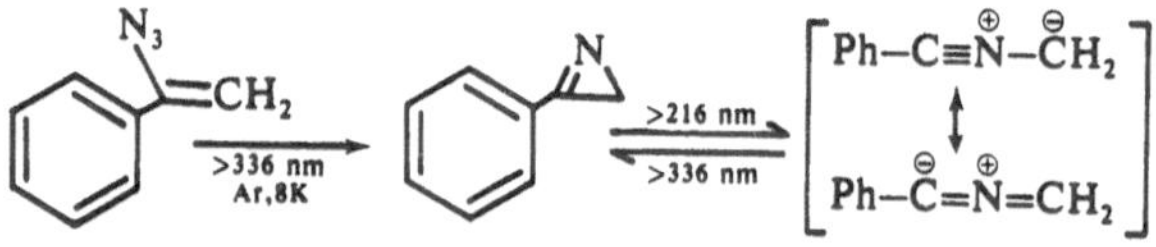

The implications for this intermediate in the mechanism of arylnitrene ring expansions will be discussed later. Only structural considerations will be dealt with here to try to help rationalize why the apparently much less stable inter-mediate 1-aza-1,2,4,6-cycloheptatetraene should be favored over benzazirine. Chapman[62] has proposed the VB (**13**) and MO (**17–18**) structures for 1-aza-1,2,4,6-cycloheptatetraene.

An *ab initio* study by Salem[64] of the opening of 2*H*-azirines gives a calculated C–N–C bond angle of 157° for **19**. Closing this C–N–C angle from 157° to 129° should not produce a large increase in energy, which accords with the observed small change in heterocumulene stretching vibrations in **16** (1930 cm^{-1}) and **13** (1895 cm^{-1}). It is still not clear whether **13** is generated directly from singlet excited phenyl azide or via phenylnitrene. Evidence for

$$H-\overset{\oplus}{C}\equiv\overset{\ominus}{N}-CH_2$$
19

benzazirine intermediacy in azepine formation has always been circumstantial. Thus comparisons have been made with vinyl azides which on photolysis are known to yield azirines.[6] Molecular orbital calculations which show **13** to have a higher energy than benzazirine were considered to support the latter as an intermediate. Chapman's elegant study of the photochemistry of phenyl azide in a matrix at 8 K throws doubt upon the intermediacy of benzazirine in the solution photochemistry of phenyl azide in the presence of nucleophiles. Clearly, this observation is going to stimulate a renewed interest in the mechanism of azepine formation from aryl azides.

II. INTERMOLECULAR REACTIONS

Some intermolecular reactions of arylnitrenes and benzazirines† are outlined in Scheme 1.

SCHEME 1

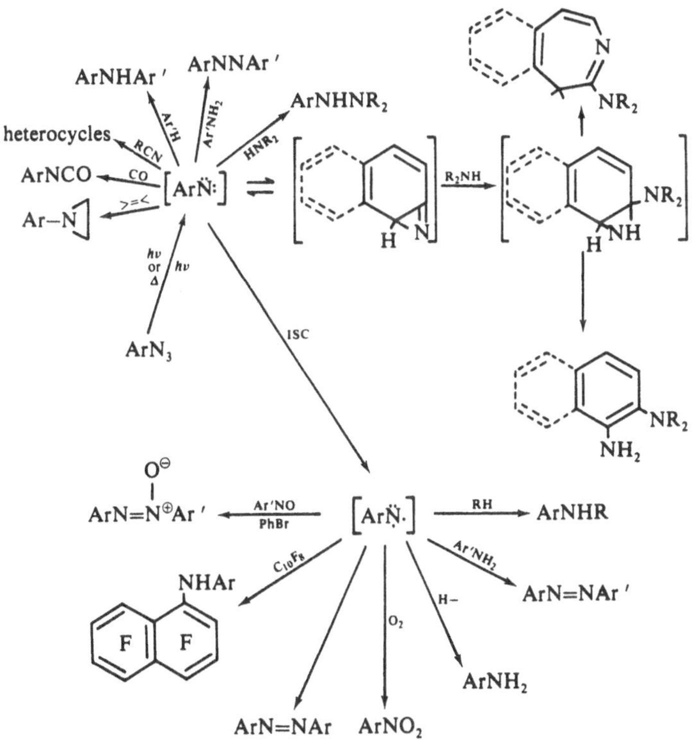

† Benzazirine intermediacy is assumed here; justification for this is presented in Section II.3.A.

Whether the nitrene generated reacts as a singlet or triplet depends upon the nature of the substituents on the aryl nucleus, the nature of the solvent, and the method of generation. These intermolecular reactions will be discussed according to the multiplicity of the nitrene thought to be involved.

1. Singlet Reactions

A. Aromatic Substitution

Intramolecular substitution by arylnitrenes has been known for a long time and the cyclization of o-azidobiphenyls to carbazoles is representative of an important series of nitrene reactions that yield heterocycles.[6,8] Only fairly recently has an intermolecular counterpart of these reactions been observed. Originally when phenyl azide was thermolyzed in benzene only azobenzene (11%), aniline (18%), and tars were obtained, and no diphenylamine was found.[66] Abramovitch[67] considered that phenylnitrene is not sufficiently elec-trophilic to attack benzene, so he decomposed a series of aryl azides (bearing strongly electron-withdrawing substituents, NO_2, CN, CF_3) in solvents activated

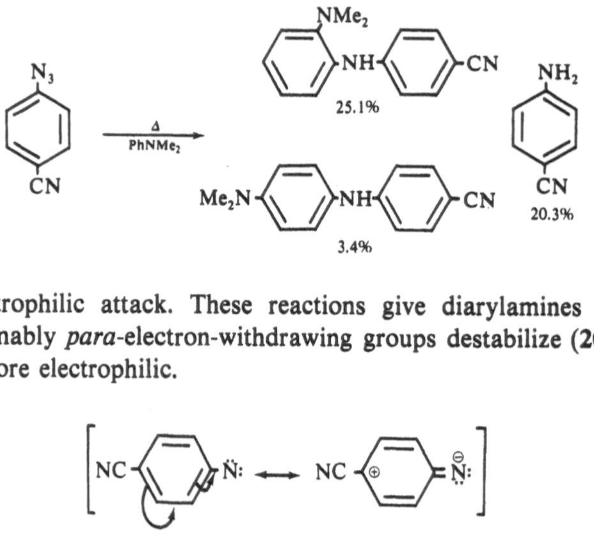

towards electrophilic attack. These reactions give diarylamines in 13–28% yields. Presumably *para*-electron-withdrawing groups destabilize (**20b**), making the nitrene more electrophilic.

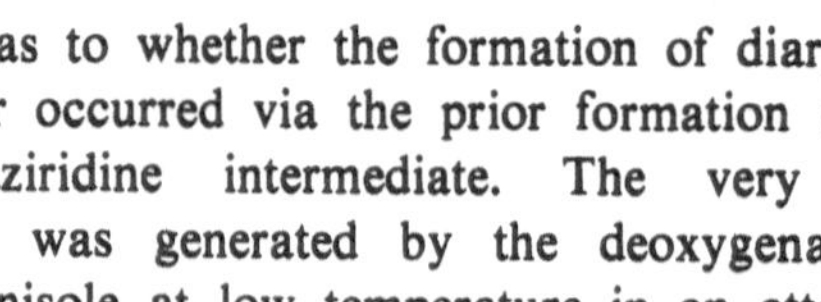

The question arose as to whether the formation of diarylamines involved a direct substitution or occurred via the prior formation and subsequent ring opening of an aziridine intermediate. The very electrophilic pen-tafluorophenylnitrene was generated by the deoxygenation of nitrosopen-tafluorobenzene in anisole at low temperature in an attempt to resolve this question[68,69]:

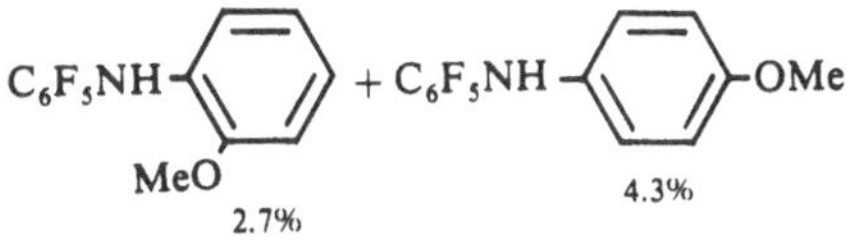

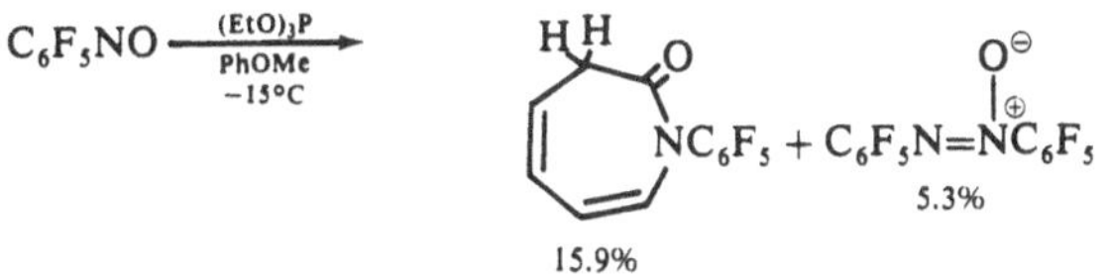

These and other results suggest that predominance of either a σ-complex (22) or an aziridine (21) after initial attack by the nitrene on the aromatic solvent depends upon the nature of the substituents on the aromatic substrates. Substituents that can stabilize positive charges should favor 22 over 21, making 23 the major product. Ring expansion to 24, the product of kinetic control, should predominate when nitrogen bears strongly electron-withdrawing substituents and the nonfluorinated aryl ring is unsubstituted.

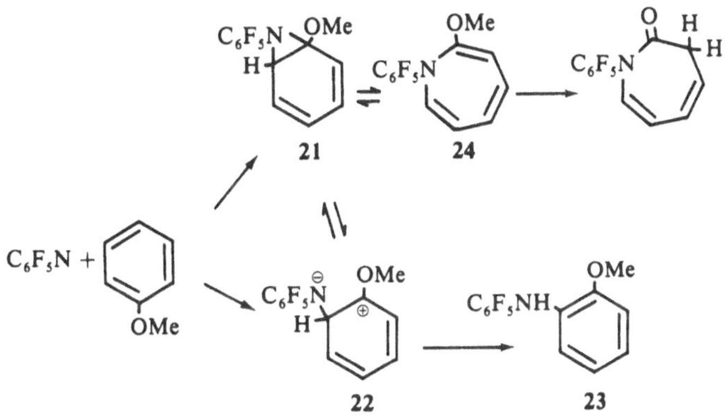

Further support for the notion that arylnitrenes need to be much more electrophilic than phenylnitrene to accomplish aromatic substitution comes from the work of Huisgen.[70] 4,5-Dimethyl-2-pyrimidinylnitrene (pyr-N) readily gives substitution with "activated" aromatics (e.g., naphthalene, anthracene, anisole, and others):

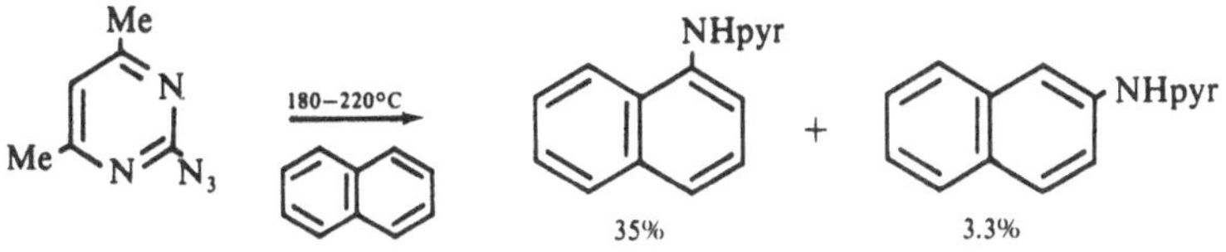

Even benzene undergoes substitution with very electrophilic arylnitrenes,[68,71] and in one case the intermediate azepine has been trapped using TCNE[68]:

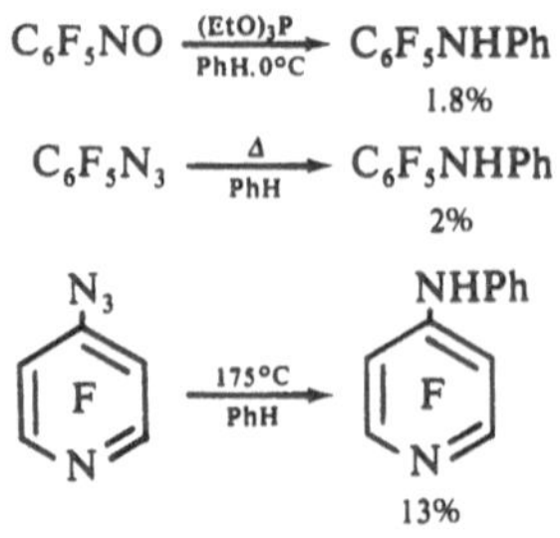

Tetrafluoropyridinylnitrene, perhaps the most electrophilic arylnitrene yet generated, gives the highest yield and its other reactions, too, are more reminiscent of carbethoxy and sulphonylnitrenes.[71]

One instance[72] of 2,4-disubstitution of an aromatic nucleus, by pentafluorophenylnitrene, has been recorded, the product so formed subsequently undergoing oxidative cyclization. Intermolecular aromatic substitution so far has been reported only on deoxygenation of nitro and nitroso aromatics and azide thermolysis:

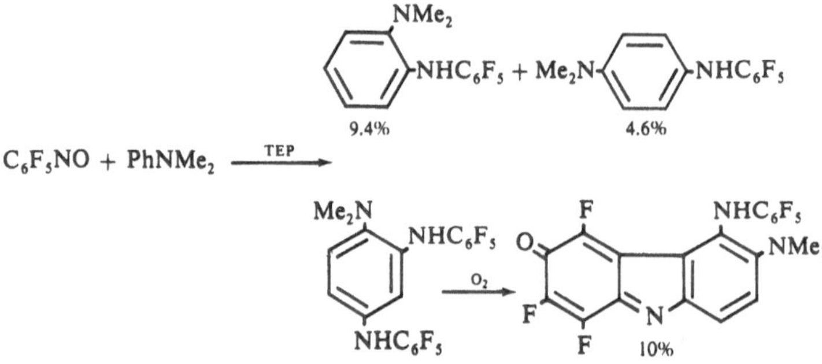

B. Addition to Alkenes

Addition of a discrete arylnitrene to an alkene was first reported[73,74] in 1971 although addition of ethoxycarbonylnitrene to olefins was well documented.[18] Photolysis (or thermolysis) of ferrocenyl azide[74] (25) in cyclohexene yields the aziridine 26 and the predictable products 27–29. The possibility of triazoline formation prior to nitrogen elimination or concerted addition and elimination from the azide were considered to be very unlikely:

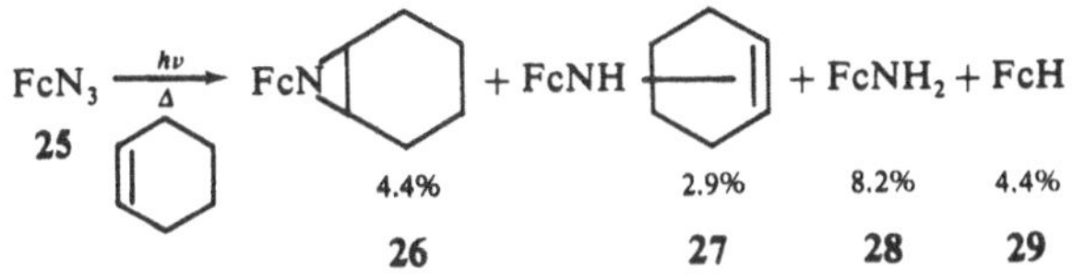

$$FcN_3 \xrightarrow[\Delta]{h\nu} FcN\text{⟨⟩} + FcNH\text{⟨⟩} + FcNH_2 + FcH$$

25

4.4%   2.9%   8.2%   4.4%

26   **27**   **28**   **29**

Further examples of this reaction have been observed[73] when highly electrophilic arylnitrenes are produced by azide decomposition or deoxygenation of
nitrosoperfluorobenzene in olefinic solvents which bear electron-donating
substituents. The highly stereospecific nature of these additions confirmed the
involvement of a singlet arylnitrene and the possibility of a 1,3-dipolar cycloaddition was ruled out.

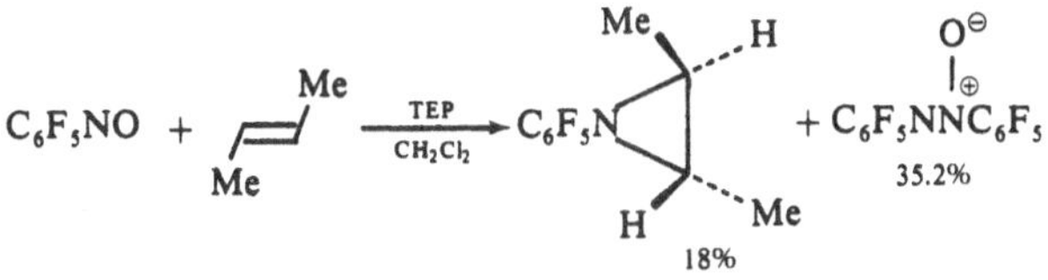

$$C_6F_5NO + \overset{Me}{\underset{Me}{=}} \xrightarrow[CH_2Cl_2]{TEP} C_6F_5N\text{⟨⟩} + C_6F_5NNC_6F_5$$

35.2%

18%

C. Reaction with Nitriles

A study[75] of the decomposition of the 2-azido-1,3,5-triazine (**30**) suggests
that the singlet triazinylnitrene so obtained is sufficiently electrophilic to react
with nitriles to give an intermediate (**31**) that subsequently undergoes a series of
rearrangements.

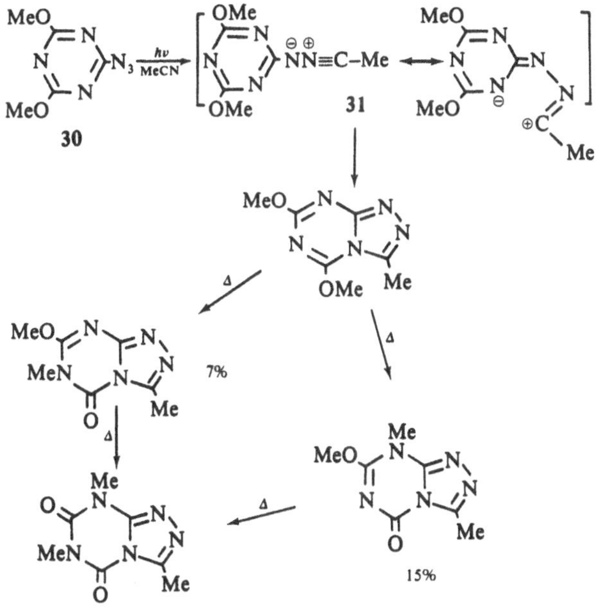

30   **31**

7%

15%

D. Trapping by Carbon Monoxide

Heating aryl azides in benzene at 160–180°C in an autoclave in the presence of carbon monoxide at 136 atm yields the corresponding isocyanates.[76] Use of a reactive solvent gives isocyanate derivatives:

$$ArN_3 \xrightarrow{160-180°C} Ar\ddot{N} \xrightarrow[\sim200\,atm]{CO} ArNCO$$

$$PhN_3 \xrightarrow[\substack{EtOH \\ CO \\ press}]{160-180°C} PhNHCO_2Et$$

E. Electrophilic Substitution at Nitrogen

Odum and Aaronson[77] found that p-cyanophenyl azide gave a substituted hydrazine (32) (70%) on direct photolysis in dimethylamine. On carrying out the photolysis in the presence of a triplet sensitizer (9-xanthenone) the yield of hydrazine dropped from 70% to 6% and the yield of amine rose to 70%, indicating that the hydrazine arose by attack of the singlet nitrene on dimethylamine:

$$p\text{-NCC}_6H_4N_3 \xrightarrow[HNMe_2]{h\nu} p\text{-NCC}_6H_4NHNMe_2 + p\text{-NCC}_6H_4NH_2$$

$$\underset{\substack{32 \\ 70\%}}{\qquad\qquad\qquad} \underset{5\%}{\qquad\qquad}$$

$$p\text{-NCC}_6H_4N_3 \xrightarrow[HNMe_2]{h\nu} p\text{-NCC}_6H_4NHNMe_2 + p\text{-NCC}_6H_4NH_2$$

$$\underset{\substack{32 \\ 6\%}}{\qquad\qquad\qquad} \underset{70\%}{\qquad\qquad}$$

+ 9-xanthenone

In a further study, Odum and Wolf[78] established that ring expansion could compete with hydrazine formation depending upon the wavelength of light used. This behevior was rationalized in terms of excess of excitation energy producing a "hot nitrene":

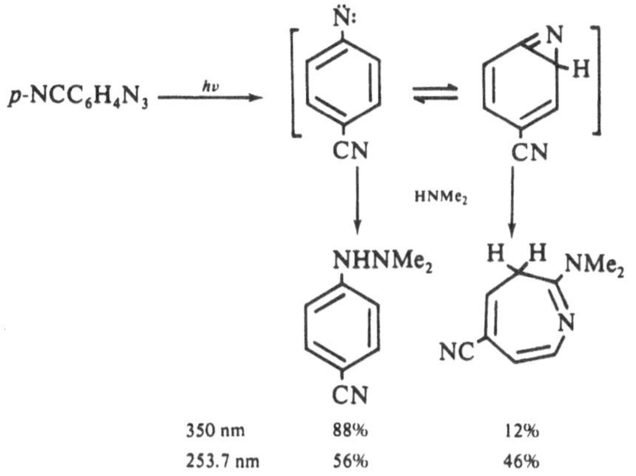

| 350 nm | 88% | 12% |
| 253.7 nm | 56% | 46% |

Thermolysis[79] of *p*-substituted aryl azides in a series of suitably substituted anilines yields mixed azo compounds as illustrated:

$$p\text{-}XC_6H_4N_3 + p\text{-}YC_6H_4NH_2 \xrightarrow{\Delta} p\text{-}XC_6H_4\text{-}N{=}N\text{-}C_6H_4Y\text{-}p$$
$$\text{(or } X{=}H)$$

$$p\text{-}YC_6H_4N_3 + p\text{-}XC_6H_4NH_2 \xrightarrow{\Delta} p\text{-}YC_6H_4\text{-}N{=}N\text{-}C_6H_4X\text{-}p$$
$$\text{(or } X{=}H)$$

X is electron withdrawing (e.g., NO_2), Y is electron donating (e.g., OMe or Me); $X = NO_2$, $Y = H$:

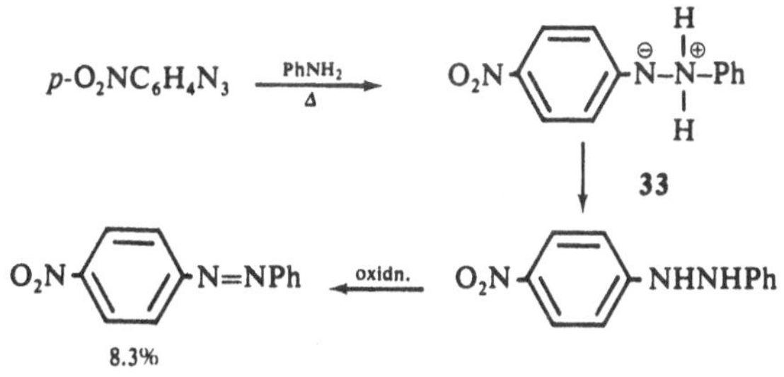

Kinetic studies[79] implicate nitrenes in all these reactions but their multiplicities are not known. It is tempting to propose electrophilic attack by a singlet nitrene at the anilino nitrogen for aryl azides bearing electron-withdrawing substituents. Ylids of the type **33** are known to be formed on the decomposition of sulphonyl azides.[80] The reactions involving the less electrophilic nitrenes such as *p*-anisylnitrene probably do not involve a singlet. These might be envisaged as insertions by triplet nitrene into an N–H bond, a process that is comparable with aliphatic C–H insertion.[58]

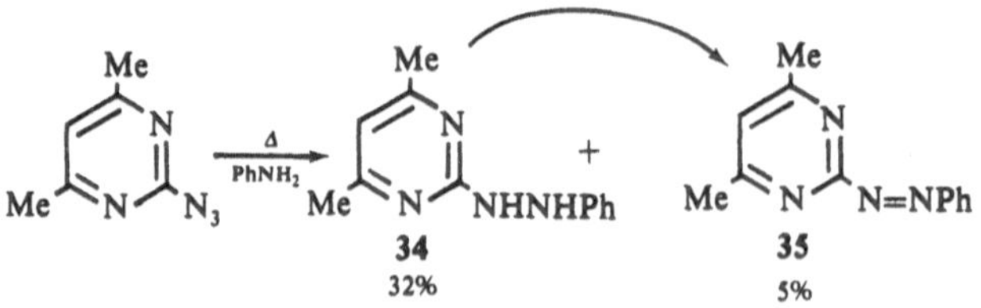

34
32%

35
5%

Several other examples[71] of azo compound formation from reaction of aryl azides with anilines are known and support for a hydrazo intermediate (**34**) comes from the work of Huisgen and von Fraunberg.[70] They showed that **34** could be converted to **35** in a separate experiment.

2. Triplet Reactions

A. Aniline Formation

Decomposition of aryl azides in aliphatic solvents is almost invariably accompanied by formation of the corresponding anilines by hydrogen abstraction from the solvent or the nitrene precursor. No comprehensive study of this reaction has been carried out probably because it is an unwanted one of no synthetic value. The yield of aniline depends upon the availability of abstractable hydrogens in the decomposition solvent. The amounts of aniline obtained on thermolysis of phenyl azide in benzene,[83] and thiophenol[84] are 25%, 44%, and 52%, respectively. In early work it was found that certain solvents undergo dehydrogenative coupling after hydrogen abstraction,[82] but this side reaction is not commonly observed:

$$PhN_3 \xrightarrow[p\text{-}MeC_6H_4Me]{\Delta} PhNH_2 + p\text{-}MeC_6H_4CH_2CH_2C_6H_4Me\text{-}p$$

This, in part, may be a reflection of the fact that an arylnitrene prefers to abstract hydrogen from starting azide rather than from the solvent.[85] In the one reaction studied, thermolysis of phenyl azide-d_5 in pentane gives two main products, aniline (containing 40% N–D) and N-pentylanilines (containing 18% N–D). Even more surprisingly, thermolysis of phenyl azide in Me_3CD affords aniline containing only 7% of N–D.

B. "Dimerization"

Formal dimerization of arylnitrenes to form azo compounds is not as ubiquitous as hydrogen abstraction (to form amines) but it frequently accompanies desired processes and occasionally predominates. Amounts of azo compound formed in these reactions vary widely; electron-donating para-substituents in the azide and solvents containing lone pairs enhance yields. Photolysis of *p*-anisyl azide in tetrahydrofuran gives azo-anisole in 82% yield, whereas only 18% of the same azo compound is obtained in benzene.[86] Electron-withdrawing substituents in the azide result in very low yields of azo compound.

The mechanism of this reaction has yet to be established. Smith (1969)[19] reviewed the possible pathways in some detail, so only the main contenders will be considered here. Attack by a thermolytically or photolytically generated nitrene on undecomposed azide is statistically much more probable than simple dimerization of two nitrenes. Experiments where arylnitrenes have been generated in the presence of an undecomposed azide lend no support for this route however. 1-Methyl-3,5-diphenyl-4-azidopyrazole was decomposed at 80°C in the presence of *p*-anisyl azide (which is stable at 80°C) in decalin but no "crossed" azo product was detected.[87] In another experiment, phenyl azide was recovered almost quantitatively from the triethylphosphite deoxygenation of nitrosobenzene in the presence of this azide.[68]

An intramolecular version of this process, the photolytic conversion of 2,2'-diazidobiphenyl (**36**) to benzo[*c*]cinnoline, is also an unfavored reaction when carried out at room temperature.[88] 4-Azidocarbazole is the main product but photolysis of **36** in a glass at 77 K gives a quantitative yield of benzo[*c*]cinnoline.[89] Presumably conformational effects play a crucial role in this reaction but it should be pointed out that nitrene intermediacy has not been confirmed.

The second mechanism worthy of consideration involves initial abstraction of one hydrogen by triplet nitrene to form an anilino radical which may subsequently either dimerize or abstract another hydrogen. This is illustrated for the thermal decomposition of *p*-anisyl azide in cumene.[27] Indirect support for the hydrazo intermediate **37** comes from the observation that thermolysis of phenyl azide in decalin affords hydrazobenzene and benzidine as well as azobenzene and aniline.[27] Anilino radical intermediates also are implicated in the oxidation of 5-aminopyrazole to azo compounds. The final step, conversion of hydrazo to azo compound, has analogies.[70,87]

Formation of a second "nitrene dimer" 2,7-dimethoxyphenazine (38) was

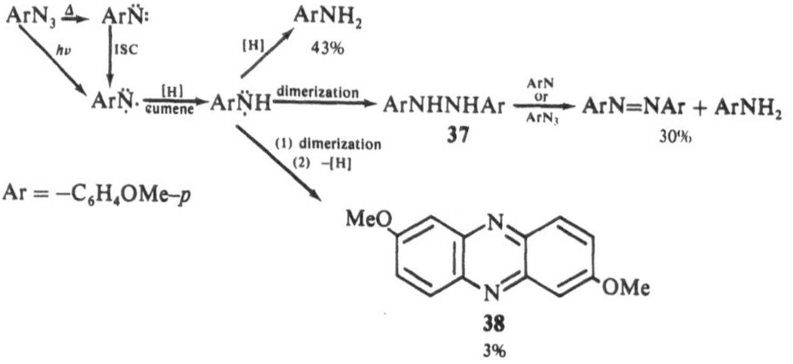

Ar = —C$_6$H$_4$OMe-p

38

3%

first reported by Waters[27] in 1962 but it remained a unique nitrene reaction until recently. Thermolysis of several bicyclic aryl and hetaryl azides gives substantial yields of phenazines and related products. Thermolysis of both α- and β-naphthyl azide in bromobenzene gives dibenzo[a,h]phenazine (39), the latter in quite good yield.[90]. The high overall conversion of these azides to

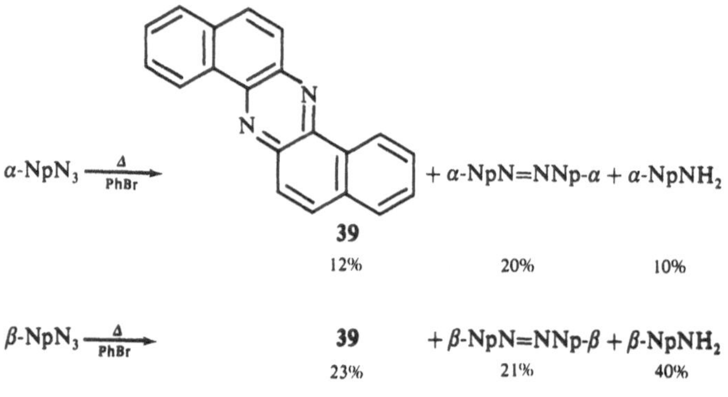

triplet products suggests reaction via a much more stable intermediate than encountered in monocyclic aryl azide decompositions. Bicyclic aromatic amino radicals are known to be much more stable than monocyclic ones, as is the case for the corresponding nitrenes.[56] Further evidence for a common intermediate(s) in the formation of anilines, azo compounds, and "phenazines" was obtained from a study of the effect of decomposition temperature on the relative yields of these products[91]:

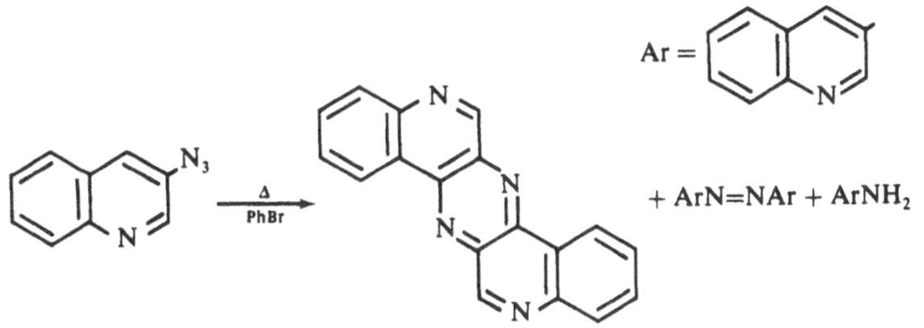

Temp.	Overall yield, %	%	%	%
130°C	80	17	31	32
140°C	81	9	26	46
150°C	80	4	17	59

It is interesting to note that whilst the overall yield of triplet derived products remains the same at higher temperature, a second hydrogen abstraction, to give amine, predominates over the more selective dimerizations. It would be bold to conclude from the above that anilino radical intermediates are involved in arylnitrene dimerizations but they do seem more likely than attack of nitrene on a molecule of starting azide.

C. Insertion into Aliphatic C–H Bonds

Insertion by arylnitrenes into aliphatic C–H bonds, which is usually an unfavored process, is now generally thought to involve the triplet state. Two obvious mechanisms are conceivable. "Classical" insertion of an arylnitrene into a C–H bond should involve a singlet species:

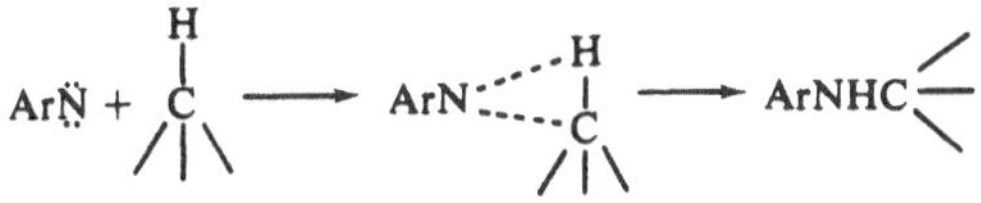

One can also envisage the same product being formed via the triplet nitrene, by an initial hydrogen abstraction followed by radical dimerization:

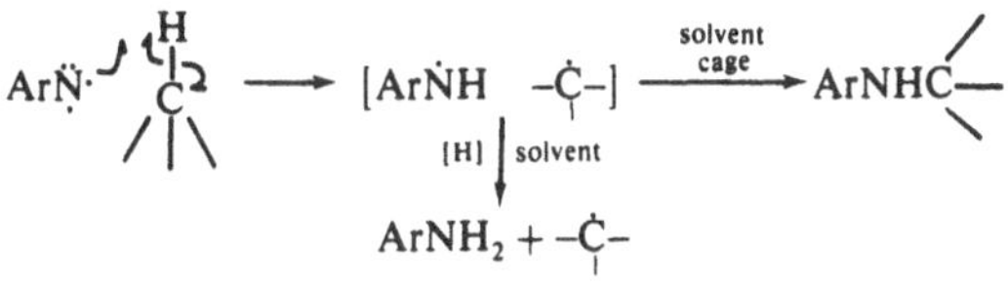

Hall and his coworkers[58] found C–H insertion by phenylnitrene to be a low-yield reaction. Thermolysis of phenyl azide in *n*-pentane yielded aniline (30%) and only 10% of a mixture of *N*-pentylanilines. Evidence for triplet intermediacy derives from a study[60] of the thermolysis of phenyl azide in a cyclohexane–neopentane mixture which gives both H abstraction and insertion products. As the concentration of cyclohexane is decreased the probability of singlet → triplet intersystem crossing should increase, leading to an increase in yield of aniline (a known triplet-derived product) at the expense of *N*- cyclohexylaniline (if it was formed from the singlet). No change in the aniline-*N*-cyclohexylaniline ratio was found with increasing the concentration of neopentane, therefore both products derive from the same intermediate, the triplet nitrene.

The extremely electrophilic tetrafluoropyridylnitrene[71] inserts into C–H bonds much more readily than does phenylnitrene, and again evidence for the triplet nitrene was presented. In other experiments the same workers discounted the possibility of a hydride abstraction mechanism. Intramolecular C–H insertions are known to proceed in high yield through the triplet nitrene (see Section III.1):

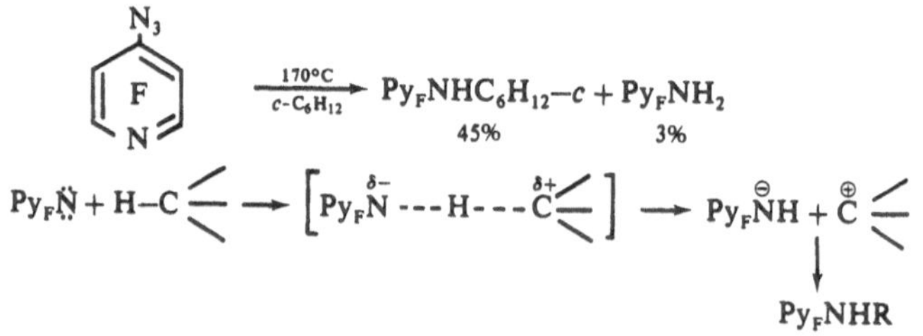

D. Reaction with Nitrosoarenes

Neiman[92] found that 2-azido-2′-nitrobiphenyl cyclizes readily to benzo[*c*]cinnoline-*N*-oxide on photolysis or thermolysis. The intermolecular counterpart of this reaction now has been found to be equally facile in some cases. Heating *α*- or *β*-naphthyl azide and *p*-nitroso-*N,N*-dimethylaniline at reflux in bromobenzene produces the azoxy compounds **40** and **41** in virtually quantitative yield[92] (Np = naphthyl).

No movement of the nitroso oxygen was found and so only one azoxy

$$\alpha\text{-NpN}_3 \xrightarrow[\text{PhBr}]{\Delta,\ p\text{-Me}_2\text{NC}_6\text{H}_4\text{NO}} \alpha\text{-NpN}{=}\overset{\oplus}{\text{N}}(\text{O}^{\ominus})\text{-C}_6\text{H}_4\text{NMe}_2\text{-}p$$

40
92%

$$\beta\text{-NpN}_3 \xrightarrow[\text{PhBr}]{\Delta,\ p\text{-Me}_2\text{NC}_6\text{H}_4\text{NO}} \beta\text{-NpN}{=}\overset{\oplus}{\text{N}}(\text{O}^{\ominus})\text{-C}_6\text{H}_4\text{NMe}_2\text{-}p$$

41
93%

isomer was produced in each case. Thermal decomposition of a series[93] of monocyclic aryl azides under the same conditions leads to yields of azoxy compounds that closely parallel those of azo compounds from decomposition of the azides in bromobenzene alone:

$$p\text{-RC}_6\text{H}_4\text{N}_3 \xrightarrow[\substack{p\text{-Me}_2\text{NC}_6\text{H}_4\text{NO} \\ \text{PhBr}}]{\Delta} p\text{-RC}_6\text{H}_4\overset{\overset{\displaystyle O^{\ominus}}{\overset{\displaystyle |}{\oplus}}}{\text{N}} = \text{NC}_6\text{H}_4\text{NMe}_2\text{-}p$$

R	% Yield
OMe	69
Me	25
H	3
NO$_2$	2

Azide decomposition in these reactions is always strictly first order, indicating nitrene intermediacy and ruling out a 1,3-dipolar cycloaddition mechanism. The substituent effects and the total suppression of azo compound and aniline formation suggests "spin trapping" of triplet nitrene by nitroso compound.[94]

o-Dinitrosobenzene, the ring open tautomer of benzofuroxan, can be trapped by *p*-anisylnitrene[95]:

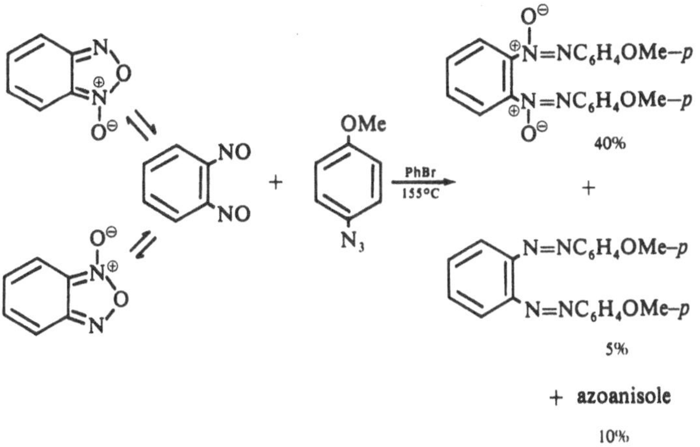

E. Reaction with Oxygen

Photolysis of aryl azides in the presence of oxygen gives nitro compounds and, in the case of substituted azides, also an azoxy compound[96]:

ArN$_3$	$\xrightarrow[O_2]{h\nu}$	ArNO$_2$	ArN=NAr	$\overset{\overset{O^{\ominus}}{\underset{\oplus}{|}}}{ArN=NAr}$	ArNH$_2$
	Solvent	%	%	%	%
	MeCN	17.6	5.3	2.02	1.7
(triplet sens.)	Me$_2$CO	19.8	1.3	4.0	a
	MeCN/				
	piperylene	5.1	1.9	1.4	a
	MeCN/Et$_3$N	18.0	2.2	3.1	a

Ar=p-MeOC$_6$H$_4$—; a, not determined.

Involvement of the triplet nitrene was ascertained by sensitization experiments. The yield of nitro compound was slightly increased by a triplet sensitizer but was much reduced when the reaction was carried out in the presence of a triplet quencher (piperylene). The possibility of singlet oxygen participation was eliminated by the observation that the presence of triethylamine (a singlet oxygen quencher) did not affect the yield of nitro compound. Further evidence for this mechanism comes from the direct observation of **42** and **43** in the esr, after irradiation of p-diazidobenzene and oxygen in a hydrocarbon glass at 77 K.[97]

$$ArN_3 \xrightarrow{h\nu} Ar\ddot{N}: \xrightarrow{ISC} Ar\dot{N}\cdot \longrightarrow Ar\dot{N}-O-\dot{O} \longrightarrow ArNO_2$$
$$\quad\quad\quad\quad\quad\quad\quad\quad \mathbf{42} \quad\quad\quad \mathbf{43}$$

Formation of the azoxy compound was ascribed to the trapping of triplet nitrene by nitroso compound, providing the first observation of this reaction (see previous section).

Trapping of nitrenes by oxygen has been observed in the deoxygenation of nitrosobenzene, thus leading to nitrobenzene.[96] Photolysis of ferrocenyl azide in benzene or cyclohexane under oxygen provides on of the best methods of preparing nitroferrocene[98]:

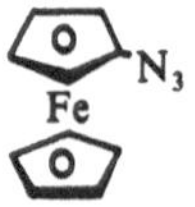

$$\text{Fc-N}_3 \xrightarrow[O_2,\ PhH]{h\nu} FcNO_2 + FcN=NFc + FcH$$
$$\quad\quad\quad\quad\quad\quad 21.2\% \quad\quad 4.5\% \quad\quad 4.8\%$$

F. Substitution of a Carbon–Fluorine Bond

Thermolysis of p-tolyl (or p-anisyl) azide in perfluoronaphthalene under nitrogen gives a low yield of the corresponding *N*-aryl-1-heptafluoronaphthylamine (**44**).[99]

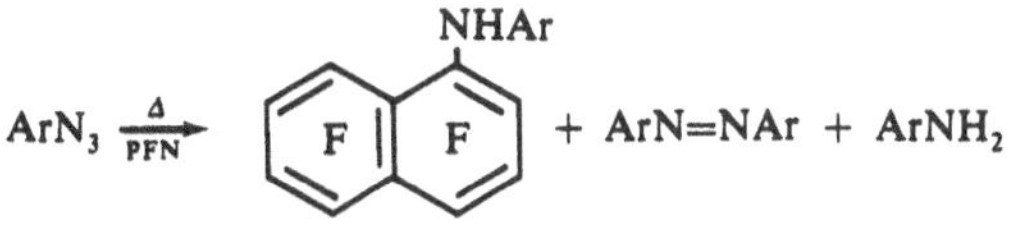

$$\text{ArN}_3 \xrightarrow[\text{PFN}]{\Delta} \quad \begin{array}{c} \text{NHAr} \\ \text{F} \quad \text{F} \end{array} \quad + \text{ArN=NAr} + \text{ArNH}_2$$

44

N$_2$	3.9%	21%	18.7%
O$_2$	—	2.9%	trace

Formation of a 1- rather than a 2-substituted perfluoronaphthalene derivative is suggestive of a radical[100] rather than simple nucleophilic attack[101] by a preformed amine. The total suppression of the formation of **44** when the experiment was carried out in the presence of oxygen lends further support to the intervention of a triplet reaction.

3. "Benzazirine" Reactions

A. With Nucleophiles

Wolff (1912)[59] was the first to observe the ring expansion of phenyl azide, on thermolysis in aniline, to a compound he called "dibenzamil" (**45**). It was many years before this compound was shown to be 2-anilino-3*H*-azepine.[60,61] This reaction was also found to take place on photolysis of aryl azides in primary and secondary aliphatic amines.[61] The mechanism that was proposed and which has since been generally accepted[10-12,14,19] involves nucleophilic attack by solvent on a benzazirine, that is believed to be in equilibrium[6] with singlet arylnitrene. Nmr evidence[102] has been advanced for the intermediacy of a 1*H*-azepine (**46**) in related expansions which, as expected, rapidly isomerizes to the more stable 3*H* tautomer. Spectroscopic evidence is still lacking[103] for the aziridine intermediate **47**, but this could simply be due to its short lifetime. On the other hand, the recent work of Chapman and Le Roux[62] discussed in Section I.4 has brought into question the intermediacy of benzazirine in phenyl

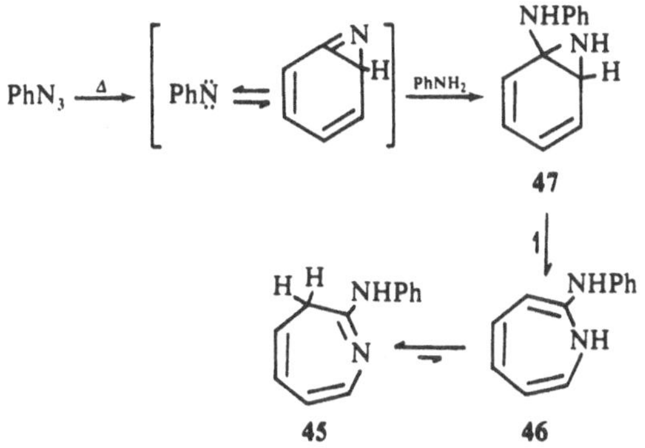

$$\text{PhN}_3 \xrightarrow{\Delta} \left[\text{Ph}\ddot{\text{N}} \rightleftharpoons \begin{array}{c}\text{N} \\ \text{H}\end{array} \right] \xrightarrow{\text{PhNH}_2} \begin{array}{c}\text{NHPh} \\ \text{NH} \\ \text{H}\end{array}$$

47

45 ⟶ **46**

13 46′ 45′

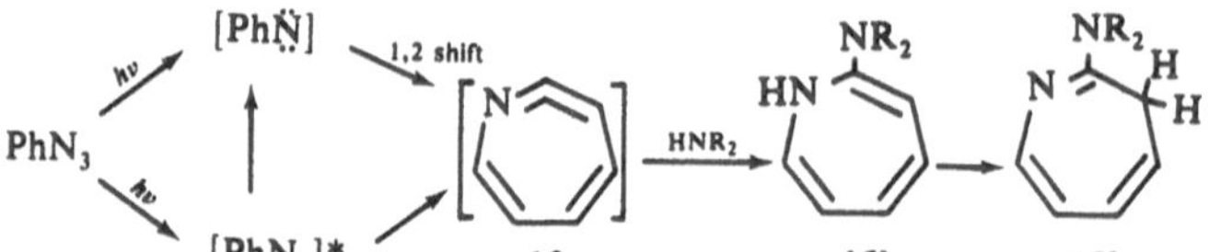

azide photolyses. Chapman envisages nucleophilic attack on 1-aza-1,2,4,6-cycloheptatetraene (**13**), which is either formed via the singlet nitrene or directly from singlet-excited phenyl azide. As a result, 1*H*-azepine (**46′**) is formed directly and the necessity to invoke an aziridine intermediate is obviated. If one can accept that results obtained from photolysis in a matrix at 8 K can be applied to solution photochemistry at room temperature this is certainly an attractive mechanism for azepine formation on phenyl azide decomposition. Also it should be borne in mind that this low-temperature spectroscopic work was carried out in the absence of nucleophiles and no azepines were isolated.

The question now arises as to whether the Chapman mechanism is applicable to the decomposition of other aryl azides. Bicyclic aryl azides, for example, have a considerable degree of double-bond fixation in their aromatic rings. This might be expected to make them behave more like vinyl azides, which are known to give azirines on photolysis. It is therefore useful to compare the nature of the products formed on photolysis in nucleophilic solvents by a series of azides that have an increasing degree of vinyl azide character:

R^1 = alkyl, R^2 = alkyl or H (Ref. 61)

(Ref. 104)

(Ref. 105)

(Ref. 106)

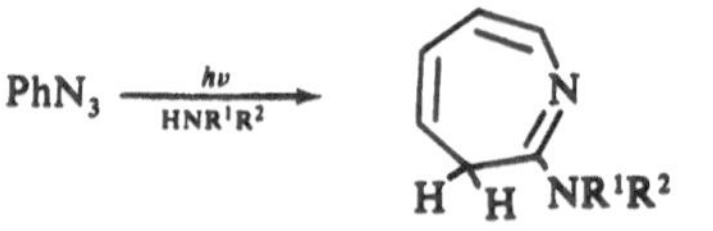

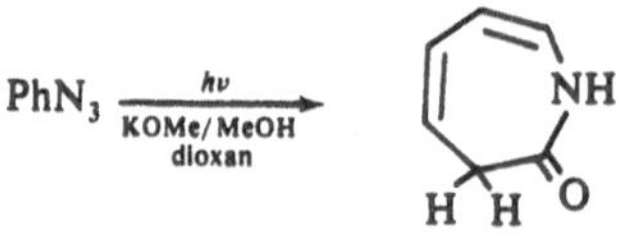

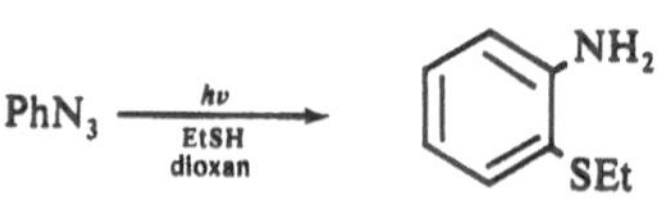

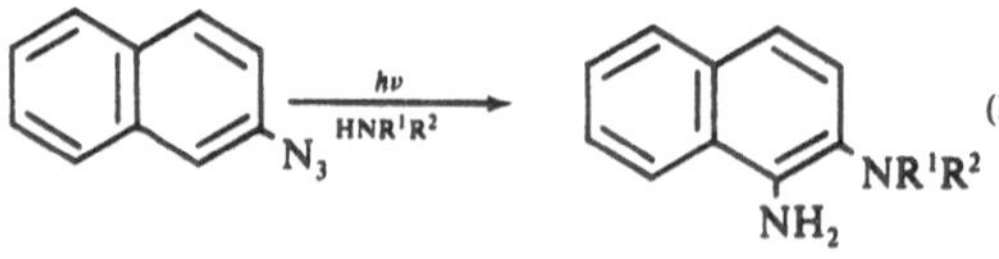

+ 47
40%

(Ref. 107)

(Ref. 108)

(Ref. 109)

(Ref. 110)

(Ref. 111)

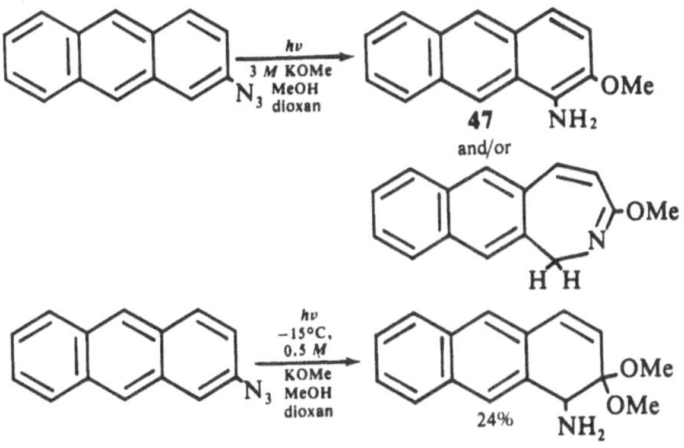

Phenyl azide itself undergoes ring expansion in the presence of nitrogen and oxygen nucleophiles, and hydrogen sulfide (in diethyl ether solution) but, in ethylmercaptan, 2-ethylmercaptoaniline, the product of rearomatization, is formed in preference to azepine. This rearomatization product can easily be accommodated by formation of an aziridine by initial attack on benzazirine and subsequent C–N bond fission of the aziridine:

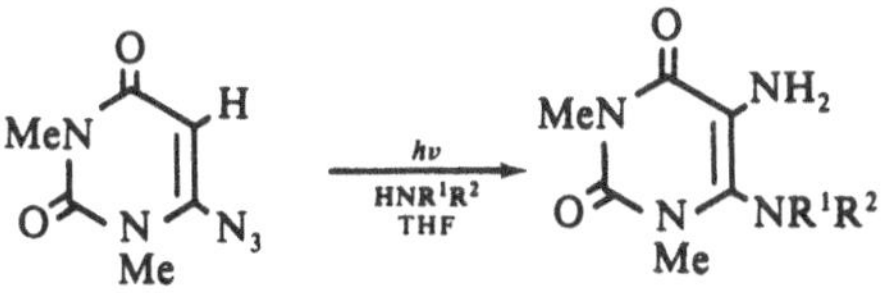

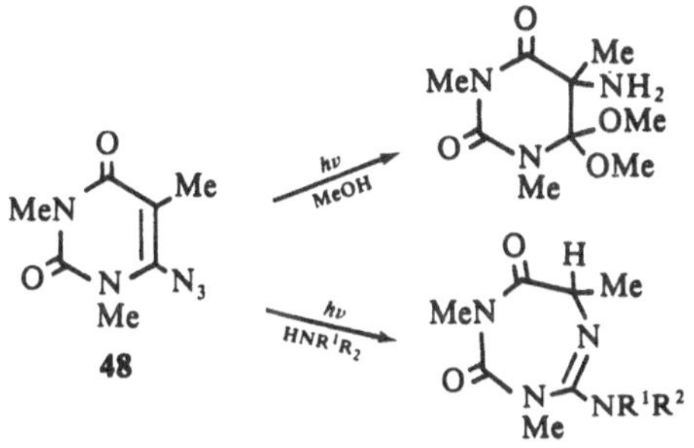

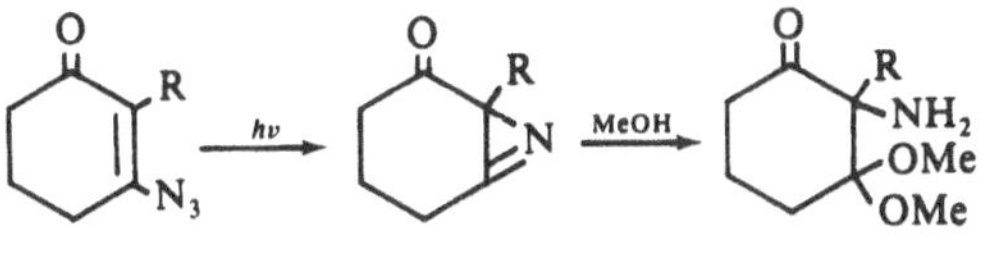

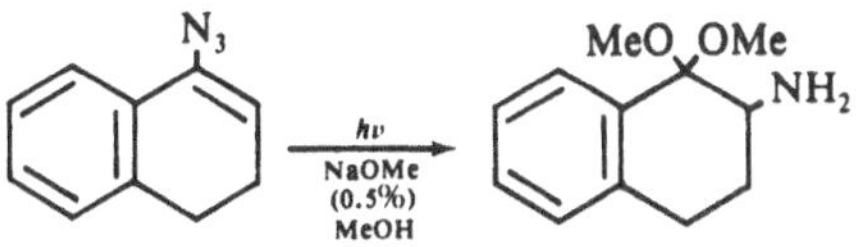

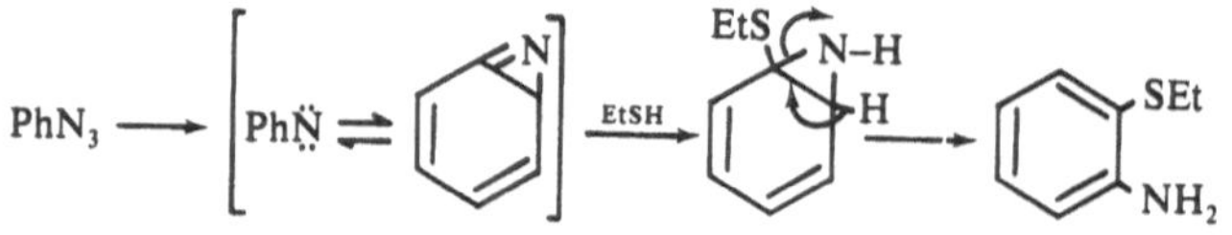

Formation of the same product by attack on 1-aza-1,2,4,6-cycloheptatetraene
(13) appears more clumsy:

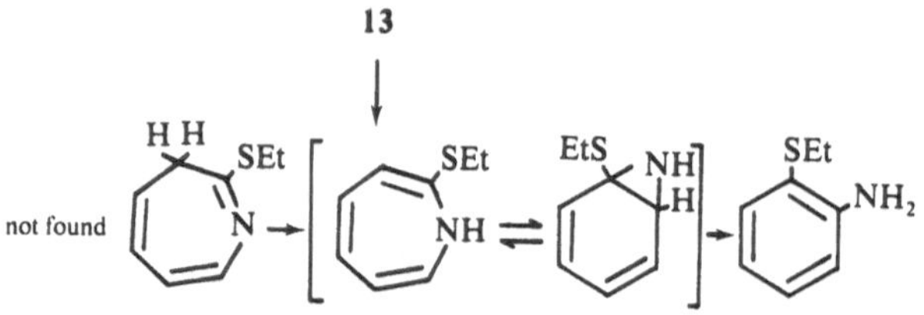

Rearomatization is much more common in bi- and tricyclic azides than
ring expansion although ring expansion may be achieved employing sodium
methoxide in methanol-dioxan as solvent. This dichotomy again may be
resolved by invoking an aziridine intermediate (*vide supra*). The formation of
aminoketals from 2-azidoanthracene and 48 is of particular interest as similar
products are commonly formed from aliphatic azirines on treatment with mildly
basic methanol.[111] At the moment, the intermediacy of benzazirine in phenyl
azide photolysis at room temperature cannot be ruled out. In the case of bi- and
tricyclic aromatic azides and azidouracil decompositions the nature of the
products strongly supports azirine involvement. Only further experimental work
will resolve this mechanistic dilemma. With this in mind, benzazirine intermediacy will be assumed for the purposes of this discussion.

Yields of azepines from photolysis and thermolysis of azides, and nitro-
and nitrosoarene deoxygenation vary greatly with the substituents on the
aromatic ring and the types of nucleophile used. Cadogan and coworkers[36] have
shown that the deoxygenation of *o*-, *m*-, and *p*-substituted nitroarenes and the
photolysis of the corresponding azides at the same temperature in diethylamine
give substituted 2-diethylamino-3*H*-azepines in almost identical yields by both
methods. This provides good evidence for a common intermediate in these reactions. Information on the wide range of aryl azides that undergo ring expansion
in primary and secondary aliphatic and aromatic amines may be found in the
above reference[36] and in the reviews already quoted. Recently, triethylamine
also has been used as a solvent for the photolytic conversion of phenyl azide to
2-diethylamino-3*H*-azepine[112]:

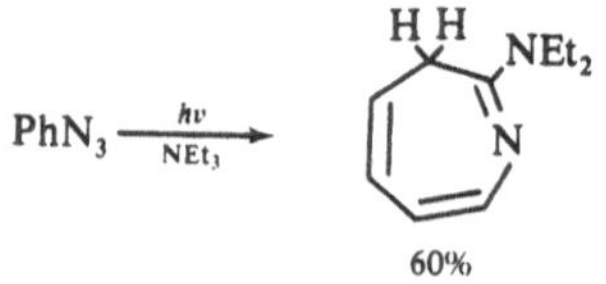

o-Substituted[102] aryl azides react via the benzazirine formed by closure onto the *ortho* carbon bearing the hydrogen and not the substituent:

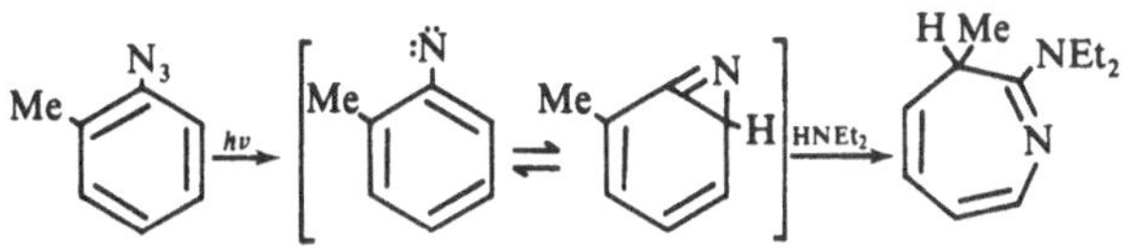

Another interesting feature of this work is the discovery of unstable oxygen-sensitive intermediates thought to be 3-alkyl-2-diethylamino-1*H*-azepines which are readily oxidized to mixtures of pyridines:

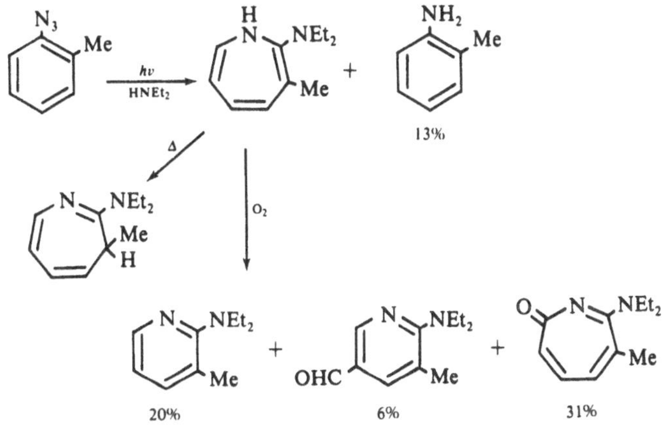

An exception to the usually observed closure of a nitrene to the less hindered *ortho*-carbon atom has been observed by Carde and Jones[113]:

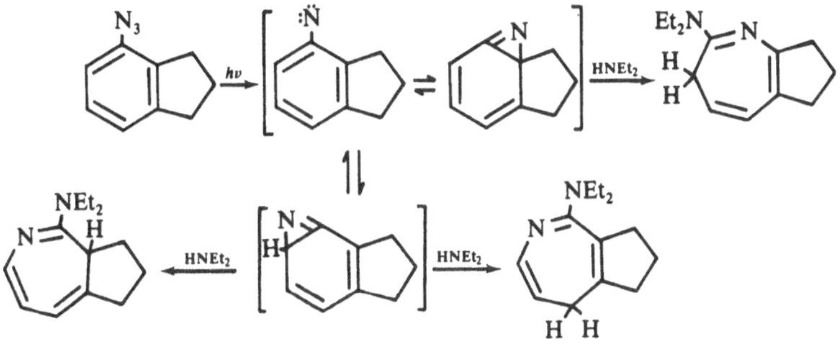

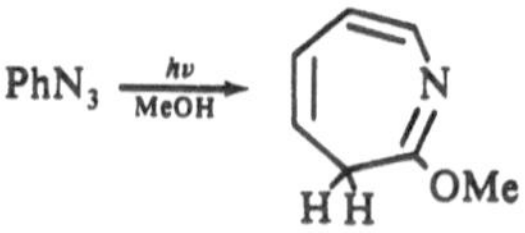

$$PhN_3 \xrightarrow[\text{MeOH}]{h\upsilon}$$

49

Phenyl azide has been claimed[114] to undergo expansion to **49** in 10% yield on photolysis in methanol, a very weak base. This reaction is much more facile when the azide bears on an *ortho* substituent containing a carbonyl group.[115] In this case, stabilization of the singlet nitrene (and suppression of intersystem crossing to triplet) can be envisaged by formation of an intermediate anthranil. These are known to give azepines on irradiation in nucleophilic solvents[47]:

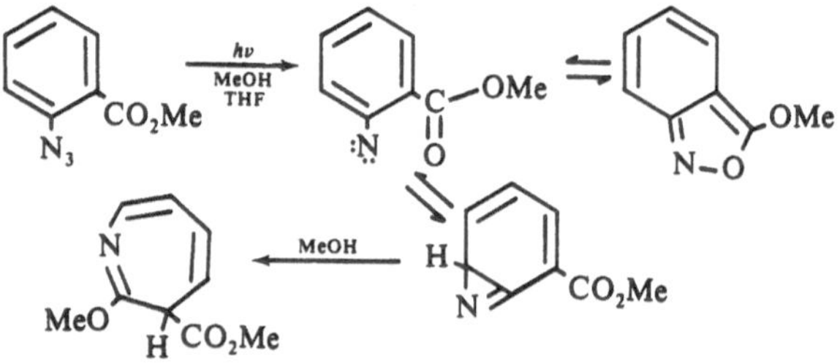

Irradiation of α-naphthyl azide in piperidine merely gives a small yield of α-naphthylamine and α-azonaphthalene attributable to typical triplet nitrene

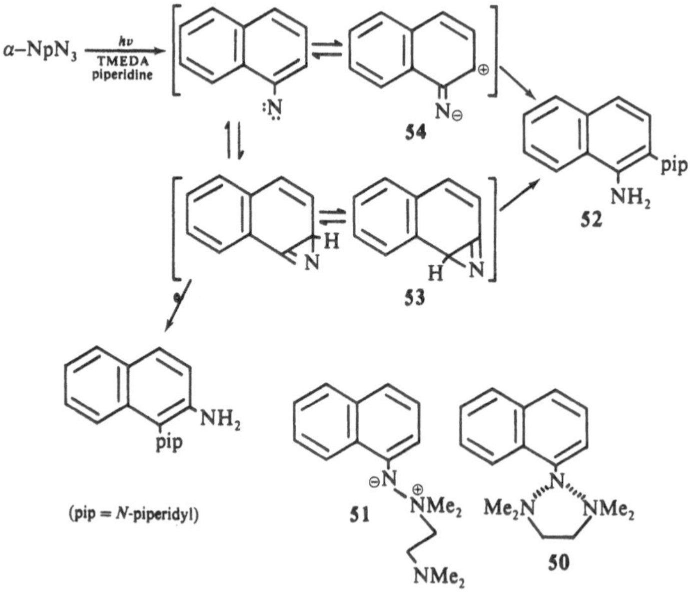

$$\alpha\text{-NpN}_3 \xrightarrow[\substack{\text{TMEDA} \\ \text{piperidine}}]{h\upsilon}$$

(pip = *N*-piperidyl)

54   **53**   **52**   **51**   **50**

reactions, but the use of TMEDA (tetramethylethylenediamine) as cosolvent changes the nature of the products dramatically.[116] It was considered that TMEDA is able to stabilize the singlet nitrene either by chelation (50) or ylid formation (51). The isomeric *o*-diamine (52) might arise by nucleophilic attack on 53 after interconversion of the azirines or by reaction with the resonance-stabilized canonical form (54).[117] Methylene chloride,[118] dioxan, [104,107] and tetrahydrofuran[108,119] also have been used as cosolvents; again, their lone pairs of electrons are thought to stabilize the singlet nitrene.

o-Diamines are usually formed on irradiation of bicyclic aromatic azides in the presence of secondary aliphatic amines[120]; only rarely have azepines been obtained[121]:

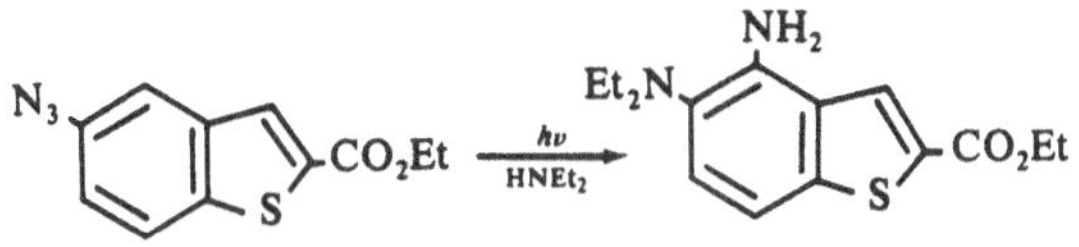

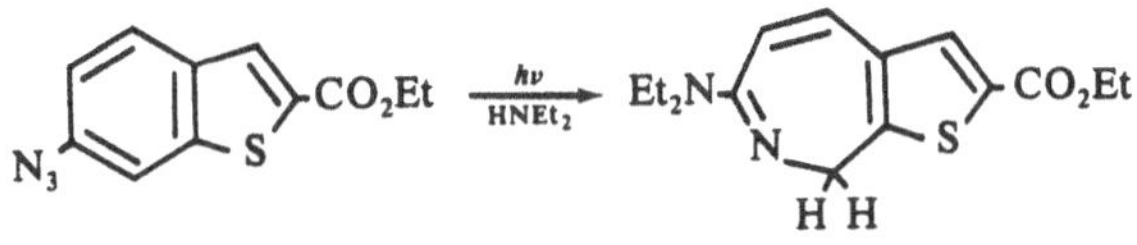

In (6-6)bicyclic aromatic systems photolysis of α-azido compounds in primary aliphatic amines favors azepine formation, while β-azido compounds form *o*-diamines. *o*-Diamine formation again predominates in photolysis of both α- and β-azido(6-6)bicyclic aromatics in secondary aliphatic amines.[106] This difference in behavior of α-azido-arenes may be accounted for by considering Scheme 2.

The first step in the formation of 55 and 56 involves nucleophilic attack on 57, giving 58, at which point the pathway bifurcates. Expansion of ring A, the conversion of 58b to 59, is concomitant with loss of aromaticity in ring B and relief of unfavorable interaction between the NR^1R^2 group and the peri substituent X. These two factors may compensate each other. Clearly, when R^1 and R^2 are alkyl, 55 should be favored; conversely when $R^2 = H$, stabilizing factors (rearomatization of ring B, introduction of an amidine structure) in the final azepine product should induce formation of the kinetic control product.

The yields of azepine and *o*-diamine may be optimized by varying the ratio of nucleophile to cosolvent used for photolysis.[123] Photolysis of α- and β-azidoarenes in methoxide–methanol–dioxan can either lead to ring expansion or

SCHEME 2

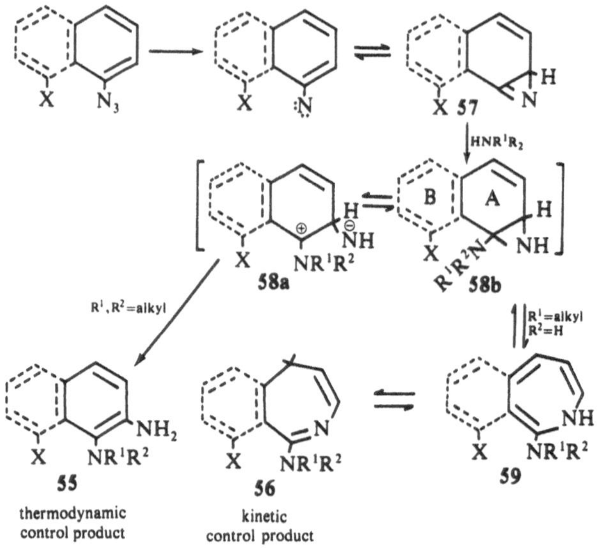

55
thermodynamic
control product

56
kinetic
control product

to *o*-methoxyamine formation depending upon workup conditions. The above reactions may also be carried out thermolytically or by deoxygenation of the appropriate nitro compounds.

Recently, phenyl azide has been found to react with "naked anions," but whether the mechanism of this reaction is similar to that of those above has not yet been established[124]:

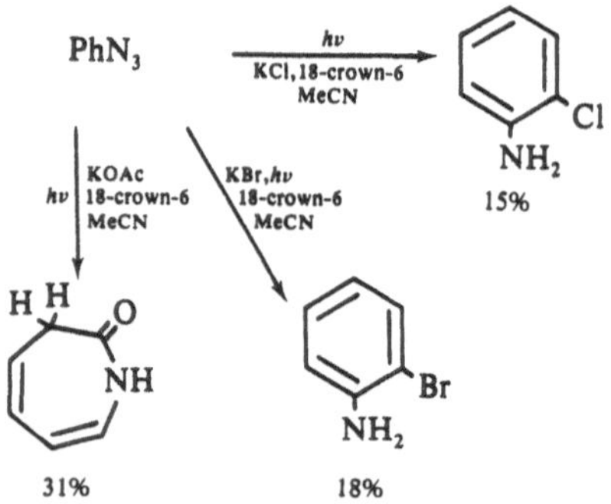

Both of these pathways have been exploited synthetically, providing novel routes to fervenulins **(60)**,[125] (involving nucleophilic attack by acyl hydrazines), lumazine **(61)**[125] and benzo-1,4-diazepines **(62)**.[126]

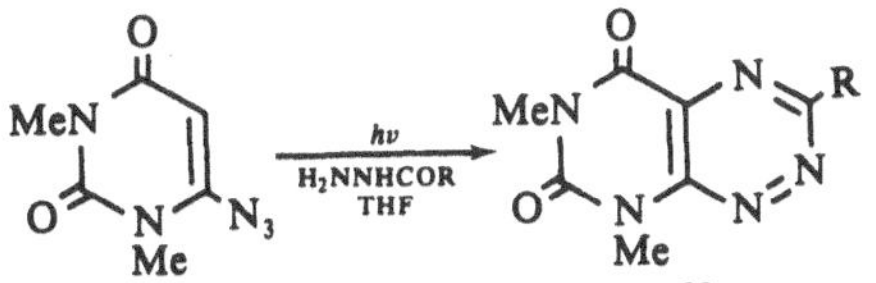

60

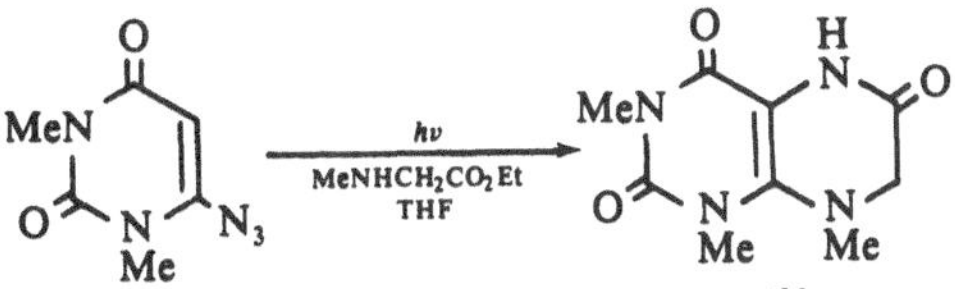

61

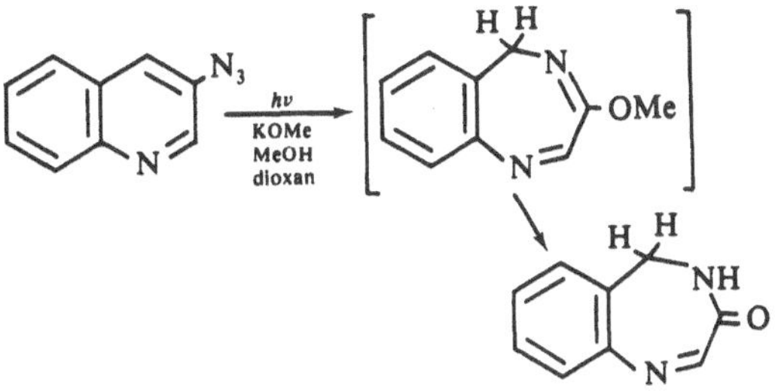

62

At this point it is useful to reflect upon the types of reaction so far observed between aryl azides (nitrenes) and nucleophiles:

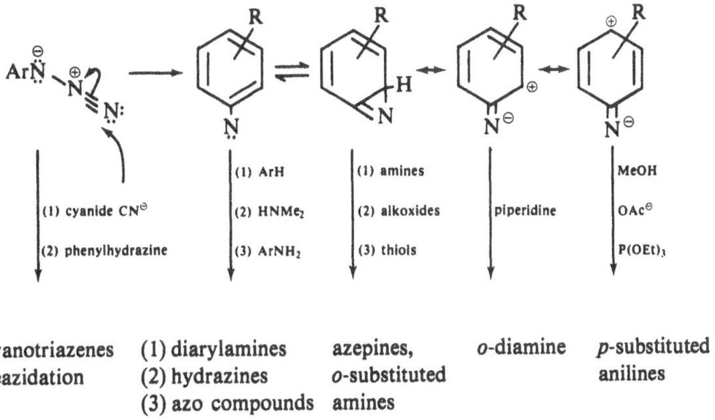

(1) cyanide CN⊖	(1) ArH	(1) amines		MeOH
(2) phenylhydrazine	(2) HNMe₂	(2) alkoxides	piperidine	OAc⊖
	(3) ArNH₂	(3) thiols		P(OEt)₃

(1) cyanotriazenes	(1) diarylamines	azepines,	*o*-diamine	*p*-substituted
(2) deazidation	(2) hydrazines	*o*-substituted		anilines
	(3) azo compounds	amines		

Aryl azides undergo nucleophilic attack at the terminal *N*-atom of the azide in the presence of "soft" nucleophiles. Heating phenyl azide in the presence of cyanide ions yields a cyanotriazene[127]:

$$PhN_3 + \overset{\ominus}{CN} \xrightarrow[\substack{(1)\ ROH-H_2O \\ (2)\ HOAc}]{\Delta} Ph-N=N-NHCN$$
$$\updownarrow$$
$$Ph-NH-N=NCN$$

Heating aryl azides at reflux in a mixture of hydrazine hydrate and ethanol leads to deazidation. This process appears to be a good method of removing azido groups from aryl systems (particularly in the case of azides bearing *ortho* substituents)[128]:

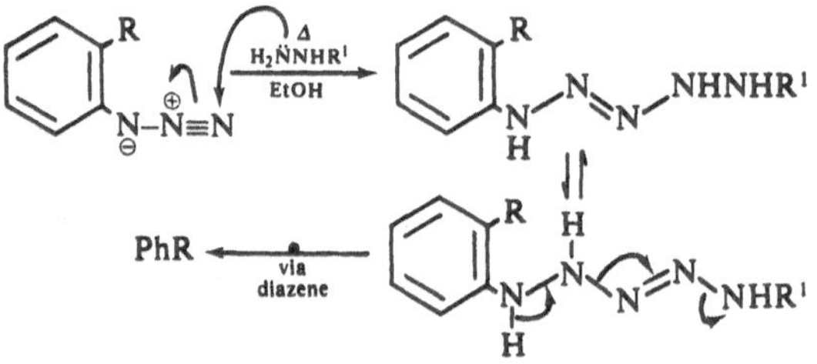

Nucleophilic attack at the para position has been observed by Sundberg[129] in the deoxygenation of nitrosobenzene, but it is thought to involve attack on a phosphorus intermediate rather than a phenylnitrene.

B. With Electrophiles

Thermolysis of aryl azides in acetic anhydride gives *N,O*-diacetyl-*o*-aminophenols plus some triplet-derived products. Yields are good for azides bearing methyl or halo substituents, but nitrophenyl azides do not undergo this reaction[130]:

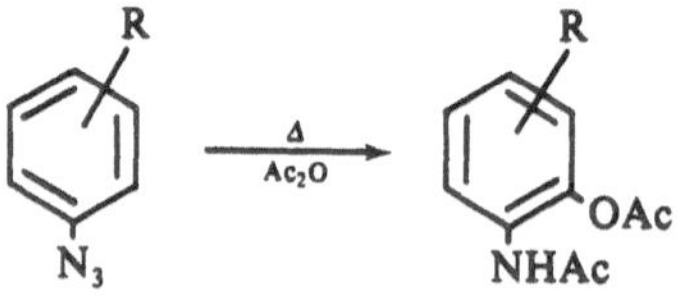

o-Chlorobenzanilides are formed on thermolysis of aryl azides in benzoyl chloride.[131] In one case 6-chloro-2-phenylbenzoxazole was also obtained. The formation of this product and the substituent effects in this reaction (decrease in *o*-chlorobenzanilide yield *p*-OMe > *p*-Me > H > *p*- Cl) prompted the authors to invoke a benzazirine intermediate, nitrene involvement having been indicated by a kinetic study:

kinetic control product

thermodynamic control product

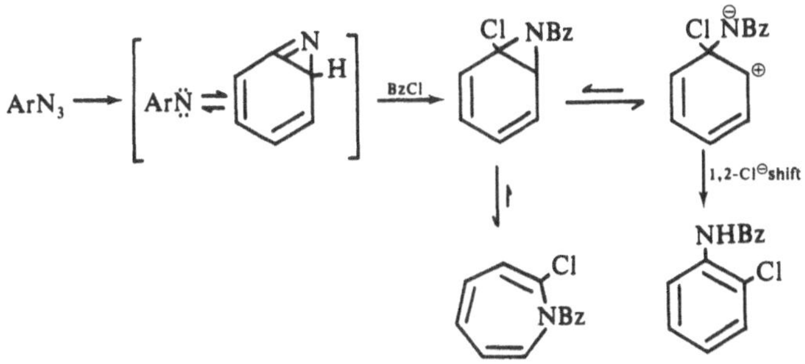

It should be noted that in none of these decompositions in acetic anhydride and benzoyl chloride was a 1,2-nitrogen movement found, as is the case for analogous nucleophilic attack. Further support for an azirine intermediate and the first observation of a 1,2 movement of nitrogen comes from Senda's work.[132] Photolysis of a 6-azidouracil (**63**) in acetyl chloride–tetrahydrofuran gives a 5-amino-6-chlorouracil (**64**). The formation of this product may be

63 **64**

70%

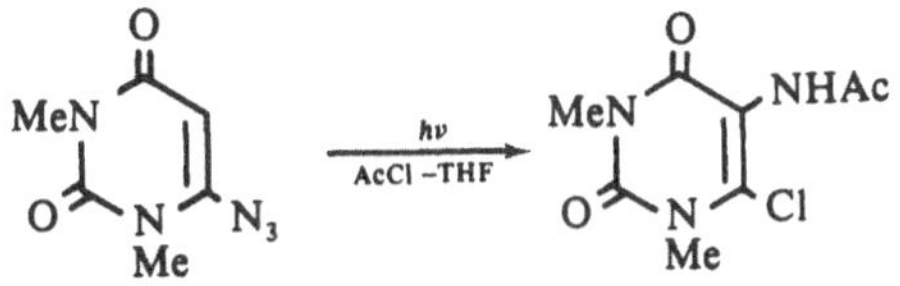

rationalized in terms of initial acetylation followed by a 1,2-H shift to give the more stable carbenium ion before attack by chloride ion to form the observed product. The photolysis of a 5-azidouracil (**65**) would be of interest, to see if **64** or **66** is formed. One would predict the formation of **64** if carbenium ions are intermediates in these reactions. This mechanism may well be followed for the decompositions in acetic anhydride and benzoyl chloride above.

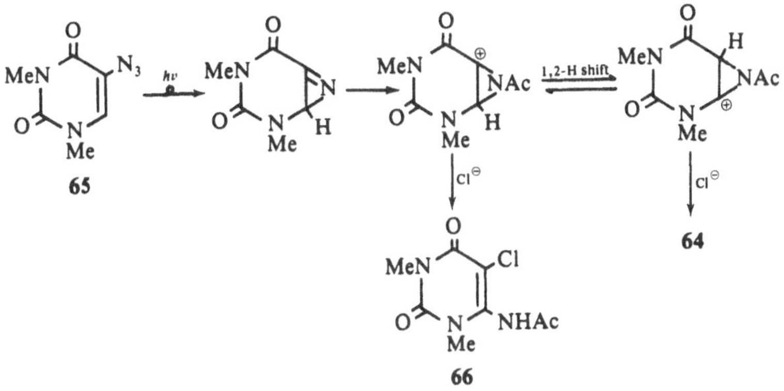

If azirine and aziridine intermediates are indeed involved in these reactions it is surprising that azepines have not been isolated, especially on photolysis in acetyl chloride, as azepines with electron-withdrawing substituents on nitrogen are known to be stabilized relative to N–H azepines.[68] Photolysis of suitable aryl azides in acetyl chloride at low temperature might permit the isolation of the product of kinetic control (azepine) rather than the thermodynamic control product which predominates at room temperature.

Heating 6-azido-1,3-dimethyluracil with alkyl halides in dimethylformamide containing potassium carbonate yields 1-alkyl-4,6-dimethyl-*v*-triazolo(4,5-*d*)pyrimidin-5,7(4*H*,6*H*)-diones[133]:

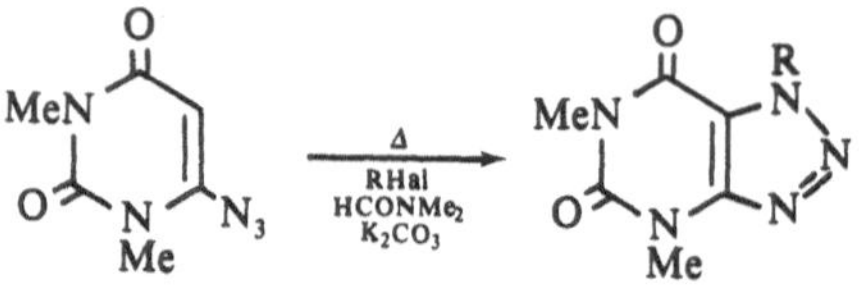

III. INTRAMOLECULAR REACTIONS

Intramolecular cyclization reactions of arylnitrenes by aromatic substitution (singlet), C–H aliphatic insertion, and attack on *ortho* nitrogen functions are usually very much more facile than their intermolecular counterparts.

1. Nitrene-Induced Cyclizations

A. To Form Five-Membered Rings

Cyclization of *o*-azidobiphenyls to carbazoles is one of the oldest and best studied arylnitrene reactions.[19] Carbazole formation (**67**) has finally been shown[134] to involve cyclization of a singlet nitrene (**68**), generated by photolysis or thermolysis of 2-azidobiphenyl, settling a controversy of some standing.[135]

The lack of pronounced substituent effects on this reaction by electron-withdrawing and -donating groups in the two aryl rings is neatly accommodated by this mechanism. The nitrene may behave as a π^2 or π^0 component, depending upon whether a full or empty orbital of the singlet nitrene is utilized. Presumably the π^6 process would be thermally allowed, the π^4 photochemically allowed.

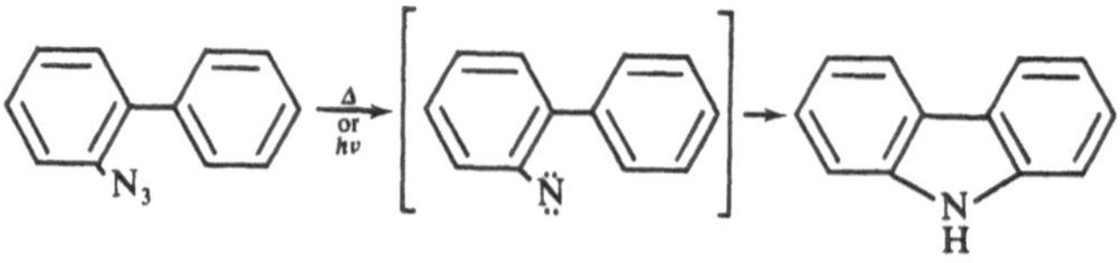

                         68   **67**

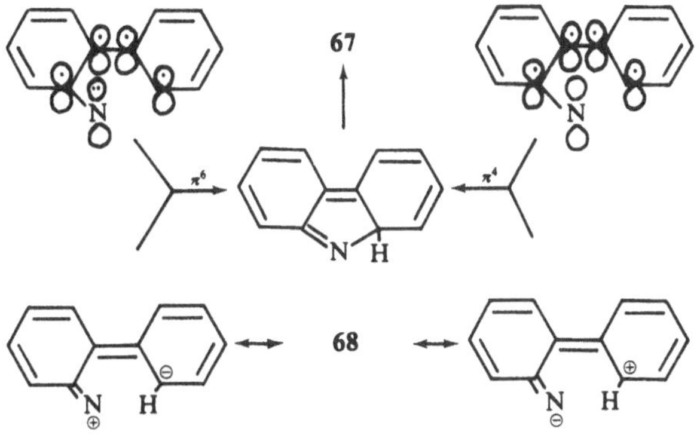

By a skillful use of sensitizers, quenchers, filters, and choice of temperature used for photolysis, the decomposition of 2-azido-2'-methyl-biphenyl can be controlled to give a predominance of either carbazole (via singlet nitrene) or phenanthridine (via triplet):

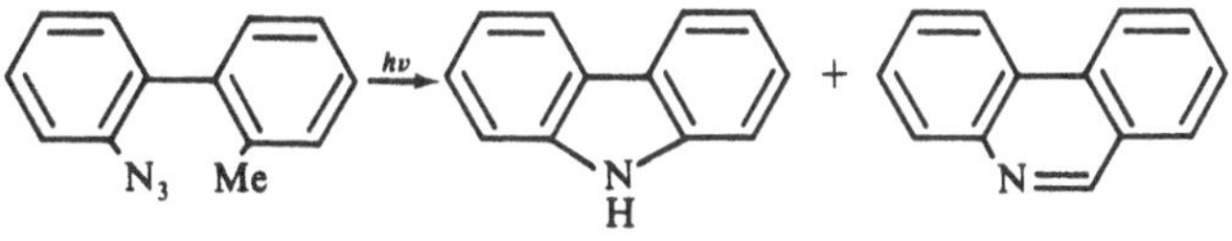

cyclohexane, RT CH₂Cl₂–pyrene or	77%	0%
10% PhCOMe-PhCl 99°C	8.5%	39.5%

In the latter example the corresponding amine (25%) and azo compound (12.5%) were also obtained.

Some recent synthetically useful examples of cyclizations that yield five-membered rings are illustrated below. They include cyclizations onto neighboring phenyl,[136] pyrone,[137] and indole[138] rings:

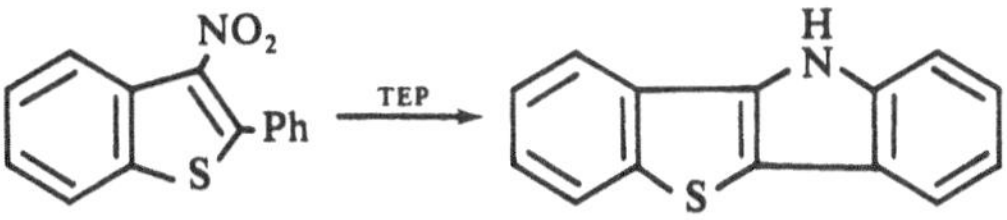

(Ref. 136)

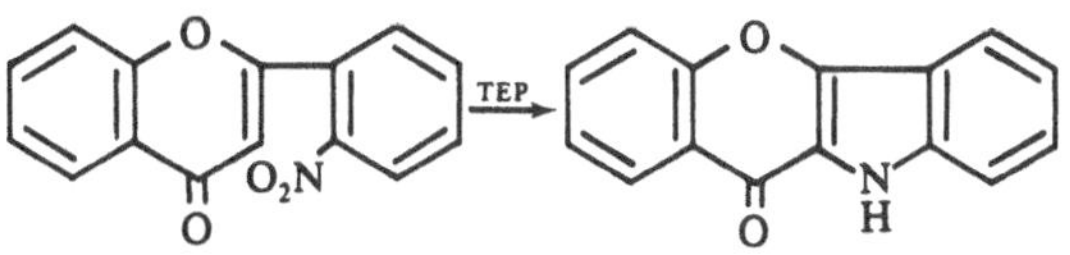

(Ref. 137)

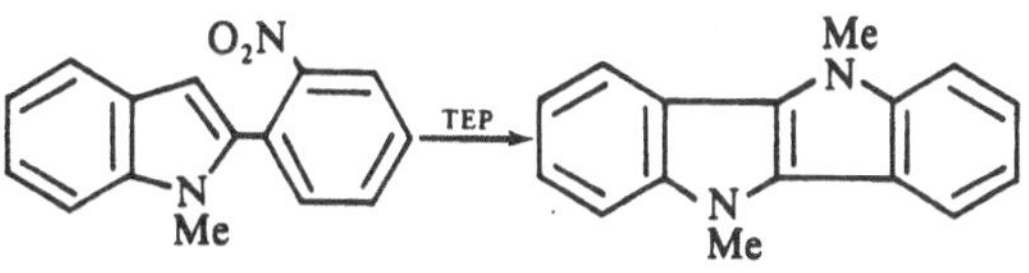

(Ref. 138)

Cyclization onto a nitrogen atom in neighboring rings[42] by deoxy-genation[139] or azide decomposition[140] yields ylids:

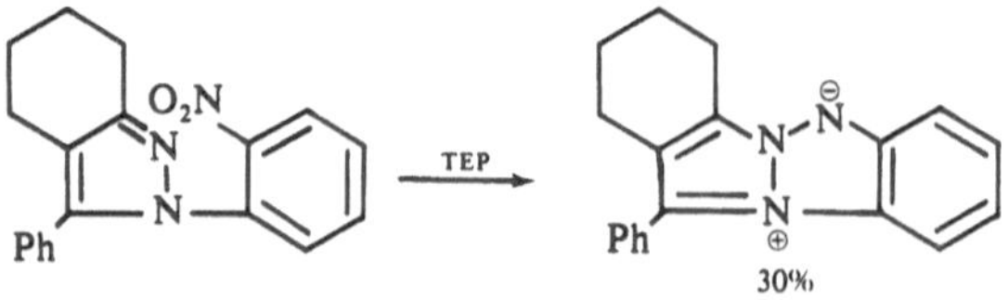

(Ref. 139)

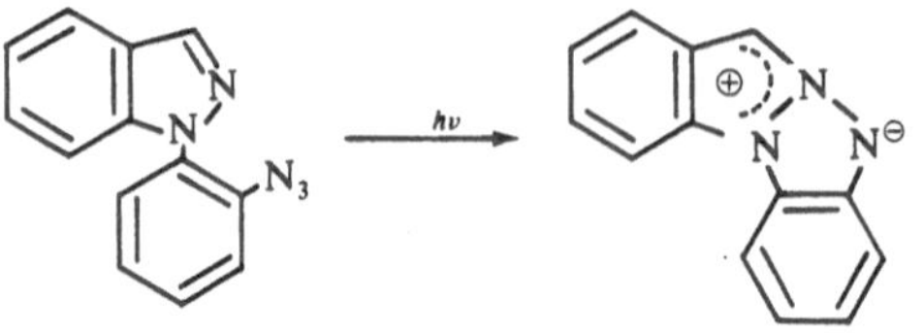

(Ref. 140)

Cyclization onto an adjacent *ortho* azo substituent can compete with closure onto carbon, but the possibility of an assisted elimination of nitrogen in the former case was not discounted[141]:

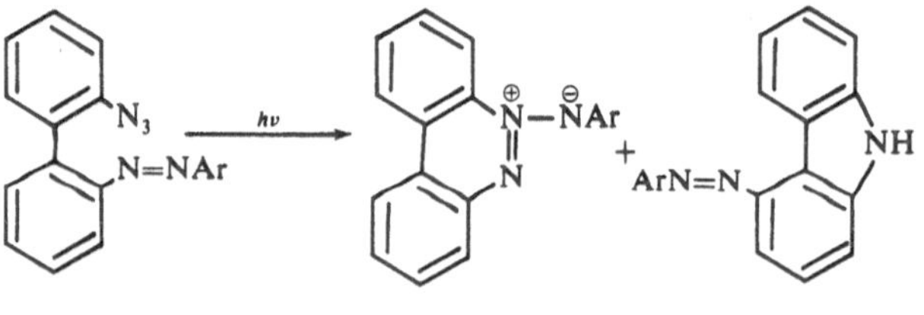

Another competition between closure onto C or N that might be expected to favor *N*-cyclization, by comparison with 2-(2-nitrenophenyl)pyridine,[42] in fact does give a substantial amount of *C*-cyclization product as well.[142]

Abramovitch and Kalinowski[143] have observed the effect of substituents *ortho* or *para* to an azido group on the competition between cyclization onto N or C:

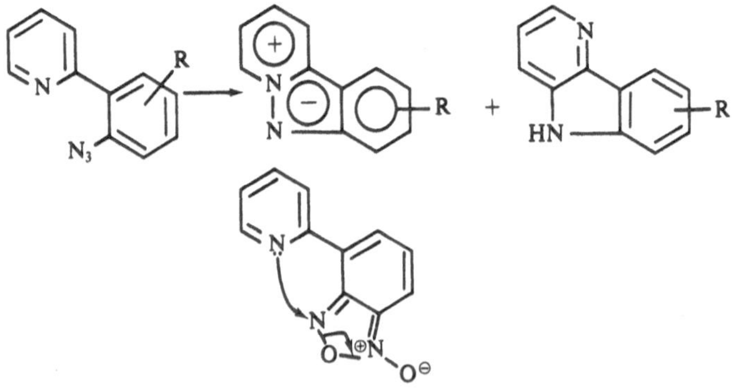

These cyclizations were effected in boiling toluene at 110°C, which suggests some participation by the pyridine lone pair in the azide decomposition, as unassisted thermal decompositions of aryl azides do not usually take place below 130°C. Cyclization onto nitrogen is the predominant mode giving a useful synthetic approach to pyrido[1,2-*b*]indazoles, except in the case with an *o*-trifluoromethyl group when the 6-trifluoromethyl-δ-carboline was formed in 29% yield. When the starting azide contained an *o*-nitro group, decomposition was very rapid, suggesting participation of a benzofuroxan intermediate which rapidly rearranged to the more stable 8-nitropyrido[1,2-*b*]indazole.

B. Closure to Six-Membered Rings

This constitutes a particularly important series of cyclizations where X in **69** may be varied and the nitrene generated from any of the usual sources. The products formed by these reactions include phenothiazines, thiazepines, phenoxazine, oxazepines, dihydrophenazines, indoloazepines, acridones, and related systems. (X = S, O, N–Ac, CH$_2$, SO$_2$, or CO).

These reactions do not involve a straightforward cyclization by electrophilic attack of the nitrene on an *ortho'* carbon atom **69** → **70**, but they occur by initial *ipso* attack to form a five-membered ring *spiro* intermediate[145] (**71**), which subsequently undergoes rearrangements. Alternatively, it has been suggested that the aziridine **72** is formed directly from **69**.[68,144] The pathway taken is probably substituent dependent.

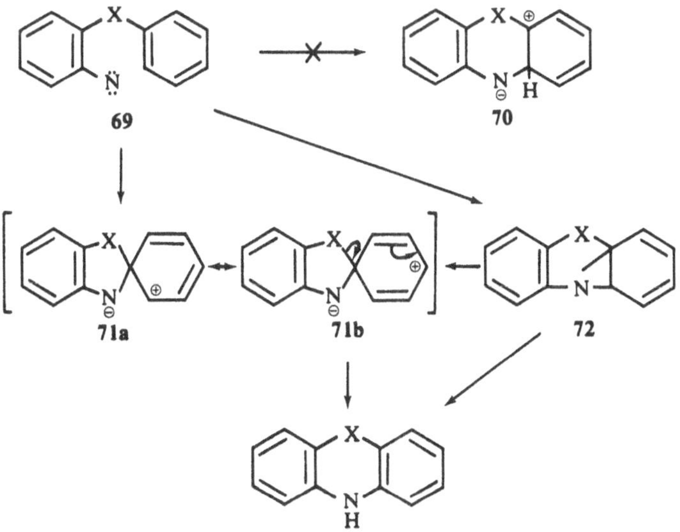

The rearrangements undergone by these systems have been reviewed by Cadogan,[13] therefore only a few examples will be quoted in this chapter. When the two *ortho*' positions are blocked by methyl groups a dibenzothiazepine (**73**) is formed.[145]

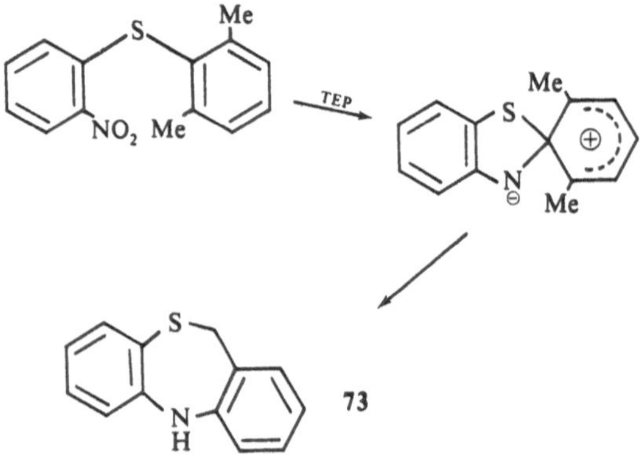

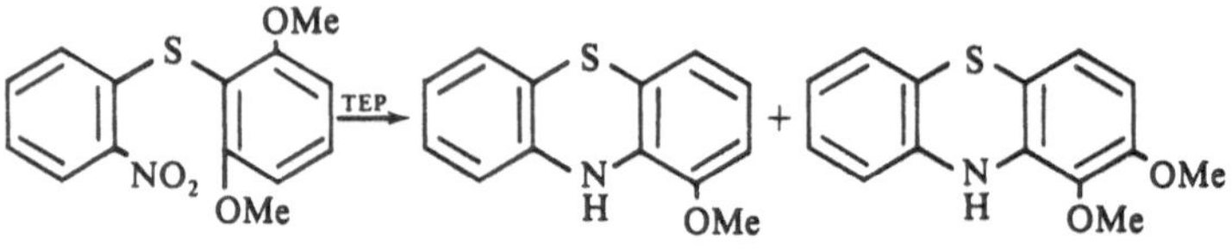

On the other hand, two *ortho*'–methoxy blocking groups lead to loss or migration of a methoxy to give **74** and **75**, respectively.[145]

Similar results have been obtained using diaryl ethers[146] and *N*-acetyldiphenylamines[147] as starting materials:

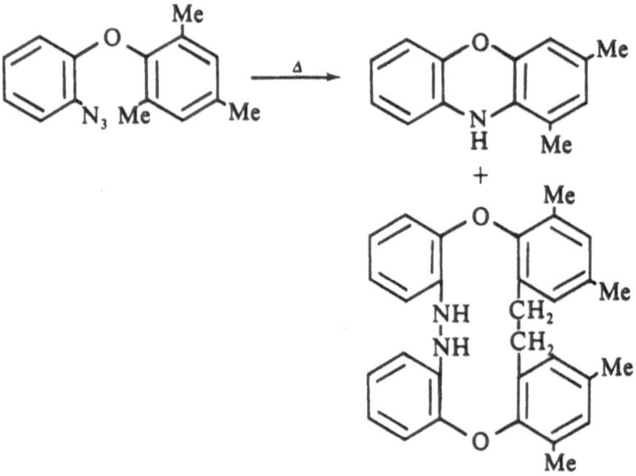

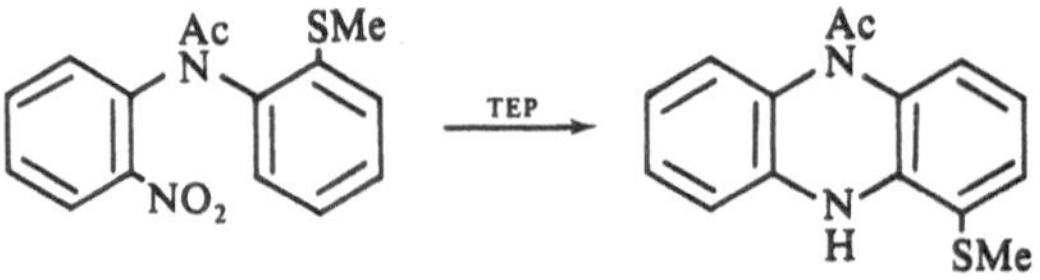

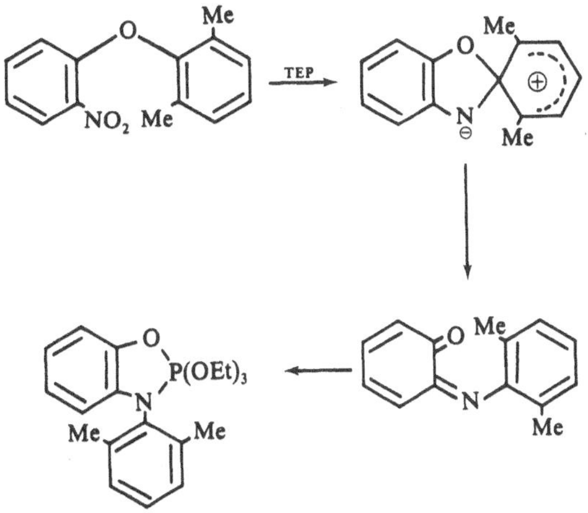

76

In the case where two *o*-methyl blocking groups are present and the spiro intermediate is generated[148] in the presence of excess triethyl phosphite, it may react with another mole of this reagent to give an oxazaphosphole **76**.

Jones has studied the decomposition of the related *o*-azidodiphenyl-methanes ($X = CH_2$) and he has found that the unsubstituted parent system undergoes thermal cyclization with ring expansion.[149] This reaction had previously been observed by Krbechek and Takimoto[150] but the structure of the product was incorrectly assigned:

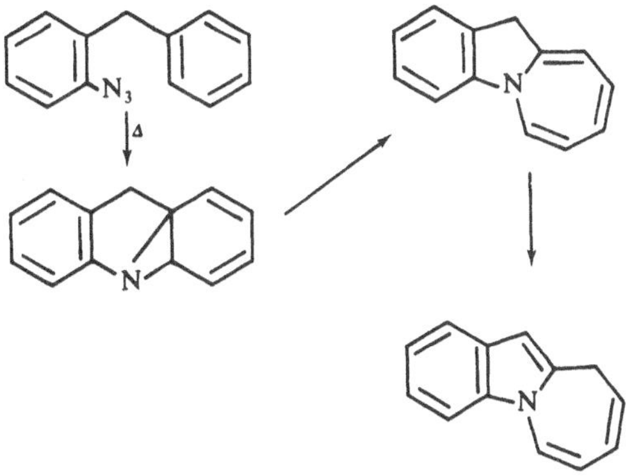

Introduction of a methoxy substituent changes the nature of the products to those predominantly of substitution. This may be the result of the *o'*-methoxy group stabilizing **77** at the expense of **78** (as has been argued in comparable situations) but it is curious that a *p'*-methoxy group does not have the same effect.[149]

Decomposition of **79** gives mainly ring substitution products rather than

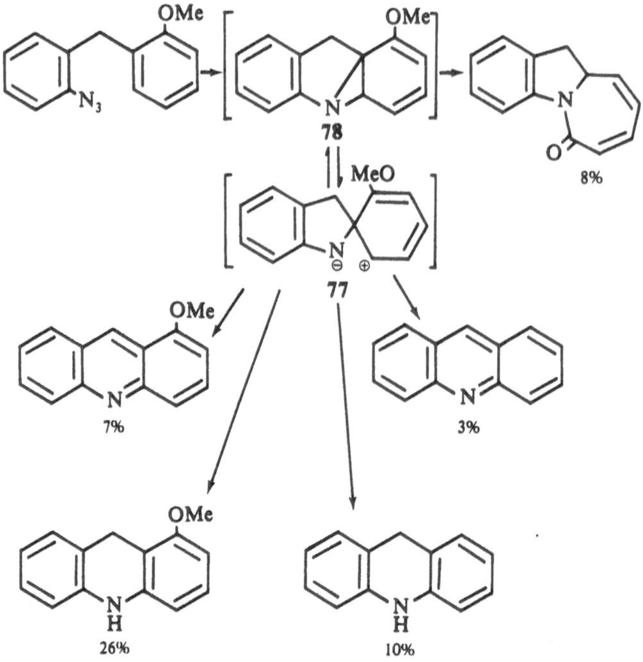

azepines probably because of loss of naphthyl resonance energy in the intermediates (80) and (81) that lead to ring expansion, e.g., 82[151]:

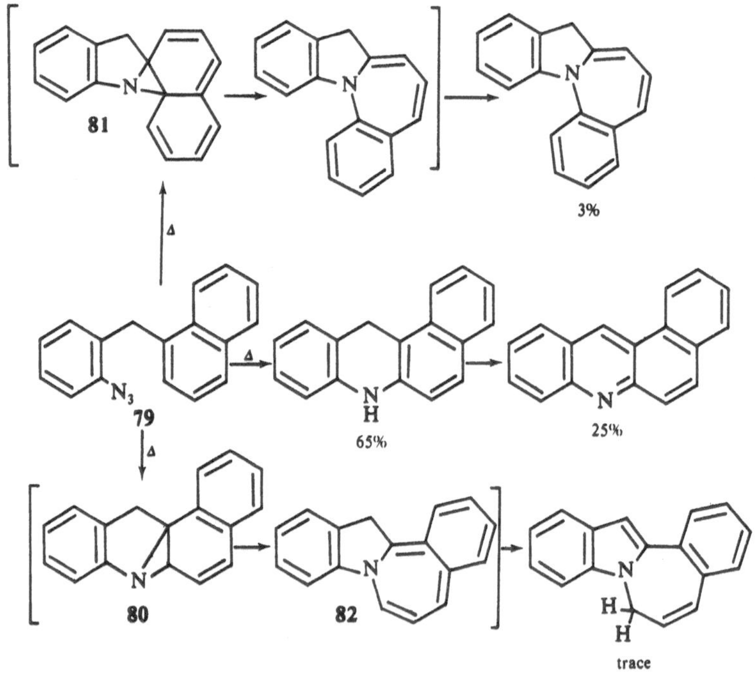

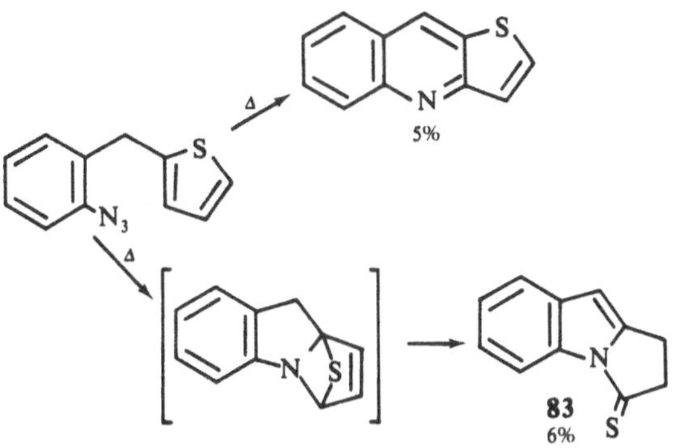

Cyclization of a nitrene onto an adjacent thiophene ring may involve a rare 1,4 addition of the nitrene to the thiophene ring to give **83**.[152]

Meth-Cohn and Suschitzky[136,153] have developed optimum conditions for the intramolecular cyclization of a triplet nitrene by insertion into an adjacent C–H bond. The cyclization of **84** is best carried out thermally at the highest convenient temperature.

The temperature at which photolysis of this azide is carried out has been shown to have a profound effect upon the nature of the products. Formation of azo compound on room temperature photolysis in preference to **85** was considered to be the reaction of a "lazy" nitrene. Raising the photolysis temperature makes the photolytically generated triplet sufficiently energetic to undergo C–H insertion to give **85**. The importance of the right choice of conditions to be used in the decomposition of aryl azides for the purpose of heterocyclic synthesis has been stressed.

Kametani[16] has reviewed the uses of nitrene cyclizations in the synthesis of heterocyclic natural products.

Many cyclizations of azides which might at first sight appear to involve nitrene intermediates have, on closer examination, been found to involve the loss of nitrogen in the rate-determining step. A consideration of these reactions is beyond the scope of this chapter.

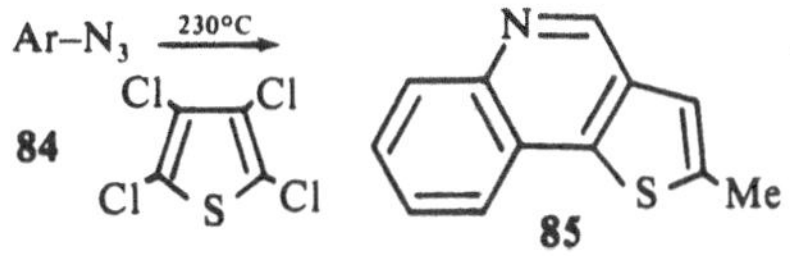

		Yields, (%)	
	64.5%	0%	22.5%
$ArN_3 \xrightarrow[\text{PhCOMe, RT}]{h\nu}$	0%	45.5%	45.5%
$ArN_3 \xrightarrow[\substack{\text{PhCOMe} \\ 107°C}]{h\nu}$	30%	0%	66.5%

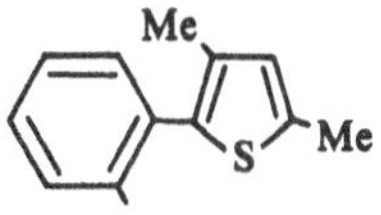

2. Heterocyclic Rearrangements

In the vapor phase, interconversions of arylnitrenes to pyridylcarbenes are well known,[154] but the corresponding reactions in solution are very rare.

Oxidation of aminopyrazoles or decomposition of azidopyrazoles usually leads to ring-opened products,[19] but recently the diaminoimidazole **86** has been found[155] to yield a mixture of triazine (**87**) and triazole (**88**) on oxidation. The intermediacy of a *C*-nitrene (**89**) was proposed but an *N*-nitrene (**90**) was not ruled out.[155] This reaction is similar to the ring expansion of triazoles to tetrazines.[156]

Another interesting ring interconversion that probably occurs via a nitrene, namely, that of a substituted thiophene to an *N*-arylpyrrole, has been reported.[157] Treatment of **91** for 14 hr with triethyl phosphite yields 3-cyano-1-

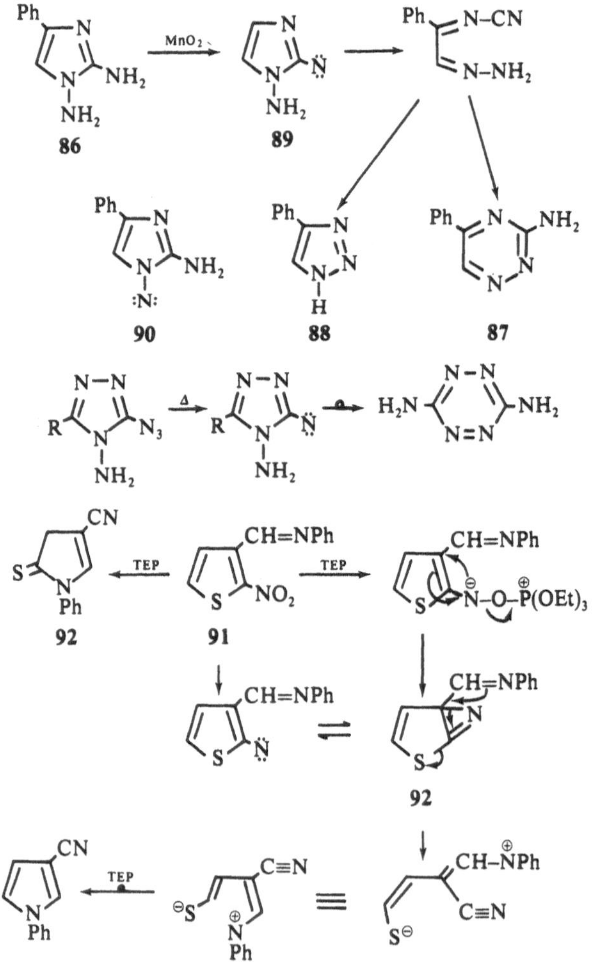

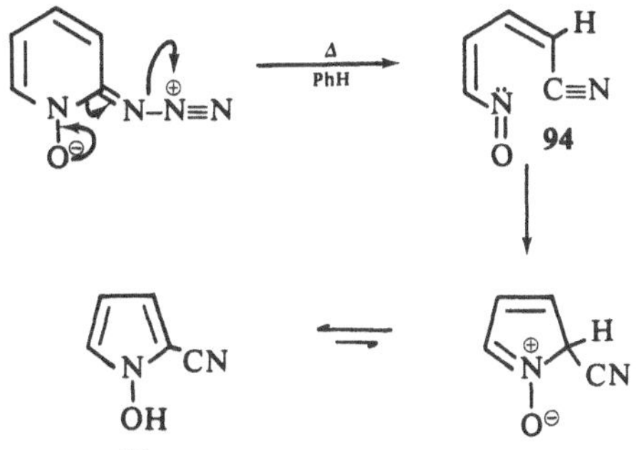

phenylpyrrole instead of the expected thienopyrazole. A mechanism was proposed (shown above), which involves a nitrenium ion, but a concerted process via **92** is equally plausible.

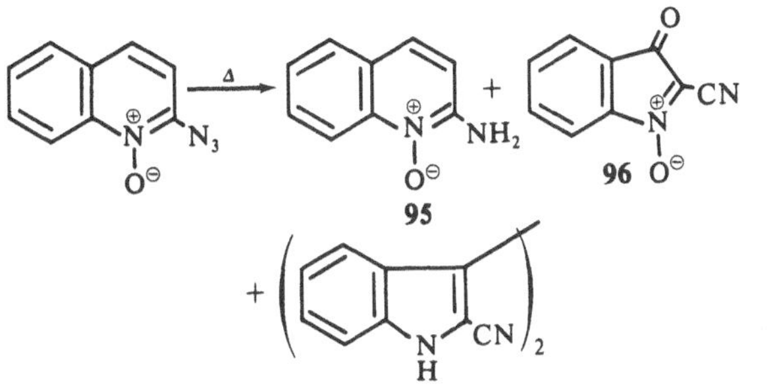

2-Azidopyridine 1-oxides undergo ring contraction on thermolysis to give the much sought after 2-cyano-*N*-hydroxypyrroles (**93**).[158] This reaction, however, is not thought to involve a nitrene.

In an inert solvent an elimination of nitrogen concerted with ring opening to give a nitrile (**94**) is proposed. The nitrile then undergoes electrocyclic ring closure and tautomerization to (**93**). This reaction appears to be fairly general for other 2-azido-*N*-oxides; for example, 2-azidopyrazine 1-oxide yields 2-cyano-1-hydroxyimidazole and 2-azidoquinoline 1-oxide gives the products **95–97**.[159]

IV. DECOMPOSITION IN THE PRESENCE OF PROTONIC AND LEWIS ACIDS

Early studies on the decompositions of aryl azides in the presence of protonic acids have been reviewed.[11] It is difficult to draw conclusions from this early work regarding nitrene involvement, as azides were heated in acids, and it is not clear if nitrenes were generated directly by thermolysis or if acid catalysis of the azide decomposition occurred.

Heating aryl azides in a mixture of a carboxylic acid and polyphosphoric acid affords a useful, fairly general method of fusing an oxazole ring to aromatics[160] and heterocycles.[161] The method suffers from the limitation that bicyclic azides in which the azido substituent is α- to the ring junction give N,O-diacetyl derivatives of the 1,4-aminohydroxy compound as, for example, the formation of 98 from α-naphthyl azide:

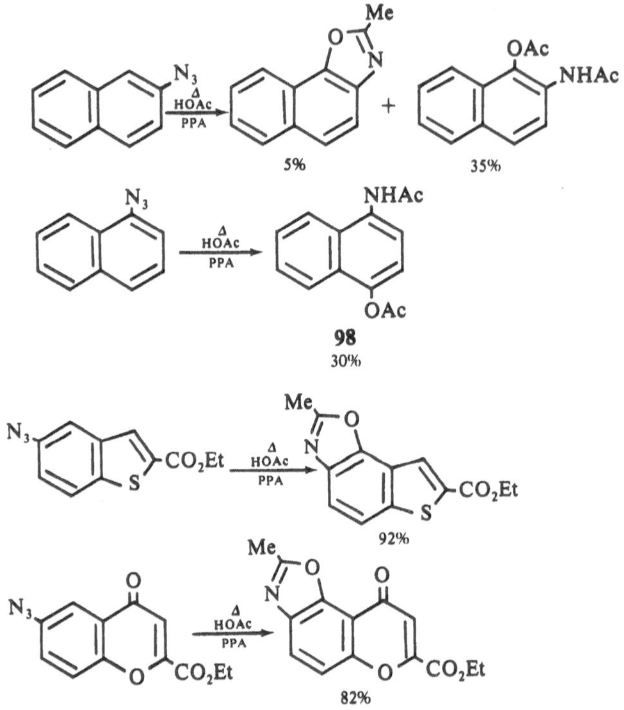

Azides that are α- to a pyridine-like nitrogen usually exist predominantly in the tetrazole form and so difficulties are often encountered in their photolysis. Photolysis of tetrazoles has been found to be facilitated by the use of trifluoroacetic acid as solvent which converts the tetrazole to azide before nitrogen is lost[162]:

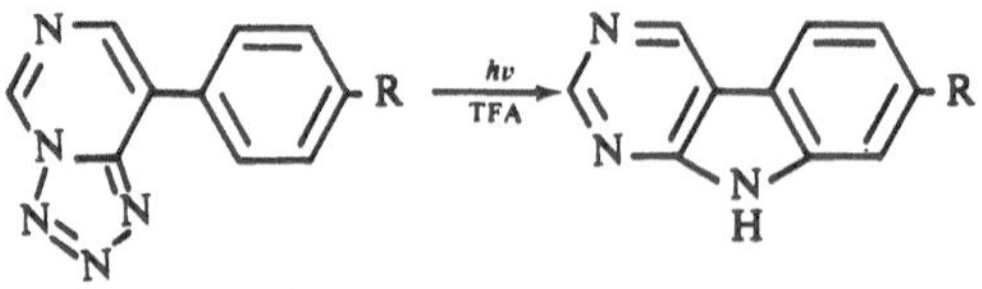

Derivatives of the new 6*H*-indolo[2,3-*b*]-1,8-naphthyridine ring system **99**, which resemble the tumor inhibitor ellipticine, have been prepared by this approach.[163]

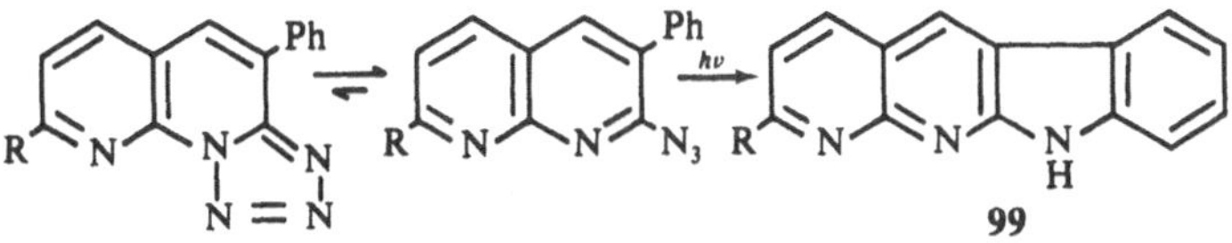

Decomposition of aryl azides in the presence of Lewis acids does not involve arylnitrenes; arylnitrenium–Lewis acid complexes are much more likely intermediates. The reaction of phenyl azide with aluminum chloride and aromatic substrates gives fair yields of diarylamines.[164] This reaction is not undergone by phenyl azide alone on thermolysis or photolysis in aromatic solvents (Section II.1.A):

$$PhN_3 \xrightarrow[PhMe]{AlCl_3} p\text{-}MeC_6H_4NHPh + PhNH_2$$

The reader is referred to a review by Gassman for an excellent treatment of the chemistry of nitrenium ions.[165]

V. CURRENT AND FUTURE APPLICATIONS

The synthetic potential of arylnitrenes has been dealt with at length in this review. The greater understanding of mechanisms of arylnitrene reactions that has been gained in recent years should lead to many further synthetic extensions.

Aryl azides have found use as photoaffinity labels in biochemistry,[166] but without the success that was at first hoped.[167] Application of some of the recent ideas of arylnitrene reactivity might lead to the development of improved labels. Aryl azides have also been exploited as labels in membrane protein studies,[168] but again the latest considerations of aryl azide chemistry, discussed in this review, have yet to be applied in this field.

Some years ago Rose[169] was the first to draw attention to the possible cytotoxic and carcinogenic role of arylnitrenes. Recently, photolysis of 2-azidofluorene *in situ* has been used as a probe for chemical carcinogenesis in mammalian cell cultures.[170] It has the advantage of bypassing metabolic activation. Clearly, one can expect to see an increase in the use of azides, which are structurally related to known carcinogens, as specific photoaffinity labels in chemical carcinogenesis studies.

APPENDIX

Recently new work has appeared that concerns the relative importance of benzazirine and didehydroazepine (azacycloheptatetraene) intermediates in aryl azide decompositions. Dunkin and Thomson[171] have published the first report of the direct observation of a benzazirine intermediate. They photolyzed α- (and β-) naphthyl azide in nitrogen or argon matrices at 12° K and observed, by ir, a naphthazirine **100** which collapsed to a more stable didehydrobenzazepine **101**. Formation of the isomers **100a** and **101a** could not be precluded in the matrix photolysis study, but products consistent with their trapping in solution photolyses have never been detected.[92,124]

Wentrup and coworkers[172] have demonstrated that a common seven membered ring carbodiimide **102** formed via nitrene intermediates is involved in gas phase, solution, and matrix decompositions of certain tetrazoles. Therefore, it appears more reasonable than formerly to apply to azide decompositions in solution information gained from gas phase and matrix studies. Photolysis of α-naphthyl azide involves formation of naphthazirine **100**, probably via singlet nitrene, that can undergo attack by a nucleophile or rearrange to a didehydroazepine **101** before nucleophilic attack. Which intermediate actually undergoes nucleophilic attack is still not clear, also of course both pathways might be operative. This will depend upon a competition between the rate of rearrangement of **100** to **101** (an intramolecular process) and the rate of nucleophilic attack on **100** (an intermolecular process). In general, three factors might be expected to control the outcome of such a competition, they are: (1) the structure of the starting aryl azide, (2) the strength of the nucleophile, and (3) the nature of the conditions and solvent used.

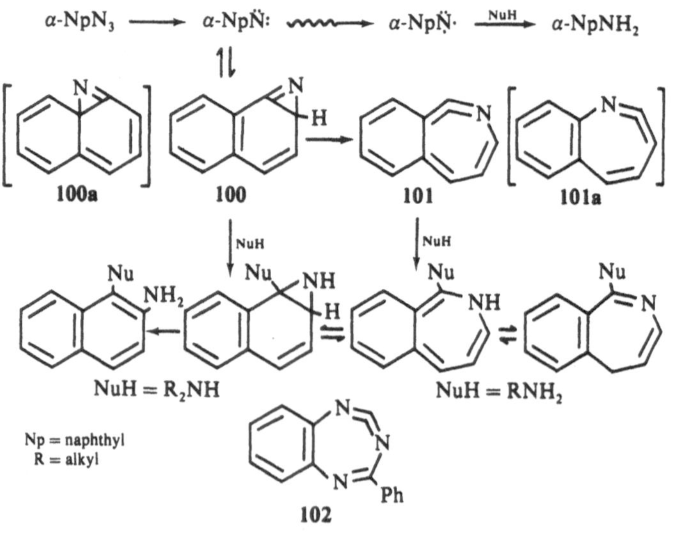

On evidence currently available, it may be said that a didehydroazepine is as likely as an azirine to be the intermediate that undergoes nucleophilic attack in bicyclic aryl azide decompositions and is probably the preferred intermediate when dealing with monocyclic aryl azides. An exception to this generalization occurs when the azide decomposed contains "vinyl azide character," for example 9-azidophenanthrene and 6-azido-1,3-dimethyluracil (Section II.3.A), in these cases azirines are almost certainly the reactive species.

ACKNOWLEDGMENTS

It remains for the author to acknowledge his gratitude to Professor R. A. Abramovitch for introducing him to nitrene chemistry, for his continuing encouragement, and to Professor H. Suschitzky and colleagues at Salford for happy and fruitful research collaboration over the last few years.

REFERENCES

1. F. Tiemann, *Ber.* **24**, 4162 (1891).
2. J. Stieglitz, *Am. Chem. J.* **18**, 751 (1896).
3. T. Curtius, *J. Prakt. Chem.* **125**, 23 (1930).
4. W. Kirmse, *Angew. Chem.* **71**, 537 (1959).
5. L. Horner and A. Christmann, *Angew. Chem.* **75**, 450 (1959).
6. R. A. Abramovitch and B. A. Davis, *Chem. Rev.* **64**, 149 (1964).
7. W. Lwowski, *Angew. Chem. Int. Ed. Engl.* **6**, 897 (1967).
8. J. I. G. Cadogan, *Q. Rev. Chem. Soc.* **22**, 222 (1968).
9. J. I. G. Cadogan, *Synthesis* **1**, 11 (1969).
10. R. A. Abramovitch, *Chem. Soc. (London) Spec. Publ.* **24**, 323 (1970); R. A. Abramovitch and R. G. Sutherland, *Top. Curr. Chem.* **16**, 1 (1970).
11. R. A. Abramovitch and E. P. Kyba, in *The Chemistry of the Azido Group*, S. Patai, Ed., Wiley New York (1971), pp. 221–329.
12. R. Belloli, *J. Chem. Ed.* **48**, 423 (1971).
13. J. I. G. Cadogan, *Acc. Chem. Res.* **5**, 303 (1972).
14. R. A. Abramovitch, in *Organic Reactive Intermediates*, Vol. 26, S. P. McManus, Ed., Academic Press, New York (1973), pp. 127–192.
15. H. Dürr and H. Kober, *Top. Curr. Chem.* **66**, 89 (1976).
16. T. Kametani, E. F. Ebetino, T. Yamanaka, and K. Nyu, *Heterocycles* **2**, 209 (1974).
17. T. L. Gilchrist and C. W. Rees, in *Carbenes, Nitrenes and Arynes*, Appleton Century Crofts, New York (1969).
18. W. Lwowski, in *Nitrenes*, W. Lwowski, Ed., Wiley, New York (1970).
19. P. A. S. Smith, in *Nitrenes*, W. Lwowski, Ed., Wiley, New York (1970), p. 99.
20. A. Lüttringhaus, J. Jander, and R. Schneider, *Chem. Ber.* **92**, 1756 (1959).
21. M. Appl and R. Huisgen, *Chem. Ber.* **92**, 2961 (1959).
22. F. O. Rice and T. A. Luckenbach, *J. Am. Chem. Soc.* **82**, 2681 (1960).
23. G. Smolinsky, *J. Am. Chem. Soc.* **83**, 2489 (1961).
24. I. L. Knunyants and E. G. Bykhovskaya, *Proc. Acad. Sci. USSR* **132**, 513 (1969).

25. P. A. S. Smith, L. O. Krbechek, and W. Resemann, in *Abstracts of the 144th National Meeting of the American Chemical Society, Los Angeles, California*, American Chemical Society, Washington, D.C. (1963), p. 35M.

26. R. A. Abramovitch, Y. Ahmed, and D. Newman, *Tetrahedron Lett.*, 752 (1961).

27. P. Walker and W. A. Waters, *J. Chem. Soc.*, 1632 (1962).

28. P. A. S. Smith and J. H. Hall, *J. Am. Chem. Soc.* **84**, 480 (1962).

29. G. L'Abbé, *Chem. Rev.* **69**, 345 (1969).

30. (a) L. K. Dyall and J. E. Kemp, *J. Chem. Soc. B*, 976 (1968); (b) L. K. Dyall, *Austral. J. Chem.* **28**, 2147 (1975); (c) L. K. Dyall, *Austral. J. Chem.* **30**, 2669 (1977).

31. J. Ashby and H. Suschitzky, *Tetrahedron Lett.*, 1315 (1971).

32. J. H. Hall and F. E. Behr, *J. Am. Chem. Soc.* **94**, 4952 (1972); J. H. Hall, F. E. Behr, and R. L. Reed, *J. Org. Chem.* **37** 4952 (1972).

33. A. Reiser, G. Bowes, and R. J. Horne, *Trans. Faraday Soc.* **62**, 3162 (1966).

34. G. Smolinsky, E. Wasserman, and W. A. Yager, *J. Am. Chem. Soc.* **84**, 3220 (1962).

35. A. Reiser, G. C. Terry, and F. W. Willets, *Nature (London)* **211**, 410 (1966).

36. T. de Boer, J. I. G. Cadogan, H. M. McWilliam, and A. G. Rowley, *J. Chem. Soc. Perkin Trans. 2*, 554 (1975).

37. J. I. G. Cadogan and A. Cooper, *J. Chem. Soc. B*, 883 (1969).

38. P. K. Brooke, R. B. Herbert, and F. G. Holliman, *Tetrahedron Lett.*, 761 (1973).

39. T. Kurihara, E. Okada, and M. Akagi, *Yakugaku Zasshi* **92**, 1557 (1972); *Chem. Abstr.* **78**, 58, 335 (1973).

40. M. Masaki, K. Fukui, and J. Kita, *Bull. Chem. Soc. Japan* **50**, 2013 (1977).

41. H. C. Waterman and D. L. Vivian, *J. Org. Chem.* **14**, 289 (1949).

42. R. A. Abramovitch and K. A. H. Adams, *Can. J. Chem.* **39**, 2516 (1961).

43. F.-P. Tsui, T. M. Vogel, and G. Zon, *J. Org. Chem.* **40**, 761 (1975).

44. E. Meyer and G. W. Griffin, *Angew. Chem.* **79**, 648 (1967); J. S. Splitter and M. Calvin, *Tetrahedron Lett.*, 1445 (1968).

45. P. A. S. Smith, B. B. Brown, R. K. Putney, and R. F. Reinisch, *J. Am. Chem. Soc.* **75**, 6335 (1953).

46. R. Kwok and P. Pranc, *J. Org. Chem.* **33**, 2880 (1968); A. Kliegl, *Ber.* **42**, 591 (1909); P. L. Coe, A. E. Jukes, and J. C. Tatlow, *J. Chem. Soc. C*, 2020 (1966).

47. M. Ogata, H. Matsumoto, and K. Kanō, *Tetrahedron* **25**, 5205 (1969).

48. R. Y. Ning, J. F. Blount, P. B. Madan, and R. I. Fryer, *J. Org. Chem.* **42**, 1791 (1977).

49. W. Heinzelmann, M. Märky, and P. Gilgen, *Helv. Chim. Acta* **59**, 2362 (1976).

50. O. Fischer and O. Heiler, *Ber.* **26**, 378 (1893).

51. F. Schröppel and J. Sauer, *Tetrahedron Lett.*, 2945 (1974).

52. F. P. Tsui, H. Y. Chang, T. M. Vogel, and G. Zon, *J. Org. Chem.* **41**, 3381 (1976).

53. S. Goldschmidt and L. Strohmenger, *Chem. Ber.* **55**, 2450 (1922).

54. C. A. Wilkie and D. R. Dimmel, *J. Am. Chem. Soc.* **94**, 8600 (1972).

55. A. Reiser, G. Bowes, and R. J. Horne, *Trans, Faraday Soc.* **62**, 3162 (1966).

56. A. Reiser, G. C. Terry, and F. W. Willets, *Nature (London)* **211**, 410 (1966); A. Reiser, F. W. Willets, G. C. Terry, V. Williams, and R. Marley, *Trans. Faraday Soc.* **64**, 3265 (1968); G. Smolinsky, E. Wasserman, and W. A. Yager, *J. Am. Chem. Soc.* **84**, 3220 (1962); G. Smolinsky, L. C. Snyder, and E. Wasserman, *Rev. Mod. Phys.* **35**, 576 (1963).

57. A. Reiser and L. Leyshon, *J. Am. Chem. Soc.* **92**, 7487 (1970).

58. J. H. Hall, J. W. Hill, and J. M. Fargher, *J. Am. Chem. Soc.* **90**, 5313 (1968); J. H. Hall, J. W. Hill, and Hu-Chu Tsai, *Tetrahedron Lett.*, 2211 (1965).

59. L. Wolff, *Justus Liebigs Ann. Chem.* **394**, 59 (1912).

60. R. Huisgen, D. Vossius, and M. Appl, *Chem. Ber.* **91**, 1 (1958).

61. W. von E. Doering and R. A. Odum, *Tetrahedron* **22**, 81 (1966).

62. O. L. Chapman and J.-P. Le Roux, *J. Am. Chem. Soc.* **100**, 283 (1978).

63. A. Hassner and F. W. Fowler, *J. Am. Chem. Soc.* **90**, 2869 (1968); E. F. Ullman and B. Singh, *J. Am. Chem. Soc.* **88**, 1844 (1966); D. W. Kurtz and H. Schechter, *Chem. Commun.*, 689 (1966); H. J. Bestman and R. Kunstmann, *Angew. Chem. Int. Ed. Engl.* **5**, 1039 (1966).
64. L, Salem, *J. Am. Chem. Soc.* **96**, 3486 (1974).
65. D. D. Schillady and C. Trindle, *Theor. Chim. Acta* **43**, 137 (1976).
66. A. Bertho, *Chem. Ber.* **57**, 1138 (1924).
67. R. A. Abramovitch and E. F. V. Scriven, *Chem. Commun.*, 787 (1970).
68. R. A. Abramovitch, S. R. Challand, and E. F. V. Scriven, *J. Am. Chem. Soc.* **94**, 1374 (1972).
69. R. A. Abramovitch, S. R. Challand, and E. F. V. Scriven, *J. Org. Chem.* **37**, 2705 (1972).
70. R. Huisgen and K. von Fraunberg, *Tetrahedron Lett.*, 2595 (1969).
71. R. E. Banks and G. R. Sparkes, *J. Chem. Soc. Perkin Trans. 1*, 2964 (1972); R. E. Banks and A. Prakash, *J. Chem. Soc. Perkin Trans. 1*, 1365 (1974).
72. R. A. Abramovitch and S. R. Challand, *J. Heterocycl. Chem.* **10**, 683 (1973).
73. R. A. Abramovitch and S. R. Challand, *J. Chem. Soc. Chem. Commun.*, 1160 (1972); R. A. Abramovitch, S. R. Challand, and Y. Yamada, *J. Org. Chem.* **40**, 1541 (1975).
74. R. A. Abramovitch, C. I. Azogu, and R. G. Sutherland, *J. Chem. Soc. Chem. Commun.*, 134 (1971).
75. H. Yamada, H. Shizuka, and K. Matsui, *J. Org. Chem.* **40**, 1351 (1975).
76. R. P. Bennett and W. B. Hardy, *J. Am. Chem. Soc.* **90**, 3295 (1968); G. Ribaldone, G. Capara, and G. Borsatti, *Chim. Ind. (Milan)* **50**, 1200 (1968).
77. R. A. Odum and A. M. Aaronson, *J. Am. Chem. Soc.* **91**, 5680 (1969).
78. R. A. Odum and G. Wolf, *J. Chem. Soc. Chem. Commun.*, 360 (1973).
79. E. F. V. Scriven, H, Suschitzky, and G. V. Garner, *Tetrahedron Lett.*, 103 (1973).
80. R. A. Abramovitch and T. Takaya, *J. Org. Chem.* **37**, 2022 (1972).
81. P. A. S. Smith and H. Dounchis, *J. Org. Chem.* **38**, 2958 (1973).
82. A. Bertho, *Chem. Ber.* **57**, 1138 (1934).
83. P. A. S. Smith and J. H. Hall, *J. Am. Chem. Soc.* **84**, 480 (1962).
84. G. Smolinsky, *J. Org. Chem.* **26**, 4108 (1961).
85. J. H. Hall, J. W. Hill, and J. Fargher, *J. Am. Chem. Soc.* **90**, 5313 (1968).
86. L. Horner, A. Christmann, and A. Gross, *Chem. Ber.* **96**, 399 (1963).
87. H. Dounchis, Ph.D. dissertation, University of Michigan (1967).
88. J. H. Boyer and G. J. Mikol, *Chem. Commun.*, 734 (1969).
89. A. Yabe and K. Honda, *Tetrahedron Lett.*, 1079 (1975).
90. S. E. Hilton, E. F. V. Scriven, and H. Suschitzky, *J. Chem. Soc. Chem. Commun.*, 853 (1974).
91. B. Nay, Ph.D. dissertation, University of Salford (1977).
92. L. A. Neiman, V. I. Maimind, and M. M. Shemyakin, *Izvest. Akad. Nauk S.S.S.R. Ser. Khim.*, 1357 (1964).
93. A. B. Bulacinski, B. Nay, E. F. V. Scriven, and H. Suschitzky, *Chem. Ind. (London)*, 746 (1975).
94. E. G. Janzen, *Acc. Chem. Res.* **4**, 31 (1971).
95. A. B. Bulacinski, E. F. V. Scriven, and H. Suschitzky, *Tetrahedron Lett.*, 3577 (1975).
96. R. A. Abramovitch and S. R. Challand, *J. Chem. Soc. Chem. Commun.*, 964 (1972).
97. J. S. Brinen and B. Singh, *J. Am. Chem. Soc.* **93**, 6623 (1971).
98. R. A. Abramovitch, C. I. Azogu, and R. G. Sutherland, *J. Chem. Soc. Chem. Commun.*, 134 (1971).
99. J. Ashby, E. F. V. Scriven, and H. Suschitzky, *J. Chem. Soc. Chem. Commun.*, 366 (1972).
100. V. L. Vlasova, L. S. Kabrina, and G. G. Yakobson, *Zh. Org. Khim.* **7**, 1224 (1971).
101. D. Price, H. Suschitzky, and J. I. Hollies, *J. Chem. Soc. C*, 1967 (1969).

102. R. J. Sundberg, S. R. Suter, and M. Brenner, *J.Am. Chem. Soc.* **94**, 513 (1972).

103. B. A. De Graff, D. W. Gillespie, and R. J. Sundberg, *J. Am. Chem. Soc.* **96**, 7491 (1974).

104. E. F. V. Scriven and D. R. Thomas, *Chem. Ind. (London)*, 385 (1978).

105. S. E. Carroll, B. Nay, E. F. V. Scriven, H. Suschitzky, and D. R. Thomas, *Tetrahedron Lett.*, 3175 (1977).

106. S. E. Carroll, B. Nay, E. F. V. Scriven, and H. Suschitzky, *Synthesis*, 710 (1975).

107. J. Rigaudy, C. Igier, and J. Barcelo, *Tetrahedron Lett.*, 3845 (1975).

108. S. Senda, K. Hirota, M. Suzuki, T. Asao, and K. Maruhashi, *J. Chem. Soc. Chem. Commun.*, 731 (1976).

109. S. Senda, K. Hirota, and T. Asao, *Tetrahedron Lett.*, 1531 (1978).

110. Y. Tamura, Y. Yoshimura, T. Nishimura, S. Kato, and Y. Kita, *Tetrahedron Lett.*, 351 (1973).

111. A. Hassner and F. W. Fowler, *J. Am. Chem. Soc.*, **90**, 2869 (1968).

112. B. Nay, E. F. V. Scriven, H. Suschitzky, D. R. Thomas, and S. E. Carroll, *Tetrahedron Lett.*, 1811 (1977).

113. R. N. Carde and G. Jones, *J. Chem. Soc. Perkin Trans. 1*, 519 (1975).

114. R. J. Sundberg and R. H. Smith, *J. Org. Chem.* **36**, 295 (1971).

115. A. C. Mair and M. F. G. Stevens, *J. Chem. Soc. C*, 2317 (1971); R. Purvis, R. K. Smalley, W. A. Strachan, and H. Suschitzky, *J. Chem. Soc. Perkin Trans. 1*, 191 (1978).

116. S. E. Carroll, B. Nay, E. F. V. Scriven, and H. Suschitzky, *Tetrahedron Lett.*, 943 (1977).

117. R. S. Atkinson, in *Special Periodic Report on Aromatic and Heterocyclic Chemistry*, Vol. 6, Chemical Society, London (1978), p. 237.

118. I. M. McRobbie, O. Meth-Cohn, and H. Suschitzky, *Tetrahedron Lett.*, 929 (1976).

119. R. J. Sundberg and R. W. Heintzelman, *J. Org. Chem.* **39**, 2546 (1974).

120. B. Iddon, H. Suschitzky, and D. S. Taylor, *J. Chem. Soc. Perkin Trans. 1*, 579 (1974).

121. B. Iddon, M. W. Pickering, H. Suschitzky, and D. S. Taylor, *J. Chem. Soc. Perkin Trans. 1*, 1686 (1975).

122. B. Nay, E. F. V. Scriven, H. Suschitzky, and Z. U. Khan, *Synthesis*, 757 (1977).

123. Z. U. Khan and E. F. V. Scriven, unpublished results, University of Salford, United Kingdom (1977).

124. R. Colman and E. F. V. Scriven, unpublished results, University of Salford, United Kingdom (1978).

125. S. Senda, K. Hirota, T. Asao, and K. Maruhashi, *J. Am. Chem. Soc.* **99**, 7358 (1977).

126. F. Hollywood, E. F. V. Scriven, H. Suschitzky, D. R. Thomas, and R. Hull, *J. Chem. Soc. Chem. Commun.*, 806 (1978).

127. L. Wolff and L. Lindenhayn, *Ber.* **37**, 2374 (1904); M. O. Forster and H. M. Judd, *J. Chem. Soc.* **97**, 254 (1910); H. Bretschneider and H. Rager, *Monatsh. Chem.* **81**, 981 (1950).

128. T. M. Paterson, R. K. Smalley, and H. Suschitzky, *Tetrahedron Lett.*, 3973 (1977).

129. R. J. Sundberg and R. H. Smith, *J. Org. Chem.* **36**, 295 (1971).

130. R. K. Smalley and H. Suschitzky, *J. Chem. Soc.*, 5571 (1963).

131. R. K. Smalley, W. A. Strachan, and H. Suschitzky, *Tetrahedron Lett.*, 825 (1974).

132. S. Senda, K. Hirota, T. Asao, and K. Maruhashi, *J. Chem. Soc. Chem. Commun.*, 367 (1978).

133. K. Senga, M. Ichiba, and S. Nishigaki, *Heterocycles* **6**, 1915 (1977).

134. J. M. Lindley, I. M. McRobbie, O. Meth-Cohn, and H. Suschitzky, *J. Chem. Soc. Perkin Trans. 1*, 2194 (1977); I. M. McRobbie, O. Meth-Cohn, and H. Suschitzky, *J. Chem. Research* **S**, 17 (1977); I. M. McRobbie, O. Meth-Cohn, and H. Suschitzky, *Tetrahedron Lett.*, 925 (1976).

135. G. Smolinsky, *J. Am. Chem. Soc.* **83**, 2489 (1961); A. Reiser, H. M. Wagner, and G. Bowes, *Tetrahedron Lett.*, 2635 (1966); A. Reiser, G. Bowes, and R. J. Horne, *Trans. Faraday Soc.* **62**, 3162 (1966); A. Reiser, F. W. Willets, G. C. Terry, V. Williams, and R.

Marley, *Trans. Faraday Soc.* **64**, 3265 (1968); P. A. Lehman and R. S. Berry, *J. Am. Chem. Soc.* **95**, 8614 (1973); J. S. Swenton, *Tetrahedron Lett.*, 3421 (1968); J. S. Swenton, I. J. Ikeler, and B. H. Williams, *J. Am. Chem. Soc.* **92**, 3103 (1970); R. J. Sundberg, M. Brenner, S. R. Suter, and B. P. Das, *Tetrahedron Lett.*, 2715 (1970); R. J. Sundberg and R. W. Heintzelman, *J. Org. Chem.* **39**, 2546 (1974); R. J. Sundberg, D. W. Gillespie, and B. A. De Graff, *J. Am. Chem. Soc.* **97**, 6193 (1975).

136. K. E. Chippendale, B. Iddon, and H. Suschitzky, *J. Chem. Soc. Perkin Trans. 1*, 2023 1972).

137. F. M. Dean, C. Patampongse, and V. Podimuhang, *J. Chem. Soc. Perkin Trans. 1*, 583 (1974).

138. A. H. Jackson, D. N. Johnston, and P. V. R. Shannon, *J. Chem. Soc. Chem. Commun.*, 911 (1975).

139. A. J. Nunn and F. J. Rowell, *J. Chem. Soc. Perkin Trans. 1*, 629 (1975).

140. O. Tsuge and H. Samura, *J. Heterocycl. Chem.* **8**, 707 (1971).

141. A. Spagnolo, A. Tundo, and P. Zanirato, *J. Org. Chem.* **42**, 292 (1977).

142. R. Y. Ning, P. B. Madan, and L. H. Sternbach, *J. Org. Chem.* **38**, 3995 (1973).

143. R. A. Abramovitch and J. Kalinowski, *J. Heterocycl. Chem.* **9**, 409 (1972).

144. R. A. Abramovitch, C. I. Azogu, I. T. McMaster, and D. P. Vanderpool, *J. Org. Chem.* **43**, 1218 (1978).

145. J. I. G. Cadogan and S. Kulik, *J. Chem. Soc. C*, 2621 (1971).

146. J. I. G. Cadogan and P. K. K. Lim, *J. Chem. Soc. Chem. Commun.*, 1431 (1971).

147. Y. Maki, T. Hosokami, and M. Suzuki, *Tetrahedron Lett.*, 3509 (1971).

148. J. I. G. Cadogan, D. S. B. Grace, P. K. K. Lim, and B. S. Tait, *J. Chem. Soc. Perkin Trans. 1*, 2376 (1975).

149. G. R. Cliff and G. Jones, *J. Chem. Soc. C*, 3418 (1971).

150. L. Krbechek and H. Takimoto, *J. Org. Chem.* **33**, 4286 (1968).

151. R. N. Carde and G. Jones, *J. Chem. Soc. Perkin Trans. 1*, 2066 (1974).

152. G. R. Cliff, G. Jones, and J. M. Wollard, *J. Chem. Soc. Perkin Trans. 1*, 2072 (1974).

153. I. M. McRobbie, O. Meth-Cohn, and H. Suschitzky, *Tetrahedron Lett.*, 929 (1976).

154. C. Wentrup, in *Reactive Intermediates*, Vol. 1, R. A. Abramovitch, Ed., Plenum, New York (1980), Chap. 4, p. 263.

155. R. Hisada, M. Nakajima, and J. P. Anselme, *Tetrahedron Lett.*, 903 (1976); A. V. Zeiger and M. M. Joullié, *J. Org. Chem.* **42**, 542 (1977).

156. H. H. Takimoto and G. C. Denault, *Tetrahedron Lett.*, 5369 (1966).

157. V. M. Colburn, B. Iddon, H. Suschitzky, and P. T. Gallagher, *J. Chem. Soc. Chem. Commun.*, 453 (1978); V. M. Colburn, B. Iddon, H. Suschitzky, and P. T. Gallagher, *J. Chem. Soc. Trans. 1*, 1337 (1979),

158. R. A. Abramovitch and B. W. Cue, *J. Org. Chem.* **38**, 173 (1973).

159. R. A. Abramovitch and B. W. Cue, *Heterocycles*, 227 (1974).

160. E. B. Mullock and H. Suschitzky, *J. Chem. Soc. C*, 1937 (1968).

161. M. I. Chohan, A. O. Fitton, B. T. Hatton, and H. Suschitzky, *J. Chem. Soc. C*, 3079 (1971); B. Iddon, M. W. Pickering, H. Suschitzky, and D. S. Taylor, *J. Chem. Soc. Perkin Trans. 1*, 1686 (1975); B. Iddon, H. Suschitzky, D. S. Taylor, and M. W. Pickering, *J. Chem. Soc. Perkin Trans. 1*, 575 (1974).

162. J. A. Hyatt and J. S. Swenton, *J. Heterocycl. Chem.* **9**, 409 (1972).

163. A. Da Settimo, G. Primafiore, V. Santerini, G, Biagi, and L. D'Amico, *J. Org. Chem.* **42**, 1725 (1977).

164. W. Borsche and H. Hahn, *Chem. Ber.* **82**, 260 (1949).

165. P. G. Gassman, *Acc. Chem. Res.* **3**, 26 (1970).

166. H. Bayley and J. R. Knowles, *Methods Enzymol.* **46**, 69 (1977).

167. J. R. Knowles, *Acc. Chem. Res.* **5**, 155 (1972).

168. H. Bayley and J. R. Knowles, *Biochemistry* **17**, 2414 (1978); K. M. Abu-Salah and J. B. C. Findlay, *Biochem. J.* **14**, 1736 (1976).
169. F. L. Rose, *Nature* **215**, 1492 (1967).
170. A. M. Sarrif, W. E. White, and N. Di Vito, *Biochem. Biophys. Res. Commun.* **83**, 506 (1978).
171. I. R. Dunkin and P. C. P. Thomson, *J.C.S. Chem. Commun.*, 499 (1980).
172. C. Wentrup, C. Thétaz, E. Tagliaferri, H. J. Lindner, B. Kitschke, H.-W. Winter, and H. P. Reisenauer, *Angew. Chem. Int. Ed. Engl.* **19**, 566 (1980); C. Wentrup and H.-W. Winter, *J. Am. Chem. Soc.* **102**, 6159 (1980).

2

Nitrile Ylides and Nitrenes from 2H-Azirines

Albert Padwa and Per H. J. Carlsen

2H-Azirines represent versatile substrates which can serve as useful precursors for the synthesis of other heterocyclic rings.[1-7] An unusual feature of this three-membered heterocyclic ring is that it is susceptible to attack by both electrophilic and nucleophilic reagents.[7] In addition, the 2π electrons present in the ring are known to participate in thermally allowed $[_\pi 2_s + _\pi 4_s]$ cycloadditions as the dienophile component.[8,9] Another intriguing aspect of this ring system is that it can participate as a dipolarophile in 1,3-dipolar cycloaddition reactions.[10-12] Reaction with diazoalkanes[10-12] and nitrile oxides[10] transforms the 2H-azirine into allylic azides and carbodiimides, respectively. Few reactions rival cycloadditions in the number of bonds that undergo transformation during the reaction, producing products considerably more complex than the reactants. Cycloaddition reactions utilizing 2H-azirines include thermal reactions with ketenes,[5,6] ketenimines,[5] cyclopentadienones,[13,14] cyclopentadiene,[15] and diphenylisobenzofuran[16,17] to yield a variety of unusual heterocyclic ring systems. 2H-Azirines also react photochemically with various carbon–carbon and hetero double bonds to give five-membered heterocyclic rings.[18,19] The photoreaction proceeds by way of irreversible opening of the azirine ring to form a nitrile ylide intermediate which is subsequently trapped by a suitable dipolarophile.[20] Products formed on thermal excitation of the 2H-azirine system, on the other hand, appear to involve vinyl nitrenes as intermediates.[21-32] The intent of this chapter is to provide an indication of the wide diversity and intriguing transformations that 2H-azirines undergo upon electronic and thermal excitation. It is hoped that an overview of the major developments and potentialities in this area will be attained.

Albert Padwa and Per H. J. Carlsen ● *Department of Chemistry, Emory University, Atlanta, Georgia 30322.*

I. GENERATION OF NITRILE YLIDES

The ultraviolet absorption spectra of substituted arylazirines (1) in cyclohexane exhibit strong ($\varepsilon \sim 10{,}000$) absorption at ~ 240 nm with a weak inflection on the long-wavelength side of the principal absorption band [i.e., ~ 285 nm ($\varepsilon \sim 500$)]. This weak band undergoes a hypsochromic shift with increasing polarity of the medium. By the empirical criteria of low intensity and shift of the absorption to shorter wavelengths in hydrogen-bonding solvents, this latter band can be attributed to an $n-\pi^*$ transition. The weakening of the C–C bond of the azirine ring as a result of this transition was rationalized[33-36] in terms of an electrocyclic transformation by analogy with the cyclopropyl $\rightarrow$ allyl cation rearrangement. One may envisage the $n-\pi^*$ excitation process as leading to a species resembling structure 4, where the nonbonding orbital on nitrogen contains only one electron. The electron that was promoted to the antibonding π^* orbital will partially reside on the carbon atom, and consequently the nitrogen atom will become somewhat electrophilic and begin to resemble an aziridinyl cation. The remaining nonbonding electron on nitrogen, which is in the plane of the σ bonds of the ring, will overlap with the back lobe of the saturated carbon and facilitate bond scission. Electron demotion of the ring-opened species 3 would then result in the formation of nitrile ylide 2. A similar nonbonding orbital overlap with an adjacent σ bond has been postulated to account for the facile α cleavage (i.e., Norrish type I reaction) encountered with carbonyl compounds.[37]

Schmid and coworkers[38,39] have reported that the irradiation of a number of substituted arylazirines in a rigid matrix at $-185°$C gave rise to a new maximum in the ultraviolet spectrum (~ 350 nm) which was attributed to a nitrile ylide. Their results showed that the dipole undergoes photochemical but not thermal reversion to the starting azirine. When the azirine was photolyzed at $-185°$C in the presence of a trapping agent such as methyl trifluoroacetate, the maximum at 350 nm was obtained again. This maximum vanished, however, upon increasing the temperature to $-160°C$, thereby indicating that cycloadduct formation is derived from a thermal 1,3-dipolar addition of the initially generated nitrile ylide with the added dipolarophile. The purely aliphatic 2,3-dipropyl-2H-azirine was also found to undergo efficient ring cleavage ($\Phi = 0.8$) to produce a nitrile ylide [$\lambda_{max} = 280$ nm ($\varepsilon > 15,000$)].[40]

A number of other methods have also been used to generate nitrile ylides. Access to this 1,3-dipole has been realized by (a) treatment of imidoyl halides (5) with base,[41] (b) thermal or photochemical elimination of phosphoric acid ester from 4,5-dihydro-1,3,5-oxazaphospholes (6),[42] (c) I,3-dipolar cycloreversion of 1-azetidines (7),[43] (d) loss of carbon dioxide from 3-oxazolin-5-ones (8)[44−47] and (e) treatment of boron-containing isonitriles (9) with base[48] (Scheme 1).

SCHEME 1

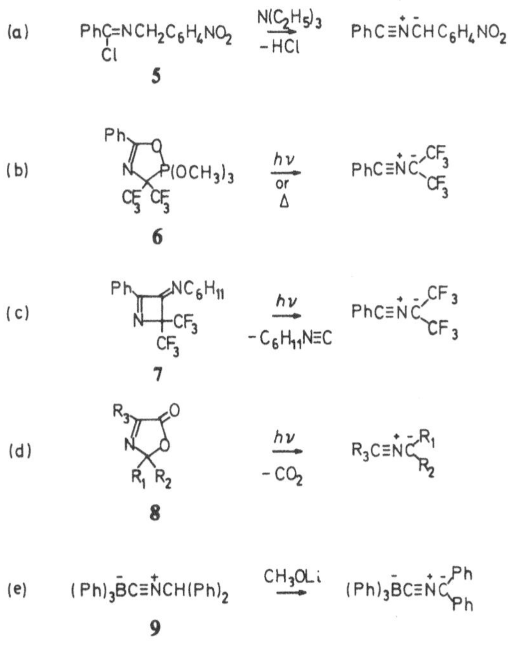

(a) $PhC(Cl)=NCH_2C_6H_4NO_2$ $\xrightarrow[-HCl]{N(C_2H_5)_3}$ $PhC\equiv\overset{+}{N}\overset{-}{C}HC_6H_4NO_2$

5

(b)   [structure **6**, Ph-substituted oxazaphosphole with $P(OCH_3)_3$ and CF_3, CF_3 groups] $\xrightarrow[\Delta]{h\nu \text{ or}}$ $PhC\equiv\overset{+}{N}\overset{-}{C}\begin{smallmatrix}CF_3\\CF_3\end{smallmatrix}$

6

(c)   [structure **7**, azetidine with Ph, NC_6H_{11}, CF_3, CF_3] $\xrightarrow[-C_6H_{11}N\equiv C]{h\nu}$ $PhC\equiv\overset{+}{N}\overset{-}{C}\begin{smallmatrix}CF_3\\CF_3\end{smallmatrix}$

7

(d)   [structure **8**, oxazolinone with R_3, R_1, R_2] $\xrightarrow[-CO_2]{h\nu}$ $R_3C\equiv\overset{+}{N}\overset{-}{C}\begin{smallmatrix}R_1\\R_2\end{smallmatrix}$

8

(e) $(Ph)_3\overset{-}{B}C\equiv\overset{+}{N}CH(Ph)_2$ $\xrightarrow{CH_3OLi}$ $(Ph)_3\overset{-}{B}C\equiv\overset{+}{N}\overset{-}{C}\begin{smallmatrix}Ph\\Ph\end{smallmatrix}$

9

Of all the methods available to generate this 1,3-dipole, the photochemical generation of nitrile ylides from 2H-azirines offers the greatest opportunity for structural variation. The 2H-azirine ring can be readily prepared in large quantities[49,50] using either the vinyl azide route developed by Hassner and coworkers[51,52] or the modified Neber reaction sequence.[53,54] 3-Aryl-2H-azirines (15) can also be obtained in good yields by reaction of alkylidene phosphoranes

(1) $\underset{\textbf{10}}{\overset{Ph}{\underset{H}{\diagdown}}\overset{H}{\underset{R}{\diagup}}}$ $\overset{\text{1. } IN_3}{\underset{\text{3. } \Delta}{\overset{\text{2. } t\text{-}\overline{B}uO^-}{\longrightarrow}}}$ $\underset{R\quad H}{\overset{Ph}{\diagup}\!\!\diagdown N}$

(2) $\underset{\textbf{11}}{\overset{N(CH_3)_2}{\underset{Ph}{\diagup}\!\!\diagdown\overset{R_1}{\underset{R_2}{\diagup}}}}$ $\underset{\text{2 Base}}{\overset{\text{1 } CH_3I}{\longrightarrow}}$ $\underset{R_1\ R_2}{\overset{Ph}{\diagup}\!\!\diagdown N}$

(3) $(Ph)_3P=CR_1R_2$ + $PhC\equiv\overset{+}{N}-\overset{-}{O}$ $\longrightarrow$ $\underset{Ph_3P\diagdown O\diagdown N}{\overset{R_1\ \overset{R_2}{\diagup}\ Ph}{\diagup}}$

       12   **13**   **14**

                                         $\underset{-\ Ph_3PO}{\overset{\Delta}{\longrightarrow}}$ $\underset{R_1\ R_2}{\overset{Ph}{\diagup}\!\!\diagdown N}$

                                                           15

(12) with nitrile oxides (13) followed by thermal extrusion of triphenylphosphine oxide.[55]

II. FEATURES OF THE PHOTOCYCLOADDITION REACTION OF 2H-AZIRINES

The photocycloaddition of 2H-azirines to electron-deficient olefins produces Δ^1-pyrrolines as primary photoproducts. Some of the dipolarophiles used include acrylic esters, acrylonitriles, fumaric and maleic esters, methyl allenecarboxylate, norbornene, and 1,2-dicyanocyclobutene.[56-60] Similarly, styrenes[58] and vinylpyridines[19] undergo smooth photocycloaddition to 2H-azirines. Addition of acetylene derivatives to the transient nitrile ylide gives 2H-pyrroles (18, 20) which rearrange to pyrroles (19) if the C-2 atom is monosubstituted.[56] 3-Phenyl-2H-azirines have been found to cycloadd to vinylphosphonium salts and to vinyl sulfones. The initial adducts (22 and 24) undergo ready loss of the phosphorous or sulfur substituent to give 2H-pyrroles (23 and 25).[60-62]

The photocycloadditions show all the characteristics of concerted reactions, including stereospecificity and regioselectivity. Concerted 1,3-dipolar additions are known to proceed via a "two-plane" orientation complex.[63] For the case of diphenylazirine and methyl acrylate, there are two possible orientation complexes (26 or 27). The interaction of substituent groups in the *syn* complex 26 can be of an attractive (π overlap, dipole–dipole interaction) or of a repulsive nature (van der Waals strain). Both effects are probably negligible in the *anti* complex 27. The ratio of the two steric courses (*syn* and *anti*) functions

16

17

18 **19**

20

21

22 **23**

24 **25**

as a probe and gives insights into the interplay of steric and electronic substituent effects in the transition state of 1,3-dipolar addition. The effect of π overlap and van der Waals strain was found to play an important role in controlling the stereochemical distribution of the products obtained.[57]

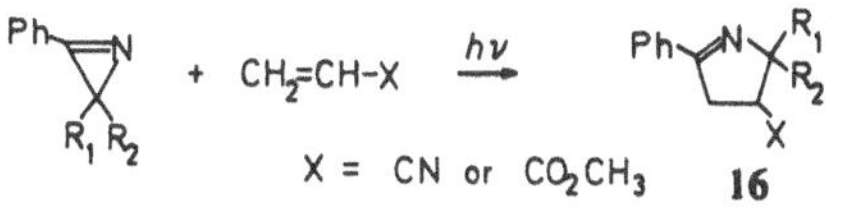

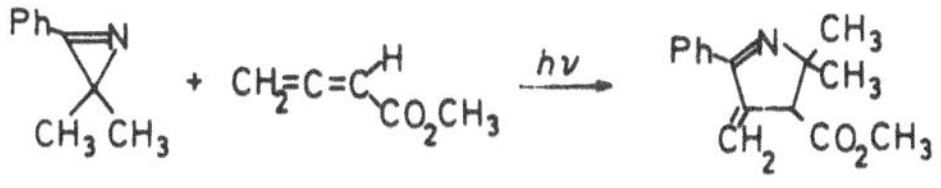

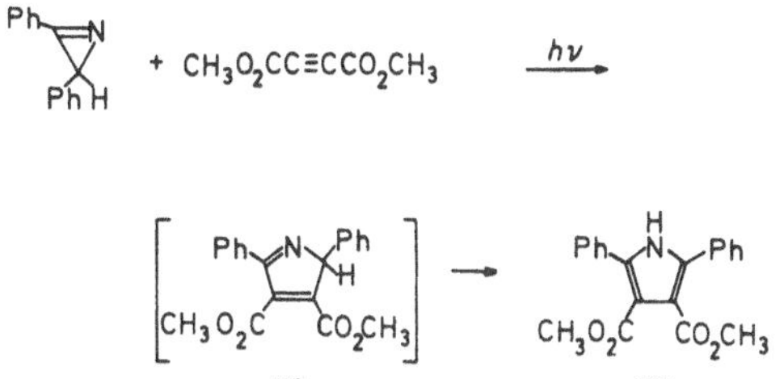

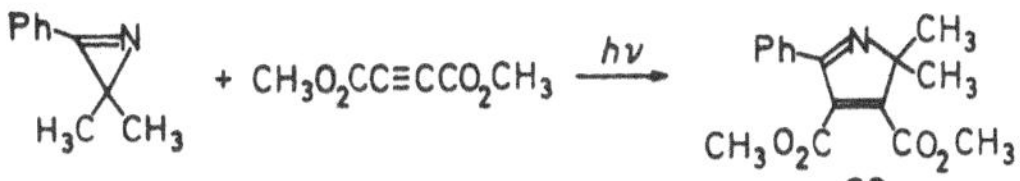

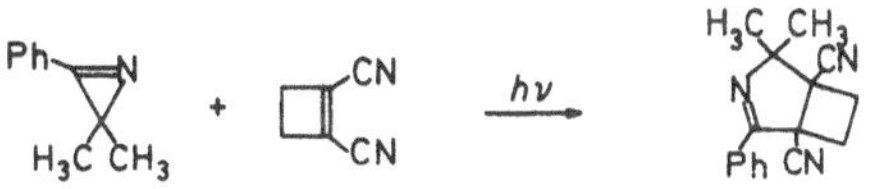

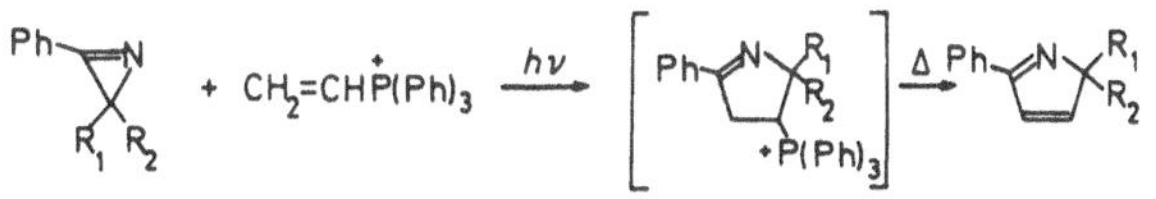

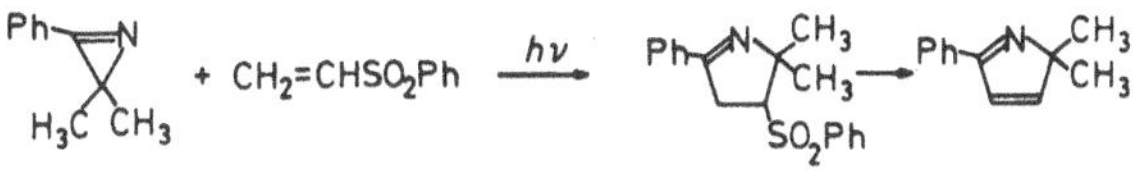

The orientation of the groups in the Δ^1-pyrrolines obtained from the photoaddition process is essentially identical to that observed by Huisgen in

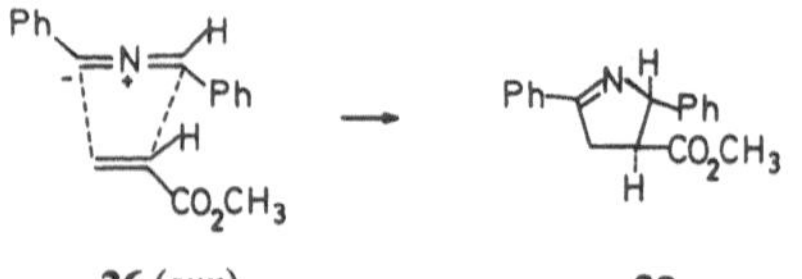

26 (*syn*)   **28**

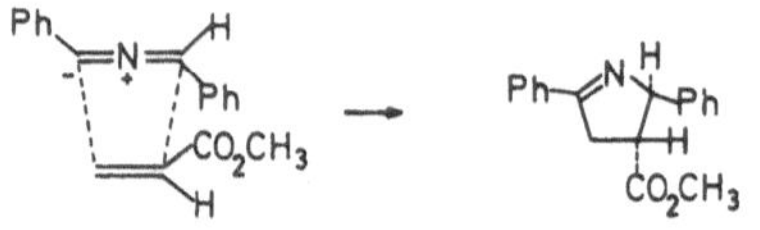

27 (*anti*)   **29**

related 1,3-dipolar additions.[63] For example, treatment of *N*-(*p*-nitrobenzyl)benzimidoyl chloride (**30**) with triethylamine in the presence of acrylonitrile has been found to give Δ^1-pyrroline, **31**.[64] This reaction has been interpreted as proceeding via a nitrile ylide intermediate, **32**.

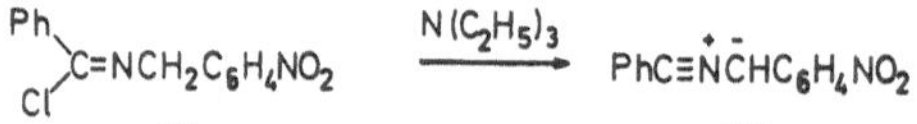

30   **32**

31

The regiospecificity of the olefin cycloaddition reaction depends on the substituent groups present on the double bond. Thus, acrylonitrile and methyl acrylate react with various nitrile ylides to give only the 4-substituted regio-isomers (i.e., **16**). Photocycloaddition of arylazirines to α-methylacrylo-nitrile and methyl methacrylate, on the other hand, give adducts of type **33** and **34** in a 3:2 ratio.[57]

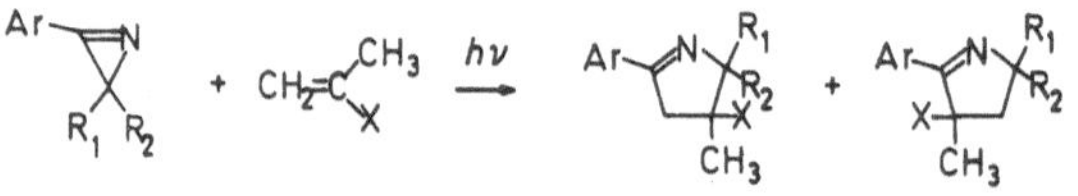

33   **34**

The regioselectivity is completely lost in the photocycloaddition of azirine **35** with diethyl vinyl phosphonate or dimethyl vinyl phosphine sulfide.[60] The two isomers, **36** and **37** as well as **38** and **39**, are found in equal quantities.

The experimentally observed regioselectivity of these and other 1,3-dipolar cycloadditions has, until recently, been a most difficult phenomenon to explain.

35

36 ; R= PO(OC$_2$H$_5$)$_2$ **37**

38 ; R= PS(CH$_3$)$_2$ **39**

Rationalizations of regioselectivity based on a concerted transition state model have invoked both electronic and steric effects.[63] A solution to the vexing problem of regioselectivity in 1,3-dipolar cycloadditions has recently been proposed by Houk and coworkers,[65] who used the frontier orbital method for rationalizing the effect of substituents on rates and regioselectivity of 1,3-dipolar cycloadditions. According to the frontier orbital treatment of 1,3-dipolar cycloadditions, the relative reactivity of a given 1,3-dipole towards a series of dipolarophiles will be determined primarily by the extent of stabilization afforded the transition state by interaction of the frontier orbitals of the two reactants.[66,67] When nitrile ylides are used as 1,3-dipoles, the dipole highest occupied (HO) and dipolarophile lowest unoccupied (LU) interaction will be of greatest importance in stabilizing the transition state. The favored cycloadduct will be that formed by union of the atoms with the largest coefficient in the dipole HO and dipolarophile LU. An electron-deficient olefin has the largest coefficient on the unsubstituted carbon in the LU orbital. In order to predict regioselectivity in the photocycloaddition of arylazirines, it is necessary to determine the relative magnitudes of the coefficients in the highest occupied orbital (HO) of the nitrile ylide. This problem was solved by carrying out the irradiation of several arylazirines in hydroxylic media.[67,68] The photoconversion of arylazirines 1 to alkoxyimines (**40**) indicates that in the highest occupied orbital of the nitrile ylide, the electron density at the disubstituted carbon is greater than at the trisubstituted carbon atom. The preferred regioisomeric transition state will be that in which the larger terminal coefficients of the interacting orbitals are united. Houk has pointed out that with all dipolarophiles except the very electron-rich, nitrile ylide reactions are HO controlled.[65] Reactions of nitrile ylides with electron-rich dipolarophiles have not been observed, indicating that the dipole LU–dipolarophile HO interaction is not very large. The photochemical addition of methanol to the nitrile ylide clearly shows that the larger HO coefficient of the nitrile ylide is, in fact, on the disubstituted carbon atom. With this conclusion, all of the regiochemical data found in the photoaddition of arylazirines[69–71] with dipolarophiles can be explained. Thus

40

the formation of the 4-substituted Δ^1-pyrroline **16**   from the irradiation of acrylonitrile or methyl acrylate with various nitrile ylides is perfectly consistent with the regioselectivity being controlled by union of the atoms with the largest coefficients in the dipole HO and dipolarophile LU. The formation of a mixture of 2*H*-azirines with α-methylacrylonitrile and methyl methacrylate can be attributed to the fact that whereas the cyano or ester group enhances the LU coefficient at the unsubstituted carbon atom of the dipolarophile, the methyl group has the opposite effect. The terminal coefficients in the LU of α-methylacrylonitrile and methyl methacrylate are more nearly the same than for the nonmethylated analogs, so that regioselectivity decreases for these dipolarophiles.

It should be noted that Schmid and coworkers[72] have reported on the photoaddition of arylazirines with other active hydrogen compounds [e.g., **35** (*hν*) → **41**] which are complementary to the observations outlined above. A

somewhat related case has also been described by Burger, who found that irradiation of azetine **7** in an alcohol solvent generated nitrile ylide **42**, which could be trapped to give *N*-(hexafluoroisopropyl)benzimidic ester, **43**.[73] It is interesting to note that the addition of alcohol to this nitrile ylide occurs in a manner opposite to that encountered in the irradiation of the arylazirines bearing alkyl groups in the 2-position of the ring. The effect of the *gem*-trifluoromethyl groups in Burger's system is apparently such that the coefficient at the trisubstituted carbon atom of the nitrile ylide is now the larger.

Dipolarophiles which contain an electron-deficient substituent undergo smooth cycloaddition reactions with nitrile ylides. The relative reactivity of the nitrile ylide toward a series of dipolarophiles is determined primarily by the extent of stabilization afforded the transition state by interaction of the dipole highest-occupied (HO) and dipolarophile lowest-unoccupied (LU) orbitals.[65,66] Substituents which lower the dipolarophile LU energy accelerate the 1,3-dipolar cycloaddition reaction.[74] For example, fumaronitrile undergoes cycloaddition at a rate which is 189,000 times faster than methyl crotonate.[74] Ordinary olefins react so sluggishly that their bimolecular rate constants cannot be measured.

44 **45** (major) **46** (minor)

47 **48** (major) **49** (minor)

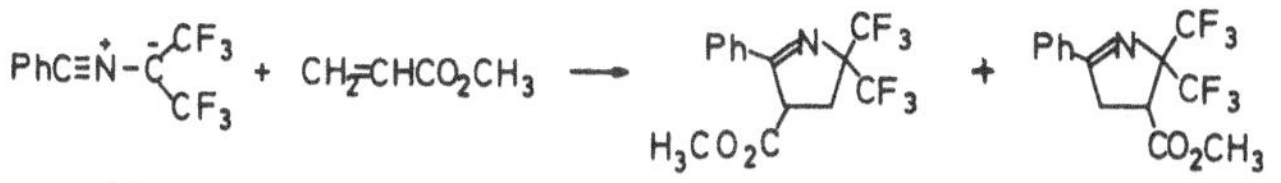

The absolute reaction rate of methyl acrylate and the nitrile ylide derived from 2,3-diphenyl-2*H*-azirine at 25°C has been estimated as 7.6×10^8 M^{-1} sec^{-1}.[39] Inductive effects exerted by substituents on the nitrile ylide also have an important effect on the regioselectivity of the cycloaddition. Benzonitrilio-hexafluoro-2-propanide (**44**) and methyl acrylate yield products with inverse regioselectivity as compared with the reactions of the related benzonitrilo-2-propanide **47**.[42] The difference in regioselectivity has been attributed to the larger coefficient at the trisubstituted carbon atom of the *gem*-trifluoromethyl substituted nitrile ylide **44**. This result parallels the different mode of addition of alcohols to nitrile ylides **44** and **47**.

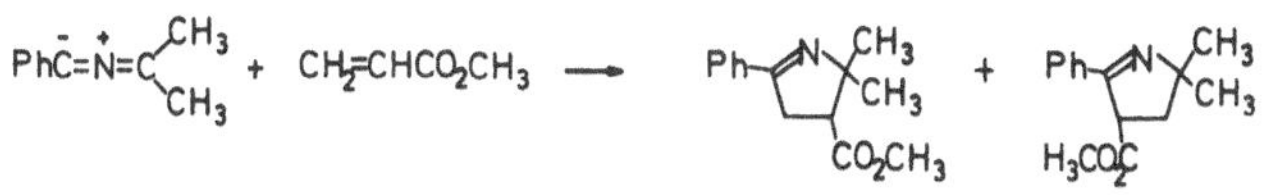

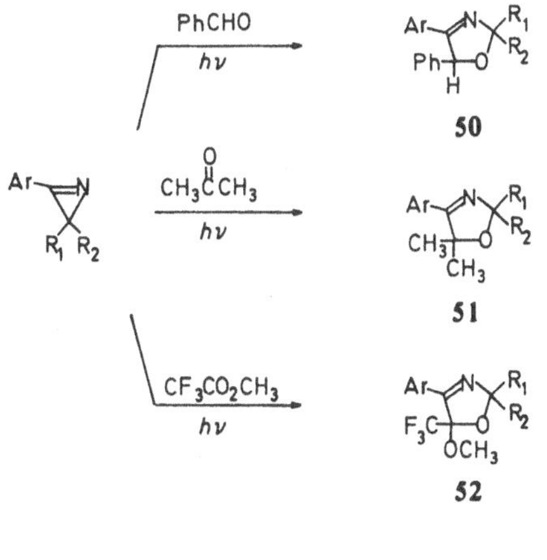

The photochemical addition of 2*H*-azirines to the carbonyl group of aldehydes, ketones, and esters also shows complete regiospecifity.[38,58,69–72,75–78] The cycloaddition of ketones with nitrile ylides proceeds much more slowly than the corresponding cycloaddition with aldehydes.[72] On the other hand, ketones with electron-withdrawing groups such as trifluoromethyl, ethoxycarbonyl, nitrile, or phosphonate at the α position react rapidly with the 1,3 dipole.[72] All of this is understandable in terms of frontier MO theory. Nitrile

50

51

52

ylides react rapidly with the more-electron-deficient carbonyl group since such a pair of addends possesses a narrow dipole HO–dipolarophile LU gap.

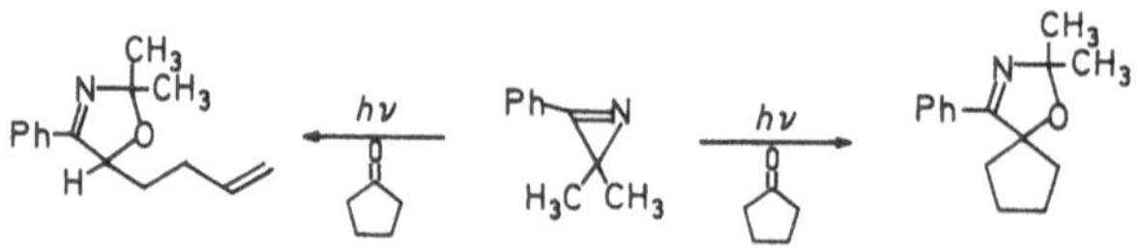

The photochemical cycloaddition of azirine **35** with cyclopentanone has been found to depend on the experimental conditions. When **35** is irradiated and cyclopentanone is slowly added, the expected *spiro*-3-oxazoline **53** is the main product.[72] However, when the cyclopentanone is irradiated first and the irradiation is continued in the presence of azirine **35**, the sole product is 3-oxazoline **54**.[72] Under the latter conditions, cyclopentanone reacts first by a Norrish type I cleavage and hydrogen transfer to yield 4-pentenal. This aldehyde reacts faster with the nitrile ylide than does the cyclic ketone still present, so that only **54** is formed. Norcamphor and camphor also react with azirine **35** under photolytic conditions via the Norrish type I reaction route to give 3-oxazolines **55** and **56**.

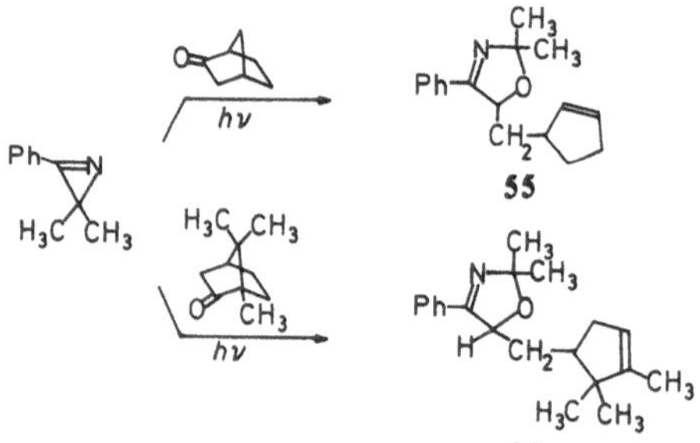

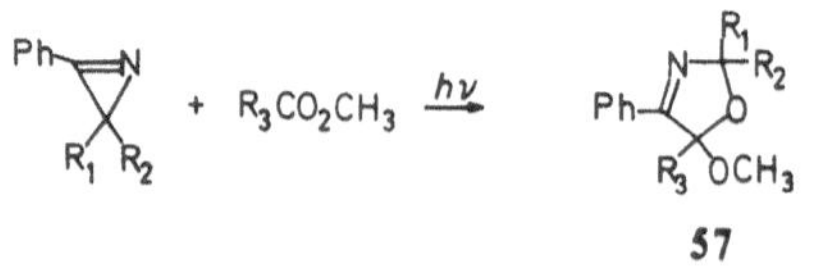

Esters of carboxylic acids which are activated by electron-withdrawing groups in the acyl or alkyl portion of the molecule also react with photochemically generated nitrile ylides to produce 5-alkoxy-3-oxazolines in high yield.[72] The addition to the carbonyl group occurs with the same regioselectivity as observed with aldehydes and ketones.[76] Esters which are not sufficiently activated, such as methyl acetate or benzoate, do not undergo cycloaddition. Schmid and coworkers have reported that the ester carbonyl

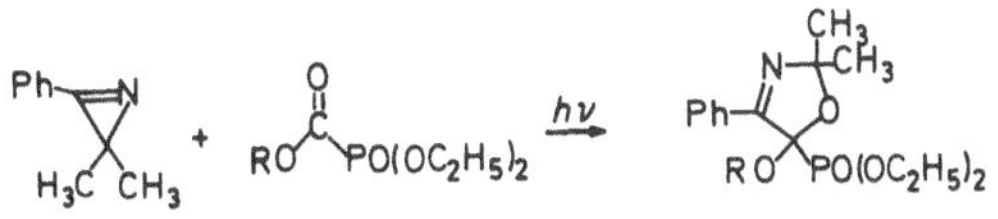

35 **58** **59**

R = C_2H_5; CH_2Ph

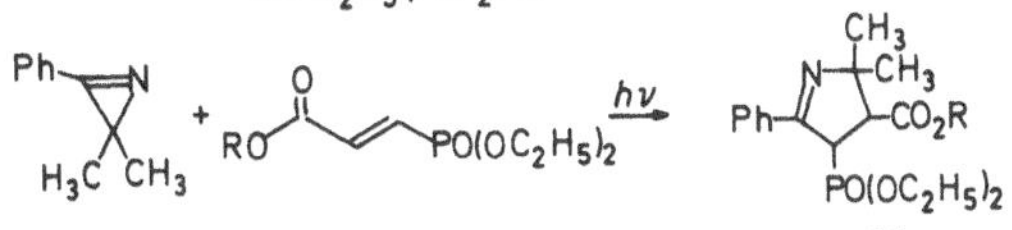

35 **60** **61**

group can also be activated by a diethyl phosphonate residue.[60] Thus irradiation of **35** with alkoxycarbonyl phosphate **58** produced oxazoline **59** in over 90% yield.[60] On the other hand, photocycloaddition of **35** to the vinylogous phosphonate **60** occurred only across the C=C bond to give cycloadduct **61**.

The Zurich group was also able to show that activation of carboxylic esters on the alkyl residue can best be achieved by esterification of the corresponding acid with 2,2,2-trifluoroethanol[72]:

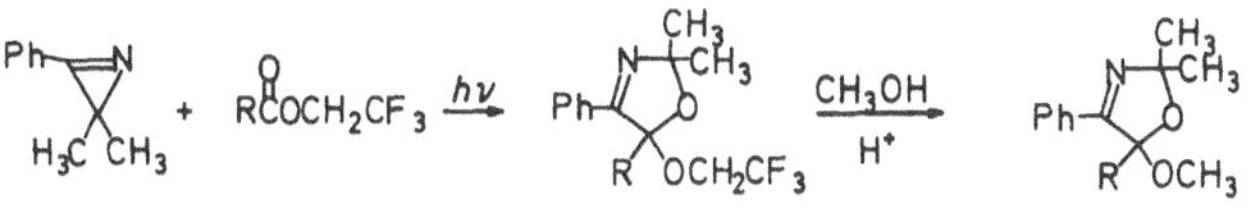

35 **62** **63**

The advantage of the trifluoroethoxy moiety is that it can easily be exchanged for other alkoxy groups under acidic conditions. This procedure allows the preparation of a variety of 5-alkoxyoxazolines which are not accessible by the direct irradiation route.

The irradiation of 2*H*-azirine **35** in the presence of ethyl acetoacetate is of interest since, in this case, protonation of the nitrile ylide competes with cycloaddition. The protonation reaction is followed by hydrolysis and elimination of ammonia to eventually give ethyl benzylideneacetoacetate **66**.[72]

The photocycloaddition reactions of arylazirines to a number of related

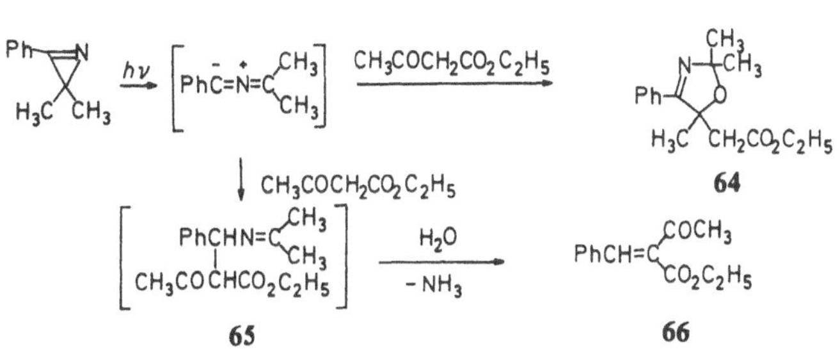

64 **65** **66**

carbonyl containing systems have also been studied.[19] Methyl thiobenzoate was found to photocycloadd with the same regiospecificity as was observed with the carboxylic ester.[72] In contrast, the reaction of methyl dithiobenzoate with azirine **68** takes place at the thiocarbonyl group with the inverse regiospecificity.[69]

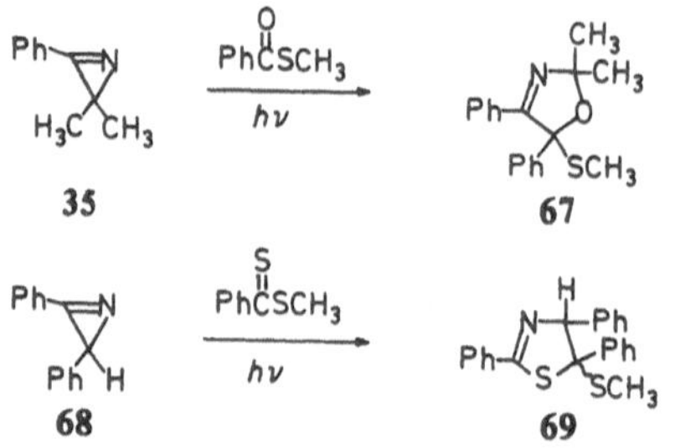

Photocycloadditions have also been observed between 2*H*-azirines and acyl chlorides.[79] The primary cycloadducts **70** are quite labile and cannot be isolated directly. The 5-chloro substituent of the photoproduct could readily be exchanged for an alkoxyl group. Cycloadducts **70**, which are monosubstituted at C-2, undergo a 1,4 elimination of hydrogen chloride on treatment with tertiary amines to form oxazoles in moderate yield.[79]

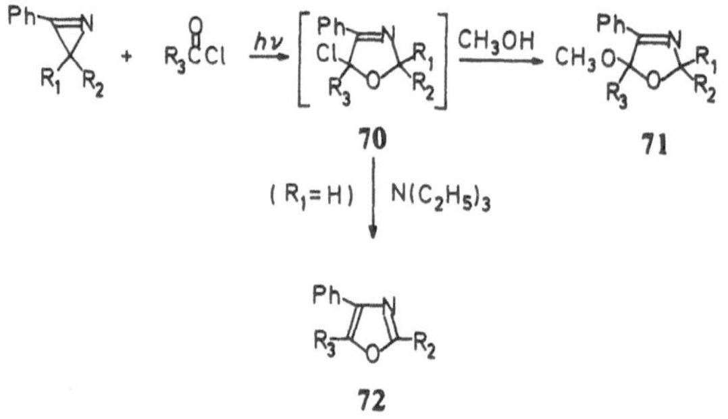

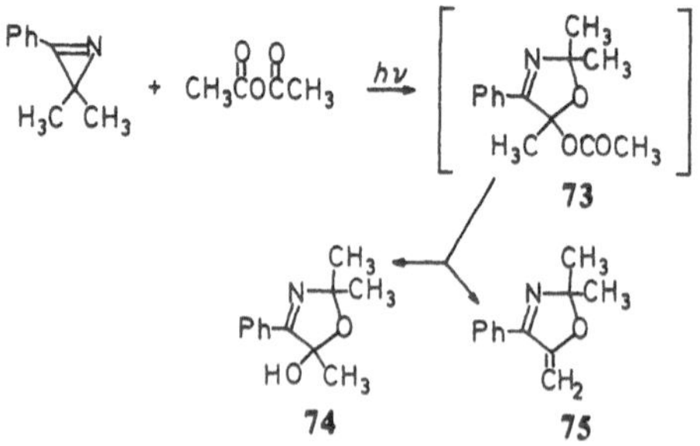

A photoreaction similar to that observed with 2*H*-azirines and acyl chlorides also takes place with anhydrides.[79] Irradiation of azirine **35** and acetic anhydride, for example, initially produces the nonisolable cycloadduct **73**, which is subsequently converted to oxazolines **74** and **75** in 50% and 25% yields, respectively. Oxazoline **75** is formed as the sole product when **35** is irradiated in the presence of ketene.[77]

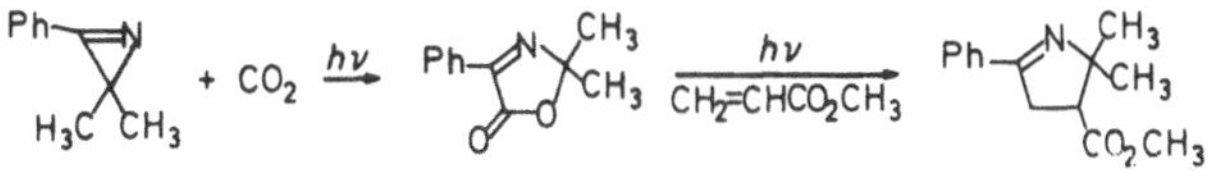

One of the more interesting photocycloaddition reactions in this series involves the photoinduced combination of 3-aryl-2H-azirines with carbon dioxide.[47,80] This reaction is carried out by passing a finely dispersed carbon dioxide stream through a benzene solution of the 2H-azirine during irradiation and results in the formation of 3-oxazolin-5-ones **76** in good yield.[81] The carbon dioxide cycloaddition with the photochemically generated nitrile ylide is reversible. Irradiation of the 3-oxazolin-5-one system (**76**) with 250–350-nm light regenerates the nitrile ylide, which can be trapped in the usual way.[40,47] A kinetic investigation involving Stern–Volmer plots and relative reactivity studies showed that the nitrile ylide generated from the photolysis of the 3-oxazolinone system is identical to that generated from the corresponding 2H-azirine.[47] Evidence was obtained which showed that the photochemically generated nitrile ylide derived from **76** did not collapse thermally back to the azirine ring.

When 2,2-dimethyl-3-phenyl-2H-azirine (**35**) was irradiated in the presence of carbon disulfide, a 2:1 adduct, 5,5-spirobis-(4,4-dimethyl-2-phenyl-2-

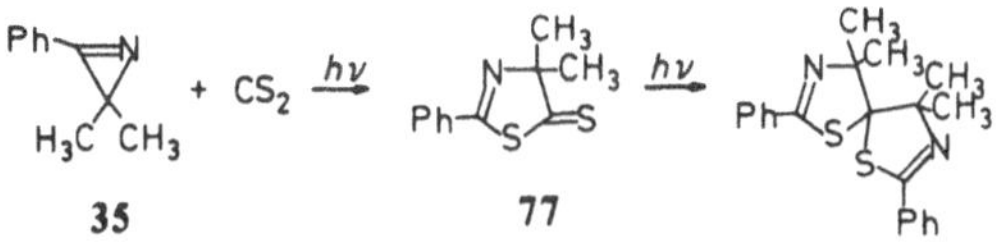

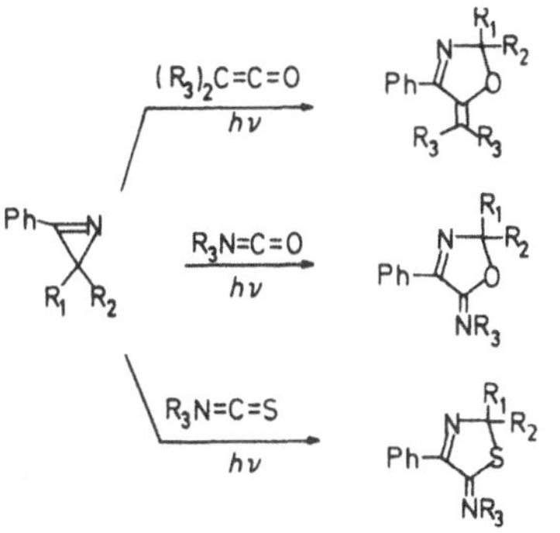

thiazoline) (**78**) was isolated.[47] Undoubtedly, the C–S double bond of the initially formed 1:1 adduct (i.e., **77**) reacts much faster than does the C–S double bond of carbon disulfide.[47,82]

Photoreactions have also been observed between 3-phenyl-2*H*-azirines and ketenes, carbodiimides, isocyanates as well as isothiocyanates. The isocyanates and their thio analogs react with the C–O and C–S double bonds, respectively, to give **79**, **80**, and **81**, and not with the C–N double bond.[77,80]

The photocycloaddition of 2*H*-azirines with a variety of other multiple π bonds proceeds in high yield and provides a convenient route for the synthesis of a variety of five-membered heterocyclic rings.[83–87] Some of the dipolarophiles used include nitriles,[71] azodicarboxylates[86] and *p*-quinones[40,87]:

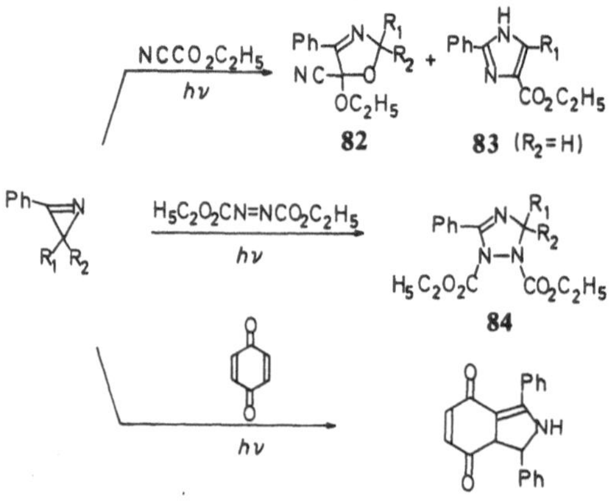

III. PHOTOCHEMICAL DIMERIZATIONS OF 2H-AZIRINES

Irradiation of arylazirines with olefins of low dipolarophilic activity or in inert solvents produced no photoadduct but instead gave dimers.[57,71,84] It was originally reported that photolysis of phenylazirine (**86**) gave azabicyclopentane (**87**).[88,89] However, in the light of the foregoing mechanistic rationale developed for the photolysis of azirines, Padwa and Schmid's groups were able to show that the photodimer isolated is actually diazabicyclohexane (**88**). Thus, in the absence of a dipolarophile, the nitrile ylide generated photochemically simply adds to a ground state azirine molecule.[57] A cross dimerization of **89** to the ground state of **86** has also been realized.[90] Additional work showed that the photodimerization of arylazirines to 1,3-diazabicyclo[3.1.0]hex-3-enes is a general reaction which is independent of the nature of the substituent groups attached to the C atom of the azirine ring.[58,78] Care is required in the choice of solvent, photolysis time, and substituents since the 1,3-diazabicyclohexanes are themselves photochemically labile.[91] Indeed, extended photolysis leads to complete ring opening and the formation of diazahexatrienes **92**. These may be isolated when $R_1 = R_2 = CH_3$. Further photolysis of *trans*-**93** ($R_1 = R_2 = CH_3$) gives the diazachrysene **96**.[92] The *cis*-enediimine **92** ($R_1 = R_2 = CH_3$) very readily cyclizes to the diazepine **95** on standing. On the other hand, if $R_1 = H$ in *cis*-**92**, then cyclization gives first the dihydropyrazine **94** and eventually the pyrazine **97**. The results indicate that the secondary photoproducts formed from the irradiation of the diazabicyclohexenes (**91**) depend on the substituent groups, the time of irradiation, and the particular solvent employed.[58,92,93]

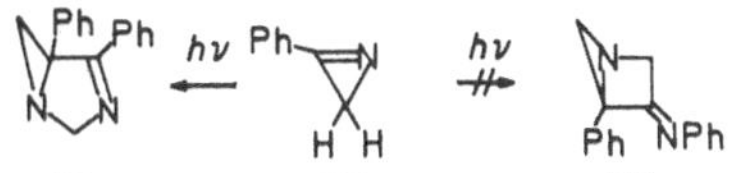

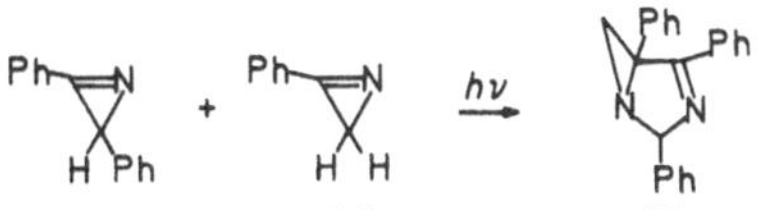

Work by Schmid's group[39] has shown that the nitrile ylide derived from diphenylazirine (**68**) also undergoes quantitative dimerization to 1,3,5,6-tetraphenyl-2,5-diaza-1,3,5-hexatriene (**99**) at −160°C. This result indicates that **99** is not only formed by the indirect route (**98** + **68** → **100** $\xrightarrow{h\nu}$ **101** → **99**) but also by dimerization of **98** by a direct head to head coupling. The intermediacy of azomethine ylide **101** in the photochemical conversion of azabicyclohexene

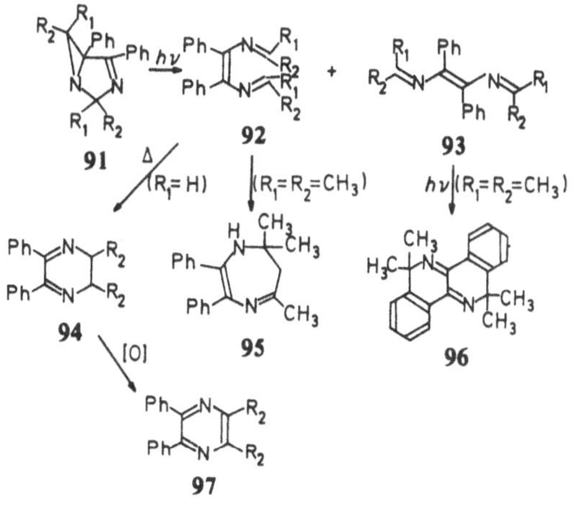

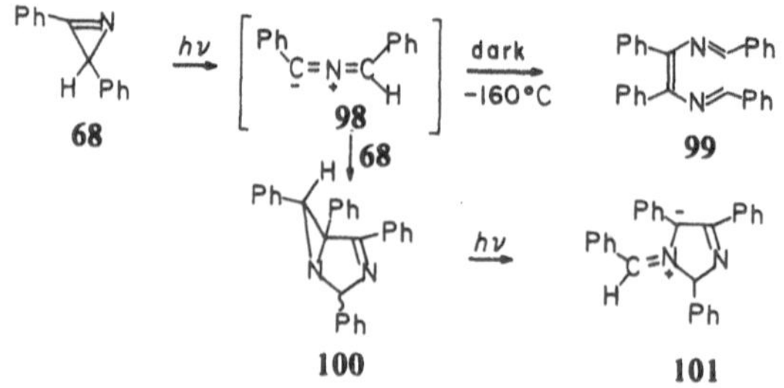

100 to diazahexatriene **99** was verified by low-temperature photolysis studies. Flash photolysis of a solution of 2,3-diphenyl-2H-azirine (**68**) in cyclohexane at room temperature allows one to monitor the disappearance of the nitrile ylide (i.e., **98**) by uv spectroscopy.[39] At high concentrations of **98** ($>10^{-7} M$), the dipolar species was found to disappear with second-order kinetics with a specific rate constant $k = 5 \times 10^7 M^{-1} \sec^{-1}$. At low concentrations of **98** ($<10^{-7} M$), the dipolar species was found to vanish with pseudo-first-order kinetics (**98** + **68** → **100**). The specific pseudo-first-order rate constant is $k = 1 \times 10^4 M^{-1} \sec^{-1}$. The direct formation of **99** has been explained in terms of a head-to-head reaction of two molecules of the dipolar intermediate in a biradicaloid or carbenoid form.[39] Alternatively, a head to tail dimerization of the nitrile ylide could occur to give **102**, which is then transformed into **99** via a subsequent hydrogen shift.[39]

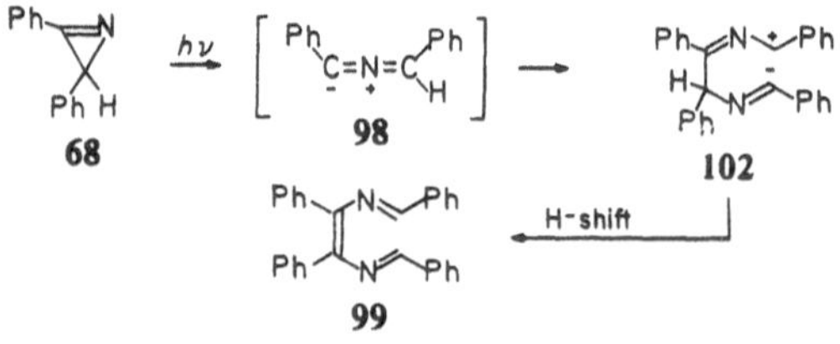

A head-to-head dimerization was also observed with benzoni-triliohexafluoro-2-propanides (**42**) generated from the thermolysis of oxaphos-phole **6**. The primary coupling product **103** was converted to structure **104** by cyclization followed by a 1,5-H shift and dehydrogenation by oxygen.

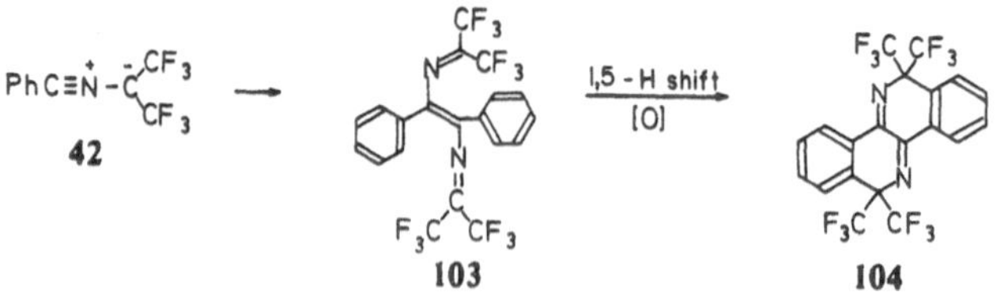

It was also shown that benzonitrilio-*p*-nitrophenyl methanide (**105**), generated chemically by the base-induced elimination of HCl from *N*-(*p*-nitrobenzyl)benzimidoyl chloride **5**, adds across the C–N double bond of 2*H*-azirines to give the 1,3-diazabicyclo[3.1.0]hexene ring system[84]:

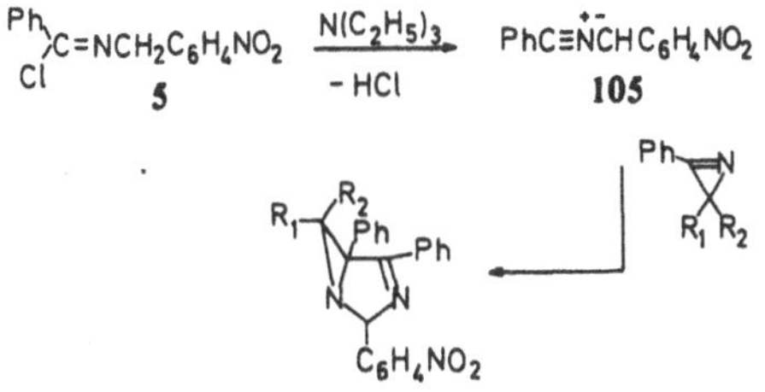

The 1,3-diazabicyclohexene ring was also prepared by treating 2-acylaziridines with aldehydes and ammonia.[94,95] This independent synthesis provides unequivocal support for the structures of the photodimers.

IV. INTRAMOLECULAR 1,5-ELECTROCYCLIZATION REACTIONS OF VINYL-SUBSTITUTED 2H-AZIRINES

Whereas the cycloaddition of arylazirines to electron-deficient olefins produces Δ^1-pyrrolines, a rearranged isomer is formed when the alkene and the azirine moieties are suitably arranged in the same molecule. This intramolecular electrocyclization reaction[96] was first observed by Padwa and Smolanoff using 2-vinyl-substituted 2*H*-azirines. Irradiation of 2*H*-azirine **107** afforded a 2,3-disubstituted pyrrole (**108**), while thermolysis gave a 2,5-disubstituted pyrrole (**109**). Photolysis of azirine **110** proceeded similarly and gave 1,2-diphenylimidazole (**111**) as the exclusive photoproduct. This stands in marked contrast to the thermal reaction of **110** which afforded 1,3-diphenylpyrazole

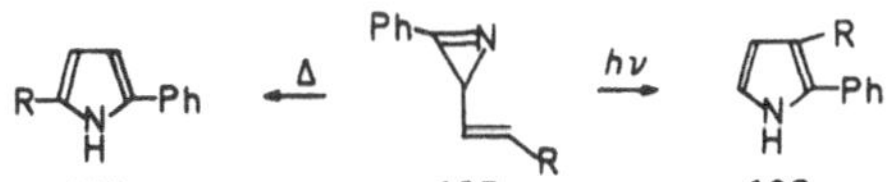

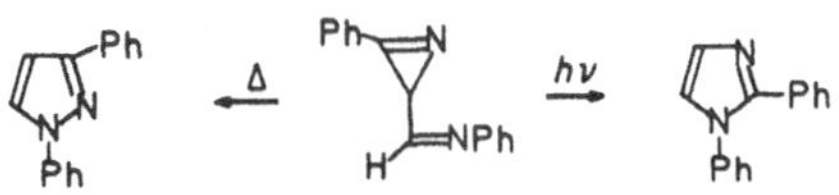

(112) as the only product. The evidence obtained clearly indicates that the
above photorearrangements proceed via a mechanism involving a nitrile ylide
intermediate, since cycloadducts could be isolated when the irradiations were
carried out in the presence of trapping agents.[97] Intramolecular electrocyclization of nitrile ylide 113 followed by a 1,3-sigmatropic hydrogen shift of the
initially formed five-membered ring readily accounts for the formation of the
final product. The thermal transformations observed with these systems has
been rationalized in terms of an equilibration of the 2*H*-azirine with a transient
vinyl nitrene which subsequently rearranges to the 2,5-disubstituted pyrrole
(*vide infra*).

In the above examples only the *E* isomers of azirine 107 could be
prepared. It is noteworthy that the photochemical conversion of the bicyclic
isoxazoline 114 to oxazepin 117 is believed to involve the *Z* isomer of azirine
115.[98,99] This species could not be isolated but goes on to give nitrile ylide 116.
1,7-Electrocyclization of 116 leads to oxazepin 117 in 80% yield. Because of
steric constraints it is clearly impossible for the *E*-vinylnitrile ylide 107
(R = CHO) to undergo 1,7-cyclization.

Analogous results were obtained from the photochemical ring opening of
Z-3-styryl-2*H*-azirine (118).[100] The *Z*-styrylnitrile ylide 119 formed in the
irradiation of 118 affords benzazepine 120 in 80% yield by a 1,7-
electrocyclization followed by a 1,5-sigmatropic shift. The photolysis of the
isomeric *E*-styrylazirine 121 followed an entirely different course and produced
2,3-diphenylpyrrole 123 as the major photoproduct. This observation requires
that opening of the azirine ring followed by intramolecular electrocyclization
proceed faster than isomerization about the C–C double bond. The formation of
120 indicates that the nitrile ylide obtained from 118 cyclizes more easily via a
seven-membered transition state and leads to the preferential formation of benzazepine 120. Cyclization of the nitrile ylide derived from the *trans* isomer to a
seven-membered ring is precluded on structural grounds, and formation of 2,3-
diphenylpyrrole occurs instead.

118 119 120 121 122 123

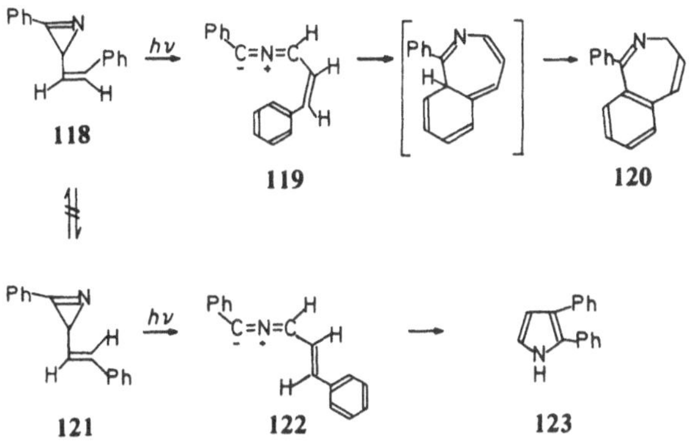

The photoconversion of the *Z*-benzoyl substituted vinylazirine **124** to tetraphenyloxazepin **125** most likely involves ring opening to a nitrile ylide followed by 1,7-electrocyclization.[101] The same product was also obtained from the thermolysis of **124**. All attempts to trap the nitrile ylide with methyl acrylate failed. This result indicates that if a nitrile ylide is formed at all during the rearrangement, it reacts much faster intramolecularly to give **125** than bimolecularly with methyl acrylate.

Irradiation of 3-vinyl-2*H*-azirine (**126**) gives, among other products, the azatriene **128**, which arises via intramolecular hydrogen transfer of vinylnitrile ylide **127**.[102] The pyrrole **129**, which would have been derived from 1,5-dipolar cyclization of **127**, was not found. When the reaction is run in the presence of acrylonitrile or methyl acrylate, the 1,3-dipolar cycloadducts **130** (X = CN or CO_2CH_3) are obtained in high yield, thus arguing for the intermediacy of the nitrile ylide.[102]

124 125 126 127 128 129 130

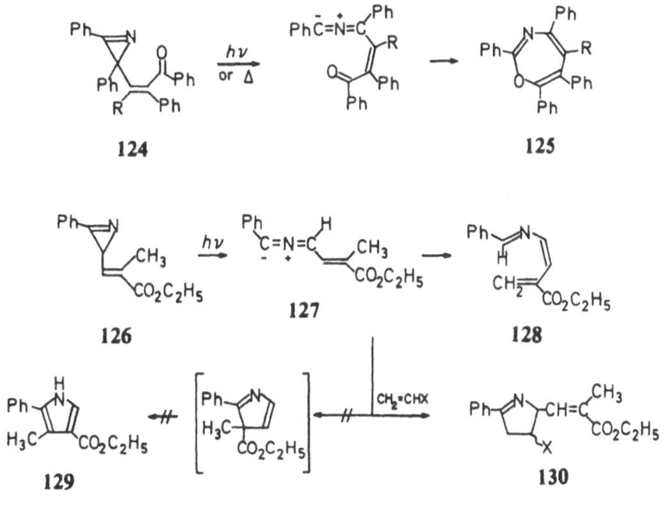

Attempts to effect a dipolar cyclization of the butadienyl nitrile ylide generated by photolysis of 2*H*-azirine **131** failed; only a complex mixture of products was obtained.[102]

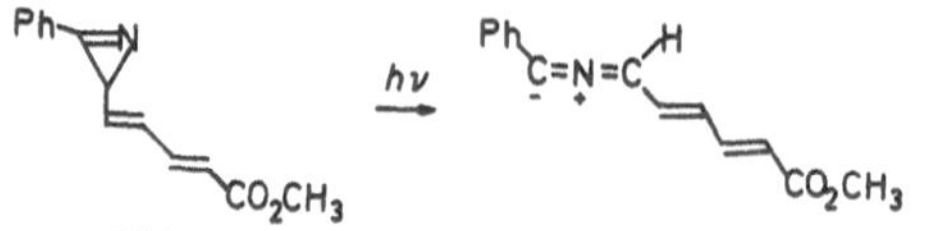

A reaction which is closely related to the above cyclizations was uncovered by Ullmann and Singh in 1966.[33,34] While studying the photochemical rearrangement of 3,5-diphenylisoxazole (**133**) to 2,5-diphenyloxazole (**134**), these workers observed the formation of an intermediate which proved to be 3-phenyl-2-benzoyl-2*H*-azirine (**135**). The photobehavior of azirine **135** was found to be controlled dramatically by the wavelength of the light used. With 3130-Å light, **135** rearranges almost quantitatively to oxazole **134**, whereas 3340-Å light causes rearrangement to isoxazole **133**. The formation of the isoxazole was suggested to occur via the $n-\pi^{*3}$ state of the carbonyl chromophore. Oxazole formation, on the other hand, was attributed to excitation of the $n-\pi^*$ state of the azirine ring. Selective $n-\pi^*$ excitation of the carbonyl group of **135** causes weakening of the C–N single bond. Cleavage of this bond leads to the vinylnitrene **136**, which collapses to isoxazole **133**. The $n-\pi^*$ excitation of the ketimine chromophore at shorter wavelengths leads to C–C bond cleavage generating the carbonylnitrile ylide **137**. 1,5-Dipolar cyclization of **137** gives oxazole **134**. This valence isomerization has subsequently aroused the interest

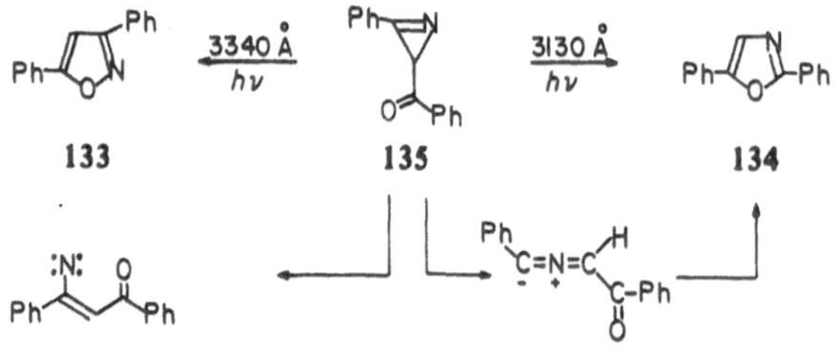

of a number of investigators and much effort has been expended in outlining the scope and mechanism of this process.[103–113] Some of the systems which have been studied in detail are outlined below:

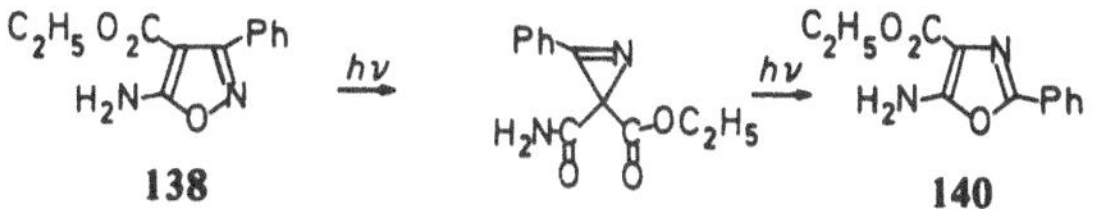

138 139 140

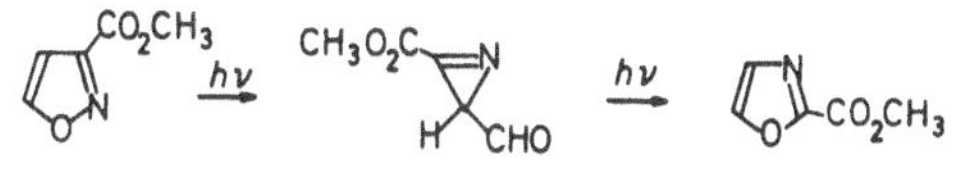

141 142 143

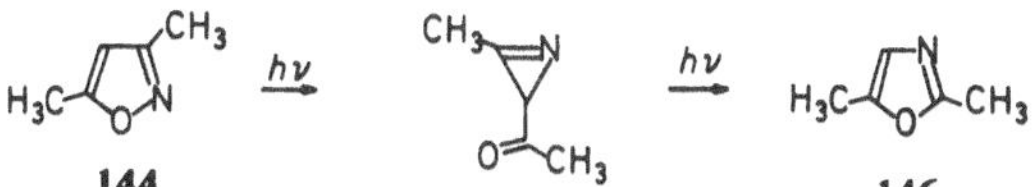

144 145 146

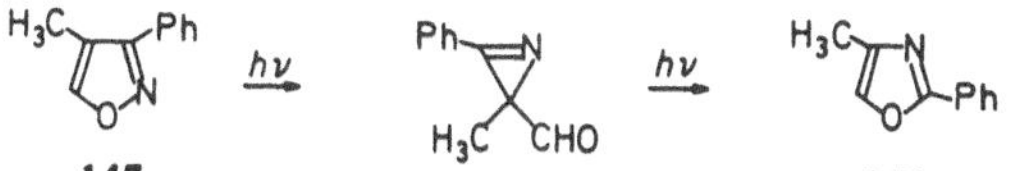

147 148 149

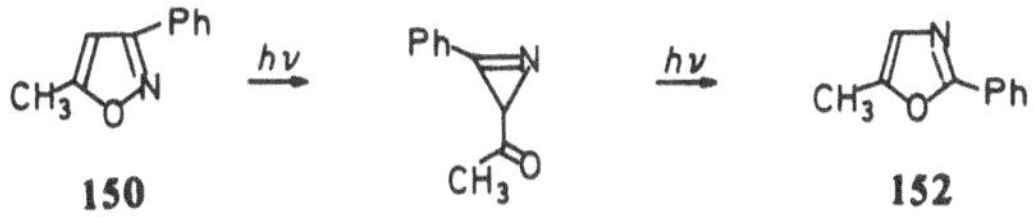

150 151 152

These reactions involve 3-carbonyl-substituted 2*H*-azirine intermediates which are not isolated but react further to give the oxazoles. The carbonylnitrile ylides do not give 1,3-dipolar cycloadducts with added dipolarophiles owing to the facility with which they undergo electrocyclic ring closure.

Although the majority of oxazole photorearrangements involve the formal interchange of the C_2 and O atoms via the intermediacy of a 2*H*-azirine intermediate, two examples of ring interchange between the C_4 and C_5 atoms have been found[113]:

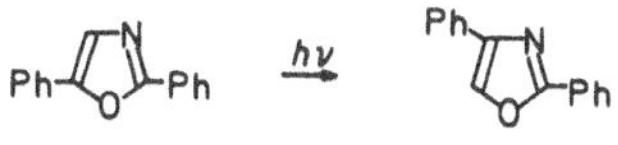

153 154

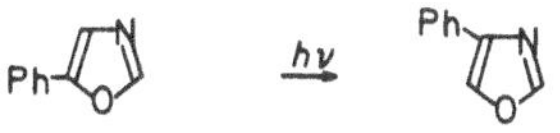

155 156

These transformations have been rationalized in terms of a 1*H*-azirine intermediate[114,115] derived by cleavage of the C_5–O bond.[113]

157

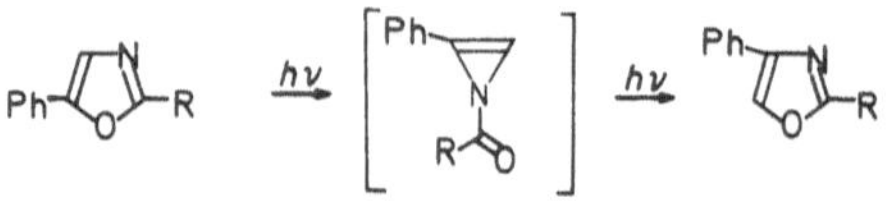

One case where an azirine intermediate is believed not to arise in the isoxazole photorearrangement has been reported by Padwa and coworkers.[116,117] When 4-benzoyl-5-methyl-3-phenyl-isoxazole (**158**) or 4-acetyl-3,5-diphenylisoxazole (**159**) was thermolyzed (230°C), a 2*H*-azirine is presumably formed which subsequently opens to the nitrile ylide **160**. Subsequent closure of **160** affords the isomeric oxazoles **161** and **162** in a ratio of 3:4.

158 **159** **159** **160** **161** **162**

160a **161** **162** **160b**

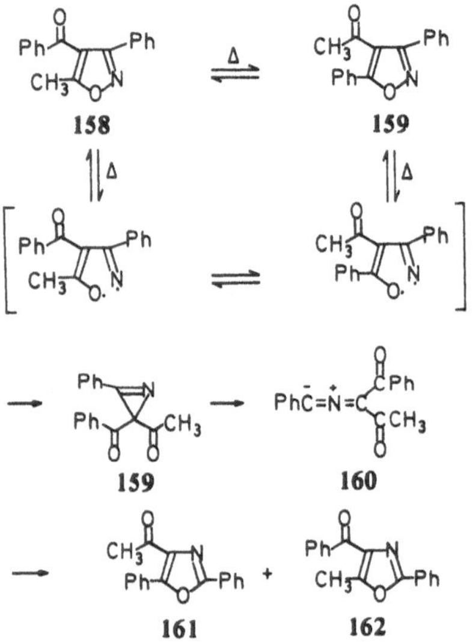

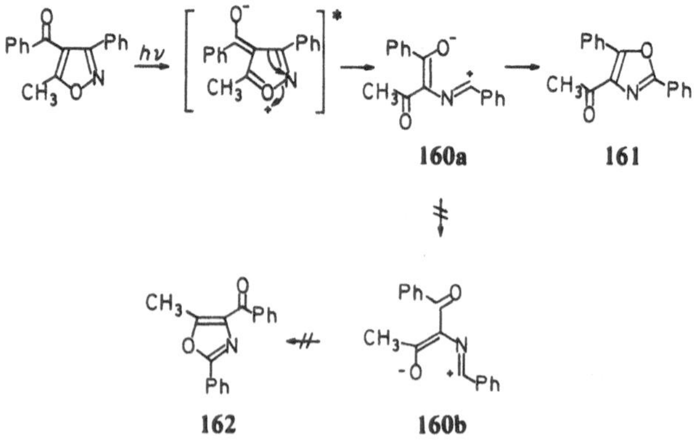

In sharp contrast to the thermal results, the photolysis of **158** affords only oxazole **161**. The authors postulated that the nitrile ylide **160** is formed directly from the excited isoxazole in the conformation **160a**.[117] Ring closure occurs before rotation around the C–N bond, yielding only oxazole **161**. Some recent MINDO/3 calculations by Dewar and Turchi indicate that the CNC angle of nitrile ylides is less than 180° and that rotation about the C–N bond in **169** is indeed slow relative to ring closure.[118]

It should be noted that Schmid and coworkers have postulated a 2H-azirine (**165**) as a transient intermediate in the photoreorganization of **163** to oxazoles **167** and **168** (1:1 mixture).[119] This reaction is also accompanied by a slower transformation of **163** to **164**.

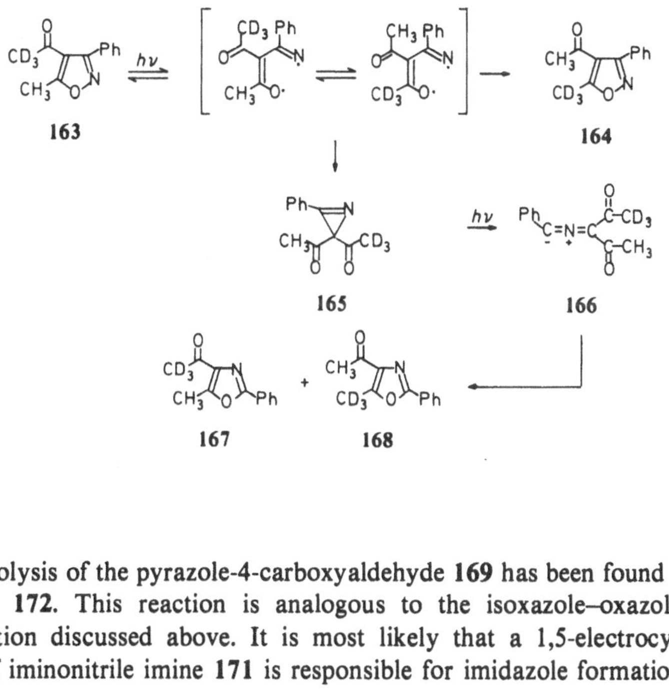

Photolysis of the pyrazole-4-carboxyaldehyde **169** has been found to afford imidazole **172**. This reaction is analogous to the isoxazole–oxazole photo-isomerization discussed above. It is most likely that a 1,5-electrocyclic ring closure of iminonitrile imine **171** is responsible for imidazole formation.[120]

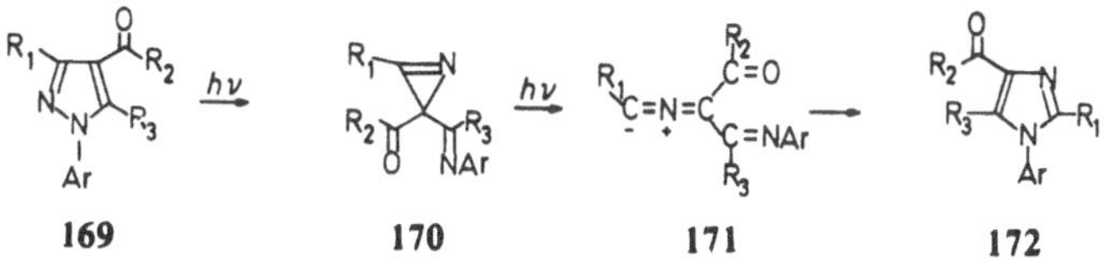

V. PROPERTIES OF NITRILE YLIDES

Nitrile ylides, generated from the photolysis of 2H-azirines, may be classified as nitrilium betaines, [121]a class of 1,3-dipoles containing a central nitrogen atom and a π bond orthogonal to the 4π allyl system. They can be intercepted with a wide variety of dipolarophiles to form five-membered heterocyclic rings. Among the possible geometric forms of a nitrile ylide, a carbene structure (**174**) can he envisaged which makes conceivable a 1,1 cycload-

dition of this 1,3-dipole. Huisgen has argued[121] that the bent geometric form (**174**) of a nitrile ylide would be less stable than the linear form (**173**), since allyl resonance would be at a maximum with the linear arrangement. The bent form would have a lone pair of electrons in an orbital with some *s* character, but this was postulated to be of lesser importance.[121] The extensive early literature dealing with nitrile ylide cycloadditions has generally been explained in terms of the linear arrangement of this 1,3-dipole.[63] Thus 1,3-dipolar cycloadditions of nitrile ylides have been suggested to proceed via a "two-plane" orientation complex in which the dipole and dipolarophile approach each other in parallel planes.[53] Formula **175** depicts the orientation complex involved in the addition of the linear nitrile ylide **173** with a dipolarophile. During the activation process, the linear bond system of the nitrile ylide must bend. This involves disruption of the orthogonal π bond at some modest energy costs but leaves the allyl anion π system undisturbed. The loss of π bond energy with **173** is partly compensated by a gain in energy resulting from rehybridization and accommodation of a lone pair of electrons in an orbital of high *s* character.

Another mechanism which could also account for the products obtained

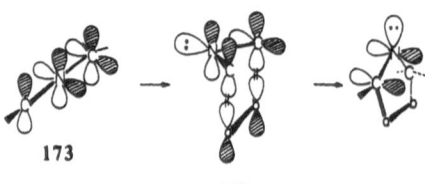

173

175

on dipolar cycloaddition of nitrile ylides with π bonds involves an initial 1,1 addition of the bent nitrile ylide (carbene form) with the dipolarophile to give a transient three-membered ring. This would be followed by a rapid intramolecular rearrangment to the five-membered heterocycle:

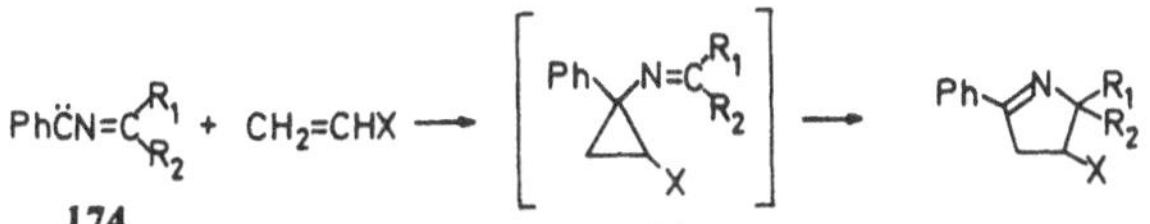

This alternate possibility was discounted by Huisgen, however, who showed that three-membered rings are not primary products in the 1,3-dipolar additions leading to five-membered heterocycles with nitrilium betaines.[122]

Recent *ab initio* LCAO–MO–SCF calculations by Salem,[123] Houk,[124,125] and Devaquet[126] indicate that the geometry of the nitrile ylide is appreciably different from that suggested by Huisgen.[121] Salem's calculations indicate that the ring-opened intermediate should be capable of dual reactivity when it is intercepted by an added dipolarophile. The behavior of the system was predicted to be dependent on the geometry of the transient intermediate generated from the photolysis. Opening of the ring to an intermediate with linear geometry will result in the formation of a 1,3-dipolar species having closed-shell zwitterionic character. Salem's calculations also indicate that if the ring is opened to give an intermediate with bent geometry, a diradical state with partial dipolar character will be obtained which may undergo reactions different from the linear species. According to Salem's calculations, the lowest-energy ground state geometry of the nitrile ylide has a HCN angle of 156.7° and is ~18 kcal/mol more stable than the linear form. A similar conclusion was reached by Houk and Caramella.[124,125] Their calculations show that the bent nitrile ylide geometry is favored over the linear, but otherwise optimized, geometry by 11.1 kcal/mol. These findings indicate that the most stable form of a nitrile ylide resembles a bent allenyl anion rather than a linear propargyl anion. The HOMO and second LUMO of the bent ylide bear a strong resemblence to the HOMO and LUMO of a singlet carbene. It should be noted that the bent nitrile ylide geometry correctly rationalizes the regioselectivity encountered with the 1,3-dipolar cycloadditions of nitrile ylides with added dipolarophiles. Houk's calculations show that the bent nitrile ylide HOMO is heavily localized at C-1, but still resembles the normal three-orbital, four-electron π system present in other 1,3 dipoles so that concerted cycloadditions can still occur. As a consequence of the bent geometry, C-1 is the nucleophilic terminus of the nitrile ylide. According to the frontier orbital treatment of 1,3-dipolar cycloadditions,[65,66] the relative reactivity of a given 1,3-dipole forward a series of dipolarophiles will be determined primarily by the extent of stabilization afforded the transition state by interaction of the frontier orbitals of the two reactants. When nitrile ylides are used as 1,3 dipoles, the dipole highest-occupied (HO) and dipolarophile lowest-unoccupied (LU) interactions will be of greatest importance in stabilizing the transition state. Regioselectivity will be controlled by union of the atoms with the largest coefficients in the dipole HO and LU. Using these generalizations, the regioselectivity of bimolecular nitrile ylide cycloadditions can readily be predicted. Protonation of the nitrile ylide is also known to occur at the C-1 carbon atom.[67] Thus all the known reactions of nitrile ylides can be adequately accounted for in terms of a bent nitrile ylide.

VI. INTRAMOLECULAR PHOTOCYCLOADDITION REACTIONS OF 2H-AZIRINES

As was mentioned in the previous section, the HOMO and second LUMO of the bent nitrile ylide bear a strong resemblance to the HOMO and LUMO of a singlet carbene. Since carbenes are known to react readily with double bonds, a 1,1-cycloaddition reaction of nitrile ylides can be expected. Padwa and Carlsen uncovered the first example of such a process during an investigation of the photochemistry of a number of 2-allyl-substituted 2H-azirines.[127] When a thoroughly deaerated solution of 2-allyl-2-methyl-3-phenyl-2H-azirine (177) was irradiated in cyclohexane with light of wavelength >280 nm for 15 min, an extremely rapid and clean conversion to 3-methyl-1-phenyl-2-azabicyclo[3.1.0]hex-2-ene (179) was observed. When the irradiation of 177 was carried out to 20% conversion, however, a 1:1 mixture of 179 and 1-methyl-3-phenyl-2-azabicyclo[3.1.0]hex-2-ene (178) was quantitatively isomerized to 179. The photochemical behavior of the isomeric 2-allyl-3-methyl-2-phenyl-2H-azirine (180) afforded a quantitative yield of azabicyclohexene (179). A control experiment showed that 177 and 180 were not interconverted by a

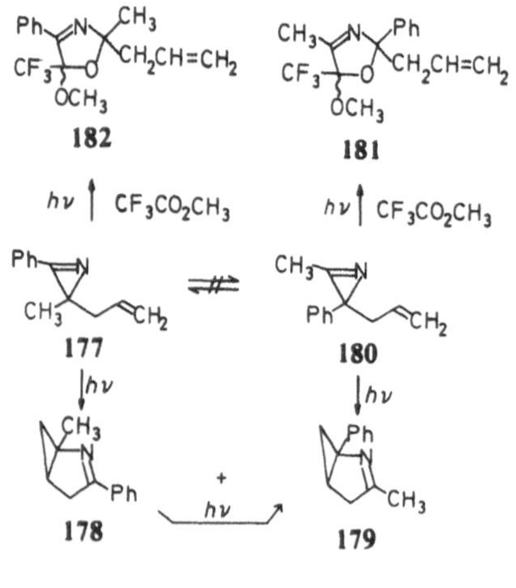

Cope reaction under the photolytic conditions. Photolysis of **180** in the presence of the very reactive dipolarophile, methyl trifluoroacetate, resulted in the trapping of a nitrile ylide and gave cycloadduct **181** in high yield. Under these conditions, the formation of **179**, which is produced in quantitative yield in the absence of a trapping agent, is entirely suppressed. Photocyclization of **177** with added methyl trifluoroacetate resulted in the formation of cycloadduct **182** in high yield. The isolation of **181** in the bimolecular trapping experiment eliminates a path by which **180** is partially isomerized to **177**, which then rearranges to **179** on further excitation. This possibility was initially considered to be a reasonable one since the extinction coefficient of **177** at 254 nm ($\varepsilon = 8700$) is much larger than that of **180** ($\varepsilon = 220$). It should be noted that no significant quantities of **178** were detected in a short-term irradiation of **180**. This is probably related to the fact that **178** possesses a much larger extinction coefficient than does **180** and is optically pumped to **179**, even at low conversions.

The formation of azabicyclohexenes **178** and **179** from the irradiation of azirine **177** clearly proceeds via a nitrile ylide intermediate since the formation of these compounds is entirely suppressed when the irradiation is carried out in the presence of an added dipolarophile. For example, when dimethyl acetylenedicarboxylate is used, cycloadduct **183** was the only product isolated. The fact that the photolysis of **177** produces a nitrile ylide intermediate which could be trapped by an added dipolarophile (i.e., as **183**) eliminates a [2+2]cycloaddition of the azirine C=N double bond with the olefin and subsequent rearrangement of a hypothetical azatricyclo[2.1.1.0^{2,5}]hexane intermediate (**184**) as the mechanism for the formation of the azabicyclo[3.1.0]hex-2-ene system.

As unusual aspect of the intramolecular photocyclization of 2-allyl-substituted 2*H*-azirines was uncovered during a study of the photochemistry of *E*-2-(2-butenyl)-2-methyl-3-phenyl-2*H*-azirine (**185**). Irradiation of **185** in cyclohexane gave rise to one major product (>95%) which was identified as *endo*-3,6-dimethyl-1-phenyl-2-azabicyclo[3.1.0]hex-2-ene (**186**). The formation of the thermodynamically less favored *endo* isomer corresponds to a complete inversion of stereochemistry about the π system in the cycloaddition process. The only product obtained on irradiation of the isomeric Z-substituted 2H-azirine **187** was azabicyclohexene **186**. Photoisomerization about the C–C double bond of starting azirine (**185** or **187**) did not occur during the course of the irradiation. Thus azabicyclohexene **186** is the exclusive product obtained with both the (E) and (Z) isomers. Irradiation of 2-(1-methylallyl)-3-methyl-2-phenyl-2H-azirine (**188**) in cyclohexane (100% conversion) also gave **186** as the exclusive photoproduct.

As was pointed out by Huisgen,[63] 1,3-dipolar additions generally proceed via a "two-plane" orientation complex in which the dipole and dipolarophile approach each other in parallel planes. Inspection of molecular models of the above 2-allyl-substituted nitrile ylides indicates that the normal "two-plane orientation approach" of the nitrile ylide and the allyl π system is impossible as a result of the geometric restrictions imposed on the system. Consequently, the normal mode of 1,3-dipolar cycloaddition cannot occur here. The most reasonable mechanism to account for the cycloadditions observed with these systems involves the reaction of a bent nitrile ylide intermediate (carbenelike). Attack of the carbene carbon on the terminal position of the neighboring double bond generates a six-membered ring trimethylene intermediate. Collapse of this species results in the formation of the observed azabicyclohexene system. The photoconversion of the azabicyclohexenes (i.e., **178 → 179**) has been ratio-nalized in terms of a trimethylene derivative. It is particularly important to note that the above cycloaddition sequence proceeds in a nonconcerted manner and bears a strong resemblance to the stepwise-diradical mechanism suggested by Firestone to account for bimolecular 1,3-dipolar cycloadditions.[128,129] It is evi-dent from the available data that unless the dipole and dipolarophile approach each other in parallel planes, an alternate nonconcerted mechanism for dipolar cycloadditions occurs. The possibility that other dipolar cycloadditions of nitrilium betaines occur via a stepwise process now merits serious attention. In fact, the cycloaddition of benzonitrile oxide to the arylidenic double bond of 3-phenyl-4-arylidenisoxazol-5-ones has been proposed to proceed through the phenylnitrosocarbene form.[130] Since Padwa's original report of this phenomenon appeared,[127] a related intramolecular carbene type of 1,1 cycloaddition of a nitrile imine has been reported by Garanti and coworkers.[131]

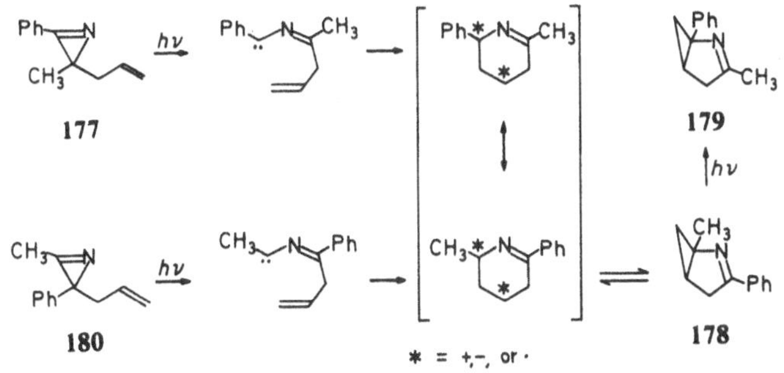

The above mechanism also accommodates the unusual stereochemical results observed with azirine **185**. As was pointed out earlier, the formation of the thermodynamically less favored *endo* isomer **186** from **185** corresponds to a complete inversion of stereochemistry about the π system in the cycloaddition process. The stereochemical results have been rationalized by assuming that collapse of the trimethylene derivative **189** to the thermodynamically more favored *exo* isomer **190** results in a severe torsional barrier on ring closure. Collapse of **189** to the thermodynamically less favored *endo* isomer **186** moves the phenyl and methyl groups increasingly further apart and accounts for the formation of the less stable product. Supporting evidence for this rationale was obtained from the irradiation of the isomeric *Z*-2-butenyl-2*H*-azirine **188**. Photolysis of this azirine resulted in the quantitative formation of the same *endo*-azabicyclohexene (**186**) and is perfectly consistent with the preferred kinetic closure of intermediate **189**. The formation of **186** from 2*H*-azirine **188** also provides convincing support for this interpretation.

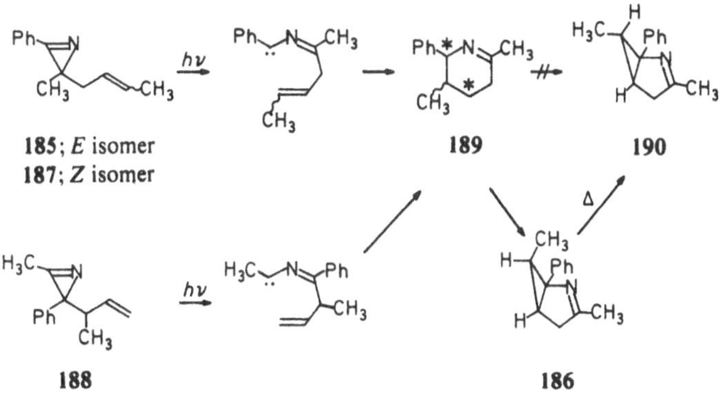

Fischer and Steglich were also able to generate an allyl-substituted nitrile ylide (**192**) by the thermal-induced 1,3-dipolar cycloreversion reaction of 3-oxazolinone **191** with elimination of carbon dioxide.[132] The resulting ylide was found to undergo 1,1 cycloaddition to give 2-azabicyclo[3.1.0]hexene (**193**). In this case the 1,1 cycloaddition of the thermally generated nitrile ylide proceeds with retention of configuration. That the nitrile ylide **192** is in fact generated in the thermolysis of **191** was demonstrated by trapping with dimethyl acetylenedicarboxylate.

The results obtained to date suggest that there are two pathways by which nitrile ylides react with multiple π bonds. The most frequently encountered path involves a bimolecular "parallel-plane approach of addends" and can be considered to be an orbital-symmetry-allowed [4+2]-concerted process. With this path, the relative reaction of dipolarophiles toward the nitrile ylide is controlled by the extent of stabilization afforded the transition state by interaction of the frontier orbitals of the two reactants. Electron-withdrawing substituents, which lower the dipolarophile LU energy, accelerate the reaction. The other path, which operates with the 2-allyl-substituted systems, occurs because the *p* orbitals of the olefinic group have been deliberately constrained to attack perpendicular to the nitrile ylide plane. Houk and Caramella have suggested that the 1,1-cycloaddition reaction is initiated by interaction of the terminal carbon of the olefin with the second LUMO of the nitrile ylide.[125] The second LUMO of the dipole is perpendicular to the ylide plane and presents a large vacancy at C_1 of the dipole for attack by the terminus of the neighboring double bond, without the possibility of simultaneous bonding at the C_3 carbon. In fact, the HOMO and second LUMO of the bent nitrile ylide bear a strong resemblance to the HOMO and LUMO of a singlet carbene. A factor which undoubtedly plays an important role in the intramolecular 1,1-cycloaddition reaction involves the interaction of the secondary orbitals of the dipole and dipolarophile. With nitrile ylides, the in-plane vacant orbital is of lower energy than the vacant π orbital. Consequently, stabilization of the transition state can be enhanced by interaction of this in-plane orbital with the dipolarophile HOMO orbital. For this to occur, a contortion away from the strictly parallel-plane approach of the dipole and dipolarophile would be necessary. With the allyl-substituted 2*H*-azirines, the transition state actually involves a geometry where the *p* orbitals of the olefinic group have been deliberately constrained to attack perpendicular to the nitrile ylide plane. Thus the 1,1-cycloaddition reaction involves appreciable interaction of both the in-plane and out-of-plane π-

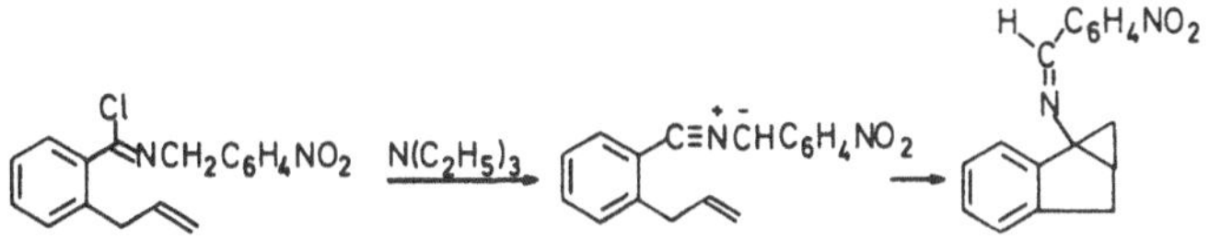

194   195   196

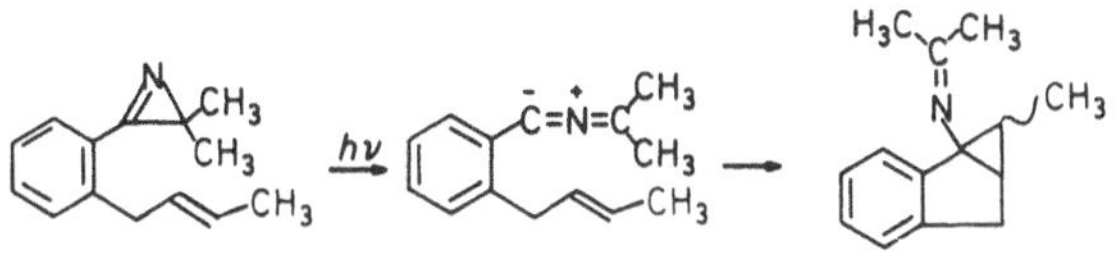

197   198   199

unoccupied orbitals of the dipole with the dipolarophile-filled orbitals. This secondary orbital interaction will significantly enhance the rate of the intramolecular 1,1-cycloaddition with unactivated olefins. Ordinary olefins react so sluggishly with nitrile ylides that their bimolecular rate constants cannot be measured. Clearly, there has been a marked enhancement of the rate of the intramolecular 1,1-cycloaddition. It should be noted, however, that while the secondary orbital effect is important, the relative reactivities of a series of π-substituted double bonds toward internal 1,1-cycloaddition is still controlled by the highest-occupied molecular orbital of the nitrile ylide.[133]

Several additional examples in the literature demonstrate the generality of the 1,1 trapping of nitrile ylides.[134-136] Thus treatment of *o*-allyl-substituted imidoyl chloride **194** with triethylamine gave benzobicyclo[3.1.0]hex-2-ene **196**. Irradiation of the closely related methyl-substituted azirine system **197** produced a mixture of *endo*- and *exo*-benzobicyclohexenes **199** in quantitative yield.[135] No detectable quantities of the isomeric 1,3-dipolar adduct were observed in either system. In the case of **197**, the major product (*exo*-**199**) is the thermodynamically more favored *exo* isomer.

Another system which was found to undergo smooth intramolecular 1,1 cycloaddition was 2*H*-azirine **200**.[135] The exclusive formation of a 1:1 cycloadduct with this system indicates that the nitrile ylide produced is capable of undergoing carbene type of addition to a vinyl group just as long as there are no considerable bond distortions involved in the transition state for the internal cycloaddition.

In contrast to the above systems, irradiation of the 3-carbomethoxy-

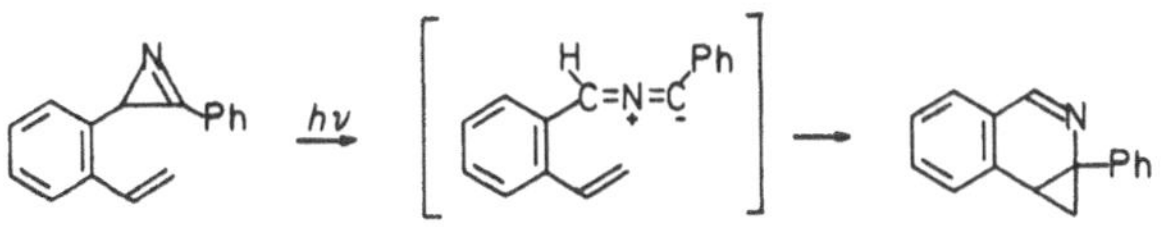

200   201   202

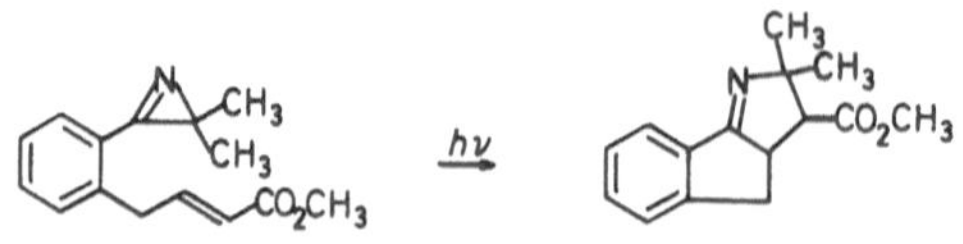

203    $\xrightarrow{h\nu}$ 204

205; X = CH₂   207   206; X = CH₂
207; X = O   208; X = O

substituted 2*H*-azirine **203** resulted in an intramolecular 1,3-dipolar cycloaddition. Related intramolecular 1,3 cycloadducts were also isolated from the
photolysis of azirines **205** and **207**. The transition state for cycloaddition with
the nitrile ylides generated from the irradiation of azirines **205** and **207** allows
easy attainment of the "parallel-plane approach" of the dipole and olefin and,
consequently, intramolecular 1,3-dipolar cycloaddition readily occurs.

In contrast to the *o*-allylphenyl substituted 2*H*-azirines (**197** and **200**)
which undergo 1,1 cycloaddition, azirine **203** cycloadds exclusively in the 1,3
sense. The formation of the 1,3-dipolar cycloadduct **204** is quite consistent with
the principles of frontier MO theory. With this system the rate of internal 1,3-
dipolar cycloaddition of the nitrile ylide derived from **203** proceeds much more
rapidly than that of the ylides derived from **197** and **200**. This is to be expected
since electron-withdrawing substituents on the double bond narrow the HO
dipole–LU dipolarophile gap. In order for the 1,3-dipolar cycloaddition of **209**
to occur, a slight distortion away from the strictly parallel-plane approach of
the dipole and dipolarophile will be necessary. The interplay of entropy and

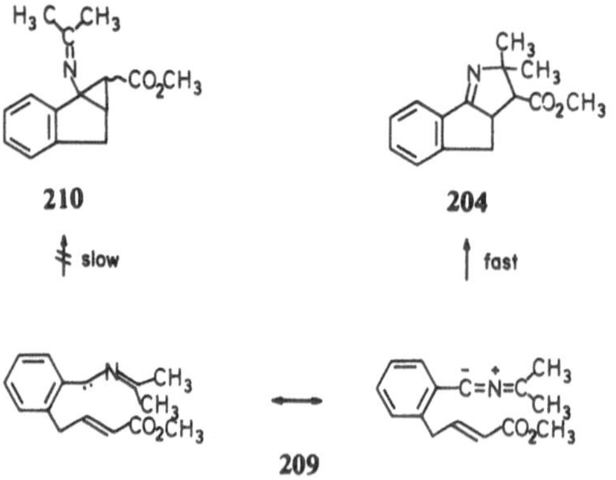

210   204

‡ slow   ↑ fast

209

enthalpy factors controls the rate-determining activation process. With azirine **203**, the enthalpy term is the dominant factor and the internal 1,3-dipolar cycloaddition process wins out.

Attempts to prepare a benzobicyclo[2.1.0]pentane from the irradiation of an *o*-substituted 2*H*-azirine **211** resulted only in the isolation of an *N*-alkylideneindene-3-amine **214**. The formation of **214** has been suggested to proceed via a nitrile ylide intermediate **212** which undergoes dipolar electrocyclization followed by a 1,5-sigmatropic shift of the initially formed isoindene ring **213**. This mechanism was supported by the trapping of nitrile ylide **212** with added dipolarophiles.

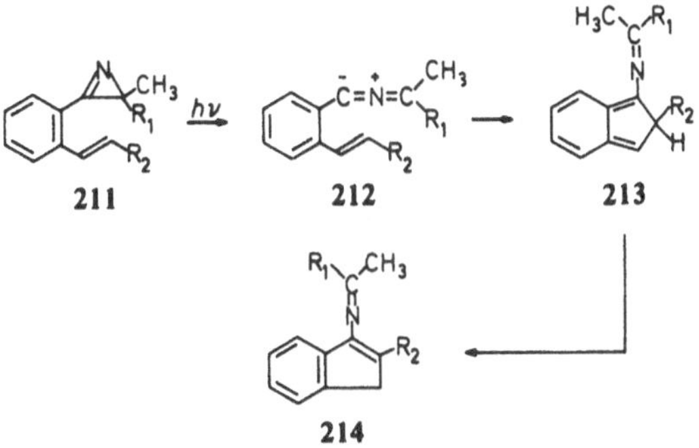

The intramolecular photocycloaddition reactions of a number of *o*-allyloxyphenyl-substituted 2*H*-azirines were also examined by Padwa's group.[136] These azirines were found to undergo intramolecular 1,1 and/or 1,3 cycloaddition depending on the substituent groups attached to the 2-position of the ring. Thus irradiation of the disubstituted 2*H*-azirine **215** resulted in the formation of a 1,1 cycloadduct, whereas the related 2*H*-azirine **217** produced the 1,3 cycloadduct **218**. The exclusive formation of a 1,3-dipolar cycloadduct from azirine **217** suggests that the mode of cycloaddition of the *o*-allyloxyphenyl system is markedly dependent on the nature of the substituent groups attached to the 2-carbon atom of the nitrile ylide. This was further borne out by the

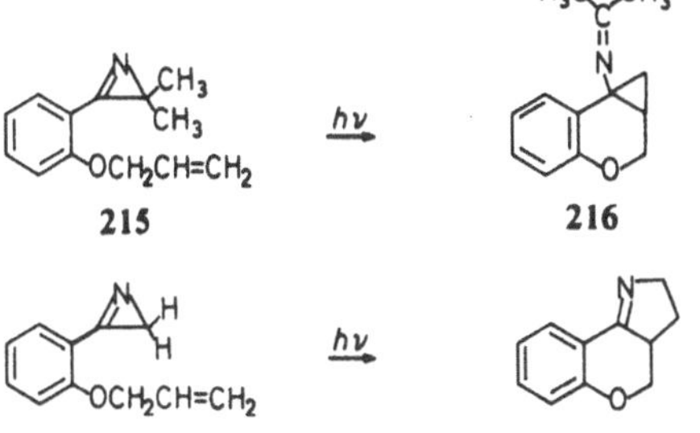

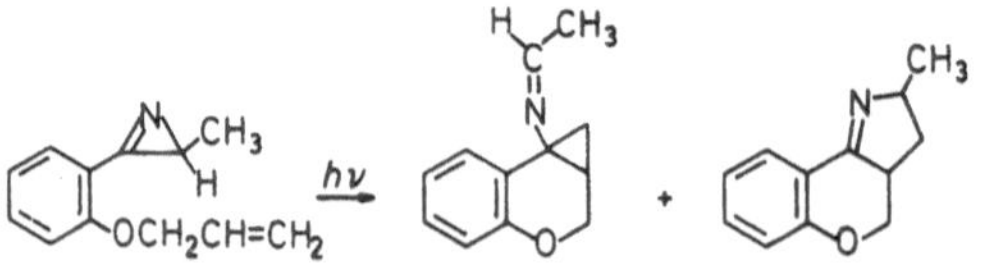

219   220   221

irradiation of the monomethyl-substituted 2H-azirine **219** which produced a 1:1 mixture of the 1,1 and 1,3 cycloadducts **220** and **221**. Another case where an oxyphenyl-substituted 2H-azirine was found to undergo both 1,1- and 1,3- dipolar cycloaddition was encountered in the photolysis of **222**. Irradiation of this compound afforded a 3:4 mixture of cycloadducts **223** and **224**. The isola-

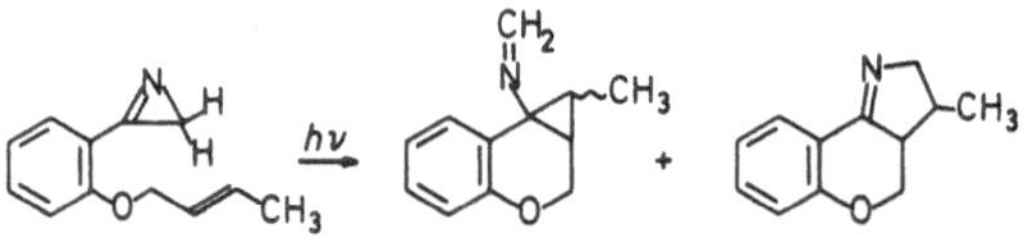

222   223   224

tion of cycloadduct **227** from the photolysis of **225** clearly establishes that the substituent effect encountered in these nitrile ylide cycloadditions is electronic rather than steric in nature. A steric effect would have been expected to produce a 1,1 cycloadduct since the trifluoromethyl group is larger than a methyl group.[137] This was clearly not the case.

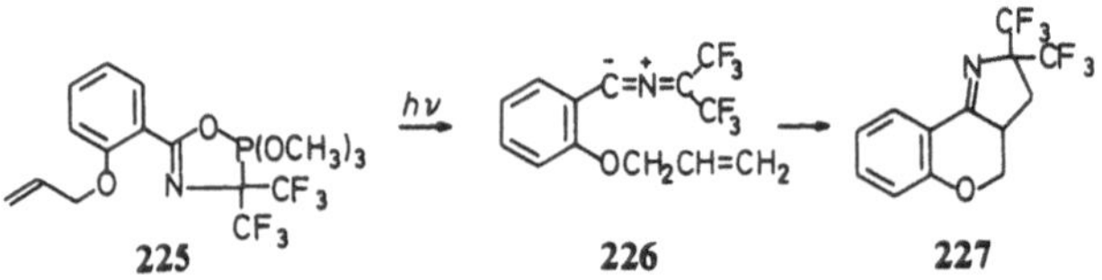

225   226   227

   Inspection of molecular models of these o-alloxyphenyl-substituted nitrile ylides shows that two paths for cycloaddition are possible depending on the geometry of the nitrile ylide. The parallel-plane approach of addends produces a 1,3 cycloadduct and occurs when the dipole possesses linear geometry. The alternate 1,1 cycloaddition process occurs when the dipole possesses bent geometry. Recent MINDO/3 calculations by Houk reveal that attachment of a phenyl group at the C_1 atom of the nitrile ylide significantly lowers the energy separation between the linear and bent forms by about 10–15 kcal/mol.[125] The calculations indicate that the o-allyloxyphenyl-substituted nitrile ylide system will be less bent and easier to make linear than the corresponding parent system (i.e., HCNCH$_2$). As the dipole becomes less bent, the C_1N bond length will shorten and NC$_3$ will lengthen, as expected for going toward a propargyl-type structure. Since the energy difference between the nonplanar bent and linear forms is quite small with these *o*-oxyallylphenyl-substituted nitrile ylides, the preferred mode of approach is strongly dependent on the substituent groups present on the nitrile ylide. The nitrile ylide species becomes more carbenelike as methyl or electron donor groups are added and is more likely to undergo the 1,1-cycloaddition reaction:

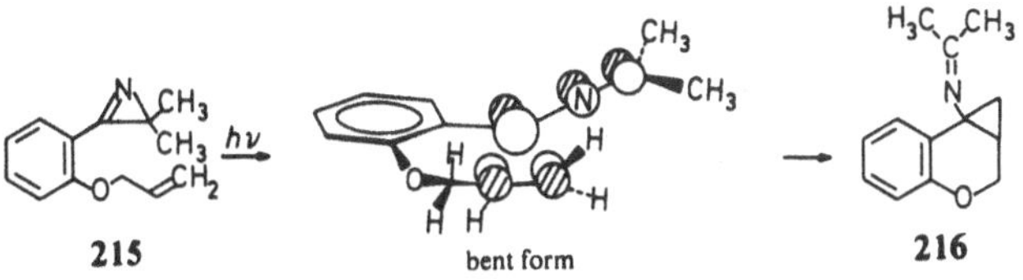

| 215 | bent form | 216 |

Placing electron-withdrawing groups at C_3 (i.e., H, CF_3, $C_6H_4NO_2$) favors linearization of the nitrile ylide and promotes the 1,3 cycloaddition:

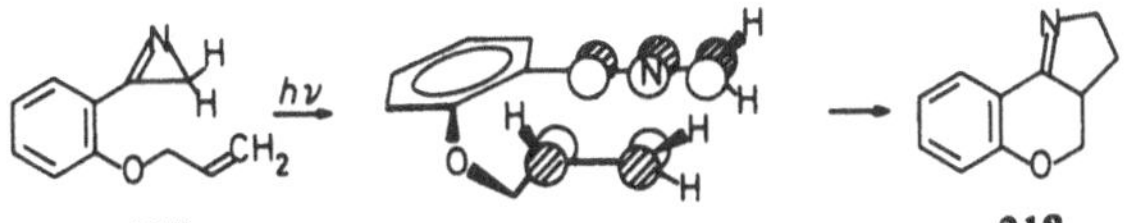

| 217 | linear form | 218 |

The results show that when the energy difference between the nonplanar bent and linear forms is small, substituent effects can play an extremely important role in determining the course of the intramolecular cycloaddition reactions of nitrile ylides. The formation of a mixture of 1,1 and 1,3 cycloadducts from the irradiation of **219** has been attributed to the fact that the energy levels of the bent and linear forms must lie very close to each other.[136] The isolation of both 1,1 and 1,3 cycloadducts from the irradiation of **222** is related to the fact that the methyl substituent on the double bond diminishes the rate of 1,3-dipolar cycloaddition and enhances the rate of the 1,1 cycloaddition.

The exclusive formation of a 1,3-dipolar cycloadduct **229** from the irradiation of azirine **228** has been attributed to a significant lowering of the activation energy associated with the 1,3-cycloaddition reaction and to a substantial increase in the activation energy for the 1,1-cycloaddition process.[136]

The isolation of 1,1 and 1,3 cycloadducts from the photolysis of 2*H*-azirines bearing unsaturated π bonds clearly indicates that the spatial

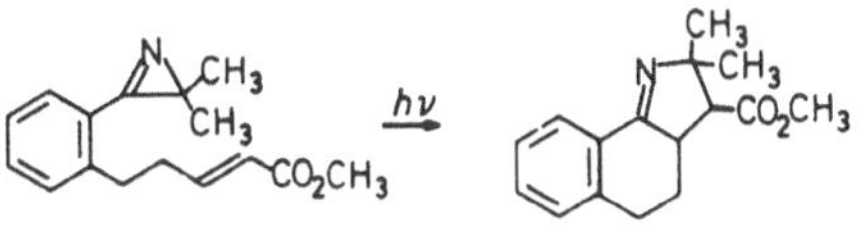

| 228 | 229 |

relationship of the dipole and dipolarophile plays an important role in controlling the intramolecular dipolar cycloaddition reactions of nitrile ylides. The primary spatial requirement for intramolecular 1,3-dipolar cycloaddition is that the distance between the two reacting centers should be short enough so that effective three-center overlap of the 1,3 dipole with the dipolarophile occurs. For concerted 1,3-dipolar cycloaddition to take place, the atoms of the dipolarophile should be arranged in such a way as to allow their p orbitals to lie in a plane parallel to the plane of the nitrile ylide. This can happen when the methylene chain between the azirine ring and the alkene end is extended to three carbon atoms. For example, irradiation of azirine 230 gives Δ^1-pyrroline 232 in quantitative yield.[138] In this case the methylene chain is sufficiently long to allow the dipole and olefinic portions to approach each other in parallel planes.

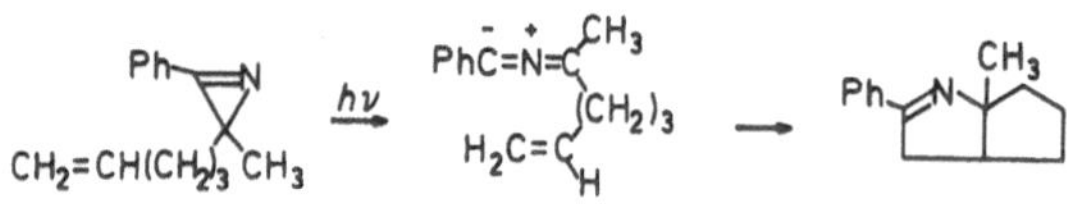

230   231   232

The exclusive orientation observed in this reaction is unusual and cannot be adequately accounted for on the basis of frontier orbital theory. A similar inconsistency was also found on irradiation of azirine 233. The orientation of the cycloaddition reaction of aldehyde 233 proceeds in an alternate regiochemical sense from that observed with related bimolecular nitrile ylide–aldehyde cycloadditions where one obtains only Δ-oxazolines 235.[69,70] In

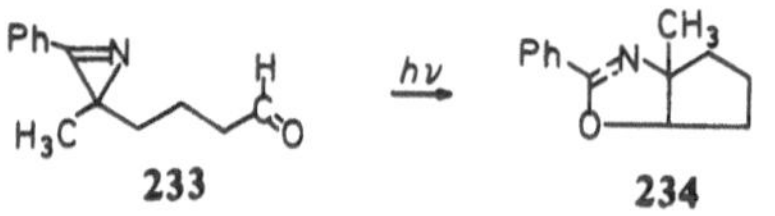

233   234

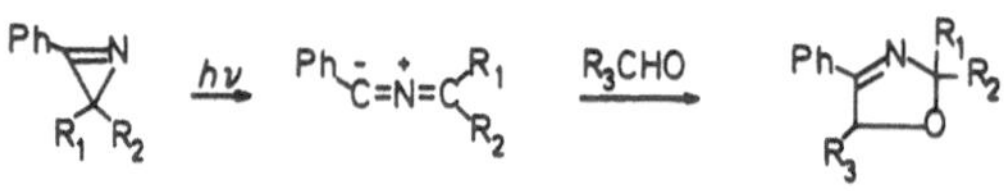

235

a somewhat related case, Schmid and coworkers reported on the photoisomerization of dihydroisoxazole 236 to dihydroxazole 239.[72] The reaction was proposed to proceed via a transient azirine (i.e., 237). This intermediate was not isolated, but was suggested to undergo rapid ring opening to nitrile ylide 238, which cyclized to the observed photoproduct via an internal 1,3-dipolar cycloaddition reaction. The intramolecular cycloaddition reactions of azirines 230 and 233 clearly indicate that geometrical factors can force the cycloaddition reaction to occur in a manner opposite to that normally encountered. The inversion of regiospecificity must be related to steric effects which destabilize the transition states for formation of the alternate bridged structures. Similar "orientation inversions" have been reported with a number of other 1,3 dipoles which undergo intramolecular dipolar cycloadditions.[139–142]

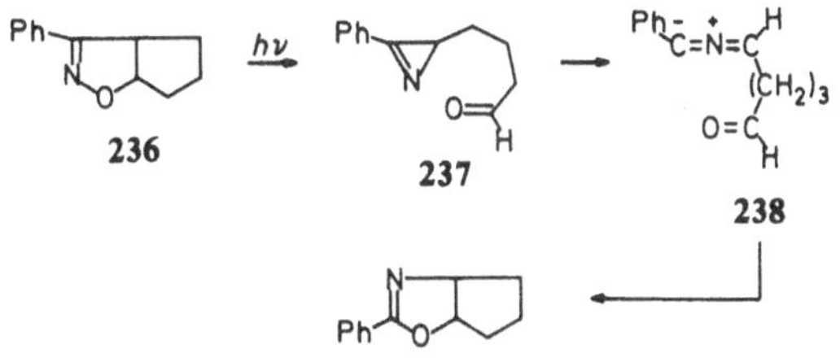

VII. FURTHER REACTIONS OF 2H-AZIRINES

All of the reactions discussed above have involved the photocycloadditions of 2H-azirines with multiple π bonds. These studies have rigorously established nitrile ylides as useful intermediates in a variety of synthetic applications. Some recent work in the literature has shown that the photolysis of certain 2H-azirines can lead to new and interesting photochemistry.

A new synthesis of cycloalkanones was devised and is based upon the photolysis of spiroazirines in alcohol followed by aqueous hydrolysis.[143] Irradiation of spiroazirine **240** in methanol resulted in the quantitative formation of imine **241**. Clean conversion to benzaldehyde and the corresponding cycloalkanone **242** was accomplished by treating the photoproduct with a 10% aqueous hydrochloric acid solution.

In contrast to the above findings, irradiation of 2-phenyl-1-azaspiro[2.2]pent-1-ene (**243**) in methanol resulted in a Griffin-type fragmentation[144] and produced ethylene and 2-phenylazirinylidene (**244**).[145] Three-membered rings are known to undergo [3→2+1] cycloelimination on irradiation.[144,146] For example, cyclopropane has been photolyzed in the vapor phase and gives methylene and ethylene,[147] while photolysis of ben-

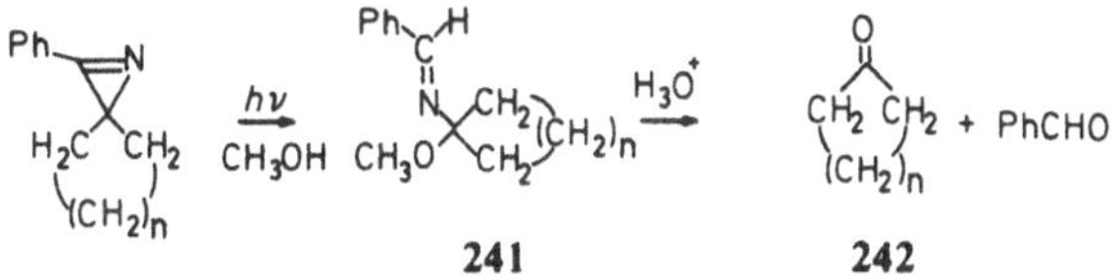

240a; $n = 1$
b; $n = 2$
c; $n = 3$

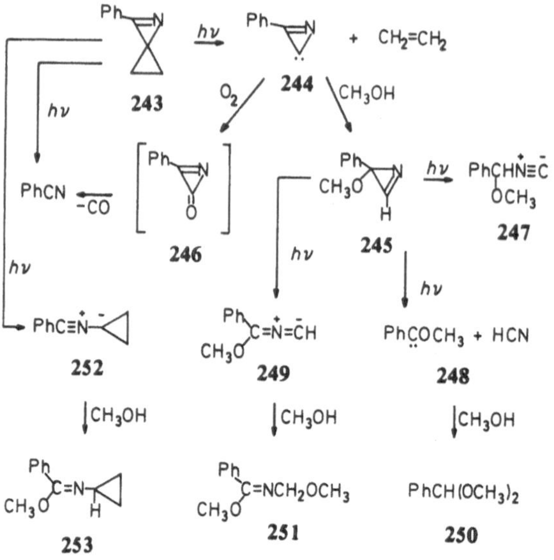

Ph—⟨N⟩ ⟶ hν ⟶ Ph—⟨N⟩ + CH₂=CH₂

243   O₂ / **244** \ CH₃OH

246   **245**   hν   PhĊHN≡C̄—OCH₃   **247**

PhCN ⟵ -CO

PhC≡N—◁   **252**

Ph\C=N=ĊH / CH₃O   **249**

PhĊOCH₃ + HCN   **248**

| CH₃OH

Ph\C=N—◁ / CH₃O H   **253**

Ph\C=NCH₂OCH₃ / CH₃O   **251**

PhCH(OCH₃)₂   **250**

zylcyclopropane leads to extensive fragmentation[148] and produces a number of hydrocarbons including ethylene and benzylcarbene. A comprehensive review of the generation of carbenes by photochemical cycloelimination from cyclopropanes has appeared.[149] The formation of the major products produced from the irradiation of spiroazirine **243** has been attributed to an initial photocycloelimination step.[145] This photocycloelimination generates ethylene and the extremely novel carbene, 2-phenylazirinylidene (**244**), which is subsequently trapped by methanol to give azirine **245**. 2-Phenylazirinylidene (**244**), by analogy with diphenylcyclopropenylidene,[150] is a carbene whose normal electrophilicity is suppressed as a result of conjugation of the double-bond electrons of the azirine ring with the vacant p orbital of the carbene. When the irradiation of **243** was carried out in the presence of oxygen, benzonitrile and carbon monoxide were formed. In the absence of oxygen, an extremely small quantity of benzonitrile was formed and is presumably derived by competitive cycloelimination from **243**. The formation of benzonitrile has been attributed to the intermediacy of 2-phenylazirinone (**246**) as a transient intermediate. Hassner and Taylor have previously shown that azirinones are unstable and readily lose carbon monoxide to form nitriles,[151] thereby providing good analogy for this reaction. On further photolysis, azirine **245** is converted into isocyanide **247**,

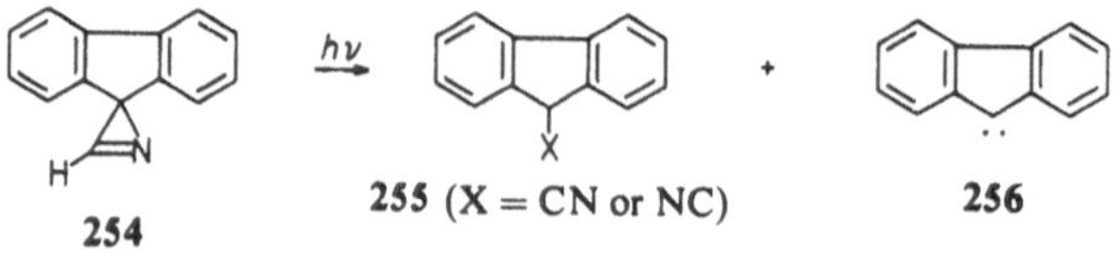

254   hν   **255** (X = CN or NC)   +   **256**

phenylmethoxycarbene **248** and nitrile ylide **249**. Both the carbene **248** and the 1,3 dipole were trapped with methanol to give **250** and **251**, respectively. Hafner and Bauer have reported that the related spiro-[2H-azirine-2,9¹-fluorene] (**254**) leads to a mixture of 9-cyano- and 9-isocyanofluorene on photolysis.[152] These workers also found that **254** undergoes loss of HCN and generates 9-fluorenylidene (**256**), thereby providing good precedent for the transformations observed with azirine **245**.

A similar fragmentation of a 2H-azirine has been reported by Shechter and Magee, who found that the irradiation of 2-cyano-2-methyl-3-phenyl-2H-azirine (**257**) gave benzonitrile and acrylonitrile. These two products are derived from carbenic collapse of **257** and hydrogen migration in the methylcyanocarbene generated.[153]

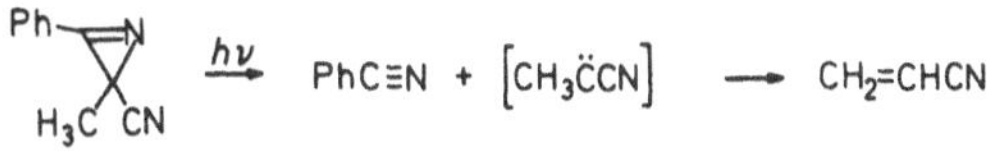

$$Ph\text{–azirine} \xrightarrow{h\nu} PhC\equiv N + [CH_3\ddot{C}CN] \longrightarrow CH_2=CHCN$$

257

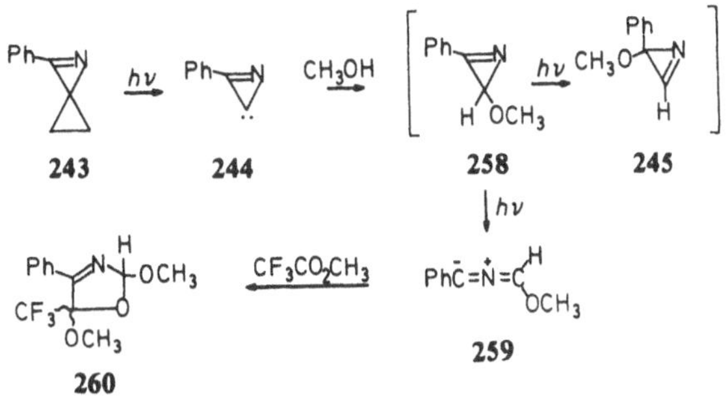

A most unusual result was encountered when the irradiation of **243** was carried out in pentane in the presence of both methanol and methyl trifluoro-acetate (excess).[145] Under these conditions, the only cycloadducts observed consisted of a mixture of the two stereoisomers of 3-oxazoline, **260**. This result suggests that 2-phenylazirinylidene (**244**) reacts with methanol to give mainly azirine **258**, which is subsequently converted to nitrile ylide **259** (and thus cycloadduct **260**) on further irradiation. The formation of **245** has been explained in terms of a photoinduced methoxy migration of **258** which competes with C–C bond cleavage of the azirine ring. Ciabattoni and Cabell[154] have previously reported that 2-chloro-2H-azirines undergo ready isomerization at room temperature via a 2π-electron azacyclopropenyl cation. A similar mechanism would rationalize the apparent photoconversion of azirines **258** and **245**. The results obtained with this spiro azirine system indicate that cyclo-elimination of ethylene from **243** is much more efficient than C–C bond scission of the azirine ring. Undoubtedly, the stability of the aromatic carbene **244** contributes to this mode of cleavage.

The formation of small quantities of methyl N-cyclopropylbenzimidate **253** from the irradiation of azirine **243** in methanol has been formulated as proceeding via nitrile ylide **252** which undergoes subsequent addition of methanol. The trapping of this ylide with methanol occurs in a different sense from that observed with most other nitrile ylides. The isolation of **253** implies that the largest HO coefficient of nitrile ylide **252** rests on the cyclopropyl carbon atom. The reluctance to develop a positive charge on the cyclopropyl carbon atom presumably contributes to this reversal in regioselectivity. The major cycloadduct **261** obtained from carrying out the irradiation of azirine **243** in pentane in the presence of methyl trifluoroacetate corresponds to the trapping of the nitrile ylide **252** in the reverse fashion from that normally observed.

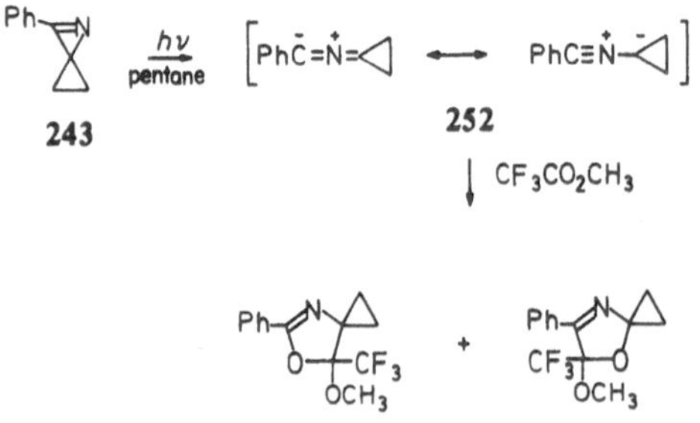

The formation of oxazoline **265** from the irradiation of azirine **263** has been formulated as proceeding via a nitrile ylide intermediate (**264**) which transfers a proton from the neighboring hydroxyl group and then collapses to the observed product.[145] Support for the nitrile ylide intermediate was provided by the irradiation of **263** in benzene which had been saturated with D_2O. A single deuterium atom was incorporated at the 2-position of the oxazoline ring, as expected for an intermediate corresponding to **264** in this reaction. The complete absence of the isomeric N-benzylidene epoxide would suggest that the zwitterion produced on transfer of the proton from the nitrile ylide prefers to collapse to a five-membered rather than a three-membered ring.

Photolysis of the hydroxyazirine with two additional methylene units afforded furanamine **267** was also explained in terms of an internal trapping of a transient nitrile ylide.[145]

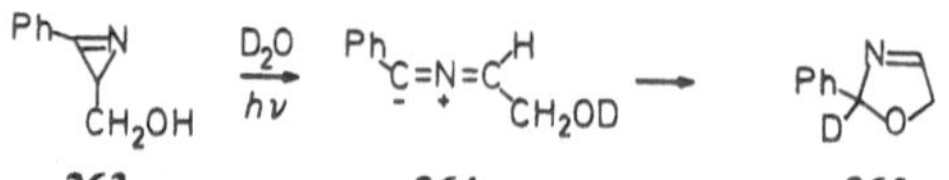

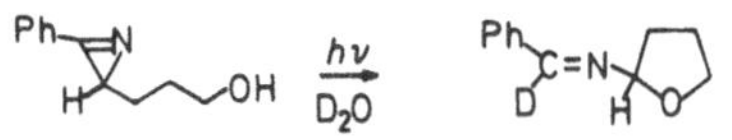

Irradiation of a series of hydroxymethyl-2*H*-azirine derivatives (**268**) which contain good leaving groups was found to give *N*-vinylimines (**270**) via a novel 1,4-substituent shift.[155] The results indicate that the migrating substituent (X) must be a reasonably good leaving group in order for the rearrangement to occur. Irradiation of a nitrile ylide intermediate (**269**) in these reactions was demonstrated by trapping experiments.

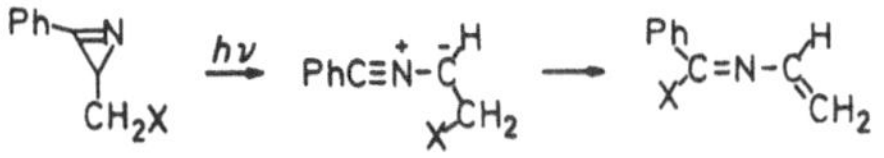

X = Cl, Br, OCOCH$_3$, OCOCF$_3$, or OCOAr

The observation that the rearrangement of the nitrile ylide derived from 3-phenyl-2-trifluoroacetoxymethyl-2*H*-azirine (**271**) proceeds at a faster rate (200 times) than that of the ylide derived from 2-acetoxymethyl-3-phenyl-2*H*-azirine (**272**) provides support for the intermediacy of an ion pair in the rearrangement

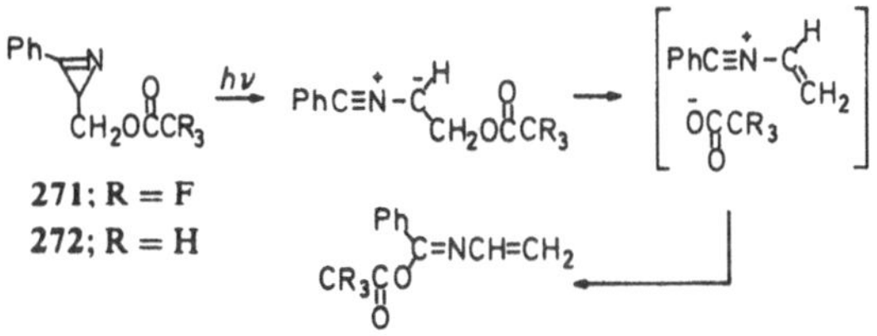

271; R = F
272; R = H

of these hydroxymethyl-2*H*-azirine derivatives. The rate of a series of substituted 3-phenyl-2*H*-azirine-2-methanol benzoate derivatives were also studied.[156] Electron-withdrawing substituents in the *para* position were found to facilitate the rearrangement while electron-donating groups retard the 1,4-substituent shift. Both the sign and magnitude of the reaction constant ($\rho = +2.15$) indicate that the transition state for the rearrangement has substantial negative charge development. Stern–Volmer treatment of the reaction rates clearly demonstrates that the rate of the rearrangement is directly related to leaving group ability and that the migrating substituent must be a reasonably good leaving group in order for the reaction to occur.

VIII. THERMAL REACTIONS OF 2H-AZIRINES

In contrast to the well-defined photochemical behavior of 2H-azirines, the thermal reactions of these compounds have been studied less thoroughly. It was not until 1968 that the first report of a thermolytic reaction of a 2H-azirine was made by Taniguchi and coworkers.[157] It should be noted that the products derived from the thermolysis of vinyl azides have, on occasion, been suggested to proceed via 2H-azirines as transient intermediates.[158] Since the first publication by Taniguchi in 1968, a large number of reports dealing with the thermal reactions of a variety of 2H-azirines have appeared in the literature over the past decade.

As was already mentioned in this chapter, the direction of opening of the 2H-azirine ring is markedly dependent on the mode of activation. Irradiation of the azirine ring results in C–C bond cleavage and formation of a nitrile ylide. In contrast to this reaction path, the products formed on thermolysis of 2H-

$$R_1\text{C}=\text{N}=\text{C}\!\begin{smallmatrix}R_2\\R_3\end{smallmatrix} \quad \xleftarrow{h\nu} \quad \begin{smallmatrix}R_1\\ \\R_2\ R_3\end{smallmatrix}\!\!\overset{N}{\underset{}{\bigtriangleup}} \quad \xrightarrow{\Delta} \quad \overset{:\text{N}:}{\underset{R_1}{\text{C}}}\!=\!\text{C}\!\begin{smallmatrix}R_3\\R_2\end{smallmatrix}$$

azirines can best be rationalized in terms of an equilibration of the heterocyclic ring with a transient vinylnitrene. Thus products formed from the thermolysis of 2H-azirines are generally consistent with C–N bond cleavage. In some recent publications, however, Ghosez[159] and Bergman[160,161] report that, under high-temperature conditions, the products obtained can be derived by C–C bond rupture. Recent MO calculations by Devaquet[162] indicate that CN bond rupture of the 2H-azirine ring is much easier than CC bond rupture. The vinylnitrene derived from CN ring opening has been calculated[162] to have no competitive reaction path other than reclosure to the azirine ring since no energy barrier exists for reclosure. Devaquet has pointed out that if thermal CC bond rupture is achieved, the intermediate generated has a longer lifetime than the species obtained by CN cleavage. This fact can explain the effect of substitution on the thermal reactivity of 2H-azirines.[160,161] According to the calculations,[162] if a stabilizing substituent is on carbon 3 of the azirine ring, the ring-opened species is always stabilized and the easiest reaction will be CN rupture. If the stabilizing substituent is on carbon 2, only the species derived from CC cleavage is stabilized. This would tend to reduce the barrier to CC cleavage and would account for the fact that this mode of opening occurs with certain systems and not others.

Strong support for the existence of an azirine–vinylnitrene equilibration comes from work carried out by Nishiwaki[25,31,163] and Padwa.[102] Nishiwaki was able to show that the vinylnitrene generated from the thermolysis of azirine 273 can actually be trapped with phosphines, resulting in the isolation of adduct

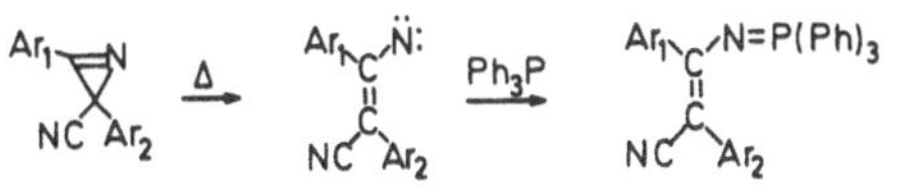

273   274   275

a; $Ar_1 = Ar_2 = Ph$
b; $Ar_1 = Ph$; $Ar_2 = p\text{-}ClC_6H_4$
c; $Ar_1 = p\text{-}ClC_6H_4$; $Ar_2 = Ph$

275. A similar product was also isolated by heating azirine **126** in the presence of tris(dimethylamino)phosphine.[102]

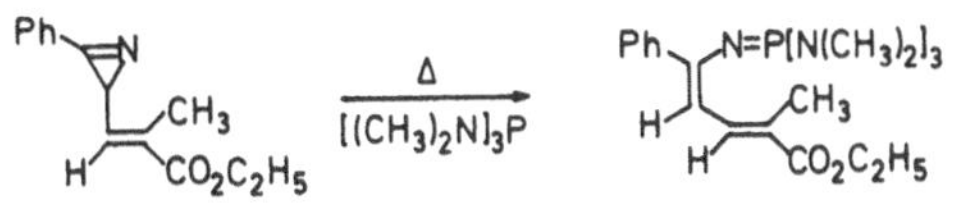

126   276

Another apparent example of the trapping of a vinylnitrene has been reported by Nomura and coworkers.[164] These workers found that the thermolysis of vinylazides in the presence of electron-deficient alkenes resulted in the formation of *N*-vinylaziridines **277** and Δ^1-pyrrolines **278**. Although this reaction was interpreted in terms of the trapping of a vinylnitrene intermediate (path **a**), it seems that concerted 1,3-dipolar cycloaddition of the azide followed by extrusion of nitrogen from a transient cycloadduct **279** could also rationalize the results (path **b**).

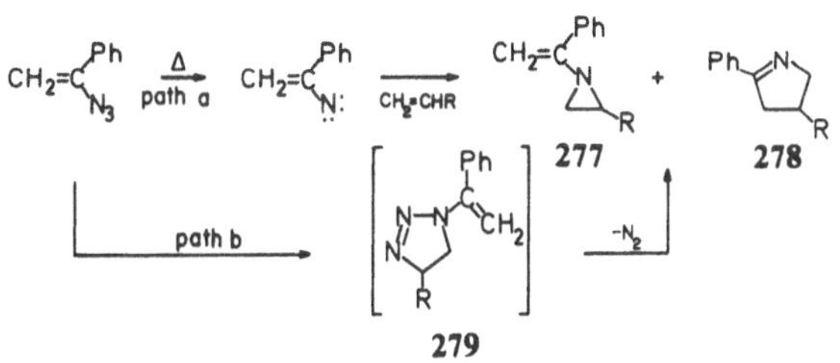

277   278

279

Taniguchi and coworkers were the first to note that the stability of the azirine ring is markedly dependent on the substituent groups present.[24] Azirines bearing a hydrogen atom in the 3-position were found to be notoriously reactive, especially in the presence of oxygen. Disubstitution with alkyl or aryl groups at the 2-position of the 2*H*-azirine ring resulted in a substantial increase in the stability of these molecules. The activation energy of a series of 2*H*-azirines towards thermal reorganization was studied by Taniguchi and coworkers.[165] They found that the reactivity of the azirine ring decreased as a function of the 3-substituent in the order $H > CH_3 > Ph$:

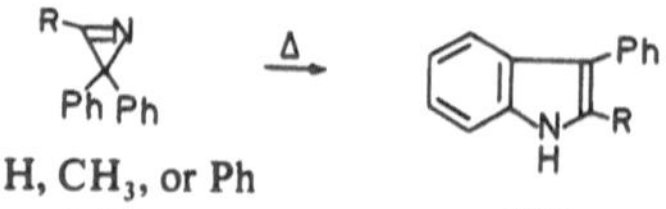

R = H, CH₃, or Ph

280   281

Polyfluoroazirines were also found to be very unstable and rapidly polymerized on standing at room temperature.[166]

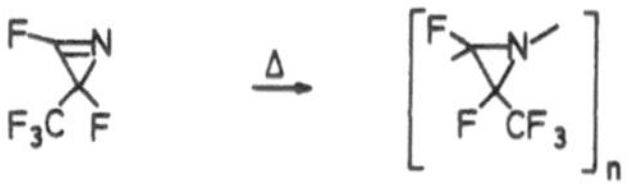

282   283

3-Unsubstituted 2H-azirines were assumed to be intermediates in the decomposition of terminal vinyl azides, but for a long time they escaped detection.[167-170] The thermal decomposition of terminal vinyl azides often results in the formation of nitriles and polymeric material without any detectable quantities of 2H-azirines. Under certain conditions, however, it is possible to detect the 2H-azirines spectroscopically and even to isolate them under controlled experimental conditions. The parent 2H-azirine **284** was recently prepared by flash vacuum pyrolysis of vinyl azide at 400°C and was characterized by its pure rotational spectrum.[171] The stability of this compound was somewhat greater than initially expected. On standing at room temperature, it decomposed to acetonitrile.

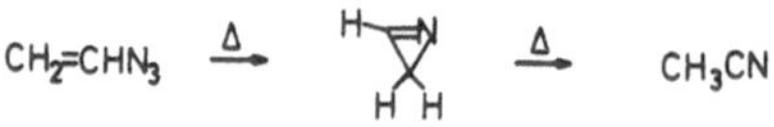

284

Hafner and Bauer were able to isolate azirine **254** in high yield from a low-temperature photolysis of the corresponding vinyl azide.[152] Thermolysis of **254** gave imine **285** and fluorenone **286**. The thermal reaction was proposed to proceed via the 9-fluorenylidene carbene, produced by elimination of HCN from **254** followed by a subsequent reaction with starting material or with oxygen.

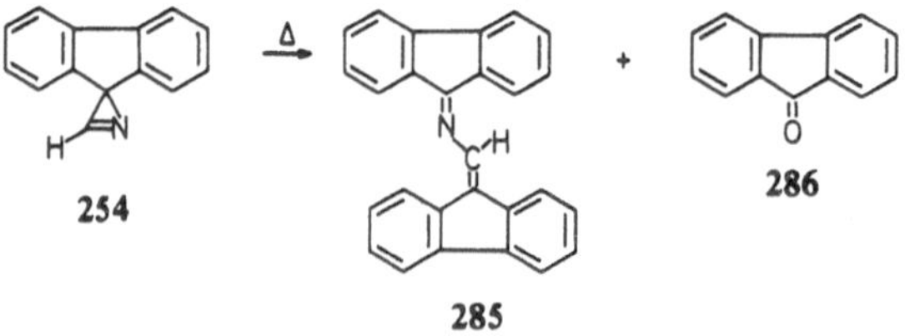

254   286

285

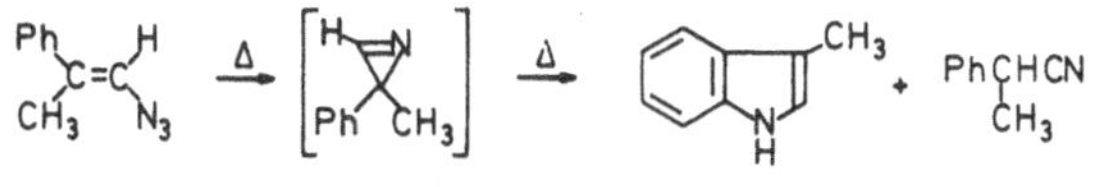

287   **288**   **289**   **290**

The products derived from the thermolysis of vinyl azide **287** have been suggested to arise via a transient 2H-azirine **288**.[158] In fact, recent studies by Taniguchi and coworkers showed that thermolysis of 2-phenyl-3-unsubstituted 2H-azirines gave indoles and phenylacetonitrile in a 1:1 ratio.[23] Similar products (**292, 293**) were produced in the same ratio from the thermolysis of the corresponding pure β-azidostyrene.

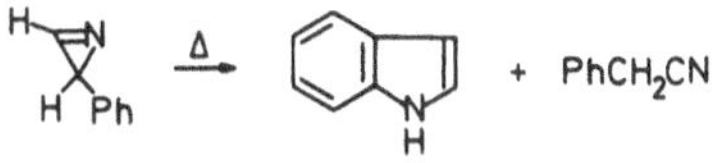

291   **292**   **293**

Heating a sample of the closely related azirine **294** gave 3-phenylindole (**297**) and dihydropyrazine **298**. The formation of **297** was suggested to proceed via a vinylnitrene intermediate which cyclizes and then undergoes a 1,5-sigmatropic shift.[165]

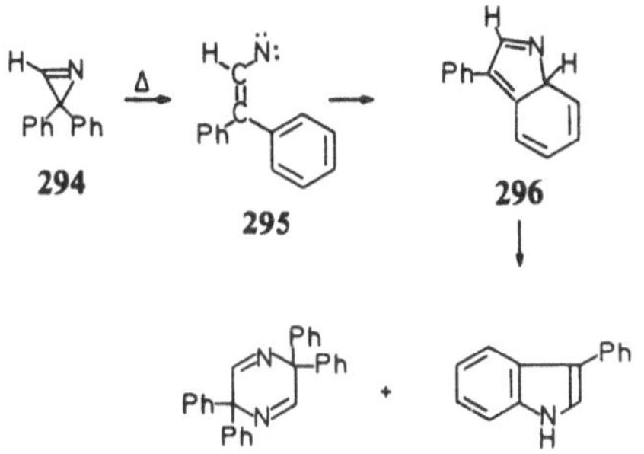

294   **295**   **296**

298   **297**

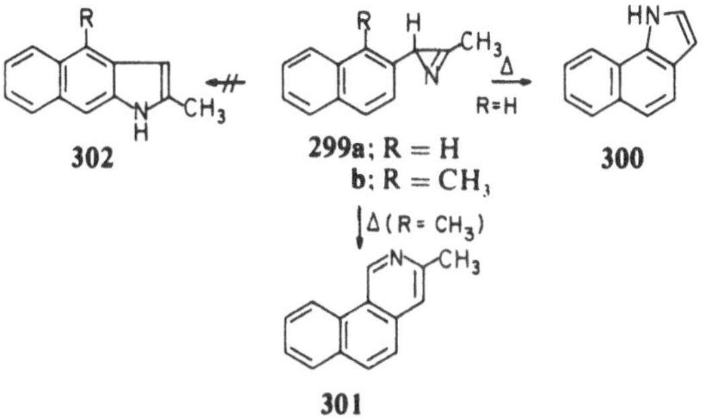

302   **299a**; R = H   **300**
                b: R = CH₃

Δ (R = CH₃)

301

The thermal chemistry of a number of related aryl-substituted 2*H*-azirines were also studied. The 2-naphthyl-substituted azirines **299** were found to rearrange to benz[*g*]indole **300**. The selectivity encountered here was attributed to the higher energy content of the intermediate leading to the alternative benz[*f*]indole system **302**. Heteroaromatic systems were also found to undergo thermal rearrangement via highly selective reaction paths[172]:

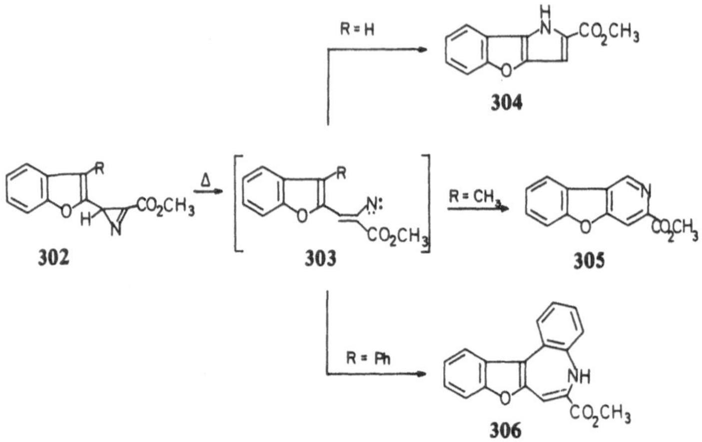

The selective formation of dihydropyrrole **308** from the thermolysis of *bis*-azirine (**307**) was rationalized on the same basis.[172]

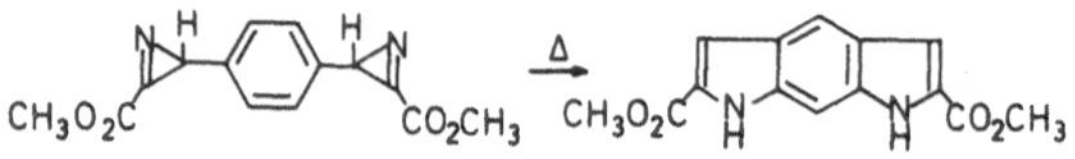

Bowier and Nussey studied the thermolysis of 2,3-di-(2,4,6-trideuterio-phenyl)-2*H*-azirine (**309**) under a variety of experimental conditions.[28] Heating a sample of **309** at 220°C produced indole **310** as the major reaction product, presumably via the intermediacy of a vinylnitrene. Similar observations were

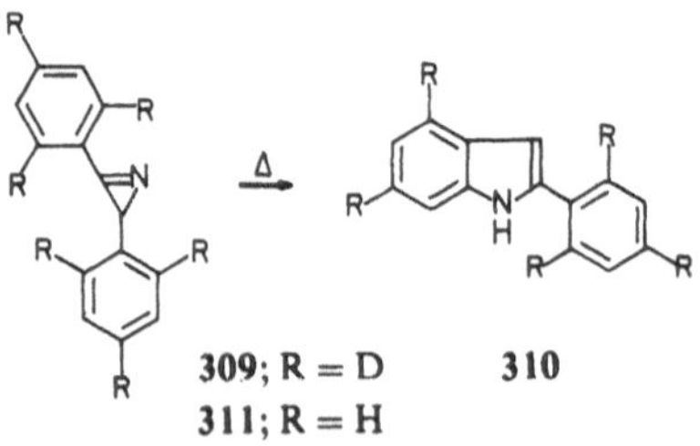

309; R = D   310
311; R = H

made by Rees and coworkers in the flash vacuum pyrolysis of **311**.[173] When **311** was heated at 250°C in a sealed tube, the reaction mixture contained a series of additional products as shown below[174]:

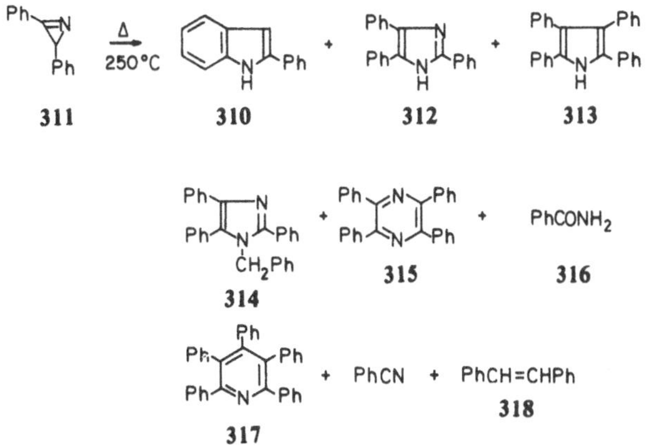

The vinylnitrene produced initially was claimed to undergo facile reorganization to indole **310**. Tetraphenylpyrrole **313**, on the other hand, was thought to be formed by a 1,3-dipolar addition of the vinylnitrene to itself.[175,176] The thermal formation of tetraphenylpyrazine (**315**) could involve the intermediacy of diazabicyclohexene **319**, which is known to rearrange to diazahexatriene **320** on thermolysis.[177] Electrocyclization of **320** to dihydropyrazine **321** followed by oxidation has been suggested to account for the formation of pyrazine **315**. Diazabicyclohexene **319** could also eliminate phenylcarbene to produce imidazole **312** or undergo a proton transfer from C_2 to C_6 to produce **314**. The formation of pentaphenylpyridine was attributed to insertion of phenylcarbene into the double bond of tetraphenylpyrrole.[174] The different products obtained from the above thermolysis reaction have been attributed to the high reactivity of the intermediates produced, their excess energy, the small energies of activation of the reactions, and the formation of products containing sufficient energy to undergo further reactions.[174]

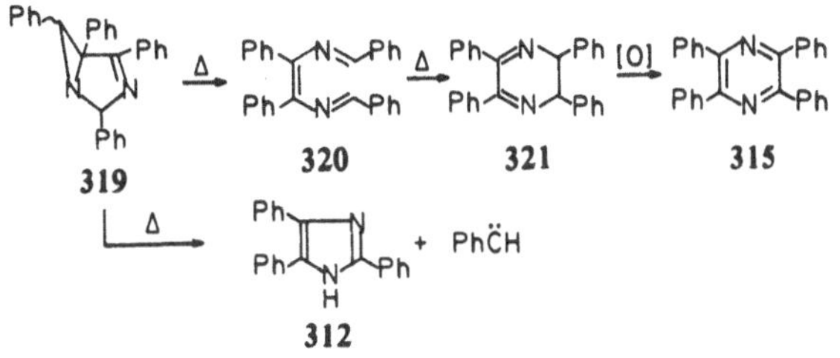

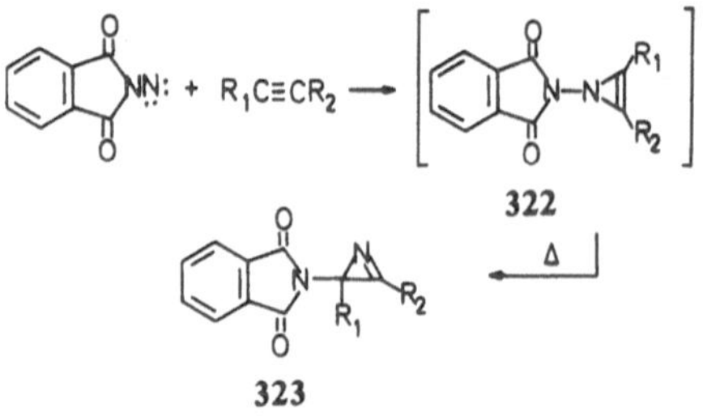

322

323

2*H*-Azirines have been reported to be the products of oxidation of *N*-aminophthalimide in the presence of alkynes.[178] The initial intermediate in this reaction is the antiaromatic 1*H*-azirine **322**. The mechanism for the 1*H*- to 2*H*-azirine rearrangement has been proposed to involve heterolytic cleavage of the N–N bond to give an azirinium cation and the phthalimide anion, followed by recombination at the ring carbon atom.[21] This mechanism is closely related to one proposed for the isomerization of 3-chloro-2*H*-azirines, which is known to take place under very mild conditions[154]:

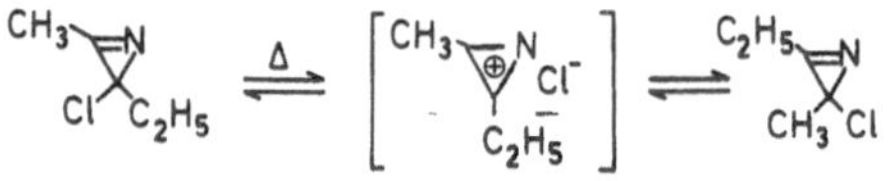

The 2*H*-azirines obtained from the vapor phase pyrolysis of 4,5-disubstituted 1-phthalimido-1,2,3-triazoles (**324**) have been found to undergo further thermal reactions.[179] Those azirines which contain a methyl group in the 2-position of the ring are cleaved to nitriles and phthalimidocarbenes, whereas those azirines which possess a phenyl substituent in the 2-position rearrange to indoles.[22,179]

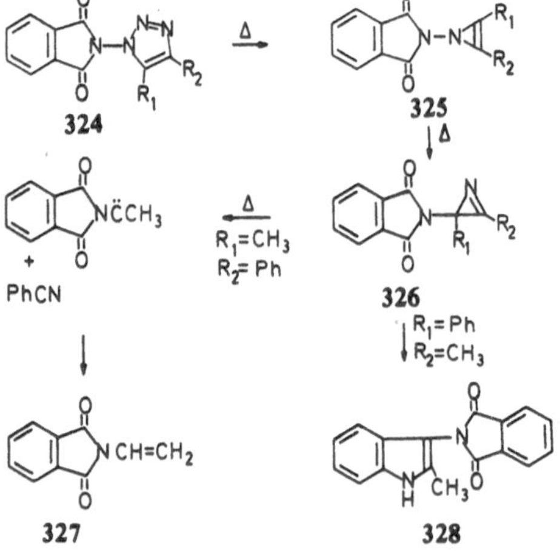

324

325

PhCN

327

326

328

As was mentioned earlier, photochemical and thermal bond cleavage preferences in 2*H*-azirines appear to be quite distinct. Products formed during photochemical isomerizations involve CC rupture, while thermal isomerization products usually arise from initial CN bond cleavage. Bergman and Wendling have recently found that the products obtained on heating azirine **329** are derived by CC bond cleavage.[160,161] The initially formed iminocarbene **330** has been suggested to undergo a 1,4-hydrogen transfer to give azabutadiene **331**. Thermal electrocyclization of this species generates a small amount of azetine **332** which fragments to the observed products. The formation of dihydroiso-

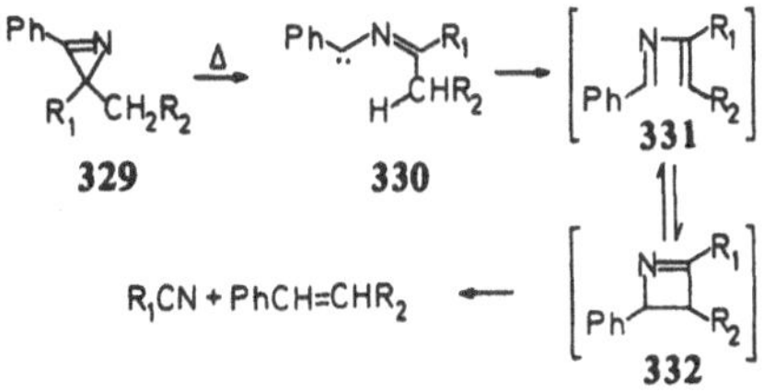

quinoline **333** by the thermolysis of 2,2-dimethyl-3-phenyl-2*H*-azirine was viewed as arising via thermal electrocyclization of the initially formed azabuta-diene intermediate followed by a subsequent 1,5-sigmatropic shift.

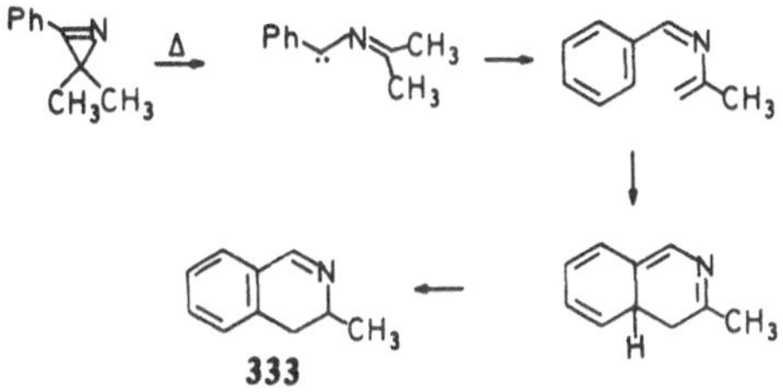

A similar thermal CC bond cleavage reaction of a 2*H*-azirine was reported by Ghosez and coworkers.[159] These workers have shown that 2-alkyl-3-amino-2*H*-azirines **334** and **336** undergo thermal reorganization to azabutadienes **335** and **337**.

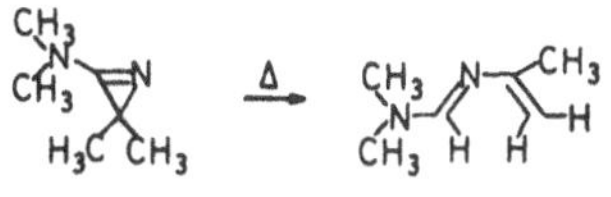

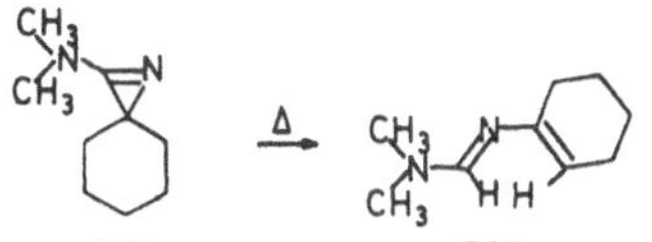

Since CN bond cleavage to form vinylnitrenes is generally the preferred bond-breaking process, it seems reasonable to suggest that CN bond rupture is also the lowest energy pathway with the above azirines. With these systems, however, no product-formation path, other than regeneration of the azirine, is available to the nitrene. Reaction products are observed only when pyrolysis temperatures high enough to cause C–C cleavage are reached. This rationale would account for the unusual regioselectivity of bond cleavage observed by Bergman[160,161] and Ghosez.[159]

The synthesis of a number of pyrrole derivatives has been effected by the thermal conversion of 3-vinyl-2H-azirines to butadienylnitrenes followed by 1,5-electrocyclization[96,97]:

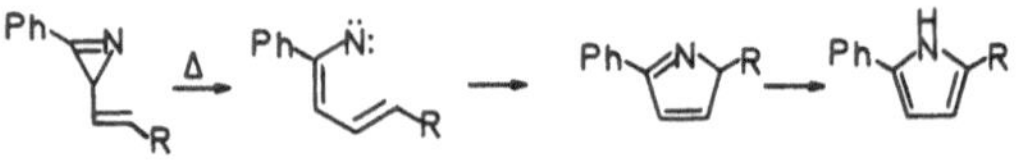

107   338   339   340

A new reaction route was discovered when the vinylazirine **126** was thermolyzed at 140°C.[102] Two major products were isolated, the pyrrole **341** and the pyridine **342**. The thermal transformations were rationalized in terms of an equilibration of the azirine with a butadienylnitrene, which subsequently rearranged to the final products. The rearrangement of **126** to pyrrole **341** was envisaged as occurring by an electrocyclic reaction followed by a [1,5]-sigmatropic ethoxycarbonyl shift and subsequent tautomerization. The presumed methoxycarbonyl shift in this reaction resembles the Van Alphen rearrangement where 3,3,4,5-tetrasubstituted pyrazolenines rearrange to N-substituted pyrazoles.[180,181] Other examples of this rearrangement have been reported to occur with ester, acyl, and cyano groups.[182,183] The formation of pyridine **342** was postulated to arise by insertion of the butadienylnitrene into the neighboring allylic methyl group followed by oxidation of the transient

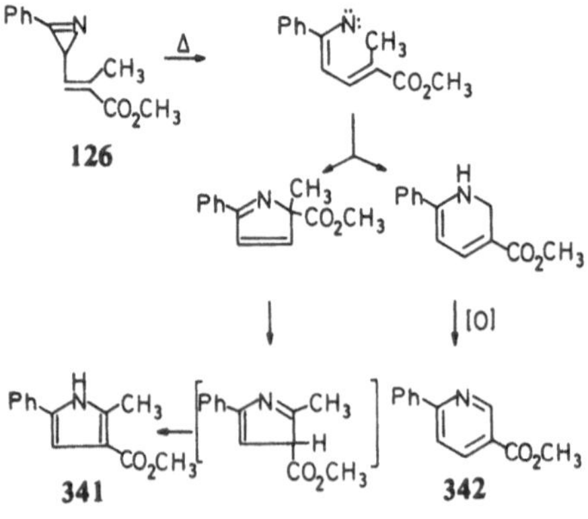

126

341   342

dihydropyridine. Support for this reaction path was obtained by carrying out the thermolysis in the presence of tris(dimethylamino)phosphine. Under these conditions, the yield of product was significantly diminished and a nitrene adduct **276** could be isolated.

A similar reaction has been observed by Friedrich and coworkers with azirine **343**.[184] In this case, the initial cycloadduct **345** did not undergo a subsequent 1,5-cyano shift.

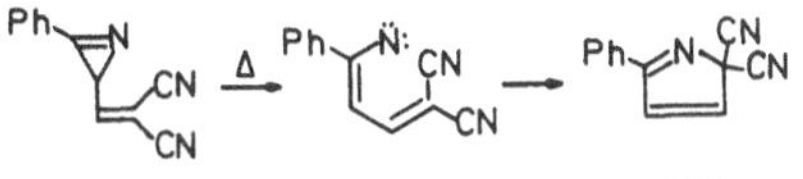

Analogous results were secured in the thermal ring-opening reaction of the corresponding diphenyl-substituted azirine **346**.[185,186] 2-Vinyl-substituted 2H-azirines have also been suggested as intermediates in the thermolysis of a number of substituted butadienyl azide systems.[187,188]

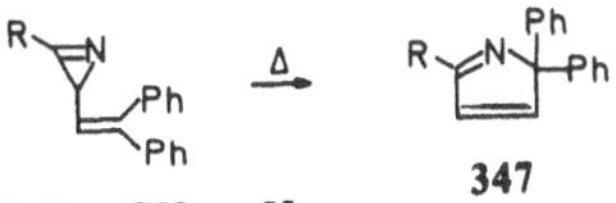

$346; R = CH_3$ or H

Further examples which illustrate the generality of the thermally induced intramolecular cyclization of 2H-azirines have been reported. When the conjugation in the 2-position was extended (**348**), two possible products can be formed.[102] In fact, however, only the 2-vinyl-substituted pyrrole **349** was detected in the reaction mixture.[172,186]

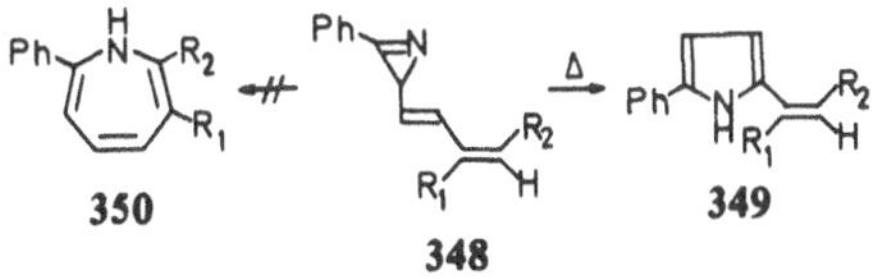

The condensation of amines with 2-formyl-3-phenyl-2H-azirine (**350**) yields the imines **352**, which undergo thermal reorganization to pyrazoles **353** via 1,5-dipolar cyclization of the resulting vinylnitrene intermediate derived from CN bond rupture.[97] The thermal reactivity of **350** and **352** parallels completely that of the 2-vinyl-substituted 2H-azirines.

The reverse reaction, where isoxazoles are thermally transformed into 2H-azirines, has also been observed by a number of investigators. Nishiwaki found that the thermolysis of a neat sample of 5-alkoxyisoxazole **354** at 200°C

350 352 351 353

354 355

resulted in the formation of 2*H*-azirine **355**.[189] Soviet workers reported a similar transformation, although catalyzed by Cu(II)-stearate.[190]

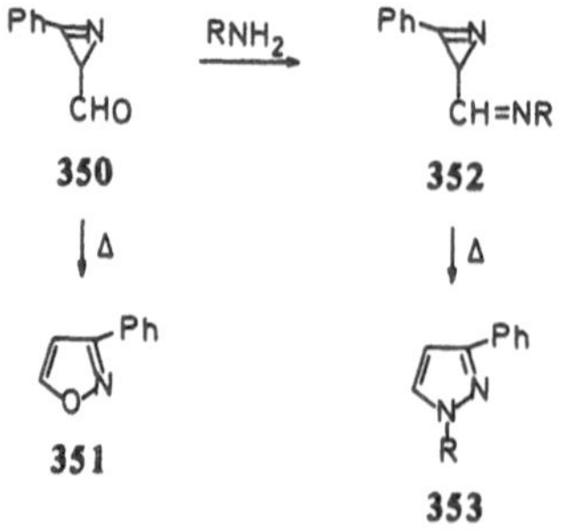

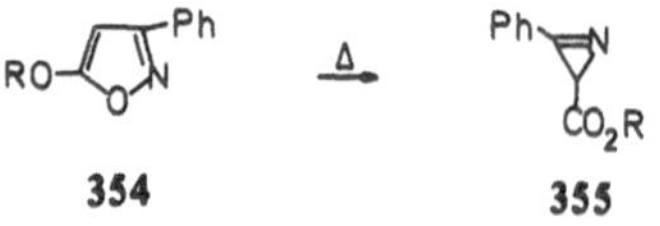

The analogous amines (**356**) were also studied and found to undergo reorganization to 2*H*-azirine-2-carboxamides (**358**).[31] Further heating of these compounds resulted in the isolation of pyrazine-2,5-dicarboxamides (**361**). The formation of **361** was suggested to proceed through either of the intermediate diradicals **358** or **359**.[27,31]

Other workers have reported the thermal rearrangement to isoxazoles where 2-acyl-2*H*-azirines are presumed to be intermediates.[191] The thermally induced fragmentation reaction of acylvinyl azides (**362**) has been reported to result in the isolation of isoxazoles (**365**) as well as nitriles (**366**), presumably via a 2-acyl-2*H*-azirine intermediate (**363**).[192,193]

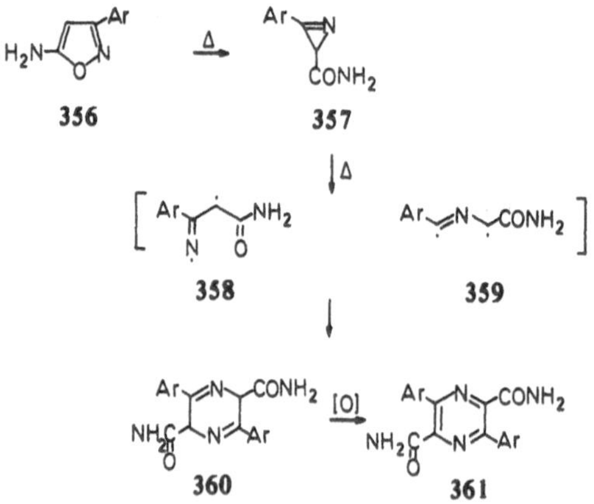

1-Aryl-2-azido-2-alken-1-ones were found to give 3-acyl-2*H*-azirines on

356 357 358 359 360 361

thermolysis.[194-196] This was confirmed by trapping these compounds with cyclopentadiene. A closely related reaction involves the thermal degradation of α-azidocinnamates (367) to indole-2-carboxylates (370).[197] It was suggested that 3-carboethoxy-2H-azirine (368) was an intermediate in this reaction. Although 368 was not isolated, nmr and ir data support the presence of this species in the crude reaction mixture.

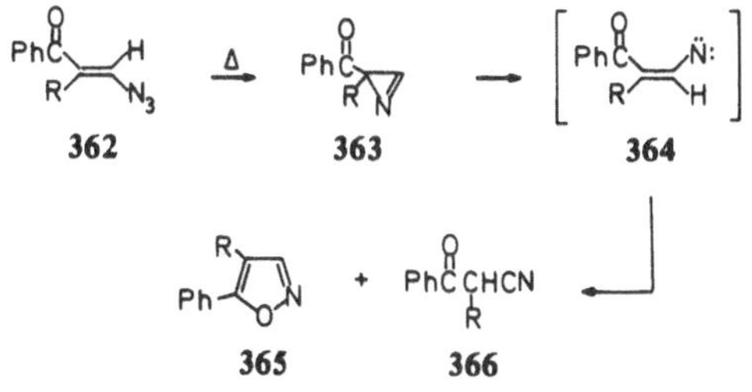

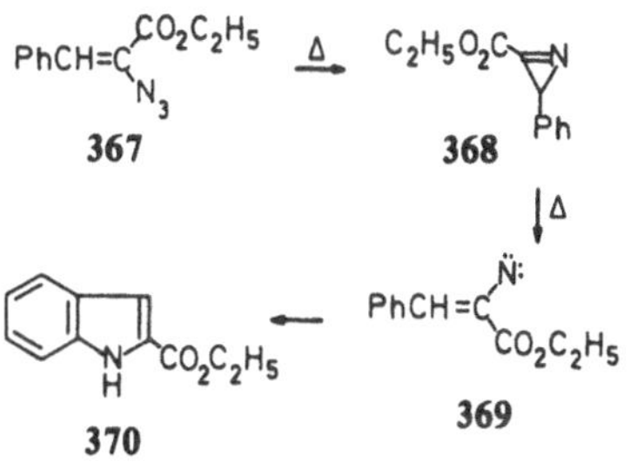

Thermolysis of the α,β-unsaturated azirinyl aldehyde 371 affords a mixture of formylpyrroles. This reaction has been suggested to proceed via the intermediacy of a vinylnitrene.[198] Cyclization of the butadienylnitrene 372 has been suggested to give pyrroline 376, which undergoes a rapid 1,5-sigmatropic formyl shift to give 373. Further heating of this pyrrole was found to produce 374 and 375.

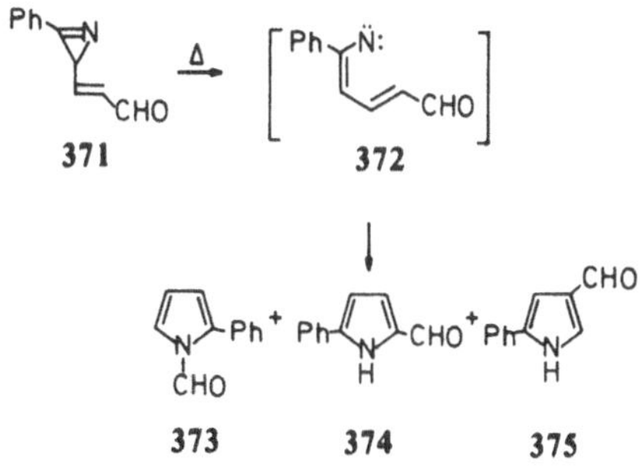

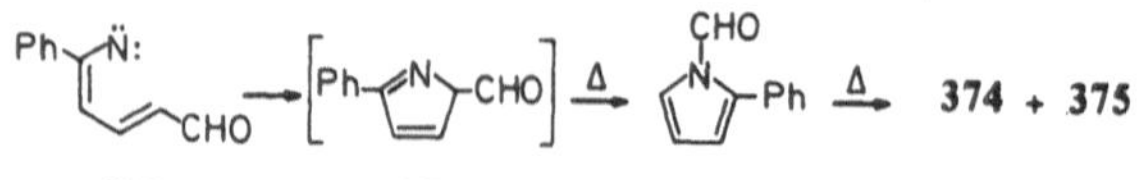

The tetraphenyloxazepins **125** (R = H or Ph) have been isolated in 90% yield from the thermolysis of Z-vinyl-2H-azirines **124**.[101] As was discussed earlier, the products formed upon thermolysis of 2H-azirines are usually derived from vinylnitrenes. In this case, CN bond rupture of azirine **124** may occur reversibly, but owing to restricted rotation around the C_2–C_3 bond of the initially formed vinylnitrene **377**, the conformation required for ring closure to **378** or **379** is inaccessible. The irreversible step then becomes the formation of vinylnitrile ylide **380**, which closes in a 1,7 fashion. Bergman and Wendling had also proposed a reversible 2H-azirine to vinylnitrene interconversion followed by irreversible opening of the azirine ring to a nitrile ylide to rationalize the formation of the products isolated in the flash vacuum pyrolysis of aralkyl 2H-azirines.[160,161]

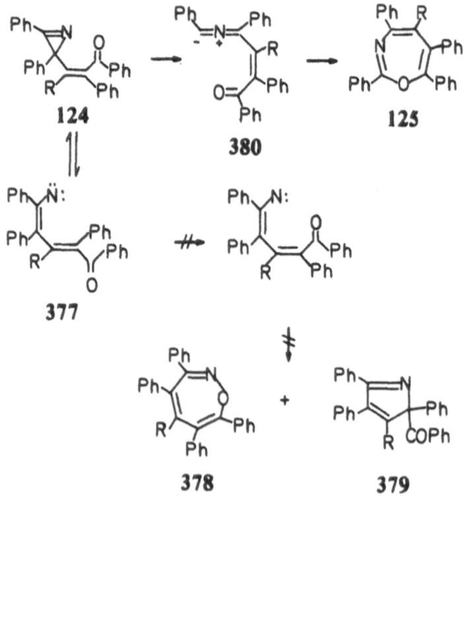

The products of the thermolysis of thiapyran **381** are pentaphenylpyridine (**387**) (37%) and tetraphenylthiophene (**388**) (12%).[199] Two pathways have been envisaged for this reaction. The first involves formation of azirine **382**, which is unstable under the reaction conditions. It undergoes CC bond cleavage affording vinylnitrile ylide **383**, which undergoes a subsequent 1,7-electrocyclization to 2,4,5,6,7-pentaphenyl-1,3-thiazepin (**384**), from which either sulfur or benzonitrile is extruded. Alternatively, **382** may produce the vinylnitrene **385**, which cyclizes to 3,4,5,6,7-pentaphenyl-1,2-thiazepin (**386**). Elimination of sulfur or benzonitrile from **386** leads to **387** or **388**.

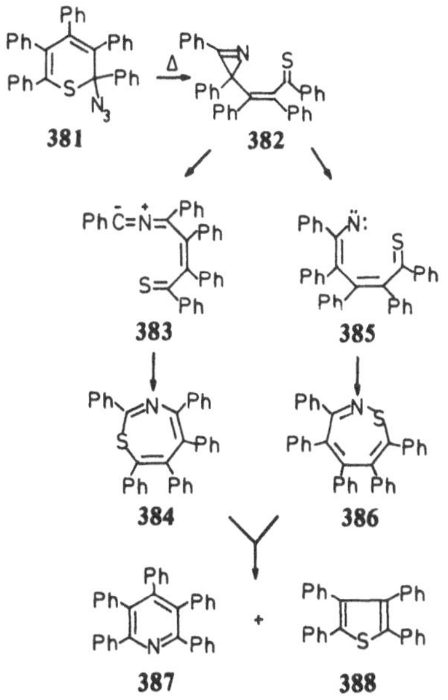

Examples of the direct addition of vinylnitrenes to olefins to give aziridines have appeared infrequently in the literature.[200] One case where this type of reaction has been suggested to occur has been uncovered by Padwa and Carlsen in a study of the thermal chemistry of a number of allyl-substituted 2H-azirines.[201–203] These compounds (**389**) were found to undergo smooth rearrangement on heating at 100°C to give 3-azabicyclo[3.1.0]hexenes (**390**). Further heating for prolonged periods of time resulted in the formation of pyridines **391**. The formation of **390** has been explained in terms of an equilibrium of the 2H-azirine ring with a transient vinylnitrene, which subsequently adds to the adjacent π bond. The initially formed bicycloaziridine rearranges to the 3-azabicyclohexene ring system by means of a 1,3-sigmatropic

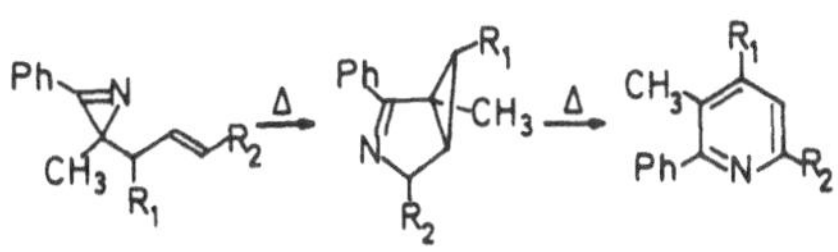

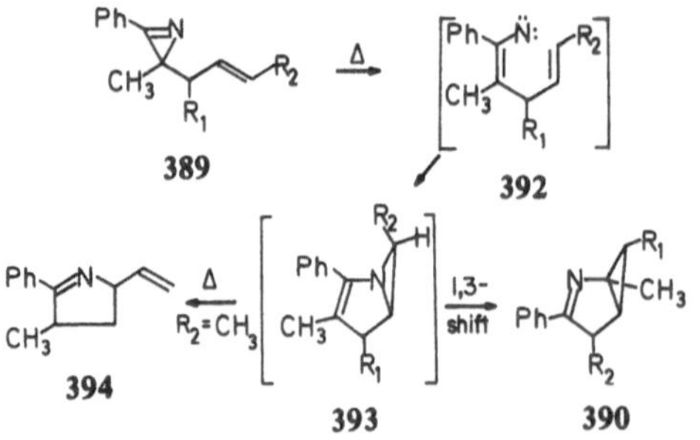

389   **392**

394   **393**   **390**

shift. Evidence favoring this pathway is provided by the isolation of 3-methyl-2-phenyl-5-vinyl-Δ^1-pyrroline (**394**) from the thermolysis of azirine **389**. The formation of the Δ^1-pyrroline ring was rationalized as proceeding via a homo[1,5]-hydrogen migration from a 6-*endo*-methyl-substituted bicycloaziridine intermediate (**393**). Several examples of intramolecular addition of nitrenes onto adjacent double bonds are available in the literature[204-206] and provide reasonable chemical analogy for the conversion of **392** to **393**

Thermolysis of 2-allyl-3-methyl-2-phenyl-substituted 2*H*-azirines has been found to afford mixtures of azabicyclohexenes (**395**) and indoles (**396**).[203] The distribution of products with this ring system is controlled by the rates of nitrene attack on the double bond vs. electrocyclization on the adjacent phenyl ring.

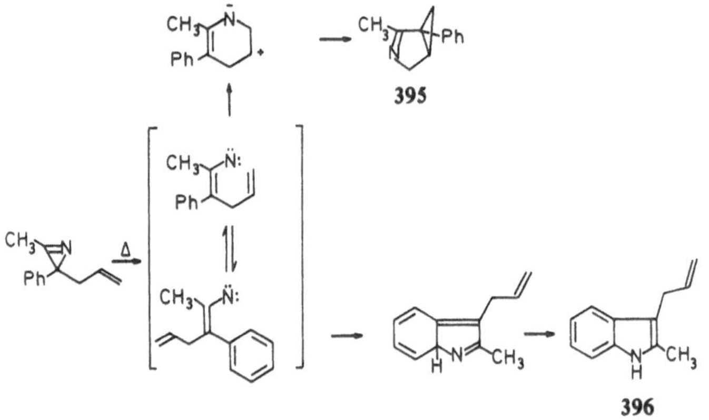

395

396

The formation of pyridine **399** from the thermolysis of azirine **397** was also suggested to be compatible with a vinylnitrene intermediate. In this case, attack by the neighboring acetylenic functionality on the nitrene has been suggested to give carbene **398**, which undergoes a facile hydrogen migration to give **399**.[203]

The thermal behavior of the carbomethoxy-substituted allyl 2*H*-azirine **400** has been explained in terms of an attack by the vinylnitrene onto the neighboring double bond to give a short-lived bicycloaziridine **401**. Heterolytic cleavage of the CN bond followed by proton reorganization furnishes pyrrole **402**. Bicycloaziridine **401** may also rearrange to the azabicyclohexene ring system **403**, which in turn gives pyridine **404** on further heating.[203]

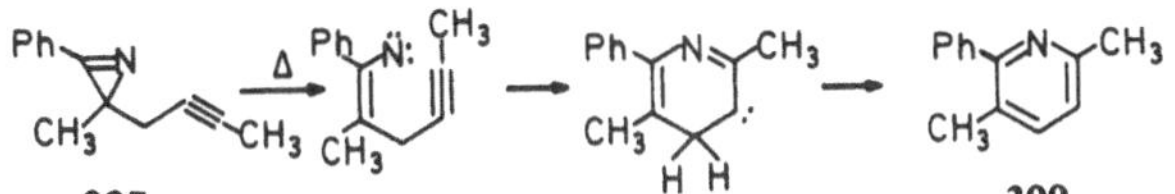

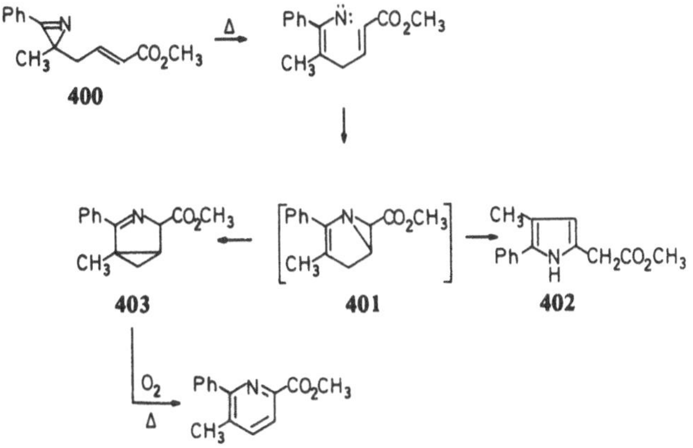

The conversion of the isomeric carbomethoxy-substituted 2H-azirine **405** to pyridine **408**, on the other hand, is thought to proceed by an entirely different pathway. This was attributed to conjugate addition of the vinylnitrene to the electron-deficient double bond.[203] The five-membered ring zwitterion **406** initially produced undergoes a subsequent fragmentation to give azatriene **407**. This would be expected to undergo a ready electrocyclic closure followed by oxidation to produce pyridine **408** ultimately. An alternate path involving nucleophilic attack by the available lone pair of electrons in the starting material onto the conjugated double bond also seems possible. The vinylnitrene formed from CN bond cleavage of **405** also cyclizes to give indole **409**. The difference in behavior between the isomeric carbomethoxy-substituted 2H-azirines **400** and **405** was attributed to the difference in nucleophilicities of the nitrogen atoms present in starting materials, or in the vinylnitrene intermediate.[203]

The thermal behavior of the closely related homoallyl-substituted 2H-azirine **410** has also been studied by Padwa and Kamigata.[138,207] Heating a solution of **410** in toluene at 195°C gave 2-methylbiphenyl (**411**) (49%) and 2,5-

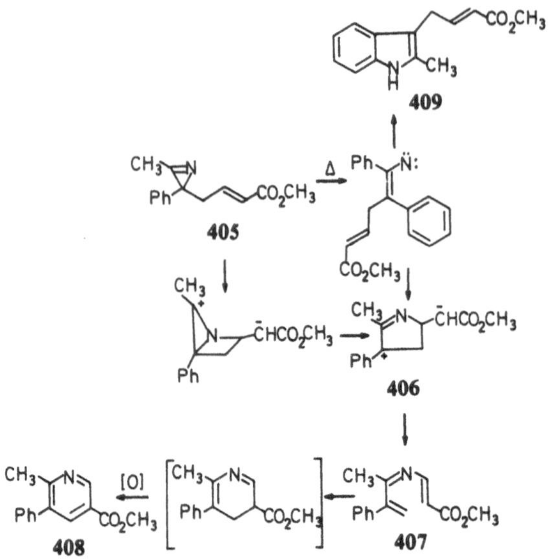

dimethyl-6-phenylpyridine **412** (15%). A similar thermolysis of azirine **413** produced pyridine **414** and methyl 6-methylbiphenyl-2-carboxylate **415**. The formation of these products was interpreted in terms of a thermal equilibration of the 2*H*-azirine with a transient vinylnitrene followed by a 1,4-hydrogen

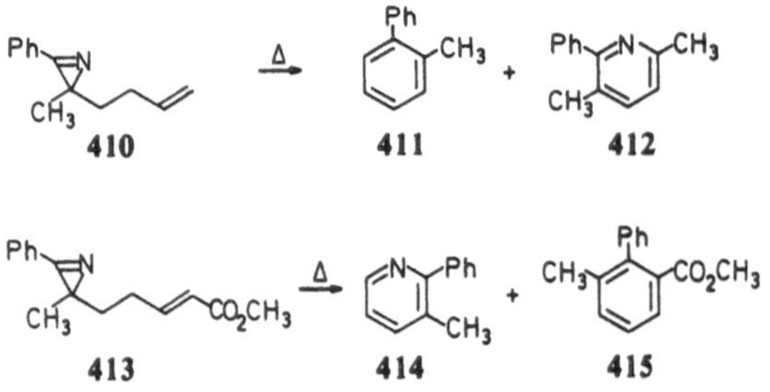

transfer from the neighboring methylene group to generate azatriene **416**. This reactive intermediate was thought to undergo a thermally allowed 1,5-sigmatropic shift to give triene **417**. Electrocyclic closure of **417** to cyclohexadiene, **418**, followed by loss of ammonia, readily accounted for the formation of the substituted biphenyl derivatives. The formation of **414** from azirine **413** involves an internal Michael addition of **417** to give **419**, followed by loss of methyl acetate. The key feature of the above reactions is the 1,4-hydrogen transfer step. Related hydrogen transfers have been observed in reactions of vinylcarbenes[208-211] and, more recently, with iminocarbenes[212] which provide reasonable chemical analogies. The proposed 1,4-hydrogen transfer was further supported by the thermal conversion of **420** to **414**. In this case, a series of 1,4- and 1,5-hydrogen transfers gives **422**, which may reasonably cyclize to 3-methyl-2-phenyl pyridine (**414**).

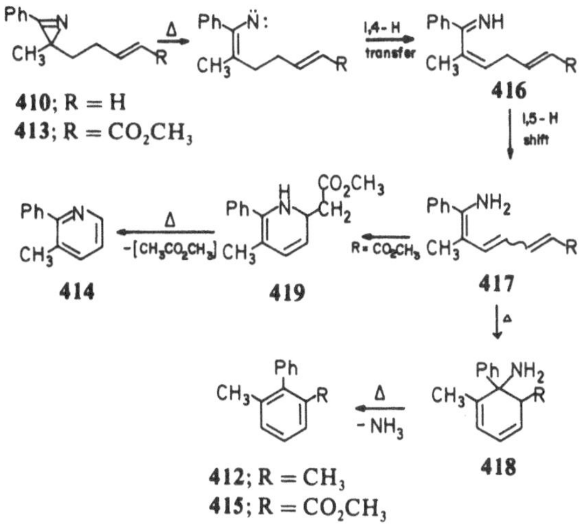

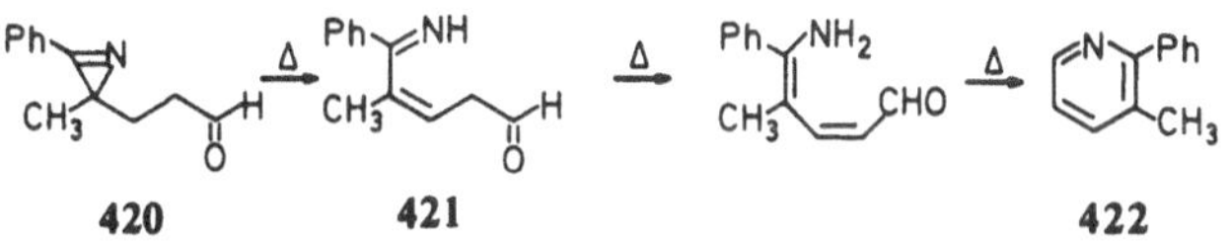

ACKNOWLEDGEMENT

The support of the National Institutes of Health and the National Science Foundation is gratefully acknowledged. A. P. would also like to thank the John S. Guggenheim Foundation for a fellowship.

REFERENCES

1. F. W. Fowler and A. Hassner, *J. Am. Chem. Soc.* **90**, 2875 (1968).
2. A. G. Hortmann and D. A. Robertson, *J. Am. Chem. Chem. Soc.* **89**, 5974 (1967).
3. S. Sato, H. Kato, and M. Ohta, *Bull. Chem. Soc. Japan.* **50**, 2936 (1967).
4. G. Smolinsky and B. Feuer, *J. Org. Chem.* **31**, 1324 (1966).
5. F. P. Woerner, H. Reiminger, and R. Merenyi, *Chem. Ber.* **104**, 2786 (1971).
6. A. Hassner, A. S. Miller, and M. J. Haddadin, *Tetrahedron Lett.*, 1353 (1972).

7. F. W. Fowler, *Adv. Heterocycl. Chem.* **13**, 45 (1971].

8. A. Hassner and D. J. Anderson, *J. Am. Chem. Soc.* **93**, 4339 (1971); *J. Org. Chem.* **38**, 2565 (1973).

9. V. Nair, *J. Org. Chem.* **37**, 802 (1972).

10. V. Nair, *J. Org. Chem.* **33**, 2121 (1968); *Tetrahedron Lett.*, 4831 (1971).

11. A. L. Logothetis, *J. Org. Chem.* **29**, 3049 (1964).

12. J. H. Bowier, B. Nussey, and A. D. Ward, *Aust. J. Chem.* **26**, 2547 (1973).

13. A. Hassner and D. J. Anderson, *J. Am. Chem. Soc.* **94**, 8255 (1972); *J. Org. Chem.* **39**, 3070 (1974).

14. D. J. Anderson, A. Hassner, and D. Y. Tang, *J. Org. Chem.* **39**, 3076 (1974).

15. H. Hemetsberger and D. Knittel, *Monatsch. Chem.* **103**, 205 (1972).

16. V. Nair, *J. Org. Chem.* **37**, 2508 (1972).

17. A. Hassner and D. J. Anderson, *J. Org. Chem.* **39**, 2031 (1974).

18. A. Padwa, *Acc. Chem. Res.* **9**, 371 (1976).

19. P. Gilgen, H. Heimgartner, H. Schmid, and J. Hansen, *Heterocycles* **6**, 143 (1977).

20. A. Padwa and J. Smolanoff, *J. Am. Chem. Soc.* **93**, 548 (1971).

21. D. J. Anderson, T. L. Gilchrist, G. E. Gymer, and C. W. Rees, *J. Chem. Soc. Perkin Trans.* 1, 550 (1973).

22. T. L. Gilchrist, G. E. Gymer, and C. W. Rees, *J. Chem. Soc. Perkin Trans.* 1, 555 (1973).

23. K. Isomura, S. Kobayashi, and H. Taniguchi, *Tetrahedron Lett.*, 3499 (1968).

24. K. Isomura, M. Okada, and H. Taniguchi, *Tetrahedron Lett.*, 4073 (1969).

25. T. Nishiwaki, *J. Chem. Soc. Chem. Commun.*, 565 (1972).

26. R. Selvarajan and J. H. Boyer, *J. Heterocycl. Chem.* **9**, 87 (1972).

27. T. Nishiwaki, A. Nakano, and H. Matsuoko, *J. Chem. Soc. C*, 1825 (1970).

28. J. H. Bowier and B. Nussey, *Chem. Commun.*, 1565 (1970).

29. D. Knittel, H. Hemetsberger, R. Leipert, and H. Weidman, *Tetrahedron Lett.*, 1459 (1970).

30. N. S. Narashimhan, H. Heimgartner, H. J. Hansen, and H. Schmid, *Helv. Chim. Acta* **56**, 1351 (1973).

31. T. Nishiwaki and F. Fujiyama, *J. Chem. Soc. Perkin Trans.* 1, 1456 (1972).

32. A. Padwa, J. Smolanoff, and A. I. Tremper, *J. Am. Chem. Soc.* **96**, 3486 (1974); *J. Org. Chem.* **41**, 543 (1976).

33. E. F. Ullman and B. Singh, *J. Am. Chem. Soc.* **88**, 1844 (1966); **89**, 6911 (1967).

34. B. Singh, A. Zweig, and J. B. Gallivan, *J. Am. Chem. Soc.* **94**, 1199 (1972).

35. I. J. Turchi and M. J. S. Dewar, *Chem. Rev.* **75**, 389 (1975).

36. T. Nishiwaki, *Synthesis*, 20 (1975).

37. H. E. Zimmerman, *Adv. Photochem.* **1**, 183 (1963).

38. W. Sieber, P. Gilgen, S. Chaloupka, H. J. Hansen, and H. Schmid, *Helv. Chim. Acta* **56**, 1679 (1973).

39. A. Orahovats, H. Heimgartner, H. Schmid, and W. Heinzelmann, *Helv. Chim. Acta* **58**, 2662 (1975).

40. A. Orahovats, H. Heimgartner, H. Schmid, and W. Heinzelman, *Helv. Chim. Acta* **57**, 2626 (1974).

41. R. Huisgen, H. Stangl, H. J. Sturm, and H. Wagenhafer, *Angew. Chem.* **74**, 31 (1962); R. Huisgen, H. Stangl, H. J. Sturm, R. Raab, and K. Bunge, *Chem. Ber.* **105**, 1258 (1972); **105**, 1296 (1972); R. Huisgen, and R. Raab, *Tetrahedron Lett.*, 649 (1966).

42. K. Burger, J. Albanbauer, and F. Manz, *Chem. Ber.* **107**, 1823 (1974); K. Burger and J. Fehn, *Chem. Ber.* **105**, 3814 (1972); K. Burger, J. Fehn, and E. Muller, *Chem. Ber.* **106**, 1 (1973); K. Burger and K. Einhellig, *Chem. Ber.* **106**, 3421 (1973).

43. K. Burger, W. Thenn, and E. Muller, *Angew. Chem. Int. Ed. Engl.* **12**, 155 (1973).

44. W. Steglich, P. Gruber, H. U. Heininger, and F. Kneidl, *Chem. Ber.* **104**, 3816 (1971).

45. P. Claus, T. Doppler, N. Gakis, M. Georgarakis, H. Geizendanner, P. Gilgen, H.

Heimgartner, B. Jackson, M. Marky, N. S. Narasimhan, H. J. Rosenkranz, A. Wunderli, H. J. Hansen, and H. Schmid, *Pure Appl. Chem.* **33**, 339 (1973).

46. H. Schmid, *Chimia* **27**, 172 (1973).
47. A. Padwa and S. I. Wetmore, Jr., *J. Am. Chem. Soc.* **96**, 2414 (1974).
48. G. Bittner, H. Witte, and G. Hesse, *Liebigs Ann. Chem.* **713**, 1 (1968).
49. A. Hassner, *Acc. Chem. Res.* **4**, 9 (1971).
50. G. L'abbé, *Angew. Chem. Int. Ed. Engl.* **14**, 775 (1975).
51. A. Hassner and L. A. Levy, *J. Am. Chem. Soc.* **87**, 4203 (1965); F. W. Fowler, A. Hassner, and L. A. Levy, *J. Am. Chem. Soc.* **89**, 2077 (1967); A. Hassner and F. P. Boerwinkle, *J. Am. Chem. Soc.* **90**, 216 (1968); *Tetrahedron Lett.*, 3309 (1969).
52. G. Smolinsky, *J. Am. Chem. Soc.* **83**, 4483 (1961); *J. Org. Chem.* **27**, 3557 (1962); M. Komatsu, S. Ichijima, Y. Ohshiro, and T. Agawa, *J. Org. Chem.* **38**, 4341 (1973); A. G. Hortmann, D. A. Robertson, and B. K. Gillard, *J. Org. Chem.* **37**, 322 (1972).
53. N. J. Leonard and B. Zwanenburg, *J. Am. Chem. Soc.* **89**, 4466 (1967).
54. R. F. Parcell, *Chem. Ind.*, 1396 (1963).
55. H. J. Bestmann and R. Kunstmann, *Chem. Ber.* **102**, 1816 (1969); J. Wulff and R. Huisgen, *Chem. Ber.* **102**, 1833 (1969).
56. A. Padwa, M. Dharan, J. Slolanoff, and S. I. Wetmore, Jr., *Pure Appl. Chem.* **33**, 269 (1973).
57. A. Padwa, M. Dharan, J. Smolanoff, and S. I. Wetmore, Jr., *J. Am. Chem. Soc.* **94**, 1395 (1972); **95**, 1945 (1973).
58. A. Padwa and S. I. Wetmore, Jr., *J. Org. Chem.* **38**, 1333 (1973); **39**, 1396 (1974); *J. Am. Chem. Soc.* **96**, 2414 (1974).
59. W. Stegmann, P. Gilgen, H. Heimgartner, and H. Schmid, *Helv. Chim. Acta* **59**, 1018 (1976).
60. N. Gakis, H. Heimgartner, and H. Schmid, *Helv. Chim. Acta* **58**, 748 (1975); **60**, 687 (1977).
61. N. Gakis, H. Heimgartner, and H. Schmid, *Helv. Chim. Acta* **57**, 1403 (1974).
62. U. Widmer, N. Gakis, B. Arnet, H. Heimgartner, and H. Schmid, *Chimia* **30**, 453 (1976).
63. R. Huisgen, *J. Org. Chem.* **33**, 2291 (1968); **41**, 403 (1976); *Angew. Chem. Int. Ed. Engl.* **2**, 565 (1963).
64. R. Huisgen, H. Stangl, H. J. Sturm, and H. Wagenhofer, *Angew. Chem. Int. Ed. Engl.* **1**, 50 (1962).
65. K. N. Houk, J. Sims, R. E. Duke, R. W. Strozier, and J. K. George, *J. Am. Chem. Soc.* **95**, 7287 (1973); **95**, 7301 (1973); **95**, 5798 (1973); **94**, 8953 (1972); K. N. Houk, *Acc. Chem. Res.* **8**, 361 (1975).
66. R. Sustmann, *Tetrahedron Lett.*, 2717 (1971).
67. A. Padwa and J. Smolanoff, *J. Chem. Soc. Chem. Commun.*, 342 (1973).
68. A. Padwa, J. K. Rasmussen, and A. Tremper, *J. Am. Chem. Soc.* **98**, 2605 (1976).
69. A. Padwa, D. Dean, and J. Smolanoff, *Tetrahedron Lett.*, 4087 (1972).
70. H. Giezendanner, M. Marky, B. Jackson, H. J. Hansen, and H, Schmid, *Helv. Chim. Acta* **55**, 745 (1972).
71. B. Jackson, M. Marky, H. J. Hansen, and H. Schmid, *Helv. Chim. Acta* **55**, 919 (1972).
72. P. Claus, P. Gilgen, H. J. Hansen, H. Heimgartner, B. Jackson, and H. Schmid, *Helv. Chim. Acta* **57**, 2173 (1974); P. Gilgen, H. J. Hansen, H. Heimgartner, W. Sieber, P. Uebelhardt, H. Schmid, P. Schonholzer, and W. E. Oberhansli, *Helv. Chim. Acta* **58**, 1739 (1975).
73. K. Burger, W. Thenn, and E. Muller, *Angew. Chem. Int. Ed. Engl.* **12**, 149 (1973).
74. A. Padwa, M. Dharan, J. Smolanoff, and S. I. Wetmore, Jr., *J. Am. Chem. Soc.* **95**, 1954 (1973).
75. H. Giezendanner, H. Heumgartner, B. Jackson, T. Winkler, H. J. Hansen, and H. Schmid, *Helv. Chim. Acta* **56**, 2611 (1973).

76. A. Orahovats, B. Jackson, H. Heimgartner, and H. Schmid, *Helv. Chim. Acta* **56**, 2007 (1973).
77. H. Heimgartner, P. Gilgen, U. Schmid, H. J. Hansen, H. Schmid, K. Pfoertner, and K. Bernauer, *Chimia* **26**, 424 (1972).
78. B. Jackson, N. Gakis, M. Maerky, H. J. Hansen, W. V. Philipsborn, and H. Schmid, *Helv. Chim. Acta* **55**, 916 (1972).
79. U. Schmid, P. Gilgen, H. Heimgartner, H. J. Hansen, and H. Schmid, *Helv. Chim. Acta* **57**, 1393 (1974).
80. N. Gakis, M. Marky, H. J. Hansen, H. Heimgartner, and H. Schmid, *Helv. Chim. Acta* **59**, 2149 (1976).
81. A. Padwa and S. I. Wetmore, Jr., *Organ. Photochem. Synth.* **2**, 87 (1976).
82. R. Huisgen, H. Stangl, H. Sturm, R. Raab, and K. Bunge, *Chem. Ber.* **105**, 1258 (1972).
83. H. Giezendanner, H. J. Rosenkranz, H. J. Hansen, and H. Schmid, *Helv. Chim. Acta* **56**, 2588 (1973).
84. N. S. Narasimhan, H. Heimgartner, H. J. Hansen, and H. Schmid, *Helv. Chim. Acta* **56**, 1351 (1973).
85. U. Schmid, H. Heimgartner, H. Schmid, and H. Heinzelmann, *Helv. Chim. Acta* **58**, 2222 (1975).
86. P. Gilgen, H. Heimgartner, and H. Schmid, *Helv. Chim. Acta* **57**, 1382 (1974).
87. P. Gilgen, B. Jackson, H. J. Hansen, H. Heimgartner, and H. Schmid, *Helv. Chim. Acta* **57**, 2634 (1974).
88. F. P. Woerner, H. Reimlinger, and D. R. Arnold, *Angew. Chem. Int. Ed. Engl.* **7**, 130 (1968).
89. F. P. Woerner and H. Reimlinger, *Chem. Ber.* **103**, 1908 (1970).
90. A. Padwa, J. Smolanoff, and S. I. Wetmore, Jr., *J. Chem. Soc. Chem. Commun.*, 409 (1972).
91. A. Padwa and E. Glazer, *J. Am. Chem. Soc.* **94**, 7788 (1972); **92**, 1778 (1970); *Chem. Commun.*, 838 (1971).
92. A. Padwa and S. I. Wetmore, Jr., *J. Chem. Soc. Chem. Commun.*, 1116 (1972).
93. R. Selvarajan and J. H. Boyer, *J. Heterocycl. Chem.* **9**, 87 (1972).
94. H. W. Heine, R. H. Weese, R. A. Cooper, and A. J. Durbetaki, *J. Org. Chem.* **32**, 2708 (1967).
95. A. Padwa, S. Clough, and E. Glazer, *J. Am. Chem. Soc.* **92**, 1778 (1970).
96. A. Padwa and J. Smolanoff, *Tetrahedron Lett.*, 29 (1974).
97. A. Padwa, J. Smolanoff, and A. Tremper, *J. Am. Chem. Soc.* **97**, 4682 (1975).
98. T. Mukai and H. Sukawa, *Tetrahedron Lett.*, 1835 (1973).
99. T. Tezuka, O. Seshimoto, and T. Mukai, *J. Chem. Soc. Chem. Commun.*, 373 (1974).
100. A. Padwa and J. Smolanoff, *Tetrahedron Lett.*, 33 (1974).
101. F. D. Bellamy, *Tetrahedron Lett.*, 4577 (1978).
102. A. Padwa, J. Smolanoff, and A. Tremper, *J. Org. Chem.* **41**, 543 (1976).
103. D. W. Kurtz and H. Schechter, *Chem. Commun.*, 689 (1966).
104. A. Padwa, *Chem. Rev.* **77**, 37 (1977).
105. A. Padwa, *Molecular Rearrangements*, P. deMayo, ed., Vol. 4, J. Wiley and Sons (1980).
106. E. C. Taylor and I. J. Turchi, *Chem. Rev.* **79**, 181 (1979).
107. P. Beak and W. Messer, *Organic Photochemistry*, O. L. Chapman, ed., Vol. 2, Marcel Dekker, New York (1969).
108. A. Lablache-Combier and M. A. Remy, *Bull. Soc. Chim. Fr.*, 679 (1971).
109. A. Lablache-Combier, *Photochemistry of Heterocyclic Compounds*, O. Buchardt, ed., J. Wiley and Sons, New York (1976), p. 123.
110. H. Wamhoff, *Chem. Ber.* **105**, 748 (1972).
111. R. H. Good and G. Jones, *J. Chem. Soc. C*, 1196 (1971).
112. T. Sato, K. Yamamoto, and K. Fukui, *Chem. Lett.*, 111 (1973).

113. M. Maeda and M. Kojima, *J. Chem. Soc. Perkin Trans.* 1, 239 (1977); *Tetrahedron Lett.*, 2379 (1969); *J. Chem. Soc. Chem. Commun.*, 539 (1973).

114. M. Ogato, M. Matsumoto, and H. Kano, *Tetrahedron* 25, 5205 (1969).

115. T. L. Gilchrist, G. E. Gymer, and C. W. Rees, *J. Chem. Soc. Perkin Trans.* 1, 555 (1973).

116. A. Padwa and E. Chen, *J. Org. Chem.* 39, 1976 (1974).

117. A. Padwa, E. Chen, and A. Ku, *J. Am. Chem. Soc.* 97, 6486 (1975).

118. M. J. S. Dewar and I. J. Turchi, *J. Chem. Soc. Perkin Trans.* 2, 724 (1977).

119. K. Dietliker, P. Gilgen, H. Heimgartner, and H. Schmid, *Helv. Chim. Acta* 59, 2074 (1976).

120. T. Nishiwaki, F. Fujiyama, and E. Minamisono, *J. Chem. Soc. Perkin Trans.* 1, 1871 (1974).

121. R. Huisgen, *Angew. Chem. Int. Ed. Engl.* 2, 633 (1963).

122. R. Huisgen, R. Sustmann, and K. Bunge, *Chem. Ber.* 105, 1324 (1972).

123. L. Salem, *J. Am. Chem. Soc.* 96, 3486 (1974).

124. P. Caramella and K. N. Houk, *J. Am. Chem. Soc.* 98, 6397 (1976).

125. P. Caramella, R. W. Gandour, J. A. Hall, C. G. Deville, and K. N. Houk, *J. Am. Chem. Soc.* 99, 385 (1977).

126. B. Bigot, A. Sevin, and A. Devaquet, *J. Am. Chem. Soc.* 100, 6924 (1978).

127. A. Padwa and P. H. J. Carlsen, *J. Am. Chem. Soc.* 97, 3862 (1975); 98, 2006 (1976); 99, 1514 (1977).

128. R. A. Firestone, *J. Org. Chem.* 33, 2285 (1968); 37, 2181 (1972).

129. R. A. Firestone, *J. Chem. Soc. A*, 1570 (1970).

130. G. LoVecchio, G. Grassi, F. Fusitano, and F. Foti, *Tetrahedron Lett.*, 3777 (1973).

131. L. Garanti, A. Vigevani, and G. Zecchi, *Tetrahedron Lett.*, 1527 (1976).

132. N. Engel, J. Fischer, and W. Steglich, *J. Chem. Res.*, 162 (1977); *Angew. Chem. Int. Ed. Engl.* 18, 167 (1979).

133. A. Padwa and P. H. J. Carlsen, *J. Org. Chem.* 43, 3757 (1978).

134. A. Padwa, A. Ku, A. Mazzu, and S. I. Wetmore, Jr., *J. Am. Chem. Soc.* 98, 1048 (1976).

135. A. Padwa and A. Ku, *J. Am. Chem. Soc.* 100, 2181 (1978).

136. A. Padwa, P. H. J. Carlsen, and A. Ku, *J. Am. Chem. Soc.* 99, 2798 (1977); 100, 3494 (1978).

137. E. W. Della, *Tetrahedron Lett.*, 3347 (1966).

138. A. Padwa and N. Kamigata, *J. Am. Chem. Soc.* 99, 1871 (1977).

139. A. Padwa, *Angew. Chem. Int. Ed. Engl.* 15, 123 (1976).

140. R. Fusco, L. Garanti, and G. Zecchi, *Tetrahedron Lett.*, 269 (1974).

141. R. Fusco, L. Garanti, and G. Zecchi, *Chem. Ind. (Milan)* 57, 15 (1975).

142. L. Garanti, A. Sala, and G. Zecchi, *J. Org. Chem.* 40, 2403 (1975).

143. A. Padwa and J. K. Rasmussen, *J. Am. Chem. Soc.* 97, 5912 (1975).

144. G. W. Griffin and N. R. Bertoniere, in *Carbenes*, M. Jones and R. A. Moss, Eds., Vol. I, Wiley, New York (1973).

145. A. Padwa, J. K. Rasmussen, and A. Tremper, *J. Am. Chem. Soc.* 98, 2605 (1976).

146. N. R. Bertoniere and G. W. Griffin, *Organic Photochemistry*, Vol. 3, O. L. Chapman, Ed., Marcel Dekker, New York (1973), p. 115.

147. C. L. Currie, H. Okabe, and J. R. McNesby, *J. Phys. Chem.* 67, 1494 (1963).

148. P. A. Leermakers and G. F. Vesley, *J. Org. Chem.* 30, 539 (1965).

149. G. W. Griffin, *Angew. Chem. Int. Ed. Engl.* 10, 537 (1971).

150. W. M. Jones, M. E. Stowe, E. E. Wells, Jr., and E. W. Lester, *J. Am. Chem. Soc.* 90, 1849 (1968).

151. A. Hassner, R. J. Isbister, R. B. Greenwald, J. T. Klug, and E. C. Taylor, *Tetrahedron* 25, 1637 (1969).

152. W. Bauer and K. Hafner, *Angew. Chem. Int. Ed. Engl.* 8, 772 (1969).

153. W. L. Magee and H. Schechter, *J. Am. Chem. Soc.* 99, 633 (1977).

154. J. Ciabattoni and M. Cabell, Jr., *J. Am. Chem. Soc.* **93**, 1482 (1971).

155. A. Padwa, J. K. Rasmussen, and A. Tremper, *J. Chem. Soc. Chem. Commun.*, 10 (1976).

156. A. Padwa, P. H. J. Carlsen, and A. Tremper, *J. Am. Chem. Soc.* **100**, 4481 (1978).

157. K. Isomura, S. Kobayashi, and H. Taniguchi, *Tetrahedron Lett.*, 3499 (1968).

158. G. Smolinsky and C. A. Pryde, *J. Org. Chem.* **33**, 2411 (1968).

159. A. Demoulin, H. Gorissen, A. M. Hesbian-Frisque, and L. Ghosez, *J. Am. Chem. Soc.* **97**, 4409 (1975).

160. L. A. Wendling and R. G. Bergman, *J. Am. Chem. Soc.* **96**, 308 (1974).

161. L. A. Wendling and R. G. Bergman, *J. Org. Chem.* **41**, 831 (1976).

162. B. Bigot, R. Ponec, A. Sevin, and A. Devaquet, *J. Am. Chem. Soc.* **100**, 6575 (1978).

163. T. Nishiwaki and T. Saito, *J. Chem. Soc. C*, 3021 (1971).

164. Y. Nomura, N. Hatanaka, and Y. Takeuchi, *Chem. Lett.*, 901 (1976).

165. H. Taniguchi, K. Isomura, and T. Tanaka, *Heterocycles* **6**, 1563 (1977).

166. C. S. Cleaver and C. G. Krespan, *J. Am. Chem. Soc.* **87**, 3716 (1965).

167. J. S. Meek and F. W. Fowler, *J. Org. Chem.* **33**, 3418 (1968).

168. J. H. Boyer, W. E. Krueger, and G. J. Mikole, *J. Am. Chem. Soc.* **89**, 5504 (1967).

169. J. S. Swenton, *Tetrahedron Lett.*, 2855 (1967).

170. A. Hassner and F. W. Fowler, *J. Am. Chem. Soc.* **90**, 2869 (1968).

171. R. G. F. Ford, *J. Am. Chem. Soc.* **99**, 2389 (1977).

172. K. Isomura, H. Tanaguchi, T. Tanaka, and H. Taguchi, *Chem. Lett.*, 401 (1977).

173. T. L. Gilchrist, C. W. Rees, and E. Stanton, *J. Chem. Soc. C*, 3036 (1971).

174. J. H. Bowier and B. Nussey, *J. Chem. Soc. Perkin Trans.* 1, 1693 (1973).

175. J. H. Boyer, W. E. Krueger, and R. Mohler, *Tetrahedron Lett.*, 6979 (1968).

176. F. P. Woerner and H. Reimlinger, *Chem. Ber.* **193**, 1908 (1970).

177. A. Padwa, S. Clough, and E. Glazer, *J. Am. Chem. Soc.* **92**, 1778 (1970).

178. D. J. Anderson, T. L. Gilchrist, and C. W. Rees, *Chem. Commun.*, 147 (1969).

179. D. J. Anderson, T. L. Gilchrist, G. E. Gymer, and C. W. Rees, *Chem. Commun.*, 1518 (1971); 1519 (1971).

180. J. van Alphen, *Rec. Trav. Chim. Pays-Bas* **62**, 485, 491 (1943).

181. R. Huttel, K. Francke, H. Martin, and J. Riedl, *Chem. Ber.* **93**, 1433 (1960).

182. H. Durr and R. Sergio, *Tetrahedron Lett.*, 3479 (1972).

183. M. Franck-Neumann and C. Buchecker, *Tetrahedron Lett.*, 937 (1972); *Angew. Chem. Int. Ed. Engl.*, 240, 259 (1973).

184. K. Friedrich, G. Bock, and H. Fritz, *Tetrahedron Lett.*, 3327 (1978).

185. K. Isomura, M. Okaka, and H. Taniguchi, *Chem. Lett.*, 629 (1972).

186. K. Isomura, M. Okaka, and H. Taniguchi, *Chem. Lett.*, 397 (1977).

187. P. Germeraad and H. W. Moore, *J. Org. Chem.* **39**, 774 (1974).

188. C. J. Sanchorawala, B. C. Subba Rao, M. K. Unni, and K. Venkataraman, *Ind. J. Chem.* **1**, 19 (1963).

189. T. Nishiwaki, *Tetrahedron Lett.*, 2049 (1969).

190. M. I. Komendantov and R. R. Bekmukhametov, *Khim. Gitero. Soed.* **9**, 1292 (1976).

191. G. L. Aldons, J. H. Bowier, and M. J. Thomson, *J. Chem. Soc. Perkin Trans.* 1, 16 (1976).

192. S. Maiorana, *Ann. Chim.* **56**, 1531 (1966).

193. M. I. Rubinskaya, A. N. Nesmeyanov, and N. K. Kochetkov, *Russ. Chem. Rev.* **38**, 961 (1969).

194. H. Hemetsberger and D. Knittel, *Monatsch. Chem.* **103**, 205 (1972).

195. D. Knittel, H. Hemetsberger, R. Leipert, and H. Weidmann, *Tetrahedron Lett.*, 1459 (1970).

196. H. Hemetsberger, D. Knittel, and H. Weidmann, *Monatsch. Chem.* **103**, 194 (1972).

197. H. Hemetsberger, D. Knittel, and H. Weidmann, *Monatsch. Chem.* **101**, 161 (1970).

198. T. Tezuka, O. Seshimoto, and T. Mukai, *Tetrahedron Lett.*, 1067 (1975).

199. J. P. LeRoux, J. C. Cherton, and P. L. Desbene, *C. R. Acad. Sci. Ser. C* **280**, 37 (1975).
200. R. A. Abramovitch, S. R. Challand, and Y. Yamada, *J. Org. Chem.* **40**, 1541 (1975).
201. A. Padwa and P. H. J. Carlsen, *Tetrahedron Lett.*, 433 (1978).
202. A. Padwa and P. H. J. Carlsen, *J. Org. Chem.* **41**, 180 (1976).
203. A. Padwa and P. H. J. Carlsen, *J. Org. Chem.* **43**, 2029 (1978).
204. A. L. Logothetis, *J. Am. Chem. Soc.* **87**, 749 (1965).
205. R. A. Abramovitch and W. D. Holcomb, *J. Am. Chem. Soc.* **97**, 676 (1975).
206. R. N. Carde and G. Jones, *J. Chem. Soc. Perkin Trans.* **1**, 2066 (1974).
207. A. Padwa and N. Kamigata, *J. Chem. Soc. Chem. Commun.*, 789 (1975).
208. E. J. York, W. Dittmar, J. R. Stevenson, and R. G. Bergman, *J. Am. Chem. Soc.* **95**, 5680 (1973).
209. R. Srinivasan, *J. Am. Chem. Soc.* **91**, 6250 (1969).
210. R. D. Streeper and P. D. Gardner, *Tetrahedron Lett.*, 767 (1973).
211. J. A. Pincock, R. Morchat, and D. R. Arnold, *J. Am. Chem. Soc.* **95**, 7536 (1973).
212. T. L. Gilchrist, G. E. Gymer, and C. W. Rees, *J. Chem. Soc. Perkin Trans.* **1**, 1 (1975).

3

Radical Cyclizations by Intramolecular Additions

J-M. Surzur

I. INTRODUCTION

In the field of free radical intermediates a lot of work has been devoted to intramolecular reactions. Probably the main reason is the very high selectivity often encountered in these reactions, which results in useful preparative methods.

The origin of the very high selectivities is not always clear, especially since the results are sometimes contrary to those which would have been anticipated from what is known about the corresponding intermolecular processes. Consequently, much speculation and experimental work has been published in an effort to obtain a better understanding of the mechanisms of these reactions.

The best known intramolecular free radical reactions probably are the 1,5-hydrogen abstractions by alkoxyl or aminium radicals used in selective functionalization. The first, known as the Barton reaction, has been reviewed,[1-6] as has the second one, the Hofmann–Löffler–Freytag reaction.[7] Synthetic uses and mechanistic aspects of intramolecular homolytic aromatic substitution have also been reviewed recently.[8-10]

This chapter will be concerned exclusively with intramolecular additions of free radicals. Here again, reviews have been published in this field but they were essentially devoted to carbon-centered radicals. Julia summarized his early work in 1964[11] and in 1967.[12] Finally, in 1971 he presented a review mainly devoted to the mechanistic aspects of intramolecular radical additions.[13] This

J-M. Surzur ● U.D.E.S.A.M., Faculté des Sciences et Techniques (St. Jérôme), 13 397 Marseille, Cédex 13, France.

subject has been covered at about the same time by Beckwith in a review principally devoted to mechanistic aspects of free radical ring-forming and ring-opening reactions.[14]

Finally, the most recent, and probably the most extensive, review was published by Wilt in 1973, in which he covered not only free radical intramolecular additions but also other free radical rearrangements.[15]

Since then, a better knowledge of intramolecular free radical additions, both from the mechanistic and synthetic points of view, has been obtained. Furthermore, these reactions are now frequently used as a probe for free radical intermediates, and important biogenetic schemes involving such reactions have been proposed. Consequently, it appeared that it would be useful to review the main principles which govern these cyclization processes. This review is nicely complemented by another written at the same time by Beckwith and Ingold, which emphasizes quantitative aspects.[628] Furthermore, interesting guidelines for free radical intramolecular additions have been summarized recently by Beckwith.[684] In our opinion, these must be handled with great care since, as is generally the case with oversimplifications, they run the risk of inducing faulty predictions. For instance, it will be seen from our review that S-, P-, and Si-centered radicals, which are also considered by Beckwith, do not obey these simple rules.

It should be noted that, although many intermolecular analogs of the reactions herein discussed are known, literature citations for such processes will rarely be provided since they are well known, and their citation would result in an undue proliferation of references.

Current abbreviations will be used: BP for benzoyl peroxide, DTBP for di-*t*-butyl peroxide, AIBN for azo-bis(isobutyronitrile), In' for initiation, esr for electron spin resonance, and CIDNP for chemically induced dynamic nuclear polarization. Furthermore, we have chosen to adopt homogeneous abbreviations for the radicals discussed. The precursor radicals will be called A', generated from AH, AX,..., and the two possible radicals resulting from intramolecular addition will be called Cy' (for cyclized) followed by a number for the size of the ring.

For instance, from the 5-hexenyl radical (A'), $CH_2{=}CH(CH_2)_3CH_2^{\cdot}$, a (Cy'5) or a (Cy'6) radical are possible (Scheme 1).

The products derived from Cy'5 are called Cy5 and those from Cy'6 are referred to as Cy6. When greater precision is desired, more specific notations, e.g., Cy5H and Cy6H, will be employed. Precisely the same notation is employed when the radical center is a heteroatom (Scheme 1).[†]

[†] For an alternative notation see Reference 16.

SCHEME 1. The Cy5–Cy6 Case

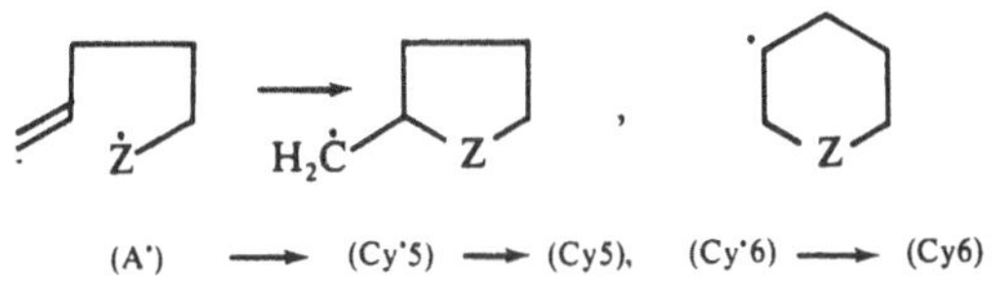

II. GENERAL PRESENTATION: THE EARLY RESULTS

1. The Unstabilized 5-Hexenyl Radicals

The first proposal of intramolecular addition of 5-hexenyl radicals was independently reported in 1957 by Marvel and by Butler. They polymerized 1,6-dienes and, finding that the products were devoid of unsaturation, assumed that the initially produced radicals had cyclized; their work will be discussed in a later section. Also to be discussed is the pioneering work of Friedlander who, in 1958, reported the cyclization of diallyl ethers and diallyl thioethers under free radical conditions.

A simpler case is due to Berson, Olsen, and Walia, who, in 1960, reported an intramolecular free radical addition[17] (Scheme 2).

SCHEME 2

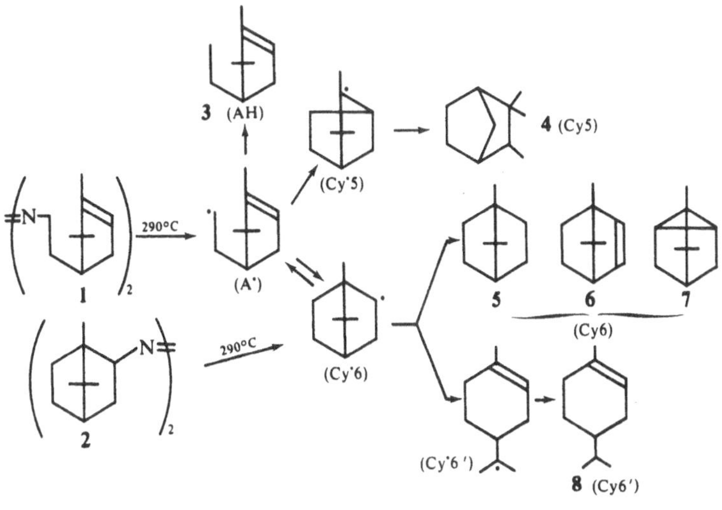

They found that the thermal decomposition of 2,2'-bis(azocamphane) (2) or of the azo compound (1) gave products AH (3), Cy5 (4), Cy6 (5, 6, 7), Cy6′ (8), assumed to have been generated from the radical (A˙) and the cyclized radicals (Cy˙5) and (Cy˙6). Ratios of products 3:4:5:6:7:8 were 2.1:0.50:1.0:trace:0.29:0.34, respectively, from 1 and 0.001:0.035:1.0:0.002: 0.001:0.17 from 2. These results were clearly in accordance with an intramolecular addition reaction Cy˙6 ⇄ A˙ → Cy˙5 complicated by a β-scission process Cy˙6 → Cy˙6′ (Scheme 2). It is, of course, impossible to discuss in this case the Cy5/Cy6 ratio since Cy˙5 could also be considered an example of a Cy˙6 process and vice versa.

Simultaneously and independently, an even more striking result was obtained from studies of the cyclohexyl radical in the gaseous phase at high temperature.[18,19a] In both cases, the presence of methylcyclopentane was established among the many decomposition products and this was rationalized as being formed by opening of the cyclohexyl radical (Cy˙6) to the 5-hexenyl radical (A˙) followed by intramolecular addition to give the methylcyclopentyl radical (Cy˙5). A direct rearrangement could not be dismissed, however, and evidence for the intermediacy of the 5-hexenyl radical was rather indirect. So the work of Lamb and coworkers[20] must be considered as the first in which the problems set up by the behavior of the 5-hexenyl radical were really considered. Furthermore, very important conclusions which are still generally accepted were reached. This work will therefore be discussed in some detail. Lamb, who was engaged in the study of the thermal decomposition of peroxides, observed that with 6-heptenoyl peroxide (AP) (0.1 M in toluene at 77°C) the products obtained (86% total yield) could be explained by Scheme 3.

The products of interest were 1-hexene (AH) (6%), methylcyclopentane (Cy5) (47%), cyclohexane (Cy6) (1%), and cyclohexene (trace). Since galvinoxyl suppressed the formation of methylcyclopentane almost completely, it was concluded that it was produced by a chain process. Furthermore, com-

SCHEME 3

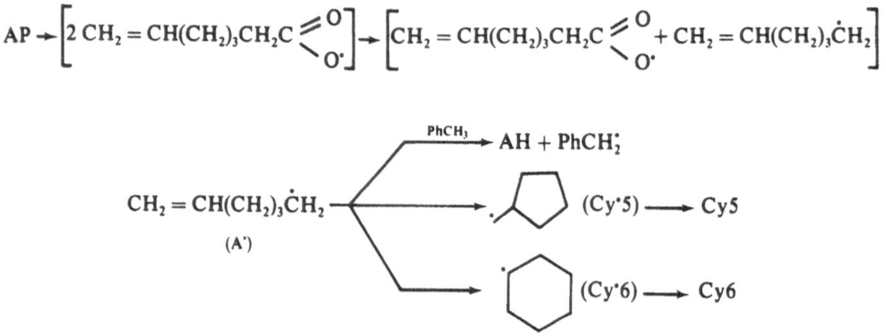

parison of the rates of thermal decomposition of heptanoyl and 6-heptenoyl peroxides (AP) led to the conclusion that there was no participation by the double bond in the decomposition of 6-heptenoyl peroxide (AP). The fact that methylcyclopentane (Cy5) was the major cyclized product was very puzzling since it would derive from anti-Kharasch intramolecular addition of A·, giving a primary rather than a secondary radical. As the cyclohexyl and the methylcyclopentyl radicals, when generated from the diacyl peroxides at 77°C in toluene, gave only cyclohexane or methylcyclopentane, respectively, Lamb concluded that intramolecular addition of the 5-hexenyl radical (A·) towards (Cy·5) and (Cy·6) radicals is irreversible and that the favored formation of the (Cy·5) radical results from kinetic control.

This conclusion was later confirmed by Walling[21] working at 130°C and by Kochi[22] working at 0°C and below by an esr study. These fundamental observations form the basis of any rationalization of the behavior of the 5-hexenyl radical. More generally, as will be seen later, depending on the radical center and the substituents, the reversibility or irreversibility of intramolecular additions is the main factor governing the selectivity of such cyclizations. But as will be seen, the matter of reversibility or irreversibility of these cyclizations must be handled with care because it depends on the experimental conditions. For instance, we have mentioned earlier that in the gas phase at temperatures of the order of 300°C the cyclohexyl radical (Cy·6) isomerizes into the 5-hexenyl radical (A·), and then to the methylcyclopentyl radical (Cy·5).[17,18] Furthermore, at temperatures as high as 420°C, methylcyclopentane disappears, and this has been ascribed to a Cy·5 → A· isomerization.[19a] Thus the generally accepted view that cyclization of the 5-hexenyl radical is irreversible is valid only at relatively low temperatures and when rather good transfer agents are used.

The essentially exclusive formation of the unexpected (Cy·5) radical from the 5-hexenyl radical in solution has been amply confirmed and, indeed, is now used as a major carbon radical probe in mechanistic studies (Section XII.1.A). The high selectivity was independently confirmed as early as 1964 by Walling,[23] who generated the 5-hexenyl radical by the SH$_2$ reaction of the corresponding thiyl radical (itself generated from the unsaturated thiol by AIBN at 60°C or DTBP at 120°C) on triethyl phosphite (Scheme 4). The Cy5H:Cy6H:AH yields were 43:0:15 at 60°C and 50:3:13 at 120°C, i.e., a result very similar to the one obtained by Lamb.

SCHEME 4

$$CH_2 = CH(CH_2)_4SH \xrightarrow{\text{In·}} CH_2 = CH(CH_2)_4S· \xrightarrow{P(OEt)_3} CH_2 = CH(CH_2)_3CH_2·$$
$$(A·)$$

$$A· \longrightarrow AH + Cy5H + Cy6H$$

SCHEME 5

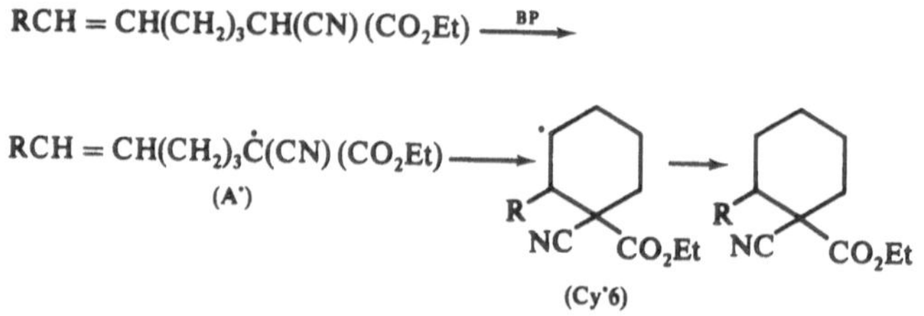

$$RCH = CH(CH_2)_3CH(CN)(CO_2Et) \xrightarrow{\ \ BP\ \ }$$

$$RCH = CH(CH_2)_3\dot{C}(CN)(CO_2Et) \longrightarrow$$
(A·)

(Cy·6)

2. The Stabilized 5-Hexenyl Radicals

In 1960, Julia, Surzur, and Katz published the first of a series of papers dealing with intramolecular free radical additions to olefins.[24] At that time, they reported the results obtained with the intramolecular analog of a reaction described two years earlier, the free radical addition of ethyl cyanoacetate to olefins.[25] The results obtained are summarized in Scheme 5. The reaction was conducted in boiling cyclohexane with BP as initiator and under high-dilution conditions to avoid polymerization; the yields of Cy6H products were 51% (R = H), 65% (R = Me), and 78% (R = Et).

At that time, products were identified by hydrolysis and comparison with authentic crystallized material, and hence it was possible that the corresponding Cy5H escaped detection. Nevertheless, two conclusions arose, the first one being that these reactions could be of synthetic value, and the second being that, in this case, the cyclized products were the Cy6 ones. Later, it was shown by chromatographic analysis that some (Cy5) products were also produced but it was confirmed that the very major products are indeed the (Cy6) type, in clear contradiction of the results obtained by Lamb and Walling described above (Section II.1).

Within a few years, a large number of papers devoted to intramolecular free radical addition reactions appeared, particular attention often being given to the matter of ring size. Before discussing these results we shall first endeavor to analyze and rationalize the behavior of 5-hexenyl radicals.

III. ATTEMPTS TO RATIONALIZE THE SELECTIVITIES OBSERVED IN THE CYCLIZATION OF 5-HEXENYL RADICALS

In a short note dealing with the cyclization of 5-hexenyl radicals[26] Julia reported that Cy·6 formation apparently was favored only when the initial

SCHEME 6

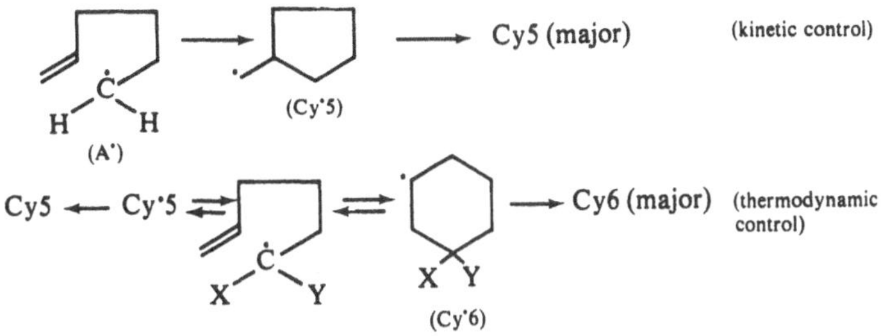

carbon radical was substituted by stabilizing groups (CN, CO_2Et, and so on). Later, Julia, Maumy, and Mion,[27] at the suggestion of deTar and Bartlett, employed a concept previously suggested by Lamb (see Section II.1) and now generally accepted, namely, that (Cy5) products arise via kinetic control. On the other hand, if the intramolecular addition is reversible, which may be expected when the initial radical (A˙) is stabilized, the main product will be derived from the thermodynamically favored (Cy˙6) radical (Scheme 6, where X and Y can be CN, CO_2Et, COMe, and so on).

These two cases will now be discussed separately.

1. The Irreversible Pathway: Why is the (Cy˙5) Radical Favored?

Even if we accept the foregoing schemes it is still necessary to explain why the (Cy5) compounds are the favored ones when the addition is an irreversible one.

Three explanations have been advanced. These are based on steric, stereoelectronic, and entropic arguments but, as will be seen, they are probably not as mutually exclusive as might at first be thought.

A. The Steric Hypothesis

The first attempt at rationalization was made by Lamb,[20] who proposed that the 5-hexenyl radical (A˙) is in the form of an intramolecular complex between the free radical site and the double bond, and that transfer of hydrogen from the solvent to this complex would be sterically easier if it were to give the (Cy5) product rather than the Cy6. The possibility of such a complex, or at least of a coiled conformation in which the terminal unsaturated linkage lies over the radical center, was later confirmed by esr.[28] Nevertheless, this proposal does not seem to have been used extensively and is not in accord with more recent ideas[37] (Section III.1.B).

A second, and rather more widely discussed, steric proposal was that

SCHEME 7

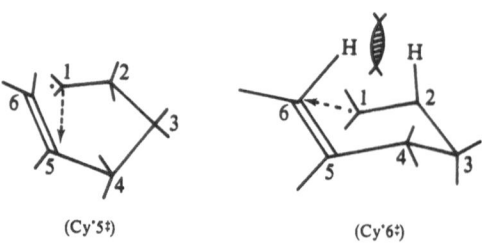

(Cy'5‡)   (Cy'6‡)

advanced by Julia following a suggestion of leBel.[29,30] Here it was assumed that if the (Cy'5‡) and (Cy'6‡) transition states resembled the products, then it could be seen by using Dreiding models that in Cy'6‡ there was severe steric interaction between the $C_{(2)}$ and $C_{(6)}$ substituents. Since such an interaction was not present in Cy'5‡, formation of the Cy'5 would be favored (Scheme 7).

This hypothesis has been discussed in the review cited[13] and it is supported by substituent effects[31] (Section IV).

B. The Stereoelectronic Hypothesis

At about the same time, Struble, Beckwith, and Gream[32] offered another rationalization of the fact that the cyclized products are formed via the thermodynamically less favorable route. They proposed that radical cyclization is largely under stereoelectronic control and that the transition state is sterically different from that which prevails in comparable cationic cyclizations reactions which usually lead to the Cy6 compound. Although little was known about the intimate structure of the transition state for homolytic addition to the double bond, Beckwith discarded the generally accepted hypothesis of a transition state involving maximum overlap of the unpaired electron with the occupied π orbital. Examination of models led him to suggest that the initial stages of the addition process involved interaction of the unpaired orbital (SOMO) with the lowest unoccupied orbital of the π system (LUMO) and that the approach of the radical occurred along a line extending almost vertically from one of the terminal carbon atoms. This view later received support from Fukui and Fleischer's studies of intermolecular radical additions.[33-36]

Beckwith concluded that bond formation occurs at the terminus of the olefinic linkage most readily approached vertically by the radical and he thus rationalized the preferred formation of the (Cy'5) radical. This view has been further discussed in a review[14] and developed in later work.[37-39] In fact, examination of models does not lead, in our hands, to such an obvious conclusion as advanced by Beckwith, unless one takes into account the same $C_{(2)}$–H / $C_{(6)}$–H interactions discussed by Julia (Scheme 7).

SCHEME 8

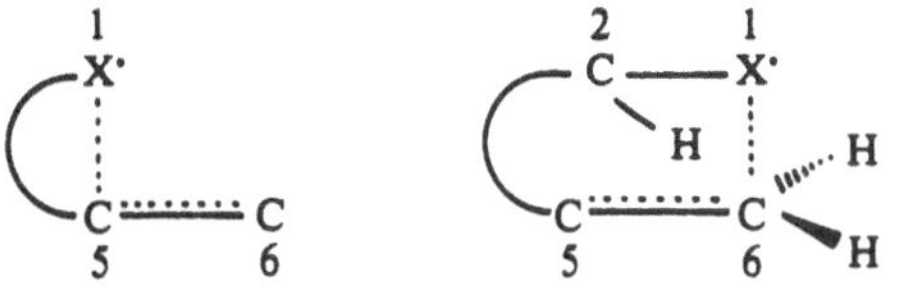

(Cy'5‡)   (Cy'6‡)

A combination of Beckwith's stereoelectronic argument and the steric hypothesis of Julia may be a fruitful approach. Rough calculations by Tordo[40] of the conformational energies E_T of the radicals were made according to Beckwith for the two situations Cy'5‡ and Cy'6‡, which lead to the (Cy'5) and (Cy'6) radicals[(Scheme 8) where X is CH_2, S, 0, NH,...], respectively. It was assumed that the energy $E(\pi, \Psi)$ of the double bond – lone electron system varied with the overlap integral S between the π^* orbital and the singly occupied orbital Ψ. The same d_{X-C} length between X and $C_{(5)}$ or $C_{(6)}$ were chosen in the two transition states and it was further assumed that for the same value of S, the lower conformational energy E_T would give the easier path toward Cy'5 or Cy'6, or that for the same E_T, the S values must be compared. The main results are summarized in Table 1.

If it is assumed that S values are nearly equivalent, it is seen that Cy'5‡ is always favored, and this bias is greater the shorter the $d_{C_{(2)}-X}$ bond length. On examining the origins of the E_T differences, it transpires that they arise mainly from the repulsive interaction between $H-C_{(2)}$ and $H-C_{(6)}$ in the (Cy'6‡) transition state, confirming the hypothesis based on Dreiding models. Thus when the $C_{(2)}-X$ bond length becomes shorter, the repulsive interaction $H-C_{(2)}/H-C_{(6)}$ increases and this disfavors (Cy'6) radical formation. From the results of Table 1 it is to be expected that when $X = CH_2$, O, or NR the Cy'5 will be

TABLE 1. Values of $S \times 10^2$ and E_T (kcal mol^{-1}) for Cy'5‡ and Cy'6‡ [a]

	S		E_T		
X	Cy'5‡	Cy'6‡	Cy'5‡	Cy'6‡	$d_{C_{(2)}-X}$, Å [b]
CH_2	4.4	3.5	0	12.4	1.54
O	2.4	2.0	0	15.4	1.43
NH	3.2	2.5	0	12.2	1.46
S	5.3	4.2	0	1	1.81

[a] Data taken from Reference 40.
[b] $d_{C_{(2)}-X} \equiv C_{(2)}-X$ bond length.

favored, but that when X = S, there will be little, if any preference between (Cy·5) and (Cy·6) radical formation.

Since Tordo's calculations are very approximate, the differences found for the E_T values are certainly too high. For instance, we will see in Section IV that introduction of alkyl substituents often modifies the ratio Cy5/Cy6 dramatically, much more than would be expected on the basis of these calculations. Nevertheless, the conclusion that the selectivity under kinetic control is mainly governed by the $C_{(2)}$–X bond length has proved to be very useful. For instance, when X = PPh, Davis, Downer, and Kirby[41] obtained what could be considered a puzzling result, namely exclusive formation of the Cy6 compound. This can be understood easily by a consideration of the large $d_{C_{(2)}-P}$ bond length (1.83 Å). Also, in the case of X = S $[d_{C_{(2)}-S} = 1.81$ Å], it will be seen later that when kinetic control is favored, selectivity toward the (Cy5) or the (Cy6) compounds is observed only in special cases. On the other hand, with X = O or NR one may expect to observe the same preference for (Cy·5) radical formation as with X = CH_2.

Finally, it should be noted that Baldwin, in a brief communication,[16] has proposed some empirical rules governing ring closure reactions and although he is mainly concerned with nucleophilic cyclizations processes[42–45] he has also considered free radical ones[16]; the origin of these rules appears similar to the one just discussed.

C. The Entropic Hypothesis

The preceding interpretations of the results are essentially based on enthalpy considerations. However, an entropic basis for the selectivity toward the Cy·5 radical in the 5-hexenyl case has also been advanced. Capon and Rees[46,47] suggested that more conformations are available for Cy5 cyclization than for Cy6 cyclization. Although sometimes used,[48,49] this way of rationalizing five-membered ring formation does not seem to have been tested.

Recently, Tedder[50] proposed that rotational entropy may be a very important factor which favors five-membered ring formation. The origin of this suggestion lies in his work in which he found small but regular variations of preexponential factors for the intermolecular addition of alkyl radicals to

SCHEME 9

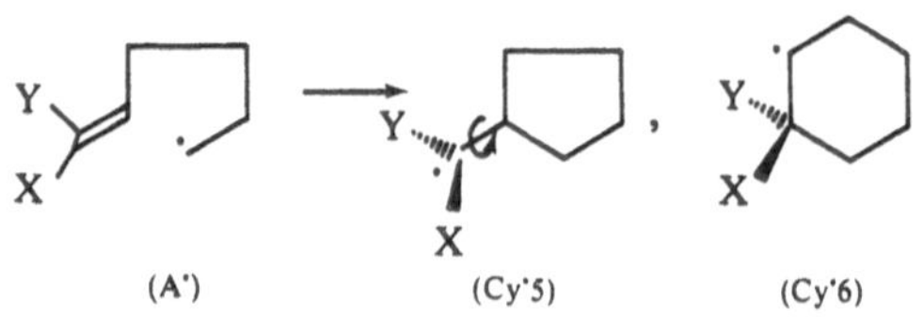

| | | |
| (A') | (Cy·5) | (Cy·6) |

substituted ethylenes. In Cy5/Cy6 intramolecular cyclizations, ring closure to give a five-membered ring would result in smaller decrease in rotational entropy than the formation of a six-membered ring (Scheme 9).

J.M. Tedder kindly informed us that Dr. Walton has calculated the entropy change for the ring closure reactions using the group parameters suggested by Benson, and that by using this approximate method it is found that the entropy change favored the smaller ring.[51]

Recent calculations suggest that (Cy'5) radical formation can be accounted for on the basis of entropy alone.[629] Since these theoretical results are not in agreement with the experimental data in both the Cy5/Cy6 and Cy6/Cy7 cases as discussed, for instance, in Table 3, they must be considered with caution.

D. Conclusion

It seems clear that none of these rationalizations is completely satisfactory and probably are not mutually exclusive. Nevertheless, it will be seen in the following sections that these approaches may be useful as qualitative guides in irreversible cyclizations processes of radicals not only in the 5-hexenyl system but also in other cases.

2. The Reversible Pathway Favoring the (Cy'6) Radical

The assumption that under thermodynamic control the more stable secondary radical (Cy'6) would be the favored one requires the demonstration that the cyclization reaction is reversible. When this hypothesis was proposed, the possibility of reversible addition of free radicals to double bonds was not a familiar one, particularly with carbon-centered radicals. This led Julia to an exhaustive study of systems such as are summarized in Scheme 10, in which X and Y were mainly CN and CO$_2$Et.

Before analyzing Julia's results, it seems worthwhile to point out two observations of some importance in reversible cyclizations. The first one is relative to the retroaddition process, i.e., the opening of the (Cy') radicals can be considered a β-scission process and to be governed by the same principles,

SCHEME 10

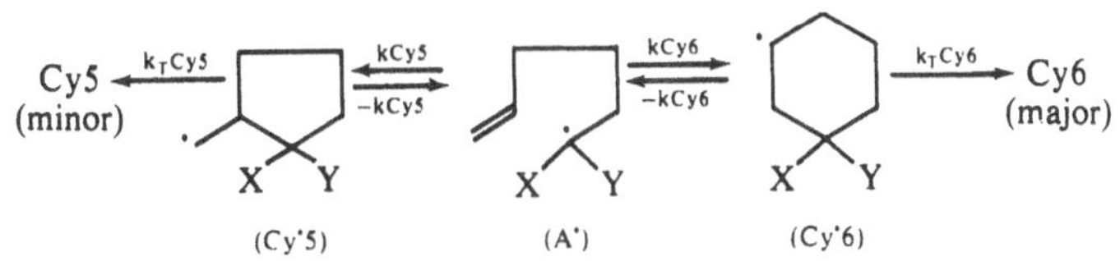

Cy5 (minor) $\xleftarrow{k_T Cy5}$ (Cy'5) $\underset{-kCy5}{\overset{kCy5}{\rightleftarrows}}$ (A') $\underset{-kCy6}{\overset{kCy6}{\rightleftarrows}}$ (Cy'6) $\xrightarrow{k_T Cy6}$ Cy6 (major)

e.g., stabilization of the open-chain radical (A˙ in Scheme 10 with X and Y stabilizing groups) or stereoelectronic factors.[14] Of course, these principles could be used to predict opening of the (Cy˙) radical in another direction as represented in Scheme 10, but obeying the same rules and complicating the nature of the isolated products, as for example in Scheme 2.

Secondly, it is important to emphasize that the cyclization processes shown in Scheme 10, while they are reversible, do not represent a system in equilibrium, for each of the cyclic radicals is capable of conversion, in an essentially irreversible manner, to final products. Thus the Cy6/Cy5 ratio will be governed by the six rate constants pictured in the Scheme. Of course, under such circumstances, the product ratio will be sensitive to experimental conditions.

Notwithstanding these difficulties, the reversible interconversion of A˙, Cy˙5, and Cy˙6 has been clearly established by Julia.[27,30] Radicals (Cy˙5) and (Cy˙6) were formed by the thermal decomposition of the corresponding *t*-butyl peresters and the Cy5/Cy6 ratio was compared to that obtained when submitting AH to a free radical initiator (Scheme 11). By treating AH (X = CN, Y = CO$_2$Et) (initial concentration 3.75×10^{-2} mol liter^{-1}) in boiling cyclohexane (81°C) as hydrogen donor R–H with BP as initiator (initial concentration 7.5×10^{-2} mol liter^{-1}) the ratio Cy6H/Cy5H was 86:14. Under the same conditions, thermolysis (100 hr) of the perester P-Cy5 led to a ratio 80:20

SCHEME 11

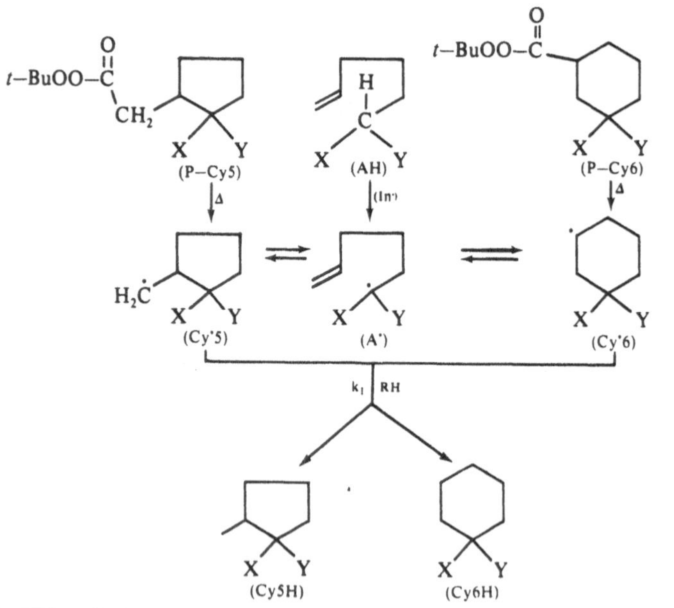

and that of the perester P-Cy6 to a ratio 85:15. It is quite clear that the near identity of these three ratios is nicely accommodated by the reversible processes as shown in Scheme 11. This near identity of ratios is not always observed. Thus Julia[30] found that when the cyclization of AH (X = CN, Y = COOEt) is initiated by a photoexcited ketone at −70°C the Cy6H/Cy5H ratio is actually the reverse (20:80) of that found at 81°C. Thus at −70°C cyclization is predominately a kinetically controlled process. In the same way, the product ratio is altered by accelerating the last, irreversible transfer steps (k_T in Scheme 11) by using as the solvent a good hydrogen donor such as toluene: at 11°C, the Cy6H/Cy5H ratio is 10:80, whereas in cyclohexane, at the same temperature, thermodynamic control begins to appear, the Cy6H/Cy5H ratio now being 33:67. Finally, it has also been observed that the nature of the initiator may have also a great influence on the ratio of products, but up to the present time no satisfactory explanation has been proposed for this effect.[30]

Cyclizations of other unsaturated stabilized radicals will be discussed in Section VII. The aim of the present section was to show that reversible cyclization processes have been clearly demonstrated and that true thermodynamic control is not always attained.

IV. ALKYL-SUBSTITUTED 5-HEXENYL RADICALS

The influence of substituents on the double bond, on the aliphatic chain, and on the carbon radical has been extensively studied in the case of irreversible cyclization processes. In this section, the results obtained by Walling, Beckwith, Julia, and their coworkers using alkyl-substituted 5-hexenyl radicals are described. The influence of heteroatoms and of stabilizing groups is discussed in another section.

Walling[52] studied the 5-hexenyl radical itself, as well as the 1- and 5-

SCHEME 12

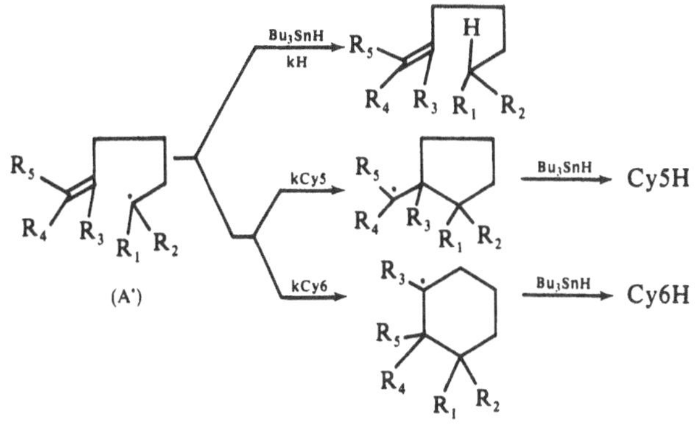

methyl substituted ones. These radicals were generated by reaction of the bromides in benzene (0.025 and 0.5 mol liter^{-1}) with tributylstannane (generally at the same concentration); as initiators he used AIBN (at 40°C and 70°C) and DTBP (at 100°C). We will not comment on these results (Table 2) because of some discrepancies in the quantitative results with those of later workers, but the main conclusion, that the Cy5/Cy6 ratio as well the facility of intra-

TABLE 2. Cyclization of
$$R_5 \diagdown \diagup (CH_2)_3-\dot{C}R_1R_2$$
$$C=C$$
$$R_4 \diagup \diagdown R_3$$

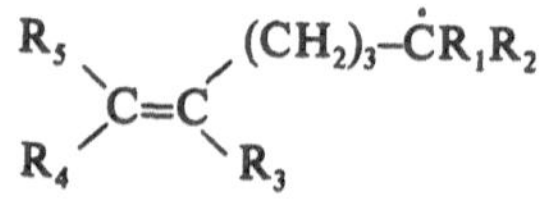

Entry	A$^{\cdot}$	kCy5/kCy6	kCy5/k°Cy5	kCy6/k°Cy5	Reference
1	R = H	>200 48 High	1.0	0.02	52 53 31
2	R$_3$ = CH$_3$	37 0.62 0.4	0.022	0.04	51 53 31
3	R$_3$ = CH(CH$_3$)$_2$	0.31	0.022	0.07	53
4	R$_4$ = CH$_3$	High			31
5	R$_4$ = (CH$_2$)$_2$CH=CH$_2$	>200	1.7	<0.009	53
6	R$_5$ = (CH$_2$)$_2$CH=CH$_2$	>200	1.4	<0.007	53
7	R$_3$ = R$_4$ = CH$_3$	0.58			31
8	R$_3$ = R$_5$ = CH$_3$	High			31
9	R$_4$ = R$_5$ = CH$_3$	>200	2.4	<0.011	53
10	R$_4$,R$_5$ = cyclohexyl	>100	0.94	<0.009	53
11	R$_1$ = CH$_3$	>99 78	1.4	0.02	52 53
12	R$_1$ = (CH$_2$)$_2$CH$_3$	206	1.6	0.008	53
13	R$_1$ = R$_2$ = CH$_3$	5 68	1.4	0.02	52 53
14	R$_1$,R$_2$ = cyclopentyl	14	0.94	0.07	53
15	R$_1$ = R$_2$ = R$_3$ = CH$_3$	1.1 <0.01	<0.0002	0.02	52 53

molecular addition can be markedly affected by alkyl substituents, remains valid. More recently, Beckwith[53] and Julia[31] have published an extensive study of the effects of alkyl substituents on the 5-hexenyl radical generated in the same way as did Walling (Scheme 12).

In Table 2 are reported the main values obtained by the three authors, $k°Cy5$ being the rate constant $kCy5$ when all R = H; substituents not indicated in the Table correspond to R = H, temperatures being 65°C (Beckwith[53]), 40°C (Walling[52]), and about 70°C (Julia[31]).

Because of some discrepancies observed (mainly in entries 2 and 15) and of the difficulty of obtaining accurate values, it is probably too early to propose a detailed rationalization of these results. Nevertheless, some interesting aspects appear. If the results of Walling,[52] mainly entries 13 and 15, are neglected, one can conclude that the nature of the initial radical, primary (entries 1–10), secondary (entries 11–12), or tertiary (entries 13–15) has little influence on $kCy5$ and $kCy6$: compare, for instance, entries 1, 11, and 13 (Beckwith) in which $kCy5/k°Cy5 \sim 1$ and $kCy6/k°Cy6 \sim 0.02$. As a result little change can be expected, not only on the yield of cyclized products, but also on the Cy5/Cy6 ratio when R_1 and (or) R_2 are H or alkyl. Of course, the opposite conclusion would be reached if Walling's results were confirmed, with the Cy5/Cy6 ratio decreasing in the order primary > secondary > tertiary. In fact, according to recent kinetic studies, cyclization of a secondary radical would be slightly faster than cyclization of a primary one, $kCy5$ (secondary)/$kCy6$ (primary) being 1.18.[630]

Alkyl substituents R_4 in the *trans-6* position also seem to have a modest influence, although as can be expected from the Kharasch effect Cy˙5 is relatively favored (with $kCy5/k°Cy5$ increasing) and Cy˙6 disfavored (with $kCy6/k°Cy5$ decreasing): compare entry 1 with entries 4 and 5. Thus, one can expect to observe a good yield of cyclized products with *trans*-alkyl substituents R_4, with a Cy5/Cy6 ratio higher than that with no substituent.

An R_3 alkyl substituent on the double bond affords a dramatic effect, also in accordance with the Kharasch effect; formation of Cy˙5 by radical attack on a crowded position of the double bond to give a primary radical is now a very unfavorable process. Thus, if the results reported by Walling in entry 2 are set aside in favor of the near identity of the results obtained by Beckwith and Julia, entry 1 (R_3 = H, $kCy5/k°Cy5 = 1.0$) can be compared with entries 2 (R_3 = CH_3) or 3 (R_3 = $(CH_3)_2CH$, $kCy5/k°Cy5 = 0.022$). Of interest is the relatively slow increase of $kCy6/k°Cy5$ (0.04 in entry 2 compared to 0.02 in entry 1). On simple thermochemical grounds a larger increase might have been expected, which led Beckwith to the conclusion that steric (but also electronic) factors are very important not only in governing the selectivity but also the ease of free radical additions to a double bond. As a result not only will an R_3 substituent change the ratio Cy5/Cy6 dramatically, with formation of (Cy6) compounds now being favored, but as this change is mainly due to a decrease of kCy5 rather to an increase in kCy6, the total yield of cyclized products will be lowered. On these grounds the result described by Beckwith in entry 15 can be understood, this tendency being, of course, reinforced by the tertiary nature of the initial radical.

Also very interesting are the results obtained mainly by Julia in which a substituent R_5 is on the terminal double bond 6-position, but this time *cis* to the aliphatic chain. This effect is well illustrated by entries 7 and 8. As already discussed, the *trans* isomeric radical ($R_4 = CH_3$, R_3, $= CH_3$, entry 7) gives the low Cy5/Cy6 ratio of 0.58 because of the R_3 substituent. But now, if in place of a *trans* $R_4 = CH_3$, there is a *cis* one (entry 8, $R_5 = CH_3$, $R_3 = CH_3$) the Cy5/Cy6 ratio is completely reversed, becoming so high that no Cy6 products are detected. These results have been nicely accommodated by Julia on the basis of Scheme 7 (see the discussion in Section III). It is quite clear that replacement of a *cis*-hydrogen atom $H-C_{(6)}$ by a *cis*-alkyl group R_5 will enhance the unfavorable interaction ($R-C_{(6)}/H-C_{(2)}$) in the (Cy'$6^{\ddagger}$) transition state. Results obtained with the two substituents R_4 and R_5 on the double bond terminal (entries 9 and 10), with an obligatory R_5 *cis* substituent may be ascribed to the same reason: the very high Cy5/Cy6 ratio observed results from a decrease in kCy6 rather to an increase in kCy5, which could have been expected from the formation of a tertiary (Cy'5) radical. In conclusion, a *cis*-alkyl R_5 group increases the Cy5/Cy6 ratio mainly by decreasing kCy6, and, as a result, one obtains a decrease in the total yield of cyclized products. This is especially true with an R_3 substituent, where Cy6 compounds are the ones favored.

As already noted, it is difficult to obtain accurate values and some discrepancies exist. Thus any conclusion or rationalization must be considered as tentative. Nevertheless, it is quite clear that alkyl substituents do have a marked influence on the free radical intramolecular addition processes.

To summarize: In the 5-hexenyl system

$$\underset{R_4}{\overset{R_5}{>}}C=C\underset{R_3}{\overset{(CH_2)_3\dot{C}R_1R_2}{<}}$$

when $R = H$, Cy5 compounds are the favored ones. With $R_3 =$ alkyl, a dramatic result is obtained: there is a significant decrease in kCy5, the Cy5/Cy6 ratio becomes <1 and the overall yield is lowered. A *cis*-R_5 alkyl substituent decreases kCy6. A *trans*-alkyl R_4 substituent produces a slight increase in the kCy5 and a slight decrease of kCy6. Substituents R_1 and R_2 on the radical center apparently have little effect, but this needs to be confirmed.

Some of these conclusions are helpful in understanding the behavior of unsaturated radicals other than the 5-hexenyl ones. It should be emphasized that only modification by alkyl substituents has been discussed in this section. It will be seen in Sections VI and VII that inductive or other effects derived from heteroatoms, for instance, may have a profound influence on the cyclization process.

V. ALKENYL RADICALS OTHER THAN 5-HEXENYL

In contrast to the 5-hexenyl system, radicals such as $\diagup\!\!C{=}\overset{|}{C}(CH_2)_n\overset{\diagdown}{\overset{|}{C}}\diagup$, where $n \neq 3$, have not been studied extensively. Not surprisingly when $n = 4$ or more, cyclization occurs with relative difficulty and, indeed, except in few special cases, the yields of cyclized products are so low that the reaction is really without synthetic utility.

1. The Higher Homologs

The kinetic results obtained by Beckwith[54] summarize what may be expected from higher alkenyl radicals. These radicals were generated from the bromides by stannyl reduction and the relative rates derived from the proportions of uncyclized and cyclized products (Scheme 13).

In Table 3 are reported the relative values obtained at 65°C by taking as 100 the value $k'Cy/kH$ for the formation of the (Cy'5) radical in the 5-hexenyl case $(n = 3)$ as well as the values $\Delta\Delta H^{\ddagger} = \Delta H_{Cy}^{\ddagger} - \Delta H_{H}^{\ddagger}$ (kcal mol^{-1}) and $\Delta\Delta S^{\ddagger} = \Delta S_{Cy}^{\ddagger} - \Delta S_{H}^{\ddagger}$ (e.u.). It is clear from these data that the ease of cyclization follows the order $n = 3 \gg 4 \gg 5$.

Thus good yields of cyclized products will only be obtained with $n > 3$ if a poor transfer agent is employed in low concentration. This conclusion could be put on a more quantitative basis, by using the value $k'Cy = 1 \times 10^5$ sec^{-1} at 25–40°C obtained by Ingold in the Cy5/Cy6 case, which will be discussed in Section XII.2.[55]

With these higher homologs of the 5-hexenyl radical a new complication sets in, namely, 1,5-intramolecular hydrogen atom transfer; this is particularly important when $n = 4$, the 5-hydrogen atom being allylic in this case.[54]

The results shown in Table 3 are in accord with those obtained in the 5-hexenyl case: formation of methylcycloalkanes is again favored with the higher homologs. From the enthalpy and entropy values, it can be seen that the former makes the more important contribution to this selectivity.[54] Nevertheless, the

SCHEME 13

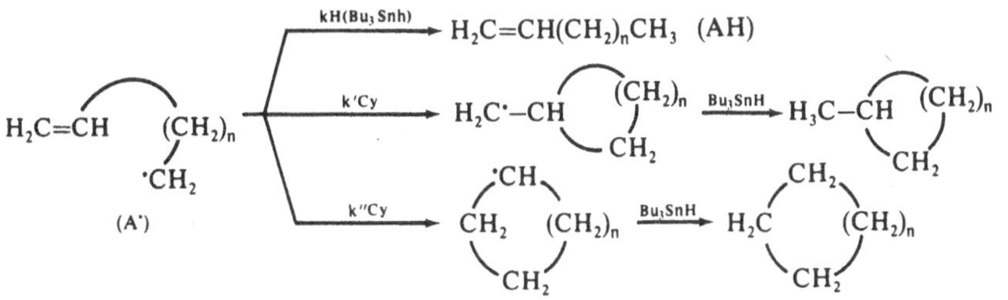

TABLE 3. Relative Kinetic Parameters Obtained in the Cyclization of
$CH_2=CH(CH_2)_nCH_2^{\cdot}$ Radicals

n	$k'Cy/kH^a$	$\Delta\Delta H^{\ddagger}$ kcal mol^{-1}	$\Delta\Delta S^{\ddagger}$ e.u.	$k''Cy/kH^a$	$\Delta\Delta H^{\ddagger}$, kcal mol^{-1}	$\Delta\Delta S^{\ddagger}$, e.u.
3 (Cy5/Cy6)	100	2.9	5.7	2.1	4.6	2.9
4 (Cy6/Cy7)	2.4	4.2	2.0	0.42	5.2	1.4
5 (Cy7/Cy8)	0.69	5.9	0.1	<0.005		

^{a} $k'Cy$ and $k''Cy$ as in Scheme 13.

Cy5/Cy6 selectivity, which is about 50, is only 6 in the Cy6/Cy7 case. These data may be rationalized on the same basis as was used in Section III.1.B (Table 1). The values obtained for the general Cy6/Cy7 case ($n = 4$), using the same hypothesis, are given in Table 4 for $CH_2=CH(CH_2)_4X^{\cdot}$ radicals.[40]

By comparing the E_T values in Tables 1 and 4, it is seen that the Cy'6‡ values for the Cy6/Cy7 case are higher than the Cy'5‡ values for the Cy5/Cy6 case, and that the differences between the E_T values for Cy'6‡ and Cy'7‡ in the Cy6/Cy7 case are much smaller than the corresponding ones for the Cy5/Cy6 case.

It would be anticipated that very low yields of cyclized products will be obtained from the Cy7/Cy8 and higher cases. In this connection a recent report of Kammerer and Steiner[56] should be noted, even though there are a number of special features to this work. They submitted a series of methacrylate esters of catechol, resorcinol, and *p*-hydroquinone, among others, to AIBN initiation in boiling benzene under high-dilution conditions. Significant yields of cyclized products were obtained in addition to polymeric material. For example, 9 gives 10 (43% yield) and 10' (22% yield), whose formation may be ascribed to addi-

TABLE 4. Values of $S \times 10^2$ and E_T (kcal mol^{-1}) for Cy'6‡ and Cy'7‡

X	S		E_T	
	Cy'6‡	Cy'7‡	Cy'6‡	Cy'7‡
CH_2	4.4	4.9	3	4.5
O	2.6	2.2	4.6	4.5
NH	3.6	3.3	1.8	4.5
S	9.1	9.4	4	5.9

SCHEME 14

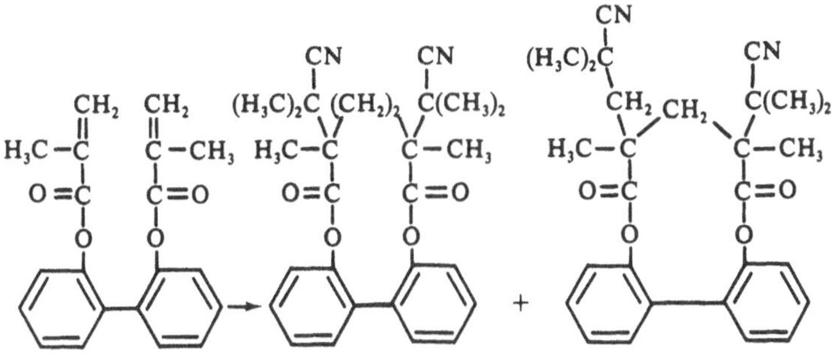

tion of the 2-cyano-2-propyl radical generated from AIBN to one or the other position of the double bond followed by intramolecular addition of the carbon radical (Scheme 14).

2. The Lower Homologs

Returning to the discussion of the behavior of the 5-hexenyl radical (Section III.1.B), if the model proposed by Beckwith (which involves the necessity for the radical to lie above the plane of the double bond) is accepted, it is clear that this will become more and more difficult as the chain length shortens. In fact, cyclized products are not observed with the 4-pentenyl and 3-butenyl radicals when these are generated from the corresponding bromides[57] under the conditions used successfully with the hexenyl bromide (Scheme 13).

A. The 4-Pentenyl Radicals

All attempts to obtain cyclized products from the 4-pentenyl radical using the same conditions under which the 5-hexenyl radical cyclizes readily failed. This was early recognized[19,23] and confirmed later.[418,419] Only in special cases, as by the use of vibrationally excited radicals in the gas phase[58-60a] or carbene triplets [60b] has cyclization been observed. In these instances, only (Cy5) and no (Cy4) products were obtained. In solution, cyclized products have been observed only from 4-pentenyl radicals possessing special features, e.g., the radical (A˙) which results from intermolecular free radical addition to *cis,cis*-1,5-cyclooctadiene (Scheme 15).

The first examples of this type of reaction were reported independently by Dowbenko[61] and by Friedman[62] with Y–X = H–CCl$_3$, H–CONR$_2$, H–COR, Cl–CCl$_3$, H–CCl$_2$CO$_2$Et, Cl–CCl$_2$CO$_2$Et, and butyrolactone,[63] yields being as high as 74%. However, with good transfer agents Y–X (such as H–SR,[64]

SCHEME 15

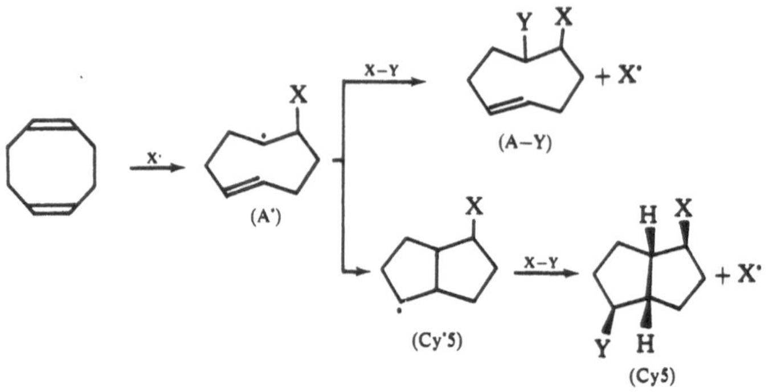

H–Br,[65] ON–NO$_2$,[66] PhICl$_2$[67]) only the 1,2 adduct A–Y, or the corresponding diadduct, was observed. Thus the cyclization process leading from A˙ to Cy˙5 must be relatively slow. Finally, it should be noted that the conversion of A˙ to Cy˙5 can be considered a Cy5/Cy6 cyclization.

In all these cases no methylcyclobutyl radicals have been observed. But even if they had been formed methylcyclobutyl radicals would open very easily. A familiar case is that of the (Cy˙4) radical resulting from intermolecular free radical addition to α- or β-pinene (Scheme 16).

Unrearranged (Cy4) products are obtained here only when Y–X is a good transfer reagent such as RSH.[68] Generally the products obtained are those resulting from β-scission of Cy˙4 to A˙, which is clearly a very fast process. A relevant example is the addition of triplet carbene (MeOCO)$_2$C: to cyclobutene: besides the expected product of biradical coupling an important competing reaction is the β-scission of the (Cy˙4) radical.[60b]

SCHEME 16

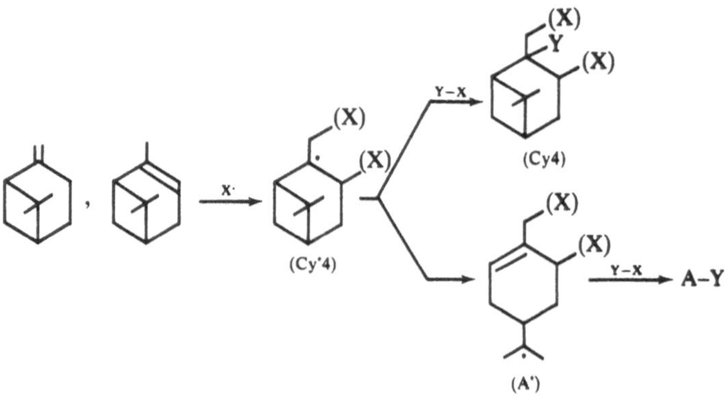

In conclusion, except in very special cases, reasonable yields of cyclized products cannot be expected in the Cy4/Cy5 case since if they are formed the (Cy'4) radical will open readily (i.e., it is the opening and not the closing process which is favored). The absence of cyclization toward the (Cy'5) radical is an excellent example of the disfavored 5-*endo*-trigonal closure in accord with Baldwin's rules.[16,683]

B. The 3-Butenyl Radicals

Extensive studies have been devoted to these radicals in connection with their structure (classical or nonclassical) and their reactivity (homoallylic rearrangement and 1,2-vinyl migration) and the main results have been reviewed.[14,15] For instance, the classical nature of the allylcarbinyl, cyclopropyl carbinyl, and cyclobutyl radicals now seems well established.

The possibility of cyclization of 3-butenyl radicals is well established, giving apparently only Cy'3, the methylcyclopropyl radicals (9.4×10^3 sec^{-1} at 40°C); but the opening of the methylcyclopropyl radical is faster (1.3×10^8 sec^{-1} at 25°C [69]) so that the cyclization process is essentially reversible ($K = 1.3 \times 10^4$ at 25°C, statistically corrected for A' and A''').[631] This is inferred mainly from products which appear to result from 1,2-vinyl migration (Scheme 17).[73,436−448]

It has been demonstrated that the β-scission process is under stereoelectronic control, involving a transition state formed by interaction of the semi-occupied orbital with a $\beta,\gamma\text{-}\sigma^*$ orbital and that in strained systems where such a situation is forbidden the opening process is too slow to be observed.[70] It must be noted that the stereoelectronic control hypothesis is more and more frequently used in order to rationalize such β-scission processes in strained systems. For instance, Agosta recently used it in order to predict the major α-cleavage products resulting from the photolysis of β,γ-cyclopropyl ketones[71a] and even as a mechanistic probe for a biradical intermediate. [71b,c] Even in unstrained systems, stereoelectronic control has been suggested to explain the unexpected regioselective ring opening of *trans*-2-alkylcyclopropylmethyl radicals to the less stable primary alkyl radicals,[632] although in these cases the selectivity is best rationalized within the framework provided by frontier

SCHEME 17

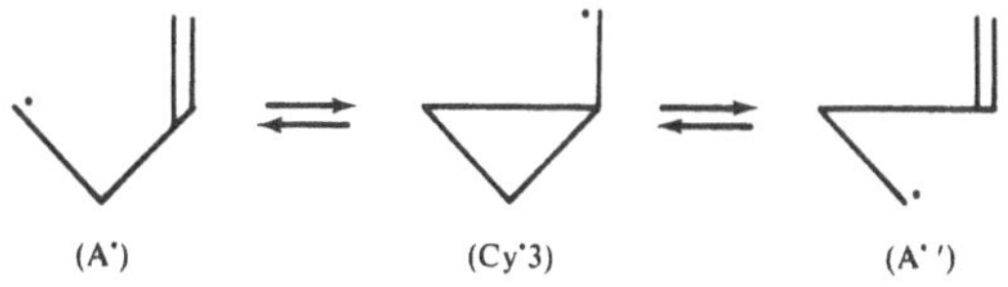

(A')   (Cy'3)   (A'')

SCHEME 18

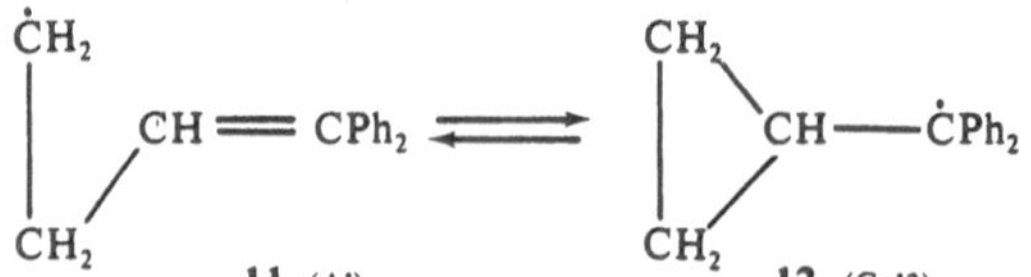

molecular orbital theory.[685] Finally, it must be pointed out that the cyclobutyl radical does not appear to form in these processes[14] and that the opening of the cyclobutyl radical is a slow process.[72] According to MINDO/3 calculations the activation energy for (Cy·3) radical formation would be about 20 kcal/mol less than that for (Cy·4) radical formation,[629,633] although Cy·4 would be more stable than Cy·3 by 11.2 kcal/mol.[633]

We are concerned here only with the possibility of observing cyclized products from 3-butenyl radicals. This is observed when the resulting methylcyclopropyl radical is stabilized by one or two phenyl groups. For instance by thermolysis of the corresponding peracetate, Roberts[73a] obtained cyclopropyl products assumed to arise from the radical 12 (Scheme 18) (although in only a 1.6% yield). Conversely, a stabilizing group such as phenyl enhances the rate of opening of the (Cy·3) radical (Scheme 17) generated, for instance, by the electroreduction of a cyclopropyl ketone; with a phenyl substituent on the radical center of Cy·3 the products result exclusively from Cy·3, whereas when the phenyl group is present on the radical center of A·, the products result exclusively from the open radical (A·).[73b]

More significant are the results described by Friedrich.[74] Tin hydride reduction (using light initiation) of 13 or of 14 gave a mixture of products resulting from the (A·) radical (15) as well from the (Cy·3) radical (16) (Scheme 19). It is of interest that the the hydrocarbon derived from 16 is among the products formed from the bromide (14). Even higher yields of hydrocarbon derived from the (Cy·3) radical (30%) were obtained starting from 2-(bromomethyl)-1,2-dihydronaphthalene, the cyclohexenyl homolog of 14.[75] It appears

SCHEME 19

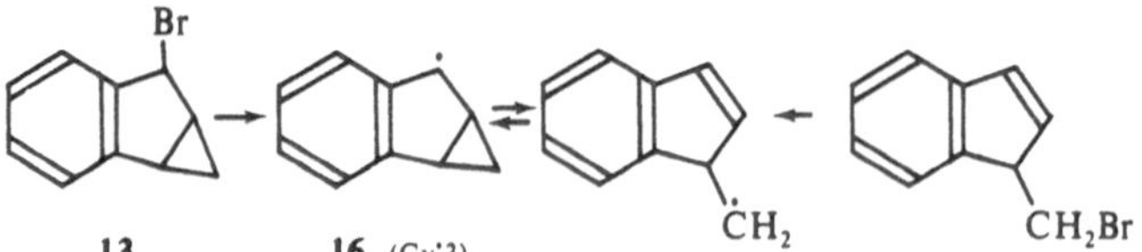

that benzylic stabilization in the (Cy·3) radical is necessary to observe products resulting from the $(A·) \rightarrow (Cy·3)$ process.[76]

A very important and much studied case of intramolecular addition of a 3-butenyl radical is that of the special case **18**; this gives high yields of cyclized (Cy·3) products (the so-called norbornenyl–nortricyclyl rearrangement).[66,155,466–472] This subject has been extensively covered by Wilt[15] and will be not discussed here except for new results of particular relevance.

A recent study by Giese[77a] affords interesting quantitative information about this reversible cyclization process (Scheme 20). By studying the inter-molecular free radical addition of $XCCl_3$ (X = Br or Cl) to norbornadiene (**17**) at different concentrations and temperatures, it was shown than the ratio **19**:[**21** + **22**] is dependent on these two factors. For instance, with X = Br this ratio is 64:36 at 0°C and 19:81 at 100°C, an increase in temperature favoring the cyclized products. Decreasing the $BrCCl_3$ concentration also favors the cyclized products resulting from the (Cy·3) radical, the ratio **19**:[**21** + **22**] being 64:36 at a concentration of 9.69 mol liter^{-1} and 21:79 at a concentration of 0.95 mol liter^{-1}. This scheme is clearly reminiscent of the one discussed earlier for the reversible cyclization of the hexenyl radical (Scheme 10) and the same precautions must, of course, be used concerning the modification of the ratio AX:Cy3X of acyclic:cyclic products depending on the addition reagent used, i.e., of the transfer rate constants k_T compared to the cyclisation k_{Cy} and opening k_{-Cy} rates constants. In the case just discussed, Giese[77a] was able to propose the following activation enthalpies (kcal mol^{-1}) in the range 0–100°C for the norbornenyl–nortricyclyl rearrangement: k_{Cy} (11.4), k_{-Cy} (9.1). When X = Br, the transfer activation enthalpies are smaller ($k_{TA} = k_{TCy} = 7.0$) than those of the rearrangement. With X = Cl, they are higher than those of the rearrangement, so that the ratio **19**:[**21** + **22**] is much smaller and less concentration sensitive than with X = Br.

It is interesting to compare this value 11.4 kcal mol^{-1} reported for the

SCHEME 20

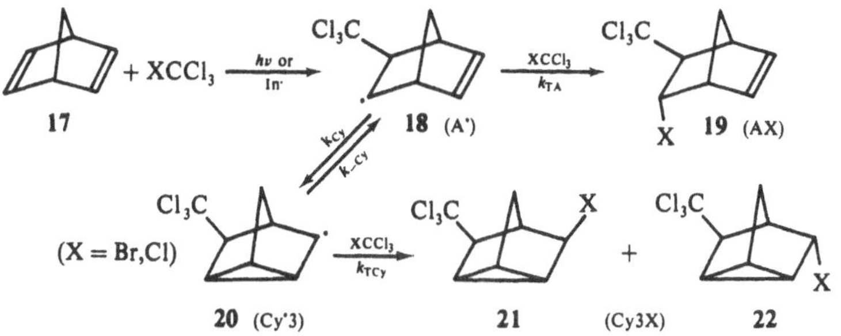

activation enthalpy of the norbornenyl radical cyclization with the similar value 7.8 kcal mol^{-1} reported by Ingold[69] for the cyclization of the 5-hexenyl radical. As a result, cyclization products may be anticipated from the norbornenyl radical even with good transfer agents, such as thiols and at relatively high concentrations, i.e., even under conditions which do not favor the establishment of equilibrium.

Further examples of the norbornenyl–nortricyclyl radical rearrangement used as a free radical probe will be discussed in Section XII.1.B.[473–480] Other methods of generating this equilibrating system than that of Scheme 20 have been employed starting with nortricyclyl,[481–483] 5-methylenenorbornene compounds,[78,431,484–486] and camphene.[78] In these cases, another complication may arise, namely, the opening of the nortricyclyl radical (Cy˙3) (Scheme 20) in the other direction. This is exemplified in the reports of Jackson[79–80] and Whitesides,[81] who chose this system as a radical probe of the borohydride reduction of organomercurials (Scheme 21). Reduction of 3-acetoxy-5-norbornen-2-yl mercuric chloride (23), 5-acetoxy-3-nortricyclyl mercuric chloride (24), or 7-*anti*-acetoxy-2-*exo*-bromo-5-norbornene (25) with tributyltin hydride gave isomeric mixtures of 2-*exo*-acetoxynorbornene (29), 3-acetoxy-nortricyclene (30), and 7-*anti*-acetoxynorbornene (31). These results were explained by the formation of the corresponding radicals (A˙) (26), (Cy˙3) (27), and (A″) (28) in equilibrium. As similar mixtures were obtained from 23 and 24 by sodium borohydride in aqueous tetrahydrofuran, it was concluded that a radical mechanism operates in the later reaction (Scheme 21).

SCHEME 21

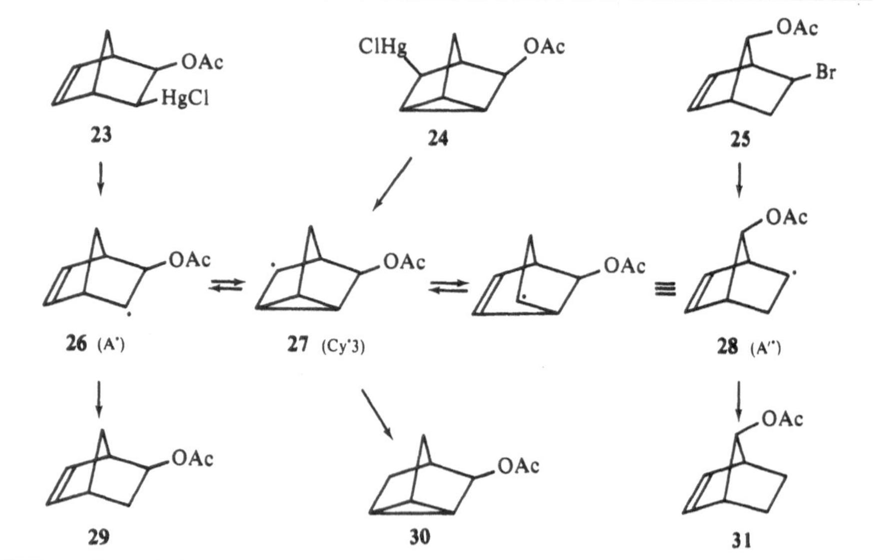

C. The Allyl Radical

Quite interestingly the behavior of the allyl radical and its cyclized counterpart the cyclopropyl radical is completely different from that of the homoallylic–methylcyclopropyl system. It does not seem that cyclized products resulting from the allyl radical have ever been observed.[82a,b,634] The opening of the cyclopropyl radical is, however, also a very slow process.[14,82c] Nevertheless, opening of substituted cyclopropyl radicals is observed when the substituents highly stabilize the allylic radical.[14,15,83,84,635] In the same way photolytic cleavage of cyclopropanes is facilitated when each of the radical centers formed is stabilized.[85]

3. Conclusion

The Cy5/Cy6 case is far and away the most facile cyclization process. In the Cy7/Cy8 case it is possible to obtain Cy7 compounds, but beyond that the cyclization is too slow to compete with other processes, except in very special cases. With the lower homologs, a sharp drop in the cyclization rate is observed in the Cy5/Cy4 case, no cyclized products being found from 4-pentenyl radicals. The Cy3/Cy4 case is a special one with high cyclization rates competing with ring opening at very high rates. As a consequence, cyclized products are observed only with particular structures such as the norbornenyl radicals. An inverse situation is encountered with allyl radicals where cyclization toward the cyclopropyl radical is too slow to be observed but where the opening of the cyclopropyl radical is also a very slow process. Similar trends are observed with other unsaturated radicals as discussed in the following sections.

VI. CARBON-CENTERED ALKENYL RADICALS CONTAINING HETEROATOMS

1. Introduction

As a first approximation it can be said that a heteroatom in the chain will not dramatically alter the course of cyclization. Nonetheless, it is true that in some instances the heteroatom is capable of facilitating the cyclization process. In addition, thanks to the presence of a heteroatom, side reactions which are not observed in purely hydrocarbon systems may occur. Thus hydrogen abstraction from an allylic position may be enhanced greatly. Another possible complication is fragmentation, e.g., as in Scheme 22.

SCHEME 22

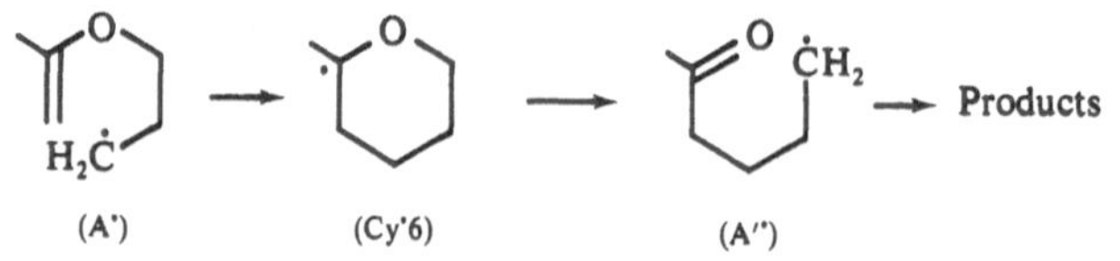

(A')   (Cy'6)   (A'')

2. Unsaturated Radicals Bearing a Heteroatom in the Chain

With these reservations in mind, the experimental results are qualitatively in good accord with what is observed with alkenyl radicals. Thus in the Cy5/Cy4 case, Walling[21] did not observe cyclized products from $CH_2=CHXCH_2CH_2^{\cdot}$ ($X = CH_2$ or O) and this is in accord with the results he obtained with the 4-pentenyl radicals (Section V.2.A). In the Cy6/Cy7 case the same author[21] observed no cyclized products from $CH_2=CHCH_2X(CH_2)_2CH_2^{\cdot}$ when $X = O$, but for $X=CH_2$, (Cy6) compounds were obtained, although generally in low yield (Section V.1). As discussed in the introduction, competitive 1,5 hydrogen abstraction, already competitive when $X = CH_2$, may become a major path when $X = O$.

Of course, the Cy5/Cy6 case has been the most extensively studied. For instance, Lamb[86] and Walling[21] have studied the behavior of the $CH_2=CHCH_2XCH_2CH_2^{\cdot}$ radical, generated by thermolysis of the corresponding peroxide[86] or by stannane reduction of the bromide.[21] In both cases, cyclized products resulted only from the (Cy'5) radical, yields of Cy5H being slightly higher (84%) when $X = O$ than when $X = CH_2$ (78%).[21]

Beckwith[87] reported that cyclization of the radicals $CH_2=CHCH_2XCH_2$-$\overset{\cdot}{C}HCH_3$ ($X = O$ or CH_2), generated by stannane reduction of the corresponding halides, gives predominantly the (Cy5) products and only a trace of the (Cy6) products. He also observed that the oxygen-bearing radical ($X = O$) cyclizes about ten times faster than does the carbon analog ($X = CH_2$). This rate enhancement has been used for mechanistic studies when 5-hexenyl cyclization is too slow a process to be detected (see Section XII.1.A).[544]

A similar rate enhancement on substituting $X = O$ for $X = CH_2$ has also been observed with o-3-butenylphenyl radicals[38,39,88,89,636] (32) which give the (Cy5) compound almost quantitatively with $X = O$, and in only 80% yield when $X = CH_2$,[89] the cyclization rate being increased about 120 times,[39] (Scheme 23, $n = 1$). With these aryl radicals, the cyclization process is also very selective toward the (Cy'6) radical in the Cy6/Cy7 case ($n = 2$) but is much slower than in the Cy5/Cy6 case. To complicate matters further, 1,5 hydrogen abstraction becomes particularly important when $X = O$. Finally, in the Cy5/Cy4 case ($n = 0$) almost no cyclized products are detected.[39]

SCHEME 23

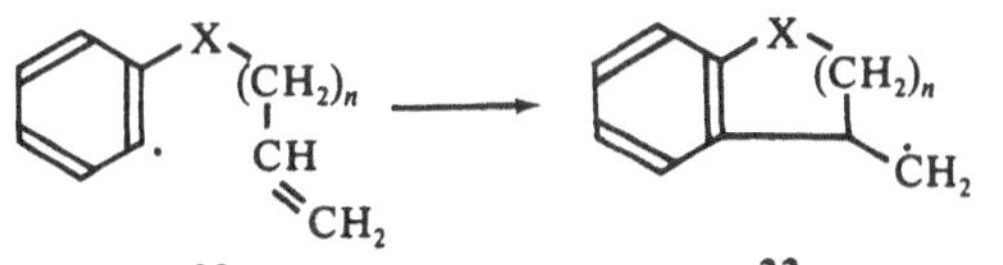

32   33

The behavior of $\overset{|}{\underset{/}{C}}=\overset{|}{C}-X-(CH_2)_n-\overset{\bullet}{\underset{\diagdown}{C}}\diagup$ radicals, the vinylic X being a heteroatom (or other stabilizing group), would be interesting to know, since one could expect, from the stabilizing effect of X, a change in the usual regioselectivity and an increase in the cyclization rate. Compared to the possible β-scission process (see discussion concerning Scheme 22) this could afford interesting data concerning the stabilizing effect of X group, a problem not yet satisfactorily quantitatively solved. Unfortunately, to our knowledge only fragmentary work[21,112] (X = O), not taking into account these possibilities, has been published until now, so that more quantitative data are needed before drawing any inference. Of course great care would be needed in the interpretation of the data since the enolic nature of the double bond may induce unexpected electrophilic reactions (see, for instance, Section VIII.3.A[176,177]). Some examples of preparative value (X = SO_2, NR, or $SiMe_2$) confirming the expected change of regioselectivity will be discussed in Section XI.2.A.a (Scheme 107).[389,637]

Thus far we have been mainly concerned with unsubstituted double bonds. It may be expected that alkyl substituent effects (discussed in Section IV) will also produce a change in the cyclic products ratio in these cases and this has been observed recently by Butler[90] in the Cy5/Cy6 case (Scheme 24). By working at different temperatures, he found what had been observed in the 5-hexenyl radical, namely, a decrease in the Cy5/Cy6 ratio with increasing temperature (43:1 at 40°C and 24:1 at 125°C for $R_3 = CH_3$), and though the dramatic effect on the Cy5/Cy6 ratio of an R_3 = Me substituent on the internal terminus of the double bond observed with the 5-hexenyl radical (Section IV) is

SCHEME 24

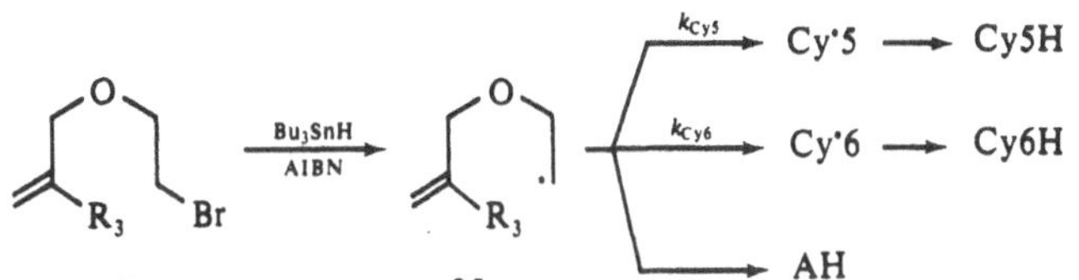

34   35 (A')

not observed in this case, it is when $R_3 = Ph$, the Cy5/Cy6 ratio now being 0.55; it has been demonstrated that the cyclization process is irreversible. Furthermore, absolute rates of cyclizations[90] are in accordance with the conclusions reached in Section IV: the decrease in the Cy5/Cy6 ratio results from a decrease of $kCy5$ rather than to an expected increase of $kCy6$. At 40°C for $R_3 = CH_3$, $kCy5 = 6.1 \times 10^4$ sec^{-1} and $kCy6 = 1.4 \times 10^3$ sec^{-1}; for $R_3 = Ph$, $kCy5 = 5.3 \times 10^4$ and $kCy6 = 9.6 \times 10^4$. These values must be compared with $kCy5 = 1.10^5$ sec^{-1} for the 5-hexenyl radical[55] and with the tenfold increase in rate expected from the presence of an oxygen atom[87] (and, of course, with the R_3 decrease; see Table 2).

The origin of the rate enhancement of the cyclization process caused by introduction of an allylic heteroatom such as oxygen in the chain is not clear. Furthermore, with other groups such as carboxylic esters, cyclization processes are greatly slowed down. Thus, even in a Cy5/Cy6 case, Walling[21] was unable to observe cyclization products from $CH_2{=}CHCH_2{-}O{-}\overset{\displaystyle O}{\underset{\displaystyle \|}{C}}{-}CH_2^\bullet$, and low yields

of what are presumed to be γ-lactones were obtained from $RCH{=}CH{-}\overset{\displaystyle O}{\underset{\displaystyle \|}{C}}{-}O{-}CH_2CH_2^\bullet$, generated by stannane reduction of 2-bromoethyl

crotonate and di(2-bromoethyl)maleate. Similarly, Hey and Cadogan[91] did not observe any cyclized products normally expected from $CH_2{=}CHCH_2{-}O{-}\overset{\displaystyle O}{\underset{\displaystyle \|}{C}}{-}\overset{\displaystyle \bullet}{C}HR$ ($R = Ph$ or $CO_2CH_2CH{=}CH_2$) when allylphenyl

acetate or diallylmalonate reacted with BP in a large excess of cyclohexane. In fact, the major products obtained resulted from cyclohexyl radical addition to the double bond, a process not usually favored. In the same way a nitrogen atom in the allylic position appears to be unfavorable to the cyclization.[502b]

At present there is no explanation for the rate enhancement or retardation of the cyclization process afforded by an allylic heteroatom in the chain. With regard to the possible interesting but more complicated effect of a vinylic heteroatom more studies are clearly needed.

3. Diallylic Substrates

We now discuss an important case of the Cy5/Cy6 cyclization. Here the process begins with the intermolecular free radical addition of a free radical ($Y^\bullet$) to a diallylic substrate with the formation of an unsaturated radical ($A^\bullet$) (Scheme 25).

Interest in this area was mainly owing to the possibility that the cyclized radical behaves as $Y^\bullet$, finally giving polymeric material. This was recognized as early as 1957 independently by Butler[92] and by Marvel.[93] Because of the wide

SCHEME 25

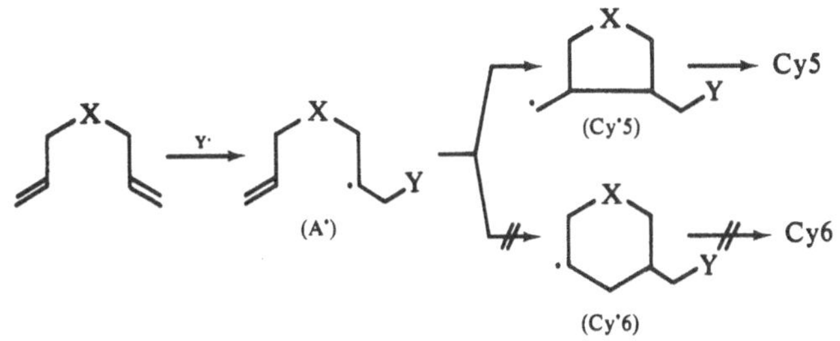

range of X groups (X = CR$_2$ or heteroatoms) possible, and of the interesting properties of the polymeric material obtained, a great deal of experimental work has been described in this area, and this process is known as *cyclo-polymerization*.[94] Because of the difficulties in identifying the polymeric materials it was assumed for some time that the cyclized unit originated from the (Cy'6) radical, according to the Kharasch orientation. In 1958, by using a large excess of the transfer reagent Friedlander[95] was able to inhibit polymerization and obtained for the first time monomeric cyclized products (Scheme 25, X = O or S). The assumed Cy6 structures proved to be in error.

The first relevant report of a selective cyclization toward the (Cy'5) radical from a diallylic substrate was made in 1964 by Brace[96] in the free radical chain reaction of 1,6-heptadiene (Scheme 25, X = CH$_2$) with 1-iodoperfluoropropane initiated by AIBN. Further work, mainly by this author, confirmed (Scheme 25) the high selectivity toward formation of the (Cy'5) radical,[490-496] in complete agreement with the results obtained with the 5-hexenyl radical (Sections II.1 and III.1). Thus all reports, particularly in the polymerization area,[94] that assumed formation of (Cy6) compounds must be regarded as suspect. For instance, Cadogan and Hey[97] reported the formation of a Cy5/Cy6 mixture by radical-initiated addition of various addenda to ethyl diallylacetate (X = CHCO$_2$Et) but had to recognize later[98] that the two compounds were in fact the two stereoisomeric (Cy5) compounds.

As discussed in the preceding section and also in Section IV, it may be expected that the introduction of alkyl substituents in the chain and more par-ticularly on the double bond would lead to a change in the behavior of the (A') radical corresponding to a decrease, or even an inversion, in the Cy'5/Cy'6 ratio. Esr results with β-monosubstituted and β,β-disubstituted *N,N*-diallyl-methylamines are in accordance with this view,[99a] the (A') radical being now of the same nature as the (A') radicals (R$_3$ = alkyl) discussed in Section IV, with competitive formation of (Cy'6) radicals as expected. Following the discussion in Section VI.3, one may also expect a change in the selectivity if a vinylic

SCHEME 26

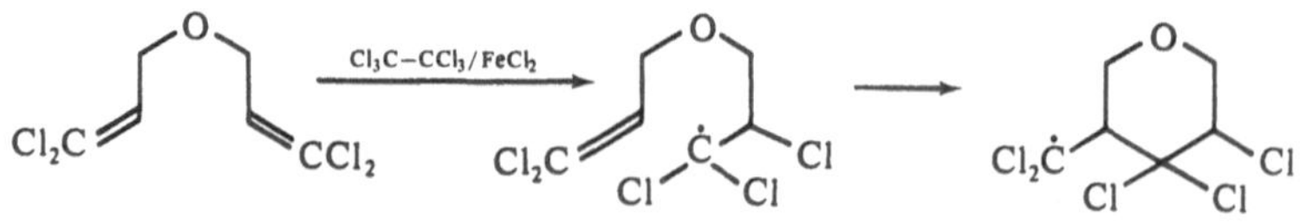

heteroatom is present in the chain. This could be the explanation of the (Cy6) products observed exclusively in the free radical addition of $(CO)_4M(PPh_2CH=CH_2)_2$ (M = Cr, Mo, or W).[99b] Nevertheless this trend is not always followed since free radical addition to $R_2C(OCH=CH_2)_2$ is reported to give only the (Cy5) compound.[492]

Another change in the cyclized products may of course be expected when stabilizing substituent on the terminal position of the double bond changes the regioselectivity of the first intermolecular addition as found by Piccardi and Modena[100] (Scheme 26). Although obtained in low yield (10.9%) because of competitive hydrogen abstraction followed by β-scission (62%) as discussed in the introduction, the only cyclized product was derived from the (Cy˙6) radical (38), as expected from the (A˙) radical (37).

Other examples of cyclizations of diallylic compounds are discussed in Section XI.3.

4. Unsaturated Radicals Bearing Fluorine Atoms on the Chain

Some very interesting results have been reported by Piccardi and Modena,[101] who employed the thermally induced radical addition (200–220°C) of pentafluoroiodoethane R_FI to 3,3,4,4-tetrafluorohexa-1,5-diene (39). They obtained a mixture containing the (Cy5) compound 42 (7% yield) and the two stereoisomeric (Cy4) compounds 44 (35% cis and 10% trans) (Scheme 27). This result takes on its full meaning when it is compared with the behavior of the 4-pentenyl radical, which does not give cyclic compounds, and with the generally very fast opening of the methylcyclobutyl radical (Cy˙4) analogous to 43 (Section V.2.A). These results appear to be exceptional, however. Thus free radical addition of carbon tetrachloride to the same substrate (39) gives poor yields of cyclized products,[102] and addition of a halogen atom gives a radical (A˙) resulting from anti-Kharasch addition, i.e., a Cy5/Cy6 case.[103] Polar effects of the fluorine atoms have been put forward in order to explain the formation of some (Cy6) products in the Cy5/Cy6 case[686] but, in fact, in the cases studied the radical is substituted and similar results are observed also with non-fluorinated olefins (see Section VII.4).

SCHEME 27

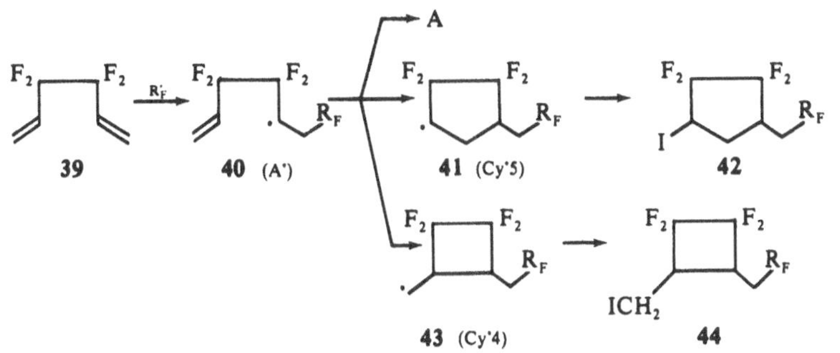

Nevertheless, these results appear attractive and it may be hoped that further studies will permit one to propose new synthetic methods, particularly in the area of small rings. One recent example may be found in the intricate mechanism ending with a Cy5/Cy6 cyclization recently proposed to rationalize some of the products obtained in the reaction of $CF_2=CCl_2$ with alkyl Grignard reagents.[638]

VII. ALKENYL RADICALS BEARING STABILIZING GROUPS ON THE CARBON RADICAL CENTER

1. Introduction

This section will be devoted to cyclization of alkenyl radicals $\overset{\diagdown}{C}=C(C)_n\dot{C}XY$ bearing groups other than alkyl ones. It must be recalled that a conclusion of Section IV was that when X and Y are alkyl groups, they have little or no influence on the yields and ratios of cyclized products. When X and Y are groups which normally stabilize carbon radicals, completely different results are obtained; for instance, in the Cy5/Cy6 case a decreased Cy5/Cy6 ratio is observed. Indeed, in some cases only the (Cy6) products are obtained. As discussed in Section III.2, the origin of this change must be sought in a cyclization process under thermodynamic control, and this has been clearly demonstrated in some cases.

These cyclization processes have been extensively studied because of the above mechanistic problems and, also for their possible use in synthesis. Indeed, under proper experimental conditions, the yields of cyclized products may be excellent. Furthermore, generation of the initial radical A˙ is generally easy.

This may be ascribed to the possibility of generating the radical selectively, e.g., by hydrogen atom abstraction, because of the stabilized nature of the generated radical.

It is of course difficult to imagine the selective generation in the same manner of an unstabilized radical. In fact, the way to obtain the unstabilized carbon-centered radicals (A·) discussed in the preceding sections is to employ a precursor (A–Z), where Z is a peroxide or perester group (thermolysis), halide (stannane reduction), or carboxylic group (anodic or metallic salt oxidation). In all these cases, the (A·) radical generated will be in a hydrogen donor medium (solvent or substrate) and a very important competitive process will be the formation of (AH) compounds, decreasing the yield of cyclized products.[†] Such a competitive pathway will not be observed if it is possible to use, as in the cases now discussed, AH itself as the A· precursor.

On the other hand, it can be assumed that because of the stabilized nature of A·, the cyclization process will be slowed when compared with unsubstituted radicals. Although kinetic measurements are scarce in these cases, this hypothesis seems confirmed by the necessity of generally using high-dilution conditions in order to minimize competitive polymerization processes. If this condition is fulfilled, high yields of cyclized compounds are often obtained. But, as a consequence of the high-dilution conditions, these reactions are seldom chain reactions, or chains will be very short. Often, the hydrogen donor to the (Cy·) radicals is the solvent (cyclohexane being a very good one) and if peroxide is used as initiator, it will have to be used in a quantity generally much higher than in classical free radical additions.[106]

We will now comment on the behavior of radicals $\diagdown\!\!C\!\!=\!\!C(\overset{|}{C})_n\overset{|}{\overset{\cdot}{C}}XY$, mainly by comparison to the case already discussed where $X = Y = H$, with particular emphasis on the ratio of cyclized radicals and on the nature of the stabilizing groups X and Y.

2. The 5-Hexenyl- and 5-Hexenylalkyl-Substituted Radicals

Under kinetic control conditions, the behavior of the $H_2C=CH(CH_2)_3\overset{\cdot}{C}XY$ radical seems to parallel the behavior of the 5-hexenyl radical ($X = Y = H$),

---

[†] For preparative purposes the method used most to generate unstabilized radicals (A·) is the reduction of the corresponding bromide by $Bu_3Sn·$, itself formed from Bu_3SnH. Since this hydride is a very good hydrogen atom donor the competitive reduction of A· to AH (Scheme 13) may be the major reaction for radicals (A·) which do not cyclize easily (substituents effects, Cy6/Cy7 case,...). In these cases the use of hexaorganoditins as initiators may be suggested in the presence of a well chosen (nature, concentration) hydrogen atom[469c] donor in the hope of facilitating the reduction Cy·5 → Cy5H rather than the reduction A· → AH.

(Cy˙5) being favored, but this is observed only at low temperatures. By increasing the temperature, the equilibrium Cy˙5/Cy˙6 favoring the secondary (Cy˙6) radical may be reached partially or totally. It will seldom be possible to obtain selectively the (Cy5) or (Cy6) compounds just by changing the experimental conditions. Nevertheless, besides changing the temperature, other alterations may be made in the experimental conditions in order to favor one or the other of the cyclized products. Thus the use of a good transfer reagent as solvent will prevent setting up the equilibrium and so will favor the (Cy5) compounds, as already discussed in Section III.2 (Scheme 11). A still clearer application of this approach is the use of what are probably the best transfer agents known, i.e., metallic salts which act as electron transfer, or ligand transfer, reagents in the last step. For instance, $CH_2=CH(CH_2)_3CH(CN)$-(CO_2Et) (45) in cyclohexane at 80°C (BP initiator) gives a product having a Cy5/Cy6 ratio of 16:84 in 58% yield.[27,104] At the same temperature, in dimethylformamide as solvent, and using cupric chloride to generate the radical by oxidation of the corresponding carbanion, only the (Cy5) compound is obtained as the chloride.[105] The nature of the X and Y groups may have some influence on the Cy5/Cy6 ratio as discussed later, but equally interesting is the influence of alkyl substituents on the chain. The influence of alkyl substituents R_3, R_4, R_5 in

$$\begin{array}{c} R_5 \\ \diagdown \\ R_4 \diagup \end{array} C=C \begin{array}{c} (CH_2)_3\dot{C}XY \\ \diagup \\ \diagdown R_3 \end{array}$$

has been discussed in Section IV when $X = Y = H$ or alkyl. One of the more pronounced effects was observed with $R_3 = $ alkyl, which gave an increased proportion of (Cy6) compounds. This was ascribed to a Kharasch effect operating even in a kinetically controlled process. One may expect to see this change of orientation amplified when the reaction is carried out under conditions of thermodynamic control. Indeed, when $R_4 = R_5 = H$, $R_3 = CH_3$, X, $Y = CN$, CO_2Et; CN, CN; or CO_2Et, CO_2Et, only the (Cy6) compound is observed (with no trace of (Cy5) compound).[27,104]

An opposite trend may be expected when $R_4 = R_5 = $ alkyl, the (Cy˙5) radical being now tertiary; with $R_4 = R_5 = $ Me and $X,Y = CN, CO_2Et$, only the (Cy˙5) product is observed, although in low yield.[24,106a]

Equally interesting, but apparently not as easy to rationalize, is the influence of a *trans* R_4 substituent. When $X = Y = H$ it was concluded in Section IV that such a substituent had little influence on the cyclization process, except to lead to a small increase of the Cy5/Cy6 ratio, this resulting mainly from a decrease in the rate of formation of Cy˙6. Under thermodynamic control conditions a greater enhancement of the amount of (Cy5) compound is expected when $R_4 = H$ is replaced by $R_4 = $ Me, as the result of the greater stabilization of the (Cy˙5) radical. In fact, very little $(X,Y = CO_2Et, COCH_3)$, or even an inverse

SCHEME 28

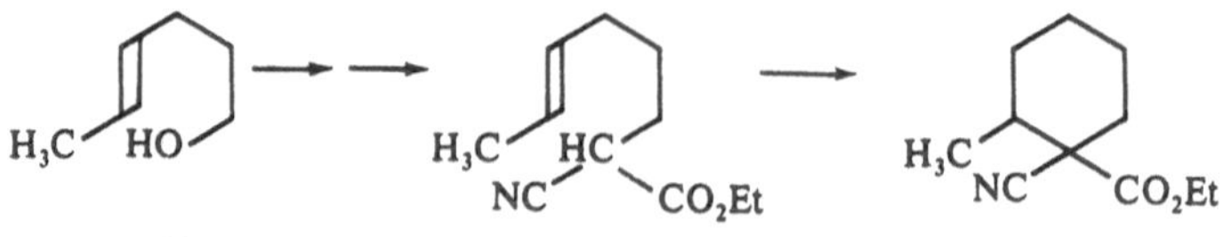

change is observed (X,Y = CO$_2$Et). More astonishing, with X,Y = CN or CN and CO$_2$Et, there is obtained under thermodynamic control a mixture of Cy5/Cy6 when R$_4$= H, but when R$_4$= CH$_3$ only the (Cy6) compound is obtained.[27,104] This last reaction is of preparative value. Thus the synthesis of ethyl 1-cyano-2-methylcyclohexanecarboxylate (47) in 75% yield from (E)-2-cyano-6-octenoate (46) has been described in *Organic Syntheses* (Scheme 28).[107]

3. Other Alkenyl Radicals

A conclusion that may be reached from Section V.2.A, is that 4-pentenyl radicals $\overset{\diagdown}{\diagup}C=\overset{|}{\overset{\bullet}{C}}(CH_2)_2\overset{\bullet}{C}XY$ are unable to cyclize when X = Y = H. With stabilized radicals (X=CN, Y=CO$_2$Et) cyclized products have been obtained, although in low yields (25%) but under more drastic conditions than with 5-hexenyl stabilized radicals.[108] Only (Cy5) and no (Cy4) compounds have been identified.

In the same way, the 6-heptenyl radicals $\overset{\diagdown}{\diagup}C=\overset{|}{C}(CH_2)_4\overset{\bullet}{C}XY$ are generally reluctant to cyclize when X = Y = H but cyclization is achieved, although in modest yields (20–40%), when X,Y = CN, CO$_2$Et [106a,109,110] or X, Y = Cl,[111] the (Cy6) products being formed exclusively.

Reports on the behavior of stabilized radicals bearing a heteroatom in the unsaturated chain are scarce. With the reservations made in Section VI.1 concerning the occurrence of competitive processes, the results parallel those obtained with purely aliphatic chains although the yields of cyclized products are generally poorer.[112,113]

4. The Stabilizing Groups

Most of the work on the behavior of stabilized radicals $\overset{\diagdown}{\diagup}C=\overset{|}{C}-(CH_2)_n\overset{\bullet}{C}XY$ has been reported using X,Y = CN,CO$_2$Et. In these instances a reversible cyclization process has been clearly demonstrated in some cases resulting mainly (Section III.2) in a decrease in the Cy5:Cy6 ratio in the Cy5/Cy6 case depending on the experimental conditions. But, as pointed out in the preceding section, other important features are also observed. The main ones are high cyclization yields in the Cy5/Cy6 case, possibility of formation of the (Cy5) products in the Cy5/Cy4 case, and of (Cy6) products in the Cy6/Cy7 case.

As these general features have been observed with other X and Y groups than CN and CO_2Et, it can be assumed that, also in these cases, a reversible cyclization process is operating, although this has not always been demonstrated. An early example was reported by Pines[114] in which $CH_2=CH(CH_2)_nCH_2Ph$ was treated by BP or DTBP, the selective formation of $CH_2=CH(CH_2)_n\dot{C}HPh$ (**48**) being assumed. In the Cy5/Cy6 case ($n = 3$) a smaller Cy5/Cy6 ratio (8.7:1) than with an alkyl substituent was observed. In the Cy5/Cy4 case ($n = 2$) some (Cy5) compound was observed, although in very low yield, and in the Cy6/Cy7 case ($n = 4$) cyclized products were also observed. Although in all the examples reported, yields were very low, these results seemed in accord with the possibility of a reversible process, particularly at the temperatures used (90–140°C). This was later confirmed independently by Walling[115] and by Julia[13] in the Cy5/Cy6 case ($n = 3$) by studying the behavior of the (Cy·5) and (Cy·6) radicals generated by stannyl reduction of the corresponding bromides,[115] and by the thermolysis of the corresponding *t*-butyl peresters[13] at 80°C and 140°C. It was observed that the 2-phenyl-cyclopentylmethyl radical (Cy·5) rearranged to some extent, depending on the reaction conditions, whereas the phenylcyclohexyl radical (Cy·6) did not. Furthermore, from the product ratio obtained, Walling[115] concluded that the Cy·5 → Cy·6 rearrangement was a direct one, without ring opening. Unfortunately, the results are obscured by the formation of large amounts of a tricyclic compound arising from efficient intramolecular aromatic alkylation of the *cis*-2-phenylcyclopentylmethyl radical. That in this case the direct Cy·5 → Cy·6 rearrangement is likely is true, but it seems unusual. Thus in the related case of the $CH_2=CH(CH_2)_3\dot{C}(Ph)(CN)$ radical, (**49**) (A·), Julia[13] obtained nearly the same mixture of products by thermolysis of the corresponding perester, as from the (Cy5) and (Cy6) peresters. As the ethylenic product is always produced this clearly proves that, at least in this case, the Cy·5 → Cy·6 rearrangement goes via the ethylenic radical (A·) (**49**) derived from ring opening. Another interesting observation reported in this work[13] is concerned with the use of photoexcited ketones to generate the (A·) radical from 2-phenyl-6-heptenonitrile. In decalin at 140°C, the Cy5/Cy6 ratio varies between 0.14 and 2.9 depending on the ketone used. There is no obvious explanation for this unexpected behavior but, from a preparative point of view, it provides a fourth variable in addition to temperature, nature and concentration of the transfer reagent for altering the Cy5/Cy6 ratio.

SCHEME 29

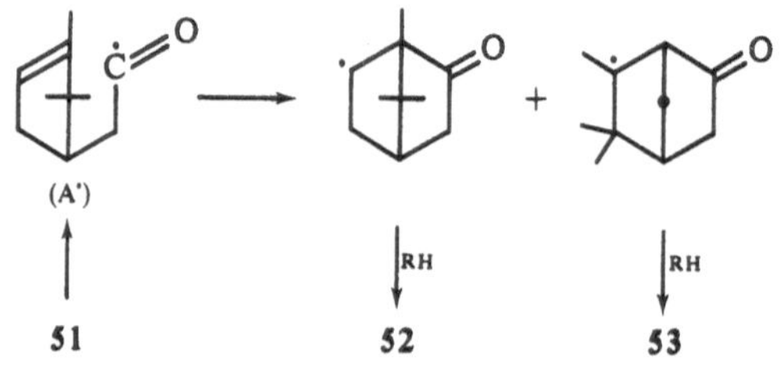

Another interesting example of cyclization which appears to be under thermodynamic control is encountered with acyl radicals $\overset{\backslash}{C}=\overset{|}{C}-(CH_2)_n-\dot{C}=O$ (50) (A'), easily generated by peroxide treatment of the corresponding aldehydes. Although promising because of the high selectivity observed, it does not appear to have been studied extensively after the early reports of Dulou, mainly in the area of terpenic compounds.[116–118] For instance, α-campholenic aldehyde (51) with acetyl peroxide in refluxing hexane gives a good yield of a mixture containing essentially camphor (52) (90–95% of total), and very little dihydrocamphenone (53) (5–10%) (Scheme 29).

In this case each of the products 52 and 53 may be considered to result from Cy'6, from Cy'5, or from both. On the basis of simpler examples it appears that a high selectivity toward the (Cy'6) radical in the Cy6/Cy5 case is the rule. For instance, 5-hexenal with BP in cyclohexane at 80°C gives cyclohexanone (41% yield) resulting from the (Cy'6) radical, with no trace of 2-methylcyclopentanone which would arise from the (Cy'5) radical.[104] A similar result has been observed by Čeković,[119] who generated the unsaturated acyl radical by tributylstannyl reduction of the corresponding acyl chloride in boiling benzene. In this way 5-hexenoyl chloride afforded cyclohexanone in 36% yield with no trace of Cy5 products.

Good yields of cyclized products are also obtained in the Cy6/Cy7 case. For instance, citronellal (54) gives a mixture (80% yield) of unreacted material (25%), menthone (55) (50%), and isomenthone (56) (25%) resulting from the (Cy'6) radical, with no products resulting from the (Cy'7) radical.[117a,639] Starting from citronellyl chloride (57), menthones (43% yield) are also obtained without (Cy7) products (Scheme 30).[119]

Although no cyclized products are obtained[117b] in the Cy5/Cy4 case the preceding results seem in accordance with a reversible cyclization process. This was demonstrated in one case by thermolysis of the cyclized t-butyl peresters corresponding to the (Cy5) and (Cy6) products expected from 5-hexenal. The (Cy6) perester was transformed exclusively into cyclohexanone but the (Cy5) perester gave a mixture of the unrearranged 2-methylcyclopentanone (30%) and,

SCHEME 30

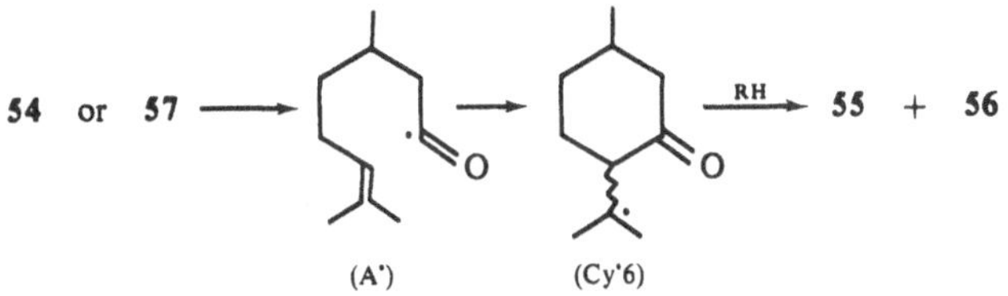

$$54 \text{ or } 57 \longrightarrow \text{(A·)} \longrightarrow \text{(Cy·6)} \xrightarrow{RH} 55 + 56$$

as the major product, cyclohexanone (70%), i.e., resulting from the Cy·5 → Cy·6 rearrangement.[30] As discussed in Section VII.1, if the reversibility results from the acyl radical (A·) being relatively stabilized, the cyclization process is expected to be slow. A recent esr study showed that at low temperatures, no ring closure of acyl radicals such as 50 ($n = 3$) or A· (Scheme 30) was apparent.[640] The above explanation accounts for the apparent contradiction between the experimental results discussed above and the esr observations.

We now turn to cases in which the X and Y substituents influence the Cy5/Cy6 ratio probably as the result of a reversible cyclization process, though this has not been demonstrated. This influence has been mainly observed with radicals

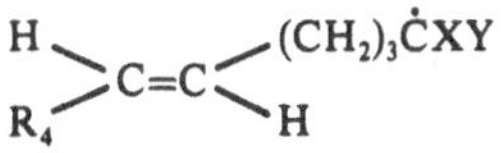

$$\underset{R_4}{\overset{H}{>}}C=C\underset{H}{\overset{(CH_2)_3\dot{C}XY}{<}}$$

58

($R_4 = H$ or Me; $X = H$; $Y = CO_2Et$, CN, CO_2H, or $COCH_3$).[26,104,120] In most of these instances, a much lower Cy5/Cy6 ratio is observed than when X, $Y = H$ or alkyl, but the change is not so dramatic as when both X and Y are stabilizing groups (Table 5). It is therefore quite tempting to conclude that in these cases also, where only one stabilizing group is present on the carbon-centered radical, a reversible cyclization process is operating but, being weaker than with two stabilizing groups, this effect is less important.

Without further evidence than just that of a lowered Cy5/Cy6 ratio, or of cyclization in other cases than Cy5/Cy6, it is probably dangerous to conclude that cyclization is under thermodynamic control. For instance, with radicals (58) ($X = H$, $Y = OH$) a low Cy5/Cy6 ratio (23:77) is observed at 140°C.[104] [Under the same conditions, with two stabilizing groups ($X = Y = CO_2Et$) the Cy5/Cy6 ratio is 60:40.[104]] The total yield of cyclized products is low (15% with $X = H$, $Y = OH$) so that any interpretation is ambiguous, and it seems difficult to assume without any clear evidence that the low Cy5/Cy6 ratio ($X = H$, $Y = OH$) results from thermodynamic control since just one hydroxyl group is certainly not as stabilizing as two carbethoxy groups. The (Cy6) compounds

SCHEME 31

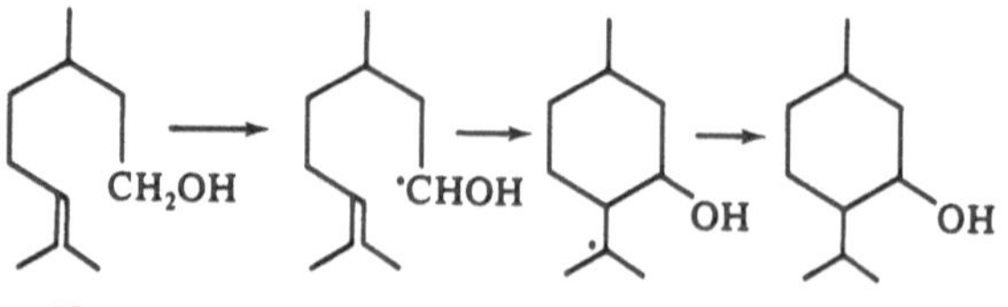

are, however, formed in the Cy6/Cy7 case, albeit in modest yield but better than those generally encountered with unstabilized radicals. This also seems in agreement with a reversible cyclization operating when X = H, Y = OH. Thus citronellol (**59**) at 110–150°C with DTBP as initiator, but without solvent, i.e., under conditions probably not optimal for cyclization, gives the isomeric menthols (**60**), (Cy6) (12%) as the only cyclized products (Scheme 31).[121]

At this time, and until more quantitative work is available on the behavior of unsaturated carbon-centered radicals bearing X and Y groups on the radical center, it seems plausible to assume that such groups can influence the ease and the selectivity of cyclization simply by favoring reversibility. But whatever may be the reasons for the observed changes, they can be very important. In Table 5 are summarized total yields of cyclized products (Cy5+Cy6), the percentage of Cy6 product with regard to the amount of cyclized products obtained from $CH_2=CH(CH_2)_3CHXY$, under the same conditions: A (one equivalent of BP, 81°C, solvent cyclohexane), or B (one equivalent of DTBP, 140°C in cyclohexane).[104]

TABLE 5. Cyclization of $CH_2=CH(CH_2)_3CHXY^a$

X	Y	Total yield, %	Cy6/Cy5+Cy6, %	Conditions[b]
O		41	100	A
CN	CO₂Et	75	86	B
CN	CN	70	80	A
H	OH	15	77	B
H	CO₂H	18	67	B
COCH₃	CO₂Et	33	50	B
H	CO₂Et	30	44	B
CO₂Et	CO₂Et	55	40	B
H	COCH₃	13	28	B
Cl	CO₂Et	15	11	A
H	CN	22	0	B

[a] Reference 104.
[b] See text.

5. Conclusion

The role of X and Y groups other than alkyl in the cyclization of $\overset{|\quad|}{\underset{/}{\overset{\backslash}{C}}=C(C)_n\overset{|}{\overset{.}{C}}XY}$ radicals may be important in making possible cyclization processes of preparative value. Manifestations of this influence are mainly the possibility of extending the reaction to cases other than the Cy5/Cy6 one, and to generally favor the (Cy6) compounds in the Cy5/Cy6 case. Furthermore, by adjusting the experimental conditions (temperature, transfer reagent, initiation mode), control over the reaction course may be exerted, i.e., toward the (Cy6) compounds. This derives generally from cyclization under thermodynamic control; but it is not certain that such a process is taking place in all the cases known.

VIII. HETEROATOM-CENTERED RADICALS

1. Introduction

Intramolecular additions of heteroatomic radicals have been studied much less than the corresponding processes involving carbon free radicals. In fact, when we began work in this area in 1964, no unambiguous example of intramolecular addition by a heteroatomic radical had been reported, although this possibility had been considered in some cases.

Since then many reports of such cyclizations have appeared, most of them directed toward preparative goals in heterocyclic chemistry. But even now much less is known about these cyclizations reactions than is known about the carbon homologs. Nevertheless, because of the synthetic potential of these cyclizations, and because of the occurrence of biogenetic processes involving this reaction, it is anticipated that this area will be more actively studied in the near future.

In general many of the features which govern the cyclization of carbon radicals are met with again in reactions involving heteroatomic radicals. But it is also true that the nature of Z may introduce subtle, or dramatic, changes in the behavior of $\overset{|}{\underset{/}{\overset{\backslash}{C}}=C(C)_n Z^{.}}$ radicals, in which $Z = O$, NR, S, or PR have been studied mainly.

It must also be pointed out that a supplementary difficulty is encountered concerning the interpretation of the results, namely, that it is a difficult task to establish that the observed cyclization really proceeds by a radical pathway. This derives from the nucleophilic character of the heteroatom Z and the possibility that cyclization proceeds via electrophilic attack of $X^{\oplus}$ on the double bond in the first step (Scheme 32).

SCHEME 32

This possibility has been clearly recognized in some cases but it is often difficult to establish, since identical Cy products resulting from both the radical and the ionic pathways are often obtained.

2. Oxygen-Centered Radicals

A. Alkoxyl Radicals

Unlike what is observed with carbon-centered radicals, the intermolecular addition of alkoxyl radicals to olefins is exceptional.[122] In fact, for reasons considered to be mysterious for a long time[123] and only recently rationalized somewhat,[124] the main reaction of alkoxyl radicals RO˙ (R = alkyl) toward olefins is allylic hydrogen abstraction. This led us to study the behavior of $\overset{\textstyle \diagdown}{\underset{\textstyle \diagup}{C}} = \overset{|}{C}(\overset{|}{C})_n O˙$ radicals in order to see if in an intramolecular, presumably favorable, situation addition could be observed.

At that time, it was already known that the lead tetraacetate oxidation of ethylenic alcohols leads to cyclized products, but it had not been ascertained that alkoxyl radicals were intermediates because of the known complexity of mechanisms in the reactions of lead tetraacetate with olefins and alcohols.[125] Indeed, as will be seen later, this oxidation of unsaturated alcohols usually follows the ionic pathway (Scheme 32).

An early example of intramolecular addition of an alkoxyl radical had been reported for a steroid bearing an allylic nitrite ester group; the possibility of intramolecular addition of an allyloxyl radical had been put forward to rationalize the formation of an epoxide.[126] But if this is the real pathway, it must be a very special one. Indeed, it is now well known that the behavior of oxyranyl alkyl radicals parallels the behavior of methylcyclopropyl radicals (Section V.2.B) and that the reverse β-scission pathway toward allyloxy radicals is very easy and very fast.[127,128,449–451] This is, in fact, faster than the other possible β-scission process involving homolysis of the carbon-carbon bond, even when stereoelectronic effects which would normally favor such C–C bond homolysis[642] are present. Carbon–carbon homolysis is only observed when the resulting carbon radical is stabilized.[449,641] Furthermore, it is conceivable that the cyclized product arose during the preparation of the nitrite ester.

These two examples exemplify the care which must be taken in assuming that cyclized products result from intramolecular addition of alkoxyl radicals. For instance, at the beginning of our studies we tried to prepare hypochlorites of ethylenic alcohols as precursors of pentenyloxyl radicals, but were unable to isolate the hypochlorites, cyclized products being obtained directly. We assumed that these products could arise from alkoxyl radicals, but further work fits the ionic pathway (Scheme 32; X = Cl) better.[129]

Fortunately, Bertrand was able to obtain nitrite esters of ethylenic alcohols as pure, characterized products. She assumed that photolysis of the nitrites would lead to the alkoxyl radical, as in the Barton reaction, with subsequent intramolecular addition. Isolation of tetrahydrofurfural oxime (62; R = H) starting from the nitrite of 4-pentene-1-ol (61; R = H) supported this hypothesis (Scheme 33).[129] This mechanism was later confirmed by esr spin-trapping techniques[130] and, independently by Rieke,[48,49] who scavenged the cyclized radical with bromotrichloromethane.

Photolyses were run at low concentration (about 0.025 M) in benzene under nitrogen, the yields of oxime (62) generally being between 50% and 60% when $HR_5= H$.[131,132] By adjusting the conditions, the highest yield of oxime obtained was 68%.[48]

Products resulting from the (Cy'6) radical have never been observed, even with $R_3= Me$, which favors the (Cy'6) radical in the 5-hexenyl case (Section IV). However, in this case the nitroxide resulting from the (Cy'6) radical was detected by an esr spin-trapping technique at room temperature. In all the other cases only the Cy5 nitroxide was detected. This seems particularly informative because of the known greater stability of secondary nitroxides derived from secondary radicals (here Cy'6,$R_3= H$) than primary nitroxides resulting from primary radicals (Cy'5, $R_5= R_4= H$). Thus the observation in all

SCHEME 33

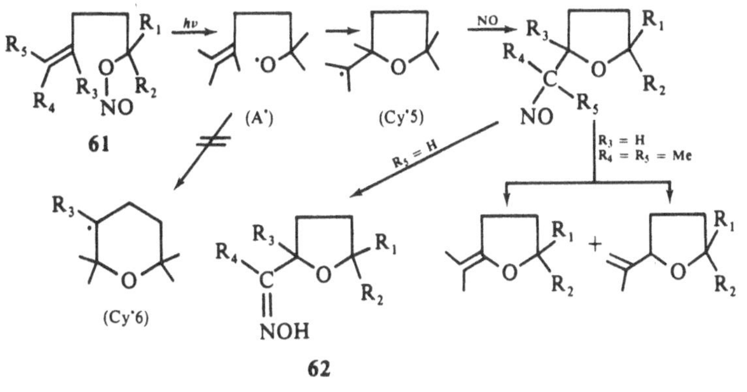

62

the cases except one, that the nitroxides result from the (Cy˙5) radical indicates a cyclization process highly selective toward the (Cy˙5) radical. Nevertheless, it has recently been reported that oxidation of 4-penten-1-ol by peroxydisulphate and a silver salt in the presence of a protonated heterocyclic base leads to alkylation of the heterocycle in high yields by the (Cy˙5) and the (Cy˙6) radicals in a 90:10 ratio, respectively. Formation of these products has been ascribed to cyclization of the intermediate 4-penten-1-oxyl radical, although oxidation of the double bond as the first step cannot be completely discarded.[133]

Photolysis of 3-buten-1-ol nitrite affords no cyclized products (Cy5/Cy4); neither does 5-hexen-1-ol nitrite (Cy6/Cy7).[131] The same result is obtained on peroxydisulphate oxidation of 5-hexen-1-ol.[133] In the Cy6/Cy7 case an important competitive pathway is probably 1,5-intramolecular allylic hydrogen abstraction and, indeed, esr spin trapping by nitrosodurene[134] provides evidence of this. Cyclization in the Cy6/Cy7 case was considered to explain the reaction products of tetrahalogeno-o-benzoquinones with 2,3-dimethylbut-2-ene but was discarded in favor of a direct cycloaddition process on the basis of spin trapping and deuteration experiments.[160] As discussed before, cyclization in the Cy3/Cy4 case must be difficult to observe because of the high β-scission rate of oxyranylalkyl radicals. Nevertheless, this pathway has been used recently to explain the formation of diepoxides in the thermal-, photochemical-, or ferrous-salt-induced decomposition of unsaturated cyclic peroxides. In view of the multistep scheme involved this conclusion must await further confirmation.[161-163] This preparatively valuable endoperoxide–diepoxide transformation has recently been employed in a number of interesting synthetic problems,[643,687] but mechanistic questions remain unanswered, although a concerted process for the bis addition of the two alkoxyl radicals rather than a stepwise intramolecular addition has been retained.[643b] Of course the possibility of a reversible cyclization-opening process, analogous to the one observed with the 3-butenyl radical (Section V.2.B), may be considered, particularly if the oxyranylalkyl radical is stabilized.[641]

All these observations are very similar to those made with unstabilized 5-hexenyl radicals and are then in agreement with an irreversible cyclization of 4-pentenyl-1-oxyl radicals. Evidence has been presented that tetrahydrofurfuryl radicals do not open and this observation has been used as a mechanistic tool to distinguish free radical from carbanionic pathways (Section XII.1.C).[135]

The relatively high yields of cyclic products, often even better than with carbon radicals, observed in the Cy5/Cy6 case is somewhat surprising when one recalls that the intermolecular addition of alkoxyl radicals to olefins is a very unfavorable process. One possible reason that this is so when the alkoxyl radicals are generated by nitrite ester photolysis is that scavenging of the cyclized (Cy˙5) radical by NO (Scheme 33) is a very fast process,[136] so that competitive reactions are minor ones. But it is also necessary to postulate that the cyclization process is very fast in the Cy5/Cy6 case. Actually, photolysis of

(61) ($R_2 = n$-Bu) (Scheme 33), where a competitive Barton reaction could be anticipated, gives only the corresponding oxime (62)[131] and no product resulting from the Barton reaction. Gilbert and Norman have recently been able to provide indirect quantitative confirmation that the cyclization of the 4-pentenyl-1-oxy radical in aqueous solution is very selective toward formation of the (Cy·5) radical, and is even faster ($>10^8$ liter mol^{-1} sec^{-1}) than the cyclization of the 5-hexenyl radical.[137]

With these results about the behavior of unsaturated alkoxyl radicals, it is now possible to compare them with the results obtained from lead tetraacetate oxidation of unsaturated alcohols. This reaction, which also leads to oxygen heterocycles, has been studied extensively since it was first reported in 1962.[138] Important differences appear. Among them are a nonselective cyclization process in the Cy6/Cy5 case and a much more general reaction, good yields of cyclized products being obtained with the higher homologs.[125,139] So it seems that cyclization by intramolecular addition of a free alkoxyl radical can be discarded as a mechanism in the lead tetraacetate cyclization of unsaturated alcohols. But the actual mechanism remains obscure because the results also are not in accord with an electrophilic cyclization process. This has led to the suggestion that a lead radical is associated with both the alkoxyl radical and the double bond.[139] Furthermore, it is now clear that apparently minor changes in experimental conditions may influence greatly the course of a lead tetraacetate oxidation. The results described above were obtained by thermal oxidation in benzene. UV photolysis of 4-pentenol with Pb(OAc)$_4$ or the thermal reaction in the presence of pyridine afford almost exclusively a (Cy5) product. This is a result in agreement with a homolytic pathway,[140] as are also the results obtained with 5-hexen-1-ol and allyloxyethanol[134] and those from esr spin-trapping experiments.[140]

B. Aryloxyl Radicals

Photolysis of 2-allylphenols (63) leads to a mixture of 2-methylcoumaran (64) and chroman (65)[141] in a Cy5/Cy6 ratio of 90:10,[142] a result nearly independent of the substituent on the ring (Scheme 34).[143] As formation of aryloxyl radicals by phenol photolysis was known, intramolecular aryloxyl

SCHEME 34

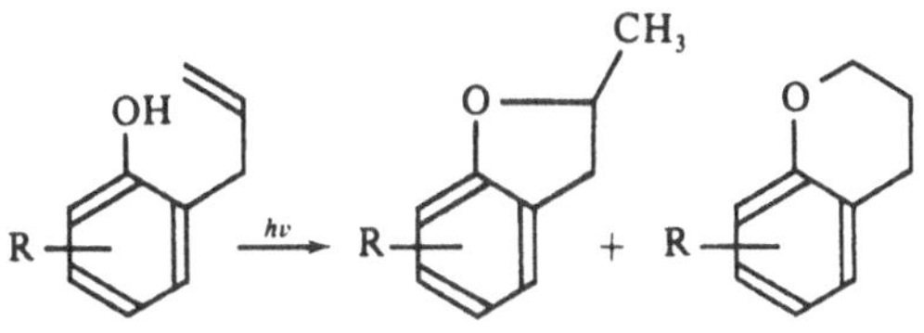

SCHEME 35

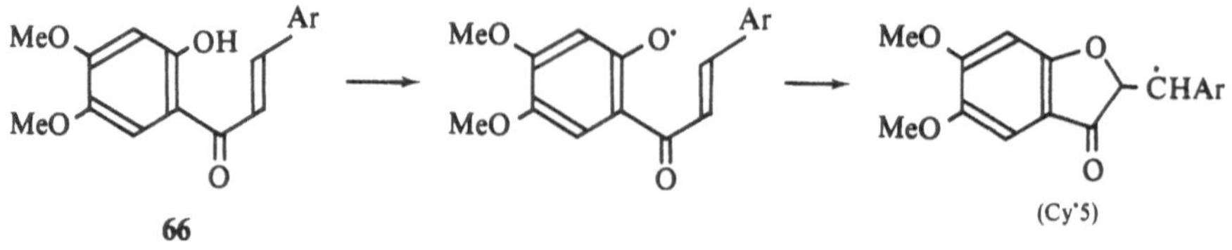

66   (Cy'5)

radical addition was invoked, but a heterolytic mechanism could not be dismissed.[141] However, the observation that o-(3-methylbut-2-enyl)phenol yielded only 2,2-dimethylchroman (the Cy6 compound) clearly spoke against a homolytic mechanism. Likewise, photolysis of cannabidiol gave the (Cy6) compounds either in the Cy5/Cy6 or the Cy6/Cy7 cases, clearly not in accordance with the results obtained with alkoxyl radicals.[144] Of course, the selectivity could be different for alkoxyl and aryloxyl radicals, but further work on the photolysis of unsaturated alcohols led to the tentative conclusion that the cyclizations observed in these cases do not involve the homolytic fission of the H–O bond.[145,146]

Furthermore, the $Mn^{(III)}$ or $V^{(V)}$ oxidation of 2-allyl-4-butylphenol (**63**; R = *p-t*-butyl), under conditions where the formation of an aryloxyl radical can reasonably be assumed does not lead to the cyclic products of Scheme 34.[147]

Thus it seems now established that the cyclized products obtained by photolysis of o-allylphenols do not result from intramolecular addition of aryloxyl radical. Furthermore, the results obtained on metallic salt oxidation lead to the conclusion that intramolecular addition of an aryloxyl radical, even in what is probably the best situation (Cy5/Cy6), is not a favored process. More work is needed before a definitive conclusion can be reached. Nonetheless, this possibility has been retained to account for the cyclized products (48.5% yield) obtained by treating the chalcone (**66**; Ar = p-HOC_6H_4) with aqueous alkali - $K_3Fe(CN)_6$ (Scheme 35).[148] This must be considered a very favorable case, however, since the (Cy'5) radical is benzylic.

In order to explain formation of a furan derivative when a perfluoromethylcyclopropenyl ketone is heated with bromine, a complicated scheme involving intramolecular addition of an enoloxy radical (Cy5/Cy4 case) has been proposed.[84c]

C. Acyloxyl Radicals

Since acyloxyl radicals add more readily intermolecularly to double bonds[13] than do alkoxyl radicals one could hope to obtain lactones from

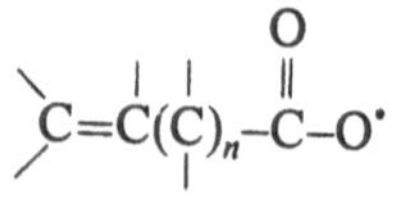

$$\text{C}=\text{C}(\text{C})_n-\overset{\text{O}}{\underset{\|}{\text{C}}}-\text{O}^\bullet$$

SCHEME 36

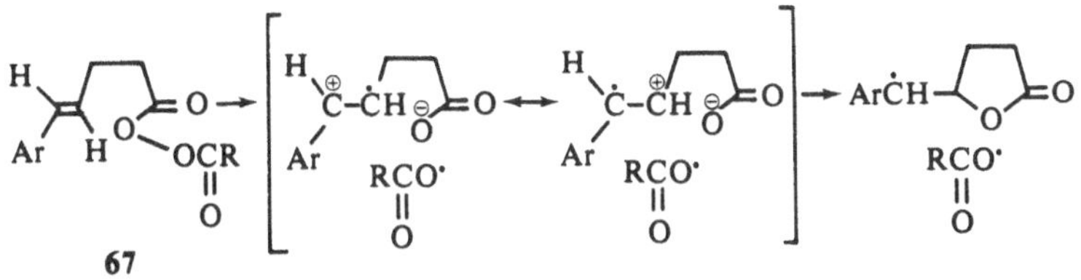

67

Of course, an important competitive reaction to be expected is the loss of carbon dioxide to give the corresponding carbon radicals $\overset{}{\underset{}{C}}=\overset{|}{\underset{|}{C}}-(\overset{|}{\underset{|}{C}})_{n-1}-\overset{\cdot}{\underset{}{C}}\overset{}{\underset{}{}}$.

The first relevant study was recorded by Lamb,[149] who thermolyzed *p*-substituted aryl peroxides **67**. Mainly on the basis of a kinetic study[150] in which the aromatic substituent and the solvent were varied, he concluded that the major pathway for the decomposition is homolytic, with neighboring group participation as pictured in Scheme 36. An analogous study, which reached similar conclusions, was described by Koenig and Martin[151] using compounds **68**. Unfortunately, products were not reported by Lamb, who assumed the presence

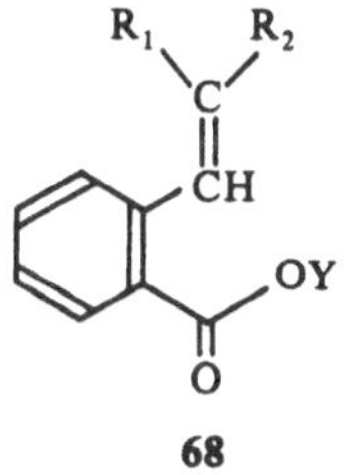

68

$(R_1, R_2 = \text{aryl or H}; Y = t - \text{BuO or ArO})$

of a lactone from an ir band at 5.7 μ. Furthermore, additional work by Lamb on the decomposition of 6-heptenoyl peroxide,[20] and by Hart and Chloupek[152] showed that unsaturated diacyl peroxides are not good substrates with which to study intramolecular addition of acyloxy radicals. Hart studied the decomposition of the diacyl peroxide (**69**) from *endo*-norbornene-5-carboxylic acid in carbon tetrachloride. A lactone (**70**) was obtained in nearly quantitative yield but could not result from path *a* (Scheme 37) because the trichloromethyl group was present instead of the expected chlorine atom. It was proposed that the lactone (**70**) arose by an induced decomposition in which a trichloromethyl radical attacked the peroxide (path *b*) and that loss of carbon dioxide from the acyloxy radical was faster than intramolecular addition (Scheme 37).

SCHEME 37

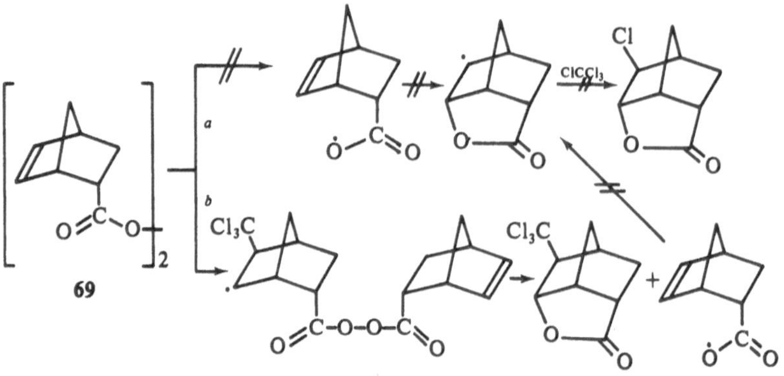

Other ways to generate unsaturated acyloxyl radicals have been used. For instance, Moriarty studied the lead tetraacetate oxidation of *endo*-norbornene-5-carboxylic acid and, after an extensive mechanistic study, he concluded that the γ-lactone observed was formed by electrophilic attack on the double bond.[153,154] The same explanation can probably be applied to other examples of lactone formation resulting from lead tetraacetate oxidation of ethylenic acids.[72b] The formation of a γ-lactone has also been reported by anodic oxidation of *endo*-norbornene-5-carboxylate, possibly by intramolecular addition of the acyloxyl radical[155] but oxidation of the double bond, as proposed for the anodic oxidation of *endo*-norbornenemethanol, could not be completely discarded.[156]

It appears then that the only unambiguous case of an intramolecular addition of an acyloxyl radical has been described by Sustmann, who used a strained system (**71**) in which β-scission of the acyloxyl radical to give a cyclopropyl radical was probably less favored than in other cases (Scheme 38).[157] The yield of Cy5 (**72**) is 2% and that of Cy6 products (**73**) 31% [22% (X = H), 5% (X = *t*-BuO) and 4% (X = PhCHMe)]. It must be noted that a very low Cy5/Cy6 ratio is observed, but that because of the special character of the system studied, it does not seem reasonable to ascribe this result to unexpected behavior by the acyloxyl radical.

In conclusion, except for this one case, there is, until now, no clear evidence for the intramolecular addition of acyloxyl radical. But this may be due to difficulty in generating unsaturated acyloxyl radicals rather than being a characteristic of these radicals.

SCHEME 38

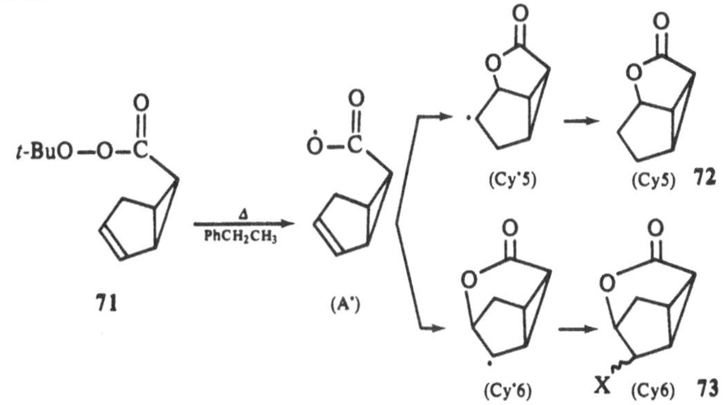

D. Peroxyl Radicals

As it is well known that peroxyl radicals add more easily to double bonds than do alkoxyl radicals, one could expect to observe easy cyclization of $\diagdown C=C(C)_n-OO^{\cdot}$. Probably because of the expected difficulties of generating these radicals without competing reactions, such additions have only been disclosed very recently.[158] Selective hydroperoxide hydrogen abstraction is readily observed on exposing unsaturated hydroperoxides (74) to the action of *t*-butoxyl radicals formed from di-*t*-butyl peroxyoxalate, or alternatively by irradiating a solution of the hydroperoxide and acetophenone in benzene. These reactions are conducted in oxygenated solutions. In the Cy5/Cy6 case, cyclized hydroperoxides (75) are obtained in good yield and in nearly pure form and are identified as the corresponding alcohols after triphenylphosphine reduction (Scheme 39). An easier way to generate peroxyl radicals analogous to A⁺ and which cyclize to 75 involves the co-oxidation of benzenethiol and 1,4-dienes[644] or even 1,3,6-trienes[688] in a remarkable regio- and stereoselective reaction. It must be noted that no products resulting from the (Cy⁺6) radical are observed, even when $R_3=CH_3(R_4=R_5=H)$. It appears, then, that the cyclization process is very easy, and this is confirmed by the results obtained in the Cy6/Cy7 case.

SCHEME 39

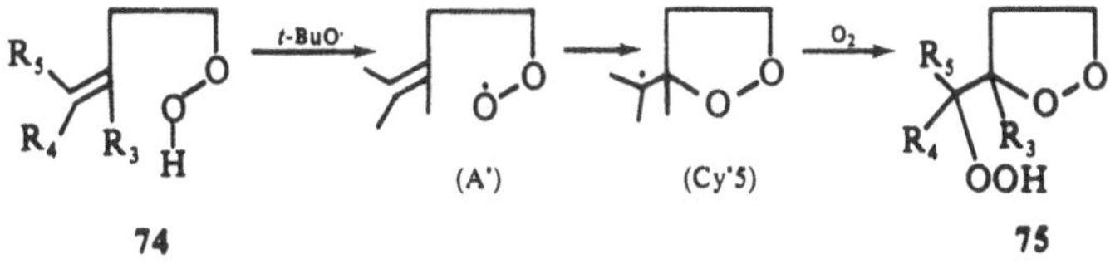

SCHEME 40

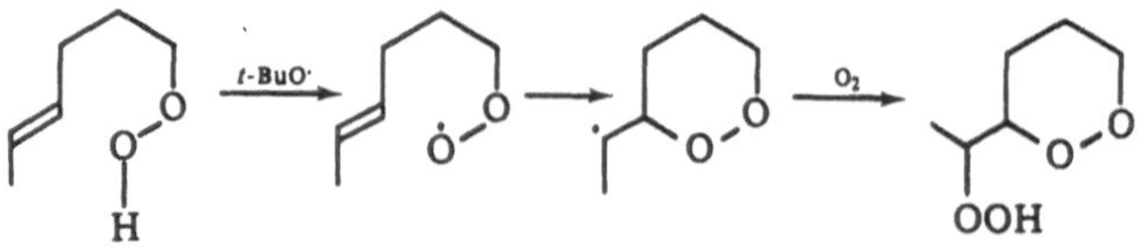

Only the Cy6 compound (77) is obtained from hydroperoxide (76), again in nearly pure form and in good yield (Scheme 40). This is a remarkable result when one considers that Cy6/Cy7 cyclizations generally give poor yields with carbon radicals and fail with alkoxyl radicals. This discovery led Porter to test the peroxyl radical cyclization mechanism leading to intermediate endoperoxides proposed for the biogenesis of prostaglandins. This process, which involves two consecutive free radical intramolecular additions, will be discussed in Section XI.4.B.a (Scheme 152). It is interesting to compare these results with those obtained on treatment of the same substrates with mercuric salts. With 74 the same cyclization selectivity is observed, but with 76 a mixture of Cy6 and Cy7 products is now obtained.[159] Undoubtedly, an understanding of the reactivity of unsaturated peroxyl radicals will lead to a better comprehension of the autoxidation of unsaturated lipids. An example is the sequential oxidation of methyl linolenate which gives rise to a single hydroperoxy-epidioxide. This reaction is assumed to proceed via the selective cyclization of a peroxyl radical.[689]

3. Nitrogen-Centered Radicals

A. Aminyl Radicals[†]

The well-known ability of aminyl radicals to add intermolecularly to double bonds suggests that one ought to observe easy intramolecular addition, and this has been confirmed; indeed, these reactions are of preparative value. Nevertheless, it was not until 1966 that Tordo[164] published the first report of such a reaction. Besides the difficulties already discussed in Section VIII.1 concerning possible competitive ionic pathways, this paper illustrated another difficulty in interpreting the results because of rearrangement of the cyclized product. Under the acidic conditions described by Neale for the intermolecular addition of aminium radicals, the *N*-chloroamine (78, R = *n*-Pr), gave 3-chloropiperidine (79) and this was assumed to arise according to path *a*

[†] See also Vol. 1, Chap. 3.

SCHEME 41

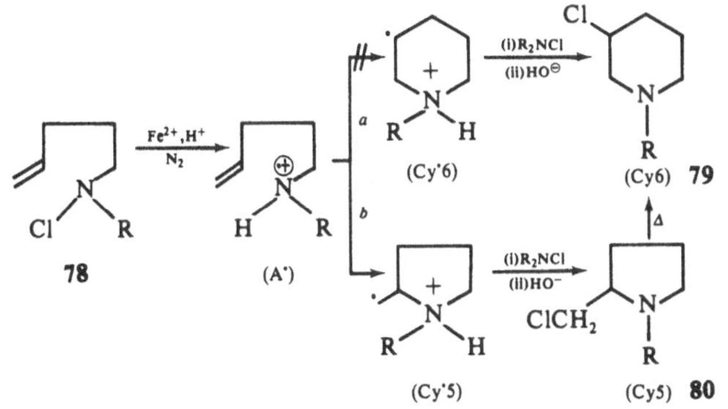

(Scheme 41). Such selectivity toward the (Cy·6) radical was not completely unexpected at that time, but it rapidly became clear that the initial product was the 2-chloromethylpyrrolidine (80) arising from the (Cy·5) radical (path *b*), which then underwent the well-known rearrangement to give 79 on workup.[165,166] An interesting facet of this reaction was that it enabled one, just as with alkoxyl radicals, to obtain functionalized cyclic compounds. This is seldom observed with carbon radicals and, consequently, Stella investigated the best experimental conditions necessary to obtain cyclic products. Using redox initiation, as proposed by Minisci for the intermolecular addition but somewhat modified by using titanous trichloride in acetic acid–water (50/50 v/v) at room temperature, pure 1-chloromethylpyrrolidine (80) (R = *n*-Pr) was isolated in 81% yield[166] (Scheme 42). The results differ from those obtained with neutral aminyl radicals, as discussed below, and therefore it was concluded that the aminyl radical (A·) is complexed with the metallic salt. The last transfer step proceeds according to *a*, i.e., as a ligand transfer when good oxidizing agents such as cupric salts are used.[166] On the other hand, when poor oxidizing salts

SCHEME 42

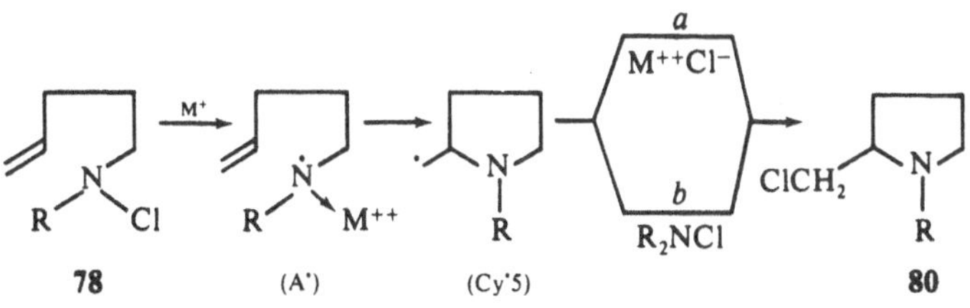

are used, e.g., Ti(IV), the reaction probably involves an atom transfer step *b*.[167] This conclusion may have important consequences as will be discussed in Section XI.4.B.b, Scheme 153.

Because of the availability of the starting materials, the high selectivity and the high yields observed, the mild conditions employed, and the fact that high-dilution conditions are not needed, the scope of the titanous chloride reduction of unsaturated *N*-chloroamines has been rather extensively studied. Some other examples of cyclization are recorded in this section as well as in Section XI.

A third way of obtaining cyclized products from unsaturated *N*-chloroamines is also available, namely, photolysis in neutral media.[168] Although the yields are generally lower (about 70%), this is an attractive cyclization process because the conditions are very mild. Here, intramolecular addition of a neutral aminyl radical is assumed, a noteworthy result, considering that inter-molecular addition is a very unfavorable process.

Finally, a fourth method of producing an unsaturated aminyl radical (in the protonated aminium form) has been described by Chow, who photolyzed unsaturated *N*-nitrosamines in acidic medium. In this case, the last step is scavenging by NO and the products are oximes[169] (see also Vol. 1, Chap. 3 in this series for more detailed discussion of this reaction).

With these four methods on hand, the main cyclization trends will be sum-marized:

As already discussed in the Cy5/Cy6 case, products resulting from the (Cy˙5) radical are generally formed exclusively. In the Cy5/Cy4 case no cyclized products have been observed from aminium radicals[170] but a low yield (13%) of 3-chloropyrrolidine, i.e., a (Cy5) product, is obtained with metal-complexed aminyl radicals.[171a]

In the Cy6/Cy7 case, intramolecular addition of aminyl radicals is a facile process. In the aminium form[170] or metal-complexed form[171b] only products resulting from the (Cy˙6) radical have been observed. In the latter case the yields are fairly high (up to 72%) when compared with the other Cy6/Cy7 cases discussed in previous sections. With neutral aminyl radicals, a low yield (14%) of cyclized products is observed as a mixture of Cy6/Cy7 products in a ratio 75:25 at −5°C and 50:50 at 45°C; the main products are dimers.[171c]

The high selectivity toward the (Cy˙5) radical in the Cy5/Cy6 case is clearly reminiscent of the behavior of unstabilized carbon-centered radicals and of oxygen radicals adding in an irreversible manner. On this basis it could be assumed that cyclization of aminyl radicals is also a process under kinetic con-trol, but is easier than in the aforementioned cases: cyclization is easy in the Cy6/Cy7 case and possible in the Cy5/Cy4 one, at least with complexed aminyl radicals.

There is, however, no real demonstration of these conclusions, which have, furthermore, been challenged recently by Michejda, at least in the case of the neutral aminyl radical.[172a] By photolysis or thermolysis of unsaturated tetrazenes in cyclohexane as a mode of generation of neutral aminyl radicals, the products were very similar to those described previously in the Cy5/Cy4 case (no cyclization) and in the Cy6/Cy7 case (mixture of Cy6 and Cy7 products in low yield). But in the Cy5/Cy6 case, this time a mixture of Cy5/Cy6 products was obtained in 19% and 34% yields, respectively, at room temperature and 41% and 16% yields at 143°C. This was the first report of the formation of a (Cy6) product in the Cy5/Cy6 aminyl case. No obvious explanation of the temperature dependence of the Cy5/Cy6 ratio is apparent since, with carbon-centered radicals, in all the cases studied the Cy6/Cy5 ratio increases with the temperature in both irreversible[23,52] and reversible[29,30] cyclizations. Furthermore during an attempt to obtain cyclizations rates of neutral aminyl radicals by esr, Ingold was unable to obtain any signals resulting from the cyclized radicals even in the Cy5/Cy6 case, at temperatures from 180 to 400°K, which would correspond to a rate constant $\leqslant 5$ sec^{-1} at 25°C.[172b] It is thus clear that further work is needed before reaching any definitive conclusion about the behavior of neutral aminyl radical.

The course of the cyclization process depends on the nature of the aminyl radical. Cyclizations with the neutral aminyl radical are smooth but do not seem general. Intramolecular addition of aminium radicals appears to be a facile process. Thus, when a competitive 1,5-hydrogen abstraction (Hofmann–Löeffler–Freytag) is possible, the only products are those resulting from intramolecular addition.[173] Nevertheless, on generation of aminium radicals from protonated N-chloroamines, an important competitive ionic reaction may occur, namely, electrophilic attack of the double bond by positive chlorine, as discussed in Section VIII.1. This pathway had been recognized by Neale in intermolecular reactions of N-chloroamines with electron-rich olefins and, later, in intramolecular reactions. This competitive reaction, although equally interesting in preparative cyclizations,[174] may be troublesome in affording other cyclized products[175] or in the interpretation of the intramolecular reactions of N-chloroamines with enol ethers which have been extensively studied by Waegell.[176,177]

It appears, therefore, that the more general way to observe intramolecular addition of aminyl radicals from unsaturated N-chloroamines is by generating the complexed aminyl radicals using reducing salts and, more specifically, by using titanous chloride.[171] But even here, competitive positive chlorine reactions may occur with substrates particularly reactive toward electrophilic reagents. For instance, titanous chloride reduction of the N-chloramine **81** (X = CH) affords the chloromethylpiperidine **82** as a mixture of *cis* and *trans* isomers in 92% yield.[178] Under the same conditions **81** (X = N) affords only products **83** resulting from o/p chlorination (30%/50%) of the aromatic nucleus and no cyclized products have been observed[179] (Scheme 43). Another complication may arise when N-chloroamines are exposed to solvolytic conditions, especially boiling methanol with, or without, silver salt catalysis. Under such conditions it

SCHEME 43

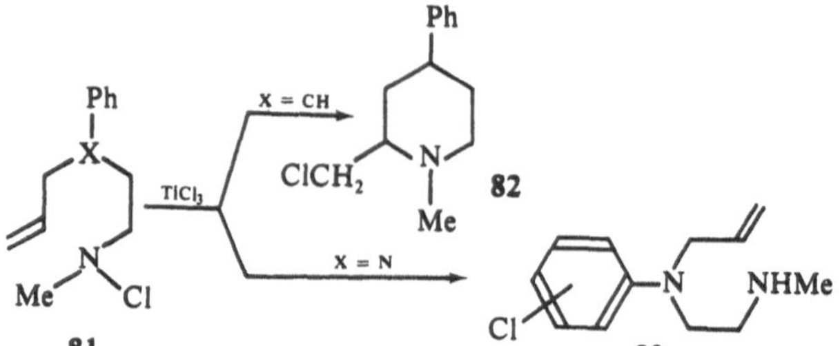

was assumed by Gassman that saturated N-chloroamines afford nitrenium ions R_2N^+.[180] With an unsaturated N-chloroamine such as **84** (X = Cl), the only cyclized product, obtained in 65% yield on boiling in methanolic solution with or without a silver salt was the (Cy6) product **85**[181] (Scheme 44). Gassman assumed that **85** was formed by the nitrenium pathway, which seems in accordance with the selectivity observed toward a (Cy6) compound rather than a selectivity toward a (Cy5) compound expected from an aminyl pathway. This conclusion was confirmed by an observation of Chow:[182] photolysis of **84** (X = NO) in acidic methanol, i.e., under aminium radical ion formation conditions, leads only to the (Cy5) products **86** and **87**.

At the same time, however, Hobson[183] obtained only (Cy5) and no (Cy6) products from N-chloro-N-methylhept-4-enamine with silver salts. His conclusion, developed by further work,[184] was that these products may arise from intramolecular free radical addition of aminyl radicals rather than from nitrenium intermediates.

However, by thermolysis or by silver-ion-catalyzed reaction of 4-(N-chloro-N-methylaminomethyl)cyclohexene Gassman and Dygos obtained a mixture of (Cy5) and (Cy6) products which they regarded as being in accord with

SCHEME 44

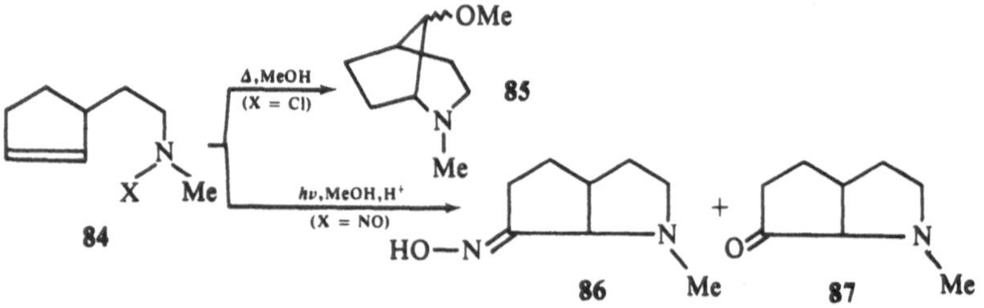

the nitrenium pathway.[185] This conclusion seemed in accordance with the results obtained independently by Chow[182] and by Nouguier[186] using the same substrate under the four conditions discussed earlier where formation of an aminyl radical (neutral, complexed, or protonated) may be assumed and where only (Cy5) products were obtained (this example is discussed fully in Section XI.2.A.b, Scheme 124).

Although these results seem to afford clear evidence for a pathway other than homolytic in cyclizations of unsaturated *N*-chloroamines by thermolysis in a polar solvent, or by silver-ion-catalyzed reaction, the nitrenium path was again challenged by Edwards in the general case of saturated *N*-chloroamines,[187] or in the unsaturated *N*-chloroamines case,[188] mainly because cyclized products are obtained only under nitrogen and with an induction period, which seems to agree with a homolytic pathway. An analogous view has been recently put forward by Furstoss, Tadayoni, and Waegell,[189] mainly on the basis that the same cyclized products are obtained under homolytic and solvolytic conditions. Unfortunately, they used a strained system in which such a behavior is not completely unexpected, as Moriarty[190] demonstrated some years ago in the cyclization of unsaturated alcohols, as also did Stella using unsaturated *N*-chloroamines.[186]

As extensive discussion will be found in the papers cited above, it was not our aim here to reach a decision about the mechanism of cyclization of unsaturated *N*-chloroamines under solvolytic conditions, but just to point out, once more, the difficulties often encountered in the interpretation of cyclization processes, and the fascinating behavior of *N*-chloroamines. As regards saturated *N*-chloroamines, the nitrenium pathway seems now clearly demonstrated by Gassman under solvolytic conditions.[191] With unsaturated *N*-chloroamines the situation is more complicated and, indeed, several pathways are proposed to account for the cyclized products. Nevertheless, some general features arise. Under conditions where an aminyl radical is generated unambiguously, the high regiospecificity toward the internal cyclization, encountered with other irreversible free radical additions, is once more observed, i.e., in the Cy5/Cy6 case the exclusive formation of the (Cy5) compounds. This high selectivity is sometimes masked by a competitive process, favored when the double bond is substituted, and apparently with strained structures, namely, cyclization initiated by positive chlorine. In fact, this cyclization process is sometimes the major one in protic media and can become the exclusive one by avoiding initiation. Although rewarding from a synthetic point of view[174a] because it is often very selective, this process may, for this same reason, introduce difficulties in the interpretation of the results when the discussion is based on the behavior of a particular substrate. A third process may be observed when submitting unsaturated *N*-chloroamines in protic media to thermolysis or silver-ion-catalyzed solvolysis. Generally, this process is not as selective and seems to involve cyclization of a nitrenium intermediate. But here also it seems that competitive processes may be involved, which are perhaps the major ones with some substrates, namely, cyclization by electrophilic addition of positive chlorine or free radical intramolecular addition of the aminyl radical.

The only secure conclusion at this time is that great care must be taken when interpreting the behavior of a particular unsaturated *N*-chloroamine, because of the possible identity of products obtained whatever the mechanism involved. This conclusion is somewhat reminiscent of the one about the behavior of unsaturated alcohols with lead tetraacetate and it leads one to ask about the behavior of unsaturated amines with lead tetraacetate. Actually, while little is known at this time about the cyclization of primary amines by lead tetraacetate it appears to be a very promising preparative cyclization method.[192-194] Finally, it must be recalled that cyclization of unsaturated amines involving, in the first step, electrophilic attack on the double bond is also an attractive cyclization process. In this connection, the aminomercuration reaction by mercuric salts has been extensively studied by Lattes, the halocyclization reaction by halogens by Staninets and Shilov, and the recent cyclofunctionalization of olefinic urethanes using benzenesulfenyl chloride by Clive and coworkers. A summary of most of these results may be found in Reference 195, but although comparison of the results (particularly of the great selectivity often observed) with the results of homolytic cyclization, would possibly be of great interest, it is beyond the scope of this review.

B. Arylaminyl Radicals

The photochemical behavior of *o*-allylanilines such as **88** has been studied.[196a] Such compounds cyclize to give the (Cy5) compounds, indolines (**89**) (Scheme 45). This result is clearly reminiscent of the photochemical behavior of *o*-allylphenols (Section VIII.3), and although the selectivity is the one expected from aminyl radical intramolecular addition, it also seems best explained by photochemical excitation of the π double bond.[196b] This is an interesting conclusion since metal-salt-induced cyclization of compounds such as **88** generally yields a mixture of (Cy5) and (Cy6) products.[197]

SCHEME 45

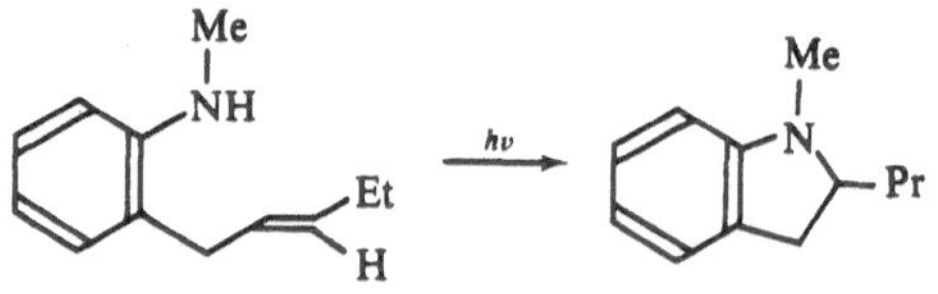

C. Amidyl Radicals

Although intramolecular hydrogen abstraction by amidyl radicals is a well-known process, the first reports of intramolecular addition of amidyl radical to double bonds were only made in 1972 independently by Chow[198] and by Flesia.[199] Unsaturated amidyl radicals were easily generated by photolysis of the corresponding *N*-nitroso or *N*-chloro-carboxamide such as **90** and the (Cy5) nature of the resulting products (**91**), obtained in yield as high as 90%, as well esr spin-trapping experiments agree with the formation of the amidyl radical (A·) followed by intramolecular addition (Scheme 46). More extensive work was described latter by Kuehne[200] and by Lessard.[202–203] Amidyl radicals were generated from olefinic *N*-chloro-*N*-methylamides with formation of

the carboxamido radical $\quad \overset{\diagdown}{\diagup}C{=}\overset{|}{\underset{|}{C}}{-}\,(\overset{|}{\underset{|}{C}})_{n-1}{-}\underset{\underset{O}{\|}}{C}{-}\overset{\cdot}{N}Me$

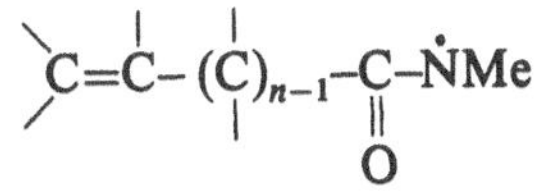

or *N*-chloroacetamides with formation of

the acetamido radical $\quad \overset{\diagdown}{\diagup}C{=}\overset{|}{\underset{|}{C}}(\overset{|}{\underset{|}{C}})_{n}{-}\overset{\cdot}{N}{-}\underset{\underset{O}{\|}}{C}{-}Me$

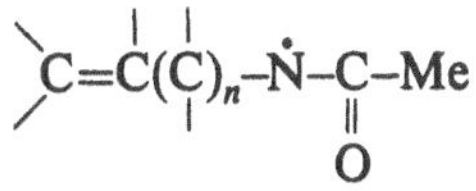

by photolysis,[200] by BP initiation,[202] or by chromous chloride reduction.[203] Most of the results obtained parallel those obtained in the other cases of irreversible intramolecular addition and, more particularly, of the aminyl radicals, with the first step being intramolecular addition of the amidyl radical followed by chlorine transfer.

Best results are generally obtained with chromous chloride initiation (yields as high as 90%) but are generally very similar whatever the initiation process used. In the Cy5/Cy6 case only the Cy5 products are observed except in a very special case in which an adamantane skeleton (Cy6) was formed rather than the more strained protoadamantane skeleton (Cy5).[203] In the Cy5/Cy4 case no cyclized products have been observed in the one case studied.[200] In the Cy6/Cy7 case only Cy6 products are observed, generally in good yield, whatever the initiation, when carboxamido radicals are generated from *N*-chloro-*N*-methyl amides. However, with acetamido radicals generated

SCHEME 46

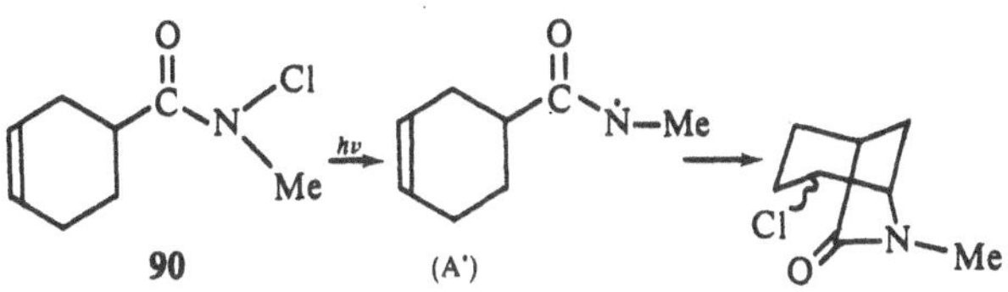

90   (A·)   **91**

SCHEME 47

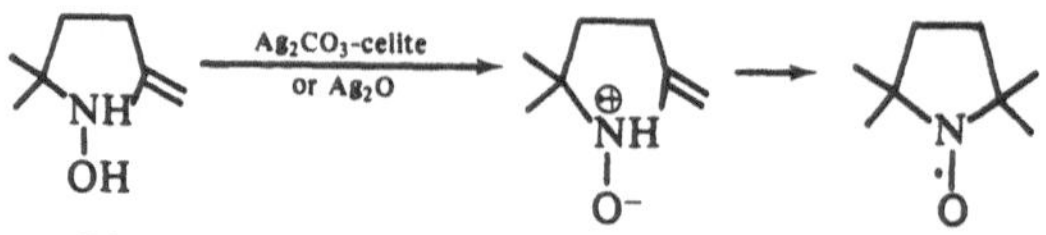

from *N*-chloroacetamides the cyclization process in the Cy6/Cy7 case appears to be very difficult.[203] This difference in behavior is difficult to rationalize, but it must be recalled that in the irreversible Cy6/Cy7 case, good yields of cyclized products are seldom observed except with complexed aminyl, carboxamido, and peroxyl radicals. In earlier work, it was observed that the acetamido radicals $CH_2=CH(CH_2)_n O\dot{N}COMe$, supposedly formed by the lead tetraacetate oxidation of the corresponding *N*-carbalkoxy-*O*-alkylhydroxylamine, did not cyclize either in the Cy6/Cy7 case ($n = 3$) or more unexpectedly, in the Cy5/Cy6 case ($n = 2$).[645] This could result from a high stabilization of the aminyl radical by a capto dative effect[380b] which would lead to the observed *N,N'*-diacyl-*N,N'*-dialkoxyhydrazines (formed by the known easy coupling reaction of these radicals, which is even easier in this case than intramolecular addition).

D. Nitroxide Radicals

Oxidation of the unsaturated hydroxylamine **92** using Ag_2CO_3-celite, or Ag_2O, with a view to obtaining the corresponding unsaturated nitroxide, gave the cyclized nitroxide (**93**) as the unexpected major product (20%). It was assumed that **93** resulted from intramolecular addition of the intermediate protonated nitroxide radical (A˙) (Scheme 47).[204] An analogous cyclization process was postulated latter by House[205] to explain the unexpected results obtained on heating 3,3-diallyl-2,4-pentanediones with excess hydroxylamine resulting from homolytic cyclization of an unsaturated hydroxylamine intermediate. This interpretation was confirmed[205] by heating the unsaturated hydroxylamine **94**; the cyclized (Cy5) product (**95**) obtained is nicely accounted

SCHEME 48

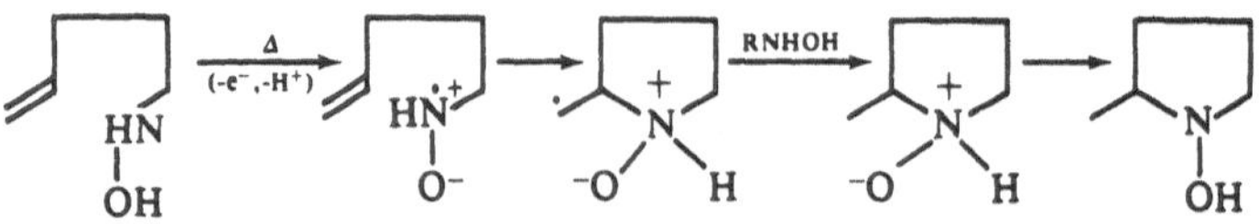

SCHEME 48A

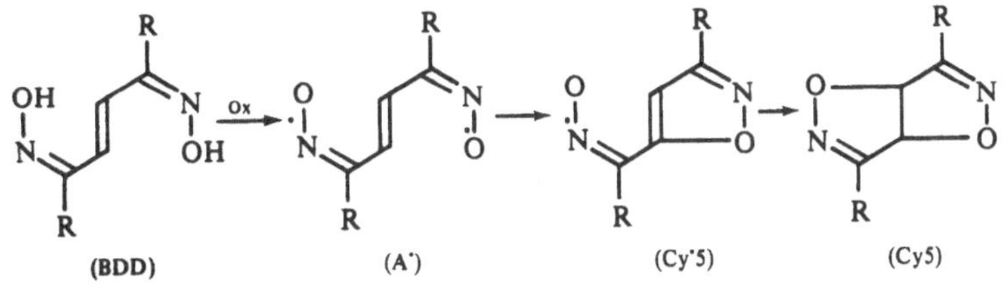

for by a homolytic cyclization (Scheme 48). According to House the initiator in the first step would be traces of oxidizing agents. The results agree with those obtained in aminyl radical intramolecular addition, namely, exclusive formation of the (Cy5) compound in 65% yield in the Cy5/Cy6 case, exclusive formation of the (Cy6) compounds in the Cy6/Cy7 case, with somewhat lower yields (30–40%) and no cyclized products in the Cy5/Cy4 case.[206]

Cyclization of unsaturated iminoxyl radicals has been proposed to rationalize the products resulting from the reaction of 2-butene-1,4-dione dioximes (BDD) with oxidizing agents such as lead tetraacetate (Scheme 48A). As yields of dihydroisoxazoloisoxazoles (Cy5) of up to 60% may be obtained (R = Ph),[646] this would be a remarkable reaction since it would involve the generally much disfavored Cy5/Cy4 cyclization. But it must be noted that the (Cy'5) radical is stabilized, particularly when R = Ph. Examples of oxidative cyclization of oximes of α,β-ethylenic ketones to dihydroisoxazoles have been reported in the steroïd series and could involve an analogous mechanism.[647]

4. Sulfur-Centered Radicals

A. Unsaturated Thiyl Radicals Generated from Unsaturated Mercaptans

In the preceding sections it was concluded that the cyclization of alkoxyl and aminyl radicals parallels that of carbon radicals under kinetic control. It will be seen in this section that the situation is much less clear as regards the behavior of unsaturated thiyl radicals.

A major difficulty is encountered when generating, unambiguously, unsaturated thiyl radicals. Most studies have been concerned with ethylenic mercaptans but these are generally difficult to handle, readily give tars, as well as cyclized products, under acidic conditions, thermolysis, photolysis and even chromatographic analysis. Thus it is not always easy to ascribe the cyclized products to an ionic or a homolytic pathway, and this could explain the dis-

SCHEME 49

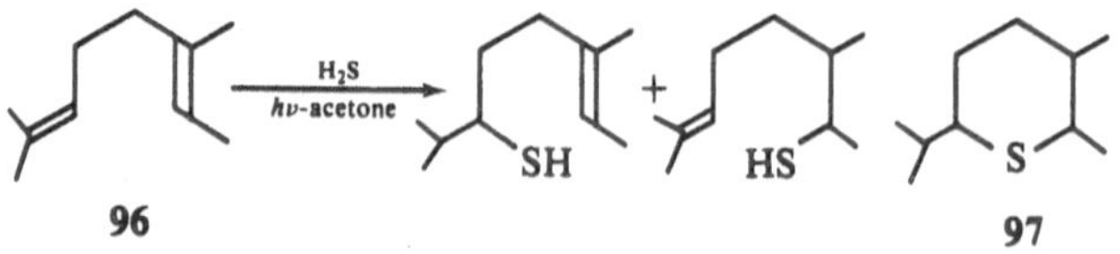

crepancies, especially as the selectivity of the cyclization processes, when comparing reactions not conducted under exactly the same experimental conditions.

Even when a free radical process can reasonably be assumed, the results are generally complicated by the possible occurrence of reversible cyclizations leading to a mixture of the two thiaheterocyclic compounds. And even when the cyclization process may be assumed to occur under kinetic control, the reaction appears to be much less selective than in the cases discussed in the preceding sections, although the ratio of the two cyclized products may be dramatically altered compared to reversible processes. With these difficulties in mind, and because the relatively small number of studies devoted to cyclizations of unsaturated thiyl radicals, any attempt at rationalization must be considered as tentative at this time.

Despite the ability of unsaturated mercaptans to give tars readily they were prepared as early as 1934 by Von Braun, who did not recognize the ability of these compounds to cyclize.[207] Likewise photolysis of hydrogen sulfide in the presence of allyloxyethanol,[208] or decomposition of the adducts of thiolacetic acid with 4-vinyl-1-cyclohexene or *d*-limonene,[209] gave only polymers, and no cyclized products, probably because high-dilution conditions were not used.

However, cyclization of unsaturated mercaptans was described in 1947 by Naylor[210] who photolyzed hydrogen sulfide in diene **96**, acetone being used as photosensitizer (Scheme 49). Since under ionic conditions other products are obtained, it can be assumed that the thiapyran **97** results from intramolecular addition of corresponding thiyl radical. Equally relevant are the results of Dyer and Osborne,[211] who, by distillation of 6-mercapto-1-hexene, obtained a mixture of tetrahydro-2-methyl-1-thiapyran (Cy6) and thiepan (Cy7). Also interesting, if resulting from thiyl radical intramolecular addition, is the result described by von Rühlmann,[212] who, by submitting a mixture of **98** and **99** to BP obtained the dihydrothiazepin (Cy7) (**100**) as the only product (43% yield) (Scheme 50). Intermolecular addition of the thiyl radical followed by intramolecular imine formation, rather than the reverse, is also an attractive pathway in this case.

It was not until 1964 that Walling proposed that 2-methyl-tetrahydrothiophene (Cy5) and thiacyclohexane (Cy6), obtained as contaminants in the preparation of 5-mercapto-1-pentene, are formed by intramolecular addition of the thiyl radical to the double bond.[23] The first studies of cyclization

SCHEME 50

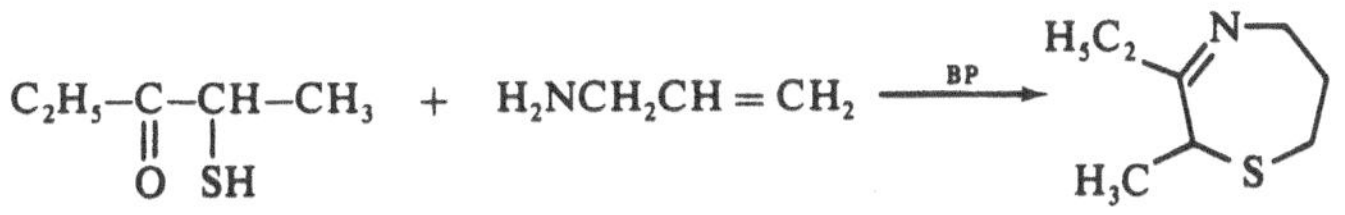

$$C_2H_5-\underset{\substack{\|\\O}}{C}-\underset{\substack{|\\SH}}{CH}-CH_3 \ + \ H_2NCH_2CH=CH_2 \ \xrightarrow{BP} \ \text{100}$$

| 98 | 99 | 100 |

of unsaturated mercaptans using the best conditions for generation of a thiyl radical (photolysis or peroxide initiation) were described in 1967 independently by Surzur, Crozet, and Dupuy[213] and by Volynskii, Galpern, and Urin.[214]

From these and later studies[215] general trends may be discerned. If conditions are chosen to minimize the competitive polymerization reaction, i.e., under high-dilution conditions, intramolecular addition of thiyl radicals, generated by peroxide initiation or by photolysis of the corresponding unsaturated mercaptans, appears to occur more easily than with the other radicals discussed before. For instance, not only in the Cy5/Cy6 case, but also in the Cy5/Cy4, Cy6/Cy7, and even in the Cy7/Cy8 case, yields of up to 90% of cyclized products may be obtained. But the selectivity is generally not as high as in the corresponding intramolecular addition of other radicals, and mixtures of the two cyclized products are obtained. For instance, although it has been reported in patents that various 5-mercapto-1-pentenes lead exclusively to thiapyrans (Cy6) by AIBN initiation,[216,217] these results are not in accord with other results also obtained in the Cy5/Cy6 case where poor selectivity is observed.[213–215]

It appears that the main reason which can be advanced to explain both the easy cyclization and the lack of selectivity is the large bond length C–S' compared to C–C', C–O', and C–N'. The orthogonal overlap of the thiyl radical with the double bond will be easier but this time in either direction since the unfavorable steric interactions which hinder addition to the terminal position of the double bond will be decreased as discussed in Section III.1.B and Table 1. A complementary molecular orbital explanation has also been proposed by Baldwin[42]: because of the presence of unoccupied $3d$ orbitals, the sulfur atom could receive electrons (back donation) from the occupied π orbital of the double bond, which would reduce the geometric constraint for the terminal ring closure.

Whatever the explanation, the lack of selectivity seems now clearly established except in special cases to be discussed later; but it is also clear that another feature complicates the interpretation of the results and modifies, often dramatically, the ratio of the two cyclized products, namely, the reversibility of intramolecular addition of thiyl radicals. Reversible intermolecular addition of thiyl radicals is well established. The following results are in accord with a reversible intramolecular addition of thiyl radicals,[218a] although the possible

SCHEME 51

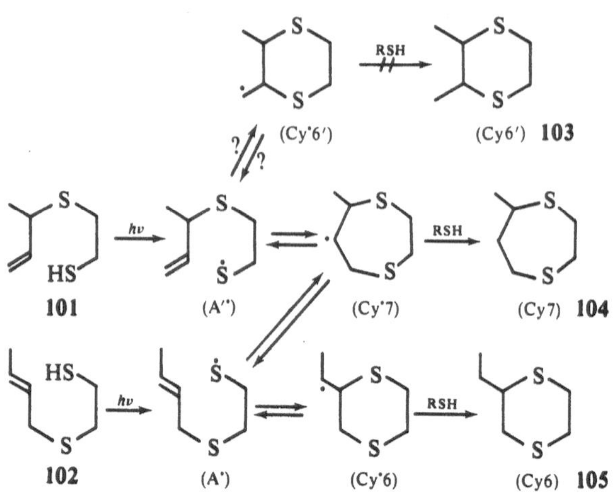

occurrence of a thiaallylic rearrangement[218b] **101** → **102** makes the demonstration less conclusive than first thought. Photolysis of **101** and **102** (0.15 mol liter^{-1} in hexane or cyclohexane) leads to mixture of cyclized products (**104**) and (**105**) in ratios very dependent on the temperature[218a] (Scheme 51). At −65°C, photolysis of **101** leads to a Cy7/Cy6 ratio, **104/105**, of 76:22 and photolysis of **102** to a 50:50 ratio. At 80°C photolysis of **101** or **102** leads to a **104/105** ratio 3:95. It appears that the best way to interpret these results, particularly the formation of **105** from **101**, may be to invoke the β-scission process of Cy·7 toward A·.

   At 80°C, the reaction would be under complete thermodynamic control but even at a temperature as low as −65°C, the formation of **105** from **101** would indicate that some of the initially formed (Cy·) radicals readily open to give A· and then Cy·6, before being trapped to give **104**. Thus the intramolecular addition is reversible, even at low temperatures. The nonoccurrence of **103** will not be discussed here because the total yields do not exceed 50–60%. Another interesting observation concerns the high Cy7/Cy6 ratio **104/105** obtained from **101** at low temperature; this would indicate that, under conditions where kinetic control is favored, formation of the (Cy·7) radical is more favored than that of the (Cy·6) radical. This is a result which is in marked contrast to those obtained with carbon or nitrogen radicals, and is confirmed by the photolysis of $H_2C=CHCH_2XCH_2CH_2SH$ (**106**) (X = CH_2,O or S), where the Cy7/Cy6 ratios increase up to 88:12 by lowering the temperature or by increasing the thiol concentration, i.e., under conditions favoring kinetic control.[219] Nevertheless, in the Cy5/Cy6 case, the (Cy5) product is favored under conditions favoring kinetic control.[215]

It is quite clear that forcasting, or even interpreting the results of intramolecular addition of thiyl radicals is complicated. One reason is the occurrence of reversible cyclization favoring thermodynamic control, especially at relatively low concentrations and high temperatures. A second is that, under conditions favoring kinetic control, a mixture of cyclized products is generally obtained, with the possibility, as in the Cy6/Cy7 case just discussed, that the major product is not the one expected from the best known behavior of carbon radicals. Thus the influence of substituents or of heteroatoms in the chain may be difficult to understand and comparisons must be made under strictly comparable experimental conditions.

One of the more unexpected results was observed on photolysis of $H_2C=CHCH_2XCH_2CH_2SH$ (106). When $X = O$ or S, the results parallel those obtained when $X = CH_2$ under the same conditions, with minor variations in the Cy6/Cy7 ratio.[219] But with $X = NR$ (107), the major product now is a thiazolidine (Cy5) (108), whose formation has been interpreted using an intramolecular hydrogen abstraction by thiyl radical[220] (Scheme 52). This rather unexpected behavior of a thiyl radical, which is known to add easily intermolecularly to double bonds rather than to give rise to hydrogen abstraction, may be ascribed to three causes: the hydrogen is allylic, α to a nitrogen atom, and in the favored 1,5 position analogous to that in the Barton reaction. Nevertheless, the fact that this behavior is observed only with $X = NR$ and only with thiyl radical is rather mysterious.

While this is a very general reaction regardless of the nature of the chain, or the presence of alkyl substituents,[221] the formation of the thiazolidine is completely suppressed in compound 109 presumably because of the stabilization provided by the chlorine atoms on the double bond. Furthermore, in this case cyclization is completely selective toward the (Cy6) compound 110 (yield 55%, $X = CH_2$;[222] 14%, $X = NMe$[223]) (Scheme 53).

Very selective cyclization has also been observed in a case where the cyclized radical appears not to be so stabilized. For example, Tanaka[224]

SCHEME 52

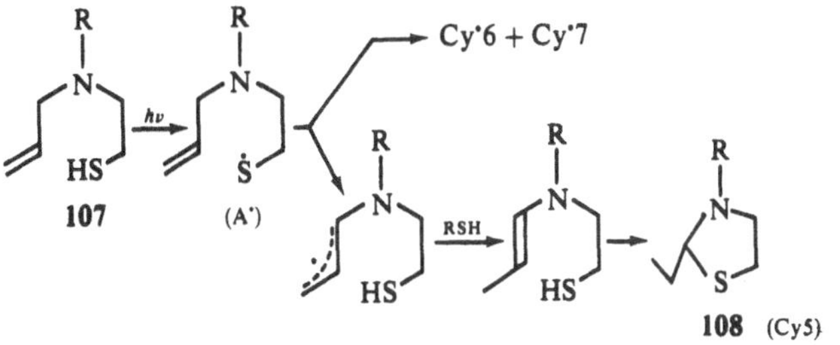

108 (Cy5)

SCHEME 53

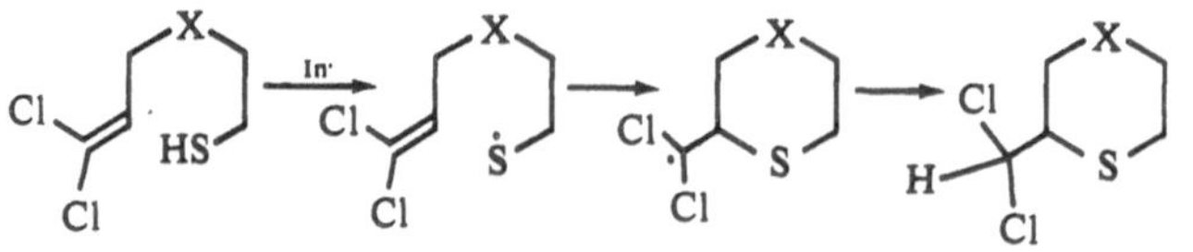

109   110

observed that uv irradiation of prenyl mercaptan (**111**) in *n*-hexane leads exclusively and quantitatively to **112** (*trans/cis*: 79:21) (Scheme 54). This is a very striking result for two reasons. A quantitative yield of **112** could hardly be expected knowing how easily allyl mercaptans polymerize, and considering that high-dilution conditions were not employed. Secondly, the high selectivity toward the (Cy6) compound is interesting. Since with crotyl mercaptan a mixture of Cy6/Cy7 is obtained,[225] this high selectivity toward the (Cy6) compound **112** may be ascribed to the stabilizing effect of the two methyl groups in the (Cy˙6) radical.

Another striking result, considering the high selectivity and the size of the cycle obtained, has been described by Weber.[226,227] Upon irradiation in dilute pentane solution at −78°C, hydrogen sulfide adds to dimethyldiallylsilane (**113**) giving only the (Cy8) compound **114** (25% yield) (Scheme 55). An analogous result has been obtained, although the yield is lower (10%), with diphenyldiallylsilane.[227] This is clearly a remarkable result if one considers the various pathways available: no cyclized products have been observed resulting from the cyclization of the carbon radical (A˙₁) and cyclization of the thiyl radical (A˙₂) leads exclusively to a product resulting from the (Cy˙8) radical with no products resulting from the (Cy˙7) radical. Furthermore, this is the first report of the formation of an eight-membered ring by radical intramolecular addition. If it is assumed that, at −78°C, the reaction is essentially under kinetic control, this result seems in accord with results obtained in the Cy6/Cy7 case where

SCHEME 54

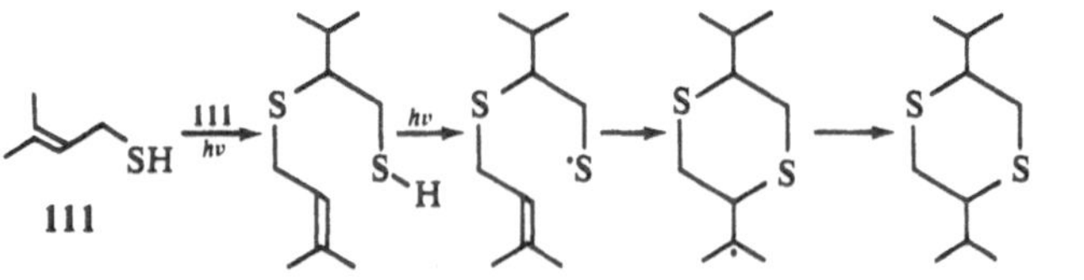

(A)   (A')   (Cy˙6)   (Cy6) **112**

SCHEME 55

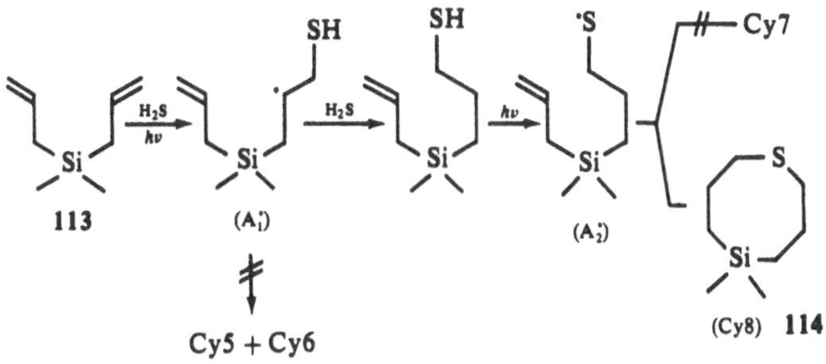

predominant formation of the (Cy7) compound under kinetic control had also been assumed. Unfortunately, it is not known at this time if this very high selectivity toward the (Cy8) compound **114** will generally be observed with aliphatic chains in the Cy7/Cy8 case or if the Si atom, perhaps because of the C–Si bond length (1.87 Å) — larger than in the cases studied up to now—will reinforce the propensity of thiyl radicals to give the larger cycle.

Other factors are able to enhance the regioselectivity of the cyclizations. For instance, in the Cy5/Cy6 case, photolysis of compounds **115** (R,R ' = H or Me) in pentane at 30–35°C gave, after careful analysis of the products, only one cyclized product,[229] identified as a [3.2.1] compound (**116**) (Cy5) (90–96% yield was obtained), rather than two as initially reported[228] (Scheme 56). In this case, the selectivity could be the result of thermodynamic control, as the (Cy˙6) radical formed unambiguously leads only, under the same conditions, to products resulting from the (Cy˙5) radical.[230]

We will comment now on the Cy5/Cy4 case. Thermolysis of 3-butenethiol

SCHEME 56

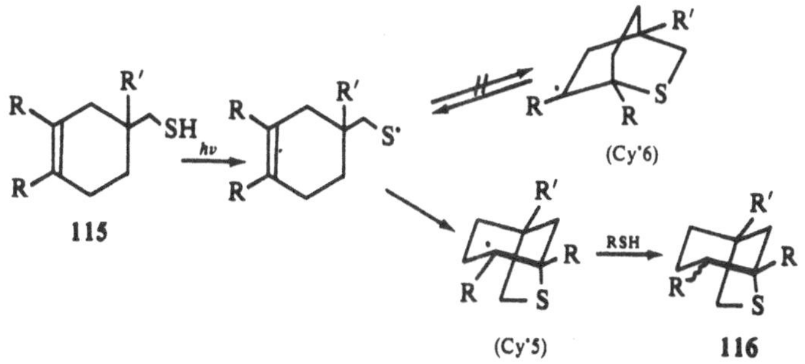

in boiling cyclohexane leads to a 43% yield of thiolane containing less than 1% of 2-methylthietane.[213,215] Analogous results have been obtained from alkyl-substituted 3-butenethiols[231a] and during one-electron oxidation of tetrahydrothiophen[231b]: it has been assumed that the cyclized product arises from intramolecular addition of the thiyl radical. If formed, the (Cy·4) radical would certainly open very rapidly as shown by Trost.[232]

With these results it is possible to draw some tentative conclusions about the major products to be expected in homolytic cyclizations of unsaturated thiols. In the Cy5/Cy4 case, the only product isolated is the Cy5 one in relatively high yield compared to the other radicals. The higher yield may be ascribed to the large C–S· bond length and the selectivity to the fact that this increase nevertheless does not permit a good overlap of the thiyl radical and the sp^2 carbon to lead to the (Cy·4) radical. In the Cy5/Cy6 case, the steric crowding which results in selectivity for (Cy·5) radical formation with carbon, alkoxyl, or aminyl radicals, is not so severe because of the increase of the C–S· bond length so that, even under conditions favoring kinetic control, a mixture of (Cy5) and (Cy6) products is obtained with an increase in the Cy6/Cy5 ratio under conditions favoring thermodynamic control. If the chain length is further increased, the reverse situation appears, the higher-membered ring is now favored, particularly under conditions favoring kinetic control. But this trend may be decreased or increased by stabilizing substituents in favor of the more stabilized (Cy·) radical. It must be recalled, however, that because of the few examples described and because cyclization depends on structural features and experimental conditions these conclusions must be considered as tentative.

It also appears that intramolecular thiyl radical addition is a very efficient process compared to other free radical cyclizations. But competitive reactions such as polymerization are sometimes very difficult to suppress. Furthermore, other easy cyclization pathways, such as ionic ones, may complicate the interpretation of the results. Dronov's work[233–236] exemplifies this possibility. Under all the experimental conditions used, 1-pentene-5-thiol led to a Cy5/Cy6 mixture of products, with the (Cy5) compound being favored in sulfuric-acid-promoted cyclization and the (Cy6) compound being favored in photolysis. Thus the claim to have observed a homolytic reaction in cyclizations of ethylenic thiols generated from fatty acids, under conditions which did not avoid acid treatment,[237,238] must be considered with care.

Finally it seems worthwhile to mention other pathways, probably ionic, for the cyclization of unsaturated mercaptans. Thus iodine addition in the presence of potassium carbonate to form a sulfenyl iodide which cyclizes via iodonium intermediates has been proposed. This cyclization reaction has been used recently by Nicolaou[239a] to prepare 6,9-thiaprostacyclin, a stable and biologically potent analog of prostacyclin PGI$_2$ in a reaction involving cyclization of a 4-pentene-1-thiol leading to the iodomethylated five-membered ring. The high selectivity toward formation of the (Cy5) compound in the Cy5/Cy6 case is again observed under other electrophilic cyclization conditions.[239b,c] An analogous result has been described by Ikegami[240] starting from the corresponding disulfide, which is easier to handle than the unsaturated mercaptan. Also, though beyond the scope of this review because an ionic mechanism is assumed, the cyclization of unsaturated sulfenic acids, which has been shown to occur in the conversion of penicillins to cephalosporins[241] and in penicillin biosynthesis,[242,243] should be mentioned.

B. Unsaturated Thiyl Radicals Generated from Unsaturated Sulfides or Disulfides

Because of the difficulties in handling unsaturated thiols and because of the competitive ionic cyclizations processes, other ways to generate unsaturated thiyl radicals were sought. Unexpected results have been obtained independently by Bastien[244,246] and by Oae[245] on thermolysis (R = SR ')[245] or photolysis (R = benzyl,[244,246] allyl,[244,246] SR '[245]) of compounds 117 (Scheme 57). The main product, obtained sometimes in quantitative yields (R = SR '), is 120, which appears to derive from the coupling of Cy6 with R. Although the possibility of double bond participation in an excited state has been suggested,[245] the fact that similar results are obtained on both thermolysis and photolysis, and the identification of the coupling product R–R when R = PhCH₂ argues for the radical pathway of Scheme 57. But the detailed mechanism is not yet clear and could be different according to the nature of the R group. It seems difficult to assume a chain mechanism, but cross experiments are in accord with a cage coupling when R = SR '[245] and an out-of-cage coupling when R = benzyl.[246b] This later conclusion seems more in accord with the unexpected high selectivity of cyclization toward the (Cy6) compound (120), which would result from the apparently more stable (Cy6) radical as discussed in Section VIII.4.A. Also, in the Cy6/Cy7 case, the only substituted cyclized product is the (Cy6) one, i.e., once more that resulting from the more stable radical.[246c] If the coupling had been a cage process one could have expected to

SCHEME 57

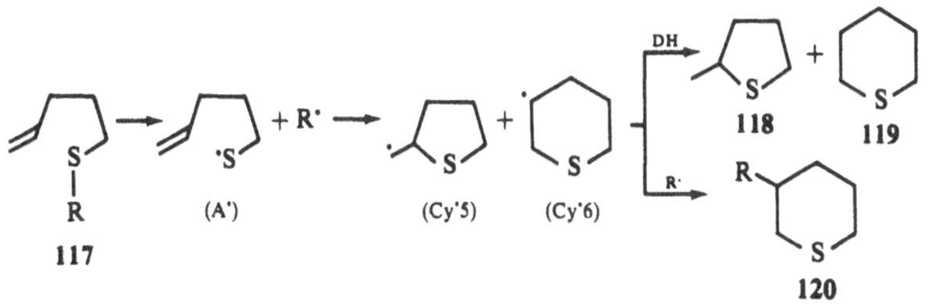

SCHEME 58

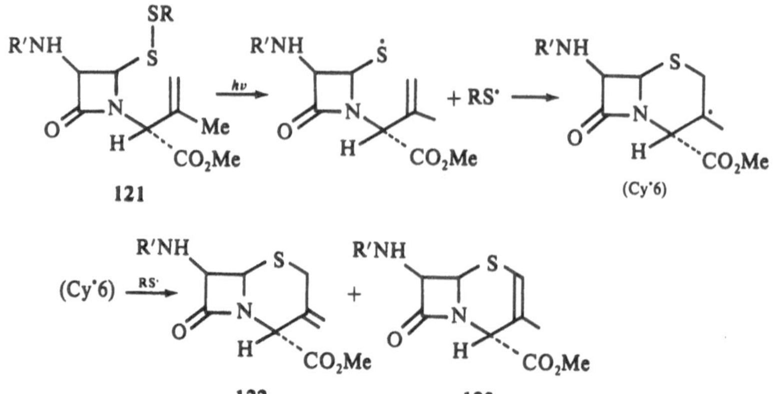

observe mainly the product formed under kinetic control, since this is a very fast process. Further work is required before a firm conclusion about the mechanism of this cyclization process is proposed.

Another unexpected and interesting result has been described by Maki[247a,b] when photolyzing the disulfide 121 (Scheme 58). The intramolecular addition of the thiyl radical to the isopropenyl bond is now followed by hydrogen abstraction to give the 3-methylenecepham 122, in high yield, (60%) and the 2-cepham 123 (15%), affording a convenient preparative method of obtaining 3-methylenecepham from penam and suggesting a possible biosynthetic route to cepham[247b] (Scheme 58). According to the results discussed above, and considering that the (Cy·6) radical is tertiary, the high selectivity toward the (Cy6) product could be reasonably expected. More troublesome, at first glance, is the fact that no product of coupling between Cy·6 and RS· is observed. This fate of Cy·6 could result from the particular nature of the R groups used in this study (2'-benzoxazolyl, 2'-benzothiazolyl, or 2'-pyridyl) but more likely is due to the tertiary nature of the (Cy·) radical: an analogous result is observed in the photolysis of unsaturated nitrite esters: when the (Cy·) radical is tertiary, one obtains unsaturated (Cy) compounds rather than the coupling products with NO.[131,132] This scheme has been confirmed by Gordon,[247c] who was able to identify the (Cy6H) product resulting from Cy·6, in addition to 122 and 123. More recently, Maki,[248a] operating at higher concentration, has obtained, in addition to 122 and 123, (Cy5) products assumed to be formed by intermolecular addition of thiyl radical to 121 followed by an SHi reaction on the S–S bond. Another possibility, namely, easier formation at the higher concentration of Cy·5 formed under kinetic control according to Section VIII.4.A, has been proposed.[246b] But quite interestingly an analogous SHi mechanism has been retained by Baldwin[248b] as a possible biosynthetic pathway to penicillins.

SCHEME 59

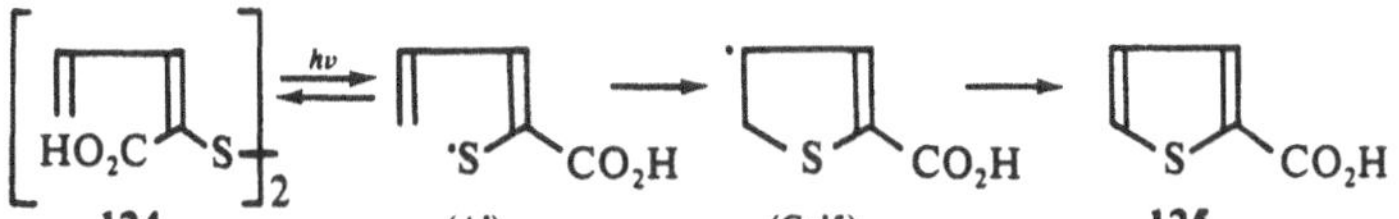

C. Enethiyl-Unsaturated Radicals, Arylthiyl-Unsaturated Radicals, and the Thio–Claisen Rearrangement

The mechanism of oxidative ring closure of compounds of the 1,3-butadiene-1-thiol type with halogens to give thiophen derivatives has been studied by Chapman.[249] A polar mechanism has been assumed, although a free radical mechanism is not completely ruled out and, indeed, is proposed for the cyclization under other conditions. Thus, photolysis of the disulfide **124** in boiling xylene gives the thiophen **125** in 40% yield; the proposed mechanism is depicted in Scheme 59. Such a mechanism involves efficient cyclization toward the (Cy5) compound in a Cy5/Cy4 case and the conversion of the (Cy˙5) radical into a thiophene.

This result could throw some light on processes which will be discussed now, even though they do not correspond exactly to the scope of this section. In connection with the mechanism of formation of volatile products from sulfur-containing amino acids (precursors of onion and garlic flavor, which are attributed to the formation of an S-1-alkenyl group), an interesting photochemical transformation of vinyl thioethers (**126**) has been observed, dialkylthiophenes (**127**) and (**128**) being the major products (Scheme 60, where R = alkyl, R$'$ = SR,[250] L-cysteine,[251,252] COMe[253]). All the mechanisms proposed begin with homolysis of the S–R$'$ bond, the initial enethiyl radical

SCHEME 60

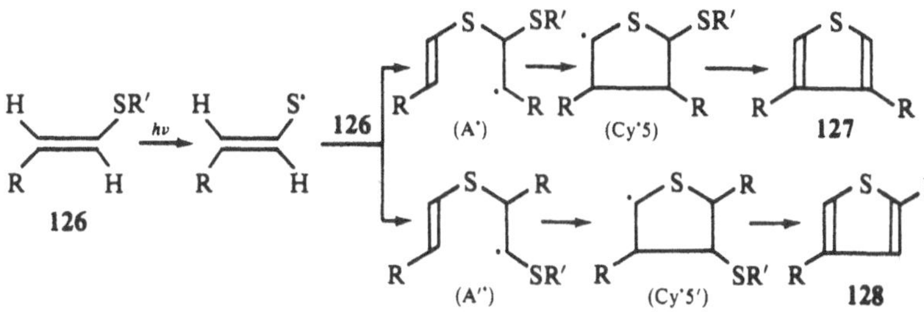

having been identified by esr.[251] Two possibilities are then proposed: one, the dimerization of the enethiyl radical, followed by sigmatropic rearrangement; the second, the addition of the enethiyl radical to the α or β position of the starting material, followed by the carbon radical intramolecular addition in this Cy5/Cy4 case toward Cy'5 and thence to the thiophenes **127** and **128**[250–253] according to Scheme 60. Although a carbon-free radical cyclization would be involved here, this result is clearly reminiscent of the thiyl radical cyclization discussed in Scheme 59. The possibility of observing carbon radical cyclization in a Cy5/Cy4 case, while generally not favored (see Section V.2.A), may be attributed to the sulfur atom in the chain acting to permit a better overlap of the radical center and the double bond and to stabilize the (Cy'5) and (Cy'5') radicals.

Many studies have been devoted to the thio-Claisen rearrangement, a [3,3] sigmatropic process involving synchronous cleavage of the C–S bond and formation of a new C–C bond in **129** and **130** giving finally the cyclized products **131–134** (Scheme 61). Discussion of the mechanistic aspects of this interesting rearrangement[254] is beyond the scope of this review and only the main conclusions concerning the possible occurrence of a free radical cyclization will be summarized. Direct cyclization pathways from the sulfide have been proposed,[255,256] but it is now generally admitted that an unsaturated mercaptan is an intermediate in the rearrangement of vinyl allyl sulfides (**129**)[257,258] as well as hetaryl or aryl allyl sulfides (**130**),[259–263] although in some cases different results have been observed starting from the sulfide or from the corresponding thiol.[264] Interpretation of the last step, the cyclization of the unsaturated thiol, is generally difficult because of the possible ionic and homolytic pathways available and the variety of experimental conditions often used to achieve the thio-Claisen rearrangement. A further difficulty is that, in certain cases, cyclized products are not observed.[261] When cyclized products have been obtained, it has been suggested that the Cy6 ones result from homolytic

SCHEME 61

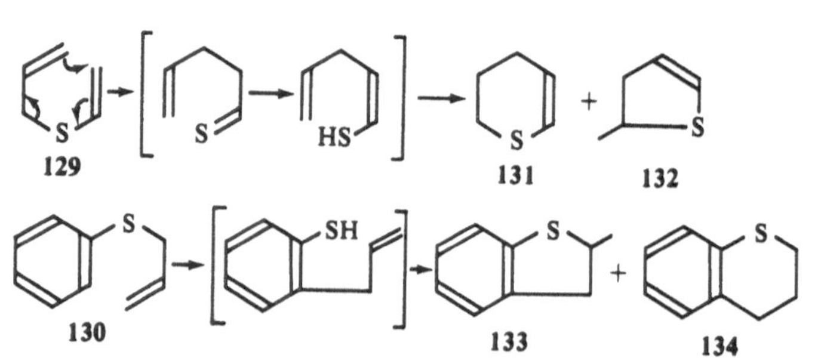

pathway and the Cy5 ones from an ionic pathway.[255] From the discussion in Section VIII.4.A, this appears to be a rather crude conclusion and it should not be generalized.

5. Other Heteroatom-Centered Radicals

Reports of the intramolecular addition of heteroatomic radicals other than alkoxyl, aminyl, and thiyl are rather scarce, although very interesting and unexpected behavior has often been reported.

A. Phosphorus-Centered Radicals

Irradiation of secondary alkenyl phosphines **135** in boiling light petroleum[41] gave only the (Cy) compounds **136** resulting from terminal attack on the double bond (Scheme 62). Besides polymeric material, the cyclic phosphines **136** were obtained in yields as high as 38% ($n = 2$, Cy5), 64% ($n = 3$, Cy6), 41% ($n = 4$, Cy7), and 0% ($n = 9$, Cy12). Although no mechanistic pathway was proposed, it seems reasonable to assume that the cyclized products arise from intramolecular addition of the phosphinyl radicals. If so, the very high selectivity toward the higher Cy compound appears astonishing at first sight. Before discussing the origin of this unusual selectivity, it would be good to have confirmation of these results, published in 1966, at a time when it was still assumed that the same rules govern the selectivity of inter- and intramolecular free radical additions and when the products were identified without the help of nmr. Nevertheless such a result is not completely unexpected if one considers the C–P· bond length (1.83 Å), which would avoid the unfavorable interaction toward the bigger (Cy·) radical, as already discussed in Section III.1.B, so that the trend toward the formation of the bigger Cy· which appeared with the thiyl radical cyclization under kinetic control would be reinforced here.

With unsaturated phosphoranyl radicals (**137**) ($X = O$ or NH), Davies[265,266] has recently described the exact opposite results as observed by

SCHEME 62

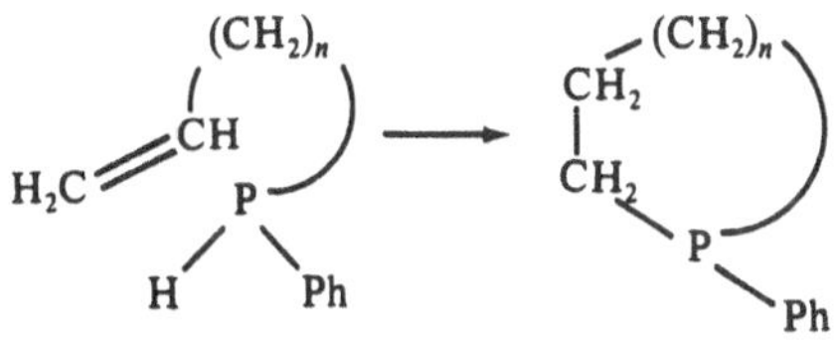

135   136

SCHEME 63

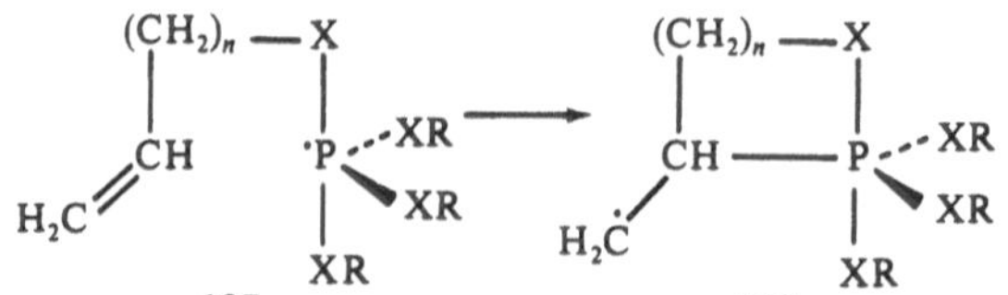

esr. This time only the smaller (Cy˙) radical (**138**) is observed (Scheme 63). Furthermore, results of these esr studies appear in total disagreement with what would be expected from other intramolecular radical additions. Thus, in the Cy5/Cy4 case ($n = 1$), the only radical observed is the (Cy˙4) one, which is never observed in other cases. Secondly, the Cy4/Cy5 process is faster than the Cy5/Cy6 process. From these results one might think that intramolecular addition of phosphoranyl radicals would be an easy process, but no cyclization is observed in the Cy6/Cy7 or Cy7/Cy8 cases. So it seems clear that with phosphoranyl radicals other features not yet considered control the cyclization process. Nevertheless, it must be recalled that these conclusions arise only from esr studies. Clearly, further studies are needed in the intramolecular addition of phosphorus-centered radicals.

B. Silicon-Centered Radicals

Very little is known of the behavior of unsaturated silyl radicals. Hydrosilylation of pentenylsilanes (**139**) has been described under ionic conditions by treatment with chloroplatinic acid: cyclized products are obtained in high yields, the (Cy5) compounds **140** being the major or exclusive, ones.[267-270] In a very brief note, Sakurai[270] described the radical cyclization of unsaturated silyl radicals (A˙) generated from the silanes (**139**) by DTBP (di-*t*-butyl peroxide) at 135°C (Scheme 64). Cyclized products **140** and **141** are obtained in low yields (7–10%) with a low Cy5/Cy6 ratio **140/141** (0.12–0.72) when X,Y = Me, Ph; Ph, Ph; iso-Pr, Cl. With X,Y = Me, Cl, the overall yield is 84% and the Cy5/Cy6 ratio becomes 5.7. The same silanes (**139**), upon treatment with chloroplatinic acid, give cyclized products in overall yield 64–100%, with the (Cy5) compound **140** much favored: the ratio **140/141** is 5.2–17.5. The dramatic change observed in the homolytic cyclization on changing the substituents on the radical center is without precedent in radical intramolecular addition except when a stabilizing substituent is replaced by a nonstabilizing one. In this case, however, it seems difficult to understand how the change from a methyl to an isopropyl group would induce such an important difference in the behavior of the silyl radical. Clearly further study is needed.

SCHEME 64

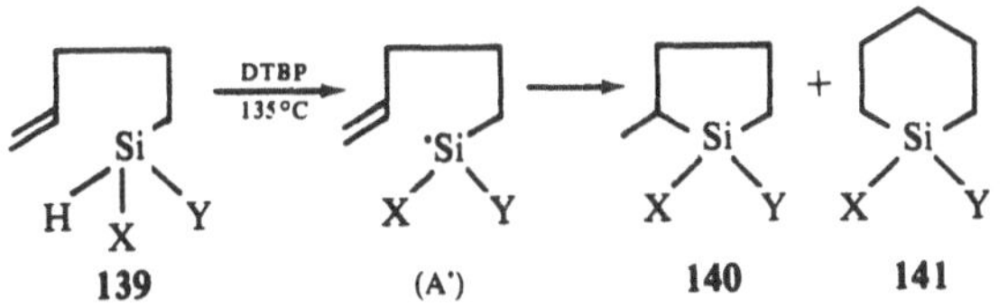

The mechanism of the thermal conversion of allyltrimethylsilane to trimethylvinylsilane at 525°C would involve homolytic cleavage of the silicon-methyl bond (rather than that of the silicon–allyl bond expected) followed by cyclization of the silyl radical to the (Cy'3) radical in the Cy3/Cy4 case, followed by the β-scission process.[648]

C. Germanium-Centered Radicals

Cyclization of unsaturated germanes under ionic conditions upon treatment with chloroplatinic acid,[271] as well under homolytic conditions[272,273] by treatment by AIBN at 100°C or upon uv irradiation has been described by Satge. Intramolecular addition of germyl radical is a very efficient process since the compound **142**, on AIBN initiation at 100°C, gives the cyclized product **143** in 70% yield (Scheme 65). This is the only example reported of intramolecular addition in the Cy11/Cy12 case. Even in the Cy3/Cy4 case, radical cyclization toward the (Cy4) compound is observed, although only in 12% yield.

D. Tin-Centered Radicals

In an esr investigation of S_H2 reaction of various radicals on stannacycloalkanes (**144**) Davies[274] assumed that a competitive process was the

SCHEME 65

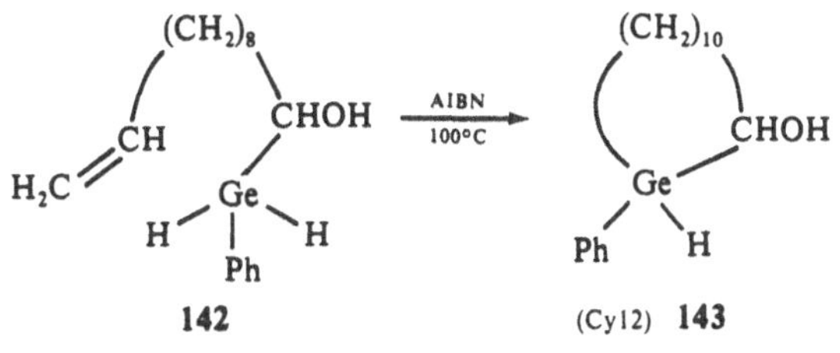

SCHEME 66

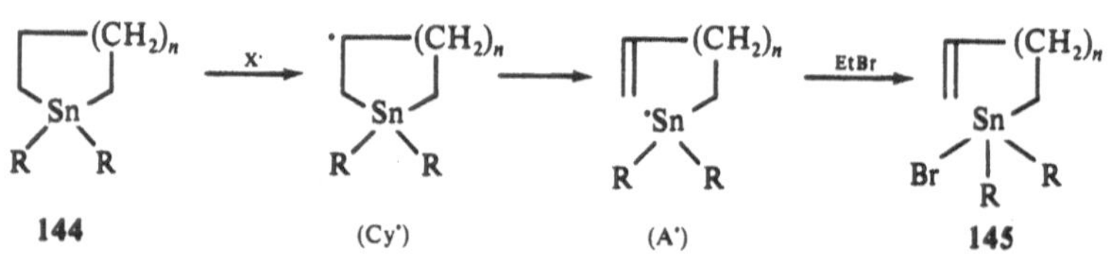

opening of the (Cy˙) radical, formed by abstraction of hydrogen from a β-methylene group, to A˙, which can be trapped by EtBr to give **145** (Scheme 66). From this result, cyclization of unsaturated stannyl radicals does not appear to be a likely process, in agreement with the well-known reversible addition of stannyl radicals to double bonds.[469b]

E. Selenium-Centered Radicals

Thermolysis of aryl allyl selenides in quinoline leads[275,276] to a mixture of (Cy5) and (Cy6) compounds, paralleling the results obtained in the thio-Claisen rearrangement. These compounds may arise by intramolecular free radical addition but, as discussed in Section 4.C for the thio-Claisen rearrangement, other pathways must be considered.

6. Conclusion

Intramolecular addition of heteroatomic radicals to a double bond is, in most of the cases studied, a more general and more efficient process than intramolecular addition of carbon radicals. It is thus possible to propose new methods for the preparation of heterocyclic compounds and to consider that these processes are involved in important biogenetic schemes such as penicillin or prostaglandin biosynthesis. It is not always easy to conclude that a free radical process is involved because of the possibility of competitive ionic cyclizations.

Finally, it must be said that while the main features concerning the cyclization of unsaturated O-, N-, or S-centered radicals are beginning to be understood, very little is known about the behavior of other heteroatomic radicals, although some very interesting features emerge from the first reports published.

IX. INTRAMOLECULAR ADDITION TO OTHER CARBON–CARBON BONDS

1. Introduction

This section will be mainly devoted to intramolecular radical addition to acetylenic bonds. Some studies with radical addition to allenic systems and cyclopropane rings will also be discussed. Compared to the well-known intermolecular free radical addition to ethylenic bonds, very little is known about free radical additions to acetylenic or allenic bonds, the same is true of intramolecular processes.

The origin of this apparent lack of interest may be found in the difficulty of devising experimental conditions which lead to addition products in good yields. This may easily be understood by recalling that the first addition product will be an olefin equally sensitive to free radical intermediates, so that competitive reactions become important.

2. Intramolecular Addition to Acetylenic Bonds of Carbon-Centered Radicals

In 1962, M. Julia[277] reported the cyclization under BP initiation of compound **146** to **147** (R = H, 27% yield; R = CH$_3$, 63% yield) and **148** to **149** as the major product (65%) of a mixture obtained in 55% yield (Scheme 67).

Hence cyclization is selective toward the (Cy5) compound in the Cy5/Cy6 case and toward the (Cy7) compound in the Cy6/Cy7 case. In spite of these promising results, the scope of the reaction and the origin of a selectivity different from that observed when the same stabilized radicals add to double bonds have not been studied.

SCHEME 67

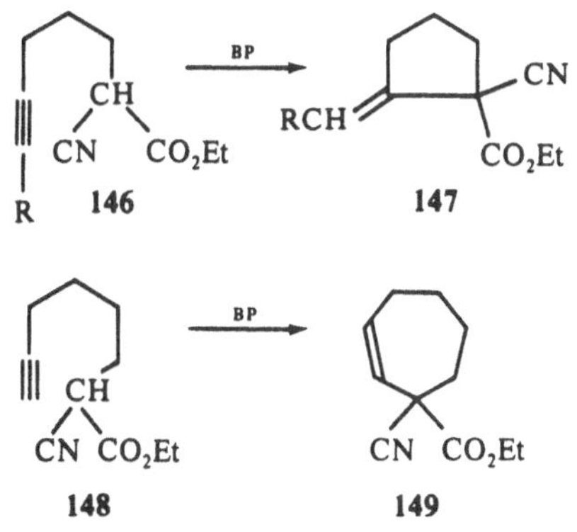

SCHEME 68

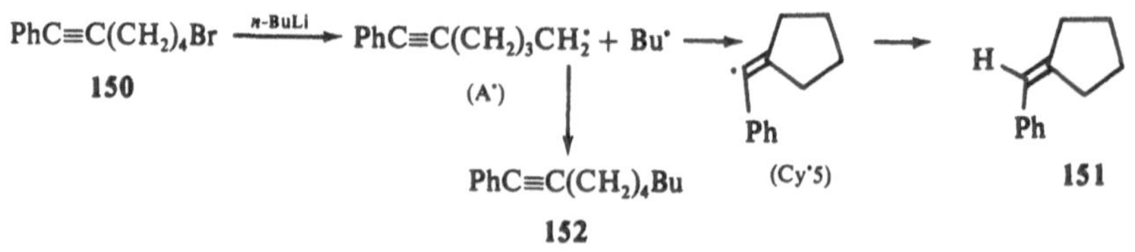

With nonstabilized carbon radicals, Dessy suggested in 1965 that when an organometallic and an acetylenic bond are put in rigid and close proximity to an aromatic substrate, ring closure to Cy5 (in the Cy5/Cy6 case) proceeds both by a carbanionic and a radical pathway.[278,279] In a simpler case, Ward proposed that steric coercion is not a requirement and that, when treated with *n*-butyllithium, the acetylenic bromide corresponding to a Cy5/Cy6 case **150** provided the cyclized (Cy5) compound **151** (60%) as the major product, along with the coupling product **152** (20%) by a homolytic pathway (Scheme 68).[280]

This mechanism was confirmed in 1969 by Crandall[281] by a comparative study of the reduction of acetylenic bromides by stannyl radical and by lithium biphenyl. This second system gives more complex results which, nevertheless, are similar to those obtained on stannane reduction, which are generally considered to proceed by a radical mechanism.[469b] In the Cy3/Cy4 and Cy4/Cy5 cases no cyclized products were obtained although the (Cy˙5) radical could have been expected to form since, according to Baldwin's rules, the 5-*endo*-dig process would be favored compared to the disfavored 5-*endo*-trig one.[16,683] In the Cy5/Cy6 case only (Cy5) compounds such as **151** were observed and in the Cy6/Cy7 case only the (Cy6) ones were, although in lower yield. These results seem to parallel those obtained by intramolecular addition of unstabilized carbon radicals to double bonds. They were confirmed later with radicals generated by stannyl reduction[282–284] and with vinyl radicals generated by thiyl radical addition to one of the acetylenic bonds of a 1,6-diyne such as dimethyl dipropargylmalonate.[285] Another interesting application of this selective cyclization has been described by Kalvoda.[286] Photolysis of the nitrite **153** in toluene or in BrCl₃C gives the radical (A˙) (Barton reaction), which cyclizes to give the (Cy5) compound (X = H or Br) in 20% yield (Scheme 69).

Electroreductive cyclization of acetylenic halides has been extensively studied by Peters, particularly in the case of 6-halo-1-phenyl-1-hexyne.[287,288] A complicated scheme was proposed for the 6-chloro compound[287] to explain the results. The origin of the cyclized products may be sought among others in the radical formed by direct reduction of the carbon–halogen bond. Another possibility is the reduction of the carbon–carbon triple bond (already recognized previously in the reductive cyclization of bis-propargylamines[289])

SCHEME 69

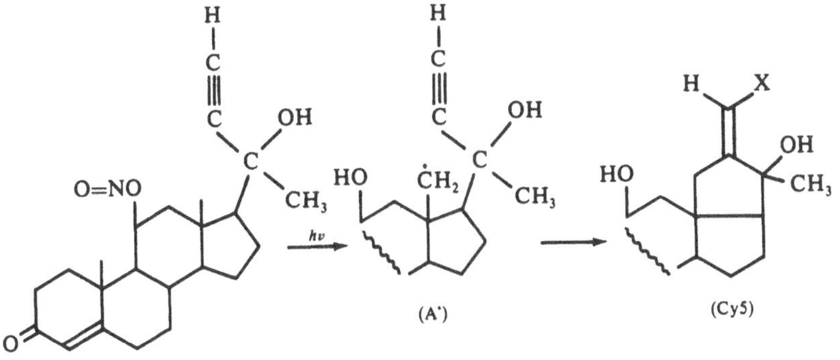

followed by cyclization of the radical anion or by an intramolecular electron transfer reaction leading to chemical reduction of the carbon–chlorine bond. Whatever the pathway involved, the conclusion seems clear in this Cy5/Cy6 case, that intramolecular addition of a carbon radical to the triple bond leads exclusively to the (Cy5) compound.

Also interesting is the conclusion of Pradhan[291] concerning the mechanism of reductive cyclization of ethynyl ketones first described by Stork.[290] The latter tentatively proposed that by using metal and liquid ammonia the acetylene radical ion is formed and that this is followed by nucleophilic attack on the ketone. In fact, by using a less powerful reducing agent such as metal naphthalenes, Pradhan[291] provided conclusive evidence that, in the first step, an electron is transferred preferentially, though reversibly, to the ketone group and that in the next, slow step the ketyl radical ion (A˙) attacks the acetylene intramolecularly as a radical and not as a nucleophile. The yield of cyclized products such as **155**, starting from **154**, may be as high as 90% ($R = C_8H_{17}$) but depends on the structure of the acetylenic ketone (for instance, with two methyl groups on the position next to the radical center in (A˙) the rate of cyclization is lowered) as well as on the counterion and the solvent. The regiospecificity is remarkable: only the (Cy5) compound is obtained in the Cy5/Cy6 case and only the (Cy6) compound in the Cy6/Cy7 case. The stereoselectivity is also remarkable since only A|B *cis* products are obtained. The mechanism of this cyclization process is summarized in Scheme 70.

In conclusion, it appears that in spite of the reservations made in the Introduction, the intramolecular addition of carbon-centered radicals to triple bonds is an efficient process in the Cy5/Cy6 case, less efficient in the Cy6/Cy7 case, and does not work with the lower homologs. The selectivity is very high toward the (Cy5) compound in the Cy5/Cy6 case, and toward the (Cy6) com-

SCHEME 70

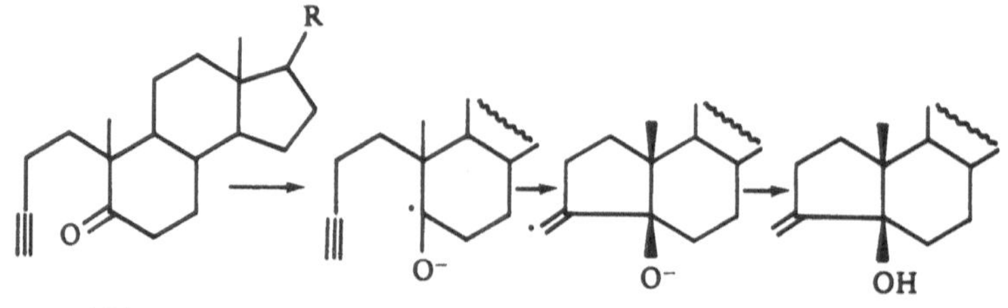

154   (A·)   (Cy·5)   155

pound in the Cy6/Cy7 case, with unstabilized radicals, but toward the (Cy7) compound in the only reported example of a stabilized radical. All these observations are reminiscent of the behavior of ethylenic carbon-centered radicals.

Comparison of these results with those of the cyclization observed with acetylenic organometallics could throw some light on the mechanism of these later processes. In fact it will be seen in Section XII.1.D.b that the mechanism of cyclization of ethylenic organometallic compounds, even though it has been studied much more, is not yet clear. Any conclusion based only on the ease and selectivity of the cyclizations must be considered as tentative, as, for instance, in the cyclization of ω-haloalkynes with nickel tetracarbonyl–potassium t-butoxide[292] or with dialkylcuprate.[293] The same is true of the cyclization of acetylenic Grignard reagents[294–296] and even during the formation of acetylenic Grignard reagents from the corresponding bromides.[296,297] Although the nature of the cyclized products accords generally with a homolytic cyclization, the observation of cyclized products also in the Cy3/Cy4 case, although in low yield,[297] as well the study of the opening process[296] led to proposal of the intervention of a "radical-like" and a four-centered cyclic reaction mechanism.[296]

3. Intramolecular Addition of Heteroatomic Radicals to Acetylenic Bonds

Cyclization of acetylenic heteroatomic radicals appears much less facile.[299] Although this has been ascribed to failure of the heteroatomic radical to undergo cyclization,[298] this is surprising if one considers that carbon-centered radicals add easily to triple bonds as discussed in the preceding section and that intramolecular addition of heteroatomic radicals to double bonds is often easier than with carbon-centered radicals. So the failure to observe cyclized products should probably be sought in the high reactivity toward radical reagents of a double bond substituted by a heteroatom which, of course, introduces new competitive pathways. Also, the cyclized products normally expected can undergo decomposition.

SCHEME 71

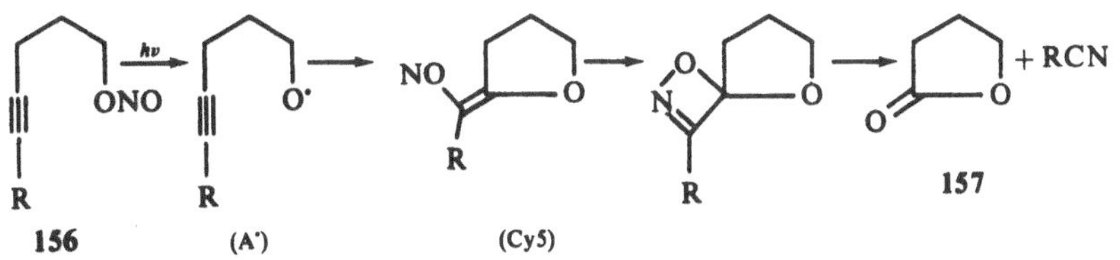

This, for instance, may be why no cyclized products have been obtained from acetylenic *N*-chloroamines under the free radical conditions discussed in Section VIII.3.A,[299a,d] whereas aminyl radicals easily added intramolecularly to double bonds.

Likewise, photolysis of 4-pentynyl nitrite (156) (R = H) led Rieke[298] to conclude that the alkoxyl radical (A˙) (Scheme 71) did not undergo free radical intramolecular addition. In fact, Dupuy[299a,d,300] was indeed able to isolate, the γ-butyrolactone (157) (although in very low yield), whose formation may be ascribed to the decomposition of a nitrosovinylic Cy5 intermediate.

In the Cy6/Cy7 case with the next homolog of 156 (R = H) no products resulting from intramolecular addition were observed, the main product identified (11%) being an isoxazole assumed to be formed in a Barton reaction in the first step.[299d]

With thiyl radicals, intramolecular addition to triple bonds has been observed by Dupuy using high-dilution conditions.[299a,b,c,301] In addition to dimeric and polymeric material, cyclized products, e.g., 159, 160, 161, are obtained in yields up to 86% (R = Ph) by photolysis of the thiol 158 (Scheme 72). The cyclization is not selective, as in the case of ethylenic thiols (Section VIII.4.A), products 159 being the major ones when R = H and 160 when R = alkyl, phenyl.[299c] The origin of compounds 161 is not clear: in some cases it has been demonstrated that compounds 160 isomerize on photolysis into 161, but prior isomerization of the thiols 158 must be considered as a possibility in other cases.[299c]

The homolytic cyclization of acetylenic thiols is general[299a,c]: the best

SCHEME 72

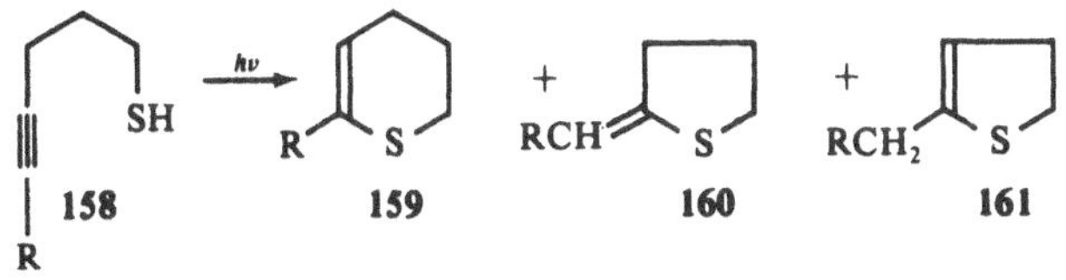

yields of cyclized products are 35% in the Cy5/Cy4 case, 89% in the Cy6/Cy7 case, 35% in the Cy7/Cy8 case, and 9% in the Cy8/Cy9 case; the presence of heteroatoms in the chain does not alter significantly the yields of cyclized products. Generally, cyclization is not regioselective, but the ratio Cy_{n-1}/Cy_n increases when R = H is replaced by R = alkyl or phenyl.

Of course ionic cyclization processes may be expected with acetylenic thiols, as discussed before with ethylenic thiols (Section VIII.4.A). The most extensive work has been described under basic conditions where intramolecular nucleophilic addition of the thiolate anion to the triple bond may be expected. Yields are generally similar to those obtained in homolytic cyclization but the cyclization is more regioselective, *exo* products such as **160** generally being the exclusive ones[299a,b,301,302]; but exceptions are known in the Cy6/Cy7 case when heteroatoms are in the chain.[299,a,b,303]

As for allyl sulfides (Section VIII.4.C), because of the possibility of competitive pathways in the cyclization of acetylenic thiols, and the small number of studies devoted to these cyclization processes, it is not possible at this time to suggest a rationale for the thio-Claisen rearrangement of propargyl sulfides.[257–259,261,264,304–308]

Very little is known of the intramolecular addition of other heteroatomic radicals to triple bonds, although very promising results have been reported by Märkl.[309] By AIBN initiation, phenylphosphine (X = P) and phenylarsine (X = As) add to hexa-1,5-diyne (**162**). The yields of cyclized products **163** are quite good (17% for X = As and 33%, for X = P), if one considers the multistep pathway probably involved (Scheme 73). It must also be emphasized that the (Cy7) compound **163** is exclusively obtained in this Cy6/Cy7 case and this is clearly reminiscent of the results obtained with ethylenic phosphines (Section VIII.5.A).

If the homolytic synthesis of unsaturated heterocyclic compounds starting from acetylenic compounds bearing a proradical group does not appear to be a general method at this time, electrophilic cyclizations of the same compounds are, unfortunately, likewise not always effective. For instance, the benzeneselenyl chloride cyclofunctionalization, which is a powerful synthetic method starting from alkenes bearing a proximate nucleophilic group,[691] fails completely with hydroxyalkynes.[692]

SCHEME 73

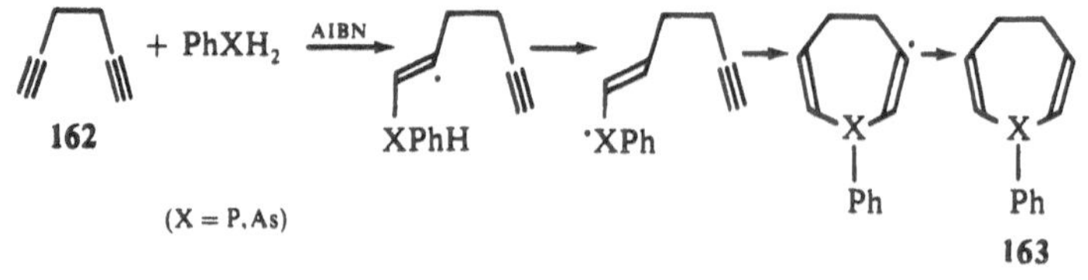

162

(X = P, As)

163

4. Intramolecular Addition to Allenes and Cyclopropanes

A very particular example of free radical intramolecular addition to an allene has been described by Gompper.[310a] It involves the free radical addition of thiophenol to **164** with opening of the vinylcyclopropyl radical (A') to A'' and cyclization of A'' (in a Cy5/Cy4 case) to the (Cy'4) radical, which gives finally **165** (Scheme 74). An analogous selectivity in the Cy5/Cy4 case has been proposed by Shellhamer[310b] but, here as well, the evidence is indirect.

Another example has been reported by Schmid[311a] during the thermolytic rearrangement of an arylpropargyl ether substituted in the *ortho* and *ortho'* positions by halogen atoms. One step of this rearrangement would involve intramolecular addition of an aryloxyl radical to the central atom of an allenic group to give the (Cy'5) radical in a Cy5/Cy6 case. The same selectivity has been observed by Gore[311b]: magnesium or lithium·reduction of 6-halo-1,2-hexadiene gives the (Cy5) product exclusively. But since the mechanism of the analogous reaction with the acetylenic isomers is not completely known[296] (Section IX.2) it is clear that more direct evidence of the possibility of intramolecular addition to the allenic group is needed before discussing these examples.

Doyle[312] thought that addition of a free radical to the cyclopropane ring, generally not a favored process in intermolecular reactions, could be favored in intramolecular reactions. He chose to study the thermolysis of *t*-butyl peresters **166** ($n = 1$, 2, 3) substituted by a phenyl group in order to favor the desired process by producing a stabilized benzylic (Cy') radical. But the yields of **167** resulting from this reaction were no better than 1.5%, the main product being **168**, which results from addition to the benzene ring (Scheme 75).

SCHEME 74

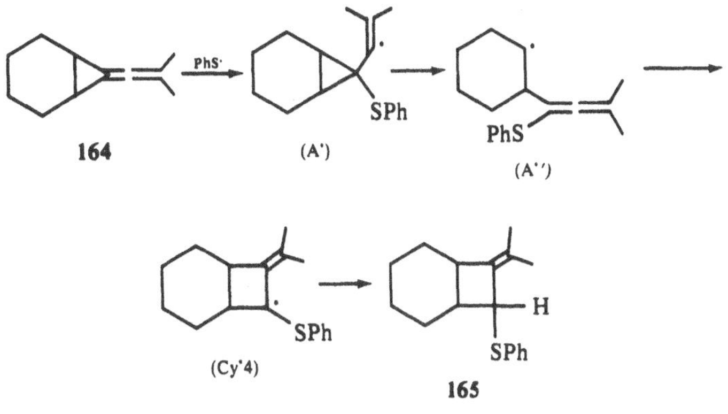

| 164 | (A') | (A'') |

| (Cy'4) | 165 |

SCHEME 75

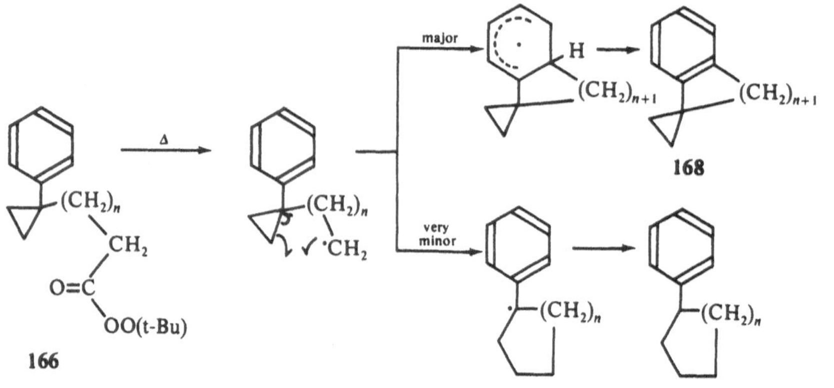

X. INTRAMOLECULAR ADDITION TO POLAR MULTIPLE BONDS SUCH AS CARBONYL OR CYANO

Free radical addition to polar bonds has not been extensively studied probably because of the interesting reactivity of the polar bonds toward nucleophilic reagents, so that study of free radical addition may appear of minor interest. Furthermore, the more familiar process in free radical addition reaction to polar bonds is the well-known β-scission process, i.e., the reverse reaction.

1. Intramolecular Addition to the Carbonyl Bond

A. The β-Scission Process

The opening of cyclic alcohols or their derivatives (**169**) under conditions generating an alkoxyl radical **170** is the well-known β-scission process and is useful in generating carbon-centered radicals **171** (Scheme 76).

This process is very fast with $n = 1$, and has been observed in the photolysis of nitrite esters[313] $X = NO$ and the oxidation of cyclopropanols,[314] with a regiochemistry of ring opening which has been compared in terms of frontier molecular orbital theory to the one observed with cyclopropylcarbinyl radicals[685] (Section V.2.B). It is also a fast process with $n = 2$ as in the oxidation of cyclobutanols,[315] although photolysis of a steroidal cyclobutyl nitrite gives the cyclobutanol in substantial amounts (unexpected), along with products resulting from the β-scission of the cyclobutanoxyl radical to tertiary and also

SCHEME 76

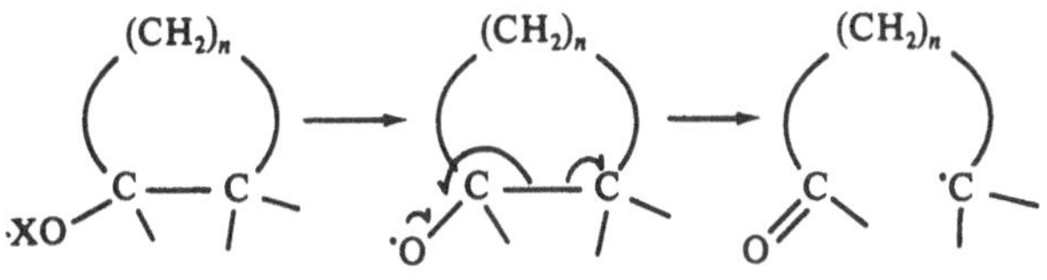

| 169 | (Cy') 170 | (A') 171 |

primary radicals.[650] This could be the result of the same stereoelectronic control proposed for the β-scission of cyclopropylmethyl radicals[632] (Section V.2.B). When $n = 3$, photolysis of nitrite esters of cyclopentanols again leads to β-scission.[316–319] With $n = 4$, the opening process has been observed in the lead tetraacetate oxidation of cyclohexanols,[320] in the ferrous salt reduction of hydroperoxides ($X = OH$),[321] and on pyrolysis of 1-nitroadamantane via the adamantyloxyl radical.[322] This process is also observed in the oxidation of cholestan-5-ols, providing an interesting way of obtaining ten-membered ring ketones.[323] With $n = 5$, the opening process has been proposed to explain the products resulting from irradiation of 2-nitrocycloheptanone.[324]

This very brief and incomplete summary of the literature shows that the opening process is very effective and may be observed under a variety of conditions. We now comment on the few cases where free radical intramolecular addition has been proposed.

B. The Intramolecular Addition

a. On the Carbon Atom of the Carbonyl Bond

The possibility of intramolecular free radical addition followed by a β-scission process has been proposed to explain some rearrangements. For instance, Reusch[325] proposed that 1,2-acyl shift from **172** to **173** proceeded by a cyclic transition state, corresponding to a Cy3/Cy4 case (Scheme 77).

SCHEME 77

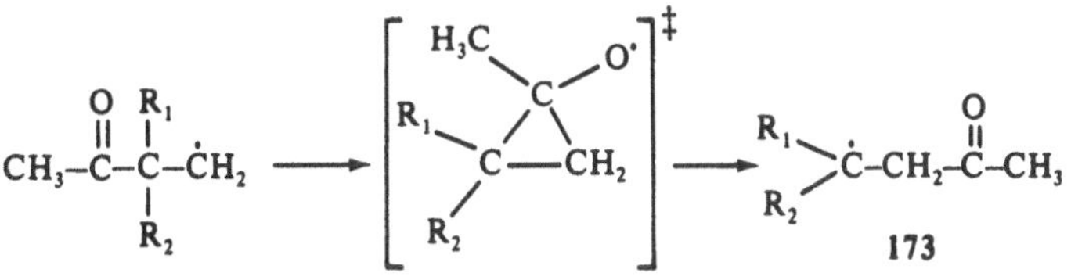

SCHEME 78

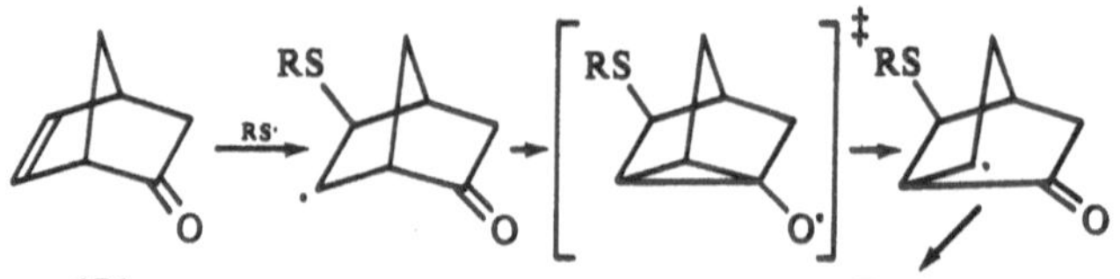

174   Products

Davies[326] suggested an analogous possibility to explain the products resulting from thiyl radical addition to 2-norbornen-5-one (174) (Scheme 78).

In the Cy5/Cy6 case a reversible addition process has been suggested by Suginome and Nickon to explain epimerization of hydroxyl group when photolyzing nitrite esters of steroïdal[327,328] or terpenic[329,330] compounds such as 175 which gives 176 (Scheme 79).[330]

An analogous scheme has been proposed to explain the formation of the inverted alcohol (in some cases in 100% yield) observed during photolysis of nitrate esters of carbohydrates.[651]

In the Cy6/Cy7 case an interesting rearrangement has been proposed by Mihailović.[331] It would involve a β-scission process from the alkoxyl radical 177 (generated by lead tetraacetate oxidation of the corresponding alcohol), followed by 1,5-hydrogen abstraction to 178, intramolecular addition to the carbonyl group to give 179, which gives 180, the expected product of lead tetraacetate oxidation (Scheme 80).

Another possibility is the observation of β-scission of a cyclic ether

SCHEME 79

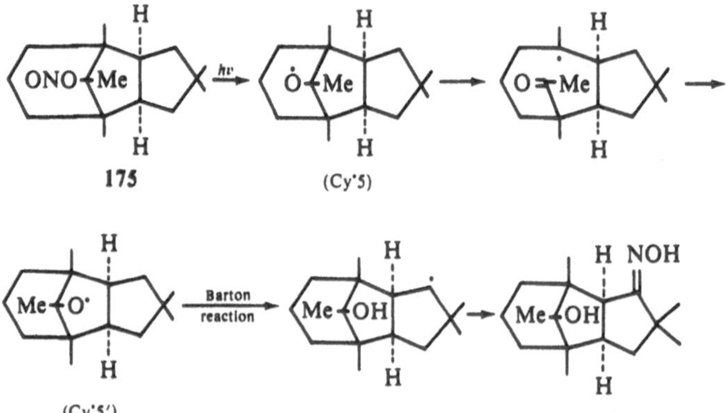

SCHEME 80

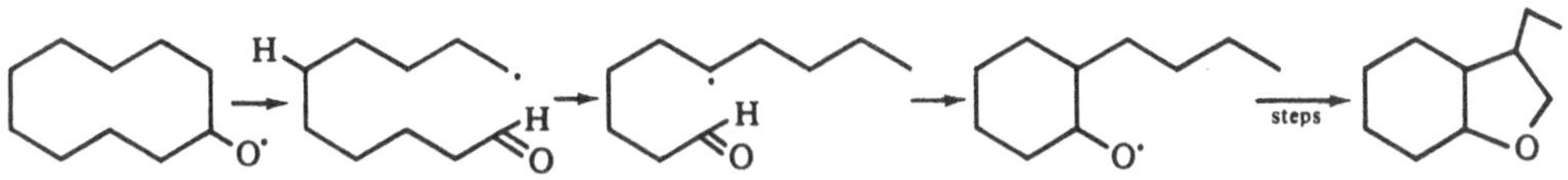

177   178   179   180

followed by intramolecular addition to the carbonyl group. This has been suggested by Huang and Lee[332] to explain the formation of 9,10-dihydro-9-phenanthrol on submitting diphenan (**180a**) to *t*-butoxy radicals, and the

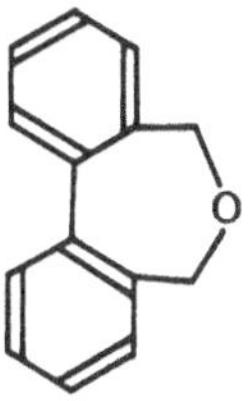

180a

process was confirmed by Flies, Lalande, and Maillard[333] using oxepan (**181**), which, when submitted to DTBP at 160°C, gave the cyclohexanol **184** (Scheme 81).

In all the cases reported, the evidence for free radical intramolecular addition on the carbonyl group was rather indirect. Nevertheless, this possibility has been confirmed in some cases by generating the radical, which then undergoes cyclization, in an unambiguous manner. Thus the mechanism shown in Scheme 81 was confirmed[333] by generating the radical **183** via stannane reduction of the corresponding ω-chloroaldehyde; the cyclohexanol **184** was obtained with no product resulting from the (Cy'7) radical **182**. Forrester, Skilling, and Thomson,[334] observed that thermolytic or photolytic persulfate oxidation of *o*-

SCHEME 81

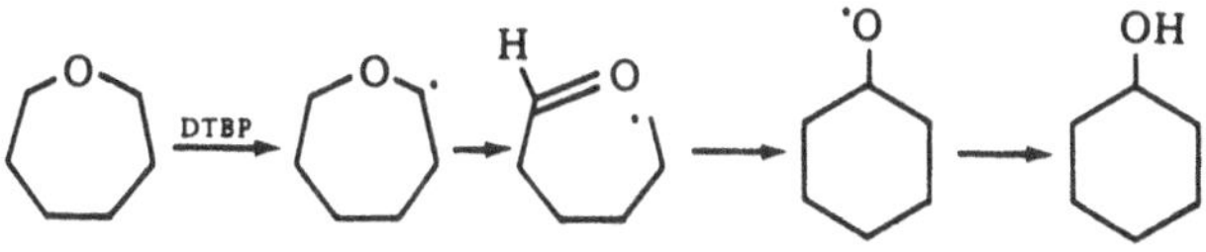

181   (Cy'7) **182**   (A') **183**   (Cy'6)   **184**

SCHEME 82

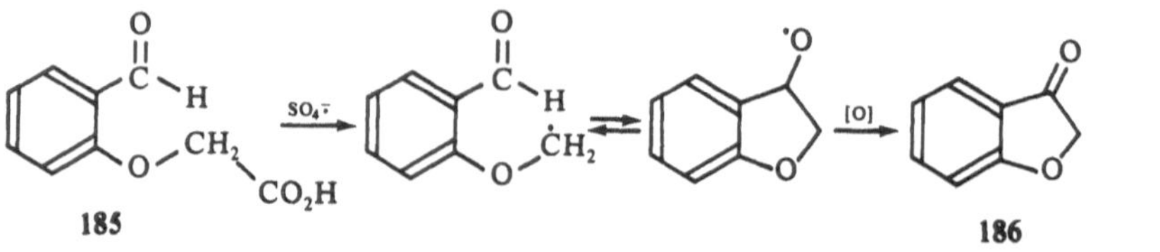

formylphenoxyacetic acid (**185**) gives benzofuran-3-one (**186**) in 55% yield (Scheme 82).

The possibility of free radical intramolecular addition to a carbonyl group has been reported in more complicated cases. For instance, hydrogenation with palladium on charcoal of the corresponding unsaturated ketone would give the adamantyloxy radical.[335] Triphenyltin hydride reduction products of 3β,12α-diacetoxy-$\Delta^{14,16}$-5α-pregnadien-20-one could result from the intramolecular addition of an alkoxyl radical to the keto group[336] in a Cy5/Cy6 case. Among other possibilities, intramolecular addition of a thiyl radical to a carbonyl bond in a Cy3/Cy4 case has been suggested to explain the formation of 2-phenylbenzo[*b*]thiophene by irradiating benzyldesyl sulfide in methanol.[337] Götschi and Eschenmoser[338a] observed easy photochemical cycloisomerization (in 89% yield) of a *seco*-ocorrinoid containing a carbonyl group (instead of the better known exomethylene group) to 1-hydroxycorrin and could not decide by which mechanism the cyclic product was formed. Another photochemical cyclization observed with diketones $RCH_2C_6H_4CO–COR$ may also be related to these reactions,[338b–d] as could the stereospecific photolytic cyclization of 3-oxobutylglycopyranosides.[652] In these last cases, 1,6-hydrogen abstraction by the excited carbonyl (rather than the better known 1,5-process, which is not feasible in these cases), followed by intramolecular radical coupling to the (Cy5) compound, is another mechanistic possibility.

b. On the Oxygen Atom of the Carbonyl Bond

Suginome proposed the intramolecular addition of an allylic radical to the oxygen atom of the carbonyl group to explain the products formed during hypoiodite oxidation of 3-hydroxy-Δ^5-steroids such as cholesterol[339] or *N*-acetyljervine.[340,341] Although the structures of the products had to be corrected later,[342] they are best accommodated by Scheme 83. The alkoxyl radical **188** generated from the alcohol **187** undergoes easy β-scission reaction to give the allylic radical **189**. This radical gives allylic iodides but would also give the radical **190** by intramolecular addition, i.e., a (Cy·7) radical in a Cy6/Cy7 case, which seems rather unexpected, especially if the behavior of the radical **183**

SCHEME 83

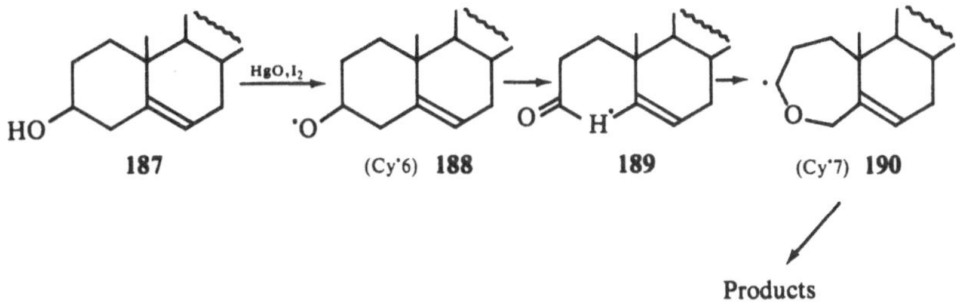

(Scheme 81), which gave only (Cy6) compounds resulting from addition to the carbon atom, is considered.

A less ambiguous example of addition to the oxygen atom, reported by Kuivila,[343a] is the stannane reduction, catalyzed by AIBN, of γ-chlorobutyrophenone (191), which gave a mixture of AH and Cy5H (ratio 20:80) in 65% yield (Scheme 84). This unexpectedly easy cyclization to the

SCHEME 84

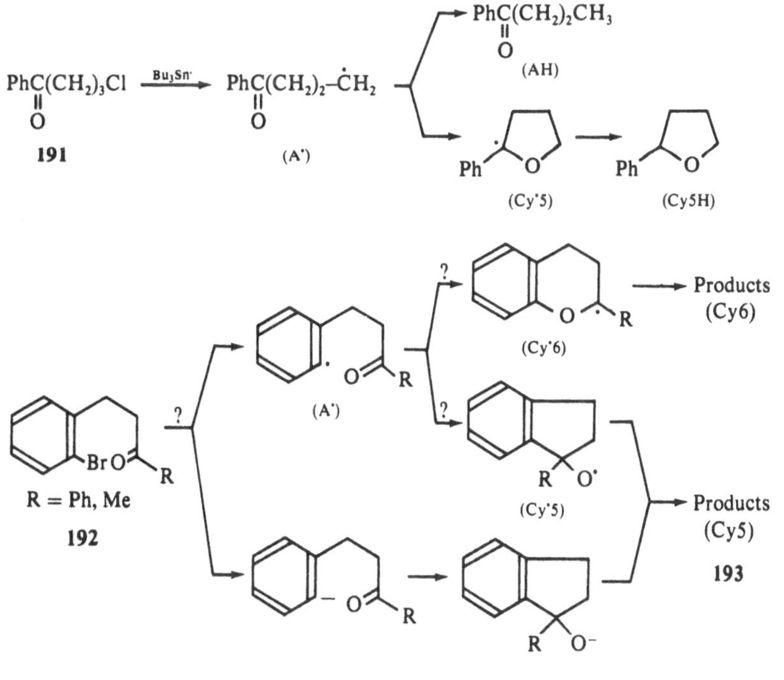

SCHEME 85

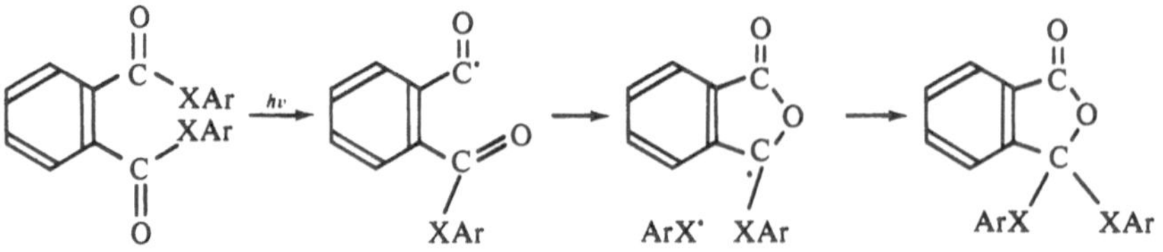

194   (A˙)   (Cy˙5) 195   196

(Cy˙5) radical in a Cy4/Cy5 case led to the suggestion[653] that this could be a general reaction and that stabilization by the phenyl group of Cy˙5 is unnecessary. Unfortunately, evidence in support of this proposal is quite indirect.[653] A negative example reported by Kuivila[343b] illustrates what is, in our opinion, the danger of generalizing such a result. In this paper, an $S_{RN}1$ mechanism between Me_3SnNa and ArBr was ruled out on the basis that 192 (Scheme 84) would give products resulting from Cy˙6, when only (Cy5) products (193) are observed, so that the mechanism would be carbanionic. We think that it is impossible to so conclude since the preceding Cy4/Cy5 case cannot be generalized without other knowledge of the behavior of A˙ in the Cy5/Cy6 case. In fact, according to the results discussed in Section X.1.B.a, cyclization of A˙ to Cy˙5 (followed by formation of 193) by addition on the carbon atom is more likely than cyclization to Cy˙6. In fact addition to the oxygen atom giving the (Cy˙6) radical, rather than addition to the carbon atom to give the (Cy˙5) radical, has been proposed, among other possibilities, to rationalize the products obtained on thermolysis of 5-(tert-butylperoxyl)-3,4,5-triphenyl-2(5H)-furanone. In this case, however, the resulting (Cy˙6) radical is stabilized not only by a phenyl group but also by another oxygen atom [641]

Other examples of addition to the oxygen atom in the Cy5/Cy4 case have been observed with acyl radicals, generated by photolysis of o-phthalaldehyde,[344] or of aryl phthalates[345] (X = O) and thiophthalates[346] (X = S) (194). Isomerized products (196) (X = O) or coupling products (X = S) from 195 have been rationalized according to Scheme 85. As in the preceding

SCHEME 86

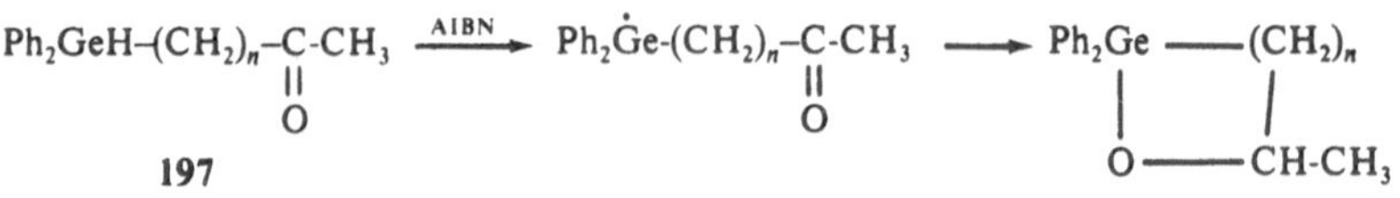

197   198

SCHEME 87

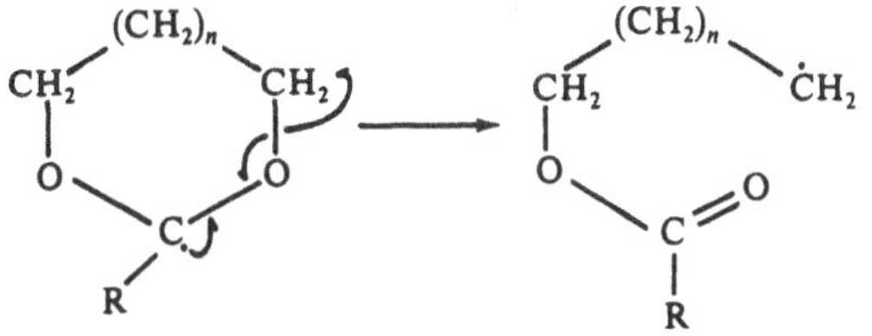

199   200

case it must be noted that the (Cy') radical is benzylic. Intramolecular addition of germyl radicals to the oxygen atom of the carbonyl group seems more general. Satge[273,347] observed this during the addition of phenylgermanes to ethylenic ketones, or by submitting germyl ketones (197) to AIBN initiation (Scheme 86). Yields of cyclized products (198) are 80% both in the Cy5/Cy4 case ($n = 2$) and in the Cy6/Cy7 case ($n = 4$).

C. Conclusion

Although the reverse process is certainly very favored, free radical intramolecular addition to the carbonyl bond has been reported in some instances. The selectivity toward the carbon or the oxygen atom is very high but the reasons for this are not clear and more exhaustive studies are required.

2. Intramolecular Addition to Carboxylic Acids and Esters

As with intramolecular addition to carbonyl bonds, the reverse reaction is the favored one (Scheme 87),[348–350,428] so that when one tries to add radicals 199 ($n = 0,1$) generated from cyclic acetals of aldehydes to ethylenic compounds, the main products are derived from the opened radical 200. Similar results are observed with radicals such as 201 which give products resulting from the open radicals 202 (Scheme 88).[351]

SCHEME 88

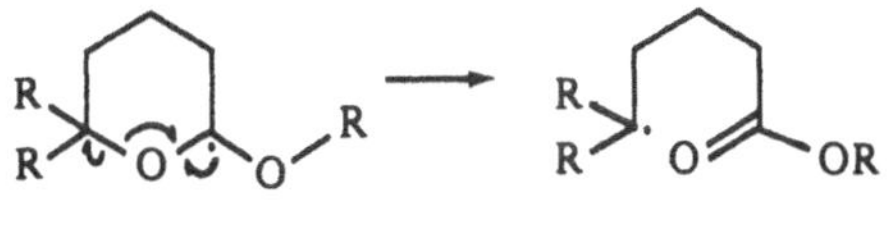

201   202

SCHEME 89

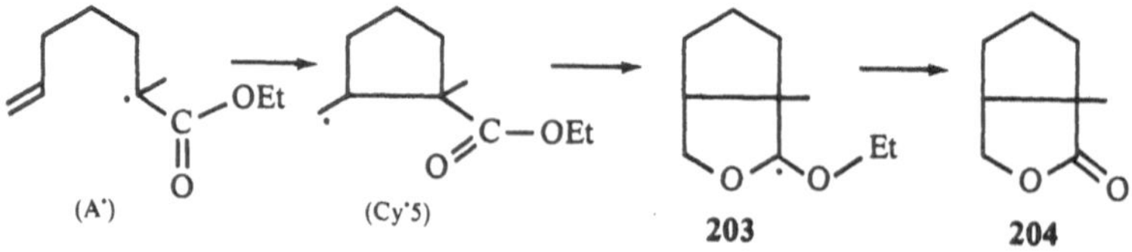

(A')   (Cy'5)   **203**   **204**

Nevertheless, the possibility of radical intramolecular addition to car-bethoxy groups has been retained in some instances, but the cyclized radical such as **203** (Scheme 89) is unstable: it may undergo a β-scission process to the lactone **204**.[352,353] This interesting β-scission in preference to reversion to the initial (Cy'5) radical has been discussed in terms of stereoelectronic control.[353]

Often the cyclized radical opens in the other direction to give rearranged products. This possibility has been used to explain the results of the pyrolysis of cyclopropyl acetates,[354] of the lead tetraacetate oxidation of α-acetoxy alcohols,[355] and of the stannane reduction of α-acetoxy chlorides.[356] In all these cases free radical addition followed by β-scission has been proposed, e.g., as in the 1,2-rearrangement of β-acyloxyl radicals **205** to **207** (Scheme 90) discovered independently by Teissier[357] and by Tanner,[358] and whose mechanism,[359-362,654] while not completely elucidated, is presumed to proceed through **206**, which must be considered as a transition state rather than an intermediate.

Since aryl radicals are less stable than alkyl radicals it was thought that 1,2 migration of an acetoxyl group in an *o*-acetoxyphenyl radical would be an easier process. Careful CIDNP studies and product analysis, however, have revealed no evidence for this migration, which has been ascribed to the inability of the sp^2 orbital of the aryl radical to become coplanar with the π orbital of the carbonyl group.[655]

SCHEME 90

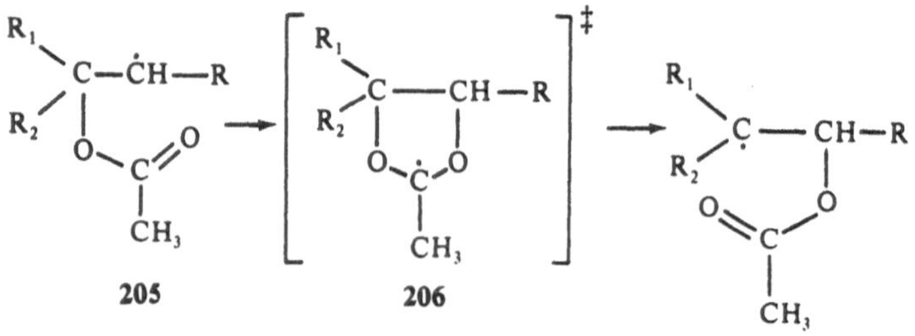

205   **206**   **207**

Of possible biological interest is the report, based on a model reaction study with hydroxocobalamine, that the coenzyme-B_{12}-dependent methyl-malonyl–CoA mutase reaction could involve intramolecular addition of a primary radical to a thioester group. The cyclopropyloxy radical so formed in a Cy3/Cy4 case would open to the carbethoxy-substituted and hence stabilized radical to give the rearranged product.[693]

Except for this last example which involves a Cy3/Cy4 case it is of interest to emphasize that the examples of intramolecular addition to carbethoxy groups so far reported deal exclusively with the formation of (Cy·5) radicals in Cy5/Cy4 cases. This cyclization was already observed with the carbonyl group (Section X.1), but other cyclization modes such as Cy6/Cy5 were also observed. What is particularly striking with the carbethoxy group is that a search for Cy5/Cy6 cyclizations, for example, with the next higher homolog of **205**, failed.[359] This is the opposite of the behavior of olefinic bonds and is in disagreement with Baldwin's rules.[16,683]

3. Intramolecular Addition to the Cyano Group

Electron spin resonance study of the 2-pyridyl radical indicates that it opens readily.[363] Nevertheless, the possibility of reversible addition to the cyano group has been nicely used by Kalvoda[5,364,365] in the oxidative cyanohydrin–cyanoketone rearrangement, an example of a "billiard reaction"; the overall result of this rearrangement is the 1,4 shift of a nitrile group (Scheme 91). In the presence of lead tetraacetate and iodine, the cyanohydrin **208** gives the alkoxyl radical **209**, which undergoes the Barton reaction to the carbon radical **210**. This radical adds intramolecularly to the cyano group to give the radical **211**, which undergoes a β-scission process toward **212** and then to the cyanoketone **213**. This reaction has been used by Kalvoda in the steroid field mainly in order to functionalize the 18-CH$_3$ group, the yields being 30–60%.

SCHEME 91

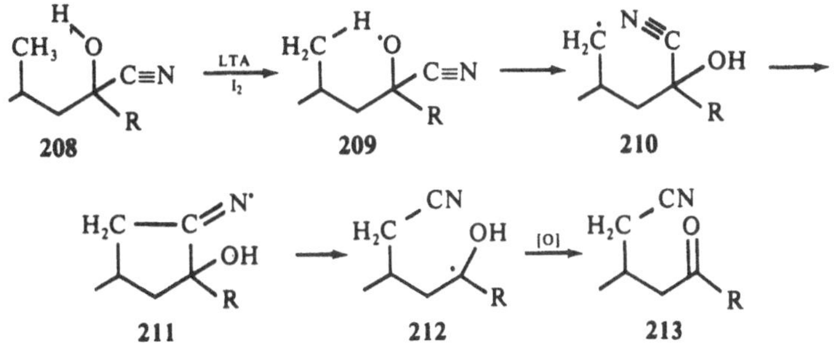

SCHEME 92

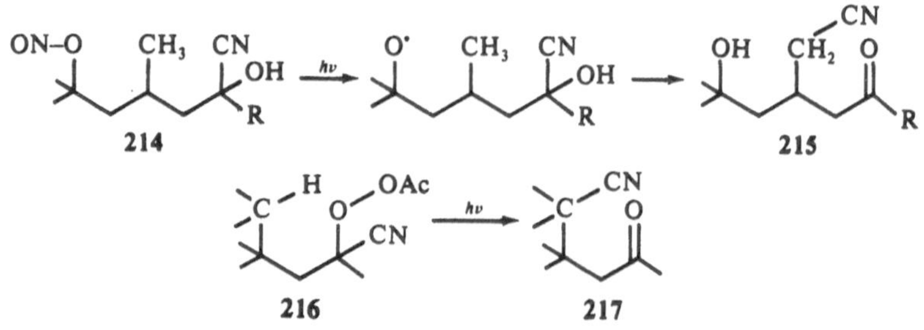

Interesting variants of this reaction have been proposed by Kalvoda, who photolyzed nitrites[366] **214** to **215**, and by Watt, who photolyzed α-peracetoxy-nitriles[367] **216** to **217** (Scheme 92) and α-azidonitriles.[368]

More recently, Nikishin and Ogibin,[369] working on aqueous solutions, have observed cyclized products resulting from intramolecular addition to the cyano group followed by hydrolysis of the imines **219** to cycloalkanones **220** (Scheme 93). The radicals **218** were obtained by thermal decomposition of the corresponding peroxides and by oxidation of the corresponding acids with $Na_2S_2O_8$–$AgNO_3$. The yields of cyclized products were 0% in the Cy3/Cy4 ($n = 1$) and Cy4/Cy5 cases ($n = 2$), 70% in the Cy5/Cy6 case ($n = 3$), 19% in the Cy6/Cy7 case ($n = 4$), and 3% in the Cy7/Cy8 case ($n = 5$), i.e., in accord with the general trends in free radical intramolecular addition.

Easy opening of the (Cy') radical ($n = 2$) has been confirmed in photochemical reactions where formation of such radicals has been proposed.[370]

Although not considered a homolytic reaction, the cyclization of ω-iodonitriles[371] by magnesium in ether at room temperature gives cycloalkanones, the yield dependence on ring size being very similar to that just discussed. Other examples[372,373] of intramolecular additions to the cyano group involving a polycyclization process will be discussed in Section XI.4.C.

In conclusion, free radical intramolecular addition to the cyano group, although not much studied, is a facile process. In all the cases reported, addi-

SCHEME 93

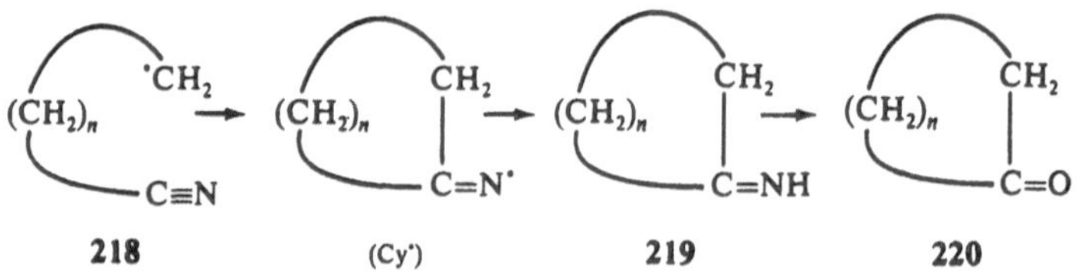

tion is exclusively to the carbon atom of the cyano group. The ease of cyclization as a function of chain length is similar to that of other known intramolecular reactions, i.e., easiest in the Cy5/Cy6 case.

4. Intramolecular Addition to Other Polar Bonds

Very few studies have been devoted to these possibilities.

A. The C=S Bond

Levesque[374] has observed that dithioacids (221) add to alkenes under AIBN initiation to give 1,3-dithiolanes (223) in high yield when the alkene is of the styrene type. This must involve intramolecular addition of the radical 222 to the sulfur atom (Scheme 94). Analogous results[375] are obtained in the free radical addition of thiols to compounds 224 [X = alkyl, OMe, NMe$_2$, or SMe; Z = CH$_2$, (CH$_2$)$_2$, O, or S] where the radical 225 adds intramolecularly to the sulfur atom to give 226.

An analogous cyclization process is observed in the free radical addition of dithioacids to 1,2- and 1,3-dienes.[656]

A less clear-cut example has been reported by Ando[376] to explain the formation of benzothiophenes (229) during the pyrolysis of styryl sulfoxides (40–50% yield) or of styryl disulfides (227) (85% yield) (Scheme 95). The radical 228 may be considered as a thiyl radical and this would be the first example of intramolecular substitution by a thiyl radical into a benzene ring, or alternatively as a carbon radical which would undergo intramolecular addition to the thiocarbonyl group.

It must be noted that most of the examples so far reported deal with the

SCHEME 94

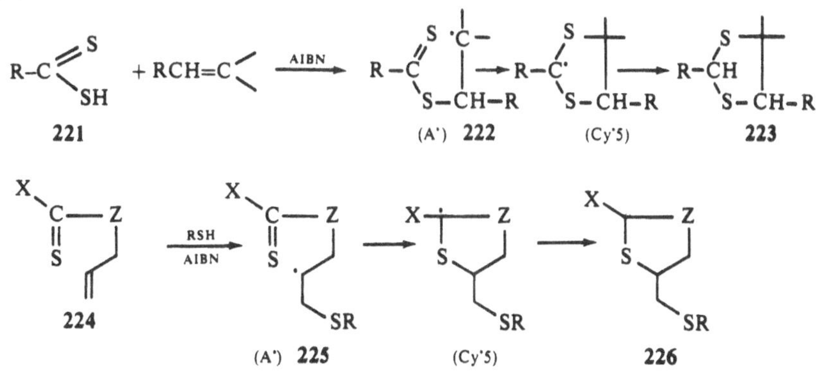

SCHEME 95

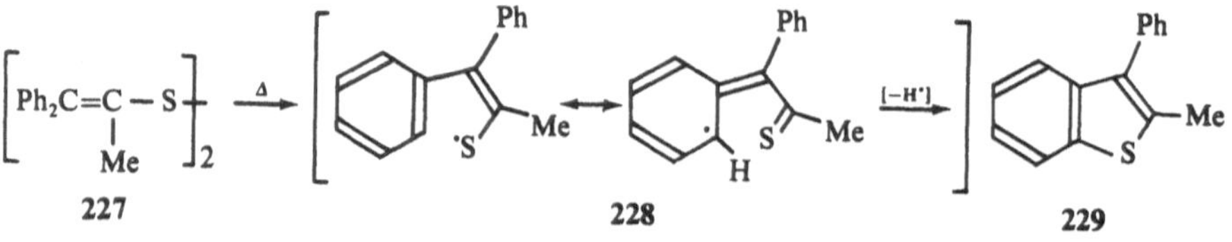

easy formation of (Cy˙5) radicals in the Cy5/Cy4 case. Possibly, in this case (as in other cases discussed in Section VIII.4.C, Scheme 60) the sulfur atom reinforces the trend to disagree with Baldwin's rules already observed and discussed in the cases of the carbonyl and carbethoxy groups (Sections X.1 and X.2).

B. The C=N Bond

From an esr study of 1-aziridylcarbinyl radical **230** at low temperature (Scheme 96), it was concluded[377a] that ring opening to **231** is a very fast process (which is not unexpected in this Cy3/Cy4 case) at temperatures above −130°C when R = H, and under all conditions when R = CH$_3$; in this latter case only the open radical **231** was observed and there was no trace of the azetidinyl (Cy˙4) radical. An analogous result was observed when the cyclopropylamino radical **232** was generated: only the opened radical **233** was observed with no evidence for the 2-azetidyl (Cy˙4) radical. A reaction involving opening of a 2-azetidyl radical has recently been advocated to explain the nature of the products[377b] but the evidence is so indirect that this interpretation must be questioned on the basis of the stability of cyclobutyl radicals (see Section V.2.B).

Nevertheless, intramolecular addition to a C=N bond has been used to explain the formation of quinolines during the pyrolysis of alkylindoles.[378] A clear example of intramolecular addition to the C=N bond of adenosines has

SCHEME 96

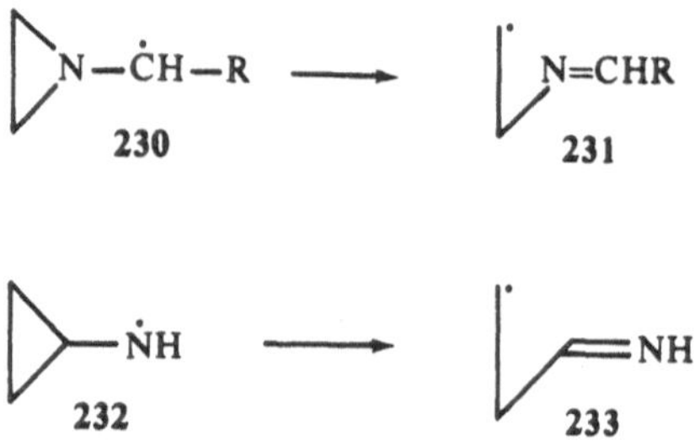

SCHEME 97

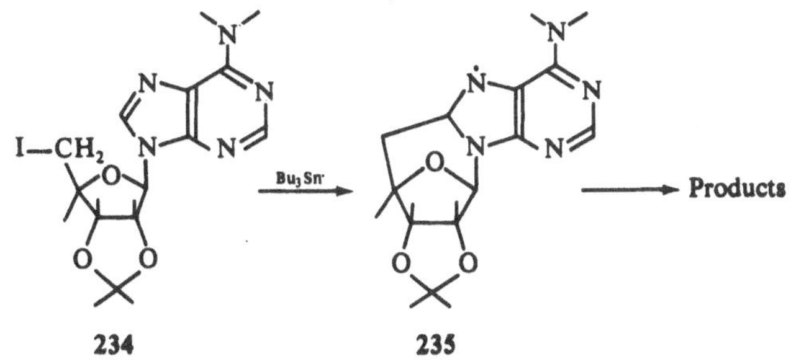

234   **235**

been reported by Gaudemer[379a] in order to explain the photolytic behavior of coenzyme B_{12}. Stannane reduction of the adenosine (**234**) leads to products resulting from the cyclized radical (**235**) in 68% yield (Scheme 97). An analogous conclusion was reached by Johnson[379b] in studies of the mechanism of action of vitamin B_{12}. Since C=O and C=N groups are expected to behave similarly, intramolecular addition on the nitrogen atom may be expected. This possibility has been proposed only recently by Tanner[380a] and, although the evidence for this process is quite indirect, it will be discussed here (Scheme 98). A mixture of methyl cyanoformate (X = MeOCO) or cyanogen (X = CN) and benzoyl peroxide heated at 90°C in 2,4-dimethylpentane (RH) leads to **236** as the major product (the yield is 22% when X = CN and 55% when X = MeOCO). The reaction would involve hydrogen abstraction from the hydrocarbon (RH) followed by intermolecular addition of R· to the C≡N group. The radical A· would isomerize in a Barton-like reaction to A·', which would give the (Cy·5) radical in a Cy4/Cy5 case, which by disproportionation and chain termination would give **236** (Scheme 98). At first sight this scheme could appear improbable, particularly if one takes into account the general dif-

SCHEME 98

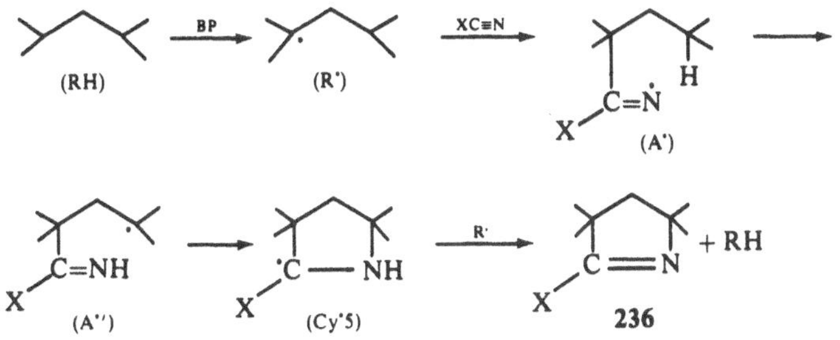

ficulty of cyclization in the Cy4/Cy5 case, reinforced in this case by an expected easy β-scission process Cy'5 → A'' instead of the reverse cyclization process assumed. Nevertheless, since X is an electron-attracting group (MeOCO, CN), the radical (Cy'5) would be highly stabilized in this case by the capto-dative effect,[380b] which would explain this easy cyclization process.

C. The –N=N– Bond

In a study of alkyl hydrazyl radicals, Ingold and coworkers[381a] observed that **237** gives only products resulting from the open radical (**238**) (Scheme 99). Nevertheless the intramolecular addition of aryl radicals to the azo group has been observed by Tundo and coworkers[381b] (Scheme 99). The radical (**239**) has been generated by reduction of the corresponding diazonium tetrafluoroborate by iodide ion in acetone. At –20°C, the hydrazine resulting from the dimerization of the radical (Cy'5) was isolated in 80% yield. At room temperature only products resulting from the open radical **239** were observed, which indicates that the opening process Cy'5 → **239** is an easy one. The exclusive formation of the (Cy'5) radical in this Cy5/Cy6 case has been confirmed by spin trapping.

D. The Nitroso and Nitrone Groups

Since these groups have been used extensively in spin-trapping experiments one would expect to observe readily nitroxides resulting from free radical intramolecular addition. Such an example has been reported by McConnell[382] in the case of caryophyllene nitrosite. According to Aurich,[694] oxidation of hydroxybenzotriazoles in aromatic solvents would involve opening of the ben-

---

SCHEME 99

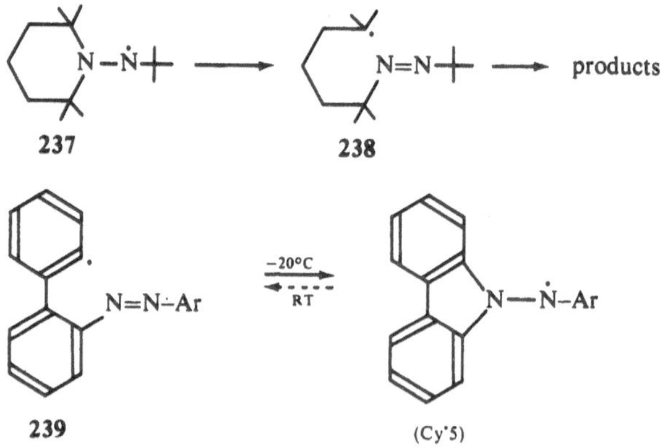

SCHEME 100

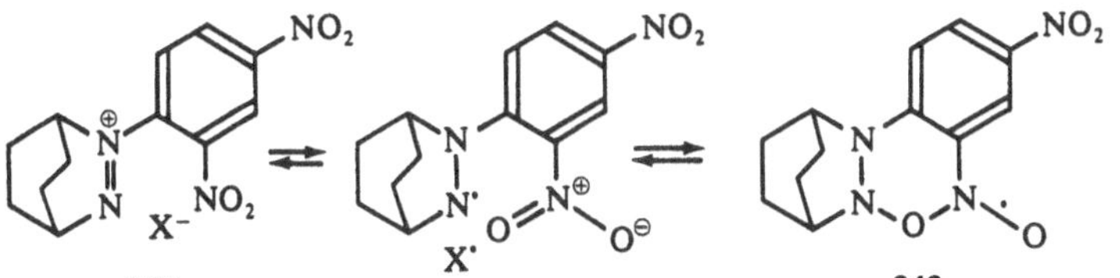

240   241   242

zotriazolyl oxide radical in a first step and intramolecular addition to the nitroso group of a cyclohexadienyl radical in a subsequent step corresponding to a Cy5/Cy6 case. As usual, the Cy4/Cy5 cyclization would be unfavorable since there is some evidence, although indirect, of the reverse opening process to the corresponding nitroso alkyl radical of the isoxazolidine radical centered on the nitrogen atom.[695]

Intramolecular addition to a nitrone, studied by Forrester,[334] has been observed only as a minor pathway because of competitive reactions.

E. The Nitro Group

An interesting example of internal spin trapping by a nitro group has been reported by Snyder[383] during a study of the diazenium cation **240**, hydrazyl radical **241**, and nitroxide radical **242** equilibrium (Scheme 100). Analogous spin trapping of alkyl radicals by the nitro group has been reported in the Cy5/Cy6[657] and Cy4/Cy5 cases.[658]

F. The Azido Group

Generation of the radical **243** (Scheme 101) by reduction of the corresponding diazonium fluoroborate by sodium iodide in acetone[384] gives

SCHEME 101

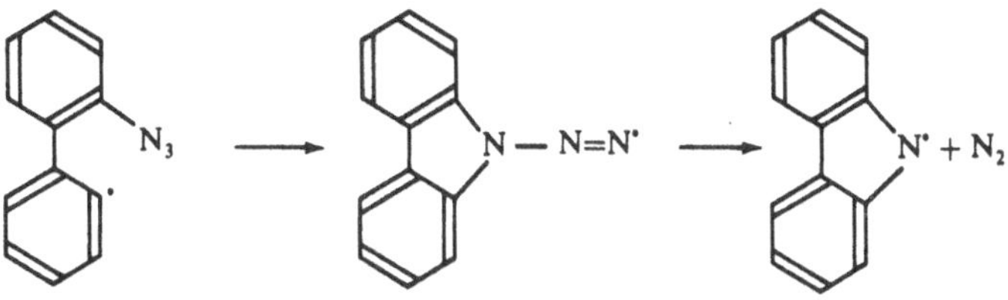

243   244   245

SCHEME 102

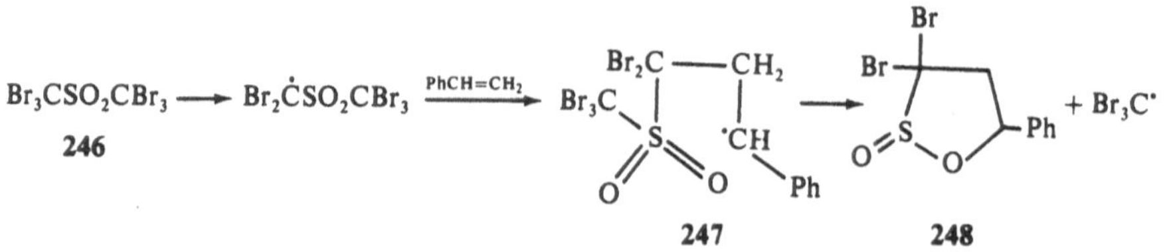

products (63% yield) resulting mainly from the coupling of the radical **245**; this latter is presumably formed by intramolecular addition to the azido group giving the triazenyl radical **244** followed by nitrogen loss.

G. The Sulfone Group

Carmack[385] has reported that thermolysis of the sulfone **246** (Scheme 102) in the presence of styrene leads to the radical **247**. This radical undergoes various transformations, one of which is intramolecular addition to the O=S bond with liberation of a tribromomethyl radical to give **248**.

5. Conclusion

The examples reported in this section provide clear evidence that free radical intramolecular addition to polar bonds occurs. These reactions may have genuine synthetic utility even when cyclization is followed by β-scission of the (Cy·) radical. As most of the examples have been reported only recently, little is known concerning the scope of these reactions or of the origin of the selectivity of addition to one or the other terminus of the polar bond. It may be expected that features such as the nucleophilic or electrophilic character of the cyclizing radical must be taken into consideration along with the factors which govern intramolecular addition to C=C double bonds.

XI. SYNTHESIS OF BI- AND POLYCYCLIC COMPOUNDS. STEREOCHEMICAL FEATURES OF FREE RADICAL INTRAMOLECULAR ADDITIONS

1. Introduction

This section will be devoted to the formation of bi- and polycyclic compounds by free radical intramolecular addition reactions which afford useful procedures for organic synthesis. It will be seen that the simple rules derived from monocyclization studies may be applied in many cases, but that exceptions are often encountered. Furthermore, mainly from these studies, interesting stereochemical features have been discovered. This interesting aspect of free radical intramolecular addition, not discussed until now, will be considered in this section.

2. Bicyclic Compounds

It could be expected that by using appropriate substrates free radical intramolecular addition could be used to synthesize bicyclic compounds and that the yields would be best in the Cy5/Cy6 case. This is why we shall first discuss examples involving the Cy5/Cy6 case.

A. The Cy5/Cy6 Case

a. Fused Rings

One of the first examples was described by Julia, Surzur, Katz, and Le Goffic[109] who used the compound **249a** (R = H) (Scheme 103); on peroxide initiation it gave the *trans*-decalin **250a** (R = H) in 40% yield. Under the same conditions compound **249a** (R = Me) does not afford a cyclized product, and compound **249b** gives **250b** in 37% yield as a *cis/trans* mixture in the ratio 82:18.[110] The isolation of a decalin or hydrindane in these cases accords with the behavior of stabilized radicals. The unstabilized radical **251** corresponding to **249a** (Scheme 104) has been extensively studied by Beckwith.[32,37,386] He obtained cyclized products derived from the *spiro* (Cy'5) radical **252** and from the (Cy'6) radical **253** in a ratio close to 1.2:1. The decalin derived from **253**

SCHEME 103

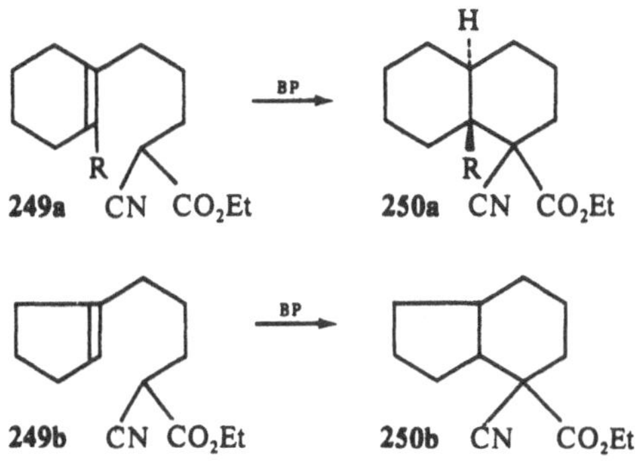

249a	CN CO₂Et	**250a**	CN CO₂Et
249b	CN CO₂Et	**250b**	CN CO₂Et

SCHEME 104

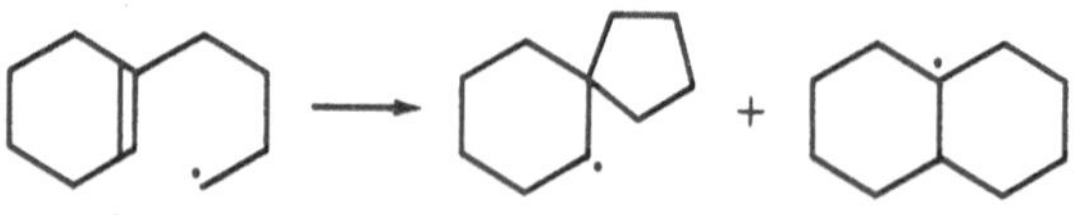

$$251 \qquad\qquad 252 \qquad + \qquad 253$$

was isolated in a *cis/trans* ratio of about 1:6. Finally, Beckwith was able to show that, under the experimental conditions used, the *spiro* radical **252** does not rearrange to **253**. It must be noted that the carboxamidyl radical corresponding to **251** cyclizes exclusively, although in low yields (13%), toward the (Cy˙5) radical analogous to **252**.[200]

Barton and coworkers[387] used free radical cyclization in the synthesis of tetracyclines (Scheme 105). Photolysis of **254** (X,Y = SR or OR) gives the corresponding radical, which cyclizes to the (Cy6) compound **255** in 80% yield when X,Y = SCH_2CH_2O. Quite remarkably, **255** is formed only in the *cis* form. Another completely stereoselective reaction toward the *cis* compound involving intramolecular addition to an acetylenic bond has been described by Pradhan[291] and was discussed in Section IX.2 (Scheme 70). An analogous reductive cyclization (K, NH_3) of ethynyl ketones has been used by Stork in the construction of a tricyclic intermediate for the synthesis of gibberellic acid.[659]

As in the tetracycline synthesis of Scheme 105 other examples have been reported in which the only cyclized products are the Cy6 ones in the Cy5/Cy6 case involving an unstabilized radical. One of these examples (Scheme 106) was reported by McLean,[388] who explained the by-products formed during the dehydrogenation of lupenyl acetate (**256**) with mercuric acetate in chloroform–acetic acid solution as deriving from the radical **257** by cyclization to a (Cy˙6) radical (**258**). It must be noted, however, that **258** is both a tertiary and allylic radical. It may also be noted that in all these cases, the situation is similar to the one discussed in Section IV (Scheme 12, Table 2, entries 1,2,7)

SCHEME 105

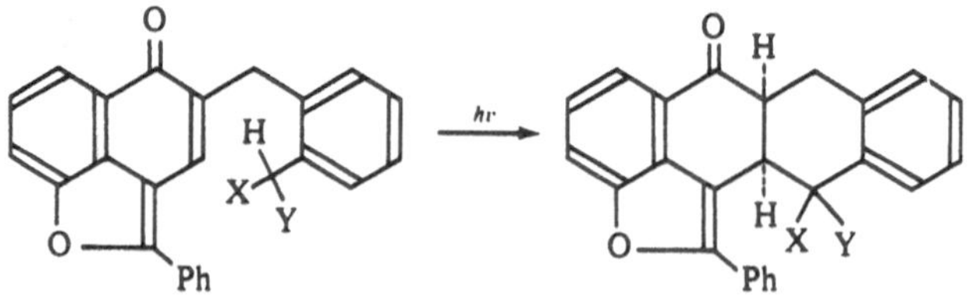

$$254 \qquad\qquad\qquad 255$$

SCHEME 106

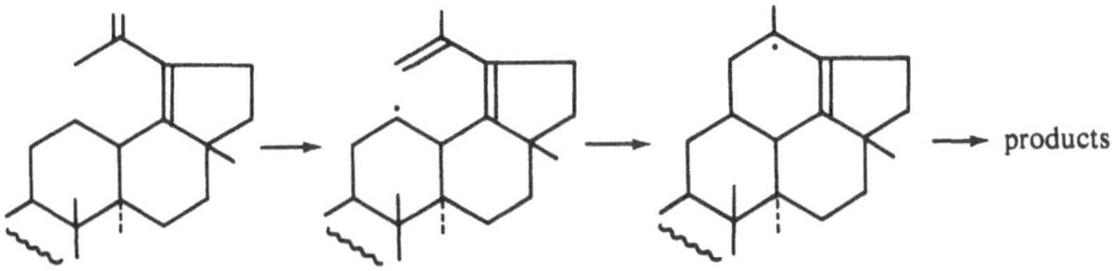

256   257   258

where an R_3 substituent favored the (Cy·6) radical formation. Another example in which only the bicyclic compounds corresponding to the (Cy·6) radical have been observed has been reported by Speckamp[389a] (Scheme 107). The radical **259**, formed by stannane reduction of the corresponding iodide, gives **260** as a mixture of epimers in 47% yield. Clearly, **260** results from the (Cy·6) radical. The fact that no cyclized products resulting from the (Cy·5) radical (which in addition is benzylic) are observed is unexpected. Nevertheless, the product (**261**) (18% yield) could result from the (Cy·5) radical. Although not studied in simpler cases, the high selectivity of **259** toward Cy·6 could result from a possible stabilizing effect of the SO_2 group in a vinylic position, as in the cases of other heteroatoms or groups (Section VI.2). Another remarkable example of very high selectivity in the cyclization, possibly due to the stabilizing effect of a vinylic nitrogen atom, was discovered by Snieckus during the photolysis of enaminoketones.[389b,c] This reaction has been exhaustively studied by Kibayashi for the synthesis in yields of up to 90% of carbazoles, phenanthridines, benzazepines, and benzazocines by a reaction considered to involve the selective intramolecular addition of an aryl radical to the enaminone group.[389d–g] One such example is given in Scheme 107: photolysis for 120 hr of **262** in dioxane–acetonitrile containing triethylamine affords the phenanthridone **263** (26% yield).[389g] The cyclization of A· in a Cy5/Cy6 case selectively to the (Cy6) product may be ascribed to the stabilization afforded by the nitrogen atom, as for analogous selective cyclizations in the Cy4/Cy5 and Cy6/Cy7 cases.[389d–g] In the silicon series the pyrolysis at 350°C of o-methylaryloxyvinyldimethylsilanes would give a benzylic radical, which would cyclize in a Cy6/Cy7 case exclusively to the (Cy·7) radical stabilized by the silicon atom.[637]

An interesting reaction has been reported by Stork.[390] Treatment of an unsaturated keto epoxide such as **264** with hydrazine in methanol gives the bicyclic compound **266** in 85% yield instead of the expected allylic alcohol from the Wharton reaction. A concerted mechanism and a free radical mechanism have been considered by Stork and, indeed, a free radical mechanism seems in accord with the other results described.[390] This mechanism

SCHEME 107

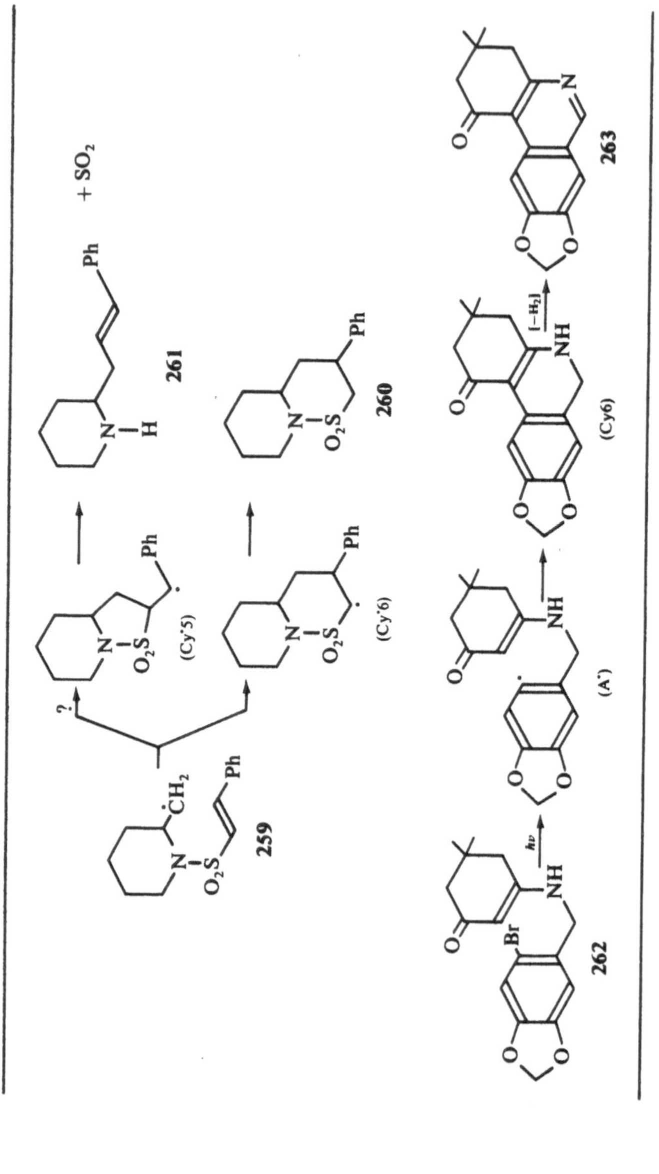

SCHEME 108

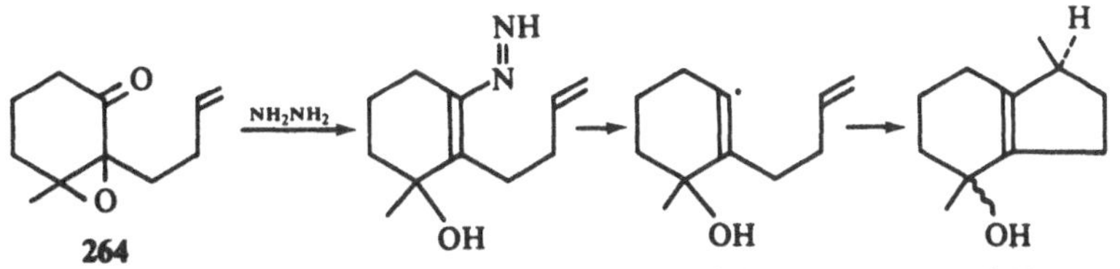

involves as the key step an intramolecular addition of the vinylic radical **265** (Scheme 108).

Similar results have been observed with heteroatomic radicals such as aminyl radical[182] (in an example already discussed in Section VIII.3.A, Scheme 44, and more recently as an important step in the total synthesis of the alkaloïd dendrobine[660]) and with the carboxamidyl radical.[200] The intramolecular addition of thiyl radicals to cyclohexenes generally gives a mixture of the (Cy5) and (Cy6) products.[214,412] In the same way, intramolecular addition to an unsaturated chain of a thiyl radical on a cyclohexane ring[233,235,412] also gives a mixture of (Cy5) and (Cy6) products. This lack of selectivity is in accord with the behavior of unsaturated thiyl radicals discussed in Section VIII.4. An interesting exception[247,248] of possible relevance to the biosynthetic route to cepham has been discussed in Section VIII.4.B (Scheme 58). The behavior of ethylenic aminothiols such as **107** under free radical initiation (Scheme 52, Section VIII.4.A) has been generalized to compounds such as **267**, which affords **268** in 87% yield[221] (Scheme 109).

In all the examples so far reported the initial ring in the substrate was a Cy5 or Cy6 ring. Other possibilities are, of course, available, such as a Cy3 ring. Such an example has been reported by Julia[391] (Scheme 110). Tributyl-stannane reduction of **269** gives the radical **270**, which cyclized exclusively toward the (Cy·5) radical **271**. The main cyclized product was the *endo* compound **272** (71% yield), and the minor one the *exo* compound **273** (14% yield). This cyclization process was not observed with compounds bearing methyl substituents on the cyclopropane ring or on the aliphatic chain.[391].

SCHEME 109

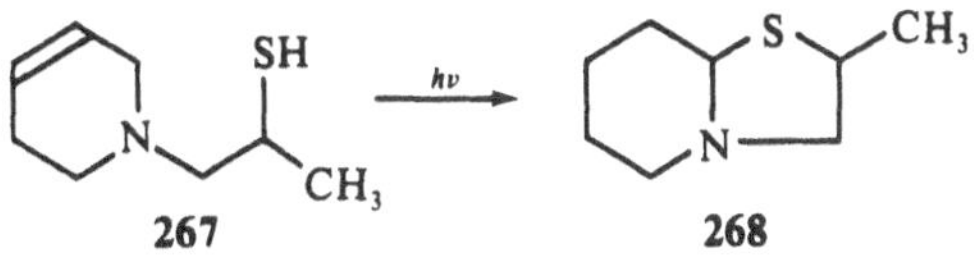

SCHEME 110

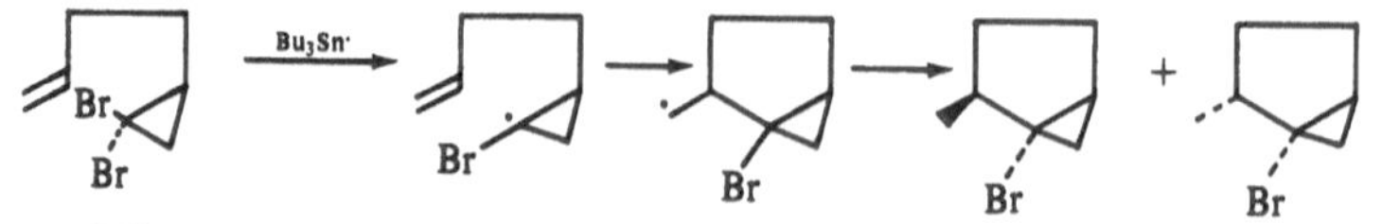

Another way of introducing a cyclopropane ring has been studied by Stein[392] (Scheme 111). The radical **274** (X = NMe), generated by titanous chloride reduction of the corresponding N-chloroamine, gives exclusively the compound **276** (X = NMe, Y = Cl) resulting from the (Cy·6) radical **275** in 55% yield.[661] The radical **274** (X = S), generated by photolysis, in pentane at −70°C, of the corresponding thiol, gives exclusively **276** (X = 3, Y = H) (65% yield). At room temperature a mixture of **276** (X = S, Y = H, 69%), **278** (X = S, 17%), and **279** (X = S, 12%) is obtained. These two last compounds clearly result from the (Cy·5) radical **277**, which, being a methylcyclopropyl radical, opens readily. With **274** (X = O) generated by photolysis of the nitrite, cyclized products have not been observed.[392]

Another way to obtain bicyclic compounds is to use free radical cycload-dition to 1,5-cyclodienes. As discussed in Section V.2.A (Scheme 15), in order to observe bicyclic compounds one must not use good transfer reagents. Thus, with cis,cis-1,5-cyclooctadiene as the substrate, the reaction is not only regioselective but also highly stereoselective toward the cis-bicyclo[3.3.0]-octane.[61-63,393] Furthermore, the X (and Y) substituent would be in the exo position,[61-63,393] as pictured in Scheme 15.

Free radical addition to cis,trans-1,5-cyclodecadiene (**280**) has been

SCHEME 111

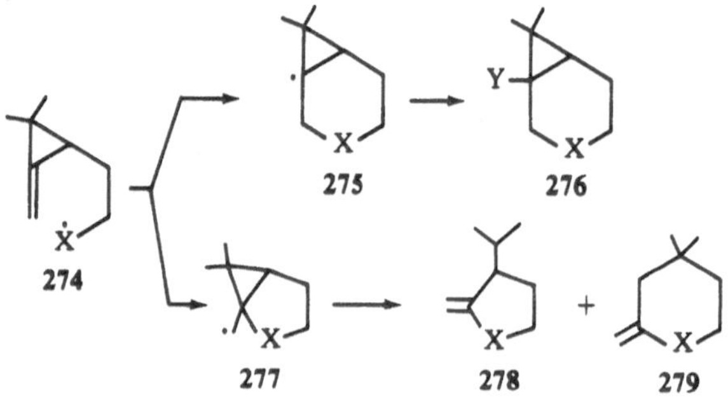

SCHEME 112

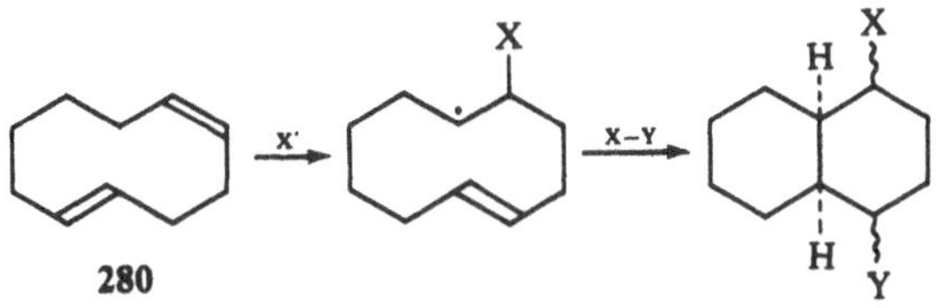

280   281

described by Traynham[394] (Scheme 112). *Cis* decalins (**281**) are obtained from
Y–X = Br–CHBr$_2$ (45% yield), Br–CCl$_3$ (61%), or Cl$_2$ (80%). It must be noted
that the same stereochemical result is obtained with cationic reagents. An
analogous result has been described by Sutherland,[395] who used the *trans,trans-*
1,5-cyclodecadiene germacrene (**282**) (Scheme 113). Free radical addition of
Y–X leads to the *trans* decalins (**283**) in 32% yield when Y–X = Cl–CCl$_3$,
34% with HSPh and 51% with Ph$_2$S$_2$. Here, too, the course of the reaction is
the same in the cation-induced cyclization.

Another interesting example of bicyclic system formation has been
observed (Scheme 114) in the attempted acyloin condensation of the diester
284.[396] The major product (50% yield) was **287** (*cis/trans*=1.0:2.3), whose formation was believed to involve transannular interaction in the radical anion **286**
formed from the dione **285**.

SCHEME 113

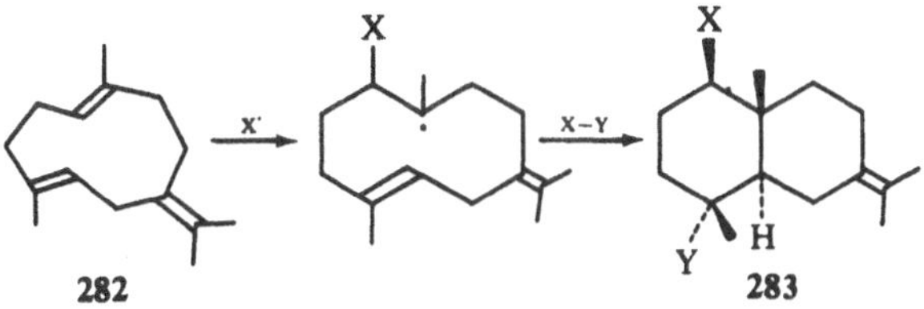

282   283

SCHEME 114

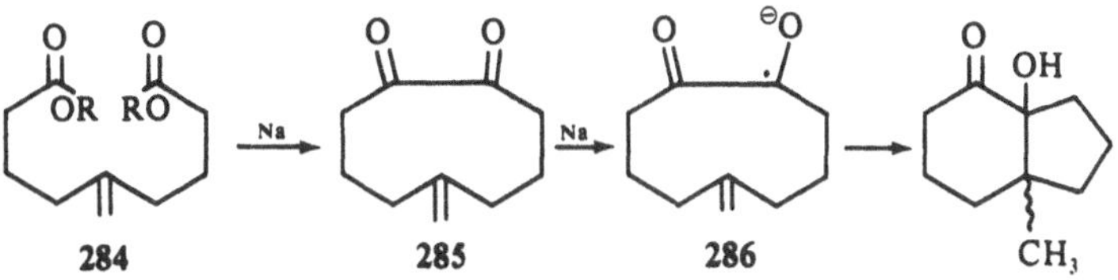

284   285   286   287

SCHEME 115

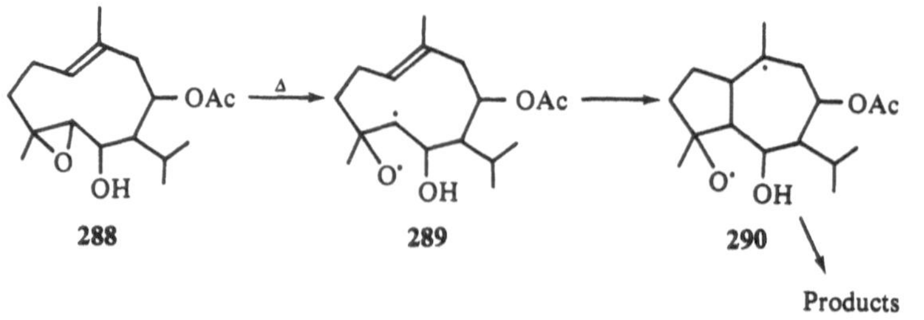

288   289   290

Products

The electroreductive cyclization of ethylenic ketones, which will be discussed fully in Section XI.3 (Scheme 138), also gives remarkable results when applied to the preparation of bicyclic compounds.[502b]

Finally, an intriguing thermal rearrangement (Scheme 115) observed during vacuum distillation of shiromodiol monoacetate (288) has been explained by postulating formation of the radical 289 followed by intramolecular addition to give the radical 290, which then undergoes cleavage or removal of hydrogen.[435] It must be noted that, in this case, the intramolecular addition would involve the formation of the (Cy˙5) radical rather than the (Cy˙6) radical as in many of the examples discussed above, and also that the cyclization is observed from a carbon radical rather than from an alkoxy radical.

In conclusion, free radical intramolecular addition in the Cy5/Cy6 case appears to be a method of preparative value for the synthesis of fused bicyclic compounds. The observed regioselectivity is often very high but it is clear that other factors than those considered for monocyclization processes are operative, for in many instances only the (Cy6) compounds are observed. A new and interesting feature appears, namely, the high stereoselectivity often observed in these cyclizations. The origin on this stereoselectivity has been discussed in the reference cited, and is often considered to result from the last step, i.e., the transfer step being favored from the less hindered side of the molecule. In other cases, such as the one discussed in Scheme 113 where four asymmetric centers are formed stereoselectively, a synchronous mechanism has been proposed. Because of the very different nature of the substrates considered in this section it does not seem worthwhile to discuss further the origin of this stereoselectivity.

SCHEME 116

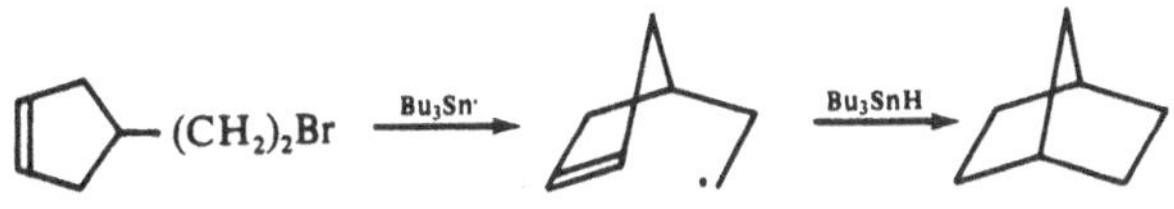

291   292

b. Bridged Rings

In the preceding section, most of the studies were devoted to carbon-centered radicals rather than to heteroatomic-centered radicals. The inverse is true for bridged ring formation where most of the known examples involve heteroatomic radicals. An example which involves a carbon-centered radical is due to Wilt,[397] (Scheme 116) who described the ring closure of 291 to 292 by tri-*n*-butyltin hydride reduction.

Although fragmentation of the 1-adamantyl radical has been recorded at high temperature,[322] Yurchenko[398] described an interesting way to obtain noradamantane derivatives (294) by the free radical addition of CCl_4 or CBr_4 to 293 (Scheme 117). This is particularly interesting when one recalls that electrophilic addition gives only adamantyl derivatives. In the free radical addition, the nor-derivatives (Cy5) are favored in the ratio of 294/295 = 3:1 with CCl_4 (total yield 64%) and 9:1 with CBr_4 (total yield 54%).

Transannular free radical addition to a polycyclic diolefin has been observed by Bird,[399a] who showed that free radical addition of *p*-chlorobenzenthiol to isodrin (296) (Scheme 118) leads to compound 297.

SCHEME 117

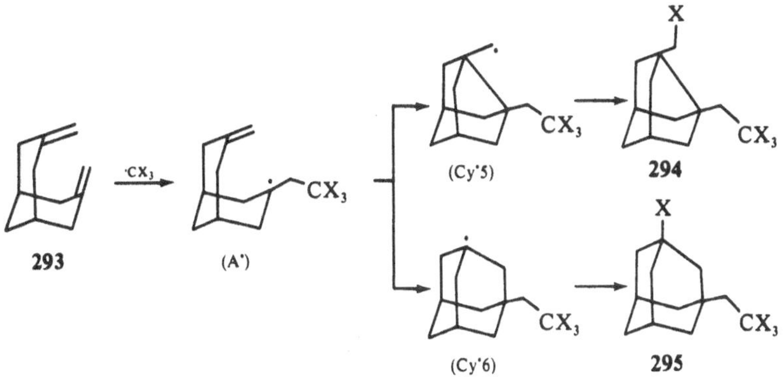

SCHEME 118

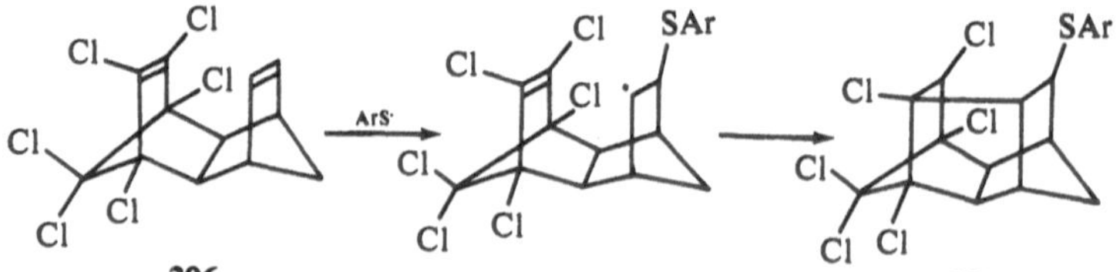

SCHEME 118

In an apparently related case (Scheme 119) Davies[400] did not observe cyclized products deriving from the radical **299** on free radical addition of BrCCl₃ or CCl₄ to **298**. It must be noted, however, that this should be considered as a Cy4/Cy5 case rather than a Cy5/Cy6 case. Finally, an interesting rearrangement observed by Jarvis, Govoni, and Zell,[401a] who assumed the direct rearrangement **301 → 303**, may be considered as proceeding through the open radical **302** (Scheme 120). Free radical addition of carbon tetrahalides to **300** (X = H, Br, or Cl) gives the rearranged compounds **304** in amounts which increase with reaction temperature and decreasing concentration of the chain transfer reagent, along with unrearranged products. The origin of these products may be ascribed to the opening of the (Cy˙6) radical **301** to **302** followed by cyclization to the (Cy˙5) radical **303**. The photolytic behavior of compounds related to **296**[399b,c] and **300**[401b,662] has been studied extensively. The formation of cyclized or rearranged products has generally been ascribed to other paths than those just discussed.

Reductive cyclization of nonconjugated enones will be discussed later (Section XI.3, Scheme 138) because of their interesting stereochemical features. Another promising behavior has been observed with suitably functionalized enones.[663] The ketone K-1 (R = Me) (Scheme 120A) on treatment with Na–THF, gives the (Cy5) compound (68% yield) expected from the cyclization of the ketyl radical (A'). Unexpectedly, the stereoisomeric ketone K-2 (R = Me) gives only the interesting patchoulenol (Cy6) (64% yield) which would result

SCHEME 119

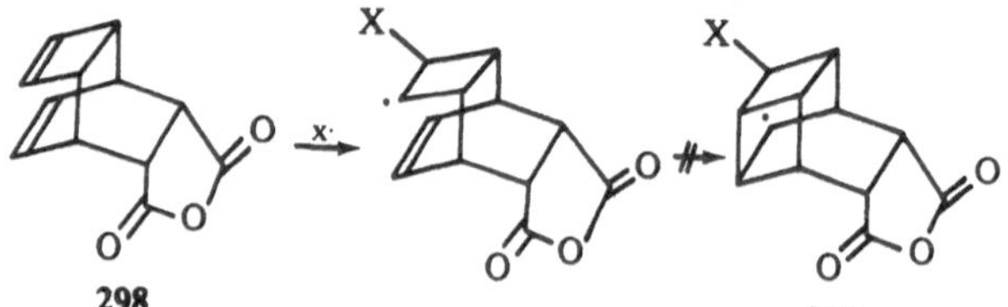

SCHEME 120

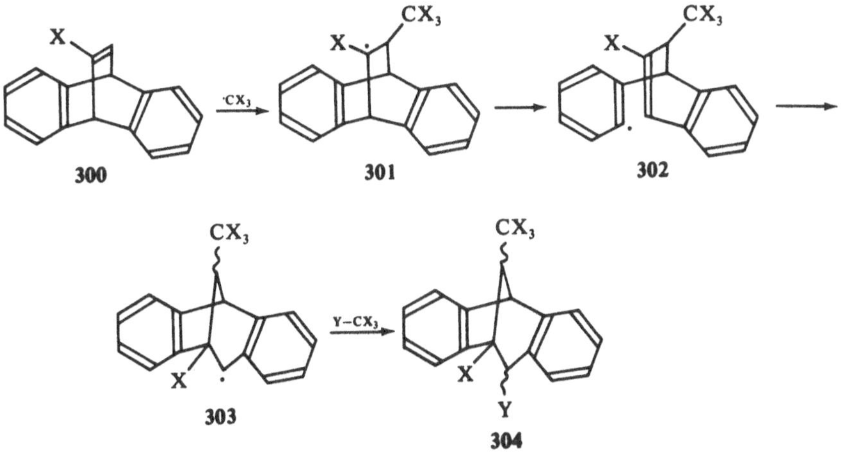

300 **301** **302**

303 **304**

SCHEME 120A

(K-1) (A') (Cy'5) (Cy5)

(A') (Cy'5) (Cy5)

(K-2)

(Cy6)

$R' = CH_2OCH_3$

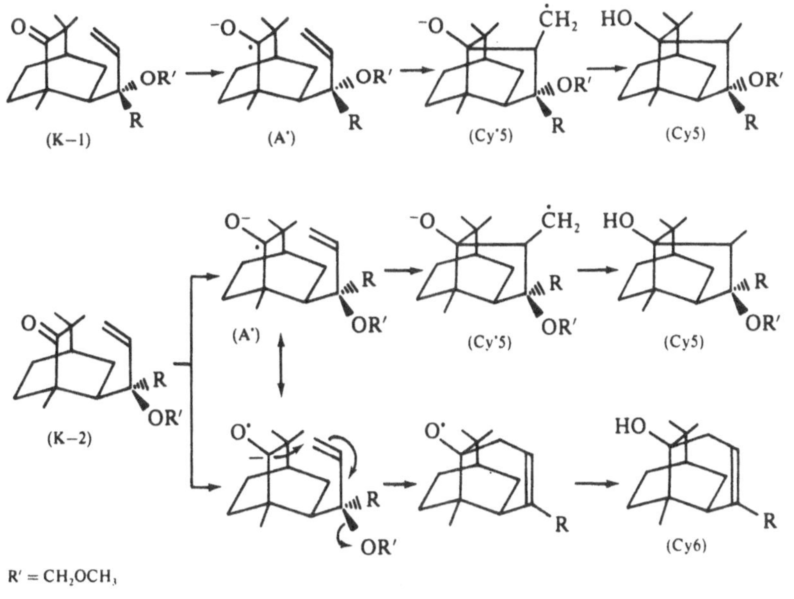

SCHEME 121

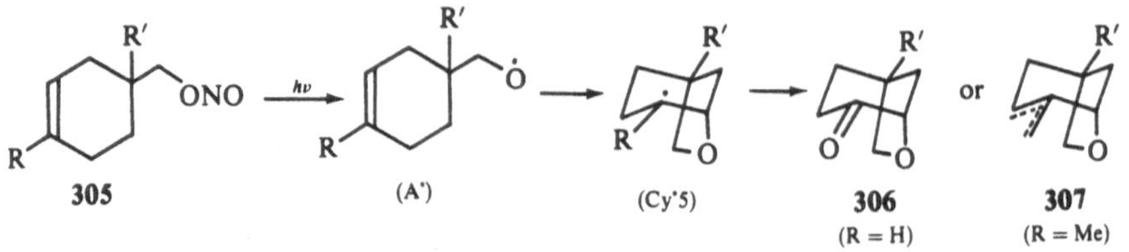

305   (A')   (Cy'5)   **306**   **307**
(R = H)   (R = Me)

from the intramolecular reaction of the ketyl radical (A') acting now as a nucleophile in a stereospecific $S_{cN'}$ reaction.[664] So the competition between the two paths is highly dependent of the configuration of the carbon atom bearing the leaving group as well as on the R group attached to this carbon. Thus the ketone K-2 (R = H) gives now a mixture of the (Cy5) compound (39%) and the most interesting norpatchoulenol (Cy6) (51%).[663]

Examples of heteroatomic radical cyclizations leading to bridged rings will now be described. Oxabicyclic compounds **306** (R = H) have been obtained by photolysis of the nitrites (**305**) (oxime not isolated) (Scheme 121).[402] When R = Me compounds **307** are obtained and, although yields are modest, this shows that intramolecular addition of an alkoxy radical is possible and that it gives exclusively the (Cy'5) radical in the Cy5/Cy6 case.

Since peroxyl radicals add so easily to double bonds (see Sections VIII.2.D and XI.4.B.a) 3-cyclopentenyl peroxyl radicals appeared to be attractive models for prostaglandin bicyclic endoperoxide formation, by analogy with the cyclization discussed in Scheme 116. Under the conditions used, however, disproportionation was preferred to intramolecular cyclization.[696]

Intramolecular addition of aminyl radicals has been more extensively studied in order to obtain azabicyclic compounds. The first report (Scheme 122) was made in 1970 by Heusler,[403a] who devised an elegant synthesis of chloroazatwistanes (**309**) (two epimers in 80% yield) by irradiating the *N*-chloroamine (**308**) in trifluoroacetic acid. While he did not propose a mechanism, it appears that the origin of the product **309** is in accord with an

SCHEME 122

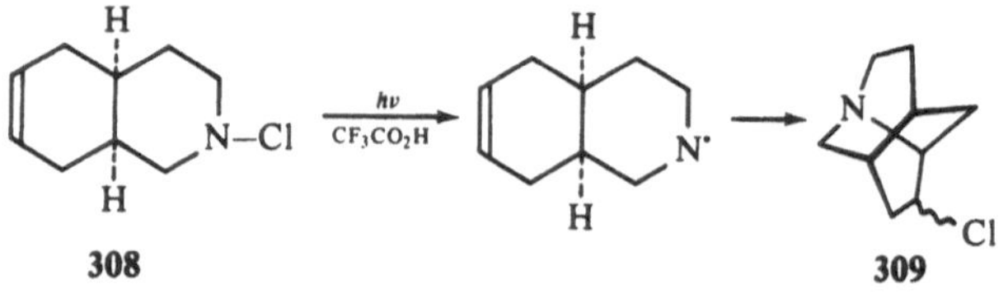

308   **309**

SCHEME 123

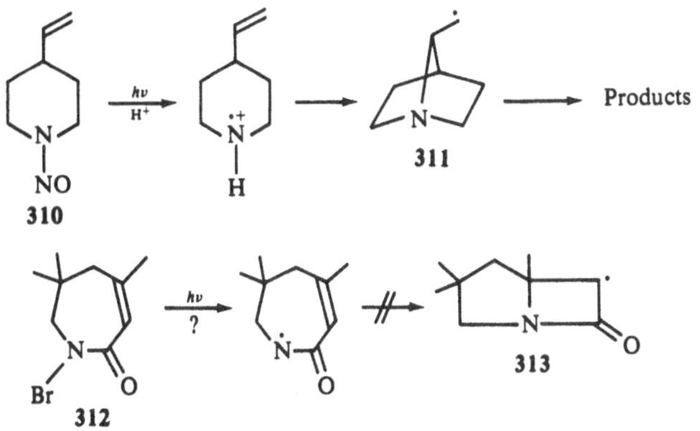

intramolecular addition of an intermediate aminium radical. Of course *cis*-decalinic systems such as **308** are *a priori* very prone to cyclize and it seems meaningless with such special structures to discuss whether the cyclization corresponds to a Cy5/Cy6 case giving directly, or after rearrangement, the (Cy6) compound, or to a Cy6/Cy7 case also giving the (Cy6) compound. Complicated cases of very efficient photochemical cyclizations involving carbon radicals may be easily rationalized in the same way.[403b]

Another example involving a bridgehead nitrogen atom has been described by Chow[169] (Scheme 123). Photolysis of **310** in methanol containing HCl gives the oxime (24%) and the hydroxylamine (22%) resulting from the (Cy˙5) radical **311**. No products resulting from the (Cy˙6) radical have been observed. In more strained systems such as **312** no cyclized products resulting from the (Cy˙6) or the (Cy˙5) radical (**313**) have been reported.[404]

More studies have been devoted to aminyl radicals carried by an aliphatic chain. The first example was described in 1971, independently by Surzur, Stella, and Nouguier[186] and by Chow[182] (Scheme 124). Under all the conditions used to generate the aminyl radical **315** (in neutral, complexed, or protonated form)

SCHEME 124

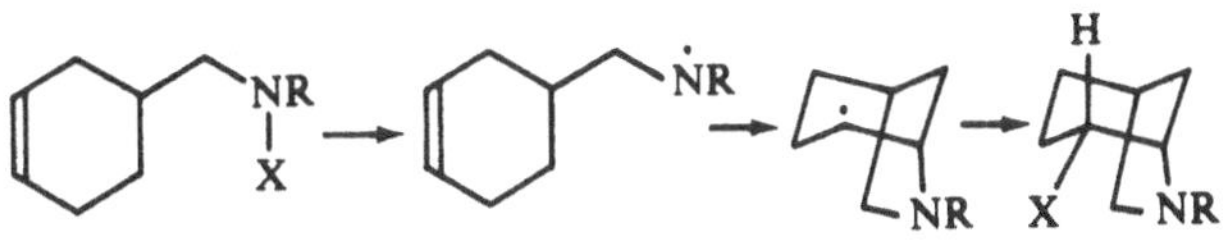

from the *N*-chloroamine[186] (**314**) (X = Cl), or from the *N*-nitrosoamine
(X = NO),[182] only the product **317** resulting from the (Cy˙5) radical **316** was
obtained. Equally interesting is the exclusive formation of the *endo*-chloro compound **317** (X = Cl) starting from the *N*-chloroamine (best yield 77%). It must
be noted that the same high regio- and stereoselectivities toward **317** (X = OAc)
are observed from the *N*-chloroamine under ionic conditions[186,411] but not under
solvolytic conditions, which favor a nitrenium pathway (Section VIII.3.A.).

Extensive study of azabicycles formation involving amidyl radicals has
been reported by Kuehne[200] and by Lessard.[201–203] Among other examples they
observed a cyclization process analogous to that shown in Scheme 124 with
carboxamidyl[198,200,202] or acetamidyl radicals[200,202] corresponding to the aminyl
radical **315**. The free radical mechanism was confirmed in this case by esr.[199]
High selectivity in the cyclization step toward the (Cy˙5) radical corresponding
to **316** was again observed but not the stereoselectivity of the transfer step, for
the *endo* isomer corresponding to **317** is obtained as well the *exo* isomer (ratio
approximatively 1:1).[198,200,202]

The cyclization of **318** [X = Cl or NO (References 405 and 406, respectively)] (Scheme 125) has been studied independently under free radical conditions as well as [407] under solvolytic conditions (X = Cl). In all cases, the only
product is **320** corresponding to selective (Cy5) formation; under free radical
conditions the yield is 85%. The cyclization of **321** (X = Cl or NO) also gives
exclusively (Cy5) compounds **323** under free radical conditions.[408a] Although
first reported to give the (Cy6) compound under solvolytic conditions,[408a] it has
now been shown that the (Cy5) compounds (**323**) are also obtained from **321**
under these conditions.[409,410] The carboxamido and acetamido radicals

SCHEME 125

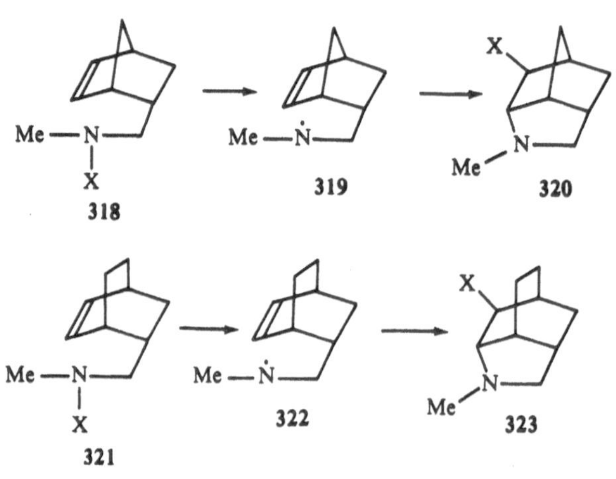

corresponding to **319** and **322** also give exclusively and in high yields the (Cy5) bicyclic compounds corresponding to **320** and **323**.[202,203] It must be noted that under all the conditions employed the X substituent and the nitrogen atom in **320** and **323** are *anti* to each other.

An interesting synthetic application of the reaction discovered by House starting from unsaturated hydroxylamines (Hy) (Section VIII.3.D, Scheme 48) has been described for the preparation of heterocyclic molecules of medicinal interest which contain a tricyclic hydrocarbon framework with a nitrogen-bridged central ring atom flanked by two aromatic rings such as Cy–H (Scheme 125A). Yields are excellent in this Cy6/Cy7 case (80%, R = H; 88%, R = Me) as well as in the Cy5/Cy6 case (yields up to 96%) and it must be noted that the attempted more classical ionic reactions failed.[665] The interesting molecules with alkyl substituents at both bridgehead positions such as Cy–Me (Scheme 125A) cannot be obtained in this way owing to the inaccessibility of the required hydroxylamine but an unusual intramolecular amine to olefin addition, generally under basic conditions, has been discovered. For instance, the (Cy–Me) compound (Scheme 125A) is obtained in 50% yield in this Cy6/Cy7 case, a similar yield being observed in the Cy5/Cy6 case. Since the ring closure is inhibited by radical scavengers, a radical chain process is suggested but one involving more probably an electron transfer to the 1,1-diphenylethylene system than an intramolecular addition of an aminyl radical.[666]

Examples of cyclization involving thiyl radicals leading to bicyclic compounds in the Cy5/Cy6 case have been also recorded. From the behavior of unsaturated aliphatic radicals one would expect to obtain a mixture of (Cy5) and (Cy6) products. In fact, the examples recorded[228,229] and discussed in

SCHEME 125A

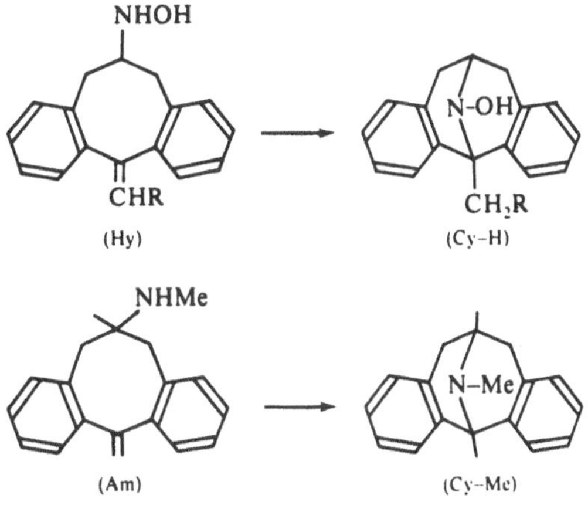

Section VIII.4.A (Scheme 56) reveal exclusive formation of the (Cy5) compound.

In conclusion, intramolecular free radical addition may be a useful method of synthesizing bridged cyclic compounds. From the examples of the Cy5/Cy6 case noted in this section, it may be concluded that a large preference for the (Cy·5) radical formation again exists and that, in some cases, a very high stereoselectivity in the last transfer step may be observed. The same stereoselectivity is often observed in cationic-induced cyclizations.

B. The Cy6/Cy7 Case

As discussed in Section V.1 for unstabilized carbon-centered radicals, free radical cyclization in the Cy6/Cy7 case must be considered as an unfavored process. As an example, the next higher homolog of **269** (Scheme 110) affords no cyclized products, whereas an 85% yield is obtained in the Cy5/Cy6 case.[391] Under these conditions cyclization of citronellyl iodide and the photochemical rearrangement of longibornyl iodide, considered a free radical process,[413] more likely involves a cationic intermediate as proposed in the original paper.[414]

Nevertheless Cy6/Cy7 cyclization is not a forbidden process. For instance, the cyclization of keto-unsaturated epoxides described by Stork and discussed in Section XI.2.A.a (Scheme 108) gives the (Cy6) compound in 60% yield in the Cy6/Cy7 case.[390] In the same way the electroreductive cyclization of ethylenic ketones (Section XI.3, Scheme 138) has been generalized for the formation in high yields of bicyclic (Cy6) compounds in the Cy6/Cy7 case.[502b] It may be expected with a stabilized (Cy·) radical to observe an easy cyclization, even in the Cy6/Cy7 case: this is verified, for instance, with enaminoketone homologs of **262** (Scheme 107) which give the (Cy7) product in yields of up to 90%.[389b–g]

An interesting application of the intramolecular free radical addition process to the synthesis of polycyclic terpenes has been described in a Cy6/Cy7 case by Bakuzis[415] (Scheme 126). Irradiation of **324** in the presence of

SCHEME 126

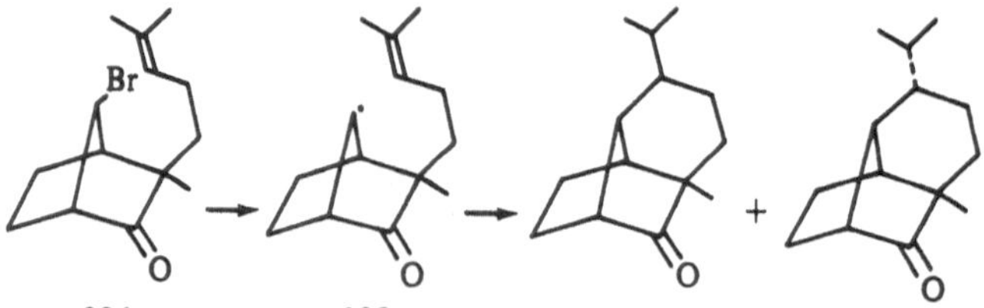

| **324** | **325** | **326** | + | **327** |

SCHEME 127

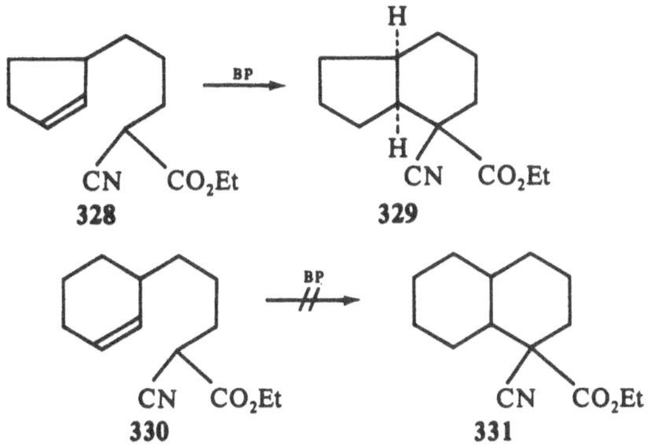

tributylstannane with *t*-butyl perbenzoate at 36°C in benzene affords the nor-sativone **326** and copacamphenilone (**327** in a 3:2 ratio (total yield 62%) by cyclization of the radical **325**. The two ketones have been separated and transformed into sativene and copacamphene, respectively.

More recently, Büchi has described the synthesis of dihydroagarofuran, an interesting fragrance of the oriental type, starting from (–)carvone: the last step involves a selective Cy6 cyclization in a Cy6/Cy7 case (67% yield).[282b]

With stabilized carbon-centered radicals, while one would expect to observe efficient cyclization reactions (Section VII.3), the yields of bicyclic compounds in the few examples reported are very low. For instance, **328** under BP initiation gives the *cis* product **329** in 22% yield[110] (Scheme 127), and **330** under the same conditions, does not give the cyclized product **331**.[109]

Bicyclic products in the Cy6/Cy7 case have been reported by Lessard in his studies of amidyl radicals[202,203] (Scheme 128). For instance, the carbox-amidyl radical generated from the corresponding *N*-chloroamide (**332**) gives the (Cy6) compound (75% *exo*, 15% *endo*), but the corresponding acetamidyl radical, generated from the amide **333**, does not give any cyclized product. The compound **334** has been cyclized under conditions in which aminium radicals are formed, to the corresponding quinuclidine derivative (Cy6) (yield 72%), a possible precursor for the total synthesis of *Cinchona* alkaloïds.[408b]

Cyclization of thiyl radicals has been reported. For instance, 5β-6-thiasteroïds (**336**) have been obtained by Speckamp[416] by photolysis of **335**; on the other hand, photolysis of **337** did not give the expected thia analog of strychnine (**338**)[417] (Scheme 129).

Thus it may be concluded that, as with monocyclization, the Cy6/Cy7 cyclization process to form polycyclic compounds is generally less favored than the Cy5/Cy6 one. Nevertheless, examples of preparative value have been

SCHEME 128

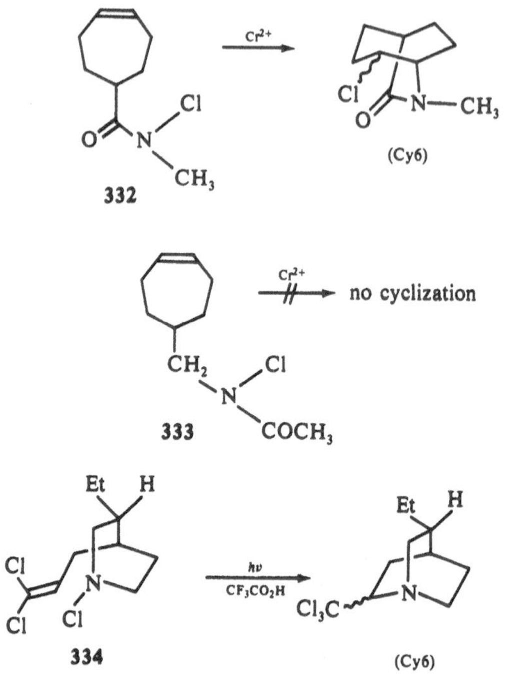

332   (Cy6)

333

334   (Cy6)

SCHEME 129

335   336

337   338

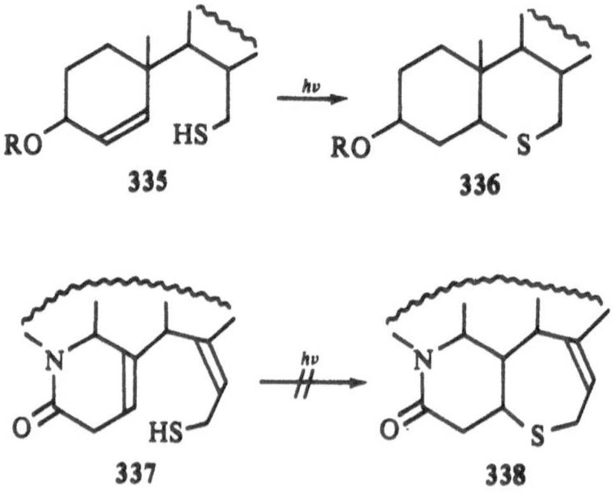

recorded and it is noteworthy that in all the cases where the cyclization has been observed only the Cy6 is produced except when the (Cy·7) radical is stabilized.

C. The Cy4/Cy5 Case

From the discussion in Section V.2.A, it is to be expected that cyclization in the Cy4/Cy5 case will not be a good way to obtain polycyclic compounds and, indeed, most of the attempts have failed. This is true, for instance, in the case of the carbon-centered radical obtained from compound **298** [see Section XI.2.A.b (Scheme 119)].[400] In the same way, the cyclization observed by Stork,[390] which gave excellent yields in the Cy5/Cy6 case (see Section XI.2.A.a, Scheme 108) and even in the Cy6/Cy7 case, failed in the Cy4/Cy5 case. The same is true for amidyl radicals, which give good results in the Cy5/Cy6 and in the Cy6/Cy7 cases.[200]

Nevertheless, with stabilized radicals, Julia[108] obtained the bicyclic product **340** in 24% yield (*cis/trans* ratio 70:30) by submitting **339** to BP initiation (Scheme 130).

Another cyclization possibility is observed now when the (Cy·5) radical is stabilized, by a nitrogen atom for instance: carbazoles have been obtained in this way from enaminoketones in yields of up to 86%.[389e,g]

The pentadienyl radical formed from piperylene is known to cyclize at 650°C to cyclopentene and cyclopentadiene[420a] although, according to an esr study, this isomerization is not observed between −130 and +180°C.[420b] A lower temperature (210°C) is sufficient to give cyclized products (Scheme 131) from the radical **343** which is assumed to form by cyclization of radical **342**, when the peroxyacetate **341** is thermolyzed, but the yield of cyclized products is less than 10%.[421]

Cyclization toward the (Cy·4) radical is never observed in a bicyclic system because of the facility with which such radicals open up. The much studied example of free radical addition to α- or β-pinene has already been reported in Section V.2.A (Scheme 16). In most of the examples reported, only products resulting from the open (A·) radical are obtained.[118,422–429] It should be

SCHEME 130

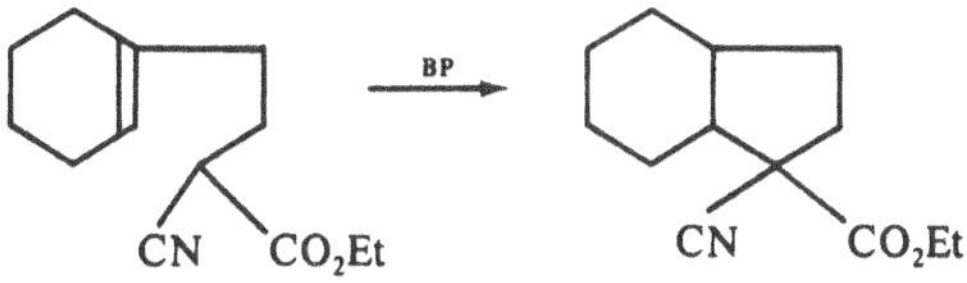

339   **340**

SCHEME 131

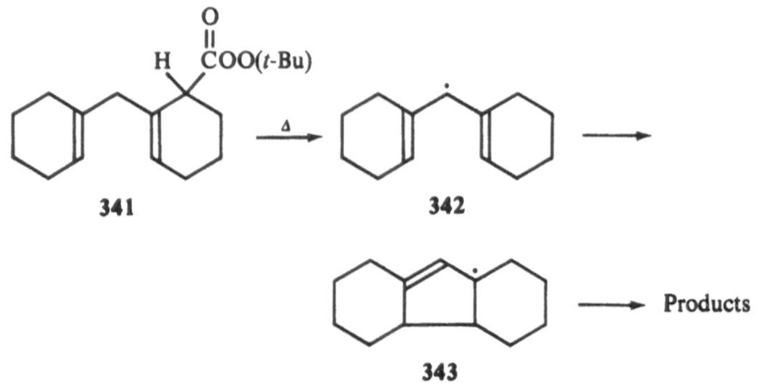

341   **342**

343   ⟶ Products

---

noted, however, that products resulting from the preformed (Cy˙4) radical may be obtained by trapping with a good transfer agent such as a thiol,[68] by using an efficient free radical scavenger such as NO,[136,430,431] or an oxidizing salt.[432-434] Even in these cases opening of the (Cy˙4) radical is a competitive reaction.

Esr investigation of bicyclo[3.2.0]heptenyl radicals[667] and 7-norbornadienyl radicals[668] in an adamantane matrix which prevents combination and disproportionation reactions allows the observation of interesting processes forbidden in solution because of their high activation barrier; using keto or formyl precursors, interesting behavior of the acyl radicals has been observed, including the reversible intramolecular addition to Cy˙4 in the Cy4/Cy5 case.[667,668]

D. The Cy3/Cy4 Case

As discussed in Section V.2.B, the 3-butenyl-alkylcyclopropyl radical rearrangement has attracted considerable interest. Generally, products resulting from the open 3-butenyl radical are observed. Stereoelectronic control in the opening process has been proposed and a great number of alkylcyclopropyl radicals which are in a rigid bicyclic system have been studied in order to test this hypothesis. The radicals studied (Scheme 132) were the bicyclo[3.1.0]hexyl (**344**),[83,452-454] the 3,5-cyclocholestan-6-yl (**345**),[70,455,456] the 2-norcaranyl (**346**),[457] and medium-sized-ring cyclopropylcarbinyl[458] or spiroalkyl radicals.[459] The selective photolytic cleavage of tricyclic ketones containing a cyclopropyl group has been explained in the same way.[71]

In some cases, reversible addition to a (Cy˙3) radical has been proposed to explain epimerization[460] or rearrangement.[76,461,667] It must be noted that in those cases (Scheme 133) products derived from the radical **348** are observed on

SCHEME 132

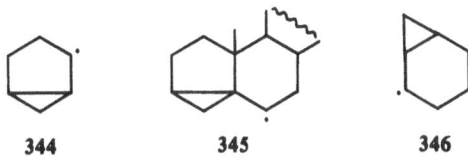

344   **345**   **346**

generation of the radical **347**; on the other hand, products resulting from the radical **347** are not observed on generation of the radical **348**. Another interesting example has been noted[462] during the photolysis of steroid nitrites (**349**) (Scheme 133) where the corresponding alkoxyl radical went to the radical **350** via the Barton reaction. This then cyclized to the radical **351**, which opened by cleavage of the C(5)–C(10) bond to give A-*nor*–B-*homo* steroids (**352**). Reversible cyclization has also been observed on free radical addition to bullvalene[463,464] and the results have been compared with the corresponding cationic rearrangement.[465]

To the best of our knowledge the only examples of bicyclic product formation from a Cy3/Cy4 cyclization (albeit in low yield) are those described by Friedrich[74,75] [see Section V.2.B (Scheme 19)], and here the resulting (Cy˙3)

SCHEME 133

'CH₂

347   **348**

ONO   HO

349   **350**

351   **352**

SCHEME 133A

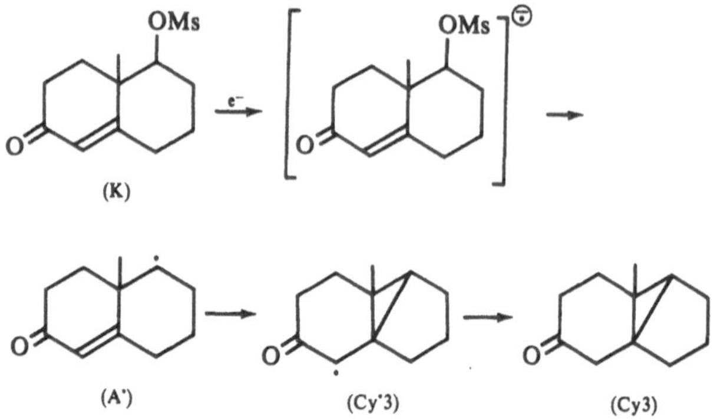

(K)

(A')   (Cy'3)   (Cy3)

radical is benzylic. The more extensively studied Cy3/Cy4 cyclization process leading to a polycyclic system is, of course, the norbornenyl–nortricyclyl radical rearrangement already discussed in Section V.2.B (Schemes 20 and 21).

Nevertheless, a cyclization very similar to the $350 \rightarrow 351$ rearrangement but not followed by β-scission has been observed recently and is another example of the synthetic potential of reductive cyclizations, particularly those which involve leaving groups. In fact, the mechanism of the cyclization now discussed[669] is assumed to be quite different of the one discussed in Section XI.2.A.b (Scheme 120A) since the displacement of the mesyloxy group would occur this time directly after the radical anion formation by electrochemical reduction (Scheme 133A). The radical (A') would give the (Cy'3) radical in a Cy3/Cy4 case. The formation of the Cy3 compound in 74% yield could result from an other reduction to the carbanion or directly by hydrogen atom abstraction from the solvent (Scheme 133A).[669]

Another interesting possibility which has been discussed involves the 7-norbornenyl radical 353 (Scheme 134) where a nonclassical structure was proposed[487] but for which no esr evidence for rearrangement to the tricyclic isomer (354) was found.[488]

SCHEME 134

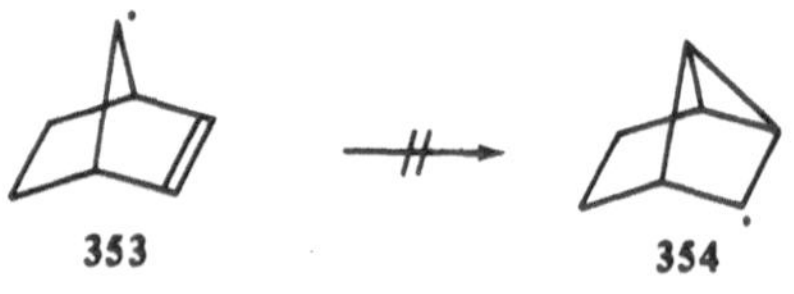

353   354

SCHEME 135

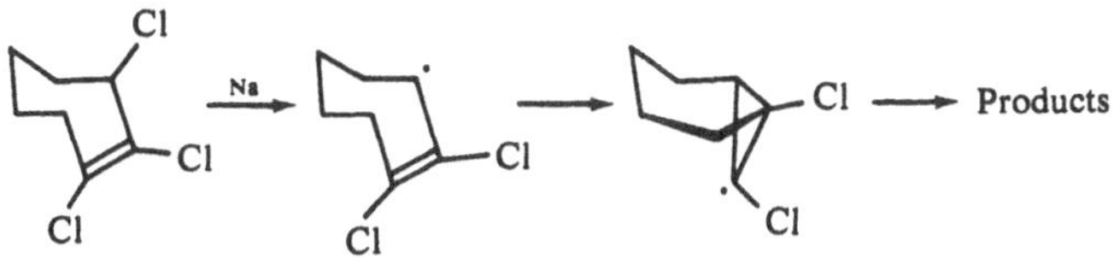

355

E. The Allyl Radical

As noted in Section V.2.C, opening of the cyclopropyl radical and closing of the allyl radical are slow processes. Nevertheless, this possibility has been used as a working hypothesis (Scheme 135) to explain the formation of products resulting from the treatment of **355** with sodium in liquid ammonia.[489]

3. Stereochemical Features of Free Radical Intramolecular Additions

In the preceding section (XI.2) it was noted that cyclization reactions leading to bicyclic products are often highly stereoselective but that a general discussion of the origin of this stereoselectivity did not seem possible at this time.

This section is devoted to the stereoselectivity often encountered in monocycle formation. First of all it must be said that asymmetric induction has been observed by Julia and Maumy[104] in the free radical cyclization of an unsaturated cyanoester similar to **46** (Scheme 28, Section VII.2) where the (−)-menthyl group replaced the ethyl group. The cyclized product analogous to **47** gave, on hydrolysis, the optically active acid in 30% optical yield. The remainder of this section will be devoted to the matter of *cis–trans* selectivity with alkyl substituted unsaturated radicals. Some of these observations will permit a better understanding of the course of polycyclization reactions discussed in the next section (XI.4). Interest in this field began with the study of free radical cyclization of 1,6-heptadiene and related diallylic substrates [Section VI.3 (Scheme 25)]. It was recognized early that **356** (Scheme 136)

SCHEME 136

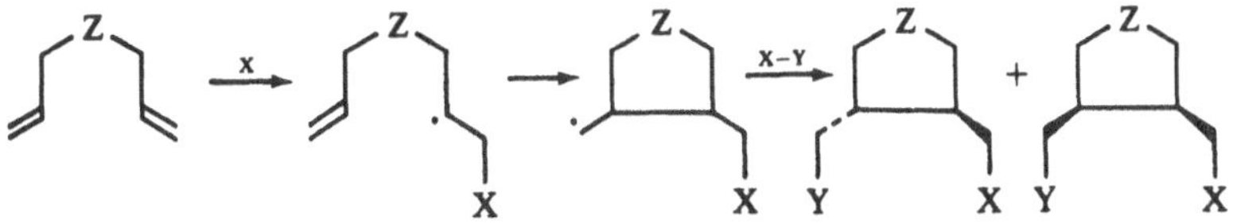

356   **357**   **358**   **359**

gives a mixture of the *trans*-**358** and the *cis* compound **359**, and somewhat later Brace reported that the *trans* compound **358** is stereoselectively formed. This trend was confirmed later, with a *cis/trans* ratio **358/359** often in the range 1:5,[67a,96,497,499,500] although in some cases equal amounts of *cis* and *trans* cyclized products have been formed.[67b,498] Quite unexpectedly, when the radical **361** (which is analogous to **357**), is generated from the corresponding halide (**360**) (Z = O or CH$_2$) by stannane reduction[87,497] (Scheme 137), the major product is the *cis* compound **362**, in a ratio *cis/trans* = **362/363** = 2.3:1. To explain why the *cis* mode of cyclization is favored, Beckwith[87] proposed a hyperconjugative mixing of the half-filled *p* orbital with adjacent σ and σ* orbitals producing a modified delocalized orbital of similar symmetry to the acceptor π* orbital as pictured in **364** (Scheme 137). This will provide a secondary attractive interaction between the alkyl substituent and the olefinic bond favoring the *cis* compound **362**. If this explanation is correct, the reason that the radical **357** gives predominantly the *trans* isomer is not clear unless one assumes that **357** is not a discrete intermediate and that the second double bond in **356** interacts with the first during the addition which leads to **357**. An analogous hypothesis, discussed in Section XI.4, has been retained to explain polycyclization results. Nevertheless this hypothesis must be regarded as tentative before an unambiguous attribution of configurations of the *cis* and *trans* isomers resulting from the cyclization of **356** has been made. It must be noted, for instance, that in a recent paper a *cis/trans* ratio 5:1, i.e., the reverse of that assumed until now has been proposed for the stereoisomers resulting from the cyclization of **356** (X = O).[501] In fact at the present time the experimental results concerning the stereochemistry appear as confusing as were the Cy5 or Cy6 nature of the products about 20 years ago.

Stereoselective cyclizations in the Cy5/Cy6 case have also been observed

SCHEME 137

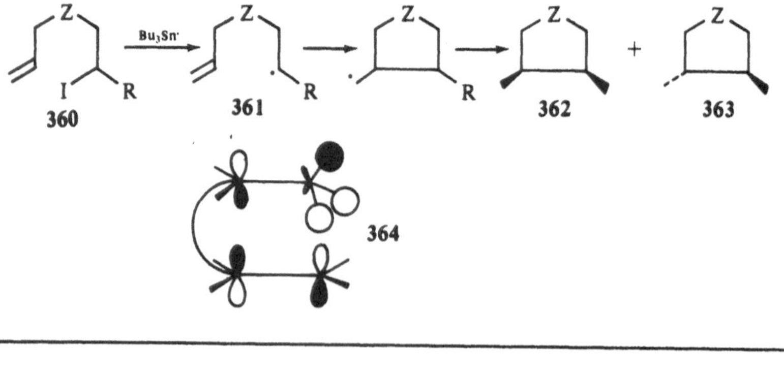

SCHEME 138

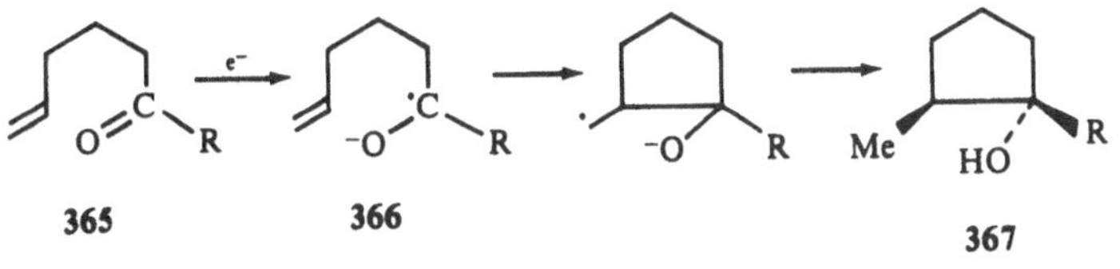

365   366   367

by Julia during the cyclization of unsaturated cyanoesters[13] and cyclopropyl radicals[391] (Scheme 110, Section XI.2.A.a).

Another interesting example of a regio- and stereoselective cyclization reaction has been reported by Shono[502] (Scheme 138), who found that electroreduction of ethylenic ketones (365) gives only the Cy5, *cis* compounds 367 (40–98% yield) in a reaction which probably involves cyclization of the radical 366 to the (Cy·5) radical. Analogous results are obtained with the next homologs in the Cy6/Cy7 case; again only the *cis*-Cy6 compounds are observed. In agreement with the discussion of the results in Table 2 (Section IV) when R_3 or $R_5 = $ Me, no cyclized products are obtained in the Cy5/Cy6 case.[502b] Although the stereochemical features have unfortunately not been reported it must be noted that nonconjugated enones may also be reductively cyclized by *N*-ethylpyrrolidone–sodium–*tert*-butyl alcohol mixtures. Alkyl radical cyclization features are again observed: a yield of 5% in the Cy4/Cy5 case, of 65% (Cy5) in the Cy5/Cy6 case, of 80% (Cy6) in the Cy6/Cy7 case, and of 6% (Cy7) and 12% (Cy8) in the Cy7/Cy8 case.[670]

It must be also noted that 1,5-ring closures of 2-, 3-, or 4-alkyl-substituted 5-hexenyl radicals are also stereoselective: 2- and 4-alkyl-substituted radicals afford mainly *trans*-disubstituted (Cy5) products,[697] while 3-substituted radicals[524] (Section XI.4.A, Scheme 149)[697] afford mainly the *cis*-Cy5 products with ratios of up to 4:1. This observation has been generalized to the Cy5/Cy6 cyclization of peroxyl radicals.[644,688] These results have been ascribed to a chair-like conformation of the transition state for the cyclizing (A·) radical with substituents preferentially occupying *pseudo*-equatorial positions.[697]

Modena and coworkers[101] reported an interesting stereoselective cyclization towards the (Cy4) compound, mainly *cis*-44 by free radical addition to the tetrafluorohexa-1,5-diene 39 (Scheme 27, Section VI.4); however, in the Cy5/Cy6 case a *cis/trans* ratio 1:1 was obtained.[503]

In most of the studies of the Cy6/Cy7 cyclization with the next higher homolog of 356 (Scheme 136) the stereochemistry of the (Cy6) compounds obtained has not been reported.[492,504–507] Nevertheless, during free radical addition of isobutyric acid to 1,7-octadiene (368) (Scheme 139) Forbes[500] obtained the three products 369 (9%), 370 (17%), and 371 (6%), after methanol

SCHEME 139

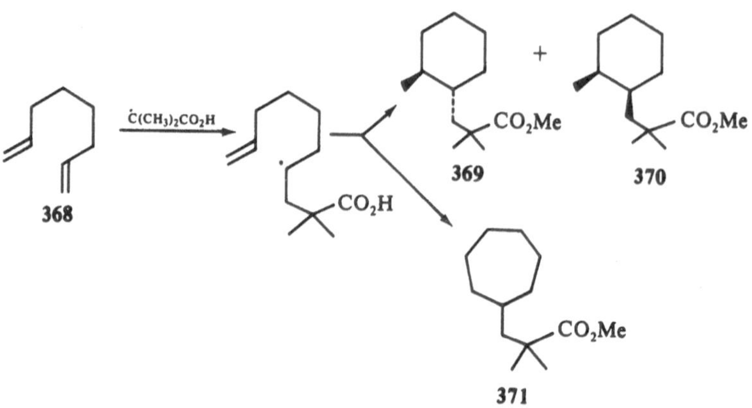

esterification. It should be noted that the major product isolated is the (Cy6) *cis* product **370**. In conclusion, intramolecular free radical addition is generally a stereoselective reaction, the *cis* compound being the major one formed in many cases.

4. Polycyclization Reactions

A. Carbon-Centered Radicals

In 1962 Breslow[508] proposed that many of the mechanistic requirements of squalene cyclization are better met by a free radical process. However, free radical initiation failed to cyclize the polyene appreciably,[508] and more recent studies showed that squalene epoxide is an intermediate in the biosynthetic pathway, probably cyclizing by a carbenium ion mechanism. Nevertheless, the free radical hypothesis generated extensive work in free radical polycyclizations and gave rise to new synthetic reactions, often remarkably specific. The first successful attempts were described by Julia.[13,509,510] The radical **372** (X,Y = CN, CO$_2$Et) (Scheme 140), generated from the corresponding cyanoester by BP in boiling cyclohexane, gave the *trans*-decalin compound **375** (X,Y = CN, CO$_2$Et) (41% yield)[509] by a pathway which presumably first involves a cyclization to **373** followed by cyclization to **374**. The first cyclization to a (Cy·6) radical **373** is in accord with the behavior of a stabilized radical **372** (X,Y = CN, CO$_2$Et). Indeed the cyclization of **372** (X,Y = H) generated by stannane reduction of the corresponding bromide affords only the product resulting from the (Cy·5) radical, now unable to cyclize since the second cyclization would correspond to a Cy4/Cy5 case.[511] The second cycliza-

SCHEME 140

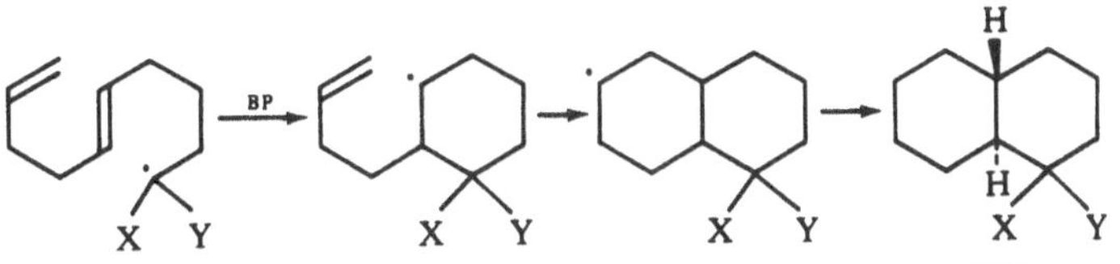

372   373   374   375

tion process to a (Cy˙6) radical (**374**), when X,Y = CN, CO₂Et is more unexpected since the radical **373** is now unstabilized and would have given a (Cy˙5)
radical. Indeed when the radical **373** (X,Y = H) is generated from the
corresponding chloride by stannane reduction, the major products (80%) are the
various stereoisomers of 1-methylhydrindane resulting from the (Cy˙5)
radical.[511]

Actually, such a Cy6 selectivity is often observed with dienic systems as
discussed in Section XI.2.A.a and is also observed with the compound **376**
(Scheme 141), which gives the compound **379** in 26% yield via the intermediates **377** and **378**.[510]

Before discussing Breslow's results, which involve remarkable specificity, it
seems worthwhile to discuss a monocyclization reaction described by Julia[512]
(Scheme 142). Treatment of the diene **380** with benzoyl peroxide, cuprous
chloride, and cupric benzoate in acetonitrile at 80°C affords the (Cy5) product
383 in ∼35% yield and less than 2% of the (Cy6) product. Thus a significant
proportion of the benzoyloxyl radicals add to **380** at the more substituted
double bond giving radical **381**, which then cyclizes to the (Cy˙5) radical **382**.

SCHEME 141

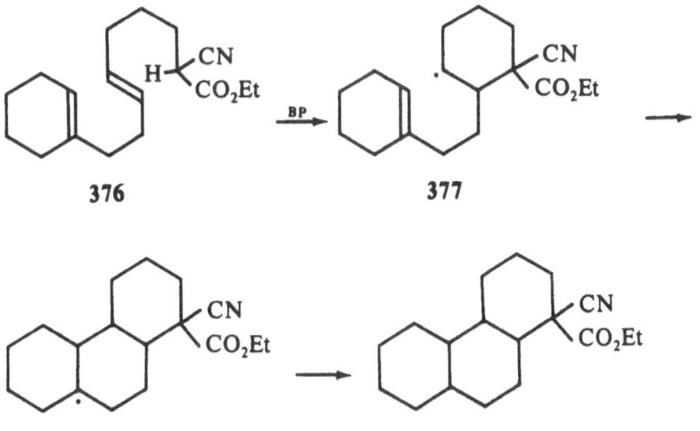

376   377

378   379

SCHEME 142

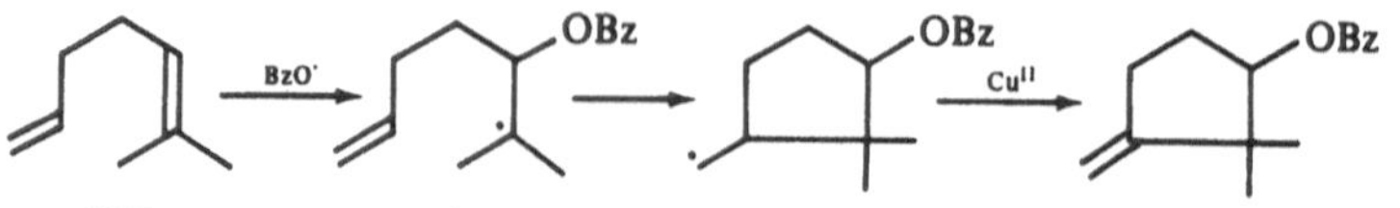

380   381   382   383

It is of interest to compare this result with the results obtained by McQuillin (Scheme 143), who studied manganese(III)[513] or photoinduced[514] radical additions to **384**, a homolog of **380**. The addition of R· to the more highly substituted double bond to give **385** is again the more favored process, but the radical **385** (a Cy6/Cy7 case) does not cyclize and gives, instead, the corresponding monoaddition products or it isomerizes to **386**, which then cyclizes to the (Cy·5) radical **387** according to the nature of R. A similar result is obtained with linalyl acetate instead of **384**.[698]

The high selectivity of free radical addition to the less hindered carbon of the more substituted double bond enables us to understand a part of the specificity observed by Breslow[515] in a reaction of preparative value[516] (Scheme 144). Treatment of geranyl acetate (**388**) with benzoyl peroxide, cuprous

SCHEME 143

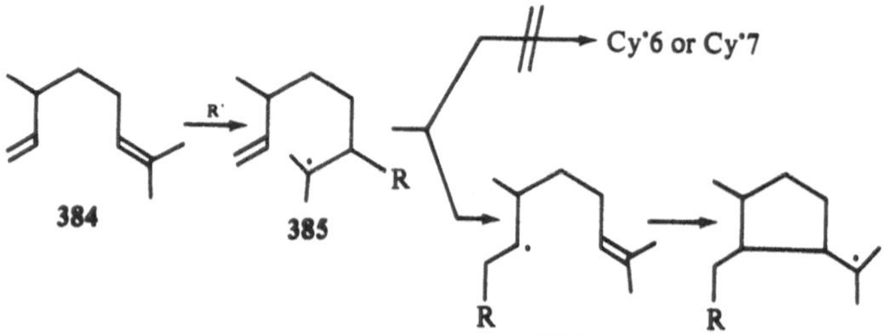

384   385   386   387

SCHEME 144

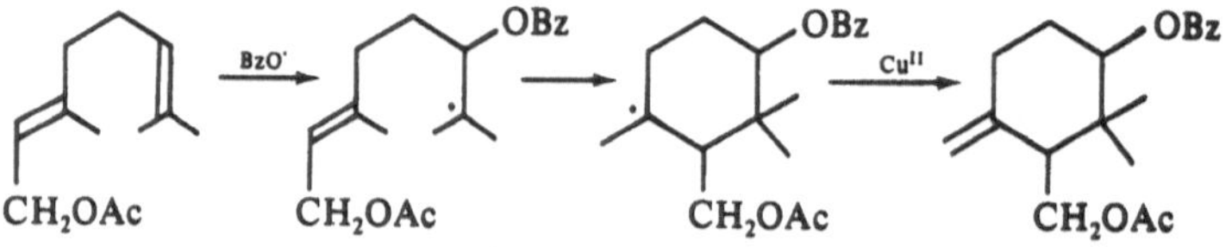

388   389   390   391

SCHEME 145

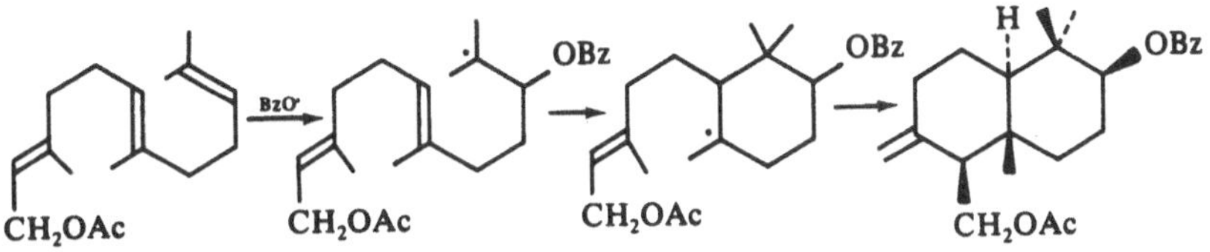

392   393   394   395

chloride, and cupric benzoate in benzene or acetonitrile at 70°C affords a
55–60% yield of the oxidatively cyclized product **391**. As in the cases discussed
above, the reaction could proceed by the selective addition of the benzoyloxy
radical to give **389**, but contrary to the fate of the radical **381**, radical **389**
cyclizes to the (Cy˙6) radical **390**. This selective cyclization to a (Cy˙6) radical
is in fact not completely unexpected according to the discussion in Section IV
(Scheme 12, Table 2, entries 2,3, and 7) where it was concluded that an R_3
substituent (here methyl) favored somewhat the Cy6 cyclization mode.

Even if this explanation is correct, the specificity observed by Breslow[517] in
the cyclization of *trans,trans*-farnesyl acetate (**392**) remains remarkable
(Scheme 145). Indeed, under the same conditions as above, **392** gives **395** in
20–30% yield. Considering the effect of the methyl substituent, as discussed
above, and the trend to often observe a Cy6 rather than a Cy5 cyclization
giving *trans*-decalin systems in bicycle formation (Section XI.2.A.a), the
regioselectivity is not completely unexpected. More remarkable is the highly
stereoselective reaction observed, since four asymmetric centers have been

SCHEME 146

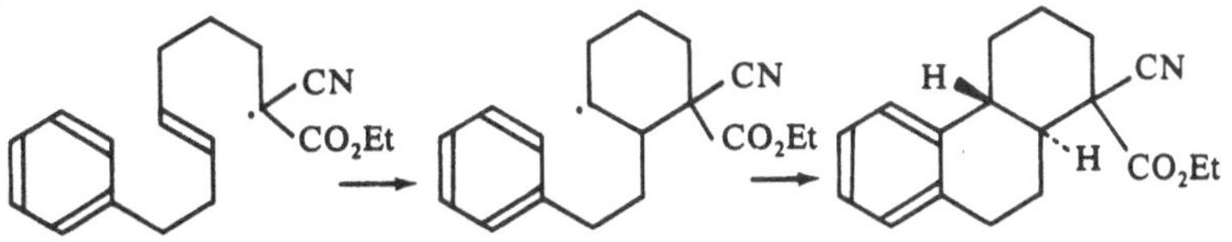

396   397   398

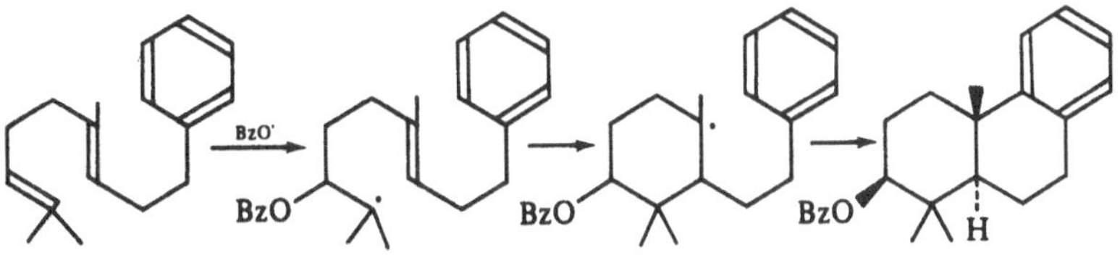

399   400   401   402

established in specific relative configurations. This highly stereoselective reaction is reminiscent of the one observed during the free radical cyclization of germacrene (Section XI.2.A.a, Scheme 113). As for the carbenium ion cyclization mechanism of squalene, a concerted mechanism, rather than a stepwise mechanism involving radicals 393 and 394, has been proposed to explain the results. Calculations are in accord with an intermediate situation, namely, not a concerted pathway but one in which bond formation during the first cyclization is facilitated by the second double bond.[518]

Another polycyclization possibility has been described by Julia.[13] In this case, an aromatic ring is attacked in the last step (Scheme 146). The radical 396 generated from the corresponding cyanoester by BP initiation in cyclohexane at 80°C gives the cyclized *trans* compound 398 (42% yield) in addition to the monocyclization product.[519,520] Compound 399, when treated with BP in refluxing benzene, gives the *trans* compound 402 in 20% yield.[521]

A more striking result[522] has been observed with compound 403 (easily obtained from *trans*-geranyl acetone) (Scheme 147). Under BP initiation the only product identified is 404 (12% yield).

Polycyclization reactions involving attack on an aromatic ring by unstabilized carbon-centered radicals in the last step have also been described by Pines[114] and by Walling.[115]

Thiyl-radical-induced polycyclizations of dienes have been described by Kuehne.[523] The results are generally not as good as with benzoyloxy radical initiation. Nevertheless, this method enables one to obtain the bicyclic compound 407 (Scheme 148) nearly quantitatively by irradiating solutions of α-acoradiene (405) and dimethyl disulfide. Desulfuration of 407 gives a product identical with natural cedrene. Photolysis of dimethyl disulfide and α-bulnesene (408) gives, in low yield, the compound 410 which by desulfuration gives dihydropatchulene.

Another interesting example of free radical polycyclization leading to bridged compounds has been described by Beckwith[524] (Scheme 149). The radical 411 generated by stannane reduction of the corresponding bromide

SCHEME 147

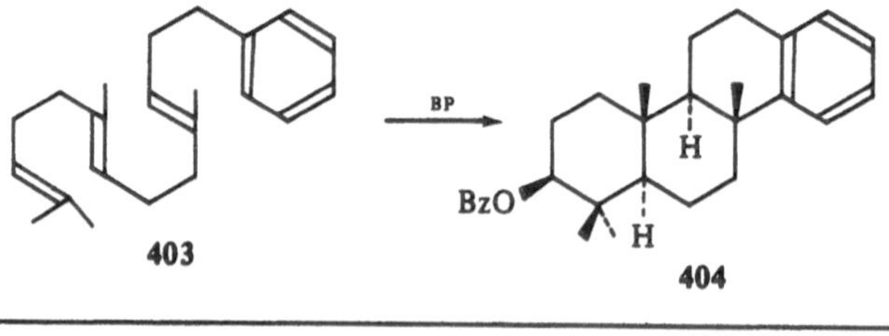

403

404

SCHEME 148

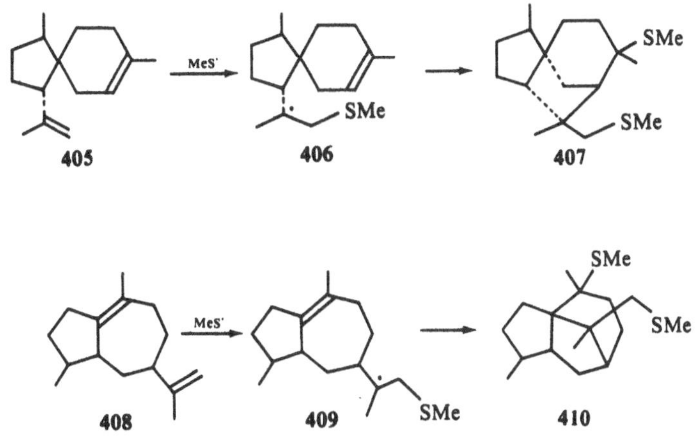

SCHEME 149

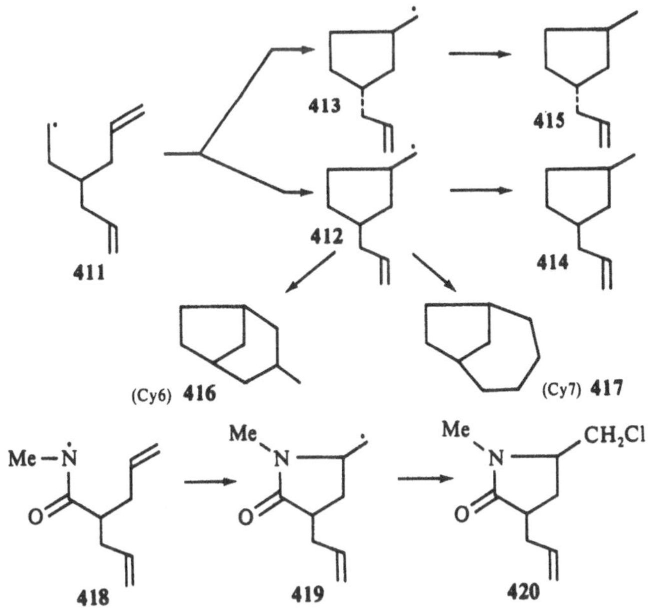

leads, via radicals **412** and **413**, to the monocyclization products *cis*-**414** and *trans*-**415** in an interesting *cis/trans* ratio of nearly 4:1. At low tributylstannane concentration (0.0027 *M*) and high temperature (100°C) the radical **412** undergoes a second cyclization in a Cy6/Cy7 case to the (Cy˙6) radical, giving **416** (22%), and to the (Cy˙7) radical, which leads to **417** (18%). This result must be compared with the fate of the carboxamido radical **418** (Scheme 149) generated from the corresponding *N*-chloroamide, which gives only the monocyclization product **420** (11%).[200]

In the same way the vinylic analog of **411** at low tributylstannane concentration affords a mixture of *exo*- and *endo*-2-methylnorbornane (71%) as the result of a double 1,5-cyclization in a Cy5/Cy6 case.[697] The obtainment of good yields of bicyclic compounds from radicals such as **411** results from the stereoselective initial cyclization to the *cis* radical **412** as expected from the behavior of 3-substituted 5-hexenyl radicals, while the corresponding 2-substituted 5-hexenyl radicals, cyclizing as they do mainly to the *trans* radicals, give poor yields of bicyclic compounds.[697]

A regiospecific and highly stereoselective approach to the synthesis of naturally occurring linearly fused tricyclopentanoids which possess antibiotic and antitumor activity has been recently devised[671]: it would involve the first example of an intramolecular 1,3-diyl trapping. By boiling the azo compounds (AZ) (Scheme 149A) the sesquiterpene (S) is obtained in 50% yield. The highly stereoselective formation of S has been explained in terms of a direct cyclization A˙′ → S involving secondary orbital interactions.[671] It may also be noted that A˙′ could cyclize to give the intermediate Cy˙5 predominantly in the *cis* form (see Section XI.3, Scheme 137) as is observed.

SCHEME 149A

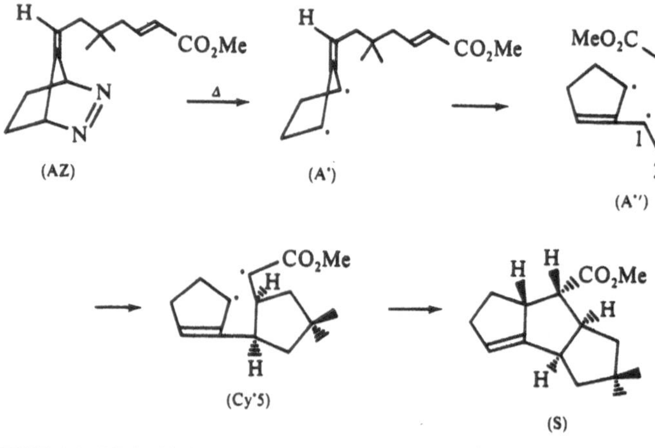

SCHEME 150

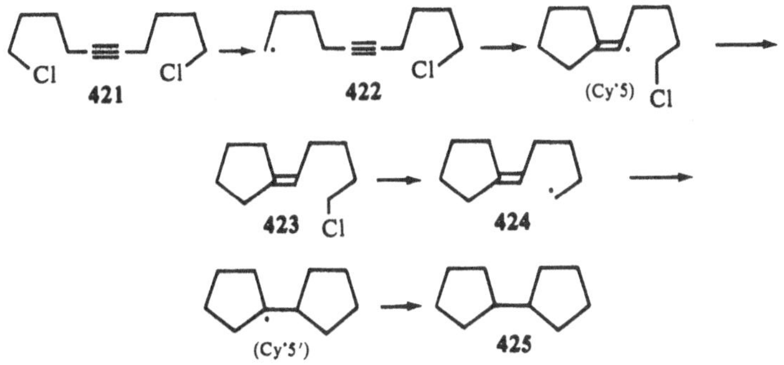

A different approach to polycyclization reactions has been reported with acetylenic compounds such as **421** (Scheme 150). By stannane reduction one obtains compound **425** in low yield.[284] The reaction involves formation of the radical **422**, which cyclizes to the (Cy·5) radical, which then gives **423**, and this, on further reduction, gives **424**, which cyclizes to the (Cy·5') radical and finally gives **425**.

Although not really a polycyclization process, but likewise involving a multistep reaction with acetylenic compounds, is an interesting free radical cyclization that has been described by Heiba and Dessau[525–527] and confirmed by Kampmeier.[528] The free radical addition of polyhalomethanes to terminal acetylenes such as heptyne **426** (Scheme 151) gives, besides the normal addition products, the cyclized compound **430** (20% yield). The vinylic radical **427**, by intramolecular 1,5-hydrogen abstraction, gives the radical **428**, which, being a Cy5/Cy6 case, cyclizes easily to the (Cy·5) radical **429** and this, on halogen elimination, gives **430**.

SCHEME 151

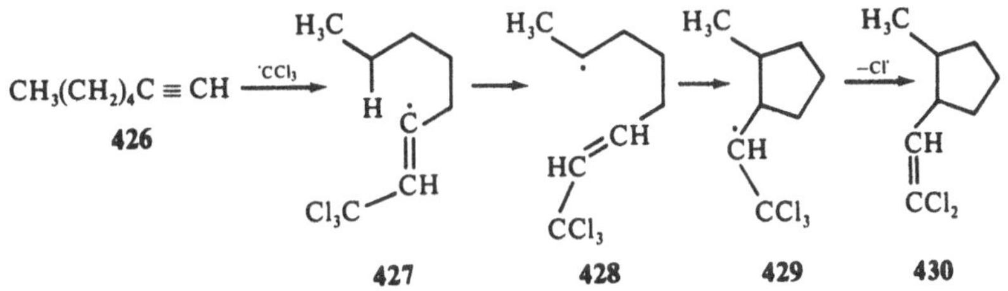

B. *Heteroatom-Centered Radicals*

a. Peroxyl Radicals

The discovery made by Porter that peroxyl radicals easily add intramolecularly to double bonds (Section VIII.2.D, Schemes 39 and 40) led this author to test the peroxyl radical cyclization mechanism proposed for the biosynthesis of prostaglandins (Scheme 152). This scheme involved formation of **432** from the unsaturated fatty acid **431**, cyclization of **432** (in a Cy5/Cy6 case) to the (Cy˙5) radical **433**, followed by a second cyclization of the carbon-centered radical **433** (also in a Cy5/Cy6 case) to the (Cy˙5) radical **434**. The radical **434** would now be oxidized to the endoperoxide **435**, which would decompose to prostaglandin $PGF_{1\alpha}$ (**436**).

Porter[529] was able to prepare a hydroperoxide from γ-linolenic acid. This hydroperoxide was transformed by his method into a peroxyl radical structurally analogous to **432** and, after $NaBH_4$ reduction, compounds structurally analogous to authentic prostaglandin $PGF_{1\alpha}$ were obtained. Autoxidation of methyl linoleate also afforded convincing evidence for a free radical mechanism in the formation of prostaglandin-like endoperoxides.[530] It is quite satisfying to see that nature uses a free radical process to synthesize important compounds like prostaglandins in the same selective manner as is observed in the laboratory, i.e., exclusive formation of (Cy˙5) radicals in the Cy5/Cy6 case. Even the *trans* stereoselectivity observed in the formation of **434** would be in accord with that observed in the cyclization of dienes as discussed in Section XI.3.

SCHEME 152

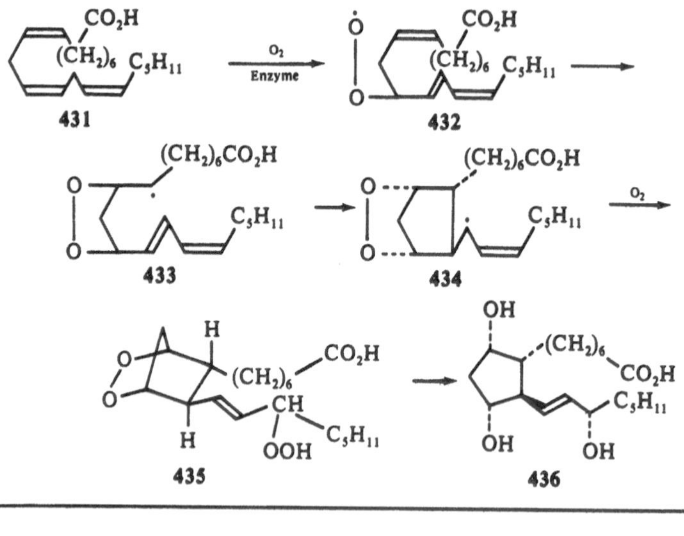

SCHEME 153

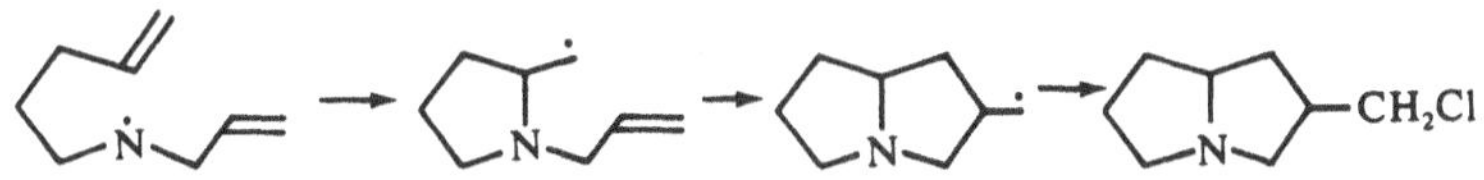

Of course, such a conclusion must not be generalized without care. For instance it has been proposed that the biosynthetic formation of the cyclopentanol ring of brefeldin A, a C_{15} macrolide antibiotic, would be analogous to the prostaglandin biosynthesis, but such a free radical pathway has recently been ruled out[531] Nevertheless, the experimental evidence is indirect, since this conclusion was arrived at using a biosynthetic path involving cyclization of an allyloxyl radical. Indeed, such a path appears to be a forbidden process (see Section VIII.2.A), and hence another pathway, ionic or radical, must be sought.

b. Aminyl Radicals

Polycyclization processes involving aminyl radicals have been described by Surzur and Stella.[167,178] The aminyl radical **437** (Scheme 153) generated by Ti(III) reduction of the corresponding N-chloroamine gives the N-chloromethyl-pyrrolizidine **440** as only product, in 66% yield by two consecutive (Cy 5) cyclizations (Cy5/Cy6 cases).[167] Cuprous halide initiation gives only monocyclic compounds by fast ligand transfer to **438**: this has been used as an argument for a radical chain mechanism in titanous initiated cyclization of unsaturated N-chloroamines[167] (see Section VIII.3.A, Scheme 42).

Alkyl-substituted radicals analogous to **437** give similar results,[532] except when the radical analogous to **438** is tertiary, when only the monocyclic product is obtained: in this case, cyclization of a tertiary radical would be less facile than cyclization of a primary radical (see Section IV, where conflicting monocyclization results have been discussed). The hexenyl homolog of **437** gives, under the same conditions, the corresponding chloromethylindolizidine as a mixture of epimers in a reaction involving, as the first step, formation of a (Cy 6) radical in a Cy6/Cy7 case. Intramolecular attack on an aromatic ring, which is the last step of the polycyclization process, has also been studied.[167] Thus the aminyl radical **441** (Scheme 154) generated by titanous salt reduction of the corresponding N-chloroamine gives a mixture of mono- and bicyclization products. The yield of benzoindolizidine (**443**) is only 35%.

That attack on an aromatic ring is less favored than attack on a double bond is confirmed by the behavior of N-chloroamine **81** ($X = CH_2$) (Section VIII.3.A, Scheme 43).[178] Only the monocyclized compound **82** is obtained

SCHEME 154

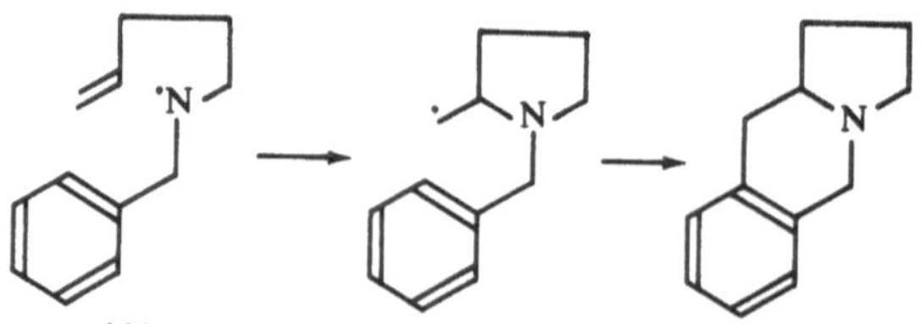

441 442 443

(92% yield) with none of the benzomorphane which would result from attack by the intermediate (Cy·6) radical on the aromatic ring. But since the benzomorphane can be obtained from **82** in 60% yield by an aluminum-chloride-induced cyclization the overall process is an attractive one for the synthesis of benzomorphan. This result may be compared to those described by Belleau,[533-535] who synthesized morphinans by cyclizing ethylenic amines or *N*-chloroamines under ionic conditions.

C. Polycyclization Reactions Involving Addition to a Polar Bond

The first step of a polycyclization reaction involving free radical addition to the oxygen atom of a carbonyl group has been described in Section X.1.B.b, Scheme 83. The second step proposed[339,341] involved cyclization of the radical **190** to the (Cy·4) radical in a Cy4/Cy5 case. According to the discussion in Section V.2.B, this is unlikely and, in fact, this proposal was later rejected.[342]

Addition to an ethylenic bond followed by addition of the radical formed to the oxygen atom of a carbethoxy group has been discussed in Section X.2, Scheme 89. But since the medium contains a cupric salt, the lactone **204** may result from an ionic cyclization, involving oxidation of the radical to the carbenium ion.[352b]

Johns and Willing[372,373] have reported an interesting example of polycyclization involving addition to a cyano group (Scheme 155). The reaction of the diallyl amine **444** with AIBN gives the ketone **448**. This would result from addition of the cyanoisopropyl radical to **444** followed by Cy5 cyclization of **445** in a Cy5/Cy6 case to the pyrrolidine radical **446**. This radical would cyclize to **447** by Cy6 intramolecular addition to the cyano group in a Cy6/Cy7 case and, finally, the intermediate imine would be hydrolyzed to the ketone **448**.

5. Conclusion

Free radical intramolecular reactions are of synthetic value for the preparation of bi- and polycyclic compounds. These reactions are often

SCHEME 155

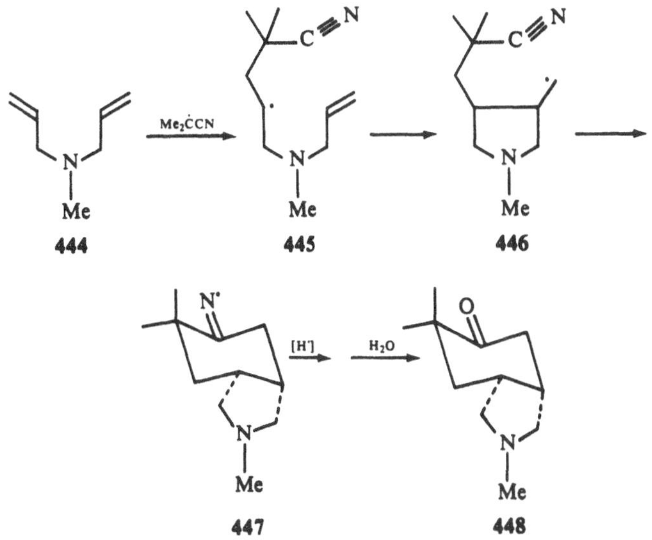

444　　　445　　　446

447　　　448

remarkably selective but the origin of the stereoselectivity is generally not obvious. Since there is good evidence for important biosynthetic pathways involving free radical polycylizations a better knowledge of these reactions is needed.

XII. FREE RADICAL INTRAMOLECULAR ADDITION AS A MECHANISTIC TOOL AND A KINETIC STANDARD

One of the most frustrating problems in the study of reactions mechanisms is the unambiguous demonstration of the occurrence of postulated intermediates on the reaction pathway. In free radical chemistry physical methods such as esr or CIDNP have made possible enormous progress in the understanding of the reactivity by often giving a precise idea of the structure of the intermediates involved. An important drawback of these techniques is their very high sensitivity, which does not permit one to assume that the observed free radical really is on the reaction pathway.

Free radical intramolecular additions, providing as they do an internal trap for the free radical in a fast and selective manner, have been one of the most popular tools used in recent years to demonstrate the occurrence of radical intermediates on the reaction pathway.

Furthermore, the cyclization rate of the 5-hexenyl radical is probably now the best known reaction rate, so that it has been used in competitive reactions to provide rather precise information concerning the rates of many other free radical reactions.

1. The Mechanistic tool

A. The 5-Hexenyl Radical

As discussed in Sections II and III, the cyclization of the 5-hexenyl radical, under usual temperature and concentration conditions, is a fast, irreversible, and selective process to the (Cy˙5) radical. Furthermore, it is well known that the corresponding 5-hexenyl carbenium ion gives rise to significant amounts of cyclohexyl products.[536-538] Therefore the occurrence of (Cy5) products and no Cy6 products from postulated precursors to the 5-hexenyl radical, is a positive proof that the postulated intermediate is really a free radical rather than a carbenium ion. Furthermore, if the yield in (Cy5) products is good, this will demonstrate unambiguously that the postulated free radical is really on the reaction pathway. Under these conditions, if it is supposed that from a starting material (RX), an intermediate (R*), free radical or carbenium ion is involved on the reaction pathway, the use of R = 5-hexenyl will enable one to decide according to the Cy5 or Cy6 nature of the products if the intermediate is a free radical or a cation (Scheme 156). Nevertheless, it will be seen at the end of this section that in a few cases, (Cy5) compounds may arise from other pathways such as concerted ones: for instance this is observed with organometallic compounds. In these ambiguous cases clearly other pieces of evidence will be needed.

The first use of 5-hexenyl radical cyclization as a mechanistic tool was proposed in 1966 by Garst and Lamb[539] for the study of alkyl halides reduction by naphthalene radical anion. Since then, Garst has extensively used this method in the study of the one-electron transfer reactions from aromatic radical anions to alkyl halides; critical discussion of the use of the method may be

SCHEME 156

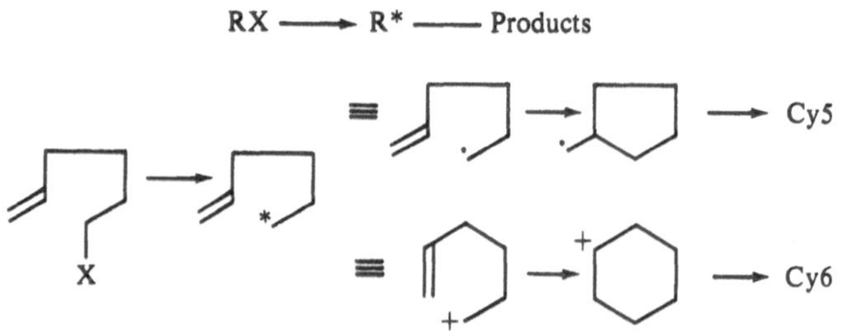

RX $\longrightarrow$ R* $\longrightarrow$ Products

found in his papers.[135,540] Other one-electron reduction processes have been studied by using the 5-hexenyl radical cyclization: reductive elimination of a cyano group on treatment with alkaline metals in HMPA (no cyclization observed)[541] naphthalene-catalyzed electrochemical reduction of the carbon–chlorine bond[542]; lithium naphthalenide reduction of quaternary ammonium salts[543]; reduction of alkyl halides by $CpV(CO)_3H^-$ [in this case cyclized products are observed from the radical $CH_2=CHCH_2XCH_2CH_2^\bullet$ (X = O) but not from the 5-hexenyl radical (X = CH_2) because of the higher cyclization rate (see Section VI.2) and would not result from an electron transfer mechanism][544]; reduction of alkyl halides by potassium graphite[545]; methoxy group displacement resulting from the reaction of o-(methoxy)-aryloxyoxazolines with organolithium or Grignard reagents (in this case an electron transfer process is discarded since the uncyclized/cyclized ratio is virtually the same as the one obtained from the direct cyclization of the ethylenic Grignard reagent: see Section XII.1.D.b)[546]; coupling and hydroxylation of lithium and Grignard reagents by oxaziridines[547]; $LiAlH_4$ reduction of aryl bromides[636] (by using o-bromophenyl allyl ethers more easily obtained and prone to cyclize: see Section VI.2, Scheme 23); and coupling reactions of silyl anions with organic halides in HMPA.[699]

Many other examples have been reported on the use of the 5-hexenyl radical cyclization as a proof that alkyl radicals are intermediates in such reactions as the following: photochemical decarboxylation of acids with thallium(III),[548] reaction of organomercury compounds with iodine where only the (Cy5) compound is obtained instead of the (Cy6) compound under acidic treatment,[549] photolysis of sulfides,[244,246b] substitution of alkyl halides by allylic organotin compounds,[550] synthesis of organocalcium compounds from iodoalkanes,[551] chlorination of olefins in gaseous phase,[552] thermolysis of methyl enol ethers of ethylenic ketones,[553] reductive dehalogenation of alkyl iodides with zinc and acid,[554] and coupling of diorganophosphides with organic halides.[555]

In other examples, the fact that no cyclized products (Cy5 or Cy6) are observed from possible 5-hexenyl intermediates has been used as proof that an S_N2 process was more likely than a free radical or carbenium ion pathway: this is the case in the solvolysis of 4-alkoxypyridinium salts.[556]

More familiar reactions but ones whose mechanism has been a controversial subject of interest for a long time have also been studied by this method. This is the case of the reductive demercuration of alkylmercuric halides by metal hydrides. In a careful study , Quirk[557,558] was able to conclude that alkyl radicals are intermediates and that the process is a radical chain mechanism. This conclusion resulted from the knowledge that the cyclization was a relatively slow process (see Section XII.2) compared with the recombination process in the solvent cage, a conclusion recently confirmed by CIDNP experiments.[559] Furthermore, the use of metal deuterides and the measurement of the kinetic isotope effect enabled Quirk to conclude that metal hydride reduction of alkylmercuric halides proceeds, in all the cases studied, via similar mechanistic pathways, i.e., hydrogen atom transfer to alkyl radicals from a common hydrogen transfer agent, presumably the alkylmercuric hydride.

The mechanisms of peroxide decomposition[559] and of Grignard compound formation[560] have been studied with the 5-hexenyl system in connection with CIDNP studies. The detailed investigation of Bickelhaupt[560] on the formation mechanism of Grignard reagents led to a better understanding of the influence of the solvent on the radical pairs involved in the process. Valuable conclusions have been drawn by Ashby concerning the radical mechanism of alkyl transfer in reactions of Grignard reagents with ketones.[561] An interesting observation was made in this study, concerning the use of the 5-hexenyl system as a radical probe. In the reaction of 5-hexenylmagnesium bromide with benzophenone no (Cy5) products were observed, while they were observed in high yield with the tertiary Grignard reagent 1,1-dimethyl-5-hexenylmagnesium chloride. This would not result from an easier cyclization of the tertiary radical (see Section IV) but from the fact that the tertiary radical, being more stable, can escape from the solvent cage and cyclize out of the cage instead of being trapped in the cage in the uncyclized form. The 5-hexenyl probe has afforded convincing evidence of a radical chain mechanism during the autoxidation of Grignard reagents to alcohols.[562,563]

The 5-hexenyl system has been used by Garst to demonstrate that the intermolecular portion of the Wittig rearrangement of aralkyl alkyl ethers and the ketyl–alkyl iodide reaction involve free radical intermediates.[564–566]

Finally the 5-hexenyl radical tool has also been used extensively to demonstrate that alkyl radicals are involved in new organometallic reactions such as the formation of platinum to carbon bonds during the reaction of alkyl halides with $Pt(PEt_3)_3$,[567a] as well as the formation of thorium to carbon bonds during the reaction of Grignard reagents with Ph_3ThCl,[567b] in the reductive demercuration of dialkylmercury compounds by heating in carbon tetrachloride,[568] in the alkylation of dinitrogen coordinated to molybdenum(0) and tungsten(0) by alkyl halides,[569] in the reaction of inorgano-Grignard reagents, i.e., compounds with transition metal–magnesium bonds such as Cp(diphos)FeMgBr with organic bromides,[570] in the reaction of alkyl cobaloximes with sulfur dioxide[571] and with oxygen,[572] and possibly in the coenzyme B_{12}-dependent methylmalonyl–CoA mutase reaction.[693]

This certainly very incomplete and fast growing list of examples was provided in order to show that the 5-hexenyl radical tool may be used in many cases. Other examples in which the 5-hexenyl cyclization also affords kinetic values will be described in Section XII.2.

B. Other Carbon-Centered Radicals

The 5-norbornenyl–3-nortricyclyl radical rearrangement Section V.2.B, Schemes 20 and 21) is also an interesting tool, because the equilibrium is quickly established, possibly in as little as $10^8 \sec^{-1}$ at $25°C$ [55] (see also Reference 77a). Since this is faster than the 5-hexenyl cyclization ($10^5 \sec^{-1}$ at $25°C$ [55]), it makes this tool an attractive one when faster reactions than the 5-hexenyl cyclization are needed. Nevertheless, an important drawback of the method is that a mixture of rearranged products, depending on the nature of the transfer agent, is obtained. An analogous mixture may be obtained from the corresponding cationic,[573,574] carbanionic,[479,575] or organometallic[639] intermediates. Under these conditions the formation of rearranged products cannot be considered as a positive proof for the intermediacy of alkyl radicals. Nevertheless, it was used as an interesting guide in reactions such as the cobaltous chloride-catalyzed reaction of organic halides with Grignard reagents,[576] the radical-anion reduction of alkyl halides,[473] the hydride reduction of organomercurials,[474,476,477] the methyllithium reaction with alkyl halides and *gem*-dihalides,[478] the radical addition of CCl_4 to olefins using the intermediate complex formed by reduction of trimethylamine oxide with iron pentacarbonyl[480] as catalyst and the bromination of olefins by bromine[486] or *N*-bromosuccinimide.[475] Once a radical mechanism has been recognised, the cyclized/uncyclized product ratio dependence on the nature of the transfer agent (see Scheme 20, Section V.2.B) may be used to give more detailed information concerning the reaction mechanism. In this way, Lessard was able to conclude that the chromous-promoted addition of *N*-halo-amides to a double bond involves a halide transfer reaction from the *N*-halo-amide in the last step rather than a ligand transfer reaction from the oxidized salt.[77b]

Allylalkyl radicals do not cyclize, with the exception of the 5-norbornenyl–3-nortricyclyl rearrangement or when the corresponding cyclopropyl radicals are highly stabilized (see Section V.2.B). In fact, the process generally observed is the opening of cyclopropylalkyl radicals to give products resulting from the allylalkyl radicals. This is also a fast process ($1.3 \times 10^8 \sec^{-1}$ at $25°C$,[69] (probably even faster when the open radical is stabilized[73b]) and hence is a useful tool. It has been used for instance to assess whether or not the peroxide-induced reaction of alkyl iodides with allylmethyl iodide leading to side-chain-substituted methylcyclopropane proceeds by an SH reaction rather than by a cyclopropylmethyl radical,[577a] to confirm the radical mechanism of the reduction of oxymercurials by sodium borohydride,[477] and to propose the intervention of an alkyl radical $R^\cdot$ in the photoreduction of esters R^1CO_2R to R^1CO_2H and RH in HMPA.[456] It has also been used to confirm the biradical mechanism for the photolytic opening of cyclopropyl ketones,[71] to distinguish between ionic and radical halogenation of double bonds,[577b] to propose that nucleophiles such as Me_3Sn^- substitute alkyl halides by an electron transfer mechanism[577c] (see below, however), to study the sodium–ammonia reduction mechanism of allenyl vinyl cyclopropanes,[577d] to propose that *t*-BuOK in DMSO is able to transfer one electron to phenylcyclopropanes,[577e] or to discard the involvement of tripletlike carbenoid or diradical intermediates in the copper-catalyzed reaction of methyl diazomalonate with olefins[†] by choosing dicyclopropylethylene as the substrate: no rearranged products are produced.[672] As with the preceding method an apparent drawback of this one is owing to the fact that the cyclopropylmethyl cation also undergoes a ring cleavage reaction to the butenyl cation. An important difference in the behavior of the two intermediates, however, results from the fact that the cyclopropyl-methyl cation is in equilibrium with the open cation and the cyclobutyl cation so that a mixture of products resulting from the three intermediates in equilibrium is obtained,[578a] although low-temperature nmr studies do not agree completely with this conclusion.[578b] On the other hand, if the methylcyclo-propyl radical is the intermediate the products obtained result exclusively from the open radical. It is possible to make a nice use of this difference by using in succession the three alkyl groups cyclobutyl, cyclopropylmethyl, and 3-butenyl. If the mixture of products obtained is the same, under the same conditions, in the three cases and derives from the three isomeric groups, it may be safely con-cluded that a cationic intermediate is involved as in the thermal decomposition of diacylperoxides.[579a] If rearranged products, in the 3-butenyl form, are observed only from the cyclopropylmethyl group when no rearranged products result from the 3-butenyl and cyclobutyl groups, it may be safely concluded that a homolytic pathway occurs (or at least that a heterolytic pathway may be ruled out) such as in the reaction of organocobalamins with oxygen.[579b] Nevertheless, on account of the particular nature of the substrates involved, this mechanistic tool must be handled with great care, as exemplified by recent studies. For instance, in a case where a nucleophilic substitution on alkyl halides seems clearly established to proceed by electron transfer, the same reac-tion with halomethylcyclopropane does not always lead to the expected open products: another mechanism would operate in this case.[544] On the other hand, open products may be formed by another mechanism[580a] as well as rearranged products resulting from a butenyl system.[580b] For instance, it was concluded that the reaction of lithium dimethyl cuprates with β-cyclopropyl-α,β-unsaturated ketones occurred by electron transfer since ring-opened products, attributed to ring opening of an intermediate radical anion, were obtained. By using stereospecifically monodeuterated compounds, however, it was observed that the products resulted from a stereospecific ring opening with inversion. This result, which provides evidence against radical anion intermediates, was interpreted in terms of direct nucleophilic attack of cuprate at the cyclopropyl

---

[†] See Vol. 1, Chapter 5 of this series.

carbon atom.[673] In the same way, the intermediacy of free radicals in the reaction of tin anionoid species with organic halides discussed above[577c] is now dubious since the use of the 5-hexenyl probe, carefully conducted by taking into account rate differences, does not give any evidence for free radical intermediates.[700]

The cyclization of alkyl-substituted cyclopentenylethyl radicals analogous to **291** (Scheme 116) has also been used as a mechanistic probe to study oxidative decarboxylation by lead tetraacetate with or without cupric salts in various solvents.[581]

In all the cases discussed above, the fate of the assumed radical R˙ (leading to cyclized products) is well known, which facilitates the interpretation of the results. It is not always possible to use such radicals as, for instance, if the assumed radical on the reaction pathway is necessarily benzylic. Nevertheless, the use of the intramolecular free radical addition probe may be fruitful. One example is found in the study of the mechanism of substitution of *p*-nitrobenzyl halides by anions.[582] Some of the best evidence for a single-electron transfer mechanism has been afforded by using 1-chloro-1-*p*-nitrophenyl-5-hexene as substrate: the formation of cyclized products, as a Cy5/Cy6 mixture, confirmed that the reaction proceeded by a chain mechanism involving the *p*-nitrobenzyl radical.[583] In the same way, it was possible to anticipate the redox-catalyzed addition of benzylic chlorides to olefins, proceeding by an analogous electron transfer mechanism.[584] The intramolecular trap has also been used to confirm the mechanism of thiazolidine formation assumed in Scheme 52 (Section VIII.4.A): the use of a 3-butenyl group as the alkyl group R on nitrogen leads to Cy5 products resulting in the trapping of the postulated allylic radical.[221] The intramolecular free radical addition probe appears also to be useful for acyl radicals, but since the behavior of unsaturated acyl radicals is relatively unknown (see Section VII.4) control experiments are needed. Bearing this in mind, it has been concluded safely that decarbonylation of aldehydes by the Rh(I) Wilkinson catalyst does not involve free radical intermediates,[639] a conclusion confirmed by the fact that Rh(I)-catalyzed intramolecular hydroacylation of unsaturated aldehydes results in high yields of cycloalkanones only in the Cy4/Cy5 case[674] (formation of cyclopentanones), which is, of course, not the result expected from a free radical cyclization process. A recent report exemplifies the care which must be exercised in using the cyclization probe with radicals whose behavior is not known. Thus it has been suggested that the formation of α-chloroketones by reaction of silyl enol ethers with an excess of FeCl$_3$ involved a vinyloxy radical since 4-chlorocyclohexanone was obtained from 2-[(trimethylsilyl)oxy]-1,5- hexadiene.[701] Although the exclusive formation of the (Cy6) compound in a Cy5/Cy6 case is not impossible from a (A˙) radical which may be considered stabilized (Section VII.4), conclusive evidence for the proposed mechanism must await further understanding of the behavior of vinyloxy radicals, unambiguously generated, on cyclization, particularly if one recalls how difficult it is to rationalize the behavior of aryloxy radicals which can be considered somewhat analogous (Section VIII.2.b, Scheme 35).

C. Heteroatomic Radicals

The problem to be examined now is as follows: is it possible to use a system such as $CH_2=CH-(CH_2)_3X^*$ analogous to the 5-hexenyl one, in order to prove that a reaction mechanism involves a heteroatomic radical X^*? In fact the answer to this question may be found in Section VIII: with oxygen- and nitrogen-centered radicals, but not with other heteroatomic radicals, the cyclization process is fast and selective to the (Cy˙5) radicals. Under these conditions the observation of rearranged (Cy5) products may be considered as evidence for oxygen or nitrogen radical intermediates. For instance, this probe has been used among others to demonstrate that the thermal decomposition of alkoxy-copper(I) reagents generates alkoxyl radicals since thermal decomposition of the copper(I) salt of 4-penten-1-ol gives only the (Cy5) product, 2-methyltetra-hydrofuran, albeit in low yield.[585] The interesting behavior of ethylenic *N*-chloroamines has been discussed at length in Section VIII.3: from the high selectivity toward the (Cy5) compounds observed under experimental conditions favouring a homolytic pathway, and from the absence of selectivity generally observed under solvolytic conditions, it was concluded that in the latter case the most probable pathway was through a nitrenium ion intermediate. Nevertheless, great care in the interpretation of heterocyclization results must be taken since a competitive and often very selective cyclization to the (Cy5) products results from the attack of a cationic species to the double bond followed by nucleophilic attack by the heteroatom (see Section VIII).

Intramolecular addition of alkoxyl radical has been used as an interesting complementary probe for alkyl radical intermediates. This has been worked out by Garst in the study of electron transfer reactions such as the reaction of halohydrocarbons with aromatic radical anions,[135,540c] the Wittig rearrangement of aralkyl ethers,[566] and the cyclization of 1,4- and 1,5-dihaloalkanes with alkali naphthalenates.[586] It was also used to study the reduction of alkyl halides by potassium graphite.[545] The principle of the method is exemplified in Scheme 157: the tetrahydrofurfuryl radical **449** does not open to the pentenyloxy radical **450**; in contrast the tetrahydrofurfuryl anion **451** does open readily to the pentenolate anion **452**. It is then possible to distinguish the products resulting directly from the radical from those obtained by further reduction into the anion.

Of course the opening of the lower homologs of **449** is a very fast process. For instance, the opening of the oxiranylmethyl radical to allyloxy radical is a fast process $>10^7 \, sec^{-1}$ (Reference 127). By using α-iodoepoxides as radical probe this permitted the conclusion to be reached that the reductive elimination of α-iodoepoxides to allylic alcohols with sodium naphthalides, as well as the

SCHEME 157

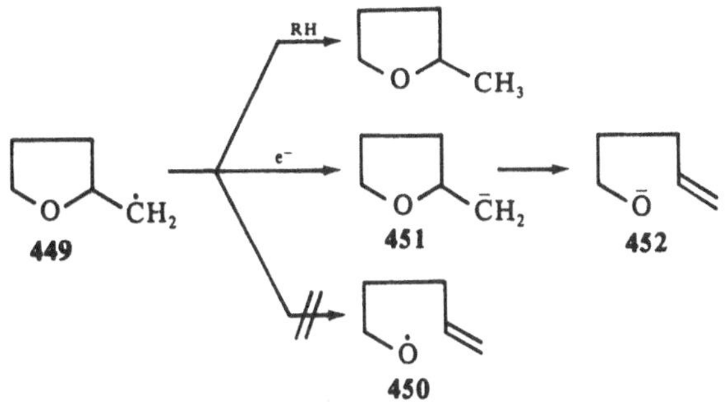

exchange reaction between butyllithium and alkyl iodides, proceeded by a single electron transfer pathway.[127]

D. Reactions related to Free Radical Intramolecular Additions

In this section we shall describe some reactions which present many features similar to intramolecular free radical additions, according to the mechanistic probes discussed in the preceding sections. These examples will permit a critical discussion of the use of the free radical intramolecular cyclization probe.

a. Photochemical Cycloaddition Reactions of Nonconjugated Olefins

By irradiation (particularly in the presence of photosensitizers) of 1,5-dienes a remarkable preference for the formation of (Cy5) products is observed: this is now known as the "rule of five."[587] This similar selectivity to the one observed during cyclization of the 5-hexenyl radical was recognized early[588,589] and considered as the result of a first intramolecular addition step under kinetic control in a Cy5/Cy6 case leading to a long-lived (Cy5) diradical intermediate.[590] Since then many other examples have been reported because of the interest of these reactions in the synthesis of bicyclic compounds. It is beyond the scope of this review to comment on all of these examples. Some of them abstracted from Dilling's review[591] are summarized in Scheme 158. The formation of the main products generally obtained may be easily rationalized by considering one of the double bonds as a triplet adding intramolecularly as does an alkyl radical to the other double bond. For instance, 1,5-hexadiene (453) gives 454, resulting only from the (Cy˙5) radical in a Cy5/Cy6 case, no products resulting from the (Cy˙6) radical or products resulting from a Cy4/Cy5 cyclization being obtained. In the same way, neryl acetate (455) gives

SCHEME 158

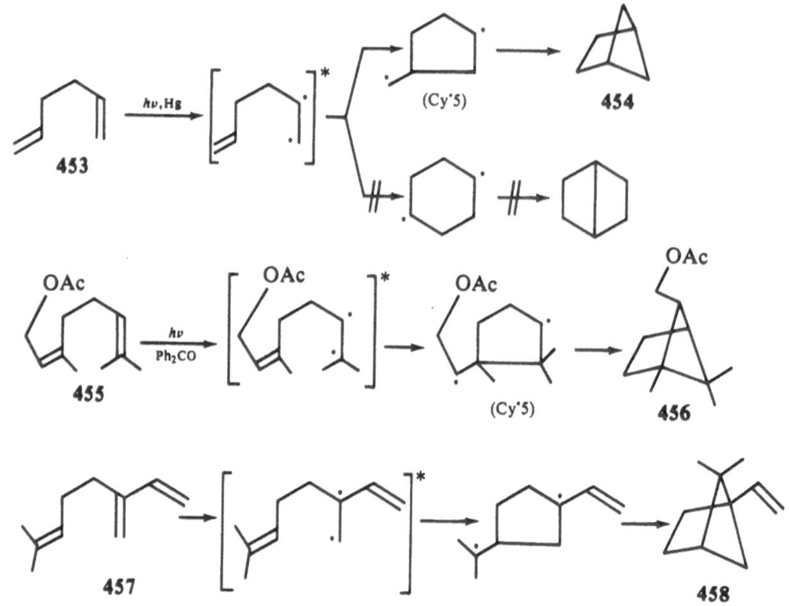

the bicyclic compound **456** in 90% yield. Even more striking is the fact that myrcene (**457**) gives **458** as the only volatile product. Furthermore not only the regio- but also the stereoselectivity toward the *cis* Cy5 compound appear to parallel those observed with 5-hexenyl radicals (Section XI.3, Scheme 137), as observed, for example, in the photolysis of geranonitrile.[370b] Raising the temperature appears to improve the analogy between photochemical (diradical) and true free radical cyclizations: photolysis of citral and related compounds such as geranonitrile gives only five-membered ring diradicals in the Cy5/Cy6 case as well as rearranged compounds. Formation of the latter is easily rationalized by postulating radical intramolecular addition onto the carbonyl or cyano groups.[702]

The same pattern is encountered with other nonconjugated olefins such as norbornadiene (**459**) (Scheme 159), which gives quadricyclane (**460**) in 95% yield by photosensitization.[591] Furthermore, this reaction is reversible, in agreement with the 5-norbornenyl–3-nortricyclyl rearrangement. The addition products obtained in the reaction of benzophenone triplets to norbonadiene and quadricyclane may also be rationalized in an analogous manner with the biradical **461** undergoing fragmentation[592a] (Scheme 159). Such considerations would have avoided faulty structure proposals such as those for the products resulting from the photocycloaddition reactions of norbornadiene and

SCHEME 159

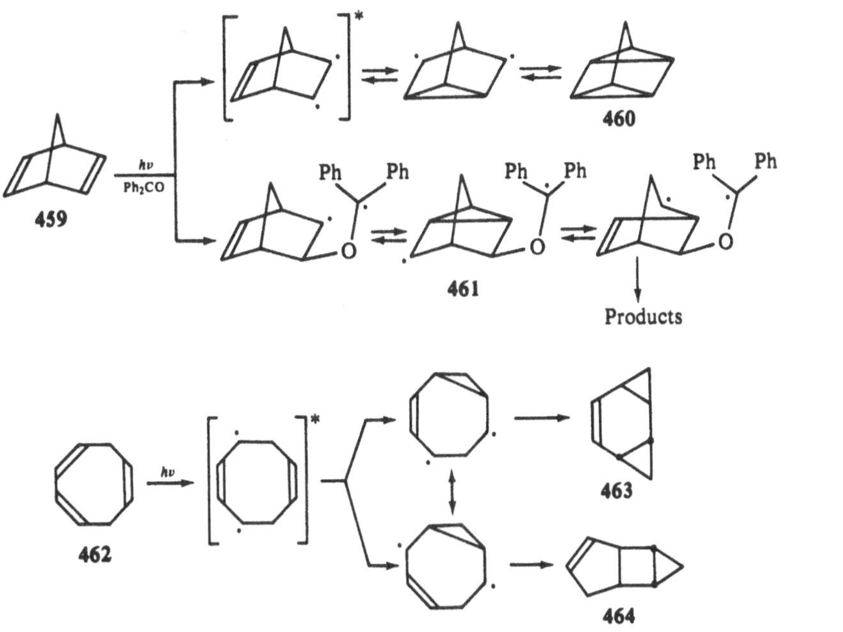

quadricyclane with *p*-benzoquinone recently reinvestigated.[675] Some of the products (463) (11%) and (464) (18%) obtained by irradiation of 1,3,6-cyclo-octatriene (462) are also easily explained by a Cy3/Cy4 cyclization giving the (Cy·3) diradical in the first step[591] (Scheme 159). From a mechanistic study of the photorearrangements of vinylcyclopropenes it appears that the free radical analogy works for part of the triplet reaction but not for the singlet rearrangement.[676]

Of course, if the analogy with free radical cyclizations is good, such cyclizations in the Cy3/Cy4 case must be exceptional. This is observed in the di-π-methane photolytic rearrangement of 1,4-dienes where the products result from the opening of a methylcyclopropyl radical,[592b] as well in the photodecarbonylation reactions of bicyclo[3.1.0]hexanones.[703] In the same manner the Cy4/Cy5 cyclization must be a difficult process. In this connection it is interesting to note that the diradical mechanism proposed for the type A photorearrangement of substituted 2-cyclohexenones to lumiketones has been recently firmly ruled out by using chiral enones[592c]: more simply, such a diradical mechanism did not appear an attractive one since it involved a Cy4/Cy5 cyclization step.

It is also very interesting to compare the photochemical behavior of 1,6-

SCHEME 160

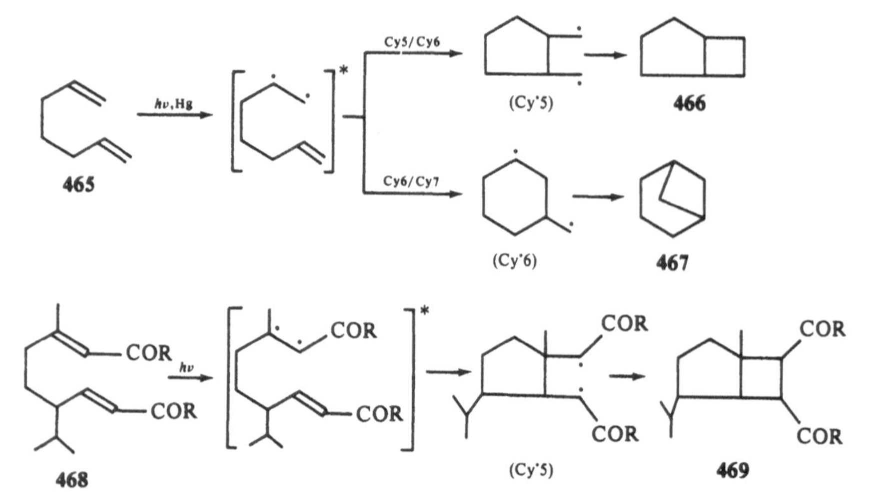

dienes with the intramolecular free radical addition. For instance, 1,6-heptadiene (**465**) (Scheme 160) gives both of the intramolecular cycloadditions products **466** and **467** by mercury-sensitized photoisomerization, **466** being the major one.[591] Such a result may be easily rationalized if **466** is considered to result from a Cy5 cyclization in a Cy5/Cy6 case, whereas **467** would result from a Cy6 cyclization in a less favored Cy6/Cy7 case (see Section V.1). Even more selective is the (usually—but not always—sensitized) photochemical cycloaddition of the 1,6-diene (**468**) (R = OEt or CH$_3$) (Scheme 160) since only **469** (53% yield) (a precursor for the synthesis of the sesquiterpene α-bourbonene) resulting from the (Cy˙5) radical is obtained.[589] Furthermore, it must be noted from these two examples that in order to observe the last step the two radical-bearing-groups in Cy˙5 must be in a relative *cis* position, i.e., the first cyclization step must follow the same *cis* stereoselectivity observed with 5-hexen-1-yl-1-alkyl radicals (see Section XI.3). In full agreement with this conclusion are the more recent results of the cyclization of 2,7-octadienones,[593a] which may be considered a generalization of the selective cyclization of 2,6-octadienals such as citral.[593b] Equally interesting is the behavior of 1,5-hexadiene-3-one corresponding to **453** which, by direct irradiation, gives the bicyclic ketone corresponding to **454** in fair yield,[677] compared to the behavior of 1,6-heptadiene-3 one, corresponding to **465**, which does not give any bicyclic product.[678] Although in these two cases a (Cy˙5) diradical resulting from a Cy5/Cy6 cyclization can be formed, this appears somewhat reminiscent of the easy Cy5/Cy6 free radical cyclization compared to the more difficult homologous Cy6/Cy7 cyclization. Nevertheless, photobicyclization of functionalized 1,6-dienes is not a forbidden process: for instance, copper(I)-catalyzed irradiation of 1,6-heptadien-3-ols affords bicyclo[3.2.0]heptan-2-ols corresponding to **466** in high yields.[679]

It is also worthwhile comparing the intramolecular photochemical cycloaddition reactions of ethylenic aldehydes and ketones with free radical intramolecular additions. For instance, irradiation of 5-hexen-2-one (**470**) (Scheme 161) in the gas phase gives the oxetane **471** as only cyclized product,[591] as expected from the known photochemical intermolecular reaction between olefins and ketones. If the irradiation is conducted in solution **470** gives **471** (26%) and **472** (18%). With other γ,δ-unsaturated ketones, the bicyclic compound analogous to **472** may become the major product.[591] With 2-allylcyclanones such as **473** (Scheme 161) bicyclic compounds are obtained (80% yield) as a mixture of **474** and **475**, with **475** being the major product, but such compounds are difficult to isolate.[594a,680] In the same manner, selective irradiation of the carbonyl group of 2-acyl-2,3-dihydro-4*H*-pyrans (**476**) leads exclusively (23% yield) to *exo*-brevicomin (**477**) (a sex attractant), neither oxetane formation nor Norrish type II reaction being observed.[594b] The formation of the compounds **472**, **475**, and **477** which was considered as unexpected

SCHEME 161

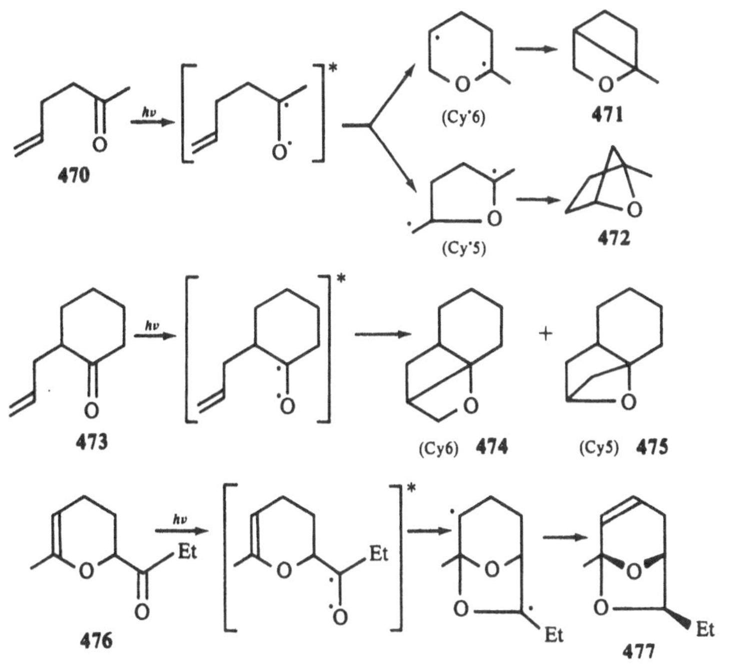

is easily rationalized in the same manner, as is the photolytic behavior of the diene **453** (Scheme 158) if one considers the excited ketone as a diradical. The first step in the cycloaddition process [followed by an intramolecular dispropor-tionation in the case of **476**] may therefore be considered as the easy selective intramolecular addition of an alkoxyl radical to the (Cy˙5) radical in a Cy5/Cy6 case (see Section VIII.2.A).

This way of rationalization is substantiated by the photolytic behavior of δ,ε-unsaturated aldehydes (**478**) (Scheme 162). Only in one of the five cases studied[595a] was **479** observed: this result is in accordance with the difficult intramolecular addition of alkoxyl radicals in the Cy6/Cy7 case. Since thiyl cyclizations appear efficient even in the Cy6/Cy7 case one might expect to observe easy photocyclizations of unsaturated thiones also in this case: the easy photochemical synthesis of quinolines from *N*-(*o*-styryl)thioamides appears to be an excellent illustration of this view. Here, the sulfur analog of **479** would be an intermediate.[681,704] To complete the analogy, one may expect the converse β-scission reaction to be efficient with radicals that readily open such as oxyranylalkyl ones (see Section VIII.2): in this way an interesting macrolide synthesis was described by Carlson.[595b] Photolysis of epoxycyclanone (**480**)

SCHEME 162

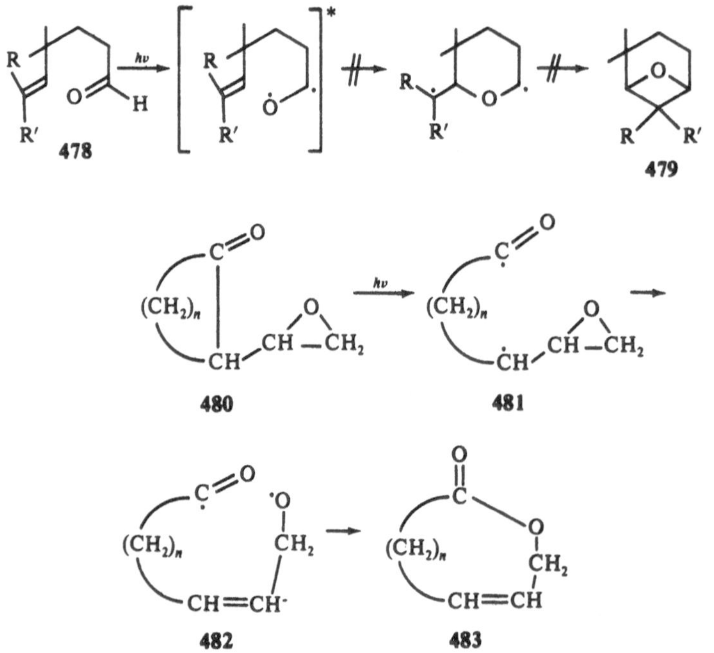

leads to the diradical **481** which, by opening of the oxyranylalkyl radical, rearranges to allyloxyl radical **482**, which couples to the macrolide **483**. Other results obtained from the photolysis of epoxyketones agree with this scheme. Of particular interest is the demonstration of a very selective homolysis of the carbon–oxygen bond rather than the carbon–carbon bond in the oxiranylmethyl radical such as **481**, even when stereoelectronic factors would favor the carbon–carbon bond homolysis.[642] Of possible interest for relative rate determinations is the conclusion that type II cyclization and elimination are apparently faster processes than epoxycarbinyl radical opening as observed in the photoreaction of epoxynaphthoquinones.[649]

Photochemical rearrangements of β,γ-unsaturated ketones have been extensively studied. Although a variety of subtle structural features may affect the mechanism and products, in some instances the oxadi-π-methane rearrangement may be considered to involve the same features as free radical cyclization. By analogy with the above examples one could predict intramolecular addition of the oxygen atom of the carbonyl group as the first step but this process would correspond to an unfavored Cy4/Cy5 cyclization, which is not observed. In fact the product formed can be readily explained by intramolecular addition of the carbon atom of the carbonyl group in a more favored Cy3/Cy4 case to give the expected (Cy·3) cyclopropyloxy radical followed by the expected β-scission process to form a new, 1,3-diradical which then closes to the product.[682]

In conclusion, many results of intramolecular photochemical cycloaddition reactions of nonconjugated olefins are easily rationalized by comparing them to the corresponding free radical intramolecular reaction. Of course such a simple rule must be used with care, having due regard for the other pathways available in photolytic reactions. From a recent study of the intramolecular photocycloaddition reactions of 3-allyl-substituted diphenylcyclopropenes which would proceed by intramolecular trapping reaction of vinylcarbenes, it appears that it would also be quite interesting to compare trapping of carbenes with radical intramolecular trapping.[690]

b. Cyclization of Olefinic Organometallic Reagents

It must be pointed out that this section is not concerned with the cyclized products obtained *during* the Grignard reagent formation from unsaturated halides[596] (or during the inverse opening reaction of strained halides) for which good evidence favors a free radical mechanism (see Section XII.1.A). The cyclizations studied here are slow reactions which necessitate boiling the preformed Grignard reagent for several hours in a solvent such as THF, the rate constants being in the range 10^{-2}–$10^{-8} \, \text{sec}^{-1}$ (to be compared with $10^5 \, \text{sec}^{-1}$ for the 5-hexenyl radical cyclization).

Since Roberts[597] observed the cyclopropylmethyl $\rightleftharpoons$ 1-but-3-enyl Grignard reagent rearrangement in 1960, the cyclization of unsaturated Grignard reagents as well as the opening of strained cyclic ones have been studied in a range of systems (main references until 1975 will be found in Reference 598). It is not the aim of this section to describe in detail these cyclization reactions but to use them as examples to show a possible limit of the cyclization probes for alkyl radicals described in Sections XII.1.A and XII.1.B. Indeed, it is only since 1975 that some agreement was reached concerning the mechanism of cyclization of unsaturated Grignard reagents, considered now as a concerted, four-center mechanism[598-600]; until that time, in addition to a concerted mechanism, three alternative mechanisms had been proposed, namely, radical, carbanion, and electron transfer.

According to Sections XII.1.A and XII.1.B, apparent positive proofs for a homolytic pathway were: a facile intramolecular cyclization in the Cy5/Cy6 case of the Grignard reagent of 6-chloro-1-heptene (**484**) (Scheme 163) (R = Me) which gives the (Cy5) compound **485** as the only cyclized product (88% yield after 5 hr boiling in THF),[601] no cyclization in the Cy6/Cy7 case,[602] rearrangement of cyclopropylmethyl to 1-but-3-enyl Grignard reagents,[597] and cyclization retarded in the Cy5/Cy6 case by an R_3 substituent[295,603] (see Section IV, Table 2). Nevertheless, the rate decrease (about 1000-fold) is much higher in this case than in the corresponding free radical cyclizations and an analogous rate decrease is observed with *trans*-R_4 substituents[295,603] when a small increase is observed during free radical cyclizations (Table 2). Other features which do not parallel the results of free radical cyclizations are: the isomerization of the Grignard reagent prepared from 5-chloro-1-hexene **486** to **487**[601]—this easy

SCHEME 163

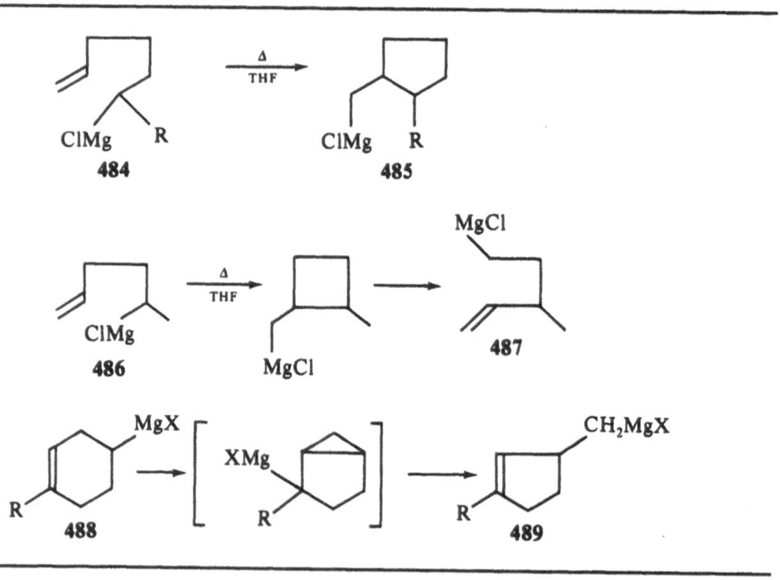

isomerization to the (Cy4) compound in a Cy4/Cy5 case is not in good accord with a radical cyclization (see Section V.2.A); the isomerization of the Grignard reagents of 4-chlorocyclohexenes **488** (R = H,Ar) to **489**,[604,605] whereas the converse is true during free radical cyclization, i.e., **347** (Scheme 133) corresponding to **489** gives **348** corresponding to **488**; and finally, the cyclization of compounds **484** which give the cyclized products **485** mainly in the *trans* form[295] [*cis/trans* ratio: 1:4, 1:9 (References 601 and 602, respectively)] whereas the corresponding free radical cyclization gives mainly the *cis* compound (*cis/trans* ratio: 2.3:1; see Section XI.3, Scheme 137). Since then, other arguments have been put forward to rule out a free radical pathway.[598–600]

Other unsaturated organometallics such as those derived from sodium, lithium,[602,606,607,609] aluminum,[608–611] gallium, indium,[609–611] zinc,[612] silver,[613] and silyllithium[614] readily give cyclized compounds. Although the cyclization is easier in the Cy5/Cy6 case and gives the (Cy5) compound, a concerted rather than an homolytic pathway is probably involved. Nevertheless, it must be noted that 1-methyl-5-hexenylaluminum gives the 1,2-dimethylcyclopentane with a *cis/trans* ratio of 2.9:1,[608] similar to the free radical cyclization ratio.

c. Other Cyclizations

Many other examples of selective formation of (Cy5) compounds in the Cy5/Cy6 case have been reported in the literature. For example, Firestone[615] used this probe among others to conclude that the orientation in 1,3-dipolar cycloadditions is better explained by a diradical mechanism than by a concerted one (though there is some disagreement with this conclusion). The sodium or potassium-catalyzed addition of benzylic and allylic hydrocarbons to olefins probably proceeds by a carbanionic intermediate.[616] Nevertheless, the selectivity observed in intramolecular additions, and the necessity of using "promoters" such as anthracene (which could favor an electron transfer process) appear as serious arguments in favor of possible free radical intermediates. The light-induced cycloisomerization of A/D–*seco*corrin complexes has been used by Eschenmoser[338a] as the key step in an elegant corrin synthesis, of value in one of the two synthetic routes to vitamin B_{12}. The exclusive formation of the (Cy5) compound in a Cy5/Cy6 case appears at first sight to favor a free radical mechanism, but since the *trans* product is exclusively obtained with alkyl-substituted compounds, the assumed concerted mechanism seems more attractive.

E. Comments on the Use of Unsaturated Systems Cyclization as a Free Radical Proof

The conclusion reached in Section C was that the cyclization of heteroatomic unsaturated systems could not be used as unambiguous proof of heteroatomic free radical formation so, this section is exclusively concerned with alkyl radicals.

The first question is: if no cyclization to the (Cy5) compounds is observed when the 5-hexenyl radical is chosen, is it possible to rule out the formation of an alkyl radical R[.] on the reaction pathway? The answer is: no, not if fast competitive intermolecular reactions are expected. In this case, it is necessary to work at low concentrations in order to favor the intramolecular process; but even under these conditions, very low yields of cyclized products are sometimes obtained. The use of the faster ring-opening processes (see Section XII.1.D) will be then a useful complementary probe. But there is a case where even these faster reactions cannot afford a positive answer, when the radical intermediate reacts, by dimerization or disproportionation, with another radical in the solvent cage, since these processes are faster than the rearrangement processes.

The second question is: if cyclization of the 5-hexenyl compound to the (Cy5) compound is observed, is it possible to conclude safely to a homolytic pathway? The answer is also no, since, as discussed in Section XII.1.D, other selective cyclizations to the (Cy5) compounds in the Cy5/Cy6 case, are known which do not involve free radical intermediates.

These warnings should not lead to the conclusion that the free radical cyclization (or ring-opening) tool must be avoided. Quite the contrary, the 5-hexenyl cyclization is certainly one of the best probes for the presence of alkyl free radical intermediates, but in dubious cases, as for instance those involving organometallic compounds, other probes are needed.

Finally it must be pointed out that if this analogous behavior can be misleading from the point of view of mechanism studies, it may be also rewarding since advances of the knowledges in one area such as photochemical cycloaddition of dienes may be of great help in the understanding of another area such as free radical intramolecular additions and vice versa.

2. The Kinetic Standard

In 1968, Carlsson and Ingold[55] published what become a useful standard in free radical rate constant determination. By using the rotating sector technique they were able to give absolute rate constants for the reduction of alkyl halides by organotin hydrides. When applied to 5-hexenyl bromide, and taking into account the cyclized/uncyclized product ratio determined by Walling[21] (see Schemes 12 and 13), they concluded that the cyclization rate constant of the 5-hexenyl radical was 1×10^5 sec^{-1} at 25–40°C.[55] When compared with intermolecular radical additions to double bonds, this value would correspond to an "effective" double bond concentration of ~40–100 M. Nevertheless, compared with many other free radical reactions rates, it is not a very fast process. This is why another absolute rate constant given in the same paper[55] is also very important. It corresponds to the opening of the 3,5-cyclocholestan-6-yl radical **345** (Scheme 132, Section XI.2.D): $k = 1.25 \times 10^8$ sec^{-1} at 30°C, 1.8×10^6 sec^{-1} at −20°C.

Some years later, Kochi[28] investigated extensively by esr the structure of alkyl radicals in nonaqueous solution. He showed that by this technique it was possible to observe the low-temperature cyclization of the 5-hexenyl radical and the opening of the cyclopropylcarbinyl radical. This led Ingold to use the esr technique in order to obtain accurate kinetic data for free radical intramolecular reactions. In this way the room temperature cyclization rate of the 5-hexenyl radical was confirmed and its temperature dependence was determined[617]: $k = 0.34 \times 10^2$ sec^{-1} at −85°C, $k = 17.0 \times 10^2$ sec^{-1} at −45°C, $k = 1.78 \times 10^5$ sec^{-1} at 40°C.[621a] Of course this high variation of the cyclization rate with the temperature must not be forgotten when this reaction is chosen as a standard for kinetic investigation. By the same esr technique, the isomerization rate of the cyclopropylcarbinyl radical to the allylcarbinyl radical was measured: $k = 1.3 \times 10^8$ sec^{-1} at 25°C.[69] In the same paper this value was compared with those of other primary alkyl radical isomerizations. The wide range in these isomerization rates (59 sec^{-1} to 1.3×10^8 sec^{-1} at 25°C) now provides one of the most useful standards for kinetic determination. Furthermore, many intramolecular addition rates to double bonds have been determined relatively to the 5-hexenyl radical. In this way absolute cyclization rates of alkyl substituted 5-hexenyl radicals may be considered as known[31,52,53,57,90] (Section IV, Scheme 12, Table 2; Section VI.2, Scheme 24) as well as cyclization rates of higher homologs[54,524] (Section V.1, Scheme 13, Table 3) and of the 4-(1-cyclohexenyl)butyl radical[37] (Section XI.2.A.a, Scheme 104).

The 5-hexenyl radical cyclization rate and other intramolecular radical isomerization rates have been used as a kinetic standard by Garst since 1969[540b] and by Kochi since 1970.[618] According to Scheme 164, a linear relationship between the concentration of the transfer reagent [XY] and the relative yields of uncyclized products [AY] and methylcyclopentane derivatives [Cy5Y] must obtain according to the following equation:

SCHEME 164

$$\frac{k_{Cy}}{k_y} = \frac{[Cy5Y]}{[AY]} \cdot [XY]$$

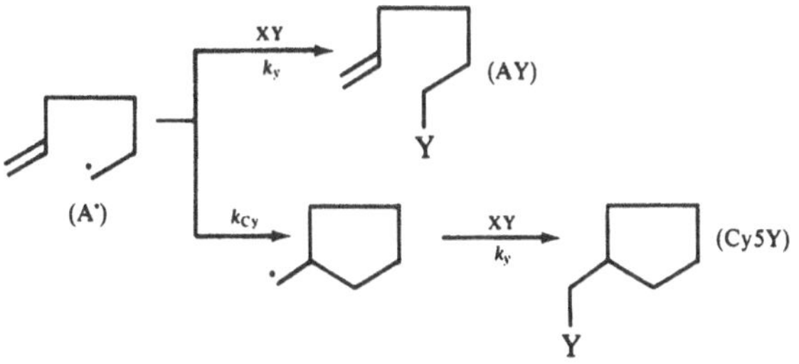

A complete discussion of the method may be found in Reference 619a. The reagent XY may be a transfer reagent, a reducing or oxidizing agent, or a radical trap (nitroso compound, heteroaromatic base,...). Since k_{Cy} is known, the rate constant k_y of the reaction studied is easily determined, if [XY] is known, according to the equation of Scheme 164.

A number of rate constants of primary alkyl radical reactions have been determined in this way: reduction by sodium naphthalenate $(2 \times 10^9 \, M^{-1} \, sec^{-1})$[540b,d] reaction with lithium benzophenone ketyl radical[565] and Wittig rearrangement[566] $(1.5 \times 10^8 \, M^{-1} \, sec^{-1})$, reaction with chromium(II) ethylene diamine $(4 \times 10^7 \, M^{-1} \, sec^{-1}$ at 25°C),[618] ligand transfer oxidation by copper(II) thiocyanate, chloride, and bromide (respectively, 3.6×10^8, 1.1×10^9, and $4.3 \times 10^9 \, M^{-1} \, sec^{-1}$ at 25°C),[72a] oxidation by copper(II) acetate $(1.2 \times 10^6 \, M^{-1} \, sec^{-1}$ at room temperature),[620] alkylation of protonated hetero-aromatic bases $(10^5–10^8 \, M^{-1} \, sec^{-1})$, of benzene $(3.8 \times 10^2 \, M^{-1} \, sec^{-1}$ at 79°C), and of anisole $(1.3 \times 10^3 \, M^{-1} \, sec^{-1}$ at 79°C),[619a] addition to quinone $(2.7 \times 10^7 \, M^{-1} \, sec^{-1}$ at 69°C), i.e., much faster than addition to butadiene $(9.2 \times 10^4 \, M^{-1} \, sec^{-1}$ at 25°C) and other conjugated olefins (but this is true only for nucleophilic radicals),[619b,653] spin trapping by phenyl-*N*-*t*-butylnitrone $(1.3 \times 10^5 \, M^{-1} \, sec^{-1}$ at 40°C) and by 2-methyl-2-nitrosopropane $(90.2 \times 10^5 \, M^{-1} \, sec^{-1}$ at 40°C).[621a] In this paper Ingold made clever use of ^{13}C-labeled compounds in order to distinguish between the uncyclized and cyclized nitroxides, both of which are primary. By this technique he was also able to measure primary alkyl radicals addition rate constants to other nitrones and nitroso compounds, to di-*t*-butyl thioketone $(7 \times 10^4 \, M^{-1} \, sec^{-1}$ at 25°C in isopentane), to 1,1-di-*t*-butylethylene $(<850 \, M^{-1} \, sec^{-1}$ at 25°C), as well as the rates of reaction of nitroxide spin adducts with primary alkyl radicals $(3–5 \times 10^8 \, M^{-1} \, sec^{-1}$ at 40°C) and of the bimolecular self-reaction of nitroxides[621b] and the solvent effects on these reaction rates.[621c]

Knowledge of the cyclization rate of the 5-hexenyl radical has also been used to study cage reactions of alkyl radicals generated in solution from diacyl peroxides[559,622] or to distinguish between intra- and intermolecular pathways in the Wittig rearrangement.[566] It has also been used to study the mechanism of the deshalogenation of 1,4-, 1,5-, and 2,3-dihaloalkanes by alkali naphthalenates.[586,623]

In many cases, particularly for the study of fast reactions $(k_y \gg k_{Cy}$ in Scheme 164), the opening of the cyclopropylmethyl radical to the isomeric allyl-carbinyl radical, one thousand times faster than the cyclization of the 5-hexenyl radical, is a useful complementary tool. It has been used mainly by Kochi for the study of cage reactions of alkyl radicals in solution[622] as well for the study of ligand and electron transfer oxidations of alkyl radicals.[72a,620,624] Further-more, in the last three studies the knowledge of the fate of the isomeric cyclobutyl intermediate was very useful in distinguishing between a homolytic (no rearrangement of the cyclobutyl radical) and a heterolytic pathway (fast rearrangement of the cyclobutyl cation). Recently, it has been shown that the reverse reaction, namely, the ring closure of the 3-butenyl radical, is a relatively slow process (9.4×10^3 sec^{-1} at 40°C).[631]

Since, it is now quite clear from these results that ligand transfer reactions are very fast, giving little chance for the 5-hexenyl radical to cyclize ($k_y \gg k_{Cy}$ in Scheme 164) unless very low concentrations of the oxidative salt are used, an interesting question arises concerning an early work of Brace's.[497] He observed that cyclization of 1,6-heptadiene (see Section VI.3, Scheme 25 and Section XI.3, Scheme 136) initiated by carbon tetrachloride free radical addition gives better yields (69%) of cyclized product using the redox transfer system (FeCl$_3$, benzoin, Et$_2$N$^+$H$_2$Cl$^-$) developed by Ascher and Vofsi rather than with AIBN (47% yield). This result does not appear in agreement with the redox mechanism of the Ascher–Vofsi intermolecular addition reaction unless it is admitted (as already discussed in Section XI.3 from the stereochemical results) that the cyclization of 1,6-heptadienes initiated by free radical addition cannot be completely explained by the cyclization of a discrete 5-hexenyl radical.

It is not always possible to use directly the unsubstituted 5-hexenyl radical as a kinetic standard. Nevertheless, provided some reasonable assumptions are made, interesting values may be obtained: the determination of 1,4-diradical lifetimes by Wagner[625] (Scheme 165) illustrates such a possibility. Irradiation of α-allylbutyrophenone (490) gives the diradical 491 by γ-hydrogen abstraction. This diradical gives the expected products of butyrophenone photolysis: 494, 495 and 496, as well the product (493) resulting from the new pathway offered to 491, i.e., the Cy5 cyclization to 492. On the basis of the critical assumption that 493 is formed quantitatively from 492 (see below, however) and that each radical site in 1,4-diradicals reacts independently of the other, the three unknown rate constants can be obtained from the relative yields and from the known k_{Cy} value. In the original paper[625] it was assumed that this value was identical with the cyclization rate of the 5-hexenyl radical since it is little affected by alkyl substituents. Since then, it has been shown that 3-alkyl-substituted radicals (such as 491) cyclize five times more rapidly than the parent 5-hexenyl radical.[57,524] With the value $k_{Cy} = 2.5 \times 10^5$ sec^{-1} for the formation of the *cis* diradical 492 (assuming that the *trans* and *cis* cyclic diradicals are formed at the same rate) the values now accepted[626] are $k_1 = 3.5 \times 10^6$ sec^{-1}, $k_2 = 2.5 \times 10^6$ sec^{-1}, and $k_3 = 9.4 \times 10^6$ sec^{-1}, and the assumption that each radical site in a 1,4-diradical is unaffected by the other has been confirmed by using the cyclopropylalkyl radical kinetic standard.[626]

In the last example, the known value of 5-hexenyl radical cyclization rate, corrected or not for alkyl substituent effects, was usefully used. A recent rate determination of secondary radical cyclizations confirms this view: the secondary 6-hepten-2-yl radical cyclizes only 1.18 times faster than does the primary 5-hexenyl radical,[630] while 2-, 3, or 4-alkyl-substituted 5-hexenyl radicals cyclize from 1.4 to 6.1 times faster than the unsubstituted radical.[697] The

SCHEME 165

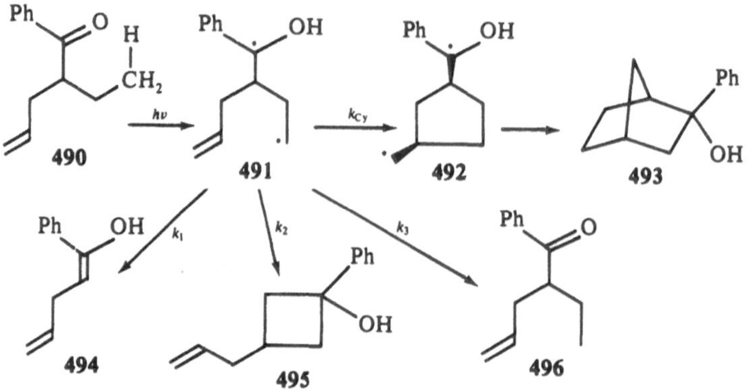

example now discussed shows clearly that this use is forbidden if important structural changes are made to the 5-hexenyl radical. One of these changes is concerned with highly stabilized radicals for which the cyclization rate is expected to be diminished but without knowing the magnitude of this rate retardation (see Section VII.1). An interesting study by House[627] demonstrates how important this rate retardation may be. He used the intramolecular tool in a effort to distinguish between reactions of an unsaturated carbonyl compound **497** (Scheme 166) with a nucleophile Nu⁻ that proceed by direct nucleophilic attack, and reactions that involve initial electron transfer to form the ion radical intermediate **498** prior to formation of the products **499** and **500**. Of course, a positive answer could only be expected if the recombination of the ion radical **498** and the radical (Nu·) proceeded in solution after diffusion, since recombination rates are higher than cyclization rates (see Section XII.1). The ketone **501** was first chosen since electron transfer would give the radical **502** and it would cyclize to the (Cy·5) radical **503**. In fact, ketone **501** subjected to three reactions believed to involve the formation of radical ion intermediates, namely, reduction with Li and *t*-BuOH in liquid NH_3, reduction with the chromium(II) complex $Cr(en)_2(OAc)_2$ and addition of $Me_2Cu\,Li$, gives products resulting only from **502** and no rearranged products resulting from the cyclized radical **503** are formed.[627] Since this result could be attributed to a very slow cyclization process, the cyclopropyl ketone **504** was then chosen in the reasonable hope that the opening of the alkylcyclopropyl radical **505** to **506** would be a faster process; in this case as well no products resulting from the rearranged radical **506** were observed.[627] Electrochemical reduction of the enones **501** and **504** afforded evidence for a remarkable stability of the corresponding radical anions **502** (half-life of the order of 10 sec) and **505** (half-life of $\sim 10^{-2}$ sec).[627] This would correspond to a rearrangement rate for these radical anions (to **503** and

SCHEME 166

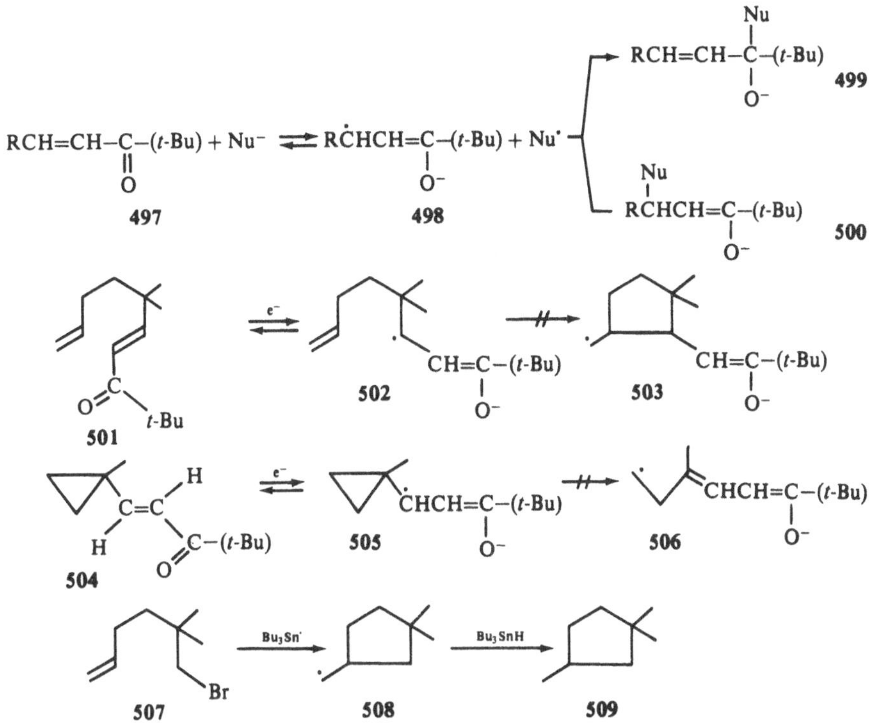

506, respectively) at least 10^6 times slower than for the unstabilized 5-hexenyl and cyclopropylmethyl radicals! Although examples of intramolecular addition of benzylic, allylic, and other stabilized radicals have been described in the preceding sections it seems that the origin of this unexpected rate retardation must be found in the allylic stabilization of the radicals studied. Indeed, the failure to observe cyclization of 502 cannot be ascribed to the ketyl radical ion structure since ketyl intramolecular addition is one of the most efficient processes known (see Section IX.2, Scheme 70; Section XI.2.A.a, Scheme 114; Section XI.3, Scheme 138). Since examples of rate retardation by *gem*-dimethyl groups are known[291] (see Section IX.2, Scheme 70) this possibility was considered but ruled out. Indeed, when the bromide 507 (Scheme 166) was submitted to stannane reduction, the cyclopentane 509 was obtained in higher yield (88% of the product) than the methylcyclopentane from the unsubstituted 5-hexenyl radical generated under the same conditions. In fact, it has been confirmed recently that the 2,2-dimethylhexenyl radical cyclizes to 508 12 times more rapidly[57] than the parent hexenyl radical, a result which has been considered[57,627] as another example of the Thorpe–Ingold effect.

In conclusion, a knowledge of the absolute rate constant of 5-hexenyl radical cyclization (and other free radical rearrangements) is of great value for kinetic investigations. Cyclization rates of alkyl-substituted 5-hexenyl radicals and other alkenyl radicals may be inferred from this value and known relative values. It would be of great help to synthetic proposals to have some ideas of cyclization rates of heteroatomic radicals [the cyclization rate of the 4-pentenyloxyl radical has been determined only indirectly and in aqueous solution[137] (see Section VIII.2), and only an upper limit has been proposed for the rate of cyclization of the neutral 4-pentenyl-aminyl radical[172b] (see Section VIII.3)] as well of those of stabilized alkyl radicals (high retardation possible as in the example just discussed) and alkyl radicals bearing heteroatoms such as fluorine (fast reactions possible according to some experimental results: see Section VI.4).

XIII. CONCLUSION

Since the first reports of radical cyclizations by intramolecular addition to double bonds in the 1960–1964 period, a great deal of activity has developed in this area. Since the Cy5/Cy6 cyclization case appeared rapidly as the easiest cyclization process, of importance in cyclopolymerization reactions and of mechanistic interest on account of the high specificity observed to the (Cy5) or (Cy6) products, most of the studies have been devoted to this case. A reasonable understanding of the specificity observed emerged from these studies. Furthermore, the need for quantitative results led the 5-hexenyl radical cyclization to become one of the best mechanistic tools for the detection of alkyl radical intermediates and a useful kinetic standard.

In the meantime, other intramolecular additions were studied, which afforded new preparative methods and which are probably involved in biogenetic schemes. Much less is known at this time of the scope and the quantitative aspect of these reactions. Among the many problems not satisfactorily solved at this time one may cite: the origin of the stereospecificity observed in mono- and polycyclization reactions, the scope of polycyclizations reactions and of intramolecular additions to acetylenic and polar bonds, a better knowledge of intramolecular additions of heteroatomic radicals and radicals bearing heteroatoms in the chain, and the limits of the cyclization processes applied to higher homologs than the Cy5/Cy6 case.

It was the aim of this review to be as exhaustive as possible, to try to bring out what appears to be reasonably well understood and what invites more study.

ACKNOWLEDGMENT

I am happy to record my indebtedness to Professor Marc Julia, who introduced me to this field of research. I am also grateful to my collaborators, whose names are recorded in this chapter, for their careful experimental work and stimulating discussions. Finally, I wish to thank Professor Nathan Kornblum not only for helping me improve my English but also for his critical discussion.

REFERENCES

1. O. L. Chapman, in *Advances in Photochemistry*, Vol. 1, G. S. Hammond and J. N. Pitts, Eds., Interscience, New York (1963), p. 323.
2. M. Akthar, in *Advances in Photochemistry*, Vol. 2, G. S. Hammond and J. N. Pitts, Eds., Interscience, New York (1964), p. 263.
3. R. H. Hesse, in *Advances in Free Radical Chemistry*, Vol. 3, G. H. Williams, Ed., Logos Press, London (1969), p. 83.
4. K. Heusler and J. Kalvoda, *Angew. Chem. Intern. Ed.* **3**, 525 (1964).
5. J. Kalvoda and K. Heusler, *Synthesis*, 501 (1971).
6. M. Lj. Mihailovic and R. E. Partch, in *Selective Organic Transformations*, Vol. 2, B. S. Tygarajan, Ed., Wiley, New York (1972), p. 97.
7. M. E. Wolff, *Chem. Rev.* **63**, 55 (1963).
   Y. L. Chow, in *Reactive Intermediates*, Vol. 1, R. A. Abramovitch, Ed., Plenum Press, New York (1979), p. 151.
8. R. A. Abramovitch, in *Advances in Free Radical Chemistry*, Vol. 2, G. H. Williams, Ed., Logos Press, London (1967), p. 87.
9. T. Kametani and K. Fukumoto, *Synthesis*, 657 (1972).
10. T. Kametani and K. Fukumoto, *Accounts Chem. Res.* **5**, 212 (1972).
11. M. Julia, *Rec. Chem. Progr.* **25**, 3 (1964).
12. M. Julia, *Pure Appl. Chem.* **15**, 167 (1967).
13. M. Julia, *Accounts Chem. Res.* **4**, 386 (1971).
14. A. L. J. Beckwith, *Chem. Soc. Spec. Publ.* **24**, 239 (1970).
15. J. W. Wilt, in *Free Radicals*, Vol. 1, J. K. Kochi, Ed., John Wiley & Sons, New York (1973), p. 333.
16. J. E. Baldwin, *J. Chem. Soc. Chem. Commun.*, 734 (1976).
17. J. A. Berson, C. J. Olsen, and J. S. Walia, *J. Am. Chem. Soc.* **82**, 5000 (1960).
18. S. Arai, S. Sato, and S. Shida, *J. Chem. Phys.* **33**, 1277 (1960).
19. (a) A. S. Gordon and S. R. Smith, *J. Phys. Chem.* **66**, 521 (1962); (b) R. F. Garwood, C. J. Scott, and B. C. L. Weedon, *J. Chem. Soc. Chem. Commun.*, 14 (1965).
20. R. C. Lamb, P. W. Ayers, and M. K. Toney, *J. Am. Chem. Soc.* **85**, 3483 (1963).
21. C. Walling, J. H. Cooley, A. A. Ponaras, and E. J. Racah, *J. Am. Chem. Soc.* **88**, 5361 (1966).
22. J. K. Kochi and P. J. Krusic, *J. Am. Chem. Soc.* **91**, 3940 (1969).
23. C. Walling and M. S. Pearson, *J. Am. Chem. Soc.* **86**, 2262 (1964).
24. M. Julia, J-M. Surzur, and L. Katz, *C. R. Acad. Sci. Ser. C*, **251**, 1030 (1960).
25. Henkel and Co., Brit. Pat. 792486; *Chem. Abstr.* **52**, 19957d (1958).
26. M. Julia and M. Maumy, *Bull. Soc. Chim. Fr.*, 434 (1966).
27. M. Julia, M. Maumy, and L. Mion, *Bull. Soc. Chim. Fr.*, 2641 (1967).

28. (a) D. J. Edge and J. K. Kochi, *J. Am. Chem. Soc.* **94**, 7695 (1972); (b) J. K. Kochi, *Chem. Soc. Spec. Publ.* **24**, 147 (1970), and references cited therein.

29. M. Julia and M. Maumy, *Bull. Soc. Chim. Fr.*, 1603 (1968).

30. M. Julia and M. Maumy, *Bull. Soc. Chim. Fr.*, 2427 (1969).

31. M. Julia, C. Descoins, M. Baillarge, B. Jacquet, D. Uguen, and F. A. Groeger, *Tetrahedron* **31**, 1737 (1975).

32. D. L. Struble, A. L. J. Beckwith, and G. E. Gream, *Tetrahedron Lett.*, 3701 (1968).

33. K. Fukui, *Tetrahedron Lett.*, 2427 (1965).

34. H. Fujimoto, S. Yamabe, T. Minato, and K. Fukui, *J. Am. Chem. Soc.* **94**, 9205 (1972).

35. S. Inagaki and K. Fukui, *Chem. Lett.*, 509 (1974).

36. G. Fleischer, *Z. Phys. Chem. (Leipzig)* **250**, 261 (1972).

37. A. L. J. Beckwith, G. E. Gream, and D. L. Struble, *Aust. J. Chem.* **25**, 1081 (1972).

38. A. L. J. Beckwith and W. B. Gara, *J. Chem. Soc. Perkin Trans.* **2**, 593 (1975).

39. A. L. J. Beckwith and W. B. Gara, *J. Chem. Soc. Perkin Trans.* **2**, 795 (1975).

40. P. Tordo, Ph. D. thesis, Marseilles (1971).

41. J. H. Davies, J. D. Downer, and P. Kirby, *J. Chem. Soc. C*, 245 (1966).

42. J. E. Baldwin, J. Cutting, W. Dupont, L. Kruse, L. Silberman, and R. C. Thomas, *J. Chem. Soc. Chem. Commun.*, 736 (1976).

43. J. E. Baldwin, *J. Chem. Soc. Chem. Commun.*, 738 (1976).

44. J. E. Baldwin and J. A. Reiss, *J. Chem. Soc. Chem. Commun.*, 77 (1977).

45. J. E. Baldwin and L. I. Kruse, *J. Chem. Soc. Chem. Commun.*, 233 (1977).

46. B. Capon and C. W. Rees, *Ann. Rep.* **61**, 261 (1964).

47. B. Capon, *Q. Rev. Chem. Soc.* **18**, 45 (1964).

48. R. D. Rieke and N. A. Moore, *Tetrahedron Lett.*, 2035 (1969).

49. R. D. Rieke and N. A. Moore, *J. Org. Chem.* **37**, 413 (1972).

50. J. M. Tedder, *CNRS International Colloquium on Organic Free Radicals*, Aix-en-Provence, 1977, CNRS, Paris (1978), p. 385.

51. J. M. Tedder, letter to the author, August 22, 1977.

52. C. Walling and A. Cioffari, *J. Am. Chem. Soc.* **94**, 6059 (1972).

53. A. L. J. Beckwith, I. A. Blair, and G. Phillipou, *Tetrahedron Lett.*, 2251 (1974).

54. A. L. J. Beckwith and G. Moad, *J. Chem. Soc. Chem. Commun.*, 472 (1974).

55. D. J. Carlsson and K. U. Ingold, *J. Am. Chem. Soc.* **90**, 7047 (1968).

56. H. Kämmerer and V. Steiner, *Makromol. Chem. Suppl.* **1**, 133 (1975).

57. A. L. J. Beckwith, *CNRS Internationale Colloquium on Organic Free Radicals*, Aix-en-Provence, 1977, CNRS, Paris (1978), pp. 373–385.

58. K. W. Watkins and D. K. Olsen, *J. Phys. Chem.* **76**, 1089 (1972).

59. W. P. L. Carter and D. C. Tardy, *J. Phys. Chem.* **78**, 1573 (1974).

60. (a) T. Sakai and D. Nohara, *Bull. Japan. Pet. Inst.* **17**, 212 (1975); (b) M. E. Hendrick and M. Jones, Jr., *Tetrahedron Lett.*, 4249 (1978).

61. R. Dowbenko, *J. Am. Chem. Soc.* **86**, 946 (1964).

62. L. Friedman, *J. Am. Chem. Soc.* **86**, 1885 (1964).

63. R. Dowbenko, *Tetrahedron* **20**, 1843 (1964).

64. J.M. Locke and E. W. Duck, *J. Chem. Soc. Chem. Commun.*, 151 (1965).

65. L. H. Gale, *J. Org. Chem.* **33**, 3643 (1968).

66. M. L. Scheinbaum, *J. Org. Chem.* **35**, 2785 (1970).

67. (a) M. C. Lasne and A. Thuillier, *C. R. Acad. Sci. Ser. C*, **273**, 1258 (1971); (b) W. R. Dolbier, Jr. and O. T. Garza, *J. Org. Chem.* **43**, 3848 (1978).

68. J. A. Claisse, D. I. Davies, and L. T. Parfitt, *J. Chem. Soc. C*, 258 (1970).

69. B. Maillard, D. Forrest, and K. U. Ingold, *J. Am. Chem. Soc.* **98**, 7024 (1976).

70. (a) A. L. J. Beckwith and G. Phillipou, *J. Chem. Soc. Chem. Commun.*, 658 (1971); (b) A. L. J. Beckwith and G. Phillipou, *Aust. J. Chem.* **29**, 123 (1976).

71. (a) I. M. Takakis and W. C. Agosta, *Tetrahedron Lett.*, 2387 (1978); (b) I. M. Takakis and W. C. Agosta, *J. Org. Chem.* **44**, 1294 (1979); (c) I. M. Takakis and W. C. Agosta, *J. Am. Chem. Soc.* **101**, 2383 (1979).
72. (a) C. L. Jenkins and J. K. Kochi, *J. Org. Chem.* **36**, 3103 (1971); (b) M. Lj. Mihailović, J. Bošnjak, and Ž. Čeković, *Helv. Chim. Acta* **59**, 475 (1976).
73. (a) Th. A. Halgren, M. E. H. Howden, M. E. Medof, and J. D. Roberts, *J. Am. Chem. Soc.* **89**, 3051 (1967); (b) L. Mandell, J. C. Johnston, and R. A. Day, Jr., *J. Org. Chem.* **43**, 1616 (1978).
74. E. C. Friedrich and R. L. Holmstead, *J. Org. Chem.* **36**, 971 (1971).
75. E. C. Friedrich and R. L. Holmstead, *J. Org. Chem.* **37**, 2546 (1972).
76. E. C. Friedrich and R. L. Holmstead, *J. Org. Chem.* **37**, 2550 (1972).
77. (a) B. Giese, and K. Jay, *Chem. Ber.* **110**, 1364 (1977); (b) H. Driguez and J. Lessard, *Can. J. Chem.* **55**, 720 (1977).
78. M. Julia, D. Mansuy, and P. Detraz, *Tetrahedron Lett.*, 2141 (1976).
79. V. M. A. Chambers, W. R. Jackson, and G. W. Young, *Chem. Commun.*, 1275 (1970).
80. G. A. Gray, W. R. Jackson, and V. M. A. Chambers, *J. Chem. Soc. C*, 200 (1971),
81. G. M. Whitesides and J. San Filippo, Jr., *J. Am. Chem. Soc.* **92**, 6611 (1970).
82. (a) S. J. Cristol, G. A. Lee, and A. L. Noreen, *J. Am. Chem. Soc.* **95**, 7067 (1973); (b) S. J. Cristol and R. P. Micheli, *J. Am. Chem. Soc.* **100**, 850 (1978); (c) M. R. Detty and L. A. Paquette, *J. Org. Chem.* **43**, 1118 (1978).
83. R. Sustmann and F. Lübbe, *Tetrahedron Lett.*, 2831 (1974).
84. (a) H. M. Walborsky and J.-C. Chen, *J. Am. Chem. Soc.* **92**, 7573 (1970); (b) J. C. Chen, *Tetrahedron Lett.*, 3669 (1971); (c) C. J. Boriack, E. D. Laganis, and D. M. Lemal, *Tetrahedron Lett.*, 1015 (1978).
85. J. C. Moutet and G. Reverdy, *Bull. Soc. Chim. Fr.*, II-442 (1978).
86. R. C. Lamb, J. G. Pacifici, and P. W. Ayers, *J. Org. Chem.* **30**, 3099 (1965).
87. A. L. J. Beckwith, I. Blair, and G. Phillipou, *J. Am. Chem. Soc.* **96**, 1613 (1974).
88. A. L. J. Beckwith and W. B. Gara, *J. Am. Chem. Soc.* **91**, 5689 (1969).
89. A. L. J. Beckwith and W. B. Gara, *J. Am. Chem. Soc.* **91**, 5691 (1969).
90. T. W. Smith and G. B. Butler, *J. Org. Chem.* **43**, 6 (1978).
91. J. I. G. Cadogan, D. H. Hey, and S. H. Ong, *J. Chem. Soc.*, 1939 (1965).
92. G. B. Butler and R. J. Angelo, *J. Am. Chem. Soc.* **79**, 3128 (1957).
93. C. S. Marvel and R. D. Vest, *J. Am. Chem. Soc.* **79**, 5771 (1957).
94. D. H. Solomon, *J. Macromol. Sci.-Chem.* **A9**, 95 (1975).
95. W. S. Friedlander, *Abstracts of the 133rd National Meeting of the American Chemic. Soc.*, San Francisco, California, American Chemical Society, Washington, D. C. (1958), paper 29, pp. 18–19N.
96. N. O. Brace, *J. Am. Chem. Soc.* **86**, 523 (1964).
97. J. I. G. Cadogan, D. H. Hey, and A. O. S. Hock, *Chem. Ind. (London)*, 753 (1964).
98. J. I. G. Cadogan, M. Grunbaum, D. H. Hey, A. S. H. Ong, and J. T. Sharp, *Chem. ind. (London)*, 422 (1968).
99. (a) A. L. J. Beckwith, D. G. Hawthorne, and D. H. Solomon, *Aust. J. Chem.* **29**, 995 (1976); (b) R. L. Keiter, Y. Y. Sun, J. W. Brodack, and L. W. Cary, *J. Am. Chem. Soc.* **101**, 2638 (1979).
100. P. Piccardi, P. Massardo, M. Modena, and E. Santoro, *Chim. Indust. (Milano)* **56**, 824 (1974).
101. P. Piccardi, M. Modena, and L. Cavalli, *J. Chem. Soc. C*, 3959 (1971).
102. P. Piccardi, P. Massardo, M. Modena, and E. Santoro, *J. Chem. Soc., Perkin Trans. 1*, 982 (1973).
103. P. Piccardi, P. Massardo, M. Modena, and E. Santoro, *Chim. Indust. (Milano)* **56**, 353 (1974).

104. M. Julia and M. Maumy, *Bull. Soc. Chim. Fr.*, 2415 (1969).
105. M. Julia and M. Barreau, *C. R. Acad. Sci. Ser. C*, **280**, 957 (1975).
106. (a) M. Julia, J-M. Surzur, and L. Katz, *Bull. Soc. Chim. Fr.*, 1109 (1964); (b) J. I. G. Cadogan, D. H. Hey, and S. H. Ong, *J. Chem. Soc.*, 1932 (1965).
107. M. Julia and M. Maumy, *Org. Synth.* **55**, 57 (1976).
108. M. Julia and F. Le Goffic, *Bull. Soc. Chim. Fr.*, 1550 (1965).
109. M. Julia, J-M. Surzur, L. Katz, and F. Le Goffic, *Bull. Soc. Chim. Fr.*, 1116 (1964).
110. M. Julia and F. Le Goffic, *Bull. Soc. Chim. Fr.*, 1555 (1965).
111. P. Piccardi, M. Modena, and P. Massardo, *Chim. Indust. (Milano)* **55**, 807 (1973).
112. J-M. Surzur and G. Torri, *Bull. Soc. Chim. Fr.*, 3070 (1970).
113. Yu. N. Ogibin and G. I. Nikishin, *Zh. Obsch. Khim.* **41**, 1277 (1971).
114. H. Pines, N. C. Sih, and D. B. Rosenfield, *J. Org. Chem.* **31**, 2255 (1966).
115. C. Walling and A. Cioffari, *J. Am. Chem. Soc.* **94**, 6064 (1972).
116. R. Dulou, Y. Chretien-Bessiere, and H. Desalbres, *C. R. Acad. Sci. Ser. C*, **258**, 603 (1964).
117. (a) J.-P. Montheard, *C. R. Acad. Sci. Ser. C*, **260**, 577 (1965); (b) Ch. F. Lochow and R. G. Miller, *J. Am. Chem. Soc.* **98**, 1281 (1976).
118. M. Chatzopoulos and J.-P. Montheard, *C. R. Acad. Sci. Ser. C*, **280**, 29 (1975).
119. Ž. Čeković, *Tetrahedron Lett.*, 749 (1972).
120. M. Julia and P. Dostert, *C. R. Acad. Sci. Ser. C*, **259**, 2872 (1964).
121. E. Van Bruggen, *Rec. Trav. Chim. Pays-Bas* **87**, 1134 (1968).
122. J.-M. Surzur and M. P. Bertrand, *Bull. Soc. Chim. Fr.*, 1861 (1973), and references cited therein.
123. Ch. Walling, *Bull. Soc. Chim. Fr.*, 1609 (1968), ans references cited therein.
124. M. P. Bertrand and J.-M. Surzur, *Tetrahedron Lett.*, 3451 (1976).
125. R. M. Moriarty, in *Selective Organic Transformations*, Vol. 2, B. S. Tyagarajan, Ed., John Wiley & Sons (1972), p. 183.
126. A. L. Nussbaum, R. Wayne, E. Yuan, O. Z. Sarre, and E. P. Oliveto, *J. Am. Chem. Soc.* **87**, 2451 (1965).
127. S. K. Pradhan and M. Girijavallabhan, *J. Chem. Soc. Chem. Commun.*, 591 (1975).
128. A. J. Dobbs, B. C. Gilbert, H. A. H. Laue, and R. O. C. Norman, *J. Chem. Soc. Perkin Trans.* **2**, 1044 (1976).
129. J-M. Surzur, P. Cozzone, and M. P. Bertrand, *C. R. Acad. Sci. Ser. C*, **267**, 908 (1968).
130. P. Tordo, M. P. Bertrand, and J-M. Surzur, *Tetrahedron Lett.*, 3399 (1970).
131. J-M. Surzur, M. P. Bertrand, and R. Nouguier, *Tetrahedron Lett.*, 4197 (1969).
132. M. P. Bertrand and J-M. Surzur, *Bull. Soc. Chim. Fr.*, 2393 (1973).
133. A. Clerici, F. Minisci, K. Ogawa, and J-M. Surzur, *Tetrahedron Lett.*, 1149 (1978).
134. M. P. Bertrand, J-M. Surzur, M. Boyer, and M. Lj. Mihailović, *Tetrahedron* **35**, 1365 (1979).
135. J. F. Garst, *Chem. Prepr.*, *Symp. Electron Affinities Aromatic Hydrocarbons and Chemistry of Radicals ions*, *Am. Chem. Soc. Div. Petrol Chem.*, San Francisco, California, **13**, pp. D65–D76 (1968).
136. H. H. Quon and Y. L. Chow, *Tetrahedron* **31**, 2349 (1975).
137. B. C. Gilbert, R. G. G. Holmes, H. A. H. Laue, and R. O. C. Norman, *J. Chem. Soc. Perkin Trans.* **2**, 1047 (1976).
138. H. Immer, M. Lj. Mihailović, K. Schaffner, D. Arigoni, and O. Jeger, *Helv. Chim. Acta* **45**, 753 (1962).
139. M. Lj. Mihailović, *J. Heterocycl. Chem.* **3**, S-111 (1976).
140. (a) M. Lj. Mihailović, Ž. Čeković, J. Stanković, N. Pavlović, S. Konstantinović, and S. Djokić-Mazinjanin, *Helv. Chim. Acta* **56**, 3056 (1973); (b) M. Lj. Mihailović, Ž. Čeković, J. Stanković, S. Djokić-Mazinjanin, D. Marinković, and S. Konstantinović, *Bull. Soc. Chim. Beograd* **43**, 69 (1978).

141. W. M. Horspool and P. L. Pauson, *J. Chem. Soc. Chem. Commun.*, 195 (1967).
142. G. Frater and H. Schmid, *Helv. Chim. Acta* **50**, 255 (1967).
143. S. Houry, S. Geresh, and A. Shani, *Israel J. Chem.* **11**, 805 (1973).
144. A. Shani and R. Mechoulam, *Tetrahedron* **27**, 601 (1971).
145. P. J. Kropp and H. J. Krauss, *J. Am. Chem. Soc.* **91**, 7466 (1969).
146. J. M. Hornback, *J. Am. Chem. Soc.* **96**, 6773 (1974).
147. D. T. Dalgleish, N. P. Forrest, D. C. Nonhebel, and P. L. Pauson, *J. Chem. Soc. Perkin Trans.* **1**, 584 (1977).
148. F. Du R. Volsteedt, D. Ferreira, and D. G. Roux, *J. Chem. Soc. Chem. Commun.*, 217 (1975).
149. R. C. Lamb, F. F. Rogers, G. C. Dean Jr., and F. W. Voigt Jr., *J. Am. Chem. Soc.* **84**, 2635 (1962).
150. R. C. Lamb, L. P. Spadafino, R. G. Webb, E. B. Smith, W. E. McNew, and J. G. Pacifici, *J. Org. Chem.* **31**, 147 (1966).
151. T. W. Koenig and J. C. Martin, *J. Org. Chem.* **29**, 1520 (1964).
152. H. Hart and F. J. Chloupek, *J. Am. Chem. Soc.* **85**, 1155 (1963).
153. R. M. Moriarty, H. G. Walsh, and H. Gopal, *Tetrahedron Lett.*, 4363 (1966).
154. R. M. Moriarty, H. Gopal, and H. G. Walsh, *Tetrahedron Lett.*, 4369 (1966).
155. E. Laurent and M. Thomalla, *Bull. Soc. Chim. Fr.*, 834 (1977).
156. T. Shono, A. Ikeda, and Y. Kimura, *Tetrahedron Lett.*, 3599 (1971).
157. R. Sustmann and F. Lübbe, *Chem. Ber.* **109**, 444 (1976).
158. (a) M. O. Funk, R. Issac, and N. A. Porter, *J. Am. Chem. Soc.* **97**, 1281 (1975); (b) N. A. Porter, M. O. Funk, D. Gilmore, R. Issac, and J. Nixon, *J. Am. Chem. Soc.* **98**, 6000 (1976).
159. J. R. Nixon, M. Cudd, and N. A. Porter, *J. Org. Chem.* **43**, 4048 (1978).
160. M. F. Ansell and A. J. Bignold, *J. Chem. Soc. Chem. Commun.*, 989 (1970).
161. W. Herz, R. C. Ligon, J. A. Turner, and J. F. Blount, *J. Org. Chem.* **42**, 1885 (1977).
162. (a) J. A. Turner and W. Herz, *J. Org. Chem.* **42**, 1895 (1977); (b) J. A. Turner and W. Herz, *J. Org. Chem.* **42**, 1900 (1977).
163. J. P. Hagenbuch and P. Vogel, *Tetrahedron Lett.*, 561 (1979).
164. J-M. Surzur and P. Tordo, *C. R. Acad. Sci. Ser. C*, **263**, 446 (1966).
165. J-M. Surzur, P. Tordo, and L. Stella, *Bull. Soc. Chim. Fr.*, 111 (1970).
166. J-M. Surzur, L. Stella, and P. Tordo, *Bull. Soc. Chim. Fr.*, 115 (1970).
167. J-M. Surzur and L. Stella, *Tetrahedron Lett.*, 2191 (1974).
168. J-M. Surzur, L. Stella, and P. Tordo, *Tetrahedron Lett.*, 3107 (1970).
169. Y. L. Chow, R. A. Perry, B. C. Menon, and S. C. Chen, *Tetrahedron Lett.*, 1545 (1971).
170. J-M. Surzur, L. Stella, and P. Tordo, *Bull. Soc. Chim. Fr.*, 1425 (1975).
171. L. Stella, Ph. D. thesis. Marseilles (1972): (a) p. 150; (b) p. 152; (c) p. 172.
172. (a) C. Michejda, D. H. Campbell, D. H. Sieh, and S. R. Koepke, in *Organic Free Radicals, ACS Symposium Series 69*, W. A. Pryor, Ed., American Chemical Society, Washington D. C. (1978), Chap. 18; (b) Y. Maeda and K. U. Ingold, *J. Am. Chem. Soc.* **102**, 328 (1980).
173. J-M. Surzur, L. Stella, and P. Tordo, *Bull. Soc. Chim. Fr.*, 1429 (1975).
174. (a) J-M. Surzur and L. Stella, *Bull. Soc. Chim. Fr.*, 255 (1977); (b) J. Lacrampe, A. Heumann, R. Furstoss, and B. Waegell, *J. Chem. Res.* (S) 334 (1978); (M) 4001 (1978).
175. P. G. Gassman and J. H. Dygos, *Tetrahedron Lett.*, 4749 (1970).
176. R. Furstoss, G. Esposito, P. Teissier, and B. Waegell, *Bull. Soc. Chim. Fr.*, 2485 (1974).
177. R. Tadayoni, A. Heumann, R. Furstoss, and B. Waegell, *Tetrahedron Lett.*, 2879 (1973).
178. L. Stella, B. Raynier, and J-M. Surzur, *Tetrahedron Lett.*, 2721 (1977).
179. B. Raynier, Ph. D. thesis, Marseilles (1975), p. 87.
180. P. G. Gassman, *Accounts Chem. Res.* **3**, 26 (1970).
181. P. G. Gassman, F. Hoyda, and J. Dygos, *J. Am. Chem. Soc.* **90**, 2716 (1968).

182. Y. L. Chow, R. A. Perry, and B. C. Menon, *Tetrahedron Lett.*, 1549 (1971).
183. J. D. Hobson and W. D. Riddell, *J. Chem. Soc. Chem. Commun.*, 1178 (1968).
184. J. W. Bastable, J. D. Hobson, and W. D. Riddell, *J. Chem. Soc. Perkin Trans.* 1, 2205 (1972).
185. P. G. Gassman and J. H. Dygos, *Tetrahedron Lett.*, 4745 (1970).
186. J-M. Surzur, L. Stella, and R. Nouguier, *Tetrahedron Lett.*, 903 (1971).
187. O. E. Edwards, D. Vocelle, and J. W. ApSimon, *Can. J. Chem.* **50**, 1167 (1972).
188. O. E. Edwards, G. Bernath, J. Dixon, J. M. Paton, and D. Vocelle, *Can. J. Chem.* **52**, 2123 (1974).
189. R. Furstoss, R. Tadayoni, and B. Waegell, *Nouv. J. Chim.* **1**, 167 (1977).
190. R. M. Moriarty and K. Kapadia, *Tetrahedron Lett.*, 1165 (1964).
191. P. G. Gassman, K. Uneyama, and J. L. Hahnfeld, *J. Am. Chem. Soc.* **92**, 647 (1977).
192. W. Nagata, S. Hirai, K. Kawata, and T. Aoki, *J. Am. Chem. Soc.* **89**, 5045 (1967).
193. W. Nagata, S. Hirai, T. Okumura, and K. Kawata, *J. Am. Chem. Soc.* **90**, 1650 (1968).
194. P. S. Portoghese and D. T. Stepp, *Tetrahedron* **29**, 2253 (1975).
195. (a) D. E. Horning and J. M. Muchowski, *Can. J. Chem.* **52**, 1321 (1974); (b) S. R. Wilson and R. A. Sawicki, *J. Org. Chem.* **44**, 287 (1979); (c) S. R. Wilson and R. A. Sawicki, *J. Org. Chem.* **44**, 330 (1979).
196. (a) N. Paillous and A. Lattes, *Tetrahedron Lett.*, 4945 (1971); (b) G. Trinquier, N. Paillous, A. Lattes, and J. P. Malrieu, *Nouv. J. Chim.* **1**, 403 (1977).
197. L. S. Hegedus, G. F. Allen, J. J. Bozell, and E. L. Waterman, *J. Am. Chem. Soc.* **100**, 5800 (1978).
198. Y. L. Chow and R. A. Perry, *Tetrahedron Lett.*, 531 (1972).
199. E. Flesia, A. Croatto, P. Tordo, and J-M. Surzur, *Tetrahedron Lett.*, 535 (1972).
200. M. E. Kuehne and D. A. Horne, *J. Org. Chem.* **40**, 1287 (1975).
201. Ph. Mackiewicz, Ph. D. thesis, Marseilles (1977).
202. Ph. Mackiewicz, R. Furstoss, B. Waegell, R. Cote, and J. Lessard, *J. Org. Chem.* **43**, 3746 (1978).
203. J. Lessard, R. Cote, Ph. Mackiewicz, R. Furstoss, and B. Waegell, *J. Org. Chem.* **43**, 3750 (1978).
204. W. B. Motherwell and J. S. Roberts, *J. Chem. Soc. Chem. Commun.*, 328 (1972).
205. H. O. House, D. T. Manning, G. D. Melillo, L. F. Lee, O. R. Haynes, and B. E. Wilkes, *J. Org. Chem.* **41**, 855 (1976).
206. H. O. House and L. F. Lee, *J. Org. Chem.* **41**, 863 (1976).
207. J. Von Braun and Th. Plate, *Chem. Ber.* **67**, 281 (1934).
208. W. E. Vaughan and F. F. Rust, *J. Org. Chem.* **7**, 472 (1942).
209. C. S. Marvel and L. E. Olson, *J. Polym. Sci.* **26**, 23 (1957).
210. R. F. Naylor, *J. Chem. Soc.*, 1532 (1947).
211. E. Dyer and D. W. Osborne, *J. Polym. Sci.* **47**, 349 (1960).
212. K. von Rühlmann, U. Schräpler, and D. Gramer, *J. Prakt. Chem.*, **10**, 325 (1960).
213. J-M. Surzur, M.-P. Crozet, and C. Dupuy, *C. R. C. R. Acad. Sci. Ser. C*, **264**, 610 (1967).
214. N. P. Volynskii, G. D. Galpern, and A. B. Urin, *Khim. Geterosikl. Soedin*, 1031 (1967).
215. M. P. Crozet, Ph. D. thesis, Marseilles (1972).
216. S. Kondo and A. Negishi, Japan. Pat. 72.36.386; *Chem. Abstr.* **78**, 4126 (1973).
217. H. Konda and A. Negishi, Japan. Pat. 72.47.035; *Chem. Abstr.* **78**, 111.130 (1973).
218. (a) M. P. Crozet, J-M. Surzur, and C. Dupuy, *Tetrahedron Lett.*, 2031 (1971); (b) S. Warren, *Acc. Chem. Res.* **11**, 401 (1978).
219. J-M. Surzur, M. P. Crozet, and C. Dupuy, *Tetrahedron Lett.*, 2025 (1971).
220. J-M. Surzur and M. P. Crozet, *C. R. Acad. Sci. Ser. C*, **268**, 2109 (1969).
221. M. Kaafarani, Ph. D. thesis, Marseilles (1977).
222. R. G. Petrova, V. V. Slepuskhin, and R. Kh. Freidlina, *Dokl. Akad. Nauk. S.S.S.R. Ser. Khim.* **202**, 857 (1972).

223. R. G. Petrova, T. D. Maiorana, and R. Kh. Freidlina, *Izvest. Akad. Nauk. S.S.S.R. Ser. Khim.*, 2540 (1973).
224. K. Takabe, T. Katagiri, and J. Tanaka, *Tetrahedron Lett.*, 4805 (1970).
225. K. Takabe, A. Kajikawa, T. Katagiri, and J. Tanaka, *Nippon Kagaku Kaishi*, 1366 (1973); *Chem. Abstr.* **79**, 78760 (1973).
226. K. E. Koenig and W. P. Weber, *Tetrahedron Lett.*, 3151 (1973).
227. K. E. Koenig, R. A. Felix, and W. P. Weber, *J. Org. Chem.* **39**, 1539 (1974).
228. J-M. Surzur, R. Nouguier, M. P. Crozet, and C. Dupuy, *Tetrahedron Lett.*, 2035 (1971).
229. R. Nouguier and J-M. Surzur, *Tetrahedron* **32**, 2001 (1976).
230. R. Nouguier and J-M. Surzur, unpublished results (1976).
231. (a) M. Dagonneau and J. Vialle, *Tetrahedron* **30**, 415 (1974); (b) B. C. Gilbert, D. K. C. Hodgeman, and R. O. C. Norman, *J. Chem. Soc. Perkin Trans. 2*, 1748 (1973).
232. S. D. Ziman and B. M. Trost, *J. Org. Chem.* **38**, 649 (1973).
233. V. I. Dronov, V. P. Krivonogov, and V. S. Nikitina, *Khim. Geterosikl. Soedin*, 335 (1970).
234. V. I. Dronov and V. P. Krivonogov, *Khim. Geterosikl. Soedin*, 1337 (1971).
235. V. I. Dronov and V. P. Krivonogov, *Khim. Geterosikl. Soedin*, 622 (1972).
236. V. I. Dronov and V. P. Krivonogov, *Khim. Geterosikl. Soedin*, 1186 (1972).
237. F. D. Gunstone, M. G. Hussain, and D. M. Smith, *Chem. Phys. Lipids* **13**, 71 (1974).
238. F. D. Gunstone, *Acc. Chem. Res.* **9**, 34 (1976).
239. (a) K. C. Nicolaou, W. E. Barnette, G. P. Gasic, and R. L. Magolda, *J. Am. Chem. Soc.* **99**, 7736 (1977); (b) K. C. Nicolaou, W. E. Barnette, and R. L. Magolda, *J. Am. Chem. Soc.* **100**, 2567 (1978); (c) K. C. Nicolaou, R. L. Magolda, and W. E. Barnette, *J. Chem. Soc. Chem. Commun.*, 375 (1978).
240. M. Shibasaki and S. Ikegami, *Tetrahedron Lett.*, 559 (1978).
241. D. H. R. Barton, *Pure Appl. Chem.* **33**, 1 (1973).
242. D. J. Aberhart, J. Y.-R. Chu, N. Neuss, C. H. Nash, J. Occolowitz, L. L. Huckstep, and N. De la Higuera, *J. Chem. Soc. Chem. Commun.*, 564 (1974).
243. D. N. Jones and D. A. Lewton, *J. Chem. Soc. Chem. Commun.*, 457 (1974).
244. J-M. Surzur, G. Bastien, M. P. Crozet, and C. Dupuy, *C. R. Acad. Sci. Ser. C*, **276**, 289 (1973).
245. H. Morita and S. Oae, *Heterocycles* **5**, 35 (1976).
246. (a) G. Bastien and J-M. Surzur, *Bull. Soc. Chim. Fr.*, II-601 (1979); (b) G. Bastien, M. P. Crozet, E. Flesia, and J-M. Surzur, *Bull. Soc. Chim. Fr.*, II-606 (1979); (c) G. Bastien and J-M. Surzur, unpublished results (1977).
247. (a) Y. Maki and M. Sako, *Tetrahedron Lett.*, 4291 (1976); (b) Y. Maki and M. Sako, *J. Am. Chem. Soc.* **99**, 5091 (1977); (c) E. M. Gordon and C. M. Cimarusti, *Tetrahedron Lett.*, 3425 (1977).
248. (a) Y. Maki and M. Sako, *J. Chem. Soc. Chem. Commun.*, 836 (1978); (b) J. E. Baldwin and T. S. Wan, *J. Chem. Soc. Chem. Commun.*, 249 (1979).
249. P. M. Chakrabarti and N. B. Chapman, *J. Chem. Soc. C*, 914 (1970).
250. H. Boelens and L. Brandsma, *Rec. Trav. Chim. Pays-Bas* **91**, 141 (1972).
251. H. Nishimura, T. Hanzawa, and J. Mizutani, *Tetrahedron Lett.*, 343 (1973).
252. H. Nishimura and J. Mizutani, *J. Org. Chem.* **40**, 1567 (1975).
253. J. R. Grunwell, D. L. Foerst, and M. J. Sanders, *J. Org. Chem.* **42**, 1142 (1977).
254. (a) T. Sheradsky, in *The Chemistry of the Thiol Group*, S. Pataï, ed., Part II, John Wiley & Sons (1974), p. 685, and references cited therein; (b) P. Metzner, T. N. Pham, and J. Vialle, *Nouv. J. Chim.* **2**, 179 (1978); (c) P. Metzner, T. N. Pham, and J. Vialle, *J. Chem. Res. (S)*, 478 (1978).
255. H. Kwart and E. R. Evans, *J. Org. Chem.* **31**, 413 (1966).
256. H. Kwart and M. H. Cohen, *J. Org. Chem.* **32**, 3135 (1967).
257. L. Dalgaard and S. O. Lawesson, *Tetrahedron* **28**, 2051 (1972).
258. L. Morin, D. Paquer, and S. Smadja, *Rec. Trav. Chim. Pays-Bas* **95**, 179 (1976).

259. Y. Makisumi and A. Murabayashi, *Tetrahedron Lett.*, 1971 (1969).
260. Y. Makisumi and A. Murabayashi, *Tetrahedron Lett.*, 2453 (1969).
261. B. W. Bycroft and W. Landon, *J. Chem. Soc. Chem. Commun.*, 168 (1970).
262. J. Z. Mortensen, B. Hedegaard, and S.-O. Lawesson, *Tetrahedron* **27**, 3831 (1971).
263. H. Kwart and J. L. Schwartz, *J. Org. Chem.* **39**, 1575 (1974).
264. L. Dalgaard and S.-O. Lawesson, *Acta Chem. Scand.* **28B**, 1077 (1974).
265. A. G. Davies, M. J. Parrott, and B. P. Roberts, *J. Chem. Soc. Chem. Commun.*, 27 (1974).
266. A. G. Davies, M. J. Parrott, and B. P. Roberts, *J. Chem. Soc. Perkin Trans.* 2, 1066 (1976).
267. K. I. Kobrakov, T. I. Chernysheva, and N. S. Nametkin, *Dokl. Akad. Nauk. SSSR* **198**, 1340 (1971); *Chem. Abstr.* **75**, 129908e (1971).
268. R. J. Fessenden and W. D. Kray, *J. Org. Chem.* **38**, 87 (1973).
269. R. A. Benkeser, E. C. Mozdzen, W. C. Muench, R. T. Roche, and M. P. Siklosi, *J. Org. Chem.* **44**, 1370 (1979).
270. (a) H. Sakurai, T. Hirose, and A. Hosomi, *Abstracts, 26th Annual Meeting, Chemical Society of Japan,* Hiratsuka III, 1001 (1972); (b) H. Sakurai, in *Free Radicals,* Vol. 2, J. K. Kochi, Ed., Wiley, New York (1973), pp. 793–794.
271. M. Massol, J. Satge, and J. Barrau, *C. R. Acad. Sci. Ser. C*, **268**, 1710 (1969).
272. R. Riviere and J. Satge, *Angew. Chem., Int. Ed.* **10**, 267 (1971).
273. R. Riviere and J. Satge, *J. Organometal. Chem.* **49**, 173 (1973).
274. A. G. Davies, B. P. Roberts, and M.-W. Tse, *J. Chem. Soc. Perkin Trans.* 2, 1499 (1977).
275. G. A. Chmutova, *Sb. Aspir. Rab., Kazan Gos. Univ. Khim. Geol.*, 70 (1967); *Chem. Abstr.* **70**, 87444 (1969).
276. N. Bellinger, D. Cagniant, and P. Cagniant, *Tetrahedron Lett.*, 49 (1971).
277. M. Julia and C. James, *C. R. Acad. Sci. Ser. C*, **255**, 959 (1962).
278. R. E. Dessy and S. A. Kandil, *J. Org. Chem.* **30**, 3857 (1965).
279. S. A. Kandil and R. E. Dessy, *J. Am. Chem. Soc.* **88**, 3027 (1966).
280. H. R. Ward, *J. Am. Chem. Soc.* **89**, 5517 (1967).
281. J. K. Crandall and D. J. Keyton, *Tetrahedron Lett.*, 1653 (1969).
282. (a) T. Ohnuki, M. Yoshida, and O. Simamura, *Chem. Lett.*, 797 (1972); (b) G. Büchi and H. Wüest, *J. Org. Chem.* **44**, 546 (1979).
283. T. Ohnuki, M. Yoshida, O. Simamura, and M. Fukuyama, *Chem. Lett.*, 999 (1972).
284. S. A. Dodson and R. D. Stipanovic, *J. Chem. Soc. Perkin Trans.* 1, 410 (1975).
285. M. E. Kuehne and W. H. Parsons, *J. Org. Chem.* **42**, 3408 (1977).
286. J. Kalvoda and J. Crob, *CNRS International Colloquium on Organic Free Radicals,* Aix-en-Provence, 1977, CNRS, Paris (1978), pp. 507–514.
287. (a) W. M. Moore and D. G. Peters, *Tetrahedron Lett.*, 453 (1972); (b) W. M. Moore, A. Salajegheh, and D. G. Peters, *J. Am. Chem. Soc.* **97**, 4954 (1975).
288. B. C. Willett, W. M. Moore, A. Salajegheh, and D. G. Peters, *J. Am. Chem. Soc.* **101**, 1162 (1979).
289. G. F. Hennion and R. H. Ode, *J. Org. Chem.* **31**, 1975 (1966).
290. G. Stork, S. Malhotra, H. Thompson, and M. Uchibayaschi, *J. Am. Chem. Soc.* **87**, 1148 (1965).
291. S. K. Pradhan, T. V. Radhakrishnan, and R. Subramanian, *J. Org. Chem.* **41**, 1943 (1976).
292. J. K. Crandall and W. J. Michaely, *J. Organometal. Chem.* **51**, 375 (1973).
293. J. K. Crandall, P. Battioni, J. T. Wehlacz, and R. Bindra, *J. Am. Chem. Soc.* **97**, 7171 (1975).
294. H. G. Richey, Jr., and A. Rothman, *Tetrahedron Lett.*, 1457 (1968).
295. W. C. Kossa, Jr., Th. C. Rees, and H. G. Richey, Jr., *Tetrahedron Lett.*, 3455 (1971).
296. J.-L. Derocque and F.-B. Sundermann, *J. Org. Chem.* **39**, 1411 (1974).
297. J.-L. Derocque, U. Beisswenger, and M. Hanack, *Tetrahedron Lett.*, 2149 (1969).
298. R. D. Rieke and B. J. A. Cooke, *J. Org. Chem.* **36**, 2674 (1971).

299. (a) C. Dupuy, Ph. D. thesis, Marseilles (1972); (b) C. Dupuy and J-M. Surzur, *Bull. Soc. Chim. Fr.*, II-353 (1980); (c) C. Dupuy, M. P. Crozet, and J-M. Surzur, *Bull. Soc. Chim. Fr.*, II-361 (1980); (d) C. Dupuy, and J-M. Surzur, *Bull. Soc. Chim. Fr.*, II-374 (1980).

300. J-M. Surzur, C. Dupuy, M. P. Bertrand, and R. Nouguier, *J. Org. Chem.* **37**, 2782 (1972).

301. J-M. Surzur, C. Dupuy, M. P. Crozet, and N. Aimar, *C. R. Acad. Sci. Ser. C* **269**, 849 (1969).

302. W. E. Truce and Th. C. Klingler, *J. Org. Chem.* **35**, 1834 (1970).

303. A. Bottini and E. Böttner, *J. Org. Chem.* **31**, 586 (1966).

304. H. Kwart and Th. J. George, *J. Chem. Soc. Chem. Commun.*, 433 (1970).

305. L. Brandsma, P. J. W. Schuijl, D. Schuijl-Laros, J. Meijer, and H. E. Wijers, *Int. J. Sulfur Chem. Part B* **6**, 85 (1971).

306. D. Schuijl-Laros, P. J. W. Schuijl, and L. Brandsma, *Rec. Trav. Chim. Pays-Bas* **91**, 785 (1972).

307. K. K. Balasubramanian and B. Venugopalan, *Tetrahedron Lett.*, 2643 (1974).

308. K. K. Balasubramanian and B. Venugopalan, *Tetrahedron Lett.*, 2645 (1974).

309. G. Märkl and G. Dannhardt, *Tetrahedron Lett.*, 1455 (1973).

310. (a) R. Gompper, and D. Lach, *Tetrahedron Lett.*, 2687 (1973); (b) D. F. Shellhamer and M. L. Oakes, *J. Org. Chem.* **43**, 1316 (1978).

311. (a) N. Šarčević, J. Zsindely, and H. Schmid, *Helv. Chim. Acta* **56**, 1457 (1973); (b) S. Arseniyadis, J. Goré, A. Laurent, and M.-L. Roumestant, *J. Chem. Res.* (S) 416 (1978), (M) 4616 (1978).

312. M. P. Doyle, P. W. Raynolds, R. A. Barents, Th. R. Bade, W. C. Danen, and Ch. T. West, *J. Am. Chem. Soc.* **95**, 5988 (1973).

313. C. H. De Puy, H. L. Jones, and D. H. Gibson, *J. Am. Chem. Soc.* **94**, 3924 (1972).

314. A. M. Martinez, G. E. Cushmac, and J. Roček, *J. Am. Chem. Soc.* **97**, 6502 (1975).

315. J. Roček and D. E. Aylward, *J. Am. Chem. Soc.* **97**, 5452 (1975).

316. H. Suginome, N. Sato, and T. Masamune, *Tetrahedron Lett.*, 3353 (1969).

317. H. Suginome, T. Mizuguchi, and T. Masamune, *Tetrahedron Lett.*, 4723 (1971).

318. H. Suginome, T. Mizuguchi, and T. Masamune, *J. Chem. Soc. Chem. Commun.*, 376 (1972).

319. H. Suginome, T. Tsuneno, N. Sato, and T. Masamune, *Tetrahedron Lett.*, 661 (1972).

320. J. Bošnjak, V. Andrejevič, Ž. Čekovič, and M. Lj. Mihailović, *Tetrahedron*, **28**, 6031 (1972).

321. K. Formanek, J. P. Aune, M. Jouffret, and J. Metzger, *Nouv. J. Chim.* **1**, 13 (1977).

322. A. I. Feinstein, E. K. Fields, P. J. Ihrig, and S. Meyerson, *J. Org. Chem.* **36**, 996 (1971).

323. M. Akhtar and S. Marsh, *J. Chem. Soc. C*, 937 (1966).

324. S. T. Reid and J. N. Tucker, *J. Chem. Soc. Chem. Commun.*, 1609 (1971).

325. C. L. Karl, E. J. Maas, and W. Reusch, *J. Org. Chem.* **37**, 2834 (1972).

326. D. I. Davies, D. J. A. Pearce, E. C. Dart, *J. Chem. Soc. Perkin Trans.* 1, 433 (1973).

327. H. Suginome, N. Sato, and T. Masamune, *Tetrahedron Lett.*, 1557 (1967).

328. H. Suginome, N. Sato, and T. Masamune, *Bull. Chem. Soc. Japan* **42**, 215 (1969).

329. A. Nickon, T. Iwadare, F. J. McGuire, J. R. Mahajan, S. A. Narang, and B. Umezawa, *J. Am. Chem. Soc.* **92**, 1688 (1970).

330. A. Nickon, R. Ferguson, A. Bosch, and T. Iwadare, *J. Am. Chem. Soc.* **99**, 4518 (1977).

331. M. Lj. Mihailović, V. Andrejevič, M. Jakovljevič, D. Jeremič, and R. E. Partch, *J. Chem. Soc. Chem. Commun.*, 854 (1970).

332. R. L. Huang and H. H. Lee, *J. Chem. Soc.*, 2500 (1964).

333. F. Flies, R. Lalande, and B. Maillard, *Tetrahedron Lett.*, 439 (1976).

334. A. R. Forrester, J. Skilling, and R. H. Thomson, *J. Chem. Soc. Perkin Trans.* 1, 2161 (1974).

335. J.-H. Liu and P. Kovacic, *J. Org. Chem.* **38**, 3462 (1973).

336. U. Pommerenk, H. Sengewein, and P. Welzel, *Tetrahedron Lett.*, 3415 (1972).

337. J. R. Collier and J. Hill, *J. Chem. Soc. Chem. Commun.*, 640 (1969).
338. (a) E. Götshi and A. Eschenmoser, *Angew. Chem. Int. Ed.* **12**, 912 (1973), and references cited therein; (b) R. Bishop and N. K. Hamer, *J. Chem. Soc. D*, 804 (1969); (c) R. Bishop and N. K. Hamer, *J. Chem. Soc. C*, 1193 (1970); (d) G. J. Boudreaux, E. I. Becker, and B. H. Arison, *J. Org. Chem.* **43**, 1827 (1978).
339. H. Suginome and K. Kato, *Tetrahedron Lett.*, 4139 (1973).
340. H. Suginome, K. Kato, and T. Masamune, *Tetrahedron Lett.*, 1161 (1974).
341. H. Suginome, K. Kato, and T. Masamune, *Tetrahedron Lett.*, 1165 (1974).
342. H. Suginome, A. Furusaki, K. Kato, and T. Matsumoto, *Tetrahedron Lett.*, 2757 (1975).
343. (a) L. W. Menapace and H. G. Kuivila, *J. Am. Chem. Soc.* **86**, 3047 (1964); (b) K. R. Wursthorn, H. G. Kuivila, and G. F. Smith, *J. Am. Chem. Soc.* **100**, 2779 (1978).
344. D. A. Harrison, R. N. Schwartz, and J. Kagan, *J. Am. Chem. Soc.* **92**, 5793 (1970).
345. A. S. Kende and J. L. Belletire, *Tetrahedron Lett.*, 2145 (1972).
346. K. Praefcke, *Tetrahedron Lett.*, 973 (1973).
347. J. Satge and P. Riviere, *J. Organometal. Chem.* **16**, 71 (1969).
348. I. Rosenthal and D. Elad, *J. Org. Chem.* **33**, 805 (1968).
349. R. Lalande, B. Maillard, and M. Cazaux, *Tetrahedron Lett.*, 745 (1969).
350. B. Maillard, M. Cazaux, and R. Lalande, *Bull. Soc. Chim. Fr.*, 467 (1971).
351. C. Bernasconi, L. Cottier, and G. Descotes, *Bull. Soc. Chim. Fr.*, 101 (1977).
352. (a) G. I. Nikishin, E. K. Starostin, B. A. Golovin, A. V. Kessenikh, and A. V. Ignatenko, *Izvest. Akad. Nauk. SSSR, Ser. Khim.*, 1842 (1972); (b) M. Julia, J. M. Salard, and J. C. Chottard, *Bull. Soc. Chim. Fr.*, 2478 (1973).
353. C. M. Rynard, C. Thankachan, and Th. T. Tidwell, *J. Am. Chem. Soc.* **101**, 1196 (1979).
354. C. H. De Puy, D. E. Zabel, and W. Wiedeman, *J. Org. Chem.* **33**, 2198 (1968).
355. S. Julia and R. Lorne, *C. R. Acad. Sci. Ser. C* **268**, 1617 (1969).
356. S. Julia and R. Lorne, *C. R. Acad. Sci. Ser. C* **273**, 174 (1971).
357. J-M. Surzur and P. Teissier, *C. R. Acad. Sci. Ser. C* **264**, 1981 (1967).
358. D. D. Tanner and F. C. P. Law, *J. Am. Chem. Soc.* **91**, 7535 (1969).
359. J-M. Surzur and P. Teissier, *Bull. Soc. Chim. Fr.*, 3060 (1970).
360. A. L. J. Beckwith and P. K. Tindal, *Aust. J. Chem.* **24**, 2099 (1971).
361. A. L. J. Beckwith and C. B. Thomas, *J. Chem. Soc. Perkin Trans. 2*, 861 (1973).
362. M. J. Perkins and B. P. Roberts, *J. Chem. Soc. Perkin Trans. 2*, 77 (1975).
363. P. M. Kasai and D. McLeod, Jr., *J. Am. Chem. Soc.* **94**, 720 (1972).
364. J. Kalvoda, *Helv. Chim. Acta* **51**, 267 (1968).
365. J. Kalvoda, *J. Chem. Soc. Chem. Commun.*, 1002 (1970).
366. J. Kalvoda and L. Botta, *Helv. Chim. Acta* **55**, 356 (1972).
367. (a) D. S. Watt, *J. Am. Chem. Soc.* **98**, 271 (1976); (b) R. W. Freerksen, W. E. Pabst, M. L. Raggio, S. A. Sherman, R. R. Wroble, and D. S. Watt, *J. Am. Chem. Soc.* **99**, 1536 (1977).
368. A. D. Barone and D. S. Watt, *Tetrahedron Lett.*, 3673 (1978).
369. (a) Yu. N. Ogibin, E. I. Troyanskii, and G. I. Nikishin, *Izv. Akad. Nauk. SSSR Ser. Khim.*, 1461 (1975); (b) Yu. N. Ogibin, E. I. Troyanskii, and G. I. Nikishin, *Izv. Akad. Nauk. SSSR Ser. Khim.*, 843 (1977).
370. (a) P. Baas and H. Cerfontain, *Tetrahedron Lett.*, 1501 (1978); (b) S. Wolff and W. C. Agosta, *J. Org. Chem.* **43**, 3627 (1978).
371. M. Larcheveque, A. Debal, and Th. Cuvigny, *J. Organometal. Chem.* **87**, 25 (1975).
372. D. G. Hawthorne, S. R. Johns, D. H. Solomon, and R. I. Willing, *J. Chem. Soc. Chem. Commun.*, 982 (1975).
373. S. R. Johns and R. I. Willing, *J. Macromol. Sci.-Chem.* **A9**, 169 (1975).
374. G. Levesque, G. Tabak, F. Outurquin, and J-C. Gressier, *Bull. Soc. Chim. Fr.*, 1156 (1976).
375. G. Levesque, A. Mahjoub, and A. Thuillier, *Colloq. Int. CNRS, Radicaux Libres Organiques*, Aix-en-Provence, July 1977, CNRS, Paris (1978), pp. 365–368.

376. W. Ando, T. Oikawa, K. Kishi, T. Saiki, and T. Migita, *J. Chem. Soc. Chem. Commun.*, 704 (1975).

377. (a) W. C. Danen and C. T. West, *J. Am. Chem. Soc.* **96**, 2447 (1974); (b) C. A. Whitesitt and D. K. Herron, *Tetrahedron Lett.*, 1737 (1978).

378. J. M. Patterson, C. F. Mayer, and W. T. Smith, Jr., *J. Org. Chem.* **40**, 1511 (1975).

379. (a) K. N. V. Duong, A. Gaudemer, M. D. Johnson, R. Quillivic, and J. Zylber, *Tetrahedron Lett.*, 2997 (1975); (b) A. J. Hartshorn, A. W. Johnson, S. M. Kennedy, M. F. Lappert, and A. W MacQuitty, *J. Chem. Soc. Chem. Commun.*, 643 (1978).

380. (a) D. D. Tanner and P. M. Rahimi, *J. Org. Chem.* **44**, 1674 (1979); (b) L. Stella, Z. Janousek, R. Merényi, and H. G. Viehe, *Angew. Chem. Int. Ed.* **17**, 691 (1978).

381. (a) R. A. Kaba, L. Lunazzi, D. Lindsay, and K. U. Ingold, *J. Am. Chem. Soc.* **97**, 6762 (1975); (b) L. Benati, G. Placucci, P. Spagnolo, A. Tundo, and G. Zanardi, *J. Chem. Soc. Perkin Trans.* **1**, 1684 (1977).

382. A. A. McConnell, S. Mitchell, A. L. Porte, J. S. Roberts, and C. Thomson, *J. Chem. Soc. B*, 833 (1970).

383. M. L. Heyman and J. P. Snyder, *J. Am. Chem. Soc.* **97**, 4416 (1975).

384. L. Benati, P. C. Montevecchi, and P. Spagnolo, *Tetrahedron Lett.*, 815 (1978).

385. C. J. Kelley and M. Carmack, *Tetrahedron Lett.*, 3605 (1975).

386. D. L. Struble, A. L. J. Beckwith, and G. E. Gream, *Tetrahedron Lett.*, 4795 (1970).

387. D. H. R. Barton, D. L. J. Clive, P. D. Magnus, and G. Smith, *J. Chem. Soc. C*, 2193 (1971).

388. S. P. Adhikary, W. Lawrie, J. McLean, and M. S. Malik, *J. Chem. Soc. C* **32** (1971).

389. (a) J. J. Köhler and W. N. Speckamp, *Tetrahedron Lett.*, 635 (1977); (b) H. O. Bernhard and V. Snieckus, *Tetrahedron Lett.*, 4867 (1971); (c) I. Tse and V. Snieckus, *J. Chem. Soc. Chem. Commun.*, 505 (1976); (d) H. Iida, T. Takarai, and C. Kibayashi, *J. Org. Chem.* **43**, 975 (1978); (e) H. Iida, Y. Yuasa, and C. Kibayashi, *J. Chem. Soc. Chem. Commun.*, 766 (1978); (f) H. Iida, Y. Yuasa, and C. Kibayashi, *Tetrahedron Lett.*, 3817 (1978); (g) H. Iida, Y. Yuasa, and C. Kibayashi, *J. Org. Chem.* **44**, 1236 (1979).

390. G. Stork and P. G. Williard, *J. Am. Chem. Soc.* **99**, 7067 (1977).

391. Ch. Descoins, M. Julia, and H. Van Sang, *Bull. Soc. Chim. Fr.*, 4087 (1971).

392. J. L. Stein, Ph. D. thesis, Marseilles (1974).

393. R. P. A. Sneeden, *Synthesis*, 259 (1971).

394. J. G. Traynham and H. H. Hsieh, *J. Org. Chem.* **38**, 868 (1973).

395. T. W. Sam and J. K. Sutherland, *J. Chem. Soc. Chem. Commun.*, 970 (1971).

396. P. Y. Johnson and M. A. Priest, *J. Am. Chem. Soc.* **96**, 5618 (1974).

397. J. W. Wilt, S. N. Massie, and R. D. Dabek, *J. Org. Chem.* **35**, 2803 (1970).

398. A. G. Yurchenko, L. A. Zosim, N. L. Dovgan, and N. S. Verpovsky, *Tetrahedron Lett.*, 4843 (1976).

399. (a) C. W. Bird and R. Khan, *Tetrahedron Lett.*, 2813 (1976); (b) L. A. Levy, *J. Chem. Soc. Chem. Commun.*, 574 (1978); (c) H. Parlar, M. Mansour, and S. Gäb, *Tetrahedron Lett.*, 1597 (1978).

400. G. I. Fray, G. R. Geen, D. I. Davies, L. T. Parfitt, and M. J. Parrott, *J. Chem. Soc. Perkin Trans.* **1**, 729 (1974).

401. (a) B. B. Jarvis, J. P. Govoni, and Ph. J. Zell, *J. Am. Chem. Soc.* **93**, 913 (1971); (b) S. J. Cristol, D. P. Stull, and R. D. Daussin, *J. Am. Chem. Soc.* **100**, 6674 (1978).

402. R. Nouguier and J-M. Surzur, *Bull. Soc. Chim. Fr.*, 2399 (1973).

403. (a) K. Heusler, *Tetrahedron Lett.*, 97 (1970); (b) W. K. Appel, T. J. Greenhough, J. R. Scheffer, and J. Trotter, *J. Am. Chem. Soc.* **101**, 213 (1979).

404. D. Gravel, J. Hebert, J. Bilodeau, E. Cavalieri and J.-P. Daris, *Can. J. Chem.* **52**, 645 (1974).

405. R. Nouguier, Ph. D. thesis, Marseilles (1975).

406. K. S. Pillay, R. N. Lockhart, T. Tezuka, and Y. L. Chow, *J. Chem. Soc. Chem. Commun.*, 80 (1974).

407. R. Tadayoni, J. Lacrampe, A. Heumann, R. Furstoss, and B. Waegell, *Tetrahedron Lett.*, 735 (1975).

408. (a) R. W. Lockhart, K. Hanaya, F. W. B. Einstein, and Y. L. Chow, *J. Chem. Soc. Chem. Commun.*, 344 (1975); (b) G. Grethe, H. L. Lee, T. Mitt, and M. R. Uskokovič, *J. Am. Chem. Soc.* **100**, 581 (1978).

409. R. Furstoss and B. Waegell, *Tetrahedron Lett.*, 365 (1976).

410. R. Furstoss, R. Tadayoni, and B. Waegell, *J. Org. Chem.* **42**, 2844 (1977).

411. G. Esposito, R. Furstoss, and B. Waegell, *Tetrahedron Lett.*, 899 (1971).

412. P. K. Claus and F. W. Vierhapper, *J. Org. Chem.* **42**, 4016 (1977).

413. J. L. Charlton and G. J. Williams, *Tetrahedron Lett.*, 1473 (1977).

414. P. D. Gokhale, A. P. Joshi, R. Sahni, V. G. Naik, N. P. Damodaran, U. R. Nayak, and S. Dev, *Tetrahedron* **32**, 1391 (1976).

415. P. Bakuzis, O. O. S. Campos, and M. L. F. Bakuzis, *J. Org. Chem.* **41**, 3261 (1976).

416. W. N. Speckamp and H. Kesselaar, *Tetrahedron Lett.*, 3405 (1974).

417. J. Szychowski and O. Achmatowicz, Jr., *Rocz. Chem.* **45**, 189 (1971).

418. R. C. Lamb, W. E. McNew, Jr., J. R. Sanderson, and D. C. Lunney, *J. Org. Chem.* **36**, 174 (1971).

419. W. R. Dolbier, Jr., I. Nishiguchi, and J. M. Riemann, *J. Am. Chem. Soc.* **94**, 3642 (1972).

420. (a) D. A. Hutchings, K. J. Frech, and F. H. Hoppstock, *Chem. Prepr., Symp. on New Routes to Olefins, Am. Chem. Soc. Div. Petrol. Chem.*, Boston, **17**, B37–B45 (1972); (b) D. Griller, K. U. Ingold, and J. C. Walton, *J. Am. Chem. Soc.* **101**, 758 (1979).

421. R. E. Lehr, J. M. Wilson, J. W. Harder, and P. T. Cohenour, *J. Am. Chem. Soc.* **98**, 4867 (1976).

422. G. Dupont, R. Dulou, and G. Clement, *Bull. Soc. Chim. Fr.*, 1115 (1950).

423. M. Cazaux and R. Lalande, *Bull. Soc. Chim. Fr.*, 3381 (1966), and references cited therein.

424. J. Moulines and R. Lalande, *Bull. Soc. Chim. Fr.*, 3387 (1966).

425. R. Lalande, B. Paskoff, and M. Cazaux, *C. R. Acad. Sci. Ser.* C **264**, 1083 (1967).

426. B. Paskoff, M. Cazaux, and R. Lalande, *Bull. Soc. Chim. Fr.*, 624 (1970).

427. M. Cazaux and R. Lalande, *Bull. Soc. Chim. Fr.*, 461 (1971).

428. B. Maillard, M. Cazaux, and R. Lalande, *Bull. Soc. Chim. Fr.*, 1368 (1973).

429. R. L. Kenney and G. S. Fisher, *J. Org. Chem.* **39**, 682 (1974).

430. H. H. Quon, T. Tezuka, and Y. L. Chow, *J. Chem. Soc. Chem. Commun.*, 428 (1974).

431. Y. L. Chow, S. K. Pillay, and H. H. Quon, *J. Chem. Soc. Perkin Trans.* 2, 1255 (1977).

432. M. Julia, D. Mansuy, and J.-Y. Lallemand, *Bull. Soc. Chim. Fr.*, 2695 (1972).

433. M. Julia and D. Mansuy, *C. R. Acad. Sci. Ser.* C **276**, 1049 (1973).

434. J.-Y. Lallemand, *Tetrahedron Lett.*, 1217 (1975).

435. K. Wada, Y. Enomoto, and K. Munakata, *Tetrahedron Lett.*, 3357 (1969).

436. L. K. Montgomery and J. W. Matt, *J. Am. Chem. Soc.* **89**, 6556 (1967).

437. H. Ladenheim and W. Bartok, *J. Am. Chem. Soc.* **89**, 1786 (1967).

438. L. K. Montgomery and J. W. Matt, *J. Am. Chem. Soc.* **89**, 3050 (1967).

439. E. C. Friedrich, *J. Org. Chem.* **34**, 1851 (1969).

440. J. K. Kochi, P. J. Krusic, and D. R. Eaton, *J. Am. Chem. Soc.* **91**, 1877 (1969).

441. J. K. Kochi, P. J. Krusic, and D. R. Eaton, *J. Am. Chem. Soc.* **91**, 1879 (1969).

442. J.-Y. Godet, M. Pereyre, J.-C. Pommier, and D. Chevolleau, *J. Organometal. Chem.* **55**, C-15 (1973).

443. W. J. Hehre, *J. Am. Chem. Soc.* **95**, 2643 (1973).

444. K. S. Chen, D. J. Edge, and J. K. Kochi, *J. Am. Chem. Soc.* **95**, 7036 (1973).

445. M. Ratier and M. Pereyre, *Tetrahedron Lett.*, 2273 (1976).

446. J-Y. Godet and M. Pereyre, *Bull. Soc. Chim. Fr.*, 1105 (1976).

447. A. G. Davies, J-Y. Godet, B. Muggleton, and M. Pereyre, *J. Chem. Soc. Chem. Commun.*, 813 (1976).
448. P. Blum, A. G. Davies, M. Pereyre, and M. Ratier, *J. Chem. Soc. Chem. Commun.*, 814 (1976).
449. E. L. Stogryn and M. H. Gianni, *Tetrahedron Lett.*, 3025 (1970).
450. H. C. Brown and M. M. Midland, *Angew. Chem. Int. Ed.* **11**, 692 (1972).
451. A. Suzuki, N. Miyaura, M. Itoh, H. C. Brown, G. W. Holland, and E. I. Negishi, *J. Am. Chem. Soc.* **93**, 2792 (1971).
452. P. K. Freeman, F. A. Raymond, J. C. Sutton, and W. R. Kindley, *J. Org. Chem.* **33**, 1448 (1968).
453. P. K. Freeman, M. F. Grostic, and F. A. Raymond, *J. Org. Chem.* **36**, 905 (1971).
454. G. W. Shaffer, *J. Org. Chem.* **37**, 3282 (1972).
455. S. J. Cristol and R. V. Barbour, *J. Am. Chem. Soc.* **90**, 2832 (1968).
456. H. Deshayes, J. P. Pete, and C. Portella, *Tetrahedron Lett.*, 2019 (1976).
457. R. S. Boikess, M. Mackay, and D. Blithe, *Tetrahedron Lett.*, 401 (1971).
458. R. W. Thies, and D. D. McRitchie, *J. Org. Chem.* **38**, 112 (1973).
459. M. Suzuki, S. I. Murahashi, A. Sonoda, and I. Moritani, *Chem. Lett.*, 267 (1974).
460. E. Müller, *Tetrahedron Lett.*, 1835 (1974).
461. L. H. Slaugh, *J. Am. Chem. Soc.* **87**, 1522 (1965).
462. H. Reimann and O. Z. Sarre, *Can. J. Chem.* **49**, 344 (1971).
463. H. P. Löffler and G. Schröder, *Tetrahedron Lett.*, 2119 (1970).
464. H. P. Löffler, *Chem. Ber.* **104**, 1981 (1971).
465. H. P. Löffler, *Tetrahedron Lett.*, 4893 (1971).
466. E. Tobler, D. E. Battin, and D. J. Foster, *J. Org. Chem.* **29**, 2834 (1964).
467. C. K. Alden and D. I. Davies, *J. Chem. Soc. C*, 2007 (1967).
468. V. A. Azovskaya, A. U. Stepanyants, D. Mondeshka, R. I. Shekhtman, I. Koshel'skaya, and E. N. Prilezhaeva, *Zh. Org. Khim.* **6**, 2568 (1970).
469. (a) H. G. Kuivila, J. D. Kennedy, R. Y. Tien, I. J. Tyminski, F. L. Pelczar, and O. R. Khan, *J. Org. Chem.* **36**, 2083 (1971); (b) H. G. Kuivila, *Acc. Chem. Res.* **1**, 299 (1968), and references cited therein; (c) H. G. Kuivila and C. C. H. Pian, *J. Chem. Soc. Chem. Commun.*, 369 (1974).
470. M. Oku and J. Ch. Philips, *J. Am. Chem. Soc.* **95**, 6495 (1973).
471. D. R. Adams and D. I. Davies, *J. Chem. Soc. Perkin Trans.* **1**, 246 (1974).
472. K. S. Pillay and Y. L. Chow, *J. Chem. Soc. Perkin Trans.* **2**, 93 (1977).
473. S. J. Cristol and R. W. Gleason, *J. Org. Chem.* **34**, 1762 (1969).
474. G. A. Gray and W. R. Jackson, *J. Am. Chem. Soc.* **91**, 6205 (1969).
475. C. W. Jefford and W. Wojnarowski, *Helv. Chim. Acta* **53**, 1194 (1970).
476. W. R. Jackson, V. M. A. Chambers, and G. W. Young, *J. Chem. Soc. C*, 2075 (1971).
477. H. C. Brown, Ph. J. Geoghegan, G. J. Lynch, and J. T. Kurek, *J. Org. Chem.* **37**, 1941 (1972).
478. R. Alexander and D. I. Davies, *J. Chem. Soc. Perkin Trans.* **1**, 83 (1973).
479. M. N. Paddon-Row, D. N. Butler, and R. N. Warrener, *J. Chem. Soc. Chem. Commun.*, 741 (1976).
480. J. Elzinga and H. Hogeveen, *J. Chem. Soc. Chem. Commun.*, 705 (1977).
481. Th. A. Halgren, J. L. Firkins, T. A. Fujimoto, and H. H. Suzukawa, *Proc. Natl. Acad. Sci. USA* **68**, 3216 (1971).
482. T. C. Morrill and F. L. Vandemark, *Tetrahedron Lett.*, 1811 (1971).
483. T. G. Burrowes and W. R. Jackson, *Aust. J. Chem.* **28**, 639 (1975).
484. K. Griesbaum, *Angew. Chem. Intern. Ed.* **9**, 273 (1970).
485. S. J. Cristol and R. Kellman, *J. Org. Chem.* **36**, 1866 (1971).
486. B. E. Smart, *J. Org. Chem.* **39**, 831 (1974).

487. J. Warkentin and E. Sanford, *J. Am. Chem. Soc.* **90**, 1667 (1968).

488. J. K. Kochi, P. Bakuzis, and P. J. Krusic, *J. Am. Chem. Soc.* **95**, 1516 (1973).

489. C. W. Jefford and F. Delay, *Tetrahedron Lett.*, 3639 (1973).

490. N. O. Brace, *J. Org. Chem.* **31**, 2879 (1966).

491. N. O. Brace, *J. Polym. Sci.* **8**, A-1, 2091 (1970).

492. B. A. Trofimov, A. S. Atavin, and G. M. Gavrilova, *Zh. Org. Khim.* **6**, 624 (1970).

493. N. O. Brace, *J. Org. Chem.* **36**, 3187 (1971).

494. N. O. Brace, *J. Org. Chem.* **38**, 3167 (1973).

495. A. L. J. Beckwith, A. K. Ong, and D. H. Solomon, *J. Macromol. Sci-Chem.* **A9**, 115 (1975).

496. N. V. Kruglova, Sh. A. Karapet'yan, and R. Kh. Freidlina, *Izvest. Akad. Nauk. SSSR Ser. Khim.*, 1569 (1975).

497. N. O. Brace, *J. Org. Chem.* **32**, 2711 (1967).

498. S. F. Reed, Jr., *J. Org. Chem.* **32**, 3675 (1967).

499. N. O. Brace, *J. Org. Chem.* **34**, 2441 (1969).

500. (a) M. A. M. Bradney, A. D. Forbes, and J. Wood, *Chem. Prepr., Am. Chem. Soc. Div. Petrol. Chem., Washington*, **16**, B20–B35 (1971); (b) M. A. M. Bradney, A. D. Forbes, and J. Wood, *J. Chem. Soc. Perkin Trans.* **2**, 1655 (1973).

501. N. O. Brace, *J. Org. Chem.* **44**, 212 (1979).

502. (a) T. Shono and M. Mitani, *J. Am. Chem. Soc.* **93**, 5284 (1971); (b) T. Shono, I. Nishiguchi, H. Ohmizu, and M. Mitani, *J. Am. Chem. Soc.* **100**, 545 (1978).

503. P. Piccardi and M. Modena, *J. Chem. Soc. Chem. Commun.*, 1041 (1971).

504. B. A. Trofimov, A. S. Atavin, G. M. Gavrilova, and G. A. Kalabin, *Zh. Obsch. Khim.* **38**, 2344 (1968).

505. A. S. Atavin, B. A. Trofimov, G. M. Gavrilova, and I. M. Korataeva, *Zh. Obsch. Khim.* **41**, 804 (1971).

506. A. S. Atavin, G. M. Gavrilova, and B. A. Trofimov, *Izvest. Akad. Nauk. SSSR Ser. Khim.*, 2040 (1971).

507. N. V. Kruglova and R. Kh. Freidlina, *Izvest. Akad. Nauk. SSSR Ser. Khim.*, 2277 (1973).

508. R. Breslow, E. Barrett, and E. Mohaczi, *Tetrahedron Lett.*, 1207 (1962).

509. M. Julia, F. Le Goffic, and L. Katz, *Bull. Soc. Chim. Fr.*, 1122 (1964).

510. M. Julia and F. Le Goffic, *Bull. Soc. Chim. Fr.*, 1129 (1964).

511. A. L. J. Beckwith and G. Phillipou, *J. Chem. Soc. Chem. Commun.*, 280 (1973).

512. M. Julia and D. Mansuy, *C. R. Acad. Sci. Ser. C* **274**, 408 (1972).

513. (a) F. J. McQuillin and M. Wood, *J. Chem. Soc. Chem. Commun.*, 65 (1976); (b) F. J. Mac-Quillin and M. Wood, *J. Chem. Soc. Perkin Trans.* **1**, 1762 (1976).

514. J. H. Edwards, F. J. McQuillin, and M. Wood, *J. Chem. Soc. Chem. Commun.*, 438 (1978).

515. R. Breslow, J. T. Groves and S. S. Olin, *Tetrahedron Lett.*, 4717 (1966).

516. R. M. Coates and L. S. Melvin, Jr., *J. Org. Chem.* **35**, 865 (1970).

517. R. Breslow, S. S. Olin, and J. T. Groves, *Tetrahedron Lett.*, 1837 (1968).

518. R. Gleiter and K. Müllen, *Helv. Chim. Acta* **57**, 823 (1974).

519. M. Julia, J.-C. Chottard, and J. J. Basselier, *Bull. Soc. Chim. Fr.*, 3037 (1966).

520. J.-C. Chottard and M. Julia, *Bull. Soc. Chim. Fr.*, 3700 (1968).

521. D. Mansuy and M. Julia, *Bull. Soc. Chim. Fr.*, 2689 (1972).

522. J. Y. Lallemand, M. Julia, and D. Mansuy, *Tetrahedron Lett.*, 4461 (1973).

523. M. E. Kuehne and R. E. Damon, *J. Org. Chem.* **42**, 1825 (1977).

524. A. L. J. Beckwith and G. Moad, *J. Chem. Soc., Perkin Trans.* **2**, 1726 (1975).

525. E. I. Heiba and R. M. Dessau, *J. Am. Chem. Soc.* **88**, 1589 (1966).

526. E. I. Heiba and R. M. Dessau, *J. Am. Chem. Soc.* **89**, 2238 (1967).

527. E. I. Heiba and R. M. Dessau, *J. Am. Chem. Soc.* **89**, 3772 (1967).

528. R. M. Kopchik and J. A. Kampmeier, *J. Am. Chem. Soc.* **90**, 6733 (1968).

529. N. A. Porter and M. O. Funk, *J. Org. Chem.* **40**, 3614 (1975).

530. W. A. Pryor and J. P. Stanley, *J. Org. Chem.* **40**, 3615 (1975).
531. (a) C. T. Mabuni, L. Garlaschelli, R. A. Ellison, and C. R. Hutchinson, *J. Am. Chem. Soc.* **99**, 7718 (1977); (b) C. T. Mabuni, L. Garlaschelli, R. E. Ellison, and C. R. Hutchinson, *J. Am. Chem. Soc.* **101**, 707 (1979).
532. J. Lacrampe, Ph. D. thesis, Marseilles (1973).
533. I. Monković, T. T. Conway, H. Wong, Y. G. Perron, I. J. Pachter, and B. Belleau, *J. Am. Chem. Soc.* **95**, 7910 (1973).
534. T. T. Conway, T. W. Doyle, Y. G. Perron, J. Chapuis, and B. Belleau, *Can. J. Chem.* **53**, 245 (1975).
535. I. Monković, H. Wong, B. Belleau, I. J. Pachter, and Y. G. Perron, *Can. J. Chem.* **53**, 2515 (1975).
536. P. D. Bartlett, *Justus Liebigs Ann. Chem.* **653**, 45 (1962).
537. W. S. Johnson, D. M. Bailey, R. Owyang, R. A. Bell, B. Jacques, and J. C. Crandall, *J. Am. Chem. Soc.* **86**, 1959 (1964).
538. P. D. Bartlett, W. D. Closson, and T. J. Gogdell, *J. Am. Chem. Soc.* **87**, 1308 (1965).
539. J. F. Garst, P. W. Ayers, and R. C. Lamb, *J. Am. Chem. Soc.* **88**, 4260 (1966).
540. (a) J. F. Garst and J. T. Barbas, *J. Am. Chem. Soc.* **90**, 7159 (1968); (b) J. F. Garst and F. E. Barton, II, *Tetrahedron Lett.*, 587 (1969); (c) J. F. Garst, *Acc. Chem. Res.* **4**, 400 (1971); (d) J. F. Garst and F. E. Barton, II, *J. Am. Chem. Soc.* **96**, 523 (1974); (e) J. F. Garst, R. D. Roberts, and J. A. Pacifici, *J. Am. Chem. Soc.* **99**, 3528 (1977); (f) J. F. Garst, J. A. Pacifici, C. C. Felix, and A. Nigam, *J. Am. Chem. Soc.* **100**, 5974 (1978).
541. Th. Cuvigny, M. Larcheveque, and H. Normant, *Bull. Soc. Chim. Fr.*, 1174 (1973).
542. J. W. Sease and R. C. Reed, *Tetrahedron Lett.*, 393 (1975).
543. I. Angres and H. E. Zieger, *J. Org. Chem.* **39**, 1013 (1974).
544. (a) R. J. Kinney, W. D. Jones, and R. C. Bergman, *J. Am. Chem. Soc.* **100**, 635 (1978); (b) R. J. Kinney, W. D. Jones, and R. C. Bergman, *J. Am. Chem. Soc.* **100**, 7902 (1978).
545. D. E. Bergbreiter and J. M. Killough, *J. Am. Chem. Soc.* **100**, 2126 (1978).
546. A. I. Meyers, R. Gabel, and E. D. Mihelich, *J. Org. Chem.* **43**, 1372 (1978).
547. F. A. Davis, P. A. Mancinelli, K. Balasubramanian, and U. K. Nadir, *J. Am. Chem. Soc.* **101**, 1044 (1979).
548. J. K. Kochi and T. W. Bethea, III, *J. Org. Chem.* **33**, 75 (1968).
549. M. Julia and E. Colomer, *C. R. Acad. Sci. Ser. C* **270**, 1305 (1970).
550. J. Grignon and M. Pereyre, *J. Organometal. Chem.* **61**, C-33 (1973).
551. M. Chastrette and R. Gauthier, *Bull. Soc. Chim. Fr.*, 753 (1973).
552. C. Benjannin, G. Lanchec, and B. Blouri, *Bull. Soc. Chim. Fr.*, 661 (1974).
553. U. Schirmer and J. M. Conia, *Tetrahedron Lett.*, 3057 (1974).
554. N. O. Brace and J. E. Van Elswyk, *J. Org. Chem.* **41**, 766 (1976).
555. B. W. Bangerter, R. P. Beatty, J. K. Kouba, and S. S. Wreford, *J. Org. Chem.* **42**, 3247 (1977).
556. M. Julia, H. Langhals, B. Mansour, D. Mansuy, and P. Mattei, *Tetrahedron Lett.*, 3439 (1976).
557. R. P. Quirk and R. E. Lea, *Tetrahedron Lett.*, 1925 (1974).
558. R. P. Quirk and R. E. Lea, *J. Am. Chem. Soc.* **98**, 5973 (1976).
559. J-Y. Nedelec, Ph. D. thesis, Paris, p. 192 (1977).
560. (a) F. Bickelhaupt, *Angew. Chem. Intern. Ed.* **13**, 419 (1974); (b) H. W. H. J. Bodewitz, C. Blomberg, and F. Bickelhaupt, *Tetrahedron* **31**, 1053 (1975).
561. E. C. Ashby and J. S. Bowers, Jr., *J. Am. Chem. Soc.* **99**, 8504 (1977).
562. R. C. Lamb, P. W. Ayers, M. K. Toney, and J. F. Garst, *J. Am. Chem. Soc.* **88**, 4261 (1966).
563. C. Walling and A. Cioffari, *J. Am. Chem. Soc.* **92**, 6609 (1970).
564. J. F. Garst and C. D. Smith, *J. Am. Chem. Soc.* **95**, 6870 (1973).

565. J. F. Garst and C. D. Smith, *J. Am. Chem. Soc.* **98**, 1520 (1976).

566. J. F. Garst and C. D. Smith, *J. Am. Chem. Soc.* **98**, 1526 (1976).

567. (a) A. V. Kramer, J. A. Labinger, J. S. Bradley, and J. A. Osborn, *J. Am. Chem. Soc.* **96**, 7145 (1974); (b) T. J. Marks and W. A. Wachter, *J. Am. Chem. Soc.* **98**, 703 (1976).

568. W. A. Nugent and J. K. Kochi, *J. Am. Chem. Soc.* **98**, 5405 (1976).

569. J. Chatt, R. A. Head, G. J. Leigh, and Ch. J. Pickett, *J. Chem. Soc. Chem. Commun.*, 299 (1977).

570. H. Felkin and B. Meunier, *Nouv. J. Chim.* **1**. 281 (1977).

571. C. J. Cooksey, D. Dodd, C. Gatford, M. D. Johnson, G. J. Lewis, and D. M. Titchmarsh, *J. Chem. Soc. Perkin Trans.* **2**, 655 (1972).

572. F. R. Jensen and R. C. Kiskis, *J. Am. Chem. Soc.* **97**, 5825 (1975).

573. C. C. Lee and E. C. F. Ko, *J. Am. Chem. Soc.* **96**, 8032 (1974).

574. G. A. Olah and G. Liang, *J. Am. Chem. Soc.* **97**, 1920 (1975).

575. J. K. Stille and K. N. Sannes, *J. Am. Chem. Soc.* **94**, 8494 (1972).

576. D. I. Davies, J. N. Done, and D. H. Hey, *J. Chem. Soc. Chem. Commun.*, 725 (1966).

577. (a) L. Kaplan, *J. Chem. Soc. Chem. Commun.*, 754 (1968); (b) D. F. Shellhamer, D. B. McKee, and C. T. Leach, *J. Org. Chem.* **41**, 1972 (1976); (c) J. San Filippo, Jr., J. Silbermann, and P. J. Fagan, *J. Am. Chem. Soc.* **100**, 4834 (1978); (d) L. Crombie, P. J. Maddocks, and G. Pattenden, *Tetrahedron Lett.*, 3483 (1978); (e) M. Newcomb, T. Seidel, and M. B. McPherson, *J. Am. Chem. Soc.* **101**, 777 (1979).

578. (a) H. G. Richey, Jr., in *Carbonium Ions*, Vol. 3, G. A. Olah and P. V. Schleyer, Eds., Interscience, New York (1972), p. 1201; (b) G. A. Olah, D. J. Donovan, and K. G. S. Prakash, *Tetrahedron Lett.*, 4779 (1978).

579. (a) K. G. Taylor, C. K. Govindan, and M. Kaelin, *CNRS International Colloquium on Organic Free Radicals*, Aix-en-Provence, July 1977, CNRS, Paris (1978), pp. 359–363; (b) S. Chemaly and J. M. Pratt, *J. Chem. Soc. Chem. Commun.*, 988 (1976).

580. (a) H. O. House, W. C. McDaniel, R. F. Sieloff, and D. Vanderveer, *J. Org. Chem.* **43**, 4316 (1978); (b) A. Bury, M. R. Ashcroft, and M. D. Johnson, *J. Am. Chem. Soc.* **100**, 3217 (1978).

581. G. E. Gream, C. F. Pincombe, and D. Wege, *Aust. J. Chem.* **27**, 603 (1974).

582. N. Kornblum, *Angew. Chem. Intern. Ed.* **87**, 797 (1975).

583. M. Barreau and M. Julia, *Tetrahedron Lett.*, 1537 (1973).

584. M. Julia, B. Mansour, and D. Mansuy, *Tetrahedron Lett.*, 3443 (1976).

585. G. M. Whitesides, J. S. Sadowski, and J. Lilburn, *J. Am. Chem. Soc.* **96**, 2829 (1974).

586. J. F. Garst and J. T. Barbas, *J. Am. Chem. Soc.* **96**, 3239 (1974).

587. F. Th. Bond, C-Y. Ho, and O. McConnell, *J. Org. Chem.* **41**, 1416 (1976).

588. R. Srinivasan and K. L. Carlough, *J. Am. Chem. Soc.* **89**, 4932 (1967).

589. M. Brown, *J. Org. Chem.* **33**, 162 (1968).

590. R. S. H. Liu and G. S. Hammond, *J. Am. Chem. Soc.* **89**, 4936 (1967).

591. W. L. Dilling, *Chem. Rev.* **66**, 373 (1966), and references cited therein.

592. (a) A. A. Gorman and R. L. Leyland, *Tetrahedron Lett.*, 5345 (1972); (b) Z. Goldschmidt and M. Shefi, *J. Org. Chem.* **44**, 1604 (1979); (c) D. I. Schuster, R. H. Brown, and B. M. Resnick, *J. Am. Chem. Soc.* **100**, 4504 (1978).

593. (a) R. L. Cargill, J. R. Dalton, S. O'Connor, and D. G. Michels, *Tetrahedron Lett.*, 4465 (1978); (b) F. Barany, S. Wolff, and W. C. Agosta, *J. Am. Chem. Soc.* **100**, 1946 (1978).

594. (a) B. Furth, G. Daccord, and J. Kossanyi, *Tetrahedron Lett.*, 4259 (1975); (b) P. Chaquin, J.-P. Morizur, and J. Kossanyi, *Tetrahedron Lett.*, 4259 (1975).

595. (a) B. Guiard, B. Furth, and J. Kossanyi, *Bull. Soc. Chim. Fr.*, 1552 (1976); (b) R. G. Carlson, J. H-A. Huber and D. E. Henton, *J. Chem. Soc. Chem. Commun.*, 223 (1973).

596. M. E. H. Howden, A. Maercker, J. Burdon and J. D. Roberts, *J. Am. Chem. Soc.* **88**, 1732 (1966).

597. M. S. Silver, P. R. Shafer, J. E. Nordlander, C. Rüchardt, and J. D. Roberts, *J. Am. Chem. Soc.* **82**, 2646 (1960).
598. H. G. Richey, Jr., and H. S. Veale, *Tetrahedron Lett.*, 615 (1975).
599. E. A. Hill, A. T. Chen, and A. Doughty, *J. Am. Chem. Soc.* **98**, 167 (1976).
600. E. A. Hill, R. J. Theissen, Ch. E. Cannon, R. Miller, R. B. Guthrie, and A. T. Chen, *J. Org. Chem.* **41**, 1191 (1976).
601. H. G. Richey, Jr., and Th. C. Rees, *Tetrahedron Lett.*, 4297 (1966).
602. V. N. Drozd, Yu. A. Ustynynk, M. A. Tselieva, and L. B. Dmitriev, *Zh. Obsch. Khim.* **39**, 1991 (1969).
603. E. A. Hill and M. R. Engel, *J. Org. Chem.* **36**, 1356 (1971).
604. A. Maercker and R. Geuss, *Angew. Chem., Intern. Ed.* **9**, 909 (1970).
605. E. A. Hill and G. E. M. Shih, *J. Am. Chem. Soc.* **95**, 7764 (1973).
606. E. A. Hill, H. G. Richey, Jr., and Th. C. Rees, *J. Org. Chem.* **28**, 2161 (1963).
607. P. T. Lansbury and F. J. Caridi, *J. Chem. Soc. Chem. Commun.*, 714 (1970).
608. A. Stefani, *Helv. Chim. Acta* **57**, 1346 (1974).
609. J. St. Denis, T. Dolzine, and J. P. Oliver, *J. Am. Chem. Soc.* **94**, 8260 (1972).
610. T. W. Dolzine and J. P. Oliver, *J. Organometal. Chem.* **78**, 165 (1974).
611. T. W. Dolzine and J. P. Oliver, *J. Am. Chem. Soc.* **96**, 1737 (1974).
612. J. St. Denis, J. P. Oliver, and J. B. Smart, *J. Organometal. Chem.* **44C**, 32 (1972).
613. G. M. Whitesides, D. E. Bergbreiter, and P. E. Kendall, *J. Am. Chem. Soc.* **96**, 2806 (1974).
614. T. W. Dolzine, A. K. Hovland, and J. P. Oliver, *J. Organometal. Chem.* **65**, C1 (1974).
615. R. A. Firestone, *J. Org. Chem.* **37**, 2181 (1972).
616. H. Pines, *Acc. Chem. Res.* **7**, 155 (1974), and references cited therein.
617. D. Lal, D. Griller, S. Husband, and K. U. Ingold, *J. Am. Chem. Soc.* **96**, 6355 (1974).
618. J. K. Kochi and J. W. Powers, *J. Am. Chem. Soc.* **92**, 137 (1970).
619. (a) A. Citterio, F. Minisci, O. Porta, and G. Sesana, *J. Am. Chem. Soc.* **99**, 7960 (1977); (b) A. Citterio, *Tetrahedron Lett.*, 2701 (1978).
620. C. L. Jenkins and J. K. Kochi, *J. Am. Chem. Soc.* **94**, 843 (1972).
621. (a) P. Schmid and K. U. Ingold, *J. Am. Chem. Soc.* **99**, 6434 (1977); (b) P. Schmid and K. U. Ingold, *J. Am. Chem. Soc.* **100**, 2493 (1978); (c) Y. Maeda, P. Schmid, D. Griller, and K. U. Ingold, *J. Chem. Soc. Chem. Commun.*, 525 (1978).
622. R. A. Sheldon and J. K. Kochi, *J. Am. Chem. Soc.* **92**, 4395 (1970).
623. J. F. Garst, J. A. Pacifici, V. D. Singleton, M. F. Ezzel, and J. I. Morris, *J. Am. Chem. Soc.* **97**, 5242 (1975).
624. C. L. Jenkins and J. K. Kochi, *J. Am. Chem. Soc.* **94**, 856 (1972).
625. P. J. Wagner and K. C. Liu, *J. Am. Chem. Soc.* **96**, 5952 (1974).
626. P. J. Wagner, *CNRS International Colloquium on Organic Free Radicals*, Aix-en-Provence, 1977, CNRS, Paris (1978), pp. 169–188.
627. H. O. House and P. D. Weeks, *J. Am. Chem. Soc.* **97**, 2778 (1975).
628. A. L. J. Beckwith and K. U. Ingold, in *Rearrangements in Ground and Excited States*, Vol. 1, P. de Mayo, Ed., Academic, New York pp. 161–310 (1980).
629. P. Bischof, *Tetrahedron Lett.*, 1291 (1979).
630. Y. Maeda and K. U. Ingold, *J. Am. Chem. Soc.* **101**, 4975 (1979).
631. A. Effio, D. Griller, K. U. Ingold, A. L. J. Beckwith, and A. K. Serelis, *J. Am. Chem. Soc.* **102**, 1734 (1980).
632. (a) M. Castaing, M. Pereyre, M. Ratier, P. M. Blum, and A. G. Davies, *J. Chem. Soc. Perkin Trans.* **2**, 589 (1979); (b) P. M. Blum, A. G. Davies, M. Pereyre, and M. Ratier, *J. Chem. Res.*, (S) 110 (1980).
633. M. J. S. Dewar and S. Olivella, *J. Am. Chem. Soc.* **101**, 4958 (1979).
634. S. J. Cristol and R. J. Daughenbaugh, *J. Org. Chem.* **44**, 3434 (1979).
635. M. Pomerantz and N. L. Dassanayake, *J. Am. Chem. Soc.* **102**, 678 (1980).

636. S-K. Chung and F-F. Chung, *Tetrahedron Lett.*, 2473 (1979).
637. J. Ancelle, G. Bertrand, M. Joanny, and P. Mazerolles, *Tetrahedron Lett.*, 3153 (1979).
638. K. Okuhara, *J. Am. Chem. Soc.* **102**, 244 (1980).
639. J. A. Kampmeier, S. H. Harris, and D. K. Wedegaertner, *J. Org. Chem.* **45**, 315 (1980).
640. A. G. Davies and R. Sutcliffe, *J. Chem. Soc. Chem. Commun.*, 473 (1979).
641. J. S. Weinberg and A. Miller, *J. Org. Chem.* **44**, 4722 (1979).
642. S. Ayral-Kaloustian and W. C. Agosta, *J. Am. Chem. Soc.* **102**, 314 (1980).
643. (a) R. Srinivasan, K. H. Brown, J. A. Ors, L. S. White, and W. Adam, *J. Am. Chem. Soc.* **101**, 7424 (1979); (b) W. Adam and M. Balci, *J. Am. Chem. Soc.* **101**, 7542 (1979); (c) W. Adam and M. Balci, *J. Am. Chem. Soc.* **102**, 1961 (1980).
644. A. L. J. Beckwith and R. D. Wagner, *J. Am. Chem. Soc.* **101**, 7099 (1979).
645. J. H. Cooley, M. W. Mosher, and M. A. Khan, *J. Am. Chem. Soc.* **90**, 1867 (1968).
646. A. Ohsawa, H. Arai, H. Igeta, T. Akimoto, A. Tsuji, and Y. Iitaka, *J. Org. Chem.* **44**, 3524 (1979).
647. S. Kaufmann, L. Tökés, J. W. Murphy, and P. Crabbé, *J. Org. Chem.* **34**, 1618 (1969).
648. S. M. Neider, G. R. Chambers, and M. Jones Jr., *Tetrahedron Lett.*, 3793 (1979).
649. K. Maruyama, A. Osuka, and H. Suzuki, *J. Chem. Soc. Chem. Commun.*, 323 (1980).
650. H. Suginome and T. Uchida, *J. Chem. Soc. Chem. Commun.*, 701 (1979).
651. R. W. Binkley and D. J. Koholic, *J. Org. Chem.* **44**, 2047 (1979).
652. G. Remy, L. Cottier, and G. Descotes, *Tetrahedron Lett.*, 1847 (1979).
653. A. Citterio, A. Arnoldi, and F. Minisci, *J. Org. Chem.* **44**, 2674 (1979).
654. S. N. Lewis, J. J. Miller, and S. Winstein, *J. Org. Chem.* **37**, 1478 (1972).
655. W. T. Evanochko and P. S. Shevlin, *J. Am. Chem. Soc.* **44**, 4426 (1979).
656. J. C. Gressier, G. Levesque, A. Mahjoub, and A. Thuillier, *Bull. Soc. Chim. Fr.*, II-355 (1979).
657. A. L. Beckwith and M. D. Lawton, *J. Chem. Soc. Perkin Trans.* 2, 2134 (1973).
658. E. G. Janzen, C. C. Lai, and R. V. Shetty, *Tetrahedron Lett.*, **21**, 1201 (1980).
659. G. Stork, R. K. Boeckmann, Jr., D. F. Taber, W. C. Still, and J. Singh, *J. Am. Chem. Soc.* **101**, 7107 (1979).
660. W. R. Roush, *J. Am. Chem. Soc.* **102**, 1390 (1980).
661. J-L. Stein, L. Stella, and J-M. Surzur, *Tetrahedron Lett.*, **21**, 287 (1980).
662. S. J. Cristol and R. M. Strom, *J. Am. Chem. Soc.* **101**, 5707 (1979).
663. (a) G. Pelerin, Ph. D. thesis, Marseilles (1980); (b) M. Bertrand, P. Teisseire and G. Pelerin, *Tetrahedron Lett.*, **21**, 2051 (1980); (c) M. Bertrand, P. Teisseire, and G. Pelerin, *Tetrahedron Lett.*, **21**, 2055 (1980).
664. G. Stork and A. R. Schoofs, *J. Am. Chem. Soc.* **101**, 5081 (1979).
665. M. E. Christy, P. S. Anderson, S. F. Britcher, C. D. Colton, B. E. Evans, D. C. Remy, and E. L. Engelhardt, *J. Org. Chem.* **44**, 3117 (1979).
666. B. E. Evans, P. S. Anderson, M. E. Christy, C. D. Colton, D. C. Remy, K. E. Rittle, and E. L. Engelhardt, *J. Org. Chem.* **44**, 3127 (1979).
667. D. Brandes, F. Lange, and R. Sustmann, *Tetrahedron Lett.*, **21**, 261 (1980).
668. D. Brandes, F. Lange, and R. Sustmann, *Tetrahedron Lett.*, **21**, 265 (1980).
669. R. A. J. Smith and D. J. Hannah, *Tetrahedron Lett.*, **21**, 1081 (1980).
670. A. F. Sowinski and G. M. Whitesides, *J. Org. Chem.* **44**, 2369 (1979).
671. R. D. Little and G. W. Muller, *J. Am. Chem. Soc.* **101**, 7129 (1979).
672. M. E. Alonso and M. Gómez, *Tetrahedron Lett.*, 2763 (1979).
673. C. P. Casey and M. C. Cesa, *J. Am. Chem. Soc.* **101**, 4236 (1979).
674. R. C. Larock, K. Oertle, and G. F. Potter, *J. Am. Chem. Soc.* **102**, 190 (1980).
675. E. A. Fehnel and F. G. Brokaw, *J. Org. Chem.* **45**, 578 (1980).
676. H. E. Zimmerman and M. C. Hovey, *J. Org. Chem.* **44**, 2331 (1979).
677. F. T. Bond, H. L. Jones, and L. Scerbo, *Tetrahedron Lett.*, 4685 (1965).

678. R. Srinivasan, Ed., *Organic Photochemical Syntheses*, Vol. 1, Wiley, New York (1971), p. 35.
679. R. G. Salomon, D. J. Coughlin, and E. M. Easler, *J. Am. Chem. Soc.* **101**, 3961 (1979).
680. P. Jost, P. Chaquin, and J. Kossanyi, *Tetrahedron Lett.*, **21**, 465 (1980).
681. P. deMayo, L. K. Sydnes, and G. Wenska, *J. Chem. Soc. Chem. Commun.*, 499 (1979).
682. R. L. Coffin, W. W. Cox, R. G. Carlson, and R. S. Givens, *J. Am. Chem. Soc.* **101**, 3261 (1979).
683. J. E. Baldwin, R. C. Thomas, L. I. Kruse, and L. Silberman, *J. Org. Chem.* **42**, 3846 (1977).
684. A. L. J. Beckwith, C. J. Easton, and A. K. Serelis, *J. Chem. Soc. Chem. Commun.*, 482 (1980).
685. P. S. Mariano and E. Bay, *J. Org. Chem.* **45**, 1763 (1980).
686. P. Piccardi, P. Massardo, M. Modena, and E. Santoro, *J. Chem. Soc. Perkin Trans.* **1**, 1848 (1974).
687. E. E. van Tamelen, and E. G. Taylor, *J. Am. Chem. Soc.* **102**, 1202 (1980).
688. A. L. J. Beckwith, and R. D. Wagner, *J. Chem. Soc. Chem. Commun.*, 485 (1980).
689. H. W. S. Chan, J. A. Matthew, and D. T. Coxon, *J. Chem. Soc. Chem. Commun.*, 235 (1980).
690. A. Padwa, T. J. Blacklock, R. Loza, and R. Polniaszek, *J. Org. Chem.* **45**, 2181 (1980).
691. D. L. J. Clive, V. Farina, A. Singh, C. K. Wong, W. A. Kiel, and S. M. Menchen, *J. Org. Chem.* **45**, 2120 (1980).
692. C. N. Filer, D. Ahern, R. Fazio, and E. J. Shelton, *J. Org. Chem.* **45**, 1313 (1980).
693. A. I. Scott, J. B. Hansen, and S. K. Chung, *J. Chem. Soc. Chem. Commun.*, 388 (1980).
694. H. G. Aurich, G. Bach, K. Hahn, G. Küttner, and W. Weiss, *J. Chem. Res. (S)*, 122 (1977); (M), 1544 (1977).
695. L. Lorenc, I. Juranić, and M. Lj. Mihailović, *J. Chem. Soc. Chem. Commun.*, 749 (1977).
696. A. J. Bloodworth and H. J. Eggelte, *Tetrahedron Lett.*, **21**, 2001 (1980).
697. A. L. J. Beckwith, T. Lawrence, and A. K. Serelis, *J. Chem. Soc. Chem. Commun.*, 484 (1980).
698. F. J. McQuillin and M. Wood, *J. Chem. Research, (S)*, 61 (1977); (M) 752 (1977).
699. R. J. P. Corriu and C. Guerin, *J. Chem. Soc. Chem. Commun.*, 168 (1980).
700. M. Newcomb and A. R. Courtney, *J. Org. Chem.* **45**, 1707 (1980).
701. Y. Ito, M. Nakatsuka, and T. Saegusa, *J. Org. Chem.* **45**, 2022 (1980).
702. S. Wolff, F. Barany, and W. C. Agosta, *J. Am. Chem. Soc.* **102**, 2378 (1980).
703. P. S. Mariano, E. Bay, D. G. Watson, T. Rose, and C. Bracken, *J. Org. Chem.* **45**, 1753 (1980).
704. P. de Mayo, L. K. Sydnes, and G. Wenska, *J. Org. Chem.* **45**, 1569 (1980).

4

Reactions of Silicon Atoms and Silylenes

Yi-Noo Tang

I. INTRODUCTION

The first two members of the Group IVA elements, carbon and silicon, share the common property that they can combine with four entities through covalent bonding, and therefore may be installed as the backbones of thousands of complicated molecules. The study of the various carbon-containing compounds during the past two centuries has already been developed into a major branch of chemistry. Although the investigation of silicon-containing systems is somewhat less extensive, vast amounts of information have been accumulated during the recent decades.

One of the best ways to mechanistic understanding of chemical reactions is through some thorough investigations of the chemical properties of the possible reaction intermediates. Knowledge of carbon-containing free radicals, carbenes, carbenium ions, and carbon atoms has given penetrating insight into the mechanism of organic reactions. Similarly, the study on the chemistry of silicon-containing free radicals and cations, silylenes, and silicon atoms should have a great impact on the understanding of silicon chemistry. In the present chapter, I attempted to summarize the various aspects on the last two reactive intermediates mentioned above. During the past decade, a number of good review articles devoted to various phases of silylene chemistry have appeared.[1-16] In particular, the excellent works of Margrave,[3,8,12] Timms,[4,11,14] Atwell,[6,13] Weyenberg,[6,13] and Gaspar[10,16] were extremely valuable in the writing of this chapter.

Yi-Noo Tang ● Department of Chemistry, Texas A & M University, College Station, Texas 77843.

There are certain similarities between the chemistry of carbon and silicon atoms, and between carbenes and silylenes. However, there are also vast differences, both quantitative and qualitative, in their reactivities towards a number of chemicals. Such dissimilarities could be attributed to the following different atomic properties of the two elements: (i) Silicon is much heavier and larger than carbon. The atomic radius of Si is 0.117 nm and that of C is 0.077 nm.[16] (ii) Silicon is more electropositive than carbon.[17] (iii) The ground state electronic configurations of C and Si are both of (ns^2np^2) in their occupied outermost energy levels. A valence of 2 will result when C or Si reacts in this state. However, a quadrivalence could be achieved after the promotion of an electron to give the (ns^1np^3) configuration. The fact that the relative stability of the lower oxidation state increases from C to Si has been attributed to the increasing stabilization of the ns (relative to the np) electrons of the valence energy levels, and to the observation that heavier elements of the same group form weaker covalent bonds.[18] (iv) Carbon has no d orbitals in its valence energy level, and therefore its valence is normally limited to 4. On the other hand, the availability of the $3d$ orbitals to Si enables the possible extension of its valence beyond 4 during the transient stages of certain reactions. (v) The overlap of p orbitals in C to form π bonds proceeds very readily, while similar overlap in Si is very difficult.[18]

This chapter will concentrate more on the methods of formation and the reaction modes of silicon atoms and silylenes, and less on the spectroscopic and thermodynamic aspects of these species. However, it should be mentioned that, historically, many of the silylenes were first discovered and studied by spectroscopists.

As for nomenclature, I am basically in agreement with Atwell and Weyenberg[6] that it is more appropriate to call the divalent species "silylenes" instead of "silenes." In this chapter, every divalent derivative of silicon is named as a substituted compound of the fundamental species, silylene (SiH_2). When there is only one substituent, no numeral prefixes are added, e.g., fluorosilylene, $SiHF$; methoxysilylene, $SiHOCH_3$. With two substituents, the prefix "di-" is added, e.g., difluorosilylene, SiF_2; dimethoxysilylene, $Si(OCH_3)_2$. Of course, it is recognized that inorganic names such as silicon difluoride for SiF_2 are equally good in describing the species, and have been repetitively used even by myself in publications. They are abandoned here because of the desire for uniformity in naming the divalent species of silicon. Another practice followed in this chapter is that whenever the formula of silylene is written in the text, the Si atom is always written first, with the symbols of the substituents following on the right-hand side. The two electron dots on the silylene formula are always omitted except in places where displaying them carries unusual significance. Abbreviations such as "Me" for CH_3, "Et" for C_2H_5, and "Ph" for C_6H_5 are also generally used.

II. METHODS OF GENERATION OF SILICON ATOMS

1. Nuclear Recoil Method

The majority of the information on silicon atom reactions is derived from studies involving the formation of ^{31}Si via nuclear transformation.[19-36] Silicon-31 atoms have been generated by the following two nuclear processes: (i) ^{31}P$(n,p)^{31}$Si; and (ii) ^{30}Si$(n,\gamma)^{31}$Si. Of the two, the first one is the most frequently employed.

The nuclear reaction, ^{31}P$(n,p)^{31}$Si, was first employed by Gaspar and coworkers in 1966 in their study of recoil silicon atom reactions.[19] The fast neutron required to induce this nuclear transformation was produced from the nuclear process, ^{9}Be$(d,n)^{10}$B, which, in turn, was induced by a 100-μA current of a 14-MeV deuteron beam from a cyclotron. The silicon atoms thus formed were born with a recoil energy of about 6×10^5 eV. They are initially positively charged. However, according to the calculations using the resonance rule,[22,37] the charge should be neutralized long before the atoms reach the chemical reaction range. Later in our laboratory,[30] we discovered that fast neutrons from a nuclear reactor are also energetic enough to initiate this (n,p) reaction because the threshold energy for the transformation is only 2 MeV.[38] We have also demonstrated that the results obtained were independent of whether a cyclotron or a nuclear reactor was used as the fast-neutron source.[30]

For the ^{30}Si$(n,\gamma)^{31}$Si reaction, silicon atoms with 75–750-eV recoil energy are formed. Owing to the low recoil energy, both the bond rupture in the transmuted precursor molecule and the charge neutralization of ^{31}Si ions may be incomplete. However, this process has the advantage that a phosphorus-containing precursor is not required to be present in the reaction mixture.

2. Thermal Evaporation Method

Skell and Owen[39,40] have studied the reactions of Si atoms with methylsilanes by heating solid silicon, resistively, to 1400°C *in vacuo* followed by co-condensation with the substrates on the liquid-nitrogen-cooled walls. The silicon vapor thus obtained may contain some Si$_2$ and Si$_3$ species, but the monoatomic Si atoms should be predominant at low pressures.[41]

In the literature, a number of reactions have been reported which presumably involved solid silicon and gaseous silicon tetrahalides at temperatures above 800°C. For example, Margrave and coworkers[1,3] reported the reaction Si(s) + SiF$_4$(g) → 2SiF$_2$(g), which takes place very readily with 90% conversion at 1400°C. Although it is generally believed that the above process is a heterogeneous gas–solid interaction, evaporated Si atoms may also be involved. Since silicon has a melting point of 1410°C and a boiling point of 2355°C, the concentration of evaporated Si atoms should be high enough to account for certain fractions of the observed reactions. Actually, the temperature employed by Margrave and coworkers to heat silicon in the study described above is about the same as that employed by Skell and Owen in their thermal evaporation studies.

III. REACTION MODES OF SILICON ATOMS

Silicon atoms are capable of forming four bonds, which usually requires more than a single encounter with the substrate molecules to achieve. A reaction mechanism consisting of two or more elementary steps is normally involved in the formation of the stable products from the free atom. This is different from the situation with free radicals, carbenes, and even monovalent and divalent atoms such as H, F, Cl, O, and S, where the final observable products can be formed directly in the primary interaction of the reacting species. Therefore, a clear picture of the Si atom reactions is always clouded by the interactions involved with the subsequent reaction intermediates. The information on the primary reactions of these atoms can only be deduced with less certainty because it has to be derived from the nature of the final products.

1. Si–H Bond Insertion

Skell and Owen[39,40] observed the formation of methyl-substituted trisilanes from the reactions of thermally evaporated silicon atoms with methylsilanes. The most logical mechanism involves a di-insertion process: Si atom reacts with a substrate molecule by insertion into its Si–H bonds to form a methyl-substituted silylene, wich subsequently inserts into the Si–H bond of another substrate molecule to give the final product.

$$:Si: + SiHR_3 \rightarrow SiR_3-SiH \qquad (R=CH_3, H, \text{ or } SiH_3) \qquad (1)$$

$$SiR_3-SiH + SiHR_3 \rightarrow SiR_3-SiH_2 - SiR_3 \qquad (2)$$

This work also reveals that thermally evaporated Si atoms do not insert into C–H or C–Si bonds.

As discussed below, the formation of Si_2H_6 from the reactions of recoil Si atoms with SiH_4 may also involve an initial step of Si insertion into Si–H bonds.

2. Apparent H Abstraction

Gaspar and coworkers[19,22] have studied the reactions of recoil ^{31}Si atoms in a $PH_3–SiH_4$ mixture where the PH_3 is required to function as a phosphorus atom source for the nuclear transmutation. They observed $^{31}SiH_3SiH_3$ as a major reaction product and have established that the reaction proceeds through $^{31}SiH_2$ as an intermediate. There are two possible mechanisms to account for the formation of $^{31}SiH_2$ from ^{31}Si atoms. One process involves the successive H abstraction from P–H or Si–H bonds, as shown in the two sequences, equation (3) followed by equation (4), or equation (5) followed by equation (6). However,

$$^{31}Si + PH_3 \rightarrow {}^{31}SiH + PH_2 \tag{3}$$

$$^{31}SiH + PH_3 \rightarrow {}^{31}SiH_2 + PH_2 \tag{4}$$

$$^{31}Si + SiH_4 \rightarrow {}^{31}SiH + SiH_3 \tag{5}$$

$$^{31}SiH + SiH_4 \rightarrow {}^{31}SiH_2 + SiH_3 \tag{6}$$

the apparent H abstraction can also be explained through an insertion–decomposition mechanism as proposed by Gaspar and coworkers.[26] This mechanism is illustrated in equation (7). The insertion of recoil ^{31}Si atoms into the P–H or

$$
\begin{array}{l}
^{31}Si + PH_3 \rightarrow ({}^{31}SiH - PH_2)^* \longrightarrow \\
\qquad\qquad\qquad\qquad\qquad {}^{31}SiH \rightarrow \rightarrow {}^{31}SiH_2 \\
^{31}Si + SiH_4 \rightarrow ({}^{31}SiH - SiH_3)^* \longrightarrow
\end{array}
\tag{7}
$$

Si–H bonds may give rise to excited adducts which subsequently decompose to give ^{31}SiH. The same insertion–decomposition cycle may be repeated to yield $^{31}SiH_2$, which eventually inserts into the Si–H bond of SiH_4 to give $^{31}SiH_3SiH_3$ as the final product. At present, there is no evidence favoring one of these two mechanisms over the other. If the direct H-abstraction process does prevail, there is also no guarantee that the H abstraction is a consecutive instead of a simultaneous process.

3. Apparent F Abstraction

In our laboratory, we have studied recoil ^{31}Si reactions with butadiene by using PF_3 as the ^{31}Si precursor.[27,33] From the formation of $[^{31}Si]$-1,1-difluorosilacyclopent-3-ene as a product, we reasoned that $^{31}SiF_2$ was formed through stepwise F abstraction by recoil ^{31}Si atoms as shown in equations (8) and (9). An insertion–decomposition mechanism similar to that shown in equation (7) is also possible.

$$^{31}Si + PF_3 \rightarrow {}^{31}SiF + PF_2 \tag{8}$$

$$^{31}SiF + PF_3 \rightarrow {}^{31}SiF_2 + PF_2 \tag{9}$$

4. Possible Si–Si Bond Insertion

Cetini and coworkers[20,24] have studied the reactions of [31]Si atoms with SiH$_4$, Si$_2$H$_6$, and Si$_3$H$_8$ as a function of moderator concentrations. They used the nuclear transformation, [30]Si(n,γ)[31]Si, as the hot atom source, and Ne as a moderator. In the SiH$_4$ system, the yields of all the [31]Si-labeled products decreased with moderation toward zero, indicating the total absence of thermal [31]Si reactions with SiH$_4$. This means that *thermal* [31]Si atoms do not insert into Si–H bonds of silane.[24] In the Si$_2$H$_6$ system, the [31]SiH$_3$SiH$_3$ yield does not change with moderation. Since the insertion of thermal [31]Si into the Si–H bond is not operational, it was deduced[24] that thermal [31]Si atoms do undergo insertion into Si–Si bonds.

On the other hand, Skell and Owen[39,40] have indicated that Si–Si bond insertion is unlikely for thermally evaporated silicon atoms. The contradictory conclusion invoked by Cetini and coworkers[24] as described above should be taken with caution. The (n,γ) nuclear process employed involves a number of potential complications, which will be discussed in detail in Section III.7. Overall it seems that Si–Si insertion, although a possible reaction mode, is not yet well established for silicon atom reactions.

5. Apparent 1,4 Addition to Dienes

The addition of Si atoms to the double bonds of olefins is also not a firmly established process. The reason is probably that the Si–olefin adducts are themselves very reactive species which undergo further interaction with other olefinic species in the system. However, in the case of dienes, the adducts formed with Si atoms are capable of rearranging to give stable molecules. Gaspar and coworkers[29,35] studied the reactions of [31]Si atoms with 1,3-butadiene, and deduced that a major product formed was indeed [[31]Si]-1-silacyclopenta-2,4-diene via equation (10). The product is likely to be formed in a single reactive collision followed by intramolecular rearrangement. Their deduction was based on the following observations. (i) The product contains C, H and Si but no phosphorus because the same compound was obtained when [30]Si, instead of ^{31}P, was used as the [31]Si source. (ii) The chromatographic behavior of this product suggests that it contains four carbon atoms. (iii) Its yield increases monotonically with 1,3-butadiene mole fraction. (iv) It was obtained in nearly identical ($20 \pm 3\%$) yields from 1:1 PH$_3$–butadiene and 1:1 PF$_3$–butadiene mixtures. Very recently, Tang and coworkers[42] confirmed the identity of this product through its heterogeneous catalytic hydrogenation to give [[31]Si]-silacyclopent-3-ene.

$$[31]Si + \quad\diagup\!\!\!\!\diagdown\!\!\!\!\diagup \quad \longrightarrow \quad \overset{[31]SiH_2}{\underset{}{\bigcirc\!\!\!\!\!\diagdown}} \qquad (10)$$

6. Reactions of Silicon with SiX_4 and Other Halides

As previously mentioned, the reactions between Si and SiX_4 may or may not involve free Si atoms in the gas phase. The equilibrium shown in equation (11) has been observed for a variety of silicon halides at temperatures above 800°C. Since the forward reaction represents an increase in entropy, it proceeds most readily under high-temperature and low-pressure conditions. For example, for the reaction of Si with SiF_4 to give SiF_2, the percent conversion increases from about 50% at 1150°C to about 90% at 1400°C.[3,4]

$$Si(s) + SiX_4(g) \rightleftarrows 2SiX_2(g) \qquad (X=F, Cl, Br, or I) \qquad (11)$$

Other reactions of silicon with halogen-containing compounds are listed below:

(i) Margrave and coworkers[43,44] have shown that monomeric SiF_2 can be detected mass spectrometrically through the reactions of Si with CaF_2:

$$Si + CaF_2 \rightarrow SiF_2 + Ca \qquad (12)$$

(ii) Zubkov and coworkers[45] studied the reactions of Si with CuCl at 180–200°C, and detected $SiCl_2$ as a gaseous product mass spectrometrically:

$$Si + 2CuCl \rightarrow SiCl_2 + 2Cu \qquad (13)$$

This same process was also believed to be responsible for the better yields of dialkyldichlorosilanes and alkyltrichlorosilanes in the synthesis of these compounds when Cu is added to the system.[46]

(iii) Joklik and Bozant[47] studied the reactions between Si and HCl and observed $SiHCl_3$ as the product. They proposed the formation of $SiCl_2$ as a reaction intermediate:

$$Si + 2HCl \rightarrow SiCl_2 + H_2 \qquad (14)$$

(iv) The reactions of Si and silica at 1350°C and 10^{-2} torr gave gaseous silicon monoxide as a product[48]:

$$Si(s) + SiO_2(s) \rightarrow 2SiO \qquad (15)$$

7. Reactions of Si Species from $\langle n, \gamma \rangle$ Reactions

Gaspar and coworkers[21,22] have studied the reactions of SiH_4 with ^{31}Si atoms recoiling from the two nuclear transformations: $^{31}P(n,p)^{31}Si$ and $^{30}Si(n,\gamma)^{31}Si$. The results obtained showed differences in two major features:

(i) A significant yield of $^{31}SiSi_2H_8$ was observed from the (n,γ) system, and (ii) the $^{31}SiH_4$ to $^{31}SiSiH_6$ product ratio was quantitatively much higher from the (n,γ) system. It is possible that neutral Si atoms are the reacting species in both of these cases and that the observed differences can be attributed to either the differences in electronic states or in translational energy content. However, it is also likely that the predominant process involved in the (n,γ) system does not include the interaction of neutral Si atoms with SiH_4 molecules.

There are three major differences observed when comparing the (n,γ) process with its (n,p) counterpart, and each of them may have determining effects on the nature of the reacting system[21]: (i) The recoil energy of the ^{31}Si atoms is about 10^5 eV when derived from the (n,p) reaction, but only about 10^2 eV when derived from the (n,γ) process. Because the recoil energy is below the peak maximum of the charge neutralization process, which is about 10^4 eV, charge neutralization may be incomplete in the (n,γ) system and certain observed reactions may be attributed to $^{31}Si^+$ ions instead of Si atoms. (ii) In an (n,γ) process of low recoil energy, bond rupture in the precursor molecule may not be complete owing to possible momentum cancellation in a gamma cascade. Because of this, some of the reacting species in the (n,γ) system may not be free ^{31}Si atoms but ^{31}Si-labeled species representing the remaining portion of the precursor molecule. (iii) The (n,γ) process involves ^{30}Si as the precursor of the nuclear reaction whose natural abundance is very low. Because of this, a large radiation dose is normally required to produce sufficient product for detection. A higher radiation dose means more radiation damage to the molecules in the system, and the instantaneous concentration of reactive species such as radicals and ions will be higher. This, in turn, will increase the chance of interaction between ^{31}Si atoms with such reactive secondary species instead of the primary reactants of the system. Owing to the above three reasons, we should always view the results from the (n,γ) systems with reservation unless further experiments indicate the absence of these complications.

Besides the work of Gaspar and coworkers[21] described above and the work by Cetini and coworkers[20,24] mentioned earlier, there are several other studies employing the $^{30}Si(n,\gamma)^{31}Si$ transformation. Studies of the chemical effects subsequent to the above nuclear transformation in tetramethylsilane have been carried out by Snediker and Miller[23] in 1968 in both gas and liquid phases with or without nitric oxide as a scavenger, and by Kawamoto[32] in 1975 on the effect of irradiation conditions. The formation of ^{31}Si-labeled products from neutron irradiation of tetra-, tri-, and diphenylsilane in the condensed phase have been studied by Wheeler and Trabal.[49] Most of the results obtained in these studies are likely to be complicated by various kinds of radical reactions.

IV. MODES OF FORMATION OF SILYLENES

1. Derived from Si Atoms and Elemental Silicon

A. Si Atom Abstraction—the Nuclear Recoil Method

As described in Sections III.2 and III.3, the nuclear recoil method can be used to derive silylenes such as $^{31}SiH_2$ and $^{31}SiF_2$ through apparent abstraction reactions.[22,30,33] There are three major features about the silylenes formed by this method. (i) The study is always a built-in radioactive tracer experiment. One can follow the reaction through the radioactivity of ^{31}Si atoms in the silylenes to identify and measure ^{31}Si-labeled products even in the presence of vast quantities of the nonlabeled version of the same molecule being introduced into the system as a reactant. For example, when a mixture of PH_3–SiH_4 is irradiated with neutrons, the labeled product, $^{31}SiH_4$, can be quantitatively measured regardless of the amount of SiH_4 present in the system.[22] (ii) Certain residue excitation of the recoil ^{31}Si atoms may carry over to the resultant silylenes. This means that the silylenes being formed may be translationally hot and electronically not in the ground state. In fact, it has been shown that both $^{31}SiH_2$ and $^{31}SiF_2$ thus formed react as a mixture of singlet and triplet species.[30,33] (iii) In a typical experiment, only 10^7–10^8 ^{31}Si-labeled silylenes are formed in a system containing about 10^{20} reactant molecules. The disadvantage is that this method cannot be employed for any synthetic purposes except in the formation of ^{31}Si-labeled species. The advantage derived from this fact is that dimerization or polymerization reactions of silylenes themselves can be suppressed if they are not desirable in a system. This is especially important in the case of difluorosilylenes because they dimerize very readily. In the nuclear recoil system, even if we assigned a lifetime of 100 sec to $^{31}SiF_2$, its instantaneous concentration is only about one out of every 10^{14} molecules. Because of this, essentially all of the $^{31}SiF_2$ species should react with other molecules instead of dimerizing.[33]

Besides the nuclear recoil methods, the formation of silylenes from the reactions of elemental silicon with various halides as described in Section III.6 can also be viewed as abstraction reactions. Thus equation (12) involves F-abstraction to give SiF_2;[43,44] equations (13) and (14) involve Cl-abstraction to give $SiCl_2$;[45,47] and equation (15) involves O-abstraction to give SiO.[48]

B. Si Atom Insertion–The Co-Condensation Method

As mentioned in Section II.2, Skell and Owen[39,40] have produced silylenes through Si atom insertion reactions by the co-condensation of thermally produced silicon vapor with methylsilanes. The molecules they employed are trimethylsilane, dimethylsilane, methylsilane, and disilane. Correspondingly, they have observed silylenes such as $SiHSiMe_3$, $SiHSiHMe_2$, $SiHSiH_2Me$, and $SiHSiH_2SiH_3$ through the reaction depicted in equation (1). The actual experiments were carried out by simultaneously depositing silicon from the vapor (20–30 mg) and a large excess of substrate ($\sim$10 g) on the liquid-nitrogen-cooled walls of an evacuated ($<1 \times 10^{-4}$-torr) reaction flask over a period of about 1 hr. Silicon vapor was produced by electron bombardment heating of a silicon electrode to its melting point using a Varian 4000-V electron gun. Reaction occurred in the condensed phase on the walls either at 77° K or upon warm-up. The resulting silylenes further inserted into a Si–H bond of another substrate molecule as shown in equation (2), or polymerized to form linear and cyclic polysilanes.

C. Disproportionation Involving Silicon—The High-Temperature Reaction Method

As mentioned in Section III.6, disproportionation between silicon and SiX_4 molecules at high temperatures generates halogenated silylenes, SiX_2. In practice, this type of reaction has been extensively employed in the formation of SiF_2, as shown in equation (16).[3,4] The chemistry of the difluorosilylene thus formed likely represents the most comprehensive study of all the known silylenes.

$$Si(s) + SiF_4(g) \rightarrow 2SiF_2(g) \qquad (16)$$

When SiF_4 passes over elemental silicon at 1100–1400°C, mass spectrometic analysis of the gas phase products indicates that SiF_2 accounts for about 60% of the species present. SiF_2 is extremely stable in the gas phase with a half-life of about 150 sec at a pressure of 0.2 torr. It shows no tendency to form *gas phase* dimers, and is essentially unaffected by the addition of many other gases. On the other hand, upon condensation at −196°C, SiF_2 is a yellow-brown paramagnetic solid. In a co-condensation experiment, SiF_2 usually reacts with the added substrate as a dimer either at the low temperature, or more likely on warming.[3,4]

The studies on the reactions of SiF_2 formed by this high-temperature method have been thoroughly reviewed in a number of articles by Margrave and by Timms.[1,3,4,8,11,12,14]

D. Abstraction from Single-Crystal Silicon—The Molecular Beam Method

Madix and Schwarz have formed $SiCl_2$ by using a molecular chlorine beam to abstract Si from a silicon single crystal.[50] The reaction was carried out at a low pressure of 10^{-6}–10^{-5} torr and a high temperature of 770–1500° K. The rate of formation of $SiCl_2$ at constant beam pressure was nearly constant with the temperature of the solid surface at high temperatures. But the reaction probability decreased rapidly with temperature below 1050° K.

$$Si(\text{single crystal}) + Cl_2(\text{beam}) \rightarrow SiCl_2 \qquad (17)$$

2. The Decomposition Method

The decomposition method is probably the most significant one for the formation of silylenes. From the fundamental point of view, this method involves a variety of decomposition modes with many different precursor molecules. Even from the practical point of view this method is important because it generally involves relatively less dramatic reaction conditions and can produce silylenes in relatively larger quantities. In the following sections, the classification of decomposition modes is generally according to the type of reactant molecules and the type of reactions involved rather than the kind of silylenes being formed.

A. From Monosilanes

The majority of monosilane decompositions involve the elimination of X_2 from molecules of the type SiX_4, or H_2 from molecules of the type SiH_3X and SiH_2X_2, as shown in equations (18)–(20). The HX elimination from $SiHX_3$ as shown in equation (21) has also been observed.

$$SiX_4 \rightarrow SiX_2 + X_2 \text{ (or 2X)} \qquad (X=H, F, Cl, Br, \text{ or } I) \qquad (18)$$

$$SiH_3X \rightarrow SiHX + H_2 \qquad (X=Cl, Br, \text{ or } CH_3) \qquad (19)$$

$$SiH_2X_2 \rightarrow SiX_2 + H_2 \qquad (X=F \text{ or } Cl) \qquad (20)$$

$$SiHX_3 \rightarrow SiX_2 + HX \qquad (X=Cl) \qquad (21)$$

a. SiH_4

The thermal decomposition of SiH_4 was first studied by Ogier[51] in 1880. However, after a century, the mechanism of this reaction has not yet been entirely resolved. Two plausible modes with different primary steps involved in the SiH_4 decomposition can be proposed: the silylene mechanism (I) and the silyl mechanism (II). In 1936, Hogness, Wilson, and Johnson[52] made a kinetic study of SiH_4 decomposition in the temperature range 380–490°C, and confirmed it to be a homogeneous unimolecular process. They favored the SiH_2

Mechanism I $SiH_4 \rightarrow SiH_2 + H_2$ (22)

$SiH_2 + SiH_4 \rightarrow Si_2H_6$ (23)

$SiH_2 + Si_2H_6 \rightarrow Si_3H_8$ (24)

Mechanism II $SiH_4 \rightarrow \cdot SiH_3 + H\cdot$ (25)

$\cdot H + SiH_4 \rightarrow H_2 + \cdot SiH_3$ (26)

$\cdot SiH_3 + SiH_4 \rightarrow Si_2H_6 + H\cdot$ (27)

$2SiH_3 \rightarrow Si_2H_6$ (28)

mechanism primarily because of the observation of $(SiH_2)_x$ as a polymeric species by Schwarz and Heinrich.[53] In 1950, Stokland's investigation[54] suggested that Si_2H_6 was an important intermediate in the SiH_4 pyrolysis. This was immediately confirmed by flow studies of silane decomposition by White and Rochow.[55] In 1966, Purnell and Walsh[56] performed a detailed kinetic study of the formation of H_2, Si_2H_6, and Si_3H_8 from pyrolysis of SiH_4. Their results are basically consistent with either mechanism I or mechanism II, but they are slightly in favor of the SiH_2 mechanism, both from energy considerations and by analogy with the primary decomposition of CH_4. However, they cautioned that the eventual choice had to await the availability of better thermochemical data for the silicon-containing species involved. Another detailed study of SiH_4 decomposition was by Ring and coworkers in 1970.[57] The pyrolysis of SiH_4–SiD_4 and SiD_4–H_2 mixtures was examined. From the distribution of the deuterium-labeled hydrogen and disilane products, they concluded that it is the SiH_3 radical mechanism which is consistent both qualitatively and quantitatively with most of their data. However, very recently, Newman, Ring, and O'Neal[58] studied the decomposition of SiH_4 in a single pulse shock tube in the range 1200–1300° K by using SiH_4, mixtures of SiH_4–SiD_4, and mixtures of SiD_4–toluene as reactants. Based on this study, they shifted their position to favor mechanism I. An activation energy of 259.1 kJ mol^{-1} has been calculated for the reaction in equation (22), the decomposition of SiH_4 to SiH_2 by H_2 elimination.

The same two primary decomposition modes are also involved in the photolysis of SiH_4, which was first studied by Emeleus and Stewart[59] in 1935. The mercury-sensitized photolysis of SiH_4 was investigated nearly simultaneously by Niki and Mains,[60] and by Gunning and coworkers.[61] However, they believed that two H atoms instead of a H_2 molecule were accompanying the SiH_2 release and that an extensive decomposition of SiH_4 to Si atoms, as shown in equation (30), was possible. The decomposition of SiH_4 to SiH_2 is also observed on vacuum uv photolysis of SiH_4.[62]

$$SiH_4 \rightarrow SiH_2 + 2H \qquad (29)$$

$$SiH_4 \rightarrow Si + 2H_2 \qquad (30)$$

The photolysis of deuterium-labeled silane molecules was again studied by Ring and coworkers.[63] For the 147.0-nm photolysis and the electric discharge decomposition of equimolar SiH_4–SiD_4 mixtures, they concluded from the relatively low HD, Si_2HD_5, and SiH_5D yields that the main primary process in silane photolysis is the formation of SiH_2 via equation (22). However, in a more recent article, they have cautioned that neither reaction shown in equations (25) and (29) can be ruled out.[57]

Purnell and Walsh[56] proposed earlier that if SiH_4 does decompose to SiH_2, a possible configuration for the transition state would involve a three-center bond in the following way:

$$\begin{matrix} H \\ | \\ H{-}Si{-}H \\ | \\ H \end{matrix} \longrightarrow \begin{matrix} H \diagdown \\ \quad\; Si \cdots \\ H \diagup \end{matrix} \begin{matrix} \cdots H \\ \vdots \\ H \end{matrix} \longrightarrow \begin{matrix} H \diagdown \\ \quad\; Si\!: \\ H \diagup \end{matrix} + \begin{matrix} H \\ | \\ H \end{matrix} \qquad (31)$$

The spin conservation rule requires that the silylene thus formed be in the singlet electronic state.

b. Methylsilane

Althoug the mercury-sensitized photolysis of the methylsilanes[61] was consistent with the Si–H bond scission as the primary step as shown in equation (32) to give H and CH_3SiH_2 radicals which combined to give the final products in unit quantum yields, other studies indicated the existence of silylenes as intermediates. The vacuum uv photolysis of CH_3SiH_3 produced SiHMe as shown in equation (33).[64,65] The major products from the pyrolysis of CH_3SiH_3 are H_2

$$CH_3SiH_3 \rightarrow CH_3SiH_2 + H \qquad (32)$$

$$CH_3SiH_3 \rightarrow CH_3SiH + H_2 \qquad (33)$$

and $CH_3SiH_2SiH_2CH_3$, which could arise from either equation (32) or (33) as the initial step.[66] The relative yields of H_2, HD, and D_2 from the deuterium-labeling experiments with CH_3SiH_3–CH_3SiD_3 mixtures performed by Ring and coworkers[57] also support the notion that both of these reactions are involved in the primary steps of the methylsilane decomposition.

c. SiX_4

The decomposition of SiX_4 to give SiX_2 is mainly observed in spectroscopic experiments and not in synthetic processes. The glow discharge of SiF_4 to give SiF_2,[67,68] and the pyrolysis of $SiCl_4$[69,70] or SiI_4[71,72] at temperatures of 900°C or above to give $SiCl_2$ or SiI_2 are known reactions. They are already summarized in equation (18).

d. SiH_3X, SiH_2X_2, and $SiHX_3$

The vacuum uv photolysis of SiH_2F_2 and SiH_2Cl_2 (References 73 and 74, respectively) and the flash photolysis of SiH_3Cl and SiH_3Br (Reference 75) were again studied for spectroscopic purposes. Halosilylenes such as SiF_2, $SiCl_2$, SiHCl, and SiHBr were detected as the decomposition products via H_2-elimination as summarized in equations (19) and (20). As indicated in equation (21), the thermal decomposition of $SiHCl_3$ at a temperature of 1000°C or above gave $SiCl_2$ and HCl as products.[76]

B. From Disilanes

In contrast to the monosilane system, the decompositions of disilanes to give silylenes are well-established processes. They can also be employed for synthetic purposes. The general process involved in the disilane decomposition can be represented by the following equation:

$$
\begin{array}{ccc}
\overset{\displaystyle A}{\underset{\displaystyle B}{-\,\overset{|}{\underset{|}{Si}}-\overset{|}{\underset{|}{Si}}-X}} & \longrightarrow & \underset{\displaystyle B}{-\,\overset{|}{\underset{|}{Si}}-X} \;+\; \overset{\displaystyle A}{\underset{\displaystyle B}{Si:}}
\end{array}
\tag{34}
$$

(A or B = H,F,Cl,Br,I,CH_3, or OCH_3; X = H,F,Cl,Br,I, or OCH_3)

The cleavage of Si–Si bonds in disilanes is always accompanied by a 1,2-shift of an entity such as H atom, a halogen atom, or an OCH_3 group. A monosilane and a silylene are formed from such a process. Vanderwielen, Ring, and O'Neal observed that this type of elimination reaction is only observed in high energy (vacuum uv) photochemical decompositions of the corresponding hydrocarbons.[77] They believed that the reason that 1,2-hydrogen shift reactions occur in polysilane thermal decompositions, but not in hydrocarbon pyrolyses, can probably be attributed to the availability of low-lying nonoccupied orbitals (4s and 3d) on silicon, which can provide appreciable bonding stabilization for pentavalent silicon in the transition state, and to the relative stability of divalent silicon.[78]

a. Si_2H_6

There are three plausible primary processes for the decomposition of Si_2H_6 as shown in equations (35)–(37), with the formation of SiH_2, SiH_3, and H atoms, respectively.

$$Si_2H_6 \rightarrow SiH_2 + SiH_4 \tag{35}$$

$$Si_2H_6 \rightarrow \cdot SiH_3 + \cdot SiH_3 \tag{36}$$

$$Si_2H_6 \rightarrow \cdot Si_2H_5 + \cdot H \tag{37}$$

The pyrolysis of Si_2H_6 at 314–360°C was first studied by Emeleus and Reid in 1939.[79] They observed the formation of H_2, SiH_4, and silicon as the major products. By analogy with the mechanism of C_2H_6 decomposition, they suggested that the reaction shown in equation (36) is the initial step for disilane decomposition. In 1948, Stokland[80] discovered that Si_2D_6 decomposes at a considerably lower rate than does Si_2H_6. He measured the kinetic isotope effect in terms of k_D/k_H as 0.71, which he regarded to be too large if the rate-determining step in the disilane decomposition is the cleavage of the Si–Si bond as shown in equation (36). As a result, he suggested that the SiH_2 mechanism, equation (34), was the operating mode. Interest in disilane decomposition revived in recent years. Ring and coworkers[81,82] have detected the formation of Si_3H_8 as a major product from this system and have confirmed Stokland's SiH_2 hypotheis. The evidence is as follows: (i) No tetrasilane was formed in the disilane process, which argues against the SiH_3 mechanism as shown in equation (36). (ii) In the pyrolysis of a Si_2D_6 and CH_3SiH_3 mixture, the very small hydrogen fraction obtained was all D_2. The absence of HD argues against the reaction in equation (37) as the primary step of Si_2H_6 decomposition. (iii) The formation of Si_3H_8 from disilane pyrolysis, and the formation of $CH_3SiH_2SiD_2H$ from the pyrolysis of the Si_2D_6–CH_3SiH_3 mixture definitely identify the silylene mechanism as shown in equation (35) to be the primary process for Si_2H_6 decomposition. All these findings by Ring and coworkers were confirmed almost at the same time by Bowrey and Purnell[83,84] in a detailed kinetic study of disilane pyrolysis. The latter authors have also measured the Arrhenius parameters for the reaction shown in equation (35), which presumably involves a 1,2-hydrogen shift to give the final products.

The information on the decomposition of various disilanes and trisilanes including Si_2H_6 is summarized in Table 1. The Arrhenius parameters available in the literature for the primary decomposition of various silanes are given in Table 2. For Si_2H_6, the E_a value was 205.02 kJ mol^{-1} and the log A value was 14.5 as measured by Bowrey and Purnell.[83,84]

b. Halogenated Disilanes

The thermal degradation of Si_2I_6 to give SiI_4 and polymeric $(SiI_2)_n$ was first studied by Friedel and Ladenburg[85] in 1880, and the corresponding decomposition of other hexahalodisilanes [see equation (38)] has also been noticed.[86–90] In 1964, Schmeisser and Ehlers[91] reported that pyrolysis of Si_2F_6 at 700°C in a high vacuum yielded SiF_4 together with SiF_2, and in 1971, Chernyshev and coworkers[92] indicated that the corresponding reaction proceeds at 450–550°C for Si_2Cl_6.

Ring and coworkers[93] studied the pyrolysis of Si_2F_6, Si_2H_5F, Si_2H_5Cl,

$$SiX_3SiX_3 \rightarrow SiX_2 + SiX_4 \tag{38}$$

TABLE 1. Summary of Studies on the Thermal Decomposition of Disilanes and Trisilanes

Disilanes[a]	Temp. of pyrolysis, °C	Decomposition products (silylenes/silanes)	Branching ratio[b] k_a/k_b	References
SiH_3SiH_3	350	SiH_2/SiH_4	—	79–84
SiH_3SiH_2F	390	(a) SiH_2/SiH_3F	5 ± 1	93
		(b) $SiHF/SiH_4$		
SiH_3SiH_2Cl	390	(a) SiH_2/SiH_3Cl	0.8 ± 0.1	93
		(b) $SiHCl/SiH_4$		
SiH_3SiHF_2	400	SiH_2/SiH_2F_2	—	93
SiH_3SiHCl_2	370	(a) SiH_2/SiH_2Cl_2	~1	93
		(b) $SiCl_2/SiH_4$		
$SiHMeClSiHMeCl$	395	(a) $SiMeCl/SiH_2MeCl$	(4.4 ± 0.4)	93
		(b) $SiHMe/SiHMeCl_2$		
SiF_3SiF_3	405–700	SiF_2/SiF_4	—	91, 93
$SiCl_3SiCl_3$	450–550	$SiCl_2/SiCl_4$	—	92
SiH_2MeSiH_3	270–375	(a) SiH_2/SiH_3Me	1.7–1.8	78, 94
		(b) $SiHMe/SiH_4$		
SiH_2MeSiH_2Me	380–400	$SiHMe/SiH_3Me$	—	94
$SiMe_3SiHMe_2$	300–325	$SiMe_2/SiHMe_3$	—	95
$SiH_3SiH_2SiH_3$	260–290	(a) SiH_2/Si_2H_6	0.4	78, 81, 103
		(b) SiH_3SiH/SiH_4		
$SiMe_3SiMe_2SiHMe_2$	300	$SiMe_2/SiMe_3SiHMe_2$	—	95
$Si(OMe)_3Si(OMe)_3$	175	$Si(OMe)_2/Si(OMe)_4$	—	6, 13
$SiMe(OMe)_2SiMe(OMe)_2$	200	$SiMe(OMe)/SiMe(OMe)_3$	—	99, 100
$SiMe(OMe)_2SiMe_2OMe$	225	$SiMe_2/SiMe_2(OMe)_2$	—	99, 100
$SiMe_2(OMe)SiMe_2SiMe_2(OMe)$	275	2 $SiMe_2/SiMe_2(OMe)_2$	—	6, 13
SiH_3PH_2	300	(a) SiH_2/PH_3	2.6	102
		(b) PH/SiH_4		
GeH_3SiH_3	245	(a) GeH_2/SiH_4	≪1	102
		(b) SiH_2/GeH_4		

[a] Including *pseudodisilanes* such as SiH_3PH_2, and GeH_3SiH_3.
[b] See text for definition.

TABLE 2. Arrhenius Parameters for the Primary Decomposition of Silanes

Silane decomposition reactions	$\log A$	$E_a,$ kJ mol^{-1}	$\Delta S^*,$ e.u.	$\Delta H^*,$ kJ mol^{-1}	Reference
$Si_2H_6 \rightarrow SiH_2 + SiH_4$	14.5	205.56	—	—	83,84
$SiH_3SiH_2Me \rightarrow SiH_2 + SiH_3Me$	15.28 ± 0.36	212.34 ± 1.51	8.3 ± 0.8	207.94 ± 1.51	78
$SiH_3SiH_2Me \rightarrow SiH_4 + SiHMe$	14.14 ± 0.14	208.74 ± 1.46	3.1 ± 0.8	204.43 ± 1.80	78
$Si_3H_8 \rightarrow SiH_2 + Si_2H_6$	15.69 ± 0.18	221.71 ± 1.80	10.1 ± 0.8	217.28 ± 1.80	78
$Si_3H_8 \rightarrow SiHSiH_3 + SiH_4$	14.68 ± 0.23	206.02 ± 2.30	5.5 ± 1.0	201.54 ± 2.30	78

SiH_3SiHF_2, SiH_3SiHCl_2, and $CH_3SiHClSiHClCH_3$ in a flow system in the presence of other silanes as silylene-trapping agents. The results demonstrated that all six disilanes decomposed into a silylene and silane. Both the temperatures of the pyrolysis and the decomposition products from these six systems are summarized in Table 1. Whenever more than one decomposition mode for each disilane is operative, the relative rates in terms of "branching ratios" can be determined by the ratio of the observed products. In fact, there are two different kinds of "branching ratios" for these disilanes. (i) For the four halogenated disilanes of the SiH_3SiH_2X and SiH_3SiHX_2 types where the two silicon atoms in each molecule do not have exactly equivalent substituents on them, two different modes of H shift are possible, namely, 1,2 shift and 2,1 shift:

$$SiH_3SiH_2X \xrightarrow{1,2\text{-shift}} SiH_4 + SiHX \qquad (39)$$

$$SiH_3SiH_2X \xrightarrow{2,1\text{-shift}} SiH_2 + SiH_3X \qquad (40)$$

The ratio k_{40}/k_{39} can be evaluated from the product yield ratio, SiH_3X/SiH_4. Such branching ratios are shown in Table 1 as k_a/k_b. The results indicate that for chlorinated disilanes, both kinds of H shift have about the same probability. But for the fluorinated disilanes, the 2,1-shift is much more favored to give SiH_2 and a fluorinated silane. In fact, the 1,2-shift does not occur for SiH_3SiHF_2. (ii) For the disilanes containing both H and halogen atoms, a competition between the H shift and X shift is possible. The reason that the branching ratio of 4.4 ± 0.4 is put in parenthesis for SiHMeClSiHMeCl is that it alternatively measures the relative rates of H-shift/Cl-shift. This value, coupled with the fact that X shift is not observed in SiH_3SiHF_2 and SiH_3SiHCl_2, indicates that the H shift is generally the predominant process.

c. Alkyl-Substituted Disilanes

The thermal decomposition of alkyl-substituted disilanes such as $CH_3SiH_2SiH_3$,[78,94] $CH_3SiH_2SiH_2CH_3$,[94] and $(CH_3)_3SiSiH(CH_3)_2$[95] has been studied, and the results as summarized in Table 1 are consistent with the following mode of decomposition:

$$SiR_3SiHR_2 \rightarrow SiHR_3 + SiR_2 \qquad (R=H \text{ or } CH_3) \qquad (41)$$

The first feature of the results is that the CH_3 group does not shift during the degradation. This is consistent with the observation that the pyrolysis of $Si_2(CH_3)_6$ occurs via the cleavage of the Si–Si bonds to give $Si(CH_3)_3$ radicals[96] because there are no Si–H bonds available for H shift:

$$(CH_3)_3Si–Si(CH_3)_3 \rightarrow 2Si(CH_3)_3 \qquad (42)$$

Therefore it is not surprising that the partially alkylated disilanes decompose at a rate which is similar to that of Si_2H_6 but much faster than that of $Si_2(CH_3)_6$.

The second feature is that because the CH_3 radical does not shift, $CH_3SiH_2SiH_3$ is the only one of the three alkyl-substituted disilanes studied to possess more than one decomposition channel. The results indicate that the 2,1-shift of H atoms is again somewhat favored when CH_3 groups are the substituents:

$$CH_3SiH_2SiH_3 \xrightarrow{\text{1,2-shift}} CH_3SiH + SiH_4 \tag{43}$$

$$CH_3SiH_2SiH_3 \xrightarrow{\text{2,1-shift}} CH_3SiH_3 + SiH_2 \tag{44}$$

Ring and coworkers[78] have also performed a detailed kinetic study of the methyldisilane decomposition and they obtained Arrhenius parameters which are included in Table 2. From the fact that the E_a values for disilane decomposition are all around 209.20 kJ mol^{-1}, while the $D(Si-Si)$ value in disilane is above 334.72 kJ mol^{-1},[97,98] they proposed a hydrogen-atom bridged transition state for the degradation which involves a pentavalent silicon as illustrated in equation (45). The sp^3 bonding of the Si atoms in the reactant molecules rehybridizes to something resembling dsp^3 bonding at the acceptor silicon center of the transition state. The relatively high A factors observed indicate that these transition states are quite "loose."

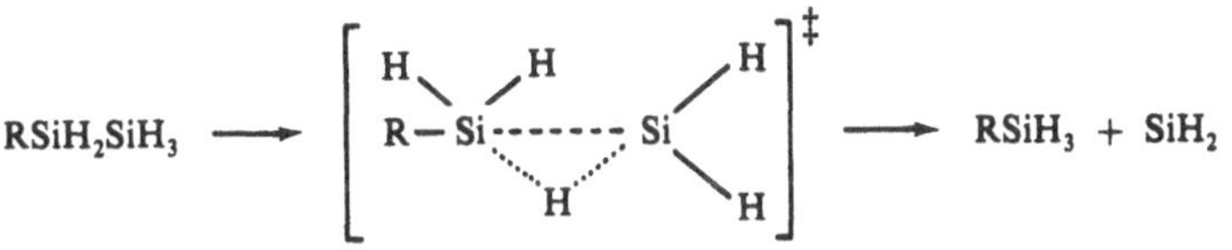

$$RSiH_2SiH_3 \longrightarrow \left[\begin{array}{c} H \quad H \qquad H \\ R-Si \text{------} Si \\ H \qquad H \end{array} \right]^{\ddagger} \longrightarrow RSiH_3 + SiH_2 \tag{45}$$

d. Alkoxy-Substituted Disilanes

The thermal decomposition of methoxy-substituted disilanes via the methoxy group shift to give a silylene and a silane has been well established by Atwell and Weyenberg.[6,13,99,100] The pyrolysis temperatures and the corresponding products are shown in Table 1. The X-elimination process involved was viewed as the reverse of the insertion of a silylene into a Si—O bond, and therefore a three-membered ring transition state with an oxygen bridge was proposed. This is illustrated below with 1,2-dimethoxytetramethyl-disilane as the starting compound[6,13]:

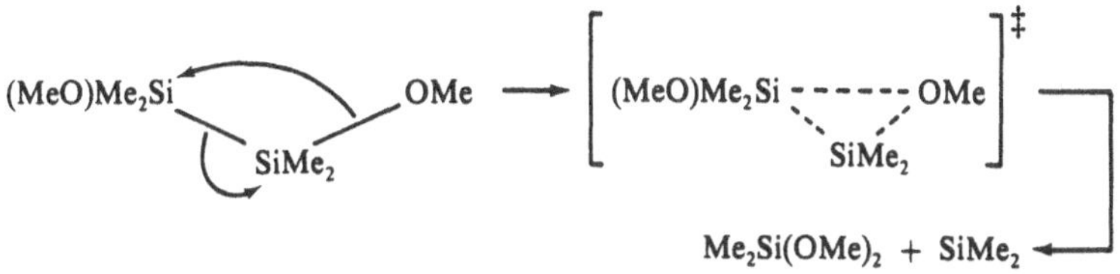

$$(MeO)Me_2Si \overbrace{}^{} OMe \longrightarrow \left[\begin{array}{c} (MeO)Me_2Si \text{------} OMe \\ SiMe_2 \end{array} \right]^{\ddagger} \tag{46}$$

$$Me_2Si(OMe)_2 + SiMe_2 \longleftarrow$$

As hinted by the temperature of the pyrolysis shown in Table 1, Atwell and Weyenberg concluded that the more highly alkoxylated compounds have less thermal stability towards degradation.[13,99] They also studied the thermolysis of a variety of other hetero-substituted disilanes and determined that the thermal stabilities in terms of the effect of substituent type follow the order:[13]

$$OR < H \simeq F < Cl \tag{47}$$

This means that the alkoxy compounds are the least stable while the analogous chloro-derivatives are the most stable.

e. Other Disilanes and Pseudodisilanes

Information in the literature shows that pentamethyldisilanyl cyanide[101] undergoes a redistribution reaction at 175°C. The low reaction temperature indicates that an α-elimination mechanism might be involved.

The term *pseudodisilane* here refers to compounds such as SiH_3PH_2 and GeH_3SiH_3 which undergo α-elimination as do the disilanes. The decomposition of this pair has been studied by Ring and coworkers[102] and the results are also included in Table 1. It is seen that a H-atom shift from Si to P is more likely in SiH_3PH_2, while in the case of GeH_3SiH_3 the H-atom shift from Ge to Si is essentially the only observed process.

C. From Trisilanes

As seen in Table 1, the thermal decomposition of trisilanes such as Si_3H_8,[72,81,103] $SiMe_3SiMe_2SiHMe_2$,[95] and $SiMe_2(OMe)SiMe_2SiMe_2(OMe)$[6,13] follows exactly the same pattern as that of the corresponding disilanes. For the two degradation modes of Si_3H_8, the shift of a primary H, as shown in equation (48), is *less* favored than the shift of a secondary H, as shown in equation (49), by a factor of 2.4.[103] However, for $CH_3SiH_2SiH_3$ the shift of a primary H is *more* favored by a factor of 1.8.[94] This contradictory picture remains the same even after the numbers are normalized for the statistical degeneracy of the reaction paths. Sefcik and Ring[103] explained this difference by recognizing that during heterolytic cleavage, the bonded electron pair should be drawn to the more acidic leaving group of the two. Since the relative acidities in Si_3H_8 are $SiH_3SiH_2 > SiH_3$, the formation of SiH_3SiH should be favored. But, on the other hand, since the relative acidities in $CH_3SiH_2SiH_3$ are $SiH_3 > CH_3SiH_2$, the formation of SiH_2 is therefore predominant.

$$SiH_3SiH_2SiH_3 \rightarrow SiH_2 + Si_2H_6 \tag{48}$$

$$SiH_3SiH_2SiH_3 \rightarrow SiH_4 + SiHSiH_3 \tag{49}$$

It is already well established that Si_2H_6 decomposes via a H shift.[79-84] Similar mechanisms can definitely account for the products formed in the Si_3H_8 system. However, the formation of SiH_2 and Si_2H_6 from trisilane is also consistent with an alternate mechanism involving a silicon bridge as the transition state, as shown in equation (50). Ring and coworkers[78] argued against this

$$SiH_3SiH_2SiH_3 \longrightarrow \left(\begin{array}{c} SiH_2 \\ H_3Si \cdots \cdots \cdots SiH_3 \end{array} \right)^{\ddagger} \longrightarrow SiH_2 + Si_2H_6 \qquad (50)$$

mechanism after comparing the kinetic data of Si_3H_8 and $CH_3SiH_2SiH_3$ decomposition as shown in Table 2. They reasoned that since the A factors for SiH_2 elimination from these two molecules differ by an amount consistent with their reaction path degeneracies, it is unlikely that these eliminations proceed via significantly different reaction paths. They ruled out the reaction in equation (50) because a similar transition state for methyldisilane would involve pentavalent carbon, and SiH_2 does not insert into either C–Si or C–H bonds in a competition with Si–H bonds. While, however, the above reasoning definitely shows that the H-shift mechanism is much more probable than the silicon-bridge mechanism for Si_3H_8, the existence of the reaction shown in equation (50) as a minor process still cannot be ruled out. Positive hints for the possible operation of such a silicon-bridge mechanism can be found in the fact that trisilanes and polysilanes decompose readily by a "breeding" mode to give silylenes as described below.[104-112]

Sakurai and coworkers[104] have synthesized 1,2,3-trisilacycloheptane derivatives and have shown that their photolysis with a low-pressure Hg lamp under a N_2 atmosphere gave substituted silylenes as products:

$$\begin{array}{c} \text{SiMe}_2 \\ \text{SiMeR} \\ \text{SiMe}_2 \end{array} \xrightarrow[N_2]{h\nu} \begin{array}{c} \text{SiMe}_2 \\ \text{SiMe}_2 \end{array} + \text{SiMeR} \qquad (R = \text{Me or } iso\text{-Pr}) \qquad (51)$$

The fact that the Si atom in the center is the one to be squeezed out to "breed" a silylene means that a silicon-bridged transition state like the one shown in equation (50) is likely. The same authors[104] have also synthesized 5,6,11-trisilaspiro-4,6-undecane, which undergoes a similar photochemical decomposition to give a cyclic silylene [equation (52)].

$$\begin{array}{c} \text{SiMe}_2 \\ \text{Si} \\ \text{SiMe}_2 \end{array} \xrightarrow{h\nu} \begin{array}{c} \text{SiMe}_2 \\ \text{SiMe}_2 \end{array} + :Si \qquad (52)$$

$$SiMe_3SiMePhSiMe_3 \xrightarrow{h\nu} SiMePh + SiMe_3SiMe_3 \qquad (53)$$

Similarly, Ishikawa, Kumada, and coworkers[105–107] formed SiMePh from the photolysis of 2-phenylheptamethyltrisilane with 253.7-nm radiation [equation (53)].

D. From Polysilanes

Earlier studies[113–115] on the thermal decomposition of permethylated polysilanes, $(SiMe_2)_n$, at 300°C showed that the yields of $SiMe_2$ are generally low. Later Ishikawa and Kumada[108–112] initiated a series of studies on the

$$(SiMe_2)_n \rightarrow nSiMe_2 \qquad (n \simeq 55) \tag{54}$$

photolysis with 253.7-nm irradiation of various permethylated derivatives of cyclic and acyclic polysilanes. They showed that dodecamethylcyclohexasilane (1)[106] and some of the permethylated linear and branched-chain polysilanes[112] readily undergo contraction of their skeletal Si–Si chain with loss of silylenes. Upon prolonged irradiation, cyclotetrasilane (2) undergoes homolytic scission of the Si–Si bond to give a diradical,[111] while octamethyltrisilane (3) is usually very resistent to photolysis.

(55)

(56)

$$Me(SiMe_2)_nMe \xrightarrow{h\nu} (n\text{-}3)SiMe_2 + \underset{\mathbf{3}}{Me(SiMe_2)_3Me} \qquad (n = 4\text{–}8) \tag{57}$$

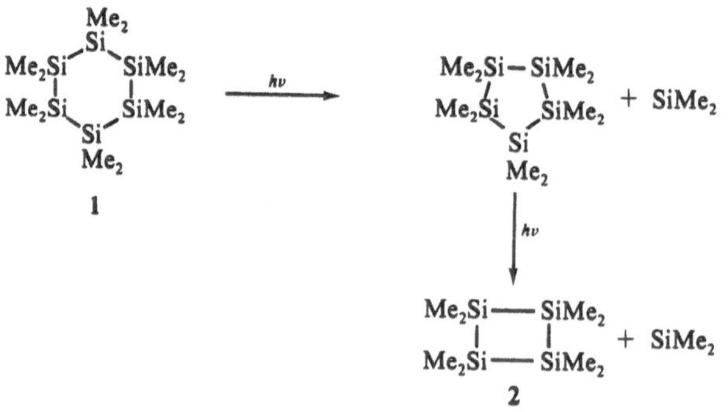

The "breeding" of a silylene from such polysilanes lends more support to the possible existence of a silicon-bridged transition state as shown in equation (50). The decomposition of the transition state shown in that equation is essentially the "breeding" of a SiH_2 unit from a cyclic silicon intermediate.

E. From Silacyclopropanes

Seyferth and coworkers recently succeeded in synthesizing a variety of the long-sought substituted silacyclopropanes.[116,117] They found that hexamethylsilirane decomposes conveniently under very mild conditions (60–80°C) to generate $SiMe_2$[118,119]:

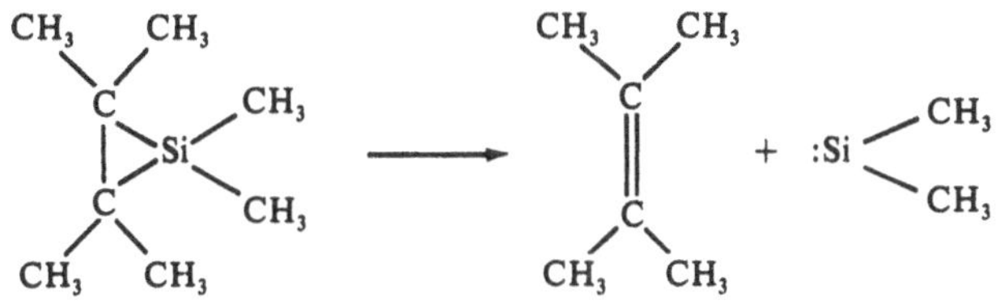

$$\tag{58}$$

However, the other silacyclopropanes synthesized are much more thermally stable than is hexamethylsilirane and do not serve as sources of $SiMe_2$ at these low temperatures.

F. From 7-Silanorbornadienes

In 1964, Gilman, Cottis, and Atwell[120,121] studied the thermal elimination of dimethyl- and diphenylsilylenes from 7-silanorbornadienes. This method was used later by other investigators to include various substituents on the basal carbon atoms.[122–125] Recently, Maruca and coworkers[125] concluded from their investigation coupled with the previous ones[120–124] that the larger the number of substituents on the basal carbon atoms, the more stable the 7-silanorbornadiene will be. They also discovered that the reaction shown in equation (59) proceeds the least readily when the substituents are bulky and electron withdrawing.

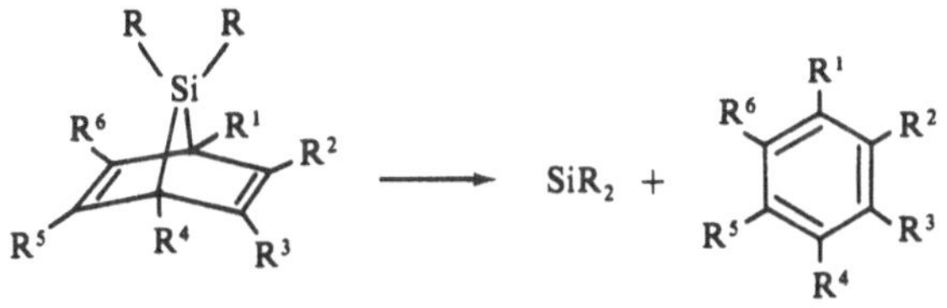

$$\tag{59}$$

G. From Metal Silyls

Wilberg and coworkers[126] studied the thermal and photolytic decomposition of metal silyls to account for the products formed they suggested, as shown in equations (60)–(63), that silylenes are involved as reaction intermediates. It is not necessary, however, that free silylenes be generated in these systems as they claimed.

More recently Kumada and coworkers[127–129] studied the thermal decomposition of disilanes such as $SiHMe_2SiHMe_2$ at 90°C in the presence of either $PtCl_2(PEt_3)_2$ or $NiCl_2(PEt_3)_2$ as catalysts and observed reactions attributable to silylene interaction. On the other hand, Yurev and coworkers[130] studied the reactions of SiH_2Me_2 and SiH_2Ph_2 with 1,3-dienes in the presence of Ni, Pd, and Co as catalysts and have observed apparent reactions of species such as $SiMe_2$ and $SiPh_2$. Again, the actual reactive intermediates involved are likely to be complexes of silylenes with metals (silylenoids) instead of free silylenes.

$$Hg(SiMe_2)_n \rightarrow Hg + nSiMe_2 \tag{60}$$

$$Al(SiMe_3)_3 \rightarrow AlMe_3 + 3SiMe_2 \tag{61}$$

$$AlMe_2(SiMe_3) \rightarrow AlMe_3 + SiMe_2 \tag{62}$$

$$Zn(SiPh_3)_2 \rightarrow Zn + SiPh_4 + SiPh_2 \tag{63}$$

3. The Reduction Method

Strictly speaking, the formation of silylenes as described in Sections IV.1.C and III.6 concerning reactions of elemental silicon with substrates can also be classified as reduction methods. Besides these, two other types of reductions to produce silylenes are observed.

A. Reduction with H_2

During the high-temperature reduction of $SiCl_4$ to Si with hydrogen, the reaction may proceed with $SiCl_2$ as an intermediate[76,131-133]:

$$SiCl_4 + H_2 \rightarrow SiCl_2 + 2HCl \tag{64}$$

B. Reduction with Metallic Elements

In 1964, Skell and Goldstein[134,135] reduced $SiCl_2Me_2$ at 280°C with Na/K vapor to give $SiMe_2$:

$$SiCl_2Me_2 + 2K \rightarrow SiMe_2 + 2KCl \tag{65}$$

However, as reasoned by Atwell and Weyenberg,[5] silylenes are not likely to be involved in the reactions of dichlorodiorganosilanes and metals in various aprotic solvents.[2]

V. SPECTROSCOPIC AND THERMODYNAMIC PROPERTIES OF SILYLENES

1. Electronic States

A. Ground Electronic States of Silylenes

Both carbon and silicon atoms have triplet (^3P) ground electronic states. For the carbenes, after long years of debate, the ground state of CH_2 has been established as a triplet.[136,135] On the other hand, halogenated carbenes such as CF_2 and CHF have been shown to possess singlet ground states.[136] Therefore, one of the most interesting questions in silylene chemistry is: what are the ground electronic states of species such as SiH_2 and SiF_2?

Although one might deduce from Skell and Goldstein's 1964 articles[134,135] that the reacting $SiMe_2$ was in the singlet ground state, the first direct confrontation with the question of ground electronic states of silylenes came from theoretical and spectroscopic studies. In 1966, Jordan[138] employed semi-empirical calculations to predict that the ground state of SiH_2 is a singlet ($^1A_1$), while the first excited state is a triplet ($^3B_1$). The energy separation between these two electronic states has been calculated to be 193 kJ mol^{-1}, and their bond angles are predicted to be 95° and 138°, respectively. The predicted structural and energetic information about the singlet state of SiH_2 were immediately confirmed spectroscopically by Dubois, Herzberg, and Verma.[139] However, these authors cautioned that it was not possible to determine what the ground electronic state of SiH_2 was from their observations.

In 1974, we studied the reaction of $^{31}SiH_2$, formed by the nuclear recoil method, with 1,3-butadiene in the presence of neon as a moderator,[30] and were able to confirm Jordan's prediction that the ground electronic state of SiH_2 is indeed a singlet. We have carried out two series of Ne-moderated $^{31}SiH_2$–butadiene studies, one in the presence and the other in the absence of nitric oxide, which is an efficient triplet scavenger. The two moderator curves decreased with the same slope, which is consistent with a singlet ground state for SiH_2. If the triplet were the ground state, the moderator curve for the well-scavenged systems should descend much more sharply in comparison with that for the unscavenged series.

In the case of SiF_2, Margrave and coworkers[140] have reasoned from esr studies that its ground electronic state is also a singlet. In their experiments, monomeric SiF_2 which was generated in the gas phase dimerized upon condensation to give $(SiF_2)_2$. The fact that no esr signal was detected in the gas phase means that the reacting monomeric SiF_2, which is most likely to be in its ground electronic state, is diamagnetic and therefore a singlet. Upon condensation without dilution, the dimeric $(SiF_2)_2$ showed paramagnetic properties with a broad esr spectrum. However, when the monomeric SiF_2 was diluted 1:50 with argon and then condensed, no esr signal was observed. It was believed that in this case the dimerization was prevented when the monomeric species was trapped in an inert matrix.

Studies of $^{31}SiF_2$ formed in the nuclear recoil systems are also consistent with the singlet being the ground state.[27,33] In fact, essentially all the kinetic studies of various silylenes are consistent with singlet ground electronic states.

B. Electronic States of Reacting Silylenes

The composition of the reacting silylenes in terms of electronic states can be quantitatively evaluated by carrying out studies at various scavenger levels. In the nuclear recoil system, [^{31}Si]-silacyclopent-3-ene was formed from the reactions of $^{31}SiH_2$ with 1,3-butadiene as shown in equation (66). Its yield as a function of nitric oxide concentration has been measured.[30] The observed product yield decreases rapidly in the 0–4% NO region and remains essentially constant thereafter. The plateau value attained is about 20% of the original unscavenged value. These results indicate that the reacting $^{31}SiH_2$ from the nuclear recoil system is about 80% triplet and 20% singlet. Similar percentages were also obtained when pentadienes were used instead of 1,3-butadiene.

$$^{31}SiH_2 \quad + \quad \diagup\!\!\!\diagdown\!\!\!\diagup \quad \longrightarrow \quad \overset{^{31}SiH_2}{\bigcirc} \qquad\qquad (66)$$

In the nuclear recoil system in which $^{31}SiF_2$ reacts with various dienes, the addition of nitric oxide actually increases the [^{31}Si]-1,1-difluorosilacyclopent-3-ene yield by a factor of about 4.[33,34] A plateau is reached with the addition of merely 0.1% NO. When O_2 is used as a scavenger, the same sharp rise in the product yield is observed. However, with further addition of oxygen, the product yield decreases and eventually declines to the same level as that of samples to which oxygen was not added. Based on the assumption that in the absence of a scavenger only singlet $^{31}SiF_2$ reacts, while the triplet $^{31}SiF_2$ can only react as $^{31}SiF_2$-donor complexes formed with the paramagnetic scavenger molecules, it is possible to assess that the singlet-to-triplet ratio of the reacting difluorosilylene in the nuclear recoil system is about 1:3.3. The O_2-scavenger results can be explained by a kinetic competition for the $^{31}SiF_2$–O_2 complexes by 1,3-butadiene and additional oxygen molecules.

Silylenes formed by the thermal decomposition processes are all likely to be in the singlet state as reasoned by Skell, Ring, Purnell, Walsh, and other investigators.[56,57,83,84,134,135]

$$^{31}SiF_2 \quad + \quad \diagup\!\!\!\diagdown\!\!\!\diagup \quad \longrightarrow \quad \overset{^{31}SiF_2}{\bigcirc} \qquad\qquad (67)$$

2. Structure

Over the years, the structural parameters of silylenes have been predicted by various types of calculations and evaluated by a number of spectroscopic studies. Among them the most thoroughly investigated one is SiF_2, which has been studied in the gas phase by microwave,[141] ultraviolet,[142] infrared,[143] and electron spin resonance spectroscopy.[140] Some infrared studies on a solid inert gas matrix have also been carried out.[144]

In Table 3 are summarized some structural parameters of silylenes in their ground electronic states. For the bond angles, essentially all the XSiY angles in the silylenes, SiXY, are less than the tetrahedral angle, and all the ground state silylenes have bent structures. The "best" values of the bond angles for various silylenes are listed below in order of increasing values: SiH_2 (92°); SiF_2 (101°); SiHX (103°), where X = Cl, Br, or I; $SiCl_2$ (105°); and $SiBr_2$ (109°). The bond lengths in silylenes are also given in Table 3. The Si–H lengths in SiH_2 and SiHX are all around 0.155 nm. The Si–X length in SiX_2 and SiHX are about 0.159 nm for Si–F, 0.205 nm for Si–Cl, and 0.223 nm for Si–Br.

There is much less structural information available for the silylenes in their excited states. For SiH_2, Dubois, Herzberg, and Verma[139] measured the following structural parameters for its $^1B_1$ state: HSiH angle = 122° and Si–H length = 0.149 nm.

3. Heats of Formation

In Table 4 are summarized the standard heats and entropies of formation of various silylenes. The methods employed for obtaining each of the values are also shown. Generally speaking, the halosilylenes have negative standard heats of formation, while for the other silylenes ΔH_f° values are positive. The most representative heat of formation values in units of kJ mol^{-1} for silylenes of the type, SiX_2, are listed below in an increasing order: SiF_2 (−582); $SiCl_2$ (−163); $SiBr_2$ (−45); SiI_2 (+76); $SiMe_2$ (+138); SiH_2 (+254). The standard entropies of formation for silylenes as shown in Table 4, mostly have values in the range of 210–320. kJ mol^{-1} deg^{-1}.

VI. CHEMICAL PROPERTIES OF SILYLENES

The overall chemical fate of a silylene, SiX_2, in a reacting system is summarized in equation (68). This scheme is a modification of the one proposed by Atwell and Weyenberg.[6] The silylene generated from its precursor may interact with three different types of species in a system. (i) It may dimerize or polymerize if such reactions are of high efficiency, and if the instantaneous concentration of the silylene is rather high. The dimers generally behave as reactive

TABLE 3. Bond Angles and Bond Lengths of Silylenes in Their Ground Electronic States

Silylenes	Bond angles	Bond length, nm		References
		Si–H	Si–X	
SiH$_2$	92.5°	0.1516	—	62,139
	91.6°	—	—	145
	97.5° (calcd.)	0.159 (calcd.)	—	146
	96° (calcd.)	—	—	147
	95° (calcd.)	—	—	138
SiF$_2$	115° (calcd.)	0.160 (calcd.)	—	148
	100°59′	—	0.159	8,12,149,150
	97.5° ± 1° (Ne matrix)	—	—	151
	102° ± 2° (Ar matrix)	—	—	151
	100° (calcd.)	—	—	152
SiCl$_2$	124° ± 2°	—	0.149 ± 0.007	68,142
	100°53′	—	0.15947	143
	106°	—	—	153
SiBr$_2$	105°	—	0.204	154
	109° ± 3°	—	—	154
SiHCl	102.8°	0.156	0.206	6,75
SiHBr	102.9°	0.156	0.223	6,75
SiHI	103°	—	—	155

TABLE 4. Heats and Entropies of Formation of Silylenes

Silylenes	ΔH_f° kJ mol⁻¹	ΔS_f° kH mol⁻¹ deg⁻¹	Methods	References
SiH₂	256 ± 5	—	Kinetic	78
	254 ± 5	—	Kinetic	78
	272 ± 38	—	Calculated	156
	243	—	Kinetic	157
	248	208	Kinetic	84,158
	339 ± 63	—	Mass Spectroscopic	159
SiF₂	−582	—	Transpiration	160
	−582	256	Spectroscopy	143
	—	—	Transpiration	44
	−582	—	Mass Spectroscopic	43
	−582	—	Transpiration	161
	−619 ± 17	—	Tensimetric	162
SiCl₂	−163 ± 2	280	Flow, static, equilibrium	153
	−167	—	Tensimetric	163
	−163 ± 3	—	Flow	160
	−160	280	Flow	160
	−164	—	Flow	164,165
	−165	289,290	Flow	166
	−171 ± 3	—	Mass Spectroscopic	160
SiBr₂	−51	305	Flow	160
	−42 ± 8	316 ± 4, 334 ± 8	Flow	164
SiI₂	75	318	Flow	160
	78	—	Flow	71
SiMe₂	138	—	Kinetic	167
SiHMe	222 ± 5	—	Kinetic	78
SiHSiH₃	279 ± 5	—	Kinetic	78

intermediates, and they may also undergo reactions with various molecules in the system. (ii) The silylene may interact with the precursor molecules if these contain reactive Si–H, Si–O, or Si–Si bonds. Insertion is normally the reaction mode because the precursor molecules are rarely unsaturated. (iii) The silylene may undergo either insertion or addition reactions with the reactant molecules of the system. The adducts formed may sometimes undergo further rearrangement to give the final products.

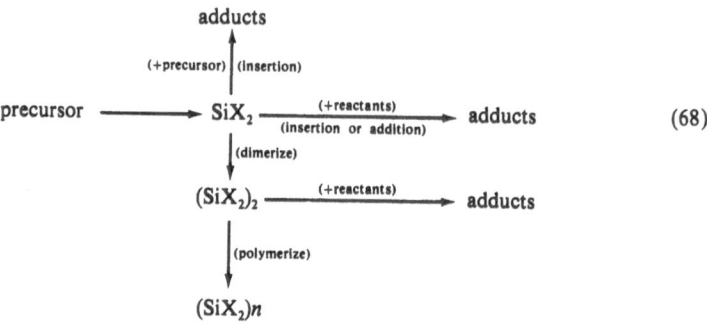

$$(68)$$

In the following sections, the chemical properties of silylenes will be described in terms of their three fundamental modes of reactions, namely, polymerization, insertion, and addition.

1. Polymerization Reactions

Monomeric silylenes may dimerize or polymerize to give species which are themselves reactive intermediates and are capable of undergoing further insertion, abstraction, and addition processes. Among the silylenes, the reactions of the monomers, dimers and polymers of SiF_2 are definitely the ones most extensively studied. The reactions of the SiF_2 species formed by the co-condensation method with various kinds of reagents are summarized in Table 5. It is seen that upon co-condensation with reagents, difluorosilylene may react either as a monomer, SiF_2, or the dimer, $(SiF_2)_2$, or the polymers, $(SiF_2)_x$.

In the table, the symbol $(SiF_2)_n$ stands for the unspecified overall spectrum of SiF_2, where $n = 1, 2,..., n$, while the symbol $(SiF_2)_x$ stands for the trimers and the higher polymeric forms of SiF_2, where $x = 3, 4,..., n$.

A. Dimerization

a. Difluorosilylenes

In Table 5 for the cocondensation of SiF_2 with reagents, it is seen that only in a few cases (such as I_2, CF_3I, C_2HF_3, or C_6F_6) do the resultant molecules

TABLE 5. Reactions of SiF_2 Formed by the High-Temperature Process, $SiF_4 + Si \rightarrow 2SiF_2$

Reactants	Reacting forms of SiF_2	Observed products	References
HF	SiF_2	SiF_3H	168
HCl	$(SiF_2)_n$	Si_2F_5H, Si_2Cl_6	4
HBr	$(SiF_2)_2$	Si_2F_5H and SiF_2BrSiF_2H, which redistributed to give SiF_3SiF_2H, $SiF_3SiFHBr$, and SiF_3SiHBr_2	169,170
H_2O	Probably SiF_2	$(SiF_2H)_2O$ and a voluminous white polymer	171
H_2S	SiF_2 and $(SiF_2)_2$	SiF_2HSH, Si_2F_5H, SiF_2HSiF_2SH, and small amounts of SiF_2HSSH and SiF_2HSiF_2SSH	172
BF–BF_3 mixture	SiF_2	$SiF_2(BF_2)_2$, $SiF(BF_2)_3$	173
BF_3	Triplet SiF_2 (perhaps)	SiF_2BF_3 (cocondensation with a collison-free path for SiF_2)	174
	$(SiF_2)_2$	$SiF_3SiF_2BF_2$, $SiF_3SiF_2SiF_2BF_2$	175
	$(SiF_2)_n$	$SiF_3(SiF_2)_nBF_2$ ($n = 2,...,13$)	175
BCl_3	$(SiF_2)_n$	SiF_nCl_{4-n}	8
B_2H_6	$(SiF_2)_n$	$1,1\text{-}(SiF_2)_nB_2H_4$ (decomposes easily)	176
B_2F_4	SiF_2	$SiF(BF_2)_3$ (1 % yield)	173,177
CO	$(SiF_2)_n$	Polymer $\xrightarrow{\Delta}$ silicon oxyflourides	144
CS_2	SiF_2 and $(SiF_2)_2$	SiF_3SSiF_3, $SiF_3SSiF_2SiF_3$	178
NH_3	$(SiF_2)_n$	Deeply colored polymeric material	8
NO	SiF_2	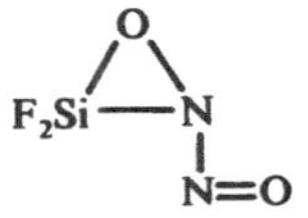 $F_2Si{-}N{-}N{=}O$ (three-membered ring with O)	144
(N_2O, NO_2, NOCl, NOF)	$(SiF_2)_n$	Deeply colored polymers; upon warming no compounds containing Si and N are formed	8
NF_3	$(SiF_2)_n$	Matrix cocondensation polymers	8
O_2	$(SiF_2)_n$	Hexafluorodisiloxane and unidentified chain and cyclic fluorinated siloxanes	3,140

(continued)

TABLE 5.—cont.

Reactants	Reacting forms of SiF_2	Observed products	References
F_2	$(SiF_2)_n$	(Reaction occurred)	8,179
Na	SiF_2	$Na^+SiF_2^-$	8
NaF, LiF	SiF_2	Low-temperature species decompose on warm-up to M_2SiF_6 and Si	8
SiH_4	$(SiF_2)_n$	Not reported	8
SiF_2	$(SiF_2)_n$	$(SiF_2)_n$ ($n = 1,...,16+$)	140,143,180
SiF_4	$(SiF_2)_n$	No reaction	11
$SiMe_4$	$(SiF_2)_n$	No reaction	181
$SiMe_3OMe$	SiF_2	$SiFMe_3$, $SiMe_3SiFMe_2$, SiF_3OMe, SiF_3SiMe_3	181
$SiClMe_3$	$(SiF_2)_n$	No reaction, only $SiFMe_3$ observed	181
PH_3	SiF_2 and $(SiF_2)_2$	SiF_2HPH_2, SiF_3PH_2, Si_2F_5H, and unstable higher molecular weight compounds	182
PF_3	$(SiF_2)_n$	A red polymer and unstable Si_nF_{2n+1}	178
PCl_3	$(SiF_2)_n$	Yellow–brown polymer	183
SO_2	$(SiF_2)_n$	Sulfur and polyfluorosiloxanes	3
SOF_2	$(SiF_2)_n$	Three series of cyclic or linear fluorosiloxanes $SiF_3(SiF_2)_nOSiF_3$ ($n = 1, 2$), SiF_3OSiF_3, $SiF_3OSiF_2OSiF_3$, $Si(OF_2)_n$ ($n = 2, 3$)	184
$(CH_3)_2SO$	SiF_2	SiF_2Me_2, $(SiF_3)_2O$	183
S_2Cl_2	$(SiF_2)_2$	SiF_3SSiF_3, perfluorosilanes, tetrafluorocyclodisilathiane	178
Cl_2	$(SiF_2)_n$	(Reaction occurred)	8,179
GeH_4	SiF_2 and $(SiF_2)_n$	$GeH_3(SiF_2)_nH$, $n = 1$(major),...,2,3	184
$Ge(CH_3)_4$	$(SiF_2)_n$	No reaction	175
Br_2	SiF_2, $(SiF_2)_n$	SiF_2Br_2, polymer (SiF_2Br_2 only when reaction is in the gas phase)	12
I_2	SiF_2	SiF_2I_2, SiF_3I (yield 3:1)	185
CF_4, CF_3, CF_2, C_2F_6, GeF_4	SiF_2	Rapid exchange reaction to give SiF_4 and C or Ge	8,168

(continued)

Reactant	SiF species	Products	Ref.
CF₃I	SiF₂, (SiF₂)₂, and (SiF₂)ₓ	(SiF₂)ₙI (n = 1 major, 2) SiF₃(SiF₂)ₙ (n = 0, 1, 2); SiF₂I₂, SiF₂ISiF₂I, SiF₂ISiF₂SiF₂I	186
CHCl₃	(SiF₂)ₙ	No reaction, only observe (SiF₂)ₓ	8
CF₃CN, CH₃CB, CNCl, (CN)₂	(SiF₂)ₙ	Perfluorosilanes and triazine derivatives	8
CH₃OH	SiF₂	(CH₃O)₂SiF₂	168
CH₂=CH₂	(SiF₂)₂	(1,2-disilacyclobutane, ring with –SiF₂–SiF₂–), and (six-membered ring with SiF₂, SiF₂)	187
CH₂=CHF	SiF₂	C₂H₄SiF₂, small amount CH₂=CHSiF₃	188
	(SiF₂)₂	CH₂=CHSiF₂SiF₃	188
CH₂=CF₂	SiF₂	CH₂=CFSiF₃, CH₂=C(SiF₃)₂	188
CF₂=CHF	SiF₂	CF₂=CHSiF₃, CHF=CFSiF₃(cis and trans)	187
CF₂=CF₂	(SiF₂)₂	CF₂=CHSiF₂SiF₃	187
CH₂=CFMe	(SiF₂)ₙ	White polymer $\xrightarrow{\Delta}$ C₂F₄(SiF₂)ₙ (n = 1, 2, 3)	188
	SiF₂	CH₂=CMeSiF₃	188
	(SiF₂)₂	CH₂=CMeSiF₂SiF₃	188
CH₂=C=CH₂	(SiF₂)₂	CH₂=C=CHSiF₂SiF₂CH₂CH=CH₂, CH≡CCH₂SiF₂SiF₂CH₂CH=CH₂	189
CH≡CH	(SiF₂)₂	(bicyclic structure with SiF₂, SiF₂ bridge), CH₂=CSiF₂SiF₂CH=CH₂,	190
CH₃C≡CH	(SiF₂)₂	(four-membered ring, CH₂=C=CHSiF₂SiF₂), CH₃CH=CH	190
CH≡CCF₃	(SiF₂)₂	(F₃C-substituted ring structures with SiF₂, SiF₂; and CF₃-substituted six-membered ring)	191

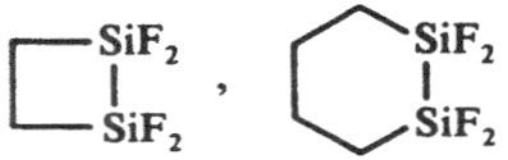

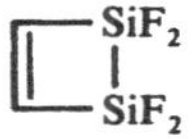

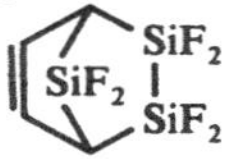

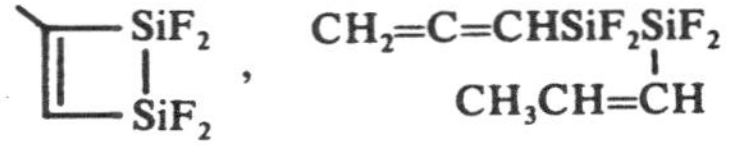

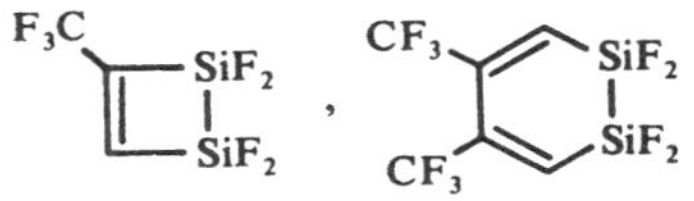

TABLE 5.—cont.

Reactants	Reacting forms of SiF$_2$	Observed products	References
CH$_2$=CHCH=CH$_2$	(SiF$_2$)$_2$	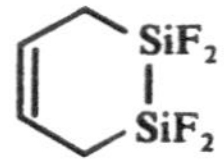 [cyclohexene ring bearing SiF$_2$ group], H$_2$C=C=CHCH$_2$SiF$_2$–SiF$_2$, H$_3$CCH=CHCH$_2$SiF$_2$	192
CH$_3$C≡CCH$_3$	(SiF$_2$)$_2$	 [ring structure with two SiF$_2$ groups]	8
c-C$_4$F$_8$	SiF$_2$	Rapid exchange reaction to given SiF$_4$ and C	8
 [cyclohexene]	SiF$_2$	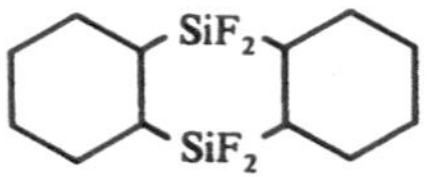 [bicyclic structure with two SiF$_2$ groups]	193
 [benzene]	(SiF$_2$)$_2$ and (SiF$_2$)$_x$	 (SiF$_2$)$_n$ ($n = 2$–8, major product: $n = 3$)	194
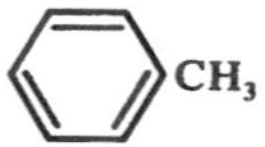 [toluene, CH$_3$]	(SiF$_2$)$_2$ and (SiF$_2$)$_x$	Analogous products as from C$_6$H$_6$ but with methyl substituents	194
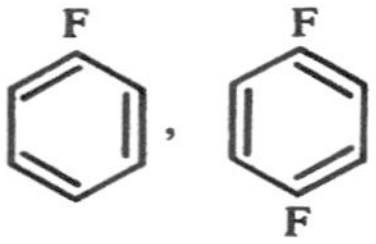 [fluorobenzenes, F]	SiF$_2$ and (SiF$_2$)$_n$	Analogous products as from C$_6$H$_6$ but with fluorine substituents	194

 (C₆F₆ hexafluorobenzene structure)	SiF_2 $(SiF_2)_2$ $(SiF_2)_x$	$C_6F_5SiF_3$ $C_6H_4(SiF_3)_2$ with *ortho*, *para*, *meta* ratio 1:6:9 Small amounts of $C_6F_3(SiF_3)_3$ and others	194
$\left[\begin{array}{c} O \\ \parallel \\ HC-Cl \end{array} \right]$	$(SiF_2)_n$	$SiF_2SiF_2-O-SiF_3$	195
$\begin{array}{c} O \\ \parallel \\ F_3CC-F \end{array}$	SiF_2	$(F_3Si-O-SiF_2)_2O$	196
$\begin{array}{c} O \\ \parallel \\ CH_3C-Cl \end{array}$	$(SiF_2)_n$	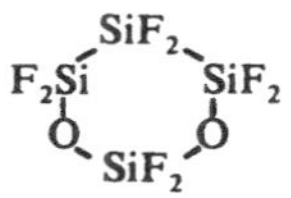 (cyclic) F_2Si—SiF_2—SiF_2—O—SiF_2—O ring $SiF_3OSiF_2OSiF_3$	195
$\begin{array}{c} O \\ \parallel \\ CF_3C-Cl \end{array}$	$(SiF_2)_2$	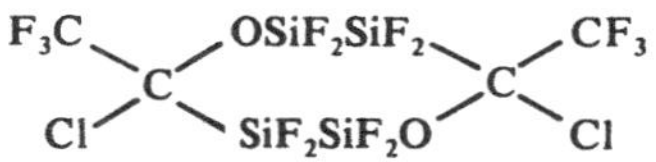 $F_3C-C(Cl)(OSiF_2SiF_2-C(CF_3)(Cl)(SiF_2SiF_2O))$	195,197
$(CF_3)_2CO$	$(SiF_2)_n$	Uncharacterized liquids	8

contain a single unit of SiF_2. The overwhelming situation is that there are two or more units of SiF_2 in each of the end products. In most of the cases, the dimerization of SiF_2 as shown in equation (69) preceded its interaction with other reagents.

$$2SiF_2 \rightarrow (SiF_2)_2 \tag{69}$$

The kinetic nature of the competition between dimerization or polymerization of SiF_2 and its interaction with other substrates can be illustrated with observations from two systems: BF_3 and 1,3-butadiene. The significant observations in the SiF_2–BF_3 system include the following[175]: (i) SiF_2 and BF_3 do not react in the gas phase. (ii) When SiF_2 is condensed alone, $(SiF_2)_x$ is formed. (iii) When SiF_2 is condensed onto BF_3, reaction occurs to give $SiF_3SiF_2BF_2$, and $SiF_3(SiF_2)_2BF_2$, but not SiF_3BF_2. (iv) When BF_3 is condensed onto a layer of SiF_2, no reaction occurs. These facts are consistent with the following scheme of reactions with the assumption that $k_d \gg k_l$ and that both k_t and k_p are very large even at $-196°C$. However, although k_{ll} and k_{lll} are less favorable than k_t and k_p because the observed yields of the two products are rather small, they do compete with the polymerization process to a certain extent.

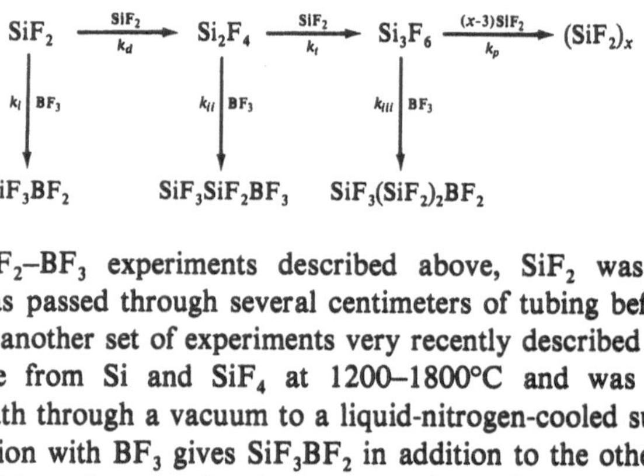

$$\begin{array}{ccccccc}
SiF_2 & \xrightarrow[k_d]{SiF_2} & Si_2F_4 & \xrightarrow[k_t]{SiF_2} & Si_3F_6 & \xrightarrow[k_p]{(x-3)SiF_2} & (SiF_2)_x \\
\downarrow{\scriptstyle k_l}{\scriptstyle BF_3} & & \downarrow{\scriptstyle k_{ll}}{\scriptstyle BF_3} & & \downarrow{\scriptstyle k_{lll}}{\scriptstyle BF_3} & & \\
SiF_3BF_2 & & SiF_3SiF_2BF_3 & & SiF_3(SiF_2)_2BF_2 & &
\end{array} \tag{70}$$

In the SiF_2–BF_3 experiments described above, SiF_2 was prepared at 1200°C and was passed through several centimeters of tubing before it reacted with BF_3.[175] In another set of experiments very recently described by Timms,[174] SiF_2 was made from Si and SiF_4 at 1200–1800°C and was passed by a collision-free path through a vacuum to a liquid-nitrogen-cooled surface. In this case, condensation with BF_3 gives SiF_3BF_2 in addition to the other products.

One plausible explanation for the two sets of experimental results is that in the gas phase SiF_2 dimerizes very readily upon collision, but its reaction with BF_3 is extremely slow. It is conceivable that the dimerization process requires the aid of a third body such as a tubing wall to remove the energy released by the recombination reaction. Therefore, in the absence of such a tubing wall or other third bodies, a certain fraction of the SiF_2 will condense as the monomer instead of as dimer or polymer. Upon condensation, the probabilities of all the SiF_2 reactions probably increase because of the increasing chances of physical contact, In particular, the SiF_2 – BF_3 reaction probability has increased to such an extent that measurable amounts of SiF_2BF_3 will be detected if enough monomeric SiF_2 survives until the condensation stage.

In the co-condensation of SiF_2 with 1,3-butadiene, the reaction shown in equation (71) was observed.[192] The nature of the end product indicates that

$$2SiF_2 \longrightarrow (SiF_2)_2 \qquad (71)$$

dimerization of SiF_2 precedes its addition. However, in our nuclear recoil experiments[27,33] when the instantaneous concentration of SiF_2 is extremely low, it is the monomeric SiF_2 which adds to butadiene, as shown in equation (72).

$$^{31}SiF_2 + \qquad \longrightarrow \qquad (72)$$

The results described above for the BF_3 and 1,3-butadiene systems, together with those summarized in Table 5, demonstrate that the dimerization of SiF_2 is efficient in both gaseous and condensed situations when the concentration of monomeric SiF_2 is sufficiently high. The observed product patterns indicate that the dimers play a major role in SiF_2 chemistry when there is no bond of sufficient lability to be attacked by monomeric SiF_2. Such reactive bonds include C–F in C_6F_6,[194] C–I in CF_3I,[186] and the O–H and S–H bonds in H_2O and H_2S (References 171 and 172, respectively). However, it should be borne in mind that an end product containing a single unit of SiF_2 is not necessarily derived from monomeric SiF_2. Some reagents may leave the Si–Si bond intact in the dimer during reaction and others may form intermediates with $(SiF_2)_2$ which eliminate one unit of SiF_2 during rearrangement. Similarly, an end product containing two units of SiF_2 is not necessarily derived from dimeric $(SiF_2)_2$. It is possible for certain reagents to acquire two separate units of monomeric SiF_2 in two consecutive steps.

Both spectroscopic and chemical evidence indicate that dimeric $(SiF_2)_2$ formed in the co-condensation experiments is a diradical species in the form of $\dot{S}iF_2\dot{S}iF_2$ instead of a molecular species in the form of $SiF_2{=}SiF_2$.[3,8] The results in Table 5 show that this diradical may undergo various apparent addition, insertion, and abstraction processes. For details of such interactions the reader is referred to the many excellent review articles on SiF_2 by Margrave,[1,3,8,12] Timms,[4,11,14] and others.[2,6] (Very recently, Margrave considered in detail the possibility for the involvement of $SiF_2{=}SiF_2$ in reactions.[198])

b. Dimethylsilylene

Recently, Conlin and Gaspar[199] demonstrated that in a flow system at 600°C, $SiMe_2$ may dimerize to give tetramethyldisilene as shown in equation (73). The evidence is derived from the fact that compounds **4** and **5** are

$$2SiMe_2 \longrightarrow Me_2Si{=}SiMe_2 \qquad\qquad (73)$$

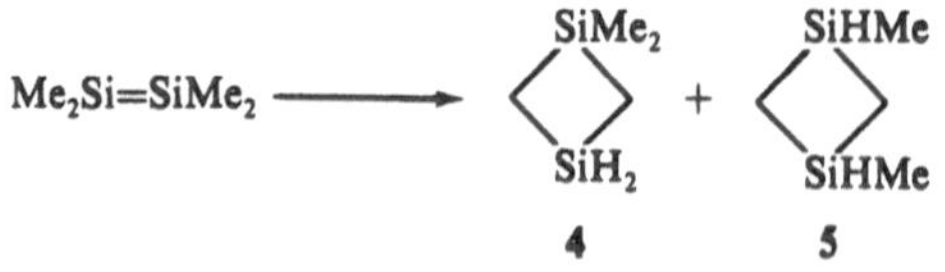

$$Me_2Si{=}SiMe_2 \longrightarrow \underset{SiH_2}{\overset{SiMe_2}{\diamondsuit}} + \underset{SiHMe}{\overset{SiHMe}{\diamondsuit}} \qquad (74)$$

$$\underset{\textbf{4}}{} \qquad \underset{\textbf{5}}{}$$

observed as products in this system, presumably formed by rearrangement of the disilene molecule as shown in equation (74). Roark and Peddle[200] obtained the same two products from tetramethyldisilene which was formed via the reaction in equation (75).

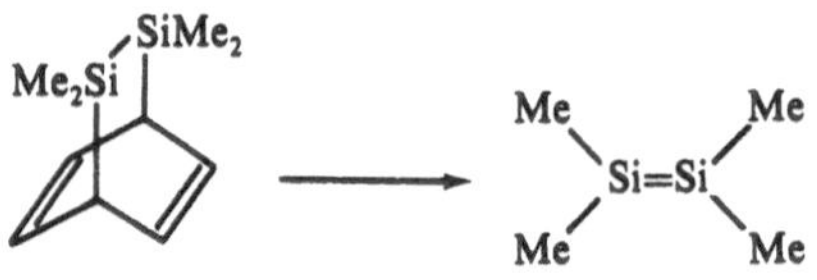

$$(75)$$

Atwell and Uhlmann[201] observed the formation of hexamethyl-1,2-disila-3-cyclobutene from the reactions of $SiMe_2$ with dimethylacetylene. As discussed by Barton and Kilgour,[202] product formation in this system may also involve an initial dimerization of $SiMe_2$ [equation (73)] followed by addition of the resulting disilylene to the acetylene:

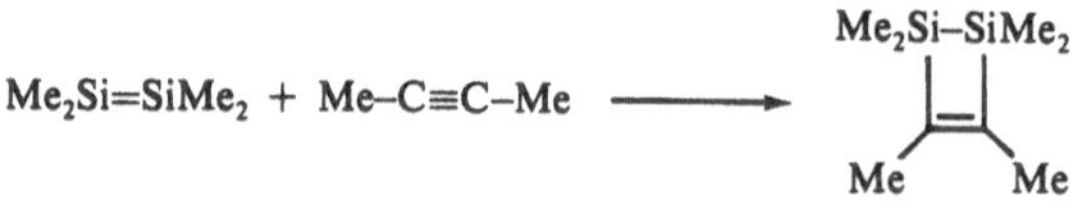

$$Me_2Si{=}SiMe_2 + Me{-}C{\equiv}C{-}Me \longrightarrow \overset{Me_2Si{-}SiMe_2}{\underset{Me \qquad Me}{\boxed{}}} \qquad (76)$$

The observation by Conlin and Gaspar[199] that the dimerization prevails even in the presence of a tenfold excess of propyne indicates that the reaction shown in equation (73) is extremely efficient. However, the structure of the dimer has not been firmly established: it may contain a silicon–silicon double bond as shown in equation (73); it also could be a diradical similar to the dimeric $(SiF_2)_2$.[3,8,198]

c. Other Silylenes

Very recently, Sakurai and coworkers[203] studied the dimerization of various silylenes to give disilene intermediates. They generated silylenes, SiR^1R^2, from either 7-silanorbornadienes or methoxypolysilanes where the R's may stand for Me, Ph, $SiMe_3$, or $SiMe_2OMe$. They trapped the dimerization products, the disilene intermediates, in the form of $R^1R^2Si = SiR^1R^2$, with either anthracene or 2,3-dimethylbutadiene. The most interesting result was the observation that methyl(trimethylsilyl)silylene dimerized stereoselectively to the disilenes, strongly in favor of one isomer which was tentatively assigned the trans geometry.

B. Polymerization

Besides SiF_2, other dihalosilylenes also polymerize very readily as shown in equation (77), where X stands for Cl, Br, and I.[2,6]

$$nSiX_2 \rightarrow (SiX_2)_n \tag{77}$$

2. Insertion Reactions

As represented in equation (78), the insertion of a silylene, SiRR', into an X–Y bond gives SiRR'XY as a product. The insertion of silylenes into single bonds is the most well-established type of reactions for these divalent species. It is the only silylene process which leads *directly* to stable products in a single step. On the other hand, the immediate reaction products of either polymerization or addition reactions of silylenes are supposed to be intermediates, which may either undergo further reactive encounters or require subsequent rearrangements to give the final stable products.

$$\text{SiRR}' + \text{X–Y} \longrightarrow \begin{matrix} \text{R} \\ | \\ \text{X–Si–Y} \\ | \\ \text{R}' \end{matrix} \tag{78}$$

Silylene insertion was first demonstrated in 1964 by Skell and Goldstein,[134,135] who showed that singlet $SiMe_2$ inserts into the Si–H bond of $SiMe_3H$:

$$SiMe_2 + Me_3Si\text{–}H \rightarrow Me_3Si\text{–}SiHMe_2 \tag{79}$$

The possible insertion of SiF_2 and $SiCl_2$ into B–F and B–Cl bonds were also known around 1965.

In Table 6, the insertion reactions of various silylenes are summarized. The reactions are classified mainly according to the type of silylenes involved. However, in the following sections, the insertion reactions are discussed according to the types of bonds undergoing insertion. As recognized by Atwell and Weyenberg,[6] the chemical bonds undergoing insertion usually contain either a hydrogen atom or one of the more electronegative elements such as a halogen or an oxygen atom.

In 1964, Skell and Goldstein[134,135] noted that $SiMe_2$ did not insert into the

TABLE 6.   The Insertion Reactions of Silylenes

Silylenes	Substrates	Bond types	References
SiH$_2$	SiH$_4$	Si–H	103
	Si$_2$H$_6$	Si–H	81,84,103,204
	Si$_3$H$_8$	Si–H	78,84,103
	SiH$_3$Me	Si–H	78,84,103,204
	SiH$_2$Me$_2$	Si–H	103,204
	SiHMe$_3$	Si–H	103,204
	SiH$_3$Et	Si–H	93
	SiH$_3$Cl	Si–H	103
	Si$_2$H$_5$Me	Si–H	78
	GeH$_3$Me	Ge–H	103
	GeH$_2$Me$_2$	Ge–H	205
	GeHMe$_3$	Ge–H	205
	CH$_3$PH$_2$	P–H	103
SiD$_2$	Si$_2$H$_6$	Si–H	206
	SiH$_3$Me	Si–H	82,83
^{31}SiH$_2$	SiH$_4$	Si–H	19,21,25
	Si$_2$H$_6$	Si–H	25
SiHF	SiH$_3$Et	Si–H	93
SiF$_2$	HF	H–F	168
	H$_2$O	O–H	171
	CH$_3$OH	O–H	168
	PH$_3$	P–H	207
	H$_2$S	S–H	172
	GeH$_4$	Ge–H	184
	BF$_3$	B–F	173,174
	BCl$_3$	B–Cl	174
	CH$_2$CHF	C–F	188
	CH$_2$CF$_2$	C–F	188
	CHFCF$_2$	C–F	188
	CH$_2$CFMe	C–F	188
	C$_6$H$_5$F	C–F	194
	C$_6$F$_6$	C–F	194
	SiMe$_3$OMe	Si–O	181
	Br$_2$	Br–Br	12
	I$_2$	I–I	185
SiHCl	SiH$_3$Et	Si–H	93
SiCl$_2$	HCl	H–Cl	47
	BCl$_3$	B–Cl	208
	CCl$_4$	C–Cl	208
	PCl$_3$	P–Cl	208
	Cl(SiCl$_2$)$_{n+1}$Cl	Si–Cl	87
SiBr$_2$	BF$_3$	B–F	209
SiHMe	SiH$_4$	Si–H	78
	SiH$_3$Me	Si–H	64
	SiH$_3$SiH$_2$Me	Si–H	78

(continued)

TABLE 6.—cont.

Silylenes	Substrates	Bond types	References
SiMe$_2$	CH$_3$OH	O–H	13
	SiHMe$_3$	Si–H	95,134,135
	SiH$_2$Et$_2$	Si–H	118
	SiHMeEt$_2$	Si–H	104,111
	SiHEt$_3$	Si–H	118
	iso-PrSiHMe$_2$	Si–H	118
	SiMe$_3$OMe	Si–O	210
	SiMe$_2$(OMe)$_2$	Si–O	113
	(SiMe$_2$OMe)$_2$	Si–O	100,199,210
	Hexamethylcyclotrisiloxane	Si–O	211
	1,1,3-3-Tetramethyl-2-oxo-1,3-disilacyclopentane	Si–O	212
 :Si⬠	SiHMeEt$_2$	Si–H	104
SiMePh	1,1,3-3-Tetramethyl-2-oxo-1,3-disilacyclopentane	Si–O	212
SiHSiH$_3$	SiH$_4$	Si–H	78
	Si$_3$H$_8$	Si–H	78
	Si$_2$D$_6$	Si–D	206
SiHSiMe$_3$	SiHMe$_3$	Si–H	39
SiHOMe	MeOH	O–H	213
SiMeOMe	MeOH	O–H	6
	SiMe(OMe)$_2$SiMe(OMe)$_2$	Si–O	100
Si(OMe)$_2$	MeOH	O–H	6

C–H bonds of ethane and trimethylsilane, which indicated that dimethyl-silylene was much less reactive than singlet CH$_2$. Subsequent studies confirmed that, in general, silylenes do not insert into the following types of bonds: (i) C–H, (ii) Si–C, and (iii) C–C sigma bonds.

A. Si–H Bonds

The Si–H bond insertion reactions have been employed very effectively as a means of trapping various kinds of silylenes. (Only SiF$_2$ and SiCl$_2$ do not insert into Si–H bonds.) The most popular trapping agents are silane, methylsilane, ethylsilane, and trimethylsilane.

a. SiH$_2$ Insertion into Si–H Bonds

The insertion of SiH$_2$ into Si–H bonds of SiH$_4$ to give Si$_2$H$_6$ [see equation (80)] was first employed in 1966 by Gaspar and coworkers[19] to account for the formation of ^{31}SiSiH$_6$ in recoil ^{31}Si reactions with silane. Simultaneously, Purnell and Walsh[56] invoked the existence of this same process in their monosilane pyrolysis system. In 1968, Strausz and coworkers[64] presented evidence of

$$SiH_2 + SiH_3-H \rightarrow SiH_3SiH_3 \tag{80}$$

SiHMe insertion into Si–H bonds during the vacuum uv photolysis of CH$_3$SiH$_3$. From the study of disilane pyrolysis, Tebben and Ring[81] established the insertion of SiH$_2$ into Si$_2$H$_6$ to give Si$_3$H$_8$, which was immediately confirmed by Bowrey and Purnell in 1970[83]:

$$SiH_2 + SiH_3SiH_2-H \rightarrow SiH_3SiH_2SiH_3 \tag{81}$$

The thermal decomposition of Si$_2$D$_6$ in the presence of CH$_3$SiH$_3$ produced SiD$_4$ and CH$_3$SiH$_2$SiD$_2$H, which is a direct mechanistic proof for both the formation of SiD$_2$ from Si$_2$D$_6$ and the subsequent insertion of SiD$_2$ into Si–H bonds.[83]

In an earlier work by Estacio, Sefcik, Chan, and Ring,[82] qualitative results from the SiH$_2$ insertions into SiH$_2$Me$_2$ and SiH$_3$Cl suggested that silylene acts as an electrophile during insertions into Si–H bonds, and that substituents do significantly affect the insertion rates into the remaining Si–H bonds. Later, Sefcik and Ring[103] studied the relative rates of SiH$_2$ insertions into several methylsilanes and methylgermane in competition with Si$_2$H$_6$ in a recirculating flow system. On a per bond basis, the following relative insertion rates were observed: SiHMe$_3$ > Si$_2$H$_6$ > SiH$_2$Me$_2$ > SiH$_3$Me > SiH$_4$ > GeH$_3$Me. No insertion products were found with CH$_3$PH$_2$ or SiH$_3$Cl in competition with disilane. The authors stated that the data could be best correlated with the negative charge on the hydrogen under attack, i.e., its hydridic character.

b. Relative Insertion Rates of Silylenes

The relative insertion rates of various silylenes, formed by the thermal decomposition of halosilanes, into the Si–H bonds of SiH$_3$Et were also examined by Ring and coworkers[93]:

$$SiXY + C_2H_5SiH_2-H \rightarrow C_2H_5SiH_2SiXYH \tag{82}$$

From the observed product ratios, coupled with reasonable assumptions of the half-lives of the reacting silylenes, they assessed the following order for silylene insertion rates: SiH$_2$ > SiHCl > SiHF $\gg$ SiCl$_2$, SiF$_2$. In fact, for the last two, which are the least reactive among the silylenes studied, no insertion of SiF$_2$ into the Si–H bonds of SiH$_2$Me$_2$ or SiH$_3$Cl, or of SiCl$_2$ into the Si–H bonds of SiH$_3$Et was detected. This fact is consistent with the previous observation by Margrave and coworkers[175] that SiF$_2$ is unreactive in the gas phase with a "half-life" of 150 sec.

On the more quantitative side, Bowrey and Purnell[84] established that the formation of Si_3H_8 from the reaction shown in equation (81) is first order with respect to Si_2H_6, and that the Arrhenius parameters for trisilane formation are: $\log A = 14.31 \pm 0.42 \text{ sec}^{-1}$, and $E_a = 205.6 \text{ kJ mol}^{-1}$. Such first-order rate constant values must be associated with the initial SiH_2 formation step from the thermal decomposition of Si_2H_6, instead of the SiH_2 insertion reaction in equation (81). Since no temperature dependence of insertion reaction rates could be detected experimentally, they concluded that only small or zero activation energies are involved. John and Purnell[157] later actually estimated the activation energy for SiH_2 insertion into SiH_4 to be $5.4 \pm 4.6 \text{ kJ mol}^{-1}$. Very recently, Vanderwielen, Ring, and O'Neal[78] evaluated the A factors for the insertion into equation (81) to be $10^{11.02} M^{-1} \text{ sec}^{-1}$, and the A factors for the following three insertion reactions as:

$$SiH_2 + SiH_3Me \rightarrow SiH_3SiH_2Me, \qquad A_{83} = 10^{10.74} M^{-1} \text{ sec}^{-1} \tag{83}$$

$$SiHMe + SiH_4 \rightarrow SiH_2MeSiH_3, \qquad A_{84} = 10^{10.08} M^{-1} \text{ sec}^{-1} \tag{84}$$

$$SiHSiH_3 + SiH_4 \rightarrow SiH_3SiH_2SiH_3, \qquad A_{85} = 10^{10.87} M^{-1} \text{ sec}^{-1} \tag{85}$$

These values indicate that the A factors for SiH_2 insertions are close to collision frequencies, while those for larger silylenes are still better than 0.1 times the collision frequencies. The various quantitative values quoted in this paragraph indicate that on both energy and entropy grounds, silylene insertion reactions such as those shown in equations (80)–(85) are extremely efficient processes.

Bowrey and Purnell[84] also evaluated the relative rate constants on a per bond basis for SiH_2 insertion into four different kinds of Si–H bonds as: SiH_3SiH_2–H, 1.0; CH_3SiH_2–H,1.4; $Si_2H_5SiH_2$–H, 2.5; and $(SiH_3)_2SiH$–H, 1.5. Two special features can be noted here. One is that the Si–H bond in methylsilane is more reactive than that of disilane, which is just the opposite of what was observed by Sefcik and Ring.[103] As pointed out by the latter authors, their results obtained with a flow system should be more dependable than Bowrey and Purnell's values which were obtained with a static system in which the primary insertion products are much more susceptible to secondary decomposition. The other feature of Bowrey and Purnell's values is that primary Si–H bonds are more susceptible to insertion than secondary Si–H bonds in Si_3H_8. This is again in contrast with some recent values of Vanderwielen, Ring, and O'Neal[78] which indicated that the insertion of silylenes such as SiH_2, SiHMe, and $SiHSiH_3$ into primary and secondary Si–H bonds in either $CH_3Si_2H_5$ or Si_3H_8 was close to statistical.

c. Insertion of Si_2H_4

A final interesting note about Si–H insertion is concerning that of Si_2H_4 formed via the thermal decomposition of Si_3H_8 as shown in equation (86). This species could exist as $SiHSiH_3$, $\dot{S}iH_2-\dot{S}iH_2$, or $SiH_2=SiH_2$. Although Sefcik and Ring[103] argued against the participation in reaction of the diradical form and favored the supposition that $SiHSiH_3$ is the sole form which inserts into Si_3H_8

$$Si_3H_8 \rightarrow SiH_3SiH + SiH_4 \qquad (86)$$

to produce the pentasilanes, they recognized that part of the $SiHSiH_3$ may isomerize to the very unstable (to polymerization) molecular species, $SiH_2=SiH_2$:

$$SiH-SiH_3 \rightarrow SiH_2=SiH_2 \qquad (87)$$

The clue is that the percentage of silylenes resulting in insertion products is about 70% for SiH_2 and SiHMe, but is only about 40% for Si_2H_4. Presumably a certain fraction of Si_2H_4 is converted from the insertive $SiHSiH_3$ form to the $SiH_2=SiH_2$ form which undergoes polymerization instead of being readily "trapped" by insertion reactions.

B. Si–Si Bonds

The possible insertion of silylenes into Si–Si bonds is still a question of debate with evidence for and against its existence. In 1970, Bowrey and Purnell[83] observed that for SiH_2 insertion into Si_3H_8, the yield ratio, n-tetrasilane/isotetrasilane, lies between the limits 4.4 and 5.2, significantly above the statistical value of 3. This result suggested that some insertion into the Si–Si bond, which can only give n-tetrasilane as shown in equation (88), was

$$SiH_2 + SiH_3SiH_2-SiH_3 \rightarrow SiH_3SiH_2SiH_2SiH_3 \qquad (88)$$

occurring. They cautioned, however, that such a proposal can only be taken as speculative. This suggestion was immediately echoed by Gaspar and Markusch[25] based on recoil ^{31}Si studies. A remarkably high ratio of $^{31}SiSi_2H_8$ to $^{31}SiH_3SiH_3$ (in the range of 4.9 ± 1.9) was obtained from $^{31}SiH_2$ reactions with Si_2H_6 and SiH_4 mixtures, which is far above the statistical ratio of 1.5. They concluded that either $^{31}SiH_2$ inserts into the Si–Si bond of Si_2H_6 to account for the additional amount of $^{31}SiSi_2H_8$ as products, or the Si–H bonds in disilane are much more susceptible to insertion than those in silane. In 1971, Bowrey and Purnell[84] reversed the position of their original proposal and concluded that SiH_2 insertion into Si–Si bonds does not occur. They stated that in order to reconcile the quantitative data obtained in their static system with the relative rate constants for SiH_2 insertion into various Si–H bonds, one had to explain them from the viewpoint that SiH_2 inserted into different Si–H bonds with different efficiencies. However, in 1973, Sefcik and Ring[103] used their flow pyrolysis results to support the operation of Si–Si bond insertion not only by SiH_2 but also by $SiHSiH_3$. Their deductions were based on the relative ratios of *n*- to *iso*-tetrasilane and *n*- to *iso*-pentasilane produced from the pyrolysis of Si_3H_8 system. Nevertheless, Ring and coworkers[78] also reversed their position in 1975 and argued that Si–Si insertion is inconsistent with their newly obtained data on the thermal decomposition of methyldisilane and trisilane. The present view seems to be in agreement that Si–Si insertion by SiH_2 does not take place.

As discussed in Section VI.2.C below, the insertion of SiMeOMe into the Si–Si bonds of alkoxy-substituted polysilanes was also not observed.[6,210] In 1972, Ishikawa, Takaoka, and Kumada[112] attempted the insertion of $SiMe_2$ into the Si–Si bonds of 1,2-diethyltetramethyldisilane. No 1,3-diethylhexamethyl-trisilane was observed, denying the insertion of $SiMe_2$ into Si–Si bonds:

$$SiMe_2 + EtMe_2Si-SiMe_2Et \nrightarrow EtMe_2SiSi(Me_2)SiMe_2Et \qquad (89)$$

Even in the presence of such negative evidence, the insertion of silylenes into Si–Si bonds is not necessarily an absolutely impossible reaction. The "breeding" modes of producing silylenes from trisilanes and polysilanes as observed by Sakurai and coworkers[104] and by Ishikawa, Kumada, and coworkers[105–112] are actually the reverse of such Si–Si bond insertion reactions, and they point to the possibility of a transition state as shown in equation (50).

C. Si–O Bonds

In 1966, Atwell and Weyenberg[99] studied the thermolysis of methoxy-substituted polysilanes to give silylenes such as $SiMe_2$ and observed products as shown in equation (90) which could be attributed to the insertion of the silylene

$$MeO(SiMe_2)_n\,OMe + SiMe_2 \rightarrow MeO(SiMe_2)_{n+1}\,OMe \qquad (90)$$

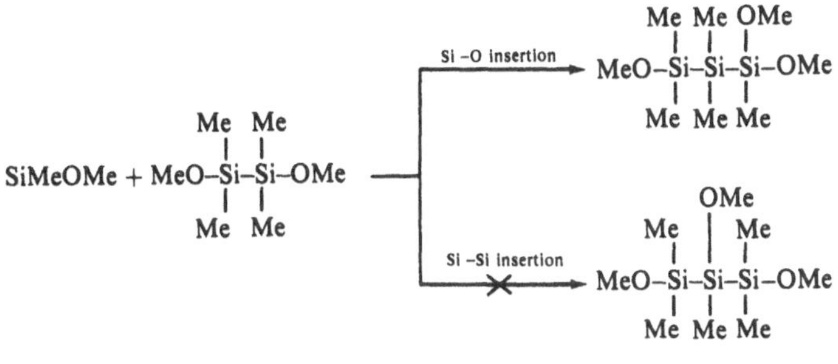

into either the Si–O or Si–Si bonds. Later, Atwell and coworkers[6,210] unambiguously demonstrated that insertion indeed occurs at the Si–O bonds of the alkoxysilanes. They showed that the reaction of methoxysilylene with *sym*-dimethoxytetramethyldisilane gives only products corresponding to Si–O insertion as shown in equation (91a) instead of Si–Si insertion as shown in equation (91b). They also demonstrated that silylene insertions into Si–O bonds of a monosilane such as trimethylmethoxysilane as shown in equation (92) are substantially less efficient than those in polysilanes.

$$SiMe_2 + SiMe_3OMe \rightarrow SiMe_3SiMe_2OMe \qquad (92)$$

In 1977, Soysa, Okinoshima, and Weber[211] demonstrated the insertion of $SiMe_2$ into the Si–O bonds of hexamethylcyclotrisiloxane. They also showed that $SiMe_2$ was capable of deoxygenating $(CH_3)_2S-O$ to give $Me_2Si=O$ as a reaction intermediate.

Very recently, Okinoshima and Weber[212] discovered that (i) SiMePh does not insert into the Si–O bonds of hexamethylcyclotrisiloxane, but into that of a 1,3-disilacyclopentane substrate, (ii) SiMePh is less reactive than $SiMe_2$, and (iii) the 1,3-disilacyclopentane substrate is a more reactive $SiMe_2$ trapping reagent than is hexamethylcyclotrisiloxane.

D. Halogen-Containing Bonds

Besides the ones discussed in the previous sections, the only other silylene insertion reaction involving silicon-containing bonds is the reaction of $SiCl_2$ with halo-substituted polysilanes as shown in equation (93).[87]

$$Cl(SiCl_2)_n Cl + SiCl_2 \rightarrow Cl(SiCl_2)_{n+1} Cl \qquad (n \geqslant 2) \qquad (93)$$

All the reported insertion reactions into halogen-containing bonds inevitably involve silylenes themselves bearing halogen substituents. As shown in Table 6, dihalosilylenes such as SiF_2, $SiCl_2$, and $SiBr_2$ may all insert into B–F and/or B–Cl bonds in BF_3 and BCl_3,[173,174,208] while $SiCl_2$ may also show an apparent insertion into P–Cl bonds of PCl_3.[208] Another feasible reaction mode for SiF_2 is its insertion into Br_2 or I_2 to give SiF_2Br_2 or SiF_2I_2.[12,185]

The apparent insertions of SiF_2 into the C–F bonds of fluoroethylenes and fluorobenzenes[194] as indicated in both Tables 5 and 6 were believed not to be direct insertion processes. Interactions of SiF_2 with the double bonds may be the primary processes followed by rearrangements to give the products finally observed.[10,11]

It is also believed that SiF_2 may insert into HF to give $SiHF_3$,[168] and $SiCl_2$ may insert into HCl to give $SiHCl_3$.[47]

E. Other Hydrogen-Containing Bonds

Both the ability of silylenes to insert into Si–H bonds and their inability to insert into C–H bonds have been described in the previous sections. Several other types of hydrogen-containing bonds besides Si–H are also accessible for silylene insertion reactions.

Both SiF_2 and SiH_2 are capable of inserting into the S–H bonds of H_2S,[172] the P–H bonds of PH_3 or CH_3PH_2,[103,207] and the Ge–H bonds of GeH_4 or methyl-substituted germanes.[103,184,205] Sefcik and Ring[103] have shown that SiH_2 insertion into Ge–H bonds is a less efficient process than its insertion into Si–H bonds of a corresponding compound.

Various kinds of silylenes including SiF_2, $SiMe_2$, $Si(OMe)_2$, and $SiMeOMe$ are noted for their ability to insert into the O–H bond of CH_3OH, as illustrated in equation (94) using dimethylsilylene as an example.

$$SiMe_2 + CH_3OH \rightarrow CH_3O\overset{\displaystyle Me}{\underset{\displaystyle Me}{\overset{\displaystyle |}{\underset{\displaystyle |}{Si}}}}H \tag{94}$$

Among the insertion reactions, the Si–H, Si–O, and O–H insertions are the three which have been experimentally employed as means to trap silylenes formed by various kinds of methods.

3. Addition Reactions

Insertion and addition are definitely two of the most fundamental reaction modes for both carbenes and silylenes. While the most convenient method for trapping carbenes involves their addition to double bonds to give stable cyclopropane derivatives, the corresponding method for trapping silylenes is generally complicated by the fact that most of the silacyclopropanes are unstable species and therefore undergo various secondary processes. In practice, with the exception of alkynes and conjugated dienes, addition reactions are rarely employed to intercept silylenes.

In Table 7, the studies on silylene addition to π-bond systems are summarized. They are classified primarily according to the types of silylenes. The addition reactions of monomeric SiF_2 formed by the high-temperature Si–SiF_4 reactions are omitted in this table because they have already been summarized in Table 5. In the following sections, the addition reactions are discussed according to the type of molecules the silylenes are adding to, namely, olefins, dienes, alkynes, and aromatic compounds.

A. Olefins

a. SiH_2

There is no report on a definite product pattern of SiH_2 addition to olefins. However, Gaspar and coworkers[22] have studied recoil [31]Si reactions in a

Table 7. Summary of Silylene Addition Reactions to π-Bond Systems

Silylenes	Substrates	Final products	References
SiH_2	1,3-Butadiene	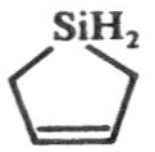 **6** (SiH₂ cyclopentene)	13,214
	Cyclopentadiene	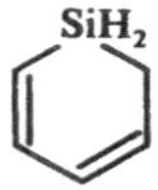 SiH₂ bicyclic	215
	trans-2-*trans*-4-hexadiene	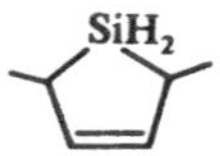 SiH₂ ring (*cis and trans* in equal amounts)	216
$^{31}SiH_2$	CH≡CH	$SiH_3C{\equiv}CH$	55,217
	$CH_2{=}CH_2$ (in the presence of PH_3)	Probably $^{31}SiH_3CH_2CH_2PH_2$	22
	1,3-Butadiene	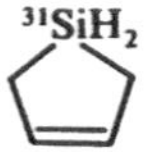 $^{31}SiH_2$ **7**	28-30
	trans- and *cis*-pentadienes	2-Methyl derivative of **7**	36
	Isoprene	3-Methyl derivative of **7**	36
SiF_2	1,3-Butadiene	Difluoro derivative of **6**	13
$^{31}SiF_2$	1,3-Butadiene	Difluoro derivative of **7**	27,31,33
	trans- and *cis*- pentadienes	2-Methyl derivative of **7**	34,36
$^{31}SiF_2$	Isoprene	3-Methyl derivative of **7**	34,36
$SiHCl$	1,3-Butadiene	Chloro derivative of **6**	214

$SiCl_2$		
$CH_3CH=CH_2$	(structure: Cl_2Si ... $SiCl_2$) (and higher homologs)	11
1,3-Butadiene	Dichloro derivative of **6**	13,218
Isoprene	3-Methyl derivative of **6**	218
Chloroprene	Analogous to above	218
Piperylene	Analogous to above	218
Cyclopentadiene	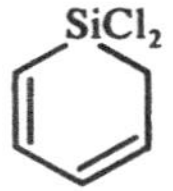 (structure: $SiCl_2$)	215,218
Furan	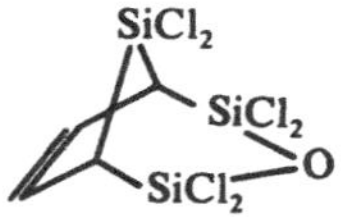 (structure with $SiCl_2$, $SiCl_2$, O)	218
CH≡CH	 (structure: Cl_2Si ... $SiCl_2$)	11,14
PhC≡CPh	Tetraphenyl derivative of above	14,218
C_6H_6	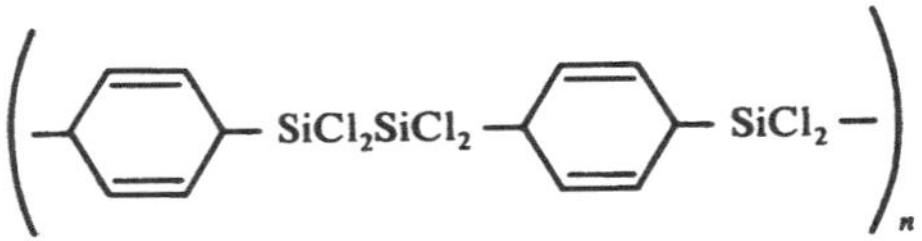 (polymer structure: $SiCl_2SiCl_2$ ... $SiCl_2$—)$_n$	4

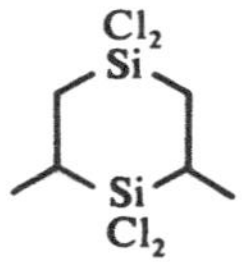

(continued)

TABLE 7.—cont.

Silylenes	Substrates	Final products	References
$SiCl_2$	1,2-Dimethylbenzene	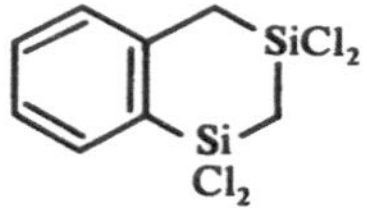 $SiCl_2$ / $Si\ Cl_2$ structure	219
	1-Methylnapthalene	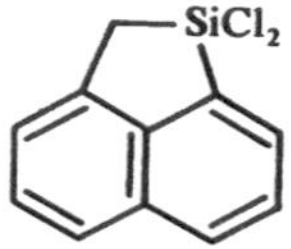 $SiCl_2$ structure	219
$SiClR$ (R = Cl, Me, or Ph)	1-Methylnaphthalene	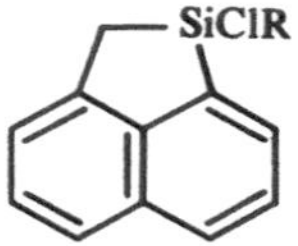 $SiClR$ structure	220
$SiHMe$ $SiMe_2$	1,3-Butadiene $CH_2=CH_2$ (gas phase) $CH_2=CH_2$ (condensed phase)	Methyl derivative of **6** $CH_2=CHSiHMe_2$ $CH_2=CHSiHMe_2$ $SiMe_2$; $SiMe_2$ / $SiMe_2$ structures $[(SiMe_2)_m-(CH_2CH_2)_n]_x$	214 134,135 114,221
	$CH_2=CMe_2$	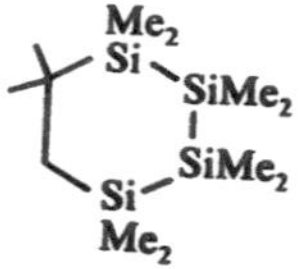 Me_2Si / $SiMe_2$ / $SiMe_2$ / $Si\ Me_2$ structure	114

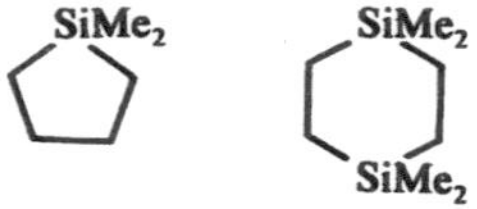

(continued)

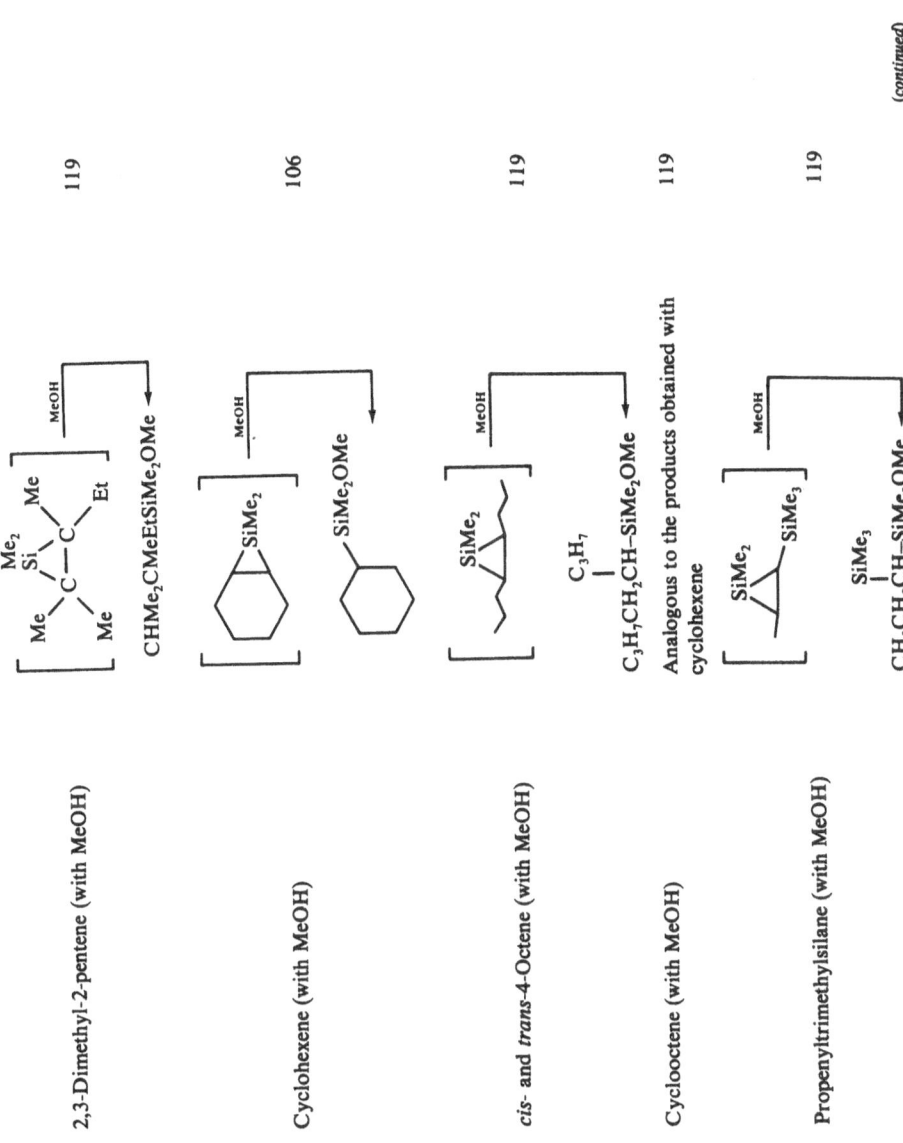

2,3-Dimethyl-2-pentene (with MeOH)   119

Cyclohexene (with MeOH)   106

cis- and *trans*-4-Octene (with MeOH)   119

Cyclooctene (with MeOH)   119

Propenyltrimethylsilane (with MeOH)   119

TABLE 7.—cont.

Silylenes	Substrates	Final products	References
SiMe₂	Stilbenes	Substituted 1,1-dimethylcyclopentanes	222,223
	1,3-Butadiene	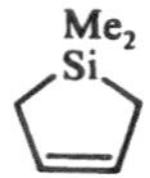 [Me₂Si cyclopentene ring structure]	5,13
	2,3-Dimethyl-1,3-butadiene Isoprene, piperylene, styrene	3,4-Dimethyl derivative of above Analogous to above	100 221
	Cyclopentadiene	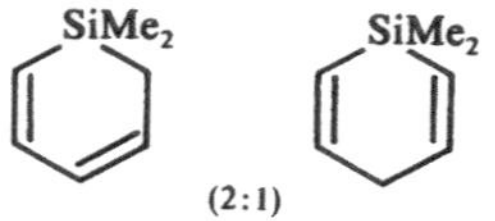 [SiMe₂ bicyclic structures] (2:1)	215
	1,3-Cyclohexadiene	 CH₂=CHSiMe₂CH=CHCH=CH₂ *and* [SiMe₂ bicyclic structure]	215
	CH≡CCH₃	Dimethylpropynylsilane	224
	2-Butyne (see CR₁≡CR₂ below)	 [SiMe₂ cyclopropene structure]	225
	2-Butyne (see CR₁≡CR₂ below)	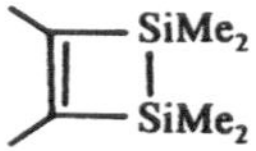 [SiMe₂–SiMe₂ disilacyclobutene structure]	201

SiHSiH₃	$R_1-C\equiv C-R_2$	$R_1 = R_2 = Me$ $R_1 = R_2 = Et$ $R_1 = Me;$ $R_2 = Ph$ $R_1 = R_2 = Ph$	202 6 5 100,113,120
	$R_1R_2C=O$	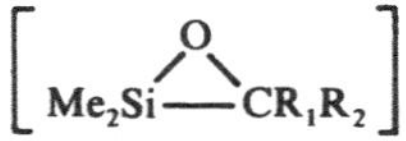 $\left[Me_2Si\!\!\overset{O}{\triangle}\!\!CR_1R_2 \right]$ as intermediates (a) $R_1 = R_2 = Et$ (b) $R_1 = Me;$ $R_2 = Ph$ (c) $R_1 = R_2 = Ph$	226
	1,3-Butadiene	 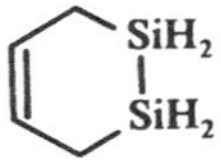	214
SiMePh	Cyclohexene		106
	2,3-Dimethyl-1,3-butadiene	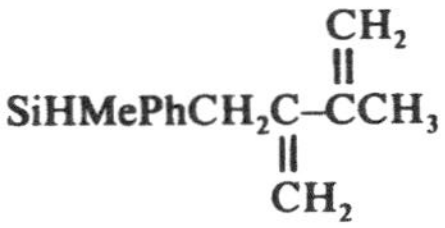 and 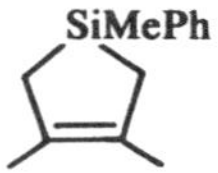	15,107

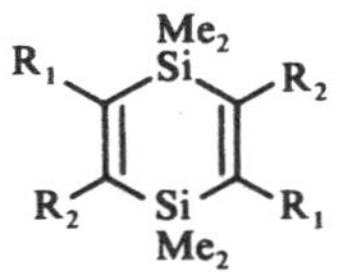

(continued)

TABLE 7.—cont.

Silylenes	Substrates	Final products	References
SiPh₂	PhC≡CPh	Analogous to 8	121
SiMeOMe	2,3-Dimethyl-1,3-butadiene	Analogous to 7	5
	1,3-Cyclooctadiene	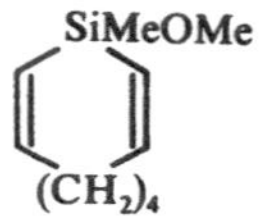 SiMeOMe (CH₂)₄	227
	1,3-Cyclododecadiene	Analogous to above	227
	2,5-Dimethylfuran	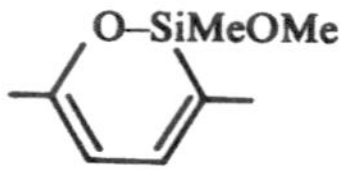 O–SiMeOMe and SiMeOMe 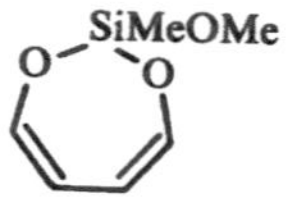	228
SiMeOMe	CH≡CH	Analogous to 8	5,6
	2-Butyne	Analogous to 8	6
	PhC≡CPh	Analogous to 8	100
Si(OMe)₂	1,3-Butadiene	Analogous to 7	201
	2,3-Dimethyl-1,3-butadiene	Analogous to 7	6

PH_3–C_2H_4 mixture and have observed a product which was tentatively iden-tified as $^{31}SiH_3CH_2CH_2PH_2$. A trace amount of $^{31}SiH_3CH_2CH_3$ was also observed. They proposed a mechanism as shown in equations (95) and (96) which involved the addition of $^{31}SiH_2$ to the double bond followed by the interaction of the intermediate silacyclopropane with a PH_3 molecule.

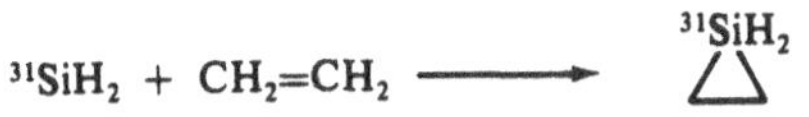

$$^{31}SiH_2 + CH_2{=}CH_2 \longrightarrow \overset{^{31}SiH_2}{\triangle} \qquad (95)$$

$$PH_3 + \overset{^{31}SiH_2}{\triangle} \longrightarrow {}^{31}SiH_3CH_2CH_2PH_2 \qquad (96)$$

b. SiF$_2$

As indicated in Table 5, the reactions of SiF_2 (formed by the high-temperature Si–SiF$_4$ interaction) with olefins such as C_2F_4 give mainly solid polymeric products.[187] The minor amount of simple products observed always contain two units of SiF_2 characterizing the interaction of dimeric $(SiF_2)_2$ instead of that of the monomeric species.[3,8]

The reactions of SiF_2 with fluorinated olefins such as $CH_2{=}CHF$, $CH_2{=}CF_2$, and $CHF{=}CF_2$ give products which apparently correspond to those expected of C–F insertion reactions. However, it is more likely that the vinylsilane derivatives obtained are actually formed via the addition of SiF_2 to the double bond followed by isomerization of the three-membered ring intermediate[187,188]:

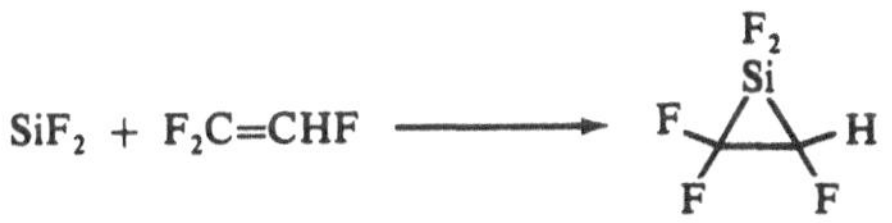

(97)

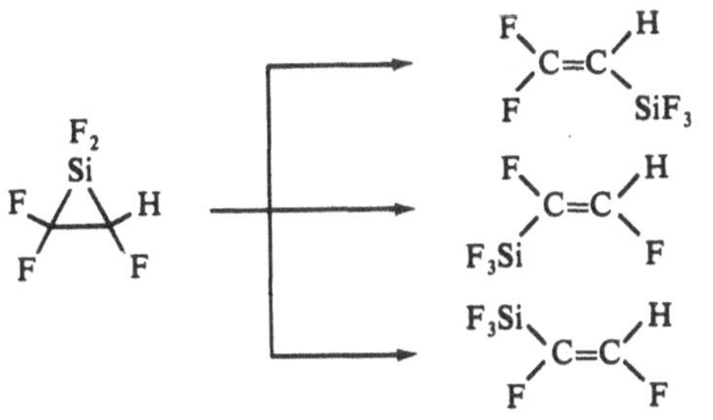

(98)

In the cyclohexane system, although the final product contains two units of SiF_2, monomeric interactions are likely to be involved followed by dimerization of the silacyclopropane intermediates[193]:

$$SiF_2 + \quad\text{[cyclohexene]} \quad \longrightarrow \quad \text{[bicyclic}\ SiF_2\text{]} \qquad (99)$$

$$2\ \text{[ring-}SiF_2\text{]} \quad \longrightarrow \quad \text{[dimer structure]} \qquad (100)$$

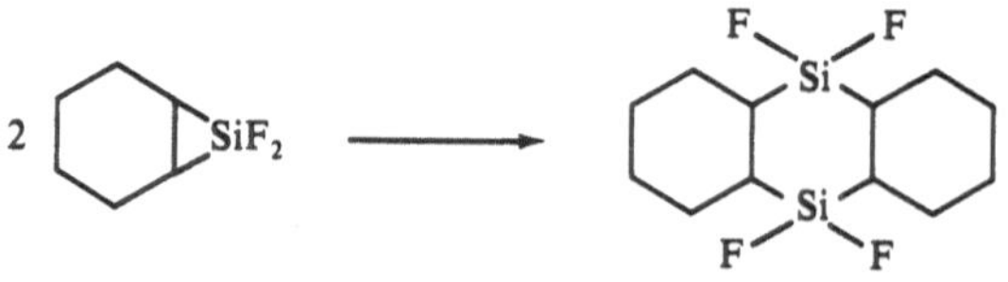

c. SiMe$_2$

Instead of 1,1-dimethyl-1-silacyclopropane, Skell and Goldstein[134,135] obtained dimethylvinylsilane as the final product from the vapor phase reactions of SiMe$_2$ with C$_2$H$_4$. They believed that the reaction proceeded as an addition–isomerization sequence rather than by a direct insertion into C–H bonds. Nefedov and coworkers[114,221] obtained products other than dimethylvinylsilane in the condensed phase; these presumably arise from secondary reactions of the silacyclopropane intermediate. As shown in the scheme in equation (101) which is a modified version of the one given by Atwell and Weyenberg,[6] the silacyclopropane intermediate may dimerize to form a heterocycle or it may undergo further reactions with C$_2$H$_4$ and/or SiMe$_2$ to form 1,1-dimethyl-1-silacyclopentane and a copolymer.

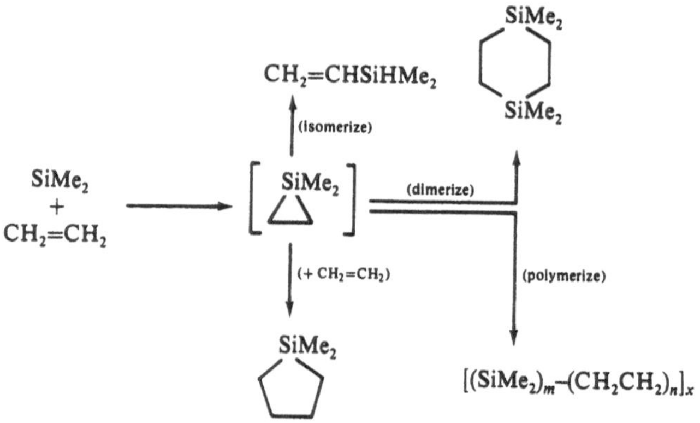

$$
\begin{array}{c}
\text{SiMe}_2 \\
+ \\
\text{CH}_2=\text{CH}_2
\end{array}
\longrightarrow
\left[
\begin{array}{c}
\text{SiMe}_2 \\
\triangle
\end{array}
\right]
\qquad (101)
$$

CH$_2$=CHSiHMe$_2$ (isomerize), (dimerize) → ring (SiMe$_2$/SiMe$_2$), (+ CH$_2$=CH$_2$), (polymerize), SiMe$_2$ cyclopentane, $[(\text{SiMe}_2)_m\text{–}(\text{CH}_2\text{CH}_2)_n]_x$

B. Conjugated Dienes

a. 1,3-Butadiene

Silylenes add to 1,3-butadiene to give silacyclopent-3-enes as shown in equation (102). This reaction was initially discovered by Atwell and

Weyenberg,[5,13,201] but most of the recent studies were carried out mainly by Tang and coworkers[27,28,30,31,33] and by Gaspar and coworkers[29] using hot [31]Si atoms formed by the nuclear recoil method, and by Ring and coworkers[214] with the recirculating flow pyrolysis method.

$$\text{SiXY} + \quad \longrightarrow \quad \overset{\text{SiXY}}{\bigcirc} \qquad (102)$$

$$(\text{SiXY} = \text{SiH}_2, \ ^{31}\text{SiH}_2, \ \text{SiF}_2, \ ^{31}\text{SiF}_2, \ \text{SiHCl},$$

$$\text{SiCl}_2, \ \text{SiHMe}, \ \text{SiMe}_2, \ \text{SiHSiH}_3, \ \text{or} \ \text{Si(OMe)}_2)$$

As mentioned in Section V.1.B, the $^{31}\text{SiH}_2$ and $^{31}\text{SiF}_2$ derived in a nuclear recoil system have both been shown to be present as about 80% triplet and 20% singlet.[30,33] For singlet silylenes, the apparent 1,4 addition was generally believed to proceed via an initial 1,2 addition step followed by isomerization to the final product. The isomerization is likely to be fast and therefore unaffected by the presence of a scavenger.

$$\text{SiXY} + \quad \longrightarrow \quad \overset{\text{SiXY}}{\triangle\!\!\!\nearrow} \qquad (103)$$

$$\overset{\text{SiXY}}{\triangle\!\!\!\nearrow} \quad \longrightarrow \quad \overset{\text{SiXY}}{\bigcirc} \qquad (104)$$

In the case of triplet silylenes, a real 1,4-addition to butadiene through a biradical intermediate is possible.

Ring and coworkers[214] have demonstrated that the insertion of SiH_2 into Si_2H_6 is about four times faster than the addition of SiH_2 to 1,3-butadiene. The rate of addition of SiHCl and SiHMe to the conjugated diene is comparable to that of SiH_2. From the reaction of 1,3-butadiene with Si_2H_4 generated from the pyrolysis of Si_3H_8, 1-silylsila-3-cyclopentene and 1,2-disila-4-cyclohexene were obtained in nearly equal amounts. These two products could be formed via the addition of SiHSiH_3 and $\text{SiH}_2{=}\text{SiH}_2$ to butadiene, respectively, as shown in equations (105) and (106), but their formation could also be explained by a mechanism involving SiHSiH_3 as the only reactant followed by ring opening and hydrogen shift to account for both products as shown by the scheme in equations (107) and (108).

b. Pentadienes

Tang and coworkers[34,36] have extended the studies of the addition of $^{31}\text{SiH}_2$ and $^{31}\text{SiF}_2$ to the pentadienes. For each of these two silylenes, the relative reac-

$$\text{SiHSiH}_3 + \text{/}\diagdown\text{/} \longrightarrow \underset{\text{SiHSiH}_3}{\bigcirc} \qquad (105)$$

$$\text{SiH}_2{=}\text{SiH}_2 + \text{/}\diagdown\text{/} \longrightarrow \underset{\text{SiH}_2{-}\text{SiH}_2}{\bigcirc} \qquad (106)$$

$$\text{SiHSiH}_3 + \text{/}\diagdown\text{/} \longrightarrow \underset{\text{SiHSiH}_3}{\triangle\diagdown\text{/}} \longrightarrow \underset{\text{SiHSiH}_3}{\bigcirc} \qquad (107)$$

$$\underset{\text{SiHSiH}_3}{\bigcirc} \longrightarrow \underset{\text{SiHSiH}_3}{\bigcirc} \longrightarrow \underset{\text{SiHSiH}_3}{\bigcirc} \qquad (108)$$

$$\underset{\text{SiH}_2{-}\text{SiH}_2}{\bigcirc} \longrightarrow \underset{\text{SiH}_2{-}\text{SiH}_2}{\bigcirc}$$

tivities of 1,3-butadienes, *trans*-pentadiene, *cis*-pentadiene, and 2-methyl-1,3-butadiene have been measured. Minor steric hindrance had to be postulated in most cases to account for the lower reactivities (about 10% lower) of the pentadienes. However, a large steric effect has been observed in *trans*- and *cis*-pentadienes as indicated by their reactivities toward the triplet $^{31}\text{SiF}_2$ species. Quantitatively, the reactivity ratio of *cis*-pentadiene to *trans*-pentadiene toward triplet $^{31}\text{SiF}_2$ is only 0.67.

c. Hexadienes

The addition of SiMe_2 and SiMeOMe to 2,3-dimethylbutadiene studied by Weyenberg and Atwell[5,100] was actually the first reported silylene addition to a conjugated diene system. In the vapor phase, the yields of silacyclopentenes were generally high and over 60%. Experimental support for the mechanism of such addition to include an initial step of 1,2-addition followed by isomerization was recently provided by Ishikawa, Ohi, and Kumada,[107] who studied the reaction as shown in equation (109) with SiMePh. They succeeded in intercepting the silacyclopropane intermediate with CH_3OH and obtained 2,3-dimethyl-4-(methylphenylmethoxysilyl)-2-butene (9), as an identifiable product.

Further mechanistic evidence for the diene addition reactions was obtained by Gaspar and Hwang[216] using *trans*-2-*trans*-4-hexadiene. The addition of SiH_2

$$\text{SiX}_2 + \diagup\hspace{-0.3em}\diagdown\hspace{-0.3em}\diagup \longrightarrow \underset{\text{SiX}_2}{\bigcirc} \qquad (\text{SiX}_2 = \text{SiMe}_2, \text{SiMeOMe, or SiMePh}) \qquad (109)$$

to this conjugated compound gave two stereoisomers in equal amounts. This means that silylene addition to dienes does not occur by a concerted 1,4-cyclo-addition process.

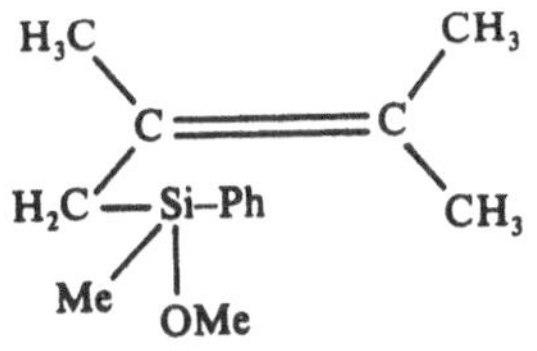

9

We have recently studied the addition of $^{31}SiH_2$ and $^{31}SiF_2$ to *trans*-2-*trans*-4-hexadiene and its *cis–cis* and *cis–trans* isomers.[229] A steric effect which decreases the reactivities of these hexadienes in comparison with that of butadiene is again observed. The relative reactivities of *trans*-2-*trans*-4-hexadiene toward silylenes in comparison with those of butadiene are around 0.6, while the corresponding values for *cis*-2-*cis*-4-hexadiene are only around 0.3.

d. Cyclic Dienes

In 1972, Chernyshev and coworkers[218] found that $SiCl_2$ reacts with cyclo-pentadiene to yield 1,1-dichloro-1-silacyclohexa-2,4-diene, a conjugated product, as shown in equation (110). On the other hand, Childs and Weber[227] studied the reactions of SiMeOMe with 1,3-cyclooctadiene and obtained an unconjugated product as shown in equation (111). Very recently, Gaspar and coworkers[215] studied the additions of SiH_2, $SiCl_2$, and $SiMe_2$ to cyclopenta-diene and the addition of $SiMe_2$ to 1,3-cyclohexadiene. They confirmed that both SiH_2 and $SiCl_2$ reacted according to equation (110) to give the conjugated product, but that $SiMe_2$ gave both the conjugated and unconjugated products as shown in equation (112) in a two-to-one ratio. In the case of $SiMe_2$ reacting

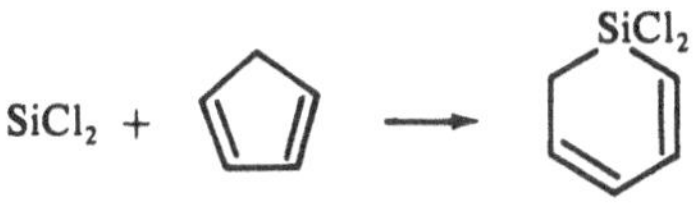

(110)

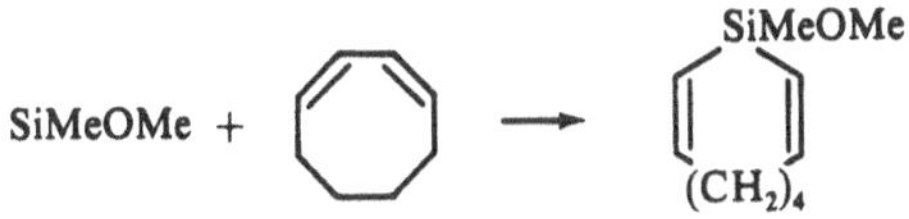

(111)

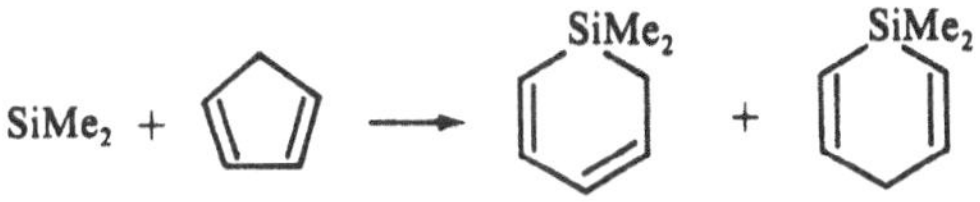

(112)

with cyclohexadiene, an open-chain compound, 3,3-dimethyl-3-sila-1,4,6-hepta-triene, and a bicyclic compound, 7,7-dimethyl-7-silanorbornene, were observed as shown in Table 7. All these products could be explained by a mechanism involving 1,2-addition of silylenes to give bicyclic vinylsilacyclopropane derivatives which underwent nonconcerted rearrangements.

C. Alkynes

a. SiH_2

The reaction of SiH_2 with acetylene was first studied by White and Rochow,[55] and later by Haas and Ring.[217] The latter investigators observed the formation of $SiH_3C{\equiv}CH$ as the major silylene product from the pyrolysis of a Si_2H_6–C_2H_2 system. Although this product could be accounted for by direct insertion of SiH_2 into a C–H bond, it is more likely to be formed via a π-bond addition process. As shown in equations (113) and (114), SiH_2 first reacts with

$$SiH_2 + CH{\equiv}CH \longrightarrow \underset{\triangle}{\overset{SiH_2}{\triangle}} \tag{113}$$

$$\underset{\triangle}{\overset{SiH_2}{\triangle}} \longrightarrow SiH_3C{\equiv}CH \tag{114}$$

C_2H_2 to form silacyclopropene, which subsequently isomerizes with H-atom migration to give the observed ethynylsilane. The work described above was actually designed to give information about the thermal decomposition of silane (see Section IV.2.A). In the corresponding SiH_4–C_2H_2 pyrolysis system, $SiH_3C{\equiv}CH$ was not formed at near-zero reaction time. Since the observed products could be accounted for by reactions of silyl radicals, these results suggest that the primary decomposition of SiH_4 is to give SiH_3 via Si–H bond scission as shown in equation (25), instead of giving SiH_2 as shown in equation (22). The authors have also shown that the apparent absence of SiH_2 from the thermolysis of SiH_4 is consistent with what would be expected from orbital symmetry considerations.[217]

b. SiF_2

As summarized in Table 5, reactions of SiF_2 with alkynes such as $CH{\equiv}CH$, $CH_3C{\equiv}CH$, and $CF_3C{\equiv}CH$ have been carried out.[190,191] A major type of product observed in these systems is tetrafluorodisilacyclobutene, which could be formed in reactions as shown in equation (115). Other linear, cyclic, or bicyclic products are also observed. It is obvious that SiF_2 dimerizes prior to interactions with alkynes when formed in high-temperature Si–SiF_4 systems.

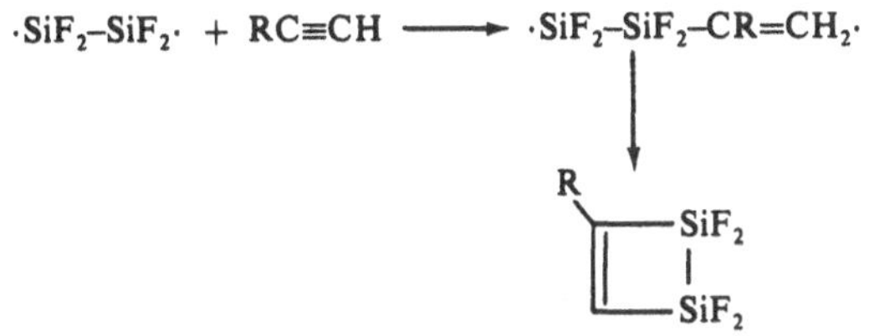

$$\cdot SiF_2-SiF_2\cdot \ + \ RC\equiv CH \ \longrightarrow \ \cdot SiF_2-SiF_2-CR=CH_2\cdot \qquad (115)$$

c. The Formation of 1,4-Disilacyclohexadiene

The study of the interaction of silylenes with alkynes to give 1,4-disila-cyclohexadiene is probably the most interesting topic in silylene chemistry. It started in 1962 when Vol'pin and coworkers[113] claimed to have obtained a substituted silacyclopropene from the reaction of $SiMe_2$ with $PhC\equiv CPh$. Soon after this report appeared, a variety of methods[230-232] were employed to show that the very stable product obtained by them was actually a substituted 1,4-disilacyclohexadiene. The net reaction, as shown in equation (116), has been subsequently found to operate with various kinds of silylenes and alkynes. These include (i) $SiMe_2$ adding to C_2Me_2,[202] C_2Et_2,[6] and C_2Ph_2[100,113,120]; (ii) $SiMeOMe$ adding to C_2H_2,[6] C_2Me_2,[6] and C_2Ph_2[100]; (iii) $SiPh_2$ adding to C_2Ph_2[121]; (iv) SiF_2 adding to $CF_3C\equiv CH$[191]; and (v) $SiCl_2$ adding to C_2H_2.[92]

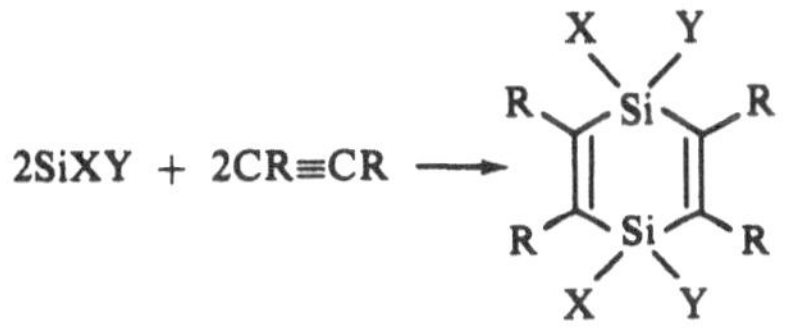

$$2SiXY + 2CR\equiv CR \ \longrightarrow \qquad\qquad (116)$$

The formation of 1,4-disilacyclohexadiene from the addition of silylene to alkynes could be explained by two separate mechanisms as indicated in the scheme in equation (117). Mechanism *a* involves the addition of SiXY to the acetylene to give a substituted silacyclopropene, which subsequently dimerizes to give the disilacyclohexadiene compound. Mechanism *b* first calls for the

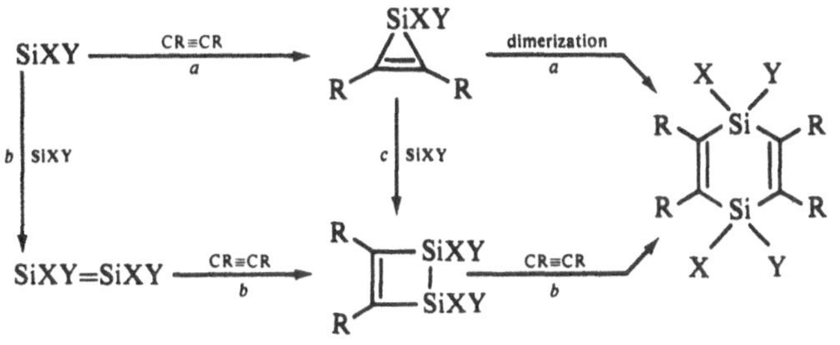

$$(117)$$

dimerization of SiXY to give a dimer which adds to two acetylene molecules in two consecutive steps to first give a disilacyclobutene and then the final product. Mechanism *c* is a semicombination of the above two mechanisms with the addition of SiXY to silacyclopropene to give disilacyclobutene. The experimental pros and cons of these three mechanisms are summarized below.

For mechanism *a*, there is no doubt that silylenes could add to alkynes to give silacyclopropenes. What is at issue is whether or not the dimerization of the silacyclopropene takes place. The evidence is as follows. (i) Stable silacyclopropenes have been recently prepared by Conlin and Gaspar[225] from the reactions of $SiMe_2$ with 2-butyne, and by Seyferth and coworkers[233] from the reactions of $SiMe_2$ with bis(trimethylsilyl)acetylene, and by Sakurai and coworkers[234] from the photolysis of (pentamethyldisilanyl) phenylacetylene:

$$SiMe_2 + RC{\equiv}CR \longrightarrow \underset{R \quad\ R}{\overset{\overset{\displaystyle Me_2}{\underset{\displaystyle Si}{}}}{\triangle}} \quad (R = Me\ or\ SiMe_3) \tag{118}$$

(ii) Atwell and coworkers first proposed that the possible dimerization of silacyclopropene might proceed via π-bond coupling, and then discarded this idea after discovering that no mixed disilacyclohexadiene of the type shown in equation (119) was obtained when $SiMe_2$ was generated in the presence of both C_2Ph_2 and 2-butyne. The possibility of silacyclopropene cleaving at the C–Si bond and then dimerizing was also excluded, since even in the presence of a large excess of C_2Ph_2, no silacyclopentadiene was formed. (iii) Recently, Atwell and Uhlmann[201] reported that a disilacyclobutene was detected as a product in the reactions of $SiMe_2$ with C_2Me_2, as shown in equation (121). They suggested that silacyclopropene may interact with another unit of $SiMe_2$ to give disilacyclobutene.

$$2SiMe_2 + CPh{\equiv}CPh + CMe{\equiv}CMe \ \nrightarrow\ \underset{Ph\quad\ \ \underset{Me_2}{Si}\quad\ Ph}{\overset{Me\quad\ \underset{Si}{Me_2}\quad\ Me}{\left|\ \ \ \ \right|}} \tag{119}$$

$$SiMe_2 + CPh{\equiv}CPh\ (large\ excess)\ \nrightarrow\ \underset{Ph\qquad\qquad Ph}{\overset{Ph\quad\ \underset{Si}{Me_2}\quad\ Ph}{}} \tag{120}$$

$$2SiMe_2 + CMe{\equiv}CMe \longrightarrow \underset{Me}{\overset{Me}{}} \begin{array}{c} SiMe_2 \\ | \\ SiMe_2 \end{array} \tag{121}$$

On the other hand, the possible operation of mechanism *b* seems to be on firmer ground. (i) The dimerization of SiF_2 is well known[8] and the dimerization of $SiMe_2$ has just been established by Gaspar and coworkers.[199] (ii) The fact that disilacyclobutenes were actually detected in alkyne systems[201] means that reactions shown in equations (115) and (121) probably take place. (iii) Barton and Kilgour[202] recently showed that disilylene formed from an alternate bicyclic precursor adds to 2-butyne to give the same type of 1,4-disilacyclohexadiene compound. (iv) Barton and Kilgour[202] also studied the pyrolysis of disilacyclobutene in the presence of excess 2-butyne, which afforded disilacyclohexadiene in 31% yield. More importantly, copyrolysis in the presence of two different alkynes gave only the unmixed isomer as product as shown in equation (122). This product is identical with that obtained from silylene reactions.

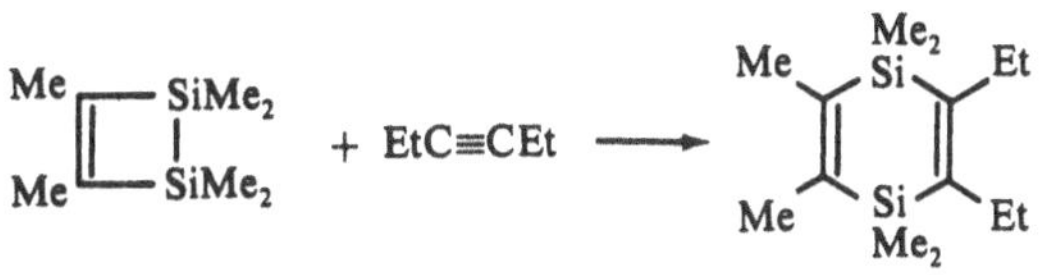

$$(122)$$

Overall, for the addition of silylenes to alkynes to give 1,4-disilacyclohexadienes, the operation of mechanism *b* seems to be firmly established. The complete operation of mechanism *a* is probably in doubt, but the conversion of silacyclopropene to disilacyclobutene (patch *c*) is still possible. However, since the primary steps of both mechanisms, the addition of silylene to π bonds and the dimerization of silylenes, are both well established, the relative importance of the two mechanisms should be kinetically controlled by the concentration of the silylenes. At some lower concentration of the reacting silylenes, the silacyclopropene route may actually prevail.

D. Aromatic Compounds

No interactions of silylenes with aromatic rings are reported except for those of SiF_2 shown in Table 5 and those of $SiCl_2$ and SiClR as shown in Table 7. For the reactions of SiF_2 with C_6H_6, it is the polymeric species which add across the ring at the 1,4 positions.[194] For C_6H_5F and C_6F_6, the apparent

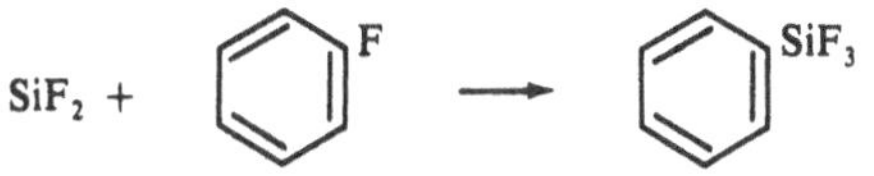

$$(123)$$

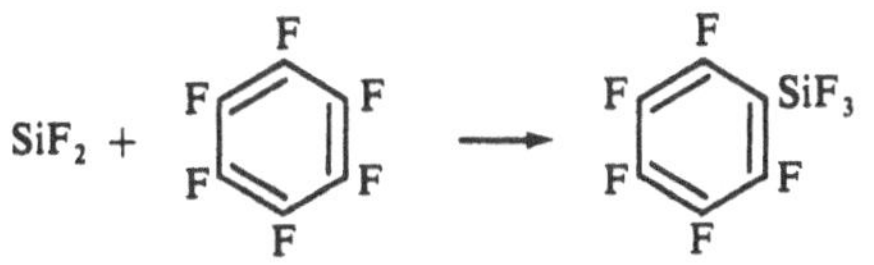

$$(124)$$

SiF_2 insertions into C–F bonds, as shown in equations (123) and (124), are likely to involve an initial attack on the aromatic π-bond systems followed by isomerization.[194]

E. Ketones

Very recently, Ando, Ikeno, and Sekiguchi[226] established that $SiMe_2$ may add to the carbonyl groups of ketones to produce oxasilacyclopropane as intermediates:

$$SiMe_2 \; + \; {>}C{=}O \; \longrightarrow \; \left[\begin{array}{c} Me \\ {>}Si{\triangle}C{<} \\ Me \end{array} \right] \tag{125}$$

Thermal cleavage of the Si–C ring bond in this intermediate gives a 1,3-diradical which may undergo either intramolecular H abstraction or addition to π bonds of the aromatic rings.

ACKNOWLEDGMENTS

This work is partially supported by U.S. Department of Energy, Contract No. EY-76-05-3898. The kind help provided by E.E. Siefert, E.B.M. Siefert, E.-C. Wu, B.J. Ezzell, and E.C.G. Tang in the preparation of this article is also appreciated.

REFERENCES

1. P. L. Timms, R. A. Kent, T. C. Ehlert, and J. L. Margrave, *Nature* 207, 187 (1965).
2. O. M. Nefedov and M. N. Manakov, *Angew. Chem. Int. Ed. Engl.* 5, 1021 (1966).
3. J. C. Thompson and J. L. Margrave, *Science* 155, 669 (1967).
4. P. L. Timms, in *Preparatve Inorganic Reactions*, Vol. 4, W. L. Jolly, Ed., Interscience, New York (1968), pp. 59-83.
5. D. R. Weyenberg and W. H. Atwell, *Pure Appl. Chem.* 19, 343 (1969).
6. W. H. Atwell and D. R. Weyenberg, *Angew. Chem. Int. Ed. Engl.* 8, 469 (1969).
7. C. H. Yoder and J. J. Zuckerman, in *Preparative Inorganic Reactions*, Vol. 6, W. L. Jolly, Ed., Wiley-Interscience, New York (1971), pp. 81-153.
8. J. L. Margrave and P. W. Wilson, *Account. Chem. Res.* 4, 145 (1971).
9. I. M. T. Davidson, *Chem. Soc. Q. Rev.* 25, 111 (1971).
10. P. P. Gaspar and B. J. Herold, Silicon, germanium, and tin structural analogs of carbenes, in *Carbene Chemistry*, 2nd ed., W. Kirmse, Ed., Academic Press, New York (1971), pp. 504–550.
11. P. L. Timms, Low temperature condensation of high temperature species as a synthesis method, in *Adv. Inorg. Chem. Radiochem.* 14, 121 (1972).
12. J. L. Margrave, K. G. Sharp, and P. W. Wilson, *Top. Curr. Chem.* 26, 1 (1972).

13. W. H. Atwell and D. R. Weyenberg, *Intra-Sci. Chem. Rep.* **7**, 139 (1973).
14. P. L. Timms, *24th Int. Congr. Pure Appl. Chem.* (1973), *Plenary Main Sect. Lect.* **4**, 25 (1974).
15. Y. Nakadaira, *Kagaku No. Ryoiki* **29**, 188 (1975); *Chem, Abstr.* **83**, 146617s (1975).
16. P. P. Gaspar, in *Reactive Intermediates*, Vol. 1, M. Jones, Jr., and R. A. Moss, Ed., John Wiley and Sons, New York (1978), pp. 229–237.
17. L. Pauling, *The Nature of the Chemical Bond and the Structure of Molecules and Crystals*, 3rd ed., Cornell University Press, Ithaca, New York (1959).
18. E. A. V. Ebsworth, *Volatile Silicon Compounds*, Pergamon Press, New York (1963).
19. P. P. Gaspar, B. D. Pate, and W. C. Eckelman, *J. Am. Chem. Soc.* **88**, 3878 (1966).
20. G. Cetini, O. Gambino, M. Castiglioni, and P. Volpe, *Atti R. Accad. Sci. Torino Cl. Sci. Fis. Mat. Nat.* **101**, 749 (1966–1967); *Chem. Abstr.,* **68**, 55657d (1968).
21. P. P. Gaspar, S. A. Bock, and C. A. Levy, *Chem. Commun.,* 1317 (1968).
22. P. P. Gaspar, S. A. Bock, and W. C. Eckelman, *J. Am. Chem. Soc.* **90**, 6914 (1968).
23. D. K. Snediker and W. W. Miller, *Radiochimica Acta* **10**, 30 (1968).
24. G. Cetini, M. Castiglioni, P. Volpe, and O. Gambino, *Ric. Sci.* **39**, 392 (1969); *Chem. Abstr.* **73**, 20515g (1970).
25. P. P. Gaspar and P. Markusch, *Chem. Commun.,* 1331 (1970).
26. P. P. Gaspar, P. Markusch, J. D. Holton, III, and J. J. Frost, *J. Phys. Chem.* **76**, 1352 (1972).
27. Y.-N. Tang, G. P. Gennaro, and Y. Y. Su, *J. Am. Chem. Soc.* **94**, 4355 (1972).
28. G. P. Gennaro, Y. Y. Su, O. F. Zeck, S. H. Daniel, and Y.-N. Tang, *J. Chem. Soc. Chem. Commun.,* 637 (1973).
29. P. P. Gaspar, R.-J. Hwang, and W. C. Eckelman, *J. Chem. Soc. Chem. Commun.,* 242 (1974).
30. O. F. Zeck, Y. Y. Su, G. P. Gennaro, and Y.-N. Tang, *J. Am. Chem. Soc.* **96**, 5967 (1974).
31. O. F. Zeck, Y. Y. Su, and Y.-N. Tang, *J. Chem. Soc. Chem. Commun.,* 156 (1975).
32. K. Kawamoto, *Ann. Rep. Res. React. Inst. Kyoto Univ.* **8**, 26 (1975).
33. O. F. Zeck, Y. Y. Su, G. P. Gennaro, and Y.-N. Tang, *J. Am. Chem. Soc.* **98**, 3474 (1976).
34. R. A. Ferrieri, E. E. Siefert, M. J. Griffin, O. F. Zeck, and Y.-N. Tang, *J. Chem. Soc. Chem. Commun.,* 6 (1977),
35. R.-J. Hwang and P. P. Gaspar, *J. Am. Chem. Soc.* **100**, 6626 (1978).
36. E. E. Siefert, R. A. Ferrieri, O. F. Zeck, and Y.-N. Tang, *Inorg. Chem.* **17**, 2802 (1978).
37. H. S. Massey and E. H. Burhop, *Electronic and Ionic Impact Phenomena*, Clarendon Press, Oxford (1952), p. 441.
38. J. A. Grundl, R. L. Henkel, and B. L. Perkins, *Phys. Rev.* **109**, 425 (1958).
39. P. S. Skell and P. W. Owen, *J. Am. Chem. Soc.* **89**, 3933 (1967).
40. P. S. Skell and P. W. Owen, *J. Am. Chem. Soc.* **94**, 5434 (1972).
41. R. E. Honig, *J. Chem. Phys.* **22**, 1610 (1954).
42. E. E. Siefert, K.-L. Loh, R. A. Ferrieri, and Y.-N. Tang, unpublished results (1979).
43. T. C. Ehlert and J. L. Margrave, *J. Chem. Phys.* **41**, 1066 (1964).
44. A. S. Kana'an and J. L. Margrave, *Inorg. Chem.* **3**, 1037 (1964).
45. W. I. Zubkov, M. V. Tichomirova, K. A. Andrianov, and S. A. Golubzov, *Dokl. Akad. Nauk SSSR* **159**, 599 (1964); *Chem. Abstr.* **62**, 4760d (1965).
46. S. A. Golubzov and K. A. Andrianov, *Dokl. Akad. Nauk SSSR* **151**, 1329 (1963).
47. J. Jolik and V. Bozant, *Collect. Czech. Chem. Commun.* **29**, 603, 834 (1964).
48. L. Brewer and R. K. Edwards, *J. Phys. Chem.* **58**, 351 (1954).
49. O. H. Wheeler and J. E. Trabal, *Rev. Latinoamer. Quim.* **1**, 61 (1970); *Chem. Abstr.* **74**, 93607u (1971).
50. R. J. Madix and J. A. Schwarz, *Surf. Sci.* **24**, 264 (1971).
51. Ogier, *Ann. Chim.* **20**, 37 (1880).

52. T. R. Hogness, T. L. Wilson, and W. C. Johnson, *J. Am. Chem. Soc.* **58**, 108 (1936).
53. R. Schwartz and F. Heinrich, *Z. Anorg. Allg. Chem.* **221**, 277 (1935).
54. K. Stokland, *Kgl. Nor. Vidensk. Selsk. Skr.* **3**, 1 (1950).
55. D. G. White and E. G. Rochow, *J. Am. Chem. Soc.* **76**, 3897 (1954).
56. J. H. Purnell and R. Walsh, *Proc. R. Soc. London* **293**, 543 (1966).
57. M. A. Ring, M. J. Puentes, and H. E. O'Neal, *J. Am. Chem. Soc.* **92**, 4845 (1970).
58. C. G. Newman, M. A. Ring, and H. E. O'Neal, *J. Am. Chem. Soc.* **100**, 5945 (1978).
59. H. J. Emeleus and K. Stewart, *Trans. Faraday Soc.* **32**, 1577 (1936).
60. H. Niki and G. J. Mains, *J. Phys. Chem.* **68**, 304 (1964).
61. M. A. Nay, G. N. C. Woodall, O. P. Strausz, and H. E. Gunning, *J. Am. Chem. Soc.* **87**, 179 (1965).
62. I. Dubois, *Can. J. Phys.* **46**, 2485 (1968).
63. M. A. Ring, G. D. Beverly, F. H. Koester, and R. P. Hollandsworth, *Inorg. Chem.* **8**, 2033 (1969).
64. O. P. Strausz, K. Obi, and W. K. Duholke, *J. Am. Chem. Soc.* **90**, 1359 (1968).
65. K. Obi, A. Clement, H. E. Gunning, and O. P. Strausz, *J. Am. Chem. Soc.* **91**, 1622 (1969).
66. J. J. Kohanek, P. Estacio, and M. A. Ring, *Inorg. Chem.* **8**, 2516 (1969).
67. J. W. C. Johns, A. W. Chantry, and R. F. Barrow, *Trans. Faraday Soc.* **54**, 1580 (1958).
68. D. R. Rao and P. Venkateswarlu, *J. Mol. Spectrosc.* **7**, 287 (1961).
69. K. Wieland and M. Heise, *Angew. Chem.* **63**, 438 (1951).
70. H. Schafer, *Z. Anorg. Allg. Chem.* **274**, 2651 (1963).
71. E. Wolf and C. Herbst, *Z. Chem.* **7**, 34 (1967).
72. D. M. Schmeiser and K. Friederrick, *Angew. Chem.* **76**, 782 (1964).
73. D. E. Milligan and M. E. Jacox, *J. Chem. Phys.* **49**, 4269 (1968).
74. D. E. Milligan and M. E. Jacox, *J. Chem. Phys.* **49**, 1938 (1968).
75. G. Herzberg and R. D. Verma, *Can. J. Phys.* **43**, 395 (1964).
76. E. Sirtl and K. Reusche, *Z. Anorg. Allg. Chem.* **332**, 113 (1964); *Chem. Abstr.* **62**, 6137f (1965).
77. K. G. Calvert and J. N. Pitts, Jr., *Photochemistry*, Wiley, New York (1966), p. 497.
78. A. J. Vanderwielen, M. A. Ring, and H. E. O'Neal, *J. Am. Chem. Soc.* **97**, 993 (1975).
79. H. J. Emeleus and C. Reid, *J. Chem. Soc.*, 1021 (1939).
80. K. Stokland, *Trans. Faraday Soc.* **44**, 545 (1948).
81. E. M. Tebben and M. A. Ring, *Inorg. Chem.* **8**, 1787 (1969).
82. P. Estacio, M. D. Sefcik, E. K. Chan, and M. A. Ring, *Inorg. Chem.* **9**, 1068 (1970).
83. M. Bowrey and J. H. Purnell, *J. Am. Chem. Soc.* **92**, 2594 (1970).
84. M. Bowrey and J. H. Purnell, *Proc. R. Soc. London* **321**, 341 (1971).
85. C. Friedel and A. Ladenburg, *Liebigs Ann. Chem.* **203**, 241 (1880).
86. C. J. Wilkins, *J. Chem. Soc.*, 3409 (1953).
87. N. W. Kohlschütter and H. Mattner, *Z. Anorg. Allg. Chem.* **282**, 169 (1955); *Chem. Abstr.* **50**, 8363d (1956).
88. A. Pflugmacher and I. Rohrmann, *Z. Anorg. Allg. Chem.* **290**, 101 (1957); *Chem. Abstr.* **51**, 7903i (1957).
89. G. Urry and A. Kaczmarczyk, *Angew. Chem.* **72**, 387 (1960).
90. G. Urry and A. Kaczmarczyk, *J. Am. Chem. Soc.* **82**, 751 (1960).
91. M. Schmeisser and K.-P. Ehlers, *Angew. Chem.* **76**, 781 (1964); *Angew. Chem. Int. Ed. Engl.* **3**, 700 (1964).
92. E. A. Chernyshev, N. G. Komalenkova, and S. A. Bashkirova, *Zh. Obshch. Khim.* **41**, 1175 (1971); *Chem. Abstr.* **75**, 76912t (1971).
93. R. L. Jenkins, A. J. Vanderwielen, S. P. Ruis, S. R. Gird, and M. A. Ring, *Inorg. Chem.* **12**, 2968 (1973).
94. R. B. Baird, M. D. Sefcik, and M. A. Ring, *Inorg. Chem.* **10**, 883 (1971).

95. H. Sakurai, A. Hosomi, and M. Kumada, *Chem. Commun.*, 4 (1969).
96. I. M. T. Davidson and I. L. Stephenson, *J. Chem. Soc. A*, 282 (1968).
97. R. E. Saalfield and H. J. Svec, *Inorg. Chem.* 3, 1442 (1964).
98. W. C. Steele and F. G. A. Stone, *J. Am. Chem. Soc.* 84, 3599 (1962).
99. W. H. Atwell and D. R. Weyenberg, *J. Organometal. Chem.* 5, 594 (1966).
100. W. H. Atwell and D. R. Weyenberg, *J. Am. Chem. Soc.* 90, 3438 (1968).
101. J. V. Urenovitch and A. G. MacDiarmid, *J. Am. Chem. Soc.* 85, 3372 (1963).
102. L. E. Elliott, P. Estacio, and M. A. Ring, *Inorg. Chem.* 12, 2193 (1973).
103. M. D. Sefcik and M. A. Ring, *J. Am. Chem. Soc.* 95, 5169 (1973).
104. H. Sakurai, Y. Kobayashi, and Y. Nakadaira, *J. Am. Chem. Soc.* 93, 5272 (1971).
105. M. Ishikawa, M. Ishiguro, and M. Kumada, *J. Organometal. Chem.* 49, C71 (1973).
106. M. Ishikawa and M. Kumada, *J. Organometal. Chem.* 81, C3 (1974).
107. M. Ishikawa, F. Ohi, and M. Kumada, *J. Organometal. Chem.* 86, C23 (1975).
108. M. Ishikawa and M. Kumada, *Chem. Commun.*, 612 (1970).
109. M. Ishikawa and M. Kumada, *J. Chem. Soc. Chem. Commun.*, 507 (1971).
110. M. Ishikawa and M. Kumada, *J. Chem. Soc. Chem. Commun.*, 489 (1971).
111. M. Ishikawa and M. Kumada, *J. Organometal. Chem.* 42, 325 (1972).
112. M. Ishikawa, T. Takaoka, and M. Kumada, *J. Organometal. Chem.* 42, 333 (1972).
113. M. E. Vol'pin, Yu. D. Koreshkov, V. G. Dulova, and D. N. Kursanov, *Tetrahedron* 18, 107 (1962).
114. O. M. Nefedov and M. N. Manakov, *Angew. Chem.* 76, 270 (1964); *Angew. Chem. Int. Ed. Engl.* 3, 226 (1964).
115. O. M. Nefedov, G. Garzo, T. Szekei, and V. I. Shiryaev, *Proc. Acad. Sci. USSR Chem. Sect. Engl. Trans.* 164, 945 (1965).
116. R. L. Lambert, Jr. and D. Seyferth, *J. Am. Chem. Soc.* 94, 9246 (1972).
117. D. Seyferth and D. C. Annarelli, *J. Am. Chem. Soc.* 97, 325 (1972).
118. D. Seyferth and D. C. Annarelli, *J. Am. Chem. Soc.* 97, 7162 (1975).
119. D. Seyferth and D. C. Annarelli, *J. Organometal. Chem.* 117, C51 (1976).
120. J. H. Gilman, S. G. Cottis, and W. H. Atwell, *J. Am. Chem. Soc.* 86, 1596 (1964).
121. J. H. Gilman, S. G. Cottis, and W. H. Atwell, *J. Am. Chem. Soc.* 86, 5548 (1964).
122. N. K. Hota and C. J. Willis, *J. Organometal. Chem.* 15, 89 (1968).
123. J. R. Maruca, *J. Organometal. Chem.* 36, 1626 (1971).
124. T. J. Barton, A. J. Nelson, and J. Clardy, *J. Organometal. Chem.* 37, 895 (1972).
125. T. J. Maruca, R. Fischer, L. Roseman, and A. Gehring, *J. Organometal. Chem.* 49, 139 (1973).
126. E. Wilberg, O. Stecker, H. J. Andrascheck, Z. Kreuzbichler, and E. Staude, *Angew. Chem.* 75, 516 (1963); *Angew. Chem. Intern. Ed. Eng.* 2, 507 (1963).
127. K. Yamamoto, H. Okinoshima, and M. Kumada, *J. Organometal. Chem.* 23, C7 (1970).
128. K. Yamamoto, H. Okinoshima, and M. Kumada, *J. Organometal. Chem.* 27, C31 (1971).
129. H. Okimoshima, K. Yamamoto, and M. Kumada, *J. Am. Chem. Soc.* 94, 9263 (1972).
130. Y. P. Yurev, I. M. Salimgaruva, O. Zh. Zhebarov, G. A. Tolstikov, and S. R. Rafikov, *Dokl. Akad. Nauk SSSR.* 224, 1092 (1075).
131. E. G. Bylander, *J. Electrochem. Soc.* 109, 1171 (1962).
132. W. Steinmaier, *Philips Res. Rep.* 18, 75 (1963).
133. R. R. Monchamp, W. J. McAleer, and P. I. Pollak, *J. Electrochem. Soc.* 111, 879 (1964).
134. P. S. Skell and E. J. Goldstein, *J. Am. Chem. Soc.* 86, 1442 (1964) (first paper).
135. P. S. Skell and E. J. Goldstein, *J. Am. Chem. Soc.* 86, 1442 (1964) (second paper).
136. W. Kirmse, *Carbene Chemistry*, 2nd ed., Academic Press, New York (1971).
137. P. P. Gaspar and G. S. Hammond, in *Carbenes*, Vol. II, R. A. Moss and M. Jones, Jr., Eds., John Wiley & Sons, New York (1975), p. 207.
138. P. C. Jordan, *J. Chem. Phys.* 44, 3400 (1966).

139. I. Dubois, G. Herzberg, and R. D. Verma, *J. Chem. Phys.* **47**, 4262 (1967).
140. H. P. Hopkins, J. C. Thompson, J. L. Margrave, *J. Am. Chem. Soc.* **90**, 901 (1968).
141. V. M. Rao and R. I. Curl, *J. Chem. Phys.* **45**, 2032 (1966).
142. V. M. Khanna, G. Besenbruch, and J. L. Margrave, *J. Chem. Phys.* **46**, 2310 (1967).
143. V. M. Khanna, R. Hauge, R. F. Curl, Jr., and J. L. Margrave, *J. Chem. Phys.* **47**, 5031 (1967).
144. J. M. Bassler, P. L. Timms, and J. L. Margrave, *Inorg. Chem.* **5**, 729 (1966).
145. C. Leibovici and J. F. Labarre, *J. Chim. Phys. Chim. Biol.* **72**, 951 (1975).
146. B. Wirsam, *Chem. Phys. Lett.* **14**, 214 (1972).
147. J. Higuchi, S. Kubota, T. Kumamoto, and I. Tolue, *Bull. Chem. Soc. Japan* **47**, 2775 (1974).
148. P. Zahradnik and J. Leska, *Collect. Czech. Chem. Commun.* **42**, 2060 (1977).
149. V. M. Rao, R. F. Curle, P. L. Timms, and J. L. Margrave, *J. Chem. Phys.* **47**, 5031 (1967).
150. J. L. Gole, R. H. Hauge, J. L. Margrave, and J. W. Hastie, *J. Mol. Spectrosc.* **43**, 441 (1972).
151. J. W. Hastie, R. H. Hauge, and J. L. Margrave, *J. Am. Chem. Soc.* **91**, 2536 (1969).
152. C. Thomson, *Theor. Chim. Acta.* **32**, 93 (1973).
153. M. W. Chase, J. L. Curnutt, A. H. Hu, H. Prophet, A. H. Syverud, and L. C. Walker, *J. Phys. Chem. Ref. Data* **3**, 408 (1974).
154. G. Maass, R. H. Hauge, and J. L. Margrave, *Z. Anorg. Allg. Chem.* **392**, 295 (1972).
155. J. Billingsley, *Can. J. Phys.* **50**, 531 (1972).
156. V. V. Dudorov, *Zh. Fiz. Khim.* **49**, 1036 (1975) (Eng. Edit. p. 607).
157. P. John and J. H. Purnell, *J. Chem. Soc. Faraday Trans. I* **69**, 1455 (1973).
158. P. John and J. H. Purnell, *J. Organomental. Chem.* **29**, 233 (1971).
159. F. G. Saalfeld and M. V. McDowell, *Inorg. Chem.* **6**, 96 (1967).
160. H. Schaefer, H. Bruderreck, and B. Morcher, *Z. Anorg. Allg. Chem.* **352**, 122 (1967); *Chem. Abstr.* **67**, 57767t (1967).
161. J. L. Margrave, A. S. Kana'an, and D. C. Pease, *J. Phys. Chem.* **66**, 1200 (1962).
162. A. D. Rusin and O. P. Yakovlev, *Vestn. Mosk. Univ. Khim.* **13**, 716 (1972); *Chem. Abstr.* **78**, 76477s (1973); **17**, 170 (1976); *Chem. Abstr.* **85**, 149768u (1976).
163. A. D. Rusin, O. P. Yakovlev, and N. A. Ereshko, *Vestn. Mosk. Univ. Khim.* **15**, 154 (1974); *Chem. Abstr.* **83**, 184421f (1975).
164. E. Wolf and C. Herbst, *Z. Anorg. Allg. Chem.* **347**, 113 (1966); *Chem. Abstr.* **66**, 41211a (1967).
165. R. Teichmann and E. Wolf, *Z. Anorg. Allg. Chem.* **347**, 145 (1966); *Chem. Abstr.* **66**, 41212b (1967).
166. M. Farber and R. D. Srivastava, *J. Chem. Soc. Faraday Trans. I* **73**, 1672 (1977).
167. I. M. T. Davidson and A. V. Howard, *J. Chem. Soc. Faraday Trans. I* **71**, 69 (1975).
168. J. L. Margrave, K. G. Sharp, and P. W. Wilson, *Inorg. Nucl. Chem. Lett.* **5**, 995 (1969).
169. J. F. Bald, K. G. Sharp, and A. G. MacDiarmid, *J. Fluorine Chem.* **3**, 433 (1973).
170. K. G. Sharp and J. F. Bald, Jr., *Inorg. Chem.* **14**, 2553 (1975).
171. J. L. Margrave, K. G. Sharp, and P. W. Wilson, *J. Am. Chem. Soc.* **92**, 1530 (1970).
172. K. G. Sharp and J. L. Margrave, *Inorg. Chem.* **8**, 2655 (1969).
173. R. W. Kirk and P. L. Timms, *J. Am. Chem. Soc.* **91**, 6315 (1969).
174. D. L. Smith, R. W. Kirk, and P. L. Timms, *J. Chem. Soc. Chem. Commun.*, 295 (1972).
175. P. L. Timms, T. C. Ehlert, J. L. Margrave, F. E. Brinckman, T. C. Farrar, and T. D. Coyle, *J. Am. Chem. Soc.* **87**, 3819 (1965).
176. D. Solan and A. B. Burg, *Inorg. Chem.* **11**, 1253 (1972).
177. R. W. Kirk, and P. L. Timms, *J. Am. Chem. Soc.* **90**, 901 (1968).
178. C. Lau and J. C. Thompson, *Inorg. Nucl. Chem. Lett.* **13**, 433 (1977).

179. D. C. Pease, U. S. Pat. 3,026,173 (1962); *Chem. Abstr.* **57**, 3081I (1962).
180. P. L. Timms, R. A. Kent, T. C. Ehlert, and J. L. Margrave, *J. Am. Chem. Soc.* **87**, 2824 (1965).
181. J. L. Margrave, D. L. Williams, and P. W. Wilson, *Inorg. Nucl. Chem. Lett.* **7**, 103 (1971).
182. K. G. Sharp and J. L. Margrave, *J. Inorg. Nucl. Chem.* **33**, 2913 (1971).
183. K. G. Sharp, *Diss. Abstr. Int.* **30**, 2072 (1969).
184. D. Solan and P. L. Timms, *Inorg. Chem.* **7**, 2157 (1968).
185. J. L. Margrave, K. G. Sharp, and P. W. Wilson, *J. Inorg. Nucl. Chem.* **32**, 1813 (1970).
186. J. L. Margrave, K. G. Sharp, and P. W. Wilson, *J. Inorg. Nucl, Chem.* **32**, 1817 (1970).
187. J. C. Thompson, J. L. Margrave, and P. L. Timms, *Chem. Commun.*, 566 (1966).
188. A. Orlando, C.-S. Liu, and J. C. Thompson, *J. Fluorine Chem.* **2**, 103 (1972–1973).
189. C.-S. Liu and J. Thompson, *J. Organometal. Chem.* **38**, 249 (1972).
190. C.-S. Liu, J. L. Margrave, J. C. Thompson, and P. L. Timms, *Can. J. Chem.* **50**, 459 (1972).
191. C.-S. Liu and J. C. Thompson, *Inorg. Chem.* **10**, 1100 (1971).
192. J. C. Thompson and J. L. Margrave, *Inorg. Chem.* **11**, 913 (1972).
193. A. G. MacDiarmid and F. M. Rabel, unpublished results quoted in O. M. Nefedov and M. N. Manakov, *Angew. Chem. Int. Ed. Engl.* **5**, 1021 (1966) and in *Preparative Inorganic Reactions*, Vol. 6, W. L. Jolly, Ed., Wiley-Interscience, New York (1971), pp. 81-153.
194. P. L. Timms, D. D. Stump, R. A. Kent, and J. L. Margrave, *J. Am. Chem. Soc.* **88**, 940 (1966).
195. F. D. Catrett, III, *Diss. Abstr. Int. B.* **33**, 1438 (1972).
196. F. D. Catrett and J. L. Margrave, *Syn. Inorg. Metal-Org. Chem.* **2**, 329 (1972).
197. F. D. Catrett and J. L. Margrave, *J. Inorg. Nucl. Chem.* **35**, 1087 (1973).
198. J. L. Margrave and D. L. Perry, *Inorg. Chem.* **16**, 1820 (1977).
199. R. T. Conlin and P. P. Gaspar, *J. Am. Chem. Soc.* **98**, 868 (1976).
200. D. N. Roark and G. J. D. Peddle, *J. Am. Chem. Soc.* **94**, 5837 (1972).
201. W. H. Atwell and J. G. Uhlmann, *J. Organometal. Chem.* **52**, C21 (1973).
202. T. J. Barton and J. A. Kilgour, *J. Am. Chem. Soc.* **96**, 7150 (1974).
203. Y. Nadadaira, T. Kobayashi, T. Oruska, and H. Sakurai, *J. Am. Chem. Soc.* **101**, 486 (1979).
204. B. Cox and H. Purnell, *Trans. Faraday Soc.* **71**, 859 (1975).
205. M. D. Sefcik and M. A. Ring, *J. Organometal. Chem.* **59**, 167 (1973).
206. J. A. Morrison and M. A. Ring, *Inorg. Chem.* **6**, 100 (1967).
207. G. R. Langford, D. C. Moody and J. D. Odom, *Inorg. Chem.* **14**, 134 (1975).
208. P. L. Timms, *Inorg. Chem.* **7**, 387 (1968).
209. P. L. Timms, unpublished results quoted in W. H. Atwell and D. R. Weyenberg, *Angew. Chem. Int. Ed. Engl.* **8**, 469 (1969).
210. W. H. Atwell, L. G. Mahone, S. F. Hayes, and J. G. Uhlmann, *J. Organometal. Chem.* **18**, 69 (1969).
211. H. S. Soysa, H. S. Dilanjan, H. Okinoshima, and W. P. Weber, *J. Organometal. Chem.* **133**, C17 (1977).
212. H. Okinoshima and W. P. Weber, *J. Organometal. Chem.* **150**, C25 (1978).
213. P. W. Owen and P. S. Skell, *Tetrahedron Lett.* **18**, 1807 (1972).
214. R. L. Jenkins, R. A. Kedrowski, L. E. Elliott, D. C. Tappen, D. J. Schlyer, and M. A. Ring, *J. Organometal. Chem.* **86**, 347 (1975).
215. R.-J. Hwang, R. T. Conlin, and P. P. Gaspar, *J. Organometal. Chem.* **94**, C38 (1975).
216. P. P. Gaspar and R.-J. Hwang, *J. Am. Chem. Soc.* **96**, 6198 (1974).
217. C. H. Haas and M. A. Ring, *Inorg. Chem.* **14**, 2253 (1975).
218. E. A. Chernyshev, N. G. Komalenkova, and S. A. Bashkirova, *Dokl. Akad. Nauk SSSR* **205**, 868 (1972); *Chem. Abstr.* **77**, 152265t (1972).

219. E. A. Chernyshev, N. G. Komalenkova, and S. A. Bashkirova, *Nov. Khim. Karbenov Mater. Vses. Soveshch Khim. Karbenov Ikh Analogov.* 1st. 1972 (Pub. 1973); *Chem. Abstr.* **82**, 57793h (1975).

220. E. A. Chernyshev, N. G. Komalenkova, T. A. Klochkova, and T. M. Kuz'mina, *Zh. Obshch. Khim.* **45**, 2223 (1975); *Chem. Abstr.* **84**, 59630t (1976).

221. O. M. Nefedov, M. N. Manakov, and A. D. Petrov, *Plaste Kautschuk* **10**, 721, 736 (1963); *Chem. Abstr.* **60**, 9304g (1964).

222. D. R. Weyenberg, L. H. Toporcer, and E. A. Bey, *J. Org. Chem.* **30**, 4096 (1965).

223. O. M. Nefedov, M. N. Manakov, and A. D. Petrov, *Dokl. Akad. Nauk SSSR* **154**, 395 (1964) (Engl. Trans., p. 76).

224. R. T. Conlin, *Diss. Abstr. Int. B.* **37**, 3955 (1977).

225. R. T. Conlin and P. P. Gaspar, *J. Am. Chem. Soc.* **98**, 3715 (1976).

226. W. Ando, M. Ikeno, and A. Sekiguchi, *J. Am. Chem. Soc.* **99**, 6447 (1977).

227. M. E. Childs and W. P. Weber, *Tetrahedron Lett.*, 4033 (1974).

228. M. E. Childs and W. P. Weber, *J. Org. Chem.* **41**, 1799 (1976).

229. E. E. Siefert, R. A. Ferrieri, and Y.-N. Tang, unpublished results (1979).

230. R. West and R. E. Bailey, *J. Am. Chem. Soc.* **85**, 2871 (1963).

231. F. Johnson, R. S. Gohlke, and W. H. Nasutavicus, *J. Organometal. Chem.* **3**, 233 (1965).

232. N. G. Bokii and Yu. T. Stuchkow, *J. Struct. Chem. USSR* **6**, 543 (1965).

233. D. Seyferth, D. C. Annarelli and S. C. Vick, *J. Am. Chem. Soc.* **98**, 6382 (1976).

234. H. Sakurai, Y. Kamiyama, and Y. Nakadaira, *J. Am. Chem. Soc.* **99**, 3879 (1977).

Five-Membered Hetarynes

Manfred G. Reinecke

I. INTRODUCTION

1. Role and Scope

Concepts as well as matter exhibit the property of inertia. Truly new concepts meet an initial resistance proportional to their novelty, which, if it is overcome, converts to a powerful force directed at extending the idea far beyond its original sphere of applicability and appropriateness. Science is no less subject to this generalization than are the social sciences and humanities but has the obvious advantage of permitting—or rather requiring—the constant testing of its concepts against experimental fact. One of the best illustrations of such intellectual inertia in chemistry is found in the concept of reactive intermediates. Free radicals, carbanions, carbenes, and carbenium ions have all gone through this cycle of postulation, proof, overextension, and retrenchment. The consequences of this process are such modifications as radicaloids, complexed carbanions, carbenoids, and the "nonclassical ion problem." The concept of the aryne intermediate has also followed this path albeit more recently and with less fanfare than some of its relatives. A central role in both the overextension and maturing of the aryne concept is played by five-membered hetarynes. It is the purpose of this chapter to critically examine this role through a detailed review of the postulation, preparation, and properties of five-membered hetarynes.

Several reviews devoted to, or containing information on, hetarynes have been published. The first category, includes an article by Kauffmann in 1965[1] and monographic chapters by den Hertog and van der Plas in 1965[2] and 1969.[3]

Manfred G. Reinecke ● Department of Chemistry, Texas Christian University, Fort Worth, Texas 76129.

The most recent review is by Kauffmann and Wirthwein in 1971[4] but intentionally slights five-membered hetarynes with one heteroatom in favor of their six-membered homologs. With the exception of Hoffmann's classic monograph[5] on arynes in 1967, reviews devoted to general aryne chemistry[6-11] tend to neglect hetarynes. Considering the recent activity in this field it seems appropriate to review five-membered hetarynes at this time both as a summary of past research and as a stimulus to future research in this area. An attempt has been made to cover all pertinent literature and unpublished observations available to the author as of December 1981.[11a]

2. Formulas and Nomenclature

The function of structural formulas and nomenclature is to unambiguously specify and characterize a molecular species without implying more than is justified by experimental fact about the structure and properties of the species. This task is particularly difficult for reactive intermediates and hence necessitates this preliminary discussion.

A. Carbocyclic Arynes

Arynes may be defined as bidentate reactive intermediates formally generated by removing two vicinal hydrogen atoms from a parent aromatic molecule. The resulting species contains two electrons and two orbitals coplanar with the aromatic ring[12] which could conceivably interact in a variety of ways. Using the aryne **1** derived by removal of adjacent hydrogen atoms from benzene as the prototypical example, the orbitals could overlap to form a bond (**1a**), or the electrons could pair in one orbital to generate dipolar species such as **1b** or **1c**, or they could distribute themselves equally between the two orbitals in either a singlet (**1d**) or a triplet (**1e**) state. If the Kékulé, rather than the sextet, notation for the benzene ring is used, then *each* of the above formulas can be represented by two forms, i.e., the "alkyne" (**1f**) and "allene" (**1g**) forms of **1a**. Naturally, all of these structures except **1e** are contributors to the same resonance hybrid, and therefore a molecular orbital picture such as **1h** or **1j** is perhaps a more accurate, if less convenient, representation.

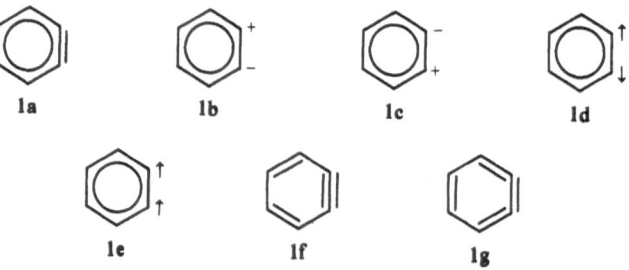

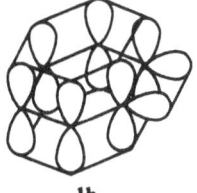

1j

1h

The question of electron distribution and molecular geometry in **1h** or **1j**, or what is equivalent, the relative importance of the valence bond structures **1a–1g** to the true structure of **1**, cannot be answered *a priori* but requires additional theoretical or experimental justification. Furthermore, since a particular aryne will adopt that geometry and electron distribution which will minimize its energy, it will not be possible to select a single type of structural formula to represent the geometry and electron distribution of all arynes. In keeping with conventional notation, however, the "alkyne" formula (**1f**) will be used for all arynes without any implication as to bonding, geometry, charge distribution, or electron multiplicity.

Although not formally arynes those species whose reactive orbitals are on nonadjacent atoms are usually considered in the same category of reactive intermediates.[5] For such species as **2** or **3**, a diradical notation, **2a** or **3a**, will be used rather than the bonded structures, **2b** or **3b**, since the latter have been shown to be distinguishably different entities.[13]

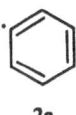

2a   **2b**   **3a**   **3b**

Generic nomenclature for arynes simply replaces the final "ene" of the parent aromatic hydrocarbon with the suffix "yne,"[14] i.e., benzene becomes benzyne, naphthalene becomes naphthalyne, etc. The ambiguity which arises with the possibility of isomeric intermediates can be clarified either with the usual prefixes or preferably by numbers. Hence **1** is called benzyne, *ortho*-benzyne, or 1,2-benzyne, while **2** becomes *meta*- or 1,3-benzyne and **3** *para*- or 1,4-benzyne.

The unwarranted structural implication associated with the suffix "yne" is eliminated by the dehydro system[5,15] of nomenclature in which **1** becomes 1,2-dehydrobenzene, **2**, 1,3-dehydrobenzene, and **3**, 1,4-dehydrobenzene. This seems to be the most popular nomenclature because of its lack of ambiguity and connotation as well as its use in the definitive monograph[5] in the field.

B. Hetarynes

The term *hetarynes* was coined by Kauffmann[16] to designate arynes formally derived from heteroaromatic compounds as defined in Section I.2.A. The generic nomenclature[17] once again uses the suffix "yne", although the possibility of isomeric hetarynes with most heterocyclic molecules makes a numbering system virtually obligatory with this method.[2,11] The popular "dehydro" system[1,3,5] also meets this need and, as in the carbocyclic case, lacks the unwarranted structural implications associated with the "yne" ending. Interestingly, after utilizing both of the above schemes,[2,3] the Dutch group switched[18] to the didehydro nomenclature, which has the further advantages of being consistent with IUPAC systematics and emphasizing the bidentate reactivity of these intermediates. This method, while not universally adopted,[11] is increasing in popularity[4] and will be used throughout this review.

The examples **4–6** illustrate these nomenclature systems as well as the

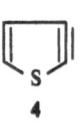 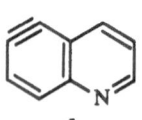 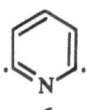

4	**5**	**6**
2,3-thiophyne	5,6-quinolyne	2,6-pyridyne
2,3-dehydrothiophene	5,6-dehydroquinoline	2,6-dehydropyridine
2,3-didehydrothiophene	5,6-didehydroquinoline	2,6-didehydropyridine

structural formulas in use. Note that bicyclic species such as **5** with the arynic bond in a carbocyclic ring are still considered hetarynes, since the heteroatom in the neighboring ring may influence the reactivity of this bond.[1] Furthermore, although only *o*-didehydrohetarenes are properly classified as hetarynes,[4,16] species such as **6** with nonadjacent reactive orbitals are, by their chemistry and analogy[5] to the carbocyclic analogs **2** and **3**, once again placed in the same category.[4] The logic of this classification is obvious from the similarity in the orbital formulations **7** and **8** of 2,3-didehydropyridine and 2,6-didehydropyridine, respectively, as well as in their mode of theoretical analysis.[19]

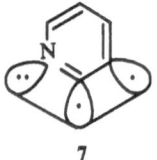 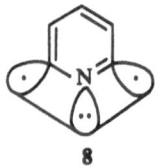

7	**8**

Finally, the *caveat* about reading too much into structural formulas is particularly to be emphasized in the hetaryne series since heteroatoms would be expected to have a pronounced effect on the relative extent to which canonical structures analogous to **1a–1d** contribute toward the true structure of the hetaryne.[19] For example, with 2,3-didehydrothiophene (**4**) and 2,3-didehydropyridine (**7**), canonical structures **4c** and **7c** might be more important owing to delocalization of the positive charge via π-type overlap with the filled, non-bonded orbital on the adjacent heteroatom **4d**[20] and **7d**.[21] Similarly, the dipolar structures **4b** and **7b** would be expected[22] to be stabilized and destabilized, respectively, based on the known effects of thiophene sulfur[23] and pyridine nitrogen[24] on adjacent carbanions. The former phenomenon also suggests the possibility of contributions from structures such as **4e** involving delocalizations of the negative charge to the sulfur atom via its empty d orbital.[25] Although these considerations serve to emphasize the inadequacy of structural formulas such as **4a** and **7a** they also preview the unique richness of chemistry and properties expected for hetarynes.

C. Related Heterocyclic Species

All the arynes considered thus far have as one of the contributing structures a form containing a carbon–carbon triple bond. Certain hetarynes have additional canonical structures such as **4d**, **4e**, and **7d** which contain triple bonds between carbon and the heteroatom. The ultimate extension of this trend leads to reactive intermediates for which *only* canonical structures containing carbon–heteroatom triple bonds can be drawn (**9–11**). Molecular orbital representations of these species (**9c–11c**) clearly show that they are isoelectronic to the analogous aryne representation **1j**. The definition given in Section I.2.A for arynes also applies if the parent aromatic systems are considered to be the pyridinium ion (**12**), the thiophenium ion (**13**),[26] and pyrrole (**14**), respectively. Consequently, in contrast to Hoffmann's choice,[27] such hetarynium ion intermediates will be covered in this review (Section III.9) strictly on structural grounds and without any prejudgement of their properties.

On the other hand the logic of including species such as **15** in this review is not nearly so compelling. Although it has been suggested that the well-known stability of carbanions adjacent to sulfur[28] may be due to d–σ overlap[29,30] as in **15b**, this view has been challenged both for closely related species[31] and as a general phenomenon.[28,32,33] Even if the contribution of **15b** to the structure of **15** were significant, however, such a species would fit neither the definition (Section I.2.A) nor the orbital representation (**15c**) of an aryne (**1j, 10c**). Accordingly, consistent with Hoffmann,[27] simple carbanions of heteroaromatic compounds[34] will not be included in this review.

Other species which will be considered within the scope of this chapter (Sections III.1.C and III.2.D) include **16** and **17**, formally derived (Section I.2.A) from the maleic acid derivatives **18** and the quinoidal compounds **19**. Since these precursors are not actually aromatic, **16** and **17** cannot be considered true hetarynes. Nevertheless, the similarity of the orbital representations

16a and 17a to those of the other arynes (1j) leads to the expectation of similar chemical properties and supports the inclusion of these intermediates in this review. A similar argument probably could be raised for heterocyclic alkynes such as 20 and 21, but since neither examples nor claims of such species were reported[586] until after this article was completed, they are not covered.

3. Historical Perspective

A. Development of the Aryne Concept

The first suggestion of a didehydroaromatic intermediate, phenylene ($C_{12}H_4$), appears to have been made by Berthelot to rationalize the destructive thermal hydrogenation of biphenyl.[35] Although this paper was published seven years after the famous Karlsruhe Congress,[36] it does not utilize the conclusions of that meeting regarding the atomic weight of carbon. If it is recognized that the correct number of carbon atoms in each molecule is actually one-half that indicated, it becomes clear from equation (1) that Berthelot meant for phenylene

$$\text{``}C_{24}H_{10}\text{''} \xrightarrow[\Delta]{H_2} \text{``}C_{12}H_6\text{''} + (\text{``}C_{12}H_4\text{''})$$
$$\text{(biphenyl)} \qquad \text{(benzene)} \quad \text{(phenylene)}$$

$$3(\text{``}C_{12}H_4\text{''}) \longrightarrow \text{``}C_{36}H_{12}\text{''}$$
$$\text{(chrysene)}$$

(1)

to represent a didehydrobenzene (C_6H_4). A similar intermediate with modern atomic weights was considered, and subsequently rejected, a few years later[37] to explain the thermolysis of diphenylmercury [equation (2)].

$$(C_6H_5)_2\,Hg \xrightarrow{\Delta} Hg + C_6H_6 + (C_6H_4)$$

$$4(C_6H_4) \longrightarrow C_6H_5-C_6H_5 + C_6H_6 + 6C$$

(2)

The first didehydroaromatic species for which a *structural* formula was proposed was the five-membered hetaryne, 2,3-didehydrobenzofuran (22).[38] This intermediate was postulated to explain the formation of 2- as well as 3-substitution products from the reaction of 3-bromobenzofuran with NaOEt or alcoholic KOH [equation (3)]. The authors recognized that such an aryne, analogous to alkynes, could add ethanol in either direction, thus accounting for the isomeric product mixture. They furthermore ascribed a very high, strain-induced instability to 22 which they held responsible for the failure of all attempts at its isolation. The precociousness of the authors' insights finally gave out, however, with the suggestion that the lachrymatory odor observed upon opening the reaction autoclave might have been due to residual **22**.

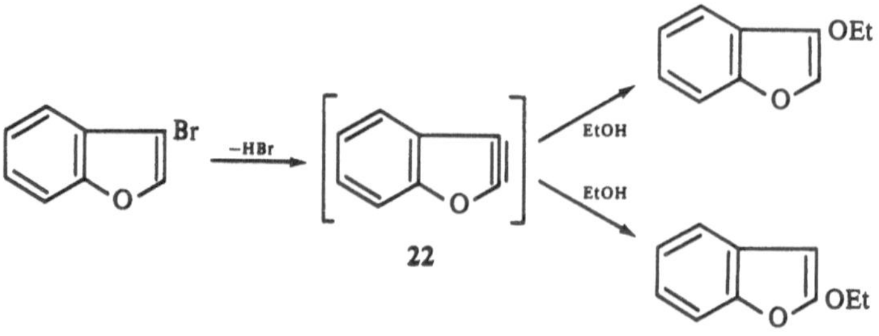

(3)

Attempts to generate a related didehydrospecies, the anhydride of acetylene dicarboxylic acid (**23**), ended in failure, although the authors[39] were sufficiently optimistic to reserve future studies in this area to themselves. A few years later the intermediacy of **23**, along with the four-membered cycloalkyne **24**,[40] was claimed[41] during the pyrolysis of acetoxymaleic acid anhydride.

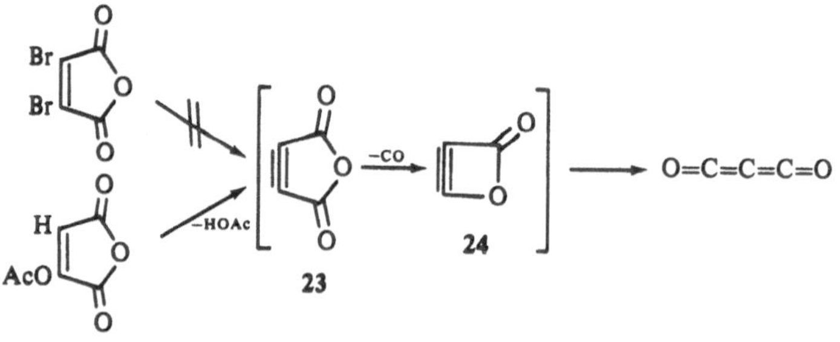

Didehydroaromatic species apparently were considered as possible inter- mediates only twice[42,43] during the next 30 years and in one of those instances[43] rejected as being too unlikely. A dipolar formulation (**1b**) of dehydrobenzene was suggested almost simultaneously by Wittig[15] and Morton[44] as an inter- mediate in the reactions of halobenzenes with alkali metal derivatives of hydrocarbons. A symmetrical structure (**1a**) for dehydrobenzene was specifically rejected by Wittig based on some earlier erroneous experimental work.[45]

The general acceptance of the existence of didehydroaromatics as we know them today began with the investigations of Roberts[14] and Huisgen[46] on the mechanism of rearrangements observed during nucleophilic aromatic substitu- tions with strong bases. These classic researches have been adequately reviewed a number of times[5–7,9,10] and will not be covered here. Recognition[47] of the dienophilic properties of dehydrobenzene as well as the development of several efficient methods of generation not requiring the use of organometallic reagents[48-50] gave additional support to the aryne concept. Particularly significant in elevating dehydrobenzene chemistry above its characterization[51] as "Chemie des 'als ob' " ("as if" chemistry) were its generation[52-54] and spectroscopic examination[52] in the gas phase and the demonstration of its identical chemical behavior from several different precursors.[55,56] The final and most recent experimental evidence for the existence of dehydrobenzene is its generation, trapping, and spectral investigation in an argon matrix at 8° K.[57] There is therefore little doubt that dehydrobenzene is a true reactive intermediate, not just a transition state, and that it can exist free of a solvent shell or complexes with residual fragments of its precursors.[58]

As the evidence for dehydrobenzene as a discreet intermediate grew stronger, analogous species were sought among other aromatic and hetero-aromatic compounds. The rigor with which the case for dehydrobenzene had been assembled usually was abandoned in these instances in favor of the much more efficient reasoning by analogy. The observation of cine-substitution (Section II.2.A) or Diels–Alder (Section II.2.B.b) products in a reaction of a potential aryne precursor (Section II.1) often was taken as presumptive evidence for an aryne or hetaryne intermediate without the time-consuming experimental elimination of alternative nonaryne pathways. In some reactions arynes were suggested as intermediates without any supporting evidence whatsoever except for the ability of the author to write down the appropriate mechanism on paper. While these shortcuts probably led to few errors in the carbocyclic series, some rather serious detours resulted in the heterocyclic area,[1-4,59] particularly with the alleged five-membered hetarynes. A brief chronological overview of this latter field will precede the more detailed consideration of the evidence to follow in Section III.

B. History of the Five-Membered Hetaryne Hypothesis

Belief in the existence of five-membered hetarynes has waxed and waned since they were first postulated[38,41] in the early part of this century. As described above, most of the development of the aryne concept took place with benzenoid compounds. Once this central idea became well established, however, it was natural to examine the scope of its application to other ring systems including five-membered heterocycles. The first experiments in this direction in the early 1960s appeared to prove the existence of didehydromaleic anhydride (23),[60] both 2,3-didehydrothiophene[61] and 3,4-didehydrothiophene,[62] 4 and 25 respectively, and even the patriarchal[38] 2,3-didehydrobenzofuran (22).[62] Older experimental results[63-65] were reinterpreted in the light of these new claims to suggest that 2,3-didehydrobenzothiophene (26)[1,66] and its dioxide (27)[67] were probably known. The first reviews[1,2] of these early results were generally optimistic about the existence of five-membered hetarynes and doubtless stimulated further claims for and consideration of both new intermediates such as 4,5-didehydroimidazole (**28**),[68] 4,5-didehydropyrazole (**29**),[69] and didehydromaleimide (**30**),[70] as well as previously reported ones such as **4** [71,72] and **26**.[72]

4   25   28   29

23: X = O
30: X = NR

22: Z = O
26: Z = S
27: Z = SO$_2$
31: Z = NR

Although Wittig[51] had warned as early as 1963 that some of the evidence on which the claims for the existence of **4**,[61] **25**,[62] and **31** [51] were based might be ambiguous, this caution was generally ignored[1,2] until the publication of Hoffmann's book.[5] After considering all of the extant data, much of it only available in the unpublished dissertations of Wittig's students, Hoffmann concluded[73] that "in fact no unambiguous evidence is as yet available for the formation of dehydro derivatives of five-membered aromatic heterocycles." The validity of this conclusion was emphasized by the timely demonstration[74] that didehydromaleic anhydride (**23**) was in fact not an intermediate in the reaction which led to its postulation.[60] Particularly significant in the further dismantling of the five-membered hetaryne hypothesis were the discoveries that both of the classical chemical methods for detecting aryne intermediates, cine-substitution (Section II.2.A)[75] and Diels–Alder cycloaddition (Section II.2.B.b),[76] could be observed in five-membered heterocyclic systems without invoking aryne intermediates. At the time of the next cycle of reviews[3,4] it was generally conceded that with the possible exception[4] of didehydroimidazole (**28**)[68] no convincing evidence for five-membered hetarynes existed. Even this lone holdout succumbed once the erroneous structure of a reaction product was corrected.[18]

Since 1970 many investigations and reinvestigations of reactions which might lead to five-membered hetarynes have been undertaken. With two possible recent exceptions from our laboratory[77–79] these have all proven negative or ambiguous. A detailed discussion of these results will be presented in Section III after a review of the chemistry of authentic arynes is presented in Section II to provide appropriate background information.

II. CHEMISTRY OF ARYNES

As will be described in Section III investigations of the existence of five-membered hetarynes have relied heavily on analogies to the chemistry of the well-authenticated six-membered carbocyclic arynes. It will therefore be the purpose of this section to briefly review and update[5] those aspects of aryne chemistry which have played an important role in this regard. For a more complete coverage of aryne chemistry the reader is referred to the several reviews previously cited[6-11] and especially to the excellent monograph by Hoffmann.[5]

1. Methods of Preparation

A. General Considerations

All practical methods of generating arynes involve either the concerted or stepwise loss of appropriately situated substituents from an aromatic precursor. Many different types of leaving groups have been utilized including those which are part of a second ring. If the elimination is stepwise several possible inter-

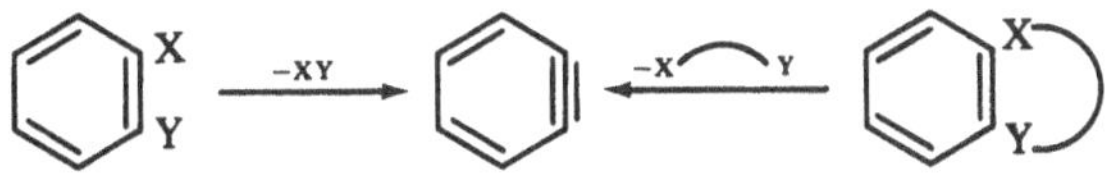

mediates might intervene depending on whether the first group to leave takes along none [equation (4)], one [equation (5)], or both [equation (6)] of its bonding electrons. Which of these pathways is followed, or whether the reaction is stepwise or concerted, is markedly dependent not only on the nature of the leaving group but also on the nature or substituents of the aromatic ring.

Since it is therefore dangerous to generalize on the mechanism of aryne

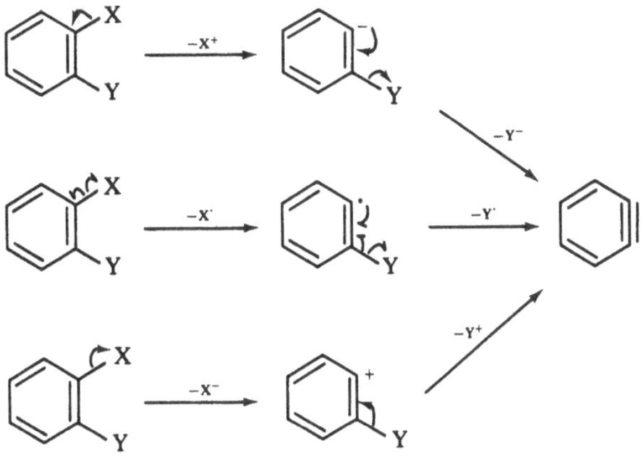

$$(4)$$

$$(5)$$

$$(6)$$

formation from a particular type of precursor, this review will indicate both the type of precursors and the most likely intermediates to be involved in an aryne generation method. Considerable overlap will exist between the classifications chosen since often several precursors will give rise to the same intermediate, or the same precursor can give rise to several intermediates.

B. Arynes from Aryl Anion Intermediates

By far the most common and historically oldest aryne generation method involves the elimination of hydrogen halides from aryl halides (**32a**) in the presence of strong bases (Table 1).[80] In many cases *o*-haloarylanions (**33**) can be trapped as intermediates, although often the elimination of halide ion to the aryne (**1**) is either too rapid to allow their detection or perhaps concerted. The effect of halogen, metal, and substituents on this reaction has been well studied.[81a]

Replacement of the ortho hydrogen of an aryl halide with another electrofugic group leads to many additional precursors of *o*-haloarylanions[81b] and in some cases arynes. Several of the more effective in the latter category (**32b–32f**) are summarized in Table 1 and in the following scheme:

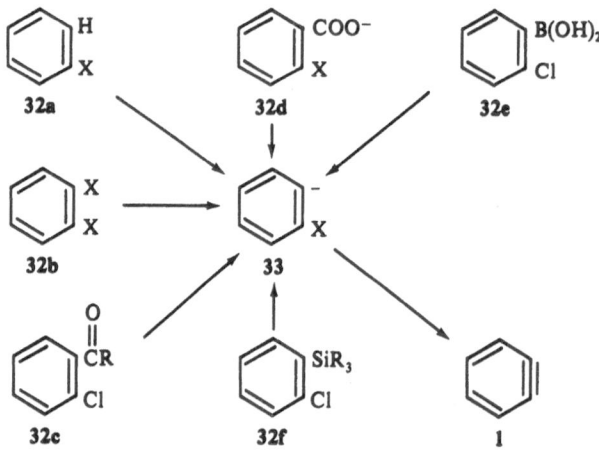

Particularly useful have been the *o*-dihalobenzenes (**32b**), which permit the generation of arynes without the use of organometallic or strongly basic reagents.[47] A recent variant traps the initially formed anion (**33**) as its TMS derivative (**32f**), which then can be converted to the aryne **1** with the nonbasic and nonnucleophilic KF.[82]

Replacement of the halogens of the above precursors with different nucleofugic groups generates another set of aryne sources (**32g–32l**) which once again probably react via the *o*-substituted anions (**34**).

Table 1. Generation of Arynes via Aryl Anion Intermediates

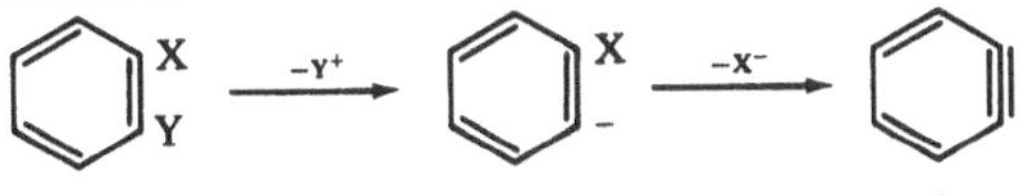

32	Y	X	Reagent	References
a	H	F,Cl,Br,I	$^-$OR,$^-$NR$_2$	80
b	Br,I	F,Cl,Br,I	Li,Mg,RLi	47
c	C(=O)R	Cl	KNH$_2$	83
d	COO$^-$	F,Cl,Br,I	300°C	84,85
e	B(OH)$_2$	Cl	BuLi,NH$_3$	86
f	SiMe$_3$	Cl,Br,I	KF, KO-*t*-Bu	82
g	H	OAr	RLi,OR$^-$	92
h	H	ClO$_3$	RLi,NR$_2$$^-$	93
i	H	$^+$IPh	*t*-BuO$^-$,$^-$OAc	94,95
j	H	OTs	LiNR$_2$	96
k	Br	OTs	RLi	97
l	COO$^-$	$^+$IPh	220°C	88,90

The mechanistically most ambiguous of the above precursors are those (**32d** and **32l**) containing carboxylate as the electrofugic group. In neither case has an anionic intermediate (**35**) been demonstrated[87,88] although it has been considered.[88–90] In fact, such evidence as is available[90] better supports[91] the intermediacy of the zwitterion **36** or one of its valence tautomers **37** or **38**.

C. Arynes by Loss of Nitrogen from Acyclic Substituents

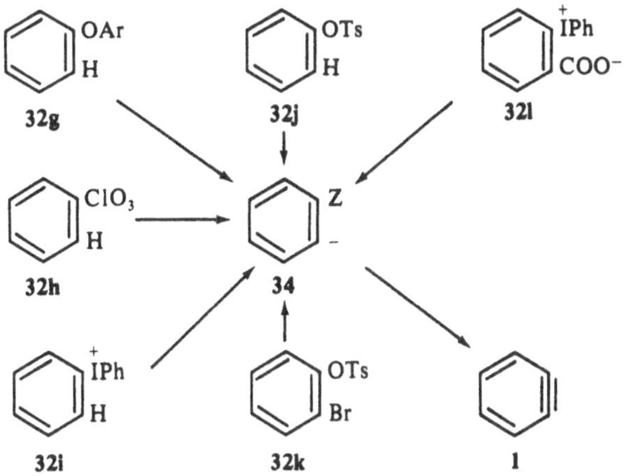

The energy requirements for generating an unstable intermediate such as an aryne can be substantially decreased if a very stable product such as

molecular nitrogen is produced simultaneously.[98] Such a process could involve either electrofugic [equation (7)], one-electron [equation (8)], or nucleofugic [equation (9)] loss of nitrogen from an appropriate precursor to generate, in the event the reaction is not concerted, anionic, radical, or cationic intermediates, respectively. Several examples of the first reaction are known as good sources of *o*-haloarylanions (**33**) but poor precursors of arynes because of rapid proton capture of **33** under the reaction conditions.[99]

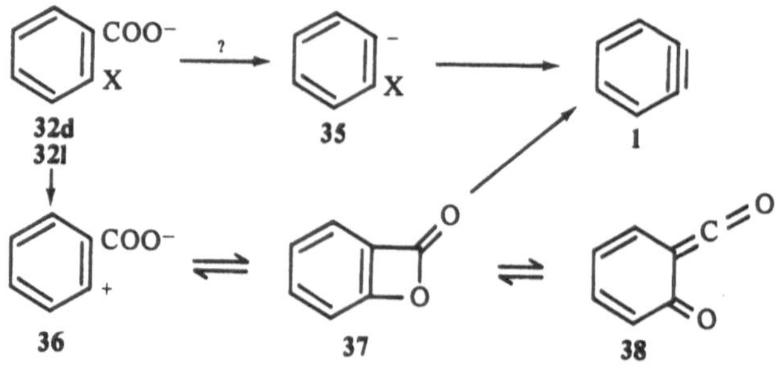

The remaining two reactions, which will be described in this section, are related through the well-known[100] formation of azo compounds **39** by the addition of nucleophiles to diazonium salts (**40**). The former compounds lose nitrogen by homolytic and the latter by heterolytic processes[100] to give, respectively, aryl radicals and sometimes, as has been recently demonstrated,[101] aryl cations. Because the equilibrium between **39** and **40** is so dependent on solvent, gegenion, and additives,[100,101] both reactions [equations (8) and (9)] often occur simultaneously.[102]

Table 2 lists several compounds which lose molecular nitrogen to produce arynes. The most useful and most studied precursors are based on ben-

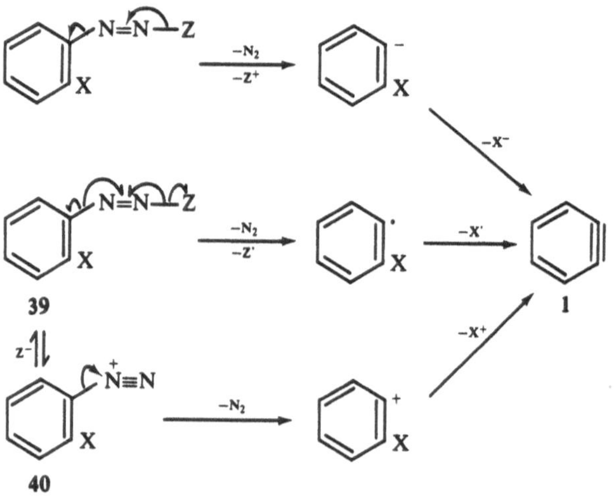

$$\text{(7)}$$

$$\text{(8)}$$

$$\text{(9)}$$

TABLE 2. Generation of Arynes by Loss of Nitrogen from o-C_6H_4XY (41)

	X	Y	Conditions	References
a	N_2^+	CO_2^-	Δ, $h\upsilon$	48
b	NH_2	COOH	RONO	103
c	N_2^+, Cl^-	COOH	Ag_2O, C_3H_6O	104,105
d	N(NO)Ac	COOH	80°C	106
e	$N=NNR_2$	COOH	150°C	107,108,109
f	N(NO)Ac	H	25°C	129
g	NH_2	H	Ac_2O/RONO/80°C	130
h	NHAc	H	BzONO/80°C	130
i	N_2^+	H	OAc^-, $(R_2N)_3PO$	133,134
j	$N=NNR_2$	H	Cl_3CCOOH, 80°C	127
k	N(NO)PO(OR)$_2$	H	80°C	95
l	N(O)=NOTs	H	PCl_3, 80°C	95
m	N(O)=NX (X=F,Ts)	COOH	190°C	136
n	N(NO)Ac	I	65°C	138
o	N_2^+	I	NaOAc	139
p	NH_2	$B(OH)_2$	RONO	141
q	SO_2N_3	SO_2N_3	135°C	142

zenediazonium carboxylates (41), which can lose *two* very stable molecules, CO_2 and N_2, to give an aryne.[48] The ease with which this decomposition takes place gives rise to the major drawback of this precursor–its tendency to detonate. In spite of their ready availability, therefore, free diazonium carboxylates are suitable aryne precursors only for very small-scale reactions. Consequently, several methods for generating 41a *in situ* have been developed, the most convenient of which is the aprotic diazotization of the related amino acid 41b.[103] It is also possible to generate 41a from its more stable salts (41c) with either silver oxide[104] or, especially, propylene oxide (C_3H_6O)[105] as acid scavengers. The thermolysis of 41d also is reported to give benzyne (1), presumably via the diazonium carboxylate 41a.[106]

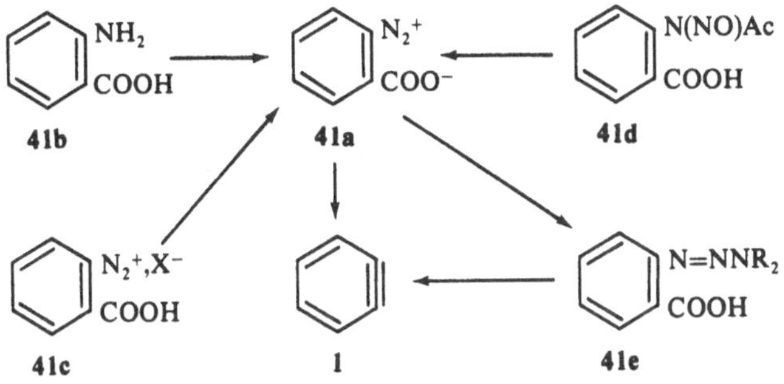

The hazard of 41a also can be avoided by reacting it with secondary amines to give the nonexplosive, easily stored triazenes 41e which thermolize smoothly to arynes.[107,108] The diazonium carboxylate **41a** can be avoided entirely by preparing the triazene **41e** from the acetylhydrazone of an *o*-nitrobenzaldehyde [equation (10)].[109]

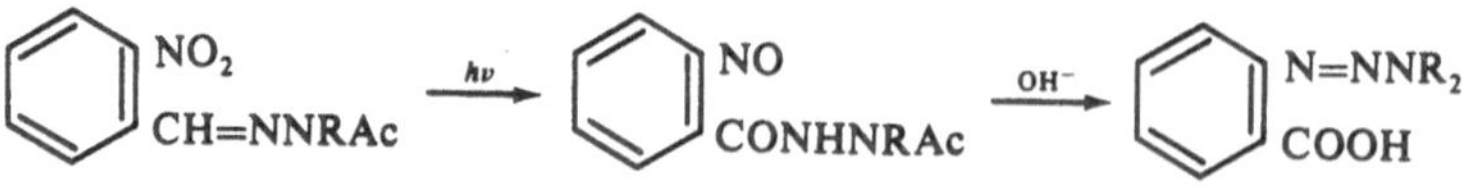

$$\text{(10)}$$

At least three species have been suggested as intermediates in the decomposition of diazonium carboxylates (**41a**) to arynes.[110] A large body of evidence, both kinetic[111] and trapping by nucleophiles,[112–115] suggests the presence of an intermediate presumably[116] similar to the zwitterion **36** considered in the decomposition of the *o*-substituted benzoic acids **32d** and **32l** discussed in Section II.1.B. It has been suggested,[118] however, that the data are equally consistent with a bimolecular displacement of nitrogen from **41a** by the various nucleophilic trapping agents[119] and hence no zwitterion intermediate **36** need be invoked. The role[110,111] of the valence tautomers **37** and **38** in the decomposition of **41a** also has been questioned,[120] although both species are intermediates in the photolysis of phthaloyl peroxide in an argon matrix at 8° K.[121] The situation is further complicated if the diazonium carboxylates **41a** are generated in the presence of their acid salts **41c**, which can either decompose to arynes via the *o*-carboxycation (**42**) or react directly with the nucleophilic trapping agents.[120,122,123]

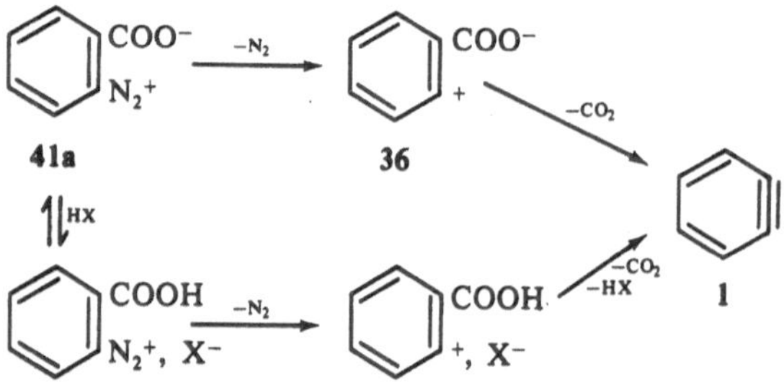

Another zwitterion intermediate (**43**), formed by preliminary loss of CO_2 from the diazonium carboxylate **41a**, also has been suggested, albeit tentatively, several times.[112,124,125] Generally this possibility has been ignored[110,118] and in at least one case[112] the evidence on which the claim was based was explained in another manner.[108]

The same intermediate (**43**) was proposed more seriously,[126] apparently substantiated,[127] and then finally rejected[128] as being involved in the generation of arynes from *N*-nitrosoacetanilides (**41f**). The precursors, which can be isolated[129] or generated *in situ*[130] from either the aniline **41g** or the acetanilide **41h**, rearrange[131] to the diazonium acetate **41i** which undergoes a concerted[128] elimination of nitrogen to the aryne **1** or homolytic decomposition to aryl radicals **44**.[129,131] The latter reaction involves a radical-induced chain mechanism which can be suppressed by the presence of certain additives[132] thereby making this method one of the simplest ways known of generating arynes. Variations on this procedure include reaction of the isolated diazonium salts **41i** directly with an appropriate base[133,134] or generation of the diazonium salts **41i** *in situ* from different precursors such as triazines (**41j**),[127] *N*-nitroso-phosphylamidates (**41k**),[95] or azoxytosylates (**41l**).[95]

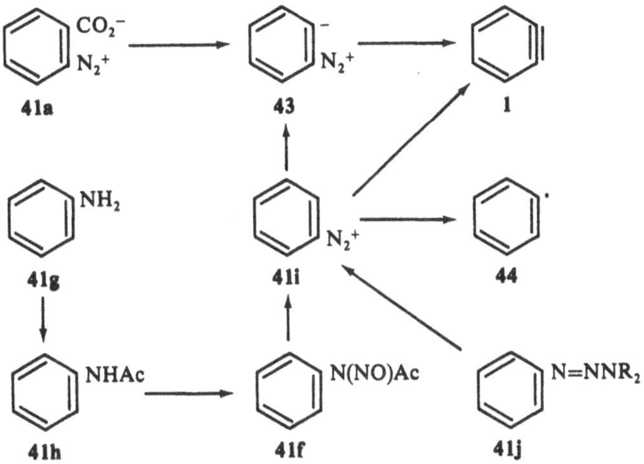

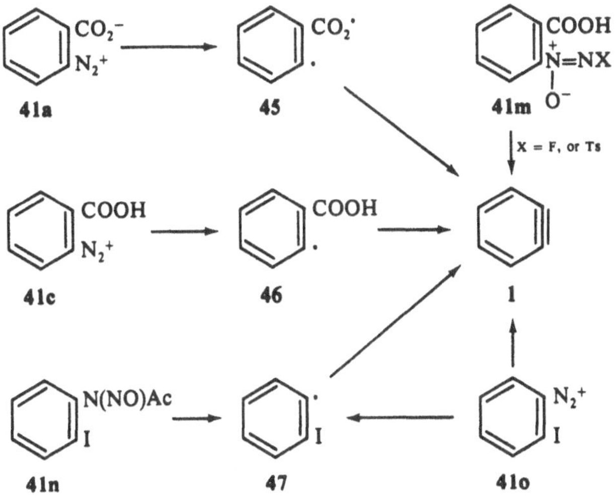

The third kind of intermediate which could intervene during the stepwise loss of molecular nitrogen to form an aryne would be a radical [equation (8)]. Although loss of N_2 and CO_2 is synchronous[54] from the flash photolysis of the *solid* diazonium carboxylate **41a**,[52] solution photolysis[135] and thermolysis[120] lead to initial loss of N_2 and eventually to products which suggest the formation of the intermediate diradical **45**. The related radical **46** also has been considered for decompositions of the salt **41c**,[105,110] although CIDNP signals could not be observed during this process.[123] Thermolysis of *o*-azoxybenzoic acids (**41m**) proceeds by elimination of CO_2 and N_2 (not N_2O!) to give benzyne (**1**) presumably via a radical process.[137] *o*-Iodophenyl radicals (**47**) appear to be implicated in the formation of arynes from *o*-iodonitrosoacetanilides (**41n**)[138] and probably from *o*-iododiazonium acetates (**41o**).[139] Curiously, some (see Section II.1.F) but not all[140] sources of *o*-iodophenyl radicals (**47**) give evidence of aryne formation.

Of the final two entries in Table 2, *o*-aminophenylboronic acid (**41p**) is a promising aryne precursor[141] whose detailed chemistry awaits further examination, and hence no information on the mechanism of its decomposition is available. Thermolysis of *o*-benzenedisulfonyl azide (**41q**) gives some evidence of benzyne formation, presumably via a nitrene and the cyclic azodisulfone **48**.[142]

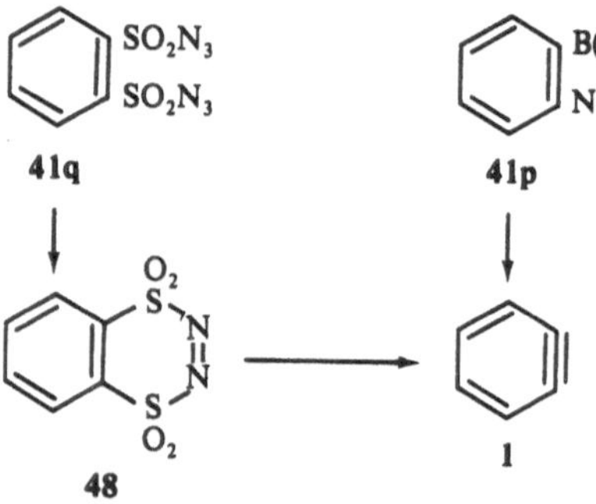

D. Arynes by Loss of Nitrogen from Cyclic Systems

Diazonium salts with a potentially divalent *o*-substituent in principle could exist as either cyclic (**49**) or open-chain (**49'**) structures. Although the former have been considered[143] for *o*-diazonium carboxylates (**49a**), solubility[104] and spectral[144] properties support the zwitterionic form (**41a**). Other cyclic azo com-

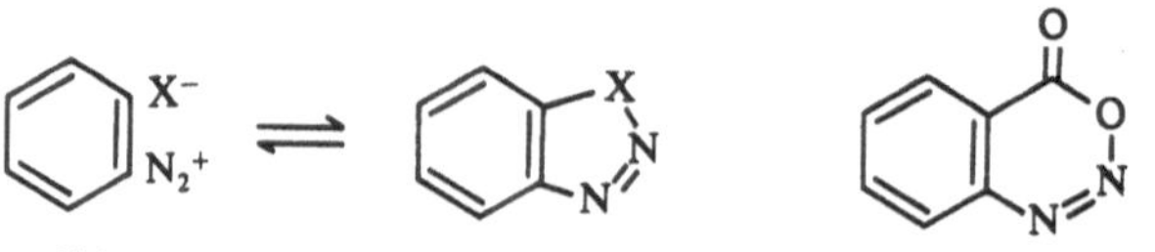

TABLE 3. Generation of Arynes by Loss of Nitrogen from 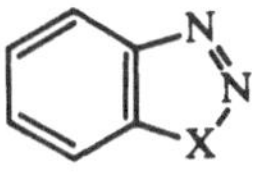 and other Cyclic Systems

49

49	X	Coproducts	Conditions	References
b	$-S(O)_2-$	SO_2, N_2	10°C	49
c	$-C(O)S-$	COS, N_2	200°C, $h\upsilon$	149,150
d	$-N=C(R)-$	RCN, N_2	850°C	151
e	$-C(R)=N-$	RCN, N_2	450°C	151
f	$-C(O)NR-$	RNCO, N_2	600°C	152
g	$-C(O)N(\bar{N}Ts)-$	N_2, CO, TsO$^-$	220°C	153
h	$-C(O)N(NH_2)-$	N_2, CO	Pb(OAc)$_4$, 80°C	155
i	$-C(O)-$	N_2, CO	400°C	157
j	$-N(NH_2)-$	N_2	Pb(OAc)$_4$	50
k	$-N(NO)-$	N_2	ROPPh$_2$	159
l	$-N(\bar{N}Ts)-$	N_2	$h\upsilon$	160
m	$-N(\bar{N}N=NNHTs)-$	N_2	25°C	160
n	$-N(N=PPh_3)$	N_2, Ph$_3$P	$h\upsilon$	161

60	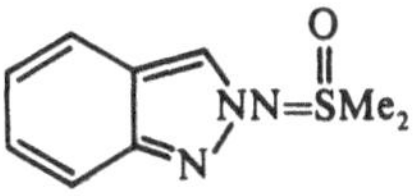	N_2, DMSO	450°C	151

61	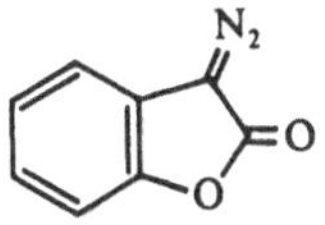	N_2, CO	$h\upsilon$	163

pounds (**49**) are known, however, and often serve as precursors of arynes. These and other cyclic systems which lose nitrogen to give arynes are summarized in Table 3.

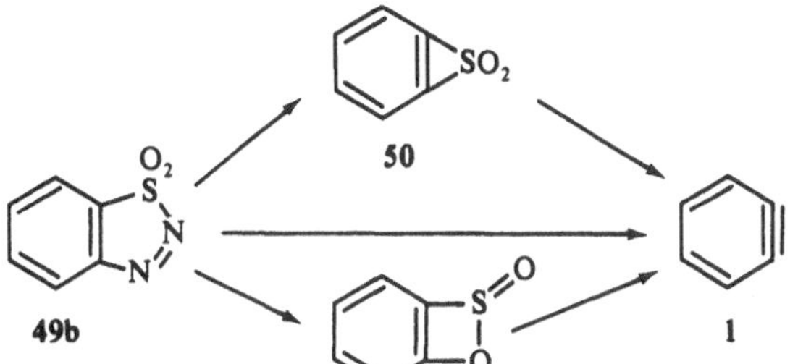

Benzothiadiazole-1,1-dioxides (**49b**) lose N_2 and SO_2 under mild conditions[49] to give arynes. Generally the major reaction path appears to be concerted[145] with only minor involvement[146] of sulfur-containing intermediates such as **50** or **51**. In certain reactions[147] or with particular precursors,[148] however, the significance of these species to the overall chemistry of **49b** may be enhanced.

In contrast to **41a**, benzenediazonium 2-thiocarboxylate (**52**) exists in the cyclic form (**49c**) and requires higher temperatures for aryne formation,[149] apparently via the thiolactone intermediate **53**[150]

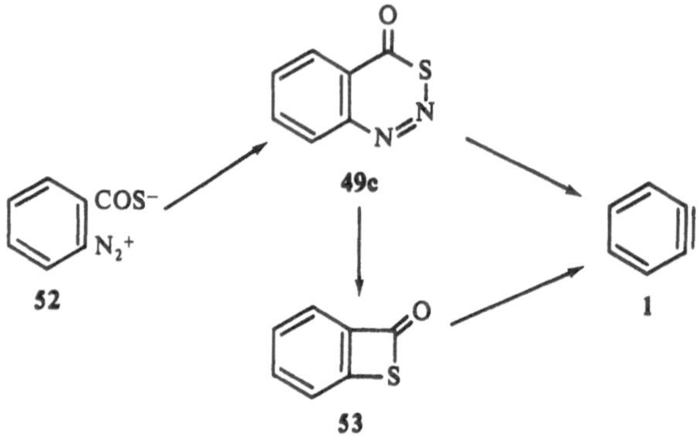

Both 1,2,4- or better 1,2,3-benzotriazines, **49d** and **49e**, respectively, loose N_2 and nitriles on thermolysis to give arynes.[151] The triazinones **49f** undergo a similar minor fragmentation with expulsion of nitrogen and isocyanates.[152] The major reaction leads to azetinones **54** which were shown not to be precursors of arynes.

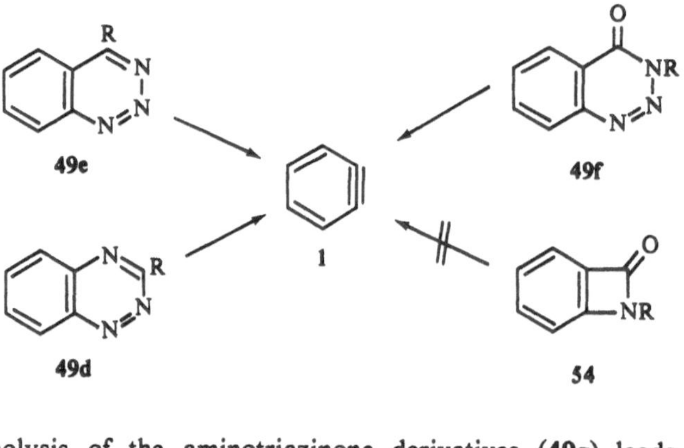

Thermolysis of the aminotriazinone derivatives (**49g**) leads to arynes, presumably via benzocyclopropenones (**55**).[153] Under the milder conditions of room temperature photolysis[153] of **49g** or oxidation[154] of the unsubstituted aminotriazinone **49h**, the benzocyclopropenone **55** can be trapped without any evidence of loss of CO to give arynes. Oxidation of **49h** at 80°C, however, leads to some aryne formation probably via benzocyclopropenone (**55**) and/or the indazolone (**49i**) formed by a simultaneous but independent route.[155] In contrast to earlier reports,[156] the indazolone **49i** will fragment to benzyne when generated by gas phase pyrolysis of its Diels–Alder adduct (**56**).[157]

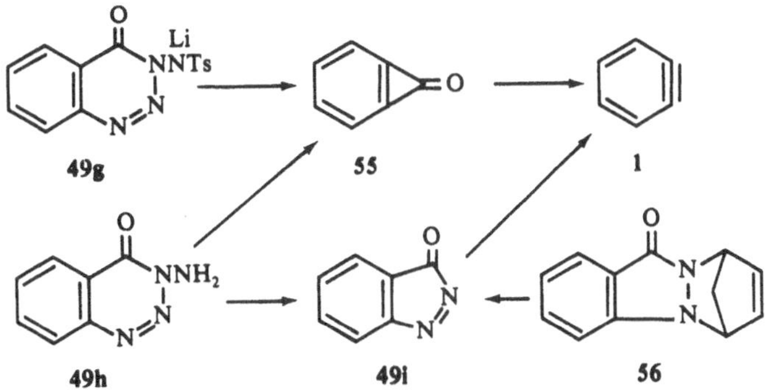

One of the best routes to arynes is the oxidation of aminotriazoles (**49j**) with lead tetraacetate.[50] Variations involve generating the intermediate nitrene **57** or nitrenoid with other oxidizing agents[158] or from other precursors (**49k–49n**).[159–161] By analogy the nitrene **58** or nitrenoid obtained by oxidation of 1-aminoindazole (**59**) might be expected to fragment to arynes by loss of N_2 and a nitrile.[162] In fact this reaction constitutes an excellent synthesis of the previously discussed aryne precursors, the 1,2,3-benztriazines (**49e**).[162] A vibrationally excited form of **49e** is presumably an intermediate in the pyrolysis of sulfoximides of 2-aminoindazoles (**60**) to arynes.[151]

In the last aryne precursor to be discussed in this section the nitrogen to be eliminated is exclusively exocyclic to the cleaving cyclic system. Photolysis of the diazobenzofuranone (**61**) gives benzyne (**1**) by the sequential loss of N_2 and CO as shown. Each of the numbered species was detected by infrared spectroscopy in an argon matrix at 8° K.[163]

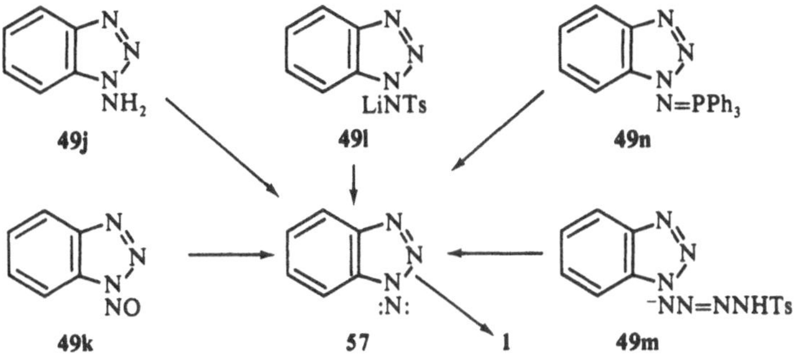

Finally, several likely aryne precursors in this category fail. Neither thermolysis nor photolysis of the tosylhydrazone derivative **64** yields any benzyne-derived products.[164] Instead, the rearranged benzoxazolone (**65**) forms and, like a: X = S;   b: X = Se

its corresponding nitrene (**66**),[165] is stable to aryne formation by loss of N_2 and CO_2. Related nitrenes in the aliphatic series fragment readily to alkenes.[166] Similarly, although 1,2,3-thiadiazoles[167] and 1,2,3-selenadiazoles[168] lose N_2 and sulfur or selenium readily to give alkynes, the benzologs **67a**[98,169] and **67b**[98,158] fragment only to a species (**68**) which retains the chalcogen and dimerizes to products such as **69**. The observation of a strong peak at $m/e=76$ in the mass spectrum of benzo-2,1,3-selenadiazoles (**71**) led to the prediction that drastic pyrolysis of **71** should produce benzyne (**1**).[170] The failure to validate this prediction in the ensuing decade was finally explained by the discovery that the peak at $m/e=76$ was due to $C_5H_2N^+$ and not $C_6H_4^+$.[171] The benzolog **70** of the successful precursor **49b** is also stable to photolysis and gives only dibenzo-thiophene dioxide on thermolysis.[53]

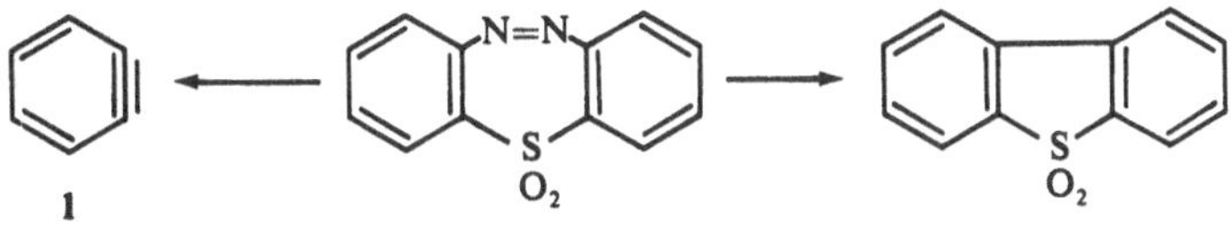

E. Arynes from Other Cyclic Systems

Fragmentation of cyclic systems to arynes without nitrogen expulsion is invariably accompanied by the loss of other low-energy molecules such as CO_2,

TABLE 4. Generation of Arynes from Cyclic Systems without N_2 Loss

Precursor	Coproducts	Conditions	References
72	CO_2	600°C	53
		hv	53,57,172
		Li/HMPT	173
73	CO_2, CO	690–800°C	175–181
		Plasmolysis	182
		hv	183
		900°C	186,187
76	CO_2, SO_2	690°C, 730°C	176,179,188
78	CO	hv	57
		800°C	189
79	CO	Plasmolysis	190
		900°C	186,187
80	CO	500°C	181,191
81	CO_2, RCN	700–860°C	192
82	CO_2, RCN	900°C	193
87	Furan	550°C	200
89	None	700°C	201
90	$R_3PO + CO_2$	R_3P, 200°C	202

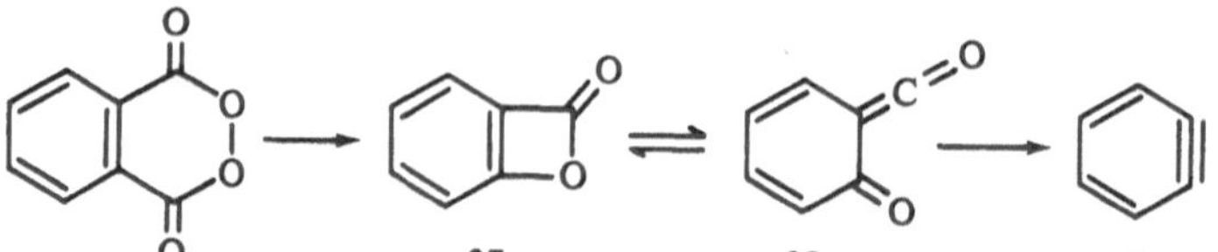

CO, or SO_2 (Table 4). Phthaloyl peroxide (72) loses CO_2 to give benzyne (1) upon thermolysis in the gas phase,[53] photolysis in solution,[53,172] or treatment with lithium in hexamethylphosphoramide.[173] The photolysis has been examined in an argon matrix at $8°$ K and the transient lactone 37 and ketene 38 detected.[57] The solution thermolysis[174] and reduction[173] of 72 also appear to proceed in a stepwise manner.

Using fragmentation behavior in a mass spectrometer as a guide to expected thermolysis behavior, three groups independently and almost simultaneously provided evidence that phthalic anhydrides (73) lose CO_2 and CO to give arynes.[175–179] This method has proven to be of synthetic value in the preparation of benzcyclobutenes[180] and biphenylenes.[181] Plasmolysis[182] or flash photolysis[183] of anhydrides (73) in the gas phase, but not in solution,[184] also generates arynes, the latter method providing independent confirmatory measurements of their uv spectra[52] and lifetimes.[54] By analogy to their mass spectral fragmentation behavior[177,185] phthalic anhydrides (73) are expected to lose CO_2 and CO sequentially when forming arynes. Some evidence for an intermediate with the composition of benzcyclopropenone (55), but written as the zwitterion 74a [175] or 74b or the diradical 75,[179,182] has been gathered from

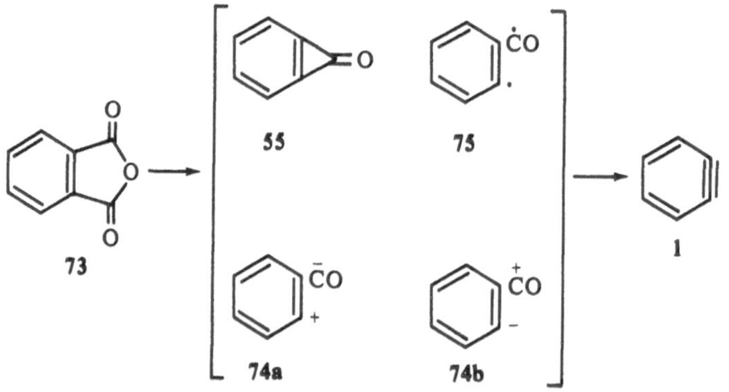

thermolysis[175,179] and especially plasmolysis[182] studies. Under flash photolysis[183] or pyrolysis–mass spectroscopy[186,187] conditions, however, the decomposition of anhydrides (73) to arynes appears to be concerted, presumably because of the severity of the conditions.

Although the major fragmentation pathway of o-sulfobenzoic anhydride (76) in its 70-eV mass spectrum involves loss of SO_3 and CO,[188] upon thermolysis SO_2 and CO_2 are lost to give benzyne (1).[176,177,188] Some evidence was found for the intermediacy of the zwitterion 77.[188]

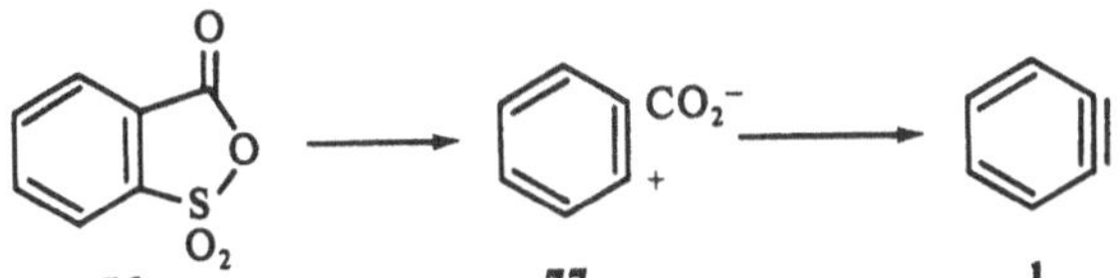

Loss of two molecules of CO leads to arynes upon photolysis[57] or severe thermolysis[189] of benzocyclobutenedione (78) and plasmolysis[190] or pyrolysis–mass spectroscopy[186,187] of acenaphthenequinone (79). With both precursors the process appears to be stepwise as shown except, once again, under pyrolysis–mass spectroscopy conditions, which is consistent with a concerted loss of CO.[186,187]

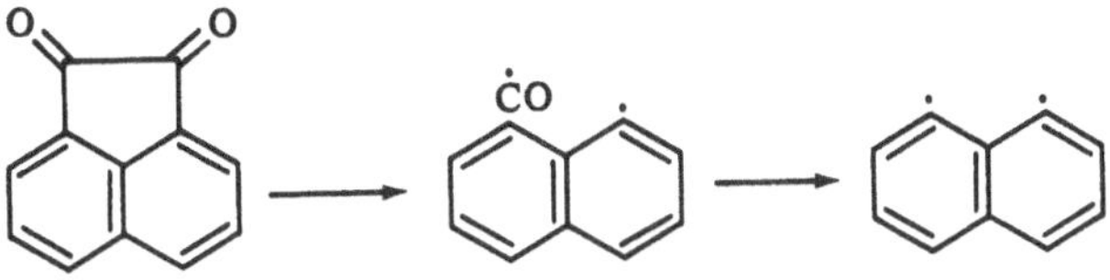

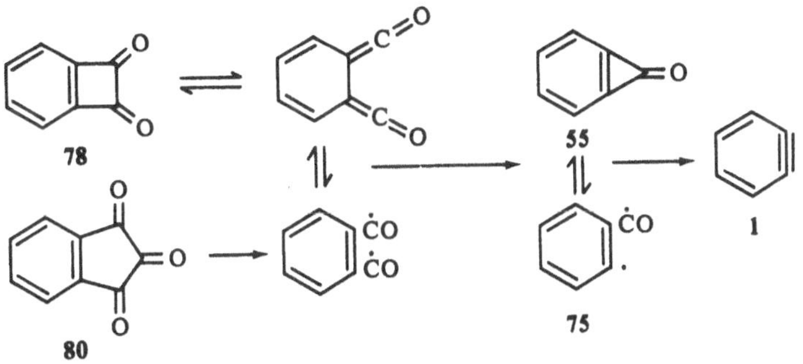

Indanetrione (80) loses three molecules of CO upon thermolysis to give arynes[191] in a preparatively useful reaction[181] for which the indicated intermediate has been suggested.

By analogy with the benzotriazines 49q and 49e (cf. Table 3), benzoxazinones might be expected to lose CO_2 and a nitrile to give arynes. Thermolysis of three of the possible isomers has been investigated. The 2,3,1-isomers (81) undergo substantial fragmentation to arynes at 700°C,[192] while the 2,4,1-

isomers (82) require a temperature of 900° C.[192,193] The 1,4,2-isomer (83) follows a different reaction course entirely, losing CO at less than 600°C to give benzoxazoles (84) in very high yield.[193] Although there has been some speculation on intermediates in these fragmentations, no compelling evidence has been presented.[193]

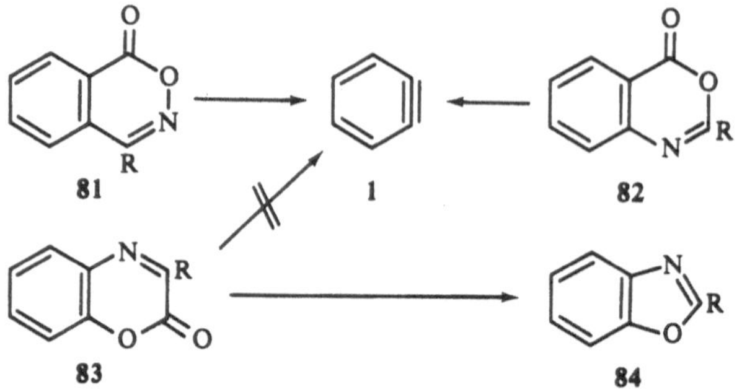

81   1   82

83   84

The retro-Diels–Alder[194] reaction has found considerable use in the preparation of reactive molecules and intermediates, particularly under flash vacuum thermolysis conditions.[195] Application of this methodology to the adducts (85) of arynes and various dienes (Section II.2.B.b) does not generally[196] regenerate the aryne and the diene but either gives the isobenzo derivative 86 and an alkyne by an alternate retro-Diels–Alder reaction,[197] or

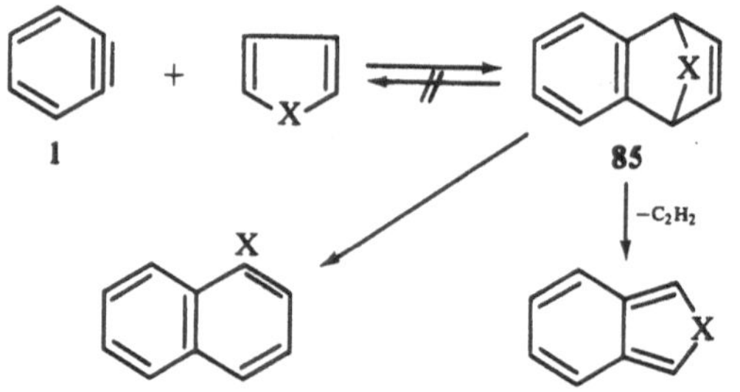

1   85

$-C_2H_2$

86

a: X = O
b: X = CH$_2$
c: X = NR

naphthalenes by ring opening.[198,199] Nevertheless, aryne formation from such a Diels–Alder adduct is sometimes postulated. An alternative explanation for the example[200] cited in Table 4 might be a concerted fragmentation of adduct 87 directly to the isolated product, diphenylbutadiyne, without the intervention of the aryne 88. If this were the case the apparent uniqueness of this example[200] could be attributed to the favorable energetics[98] of molecular nitrogen elimination in the primary cleavage step. A [2+2] cycloreversion upon thermolysis of the benzyne dimer, biphenylene (**89**), does lead to benzyne as a minor reaction pathway, however.[201]

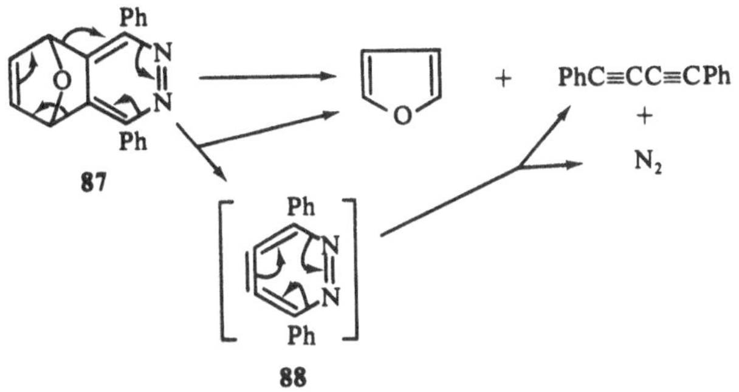

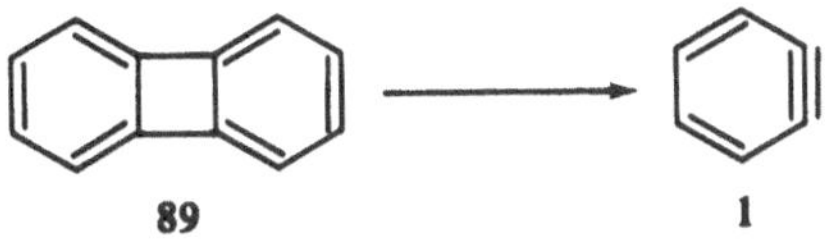

A final example of aryne formation involves the reaction of *o*-phenylene carbonate (**90**) with phosphines.[202] An analogous reaction with thiocarbonates (**91**) fails[203] as does generation of the anion of 2-phenyl-1,3-benzodioxole (**92**), which had been expected to lose benzoate ion to give benzyne.[204]

Thermolysis of phthalimides (**93**), benzofurandione (**94**), and the lactide (**95**) does not give arynes,[176] nor does photolysis of benzocyclobutadiene (**96**)[205]

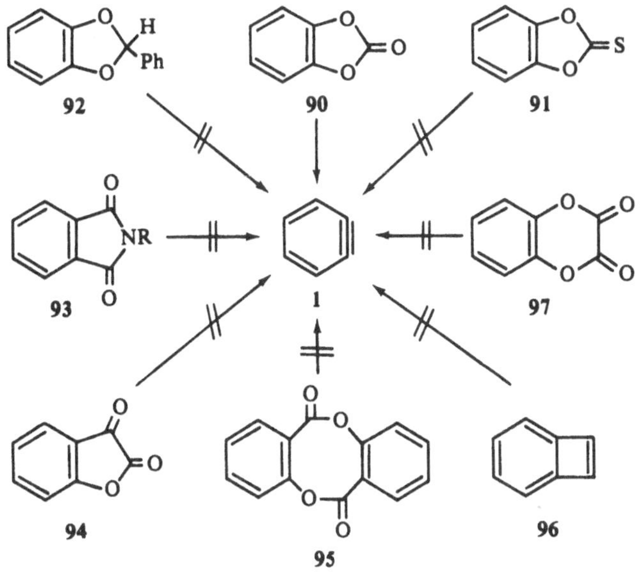

or *o*-phenylene oxalate (**97**),[206] in contrast to its isomer phthaloyl peroxide (**72**).[53,172] The failure of at least the first of these potential precursors (**93**) is consistent with a predictive mass spectral correlation proposed by Brown.[179] The use of mass fragmentation patterns as a model for thermolytic reactions has its limits, however,[207] as demonstrated by the behavior of phthalide (**98**). In the mass spectrometer **98** fragments by loss of CO and CH_2O to give what is probably the benzyne radical cation, while upon thermolysis CO_2 is lost to generate the fulvenallene **99**.[208]

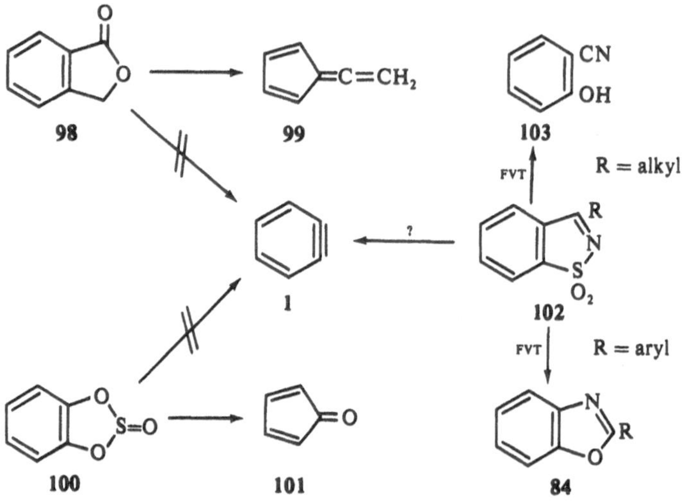

Other potential aryne precursors of the type under consideration in this section which fail include *o*-phenylene sulfite (**100**), which loses SO and CO to give the cyclopentadienone **101** instead of SO_3 to give benzyne,[209] and perhaps the benzoisothiazole dioxides (**102**). In contrast to the analogous benzo-2,3,1-oxadiazinones (**81**) which readily lose CO_2 and nitriles to give arynes,[192] the dioxides (**102**) primarily undergo deep-seated rearrangements to benzoxazoles (**84**) or *o*-cyanophenol (**103**) on flash vacuum thermolysis (FVT).[210] A minor reaction does lead to nitriles but no products derived from the expected aryne coproduct were observed. Perhaps a change in conditions or the presence of an appropriate trapping agent will provide evidence for aryne formation.

Such experimental differences and not the nature of the precursor may be responsible for the fact that certain intermediates sometimes apparently go on to arynes and sometimes do not. For example, the valence tautomers **37** and **38** are formed during the photolysis of phthaloyl peroxide (**72**)[121] but only go on to benzyne upon prolonged irradiation.[57] Similarly, the sulfur analogs **53** and **103** appear to lead to benzyne (**1**) when their precursor is the thiadiazinone **49c**[150] but not from the benzoxathianone **104**,[211] the thianaphthaquinone **105**,[212] or the thioanhydride **106b**.[213] The nitrogen analogs **107** and **108** apparently have never been shown to lead to benzyne regardless of whether the source is a triazinone (**49f**),[152] an anthranilate ester (**109**),[214] or an isatoic anhydride (**106c**).[213] Clearly some of the unproductive aryne precursors cited above might also yield arynes if the intermediates were generated under conditions where they could lose COX before undergoing a variety of bimolecular reactions.

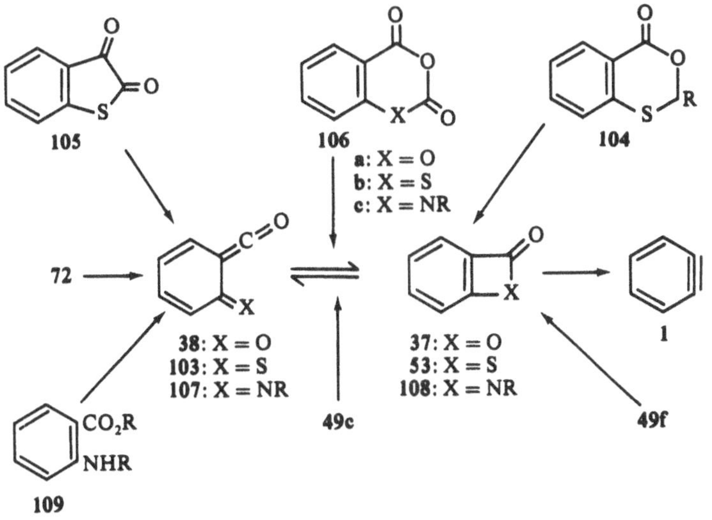

Finally, a recent review article[215] claims that the trioxolane or ozonide **110** suffers double β-scission to form benzyne (**1**) and formic anhydride analogous to similar reactions of aliphatic ozonides.[216] This is curious since unsaturated trioxolanes such as **110** derived from the reaction of singlet oxygen and isoben-zofurans (**86a**) have been well-studied and shown to fragment to *o*-diacyl-benzenes and not to benzynes (**1**).[217] Examination of the reference cited[215] for the transformation of **110** to **1** reveals that the original authors[218] were, in fact, studying the decomposition of the benzofuran ozonides **111** which fragment in the normal way to salicylic acid derivatives.

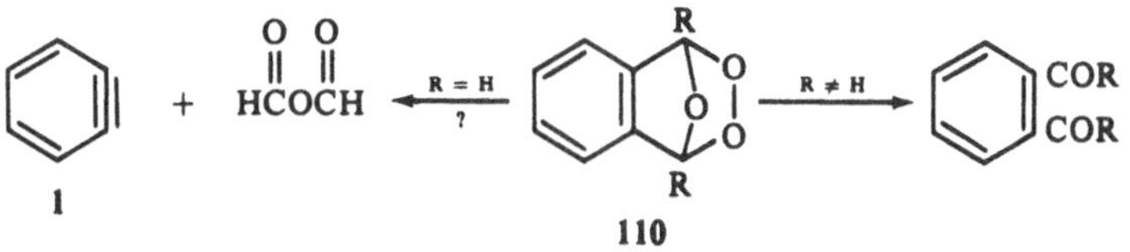

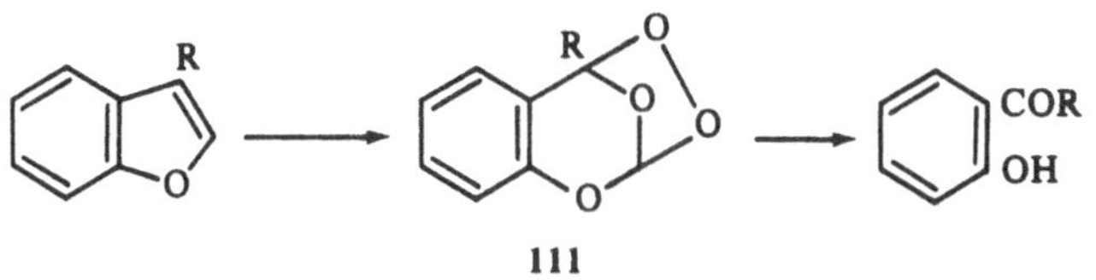

F. Arynes from Other Sources

The first two precursors in this category to be discovered (see Table 5), **112** and **113**,[53] contain the relatively weak C–I and C–Hg bonds which undergo either photolysis or thermolysis to yield benzyne (**1**). The latter process is noteworthy as one of the first generations of this intermediate in the gas phase.[219] Similar *o*-iodoorganometallic intermediates may be involved in the formation of arynes from *o*-diiodobenzene derivatives[114] in the presence of zinc[220] or copper.[221] Both direct thermolysis[186,222] and photolysis[223–226] of **114** also lead to arynes, presumably via a stepwise mechanism involving *o*-iodophenyl radicals (**47**). The extent to which the latter go on to arynes depends not only on the nature of the aryl residue[227] but, as mentioned in Section II.1.C, on the source of the radical center.[140] Pyrolysis of dibromo (**115**), iodonitro (**116**), nitrobromo (**117**), and dinitro (**118**) aromatic compounds in a mass spectrometer also produces the corresponding arynes and didehydro compounds.[187,228]

The thermolysis of heavy metal acid salts also leads to arynes. Silver *o*-halobenzoates decompose by a different,[229] apparently free radical,[230,231] pathway than their alkali metal counterparts (**32d**).[85] While the *o*-

TABLE 5.   Arynes from Other Sources o-C$_6$H$_4$XY

	X	Y	Conditions	References
112	HgI	I	*hv*	53
113	*o*-HgC$_6$H$_4$I	I	*hv*, 600°C	53
114	I	I	960°C	186,222
			Zn	220
			Cu	221
			hv	223-226
115	Br	Br	900°C	187,228
116	NO$_2$	I	900°C	228
117	NO$_2$	Br	900°C	187
118	NO$_2$	NO$_2$	900°C	228
119	COOAg	Cl	250°C	232,233
120	COOAg	Br	250°C	234
121	SO$_2$Ag	Cl	300°C	236
122	COOAg	COOAg	350°C	236,237
123	COO—Hg——OOC		300°C	238,239
	Cl	H	690°C	241
			Plasmolysis	245
	NO$_2$	H	600°C	242
	CF$_3$	H	600°C	243
	H	H	690°C	244
			Plasmolysis	245
	H	C≡CH	Plasmolysis	245

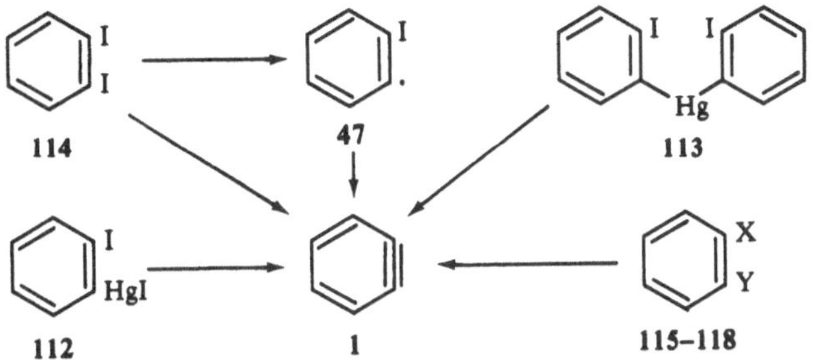

chlorobenzoates $(119)^{232}$ have been demonstrated to eventually yield arynes,[233] and the *o*-bromobenzoates (120) have been suggested[234] to behave in a like manner, neither the iodo[232] nor the fluoro analogs[235] give any evidence of aryne formation. Silver *o*-chlorobenzenesulfinates (121) also thermolize to arynes in a manner analogous to the benzoates $(119).^{236}$

Pyrolysis of silver phthalates (122) also leads to products which have suggested the intermediacy of either arynes[237] or the related diradicals $(1e).^{236}$ Mercuric phthalates (123) lose CO_2 in a stepwise manner to give the trimeric phenylmercury carboxylates (124) ($n = 3$) and the *o*-phenylenemercury trimers (125) ($n = 3$) before going on to aryne-type products.[238] Although it has been claimed that the monomeric form of 124 decomposes to benzyne $(1),^{239}$ both the trimer[238] and hexamer[240] of 125 apparently do not.

Finally, benzyne has been postulated to be formed in small amounts during the pyrolysis of chlorobenzene,[241] nitrobenzene,[242] benzotrifluoride,[243] and even benzene itself[244] by the *intra*molecular elimination of HCl, HNO_2, HCF_3, and H_2 respectively. Plasmolysis of halobenzenes, benzene, and phenylacetylene also provides arynes by a similar process.[245] *Inter*molecular dehydrogenation either with other arynes [equation $(11)]^{246}$ or nitrenes [equa-

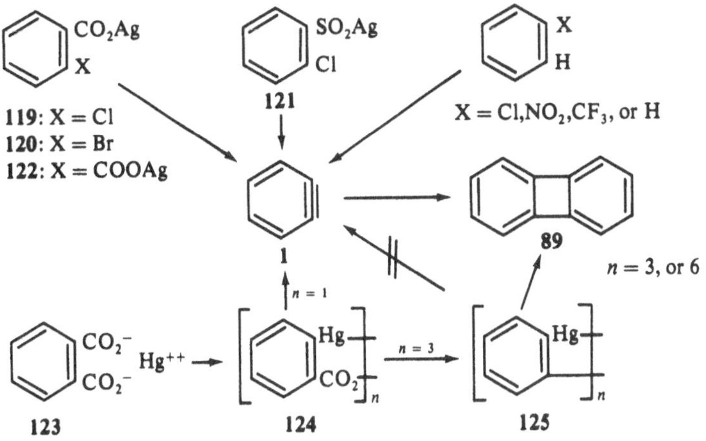

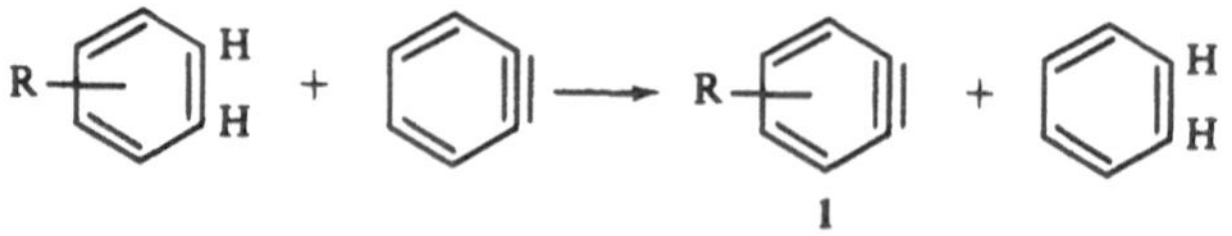

$$R + \bigcirc \substack{H \\ H} + \bigcirc \| \longrightarrow R + \bigcirc \| + \bigcirc \substack{H \\ H} \tag{11}$$

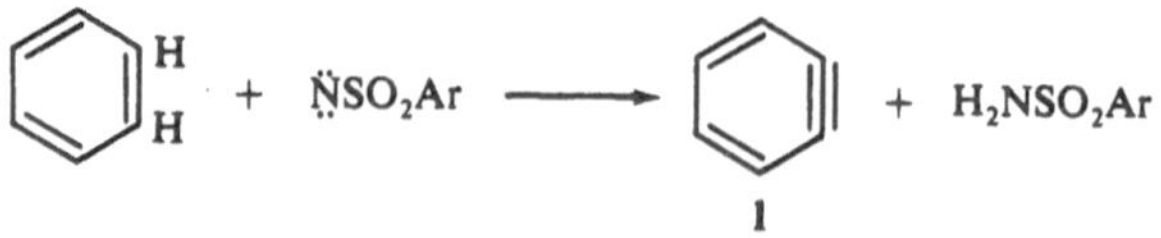

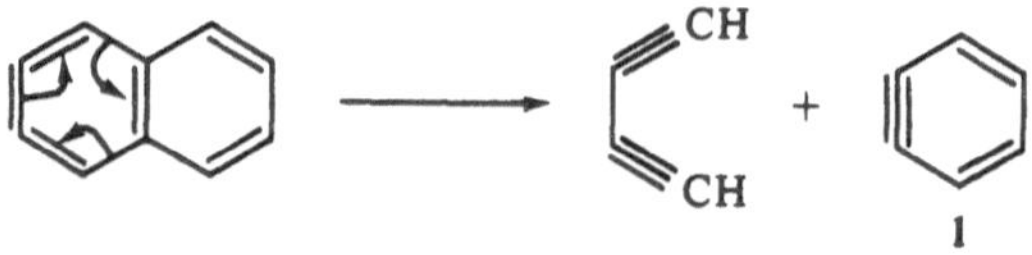

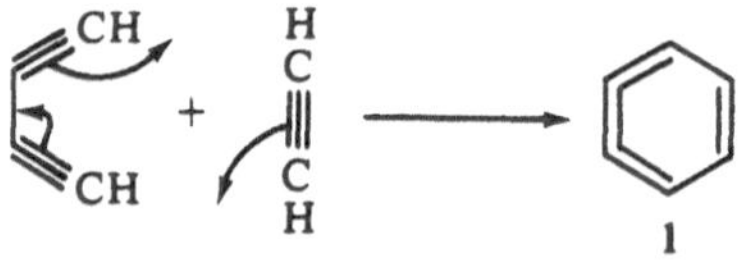

$$\text{1}$$

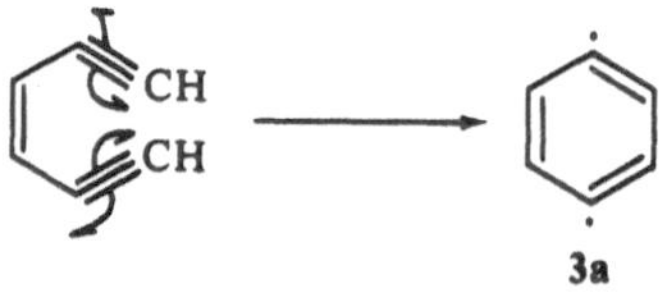

$$\bigcirc \substack{H \\ H} + \ddot{N}SO_2Ar \longrightarrow \bigcirc \| + H_2NSO_2Ar \tag{12}$$

$$\text{1}$$

$$\bigcirc\!\!\bigcirc \longrightarrow \substack{CH \\ \| \\ CH} + \bigcirc \tag{13}$$

$$\text{1}$$

$$\substack{CH \\ \| \\ CH} + \substack{H \\ C \\ \| \\ C \\ H} \longrightarrow \bigcirc \tag{14}$$

$$\text{1}$$

$$\substack{CH \\ CH} \longrightarrow \bigcirc \tag{15}$$

$$\text{3a}$$

tion (12)][247] as the hydrogen acceptors also has been suggested, although the last of these possibilities was recently disproven.[248] An additional generation of one aryne from another has been speculated to occur by a retro-Diels–Alder reaction [equation (13)],[72] the reverse of which [equation (14)] constitutes the sole claim[249] of aryne formation by a process not involving elimination. A related electrocyclic reaction [equation (15)] leads to *p*-benzyne (**3a**).[250]

2. Methods of Detection

Arynes are very reactive intermediates which enter into a large variety of bimolecular reactions including dimerization.[5] With the exception of apparently slow reactions such as equation (13),[57,72] no unimolecular decompositions of arynes are known. The *direct* detection of arynes, therefore, is possible in the gas phase or in dispersed, inert media. Utilizing such conditions the ultraviolet[52,184] and infrared[57] absorption spectra, ionization potential,[222,251] enthalpy of formation,[251] rate of dimerization,[183] and approximate lifetimes[52,54,219] of arynes have been measured. Most claims for the existence of arynes, and to date all those for five-membered hetarynes, are based on their *indirect* detection by trapping with a wide variety of reagents. In this section the criteria and reactions which have been used for this purpose will be examined. No attempt has been made to update[5] all the reactions of arynes except as they apply to the validity of claiming the existence of arynes.

A. Cine-Substitution

a. Elimination–Addition (EA) Mechanism

The first characteristic reaction of arynes to be discovered was the addition of polar species, especially nucleophiles, to the "triple bond."[252] Since arynes are bidentate intermediates, such additions could lead to two different products in the event of unsymmetrical substitution in the aryne ring. If, as is often the case for such polar additions, the aryne was generated by elimination of HX from a monosubstituted aromatic compound **126**, then the product **127** with the entering group in the position previously occupied by the substituent is called the product of normal or *ipso*-substitution while the rearranged product **128** is referred to as that of cine-substitution.[253] Should the aryne be generated from the isomeric precursor **129** then the designation of which is the normal

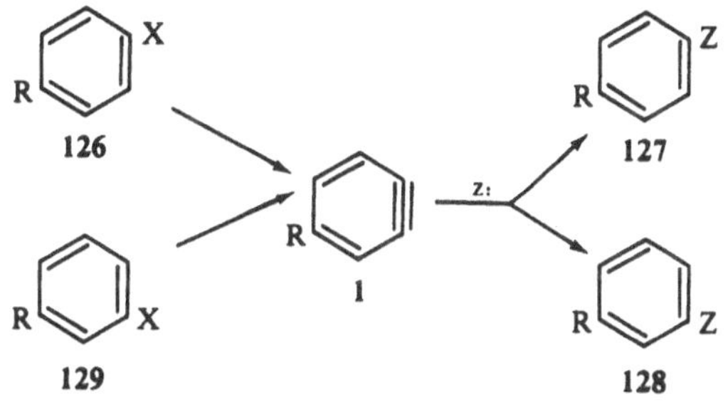

and which the cine-substitution product is of course reversed. It was the observation of such cine-substitution products in the reactions of haloaromatic compounds with strong bases which led to the first firm postulates of aryne intermediates (Section I.3.A) and the elimination–addition (EA)[1] mechanism of nucleophilic aromatic substitution.[254] It is now clear, however, that without additional information the occurrence of cine-substitution is neither a necessary nor a sufficient criterion for the existence of aryne intermediates. Examples of nucleophilic aromatic substitutions are known that occur via arynes but without cine-substitution, or with cine-substitution but without the intermediacy of arynes.

b. Normal Substitution by an EA Mechanism

The first of these possibilities can occur if the addition reaction to the aryne is regiospecific and leads only to the normal substitution product **126 → 1 → 127**. Such regiospecific reactions are known and are due to electronic and/or steric effects in either the aryne or the entering nucleophile.[252] For example, *m*-chlorobenzotrifluoride (**130**) and *m*-bromoanisole (**131**) give only the respective *m*-substituted anilines (**132**) upon amination[255] in spite of the demonstrated intermediacy of arynes.[256] Similar regiospecificity might be expected[2] even in unsubstituted hetarynes because of electronic directing effects of the heteroatom.

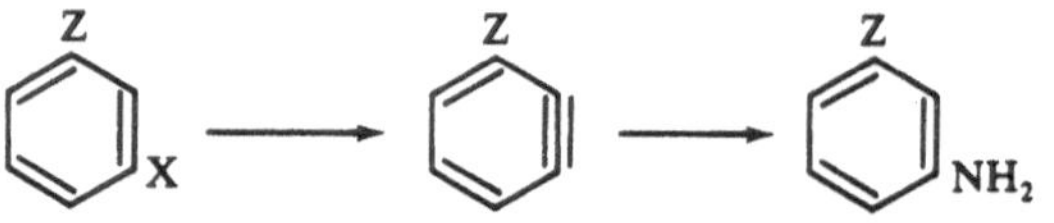

130: Z = CF$_3$; X = Cl   132
131: Z = OCH$_3$; X = Br

c. von Richter Reaction

The second possibility requires an alternative to the elimination–addition mechanism to account for the formation of the cine-substitution product. Several such reactions are known, the von Richter reaction [equation (16)] being the oldest and most thoroughly studied. Since the accepted mechanism of this reaction[257] specifically requires the presence of a nitro group, it has been suggested[8] that "if nitro groups are absent cine-substitution may be taken as good evidence of an aryne mechanism." While this statement may be valid for simple benzenoid compounds it does not necessarily hold for highly substituted or heterocyclic aromatics.

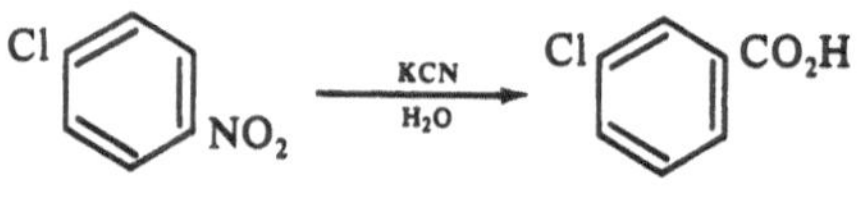

$$ (16) $$

d. Transhalogenation

In his classic paper on the mechanism of amination of halobenzenes[12] Roberts expended considerable effort and ingenuity to eliminate alternatives to the aryne mechanism. In particular he convincingly demonstrated that no rearrangement of the starting aryl halide took place. Such rearrangements do occur under aryne forming conditions, however, with polyhalobenzenes[258] and a variety of monohaloheterocycles.[75,259-266] This reaction has been labeled[267] "the base-catalyzed halogen dance" (BCHD) and shown[268] to proceed by a series of positive halogen transfers involving stable aryl anions. If such a rearranged halogen atom is eventually replaced via a normal aromatic nucleophilic substitution reaction, a cine-substitution product will be produced without intervention of an aryne intermediate [equation (17)]. Consequently the observation of cine-substitution products from aromatic compounds prone to rearrangement cannot be considered as conclusive evidence for the presence of arynes.

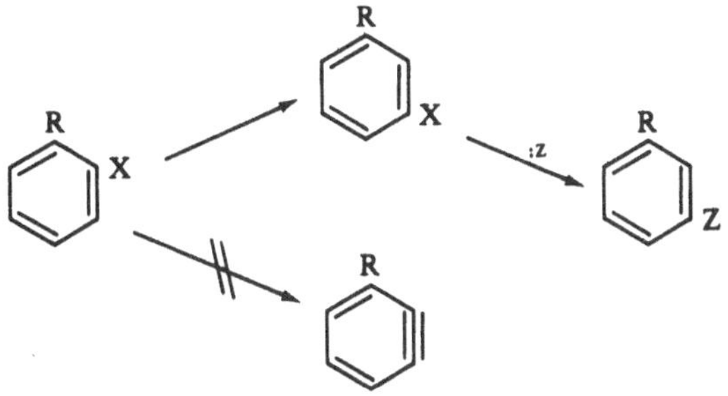

$$\text{(17)}$$

e. Abnormal Addition–Elimination (AEa)

Another nonaryne pathway to cine-substitution products is the abnormal addition–elimination (AEa)[1] mechanism [equation (18)] in which the nucleophile adds *ortho* to the leaving group which is subsequently eliminated. An essentially similar mechanism was suggested[269] and rejected[12] for the amination of halobenzenes, since the observed 1:1 ratios of normal to cine-substitution products from both chloro- and iodobenzene-1-^{14}C would require the highly unlikely, fortuitous combination of normal (AEn)[1] [equation (19)] and abnormal (AEa) [equation (18)] substitution for two different halobenzenes. Understandably, it was concluded[12] that the elimination–addition (EA)[1]

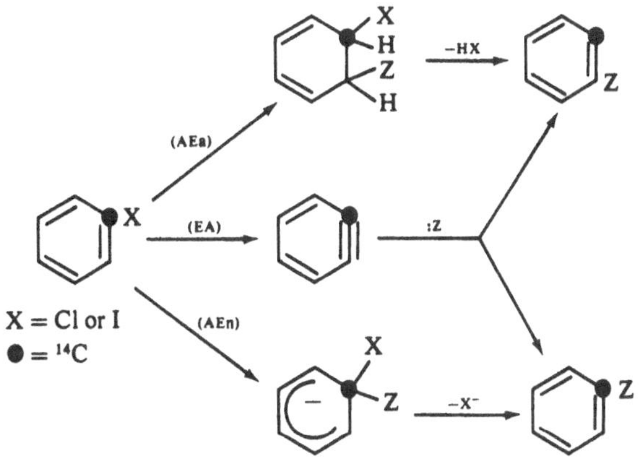

$$\text{(18)}$$

$X = \text{Cl or I}$
$\bullet = {}^{14}\text{C}$

$$\text{(19)}$$

mechanism via a benzyne intermediate was a more reasonable explanation. The AEa mechanism is not unknown, however, and has been recognized[270] as one of the mechanisms of nucleophilic aromatic substitution for several five[64] and six-membered heterocycles[270,271] as well as some carbocycles containing electron-withdrawing substituents.[272,273] The von Richter reaction [equation (16)] discussed previously also can be considered[270b] as a special case of the AEa mechanism. Even before these examples were recognized, however, the opinion was voiced that AEa mechanisms could not be *generally* excluded as possible explanations for cine-substitution.[274]

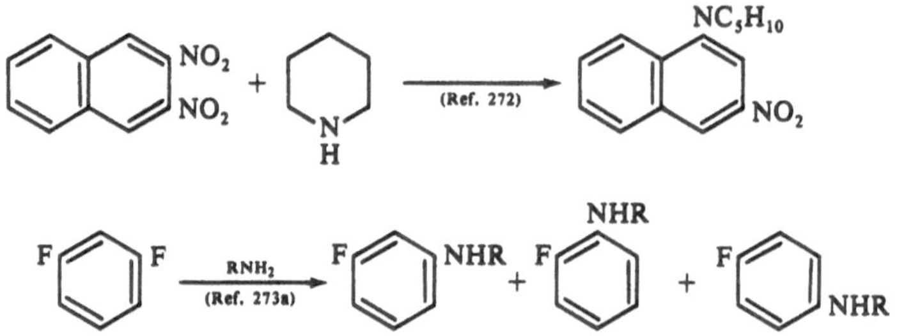

f. Addition–Substitution–Elimination (ASE)

The third mechanism considered[12] for the amination of halobenzenes, addition–substitution–elimination (ASE) [equation (20)], was deemed unlikely, since the observed ortho-hydrogen isotope effect would require the probably exothermic last step to be rate determining and the probably endothermic preceding steps to be rapid equilibria. Although these arguments were substantiated for halobenzenes by an elegant double-labeling experiment,[256] it has now been demonstrated that certain heterocycles do, in fact, react with ammonia[275,276] and other nucleophiles[277,278] rapidly and reversibly to give

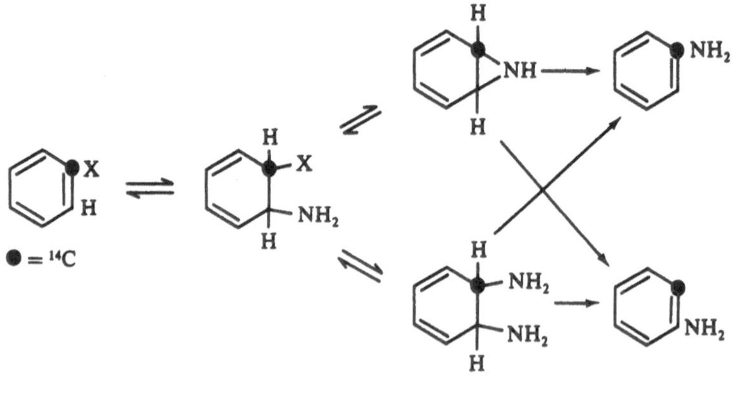

$$(20)$$

covalent addition products. Furthermore, such addition compounds have been shown to undergo substitution by a second equivalent of nucleophile at positions which are at,[279] adjacent to,[280,281] or remote from[282] the point of nucleophile attachment. In the case of the pyridazine N-oxide 133 subsequent elimination of the original nucleophile leads to an overall nucleophilic aromatic substitution by an ASE mechanism [equation (21)].[282] Recently, cine-substitution products have been isolated from such an ASE process (Section III.2.B) thereby demonstrating the plausibility of this mechanism for certain heteroaromatic compounds. In some quarters[283,284] it is still considered likely for halobenzenes.

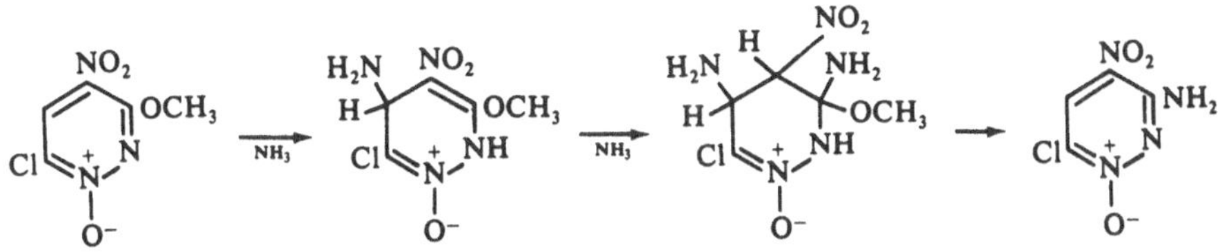

(21)

g. Addition–Ring-Opening–Elimination–Ring-Closure (ANRORC-cine)

A final nonaryne mechanism of cine-substitution is also based on the tendency of certain heteroaromatic compounds to add nucleophiles. Instead of elimination (AEa mechanism) or substitution followed by elimination (ASE mechanism), however, the initial addition product 134 undergoes a sequence of ring opening to 135, elimination to 136, and ring closure to 137. This variation of the ANRORC mechanism of nucleophilic aromatic substitution[285] was recently proposed[286] to explain at least part of the cine-amination of 4-substituted-5-halopyrimidines (138) previously[287] thought to proceed via an aryne 139.

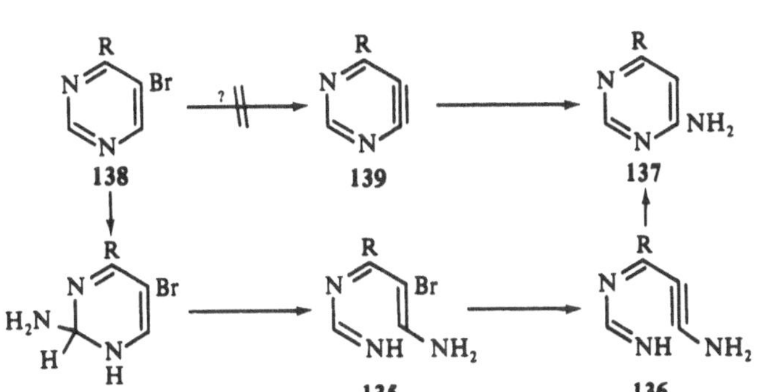

(22)

h. Precautions and Techniques

Although the validity of the aryne explanation for cine-substitution of carbocyclic aromatic compounds is generally, if not universally,[283,284] accepted, the above five examples support the prior assertion that without additional information cine-substitution is not a sufficient criterion for generally claiming an aryne intermediate, particularly in reactions of heteroaromatic compounds. It has been suggested[288a] that such sufficiency can be achieved "only if the nucleophile adds to *both* ends of the dehydro bond and the resulting isomer ratio turns out to be independent of the nature of the leaving group." In practice, the acquisition of this information may be complicated first by the simultaneous operation of several of the nucleophilic substitution mechanisms described above and second by the regiospecific addition of the nucleophile to the aryne owing to heteroatom-induced electronic effects. The "base competition method" has been used to detect an EA process in the presence of AEn and/or AEa mechanisms and also for cases where only a single substitution product is formed.[4,270,288b] These complications also can be overcome by intercepting the intermediate aryne with traps *which do not react with the aryne precursor.* Weakly basic amines,[270b] mercaptides,[289] and dienes (Section II.2.B.b) have been used for this purpose.

B. Cycloaddition

a. Dimerization and Trimerization

Their bidentate character suggests an additional characteristic reaction for arynes, namely, cycloaddition both with themselves and other reactants. During the early research on arynes this property was masked because all generation methods involved the presence of an excess of powerful nucleophiles which preferentially attacked the aryne. The introduction[47,290] of the *o*-dihalobenzene/lithium almalgam method for generating arynes (Table 1) circumvented this problem and led to the first observation of an apparent aryne cycloaddition to give the dimer and trimer of dehydrobenzene (1), biphenylene (89) and triphenylene (140), respectively. The ambiguity of this result with respect to the intermediacy of 1, and hence the occurrence of the cycloaddition reaction, was immediately recognized, however,[47,290] since both biphenylene (89) and triphenylene (140) could arise by stepwise coupling of the *o*-halophenyllithium 141 via the biphenyl 142 [291,292] and the *o*-terphenyl 143,[292] respectively. The formation of 140 during the photolysis of *o*-diiodobenzene (114) has been shown[223,293] to follow a similar stepwise path. Although it has now been demonstrated that in the gas phase, dehydrobenzenes do in fact dimerize to biphenylenes (89),[52,54,181] in solution, particularly in the presence of organometallic species, a stepwise process must be considered possible for the formation of **89** and probable for the formation of triphenylene (**140**).[294] Even if the intermediacy of an aryne can be demonstrated by other methods it does not necessarily follow that **89** and **140** arise from its dimerization or trimerization. A much more likely process for a reactive intermediate such as an aryne would be its reaction with organometallic species such as **141** or **142** to produce **89** and **140** via the biphenyl **142** and the terphenyl **143**, respectively.[294] These facts in conjunction with the expected instability of five-membered heterocyclic analogs of biphenylene such as **144**,[295,296] but not necessarily **145**,[297,298] compromise the value of observing (or not observing) such dimers as a criterion for the intermediacy of arynes and especially five-membered hetarynes.

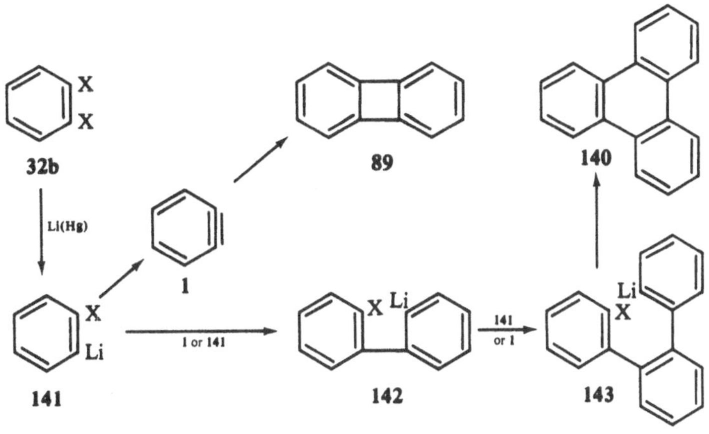

b. Diels–Alder Reactions

Wittig reasoned[47] that if the formation of biphenylene (**89**) and/or triphenylene (**140**) did in fact arise from the cycloaddition of benzyne (**1**) with itself then this intermediate ought to be sufficiently long lived to be intercepted by other cycloaddends. The validity of this hypothesis was soon verified for a variety of dienes such as furan (**146a**),[47,390] anthracene (**147**),[299] and many others (**146b–146e**).[300] It is now well known[301] that arynes undergo cycloaddition with both polar and nonpolar partners to form rings of all common sizes. The most characteristic of these reactions, however, remains the [4+2] cycloaddition or Diels–Alder reaction first explored by Wittig.[47,300] Because of the stability and availability of the diene traps, the stability and consequent good

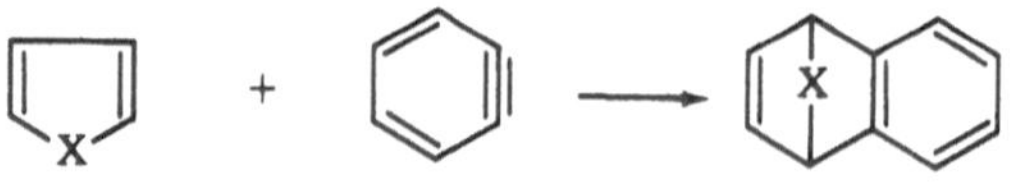

146   **1**   **148**

a: X = O;   d: X = NCH₃;
b: X = CH₂;   e: X = H and H
c: X = CH₂CH₂;

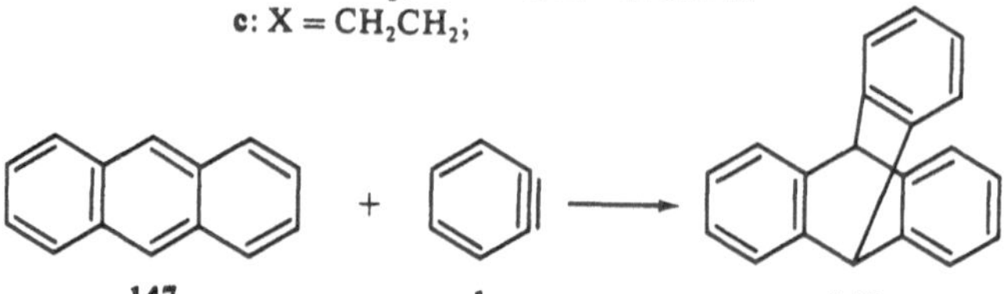

147   **1**   **149**

yield of the adducts, and finally the presumably concerted nature of the mechanism,[302] the formation of Diels–Alder adducts has become[301] the second major criterion for claiming the presence of an aryne intermediate in a reaction.

In spite of the general acceptance of this method of detection of arynes it suffers from the same limitation as does the cine-substitution criterion, i.e., the Diels–Alder adduct **148** or its transformation products might arise by an addition–elimination via **150** [equation (23)] rather than an elimination–addition process via benzyne [equation (24)].[303] The tendency for such an addition–elimination reaction to occur will depend on the nature of the aryne precursor, the existence of any intermediates during its decomposition to the aryne which might add to the diene, and finally the nature of the diene trapping agent itself. Potential precursors of five-membered hetarynes will be more prone to undergo addition reactions than benzyne precursors since the resonance energy of five-membered heterocycles is lower than that of benzene.[304] As already described in Section II.1, depending on the precursor, either carbanions, carbenium ions, free radicals, or other reactive species may be present during the formation of an aryne. Finally, there are certain dienes that have shown a

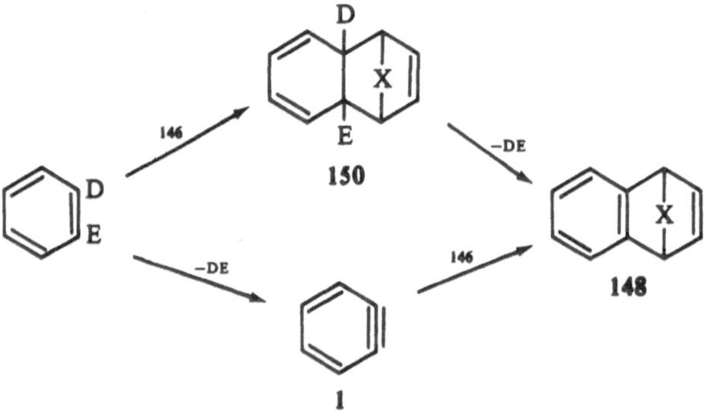

(23)

(24)

particular tendency to participate in the addition–elimination reaction, and must therefore be used with great caution when claiming the presence of an aryne.

c. Ambiguous Aryne Traps

The most prominent among these is tetraphenylcyclopentadienone (tetra-cyclone) (**151**), whose popularity as an aryne trap arises because its thermal stability[305] permits its use at high temperatures. Furthermore, although the initial adduct **152** is readily decarbonylated, the ultimate product **153** is very

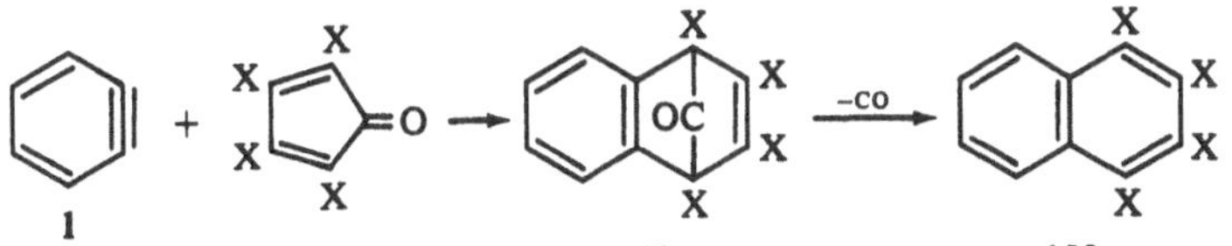

X = phenyl

stable and easily detected even in small quantities.[306] The earliest example of the formation of an "aryne adduct" from a cyclopentadienone is actually the very first example of the Diels–Alder reaction.[307] The dimerization of tetra-chlorocyclopentadienone (**154**) proceeds by an addition–elimination mechanism via **155** to give the same adduct **156** [308] which would have been expected from an elimination–addition process via the aryne **157**. Subsequently, many examples have been found[309] of the addition of alkenes to cyclopentadienones with the simultaneous evolution of CO and aromatization to give the same products as would have arisen from the corresponding alkynes. Although none of these examples involves the double bond of a benzene ring, several cycloalkenes[309] and, of special significance, several potential acenaphthyne (**158**) precursors,[310] have been shown to undergo this reaction.

A second common aryne trap which may lead to an erroneous inference is the *o*-quinoidal 1,3-diphenylisobenzofuran **159**. With benzyne precursors the

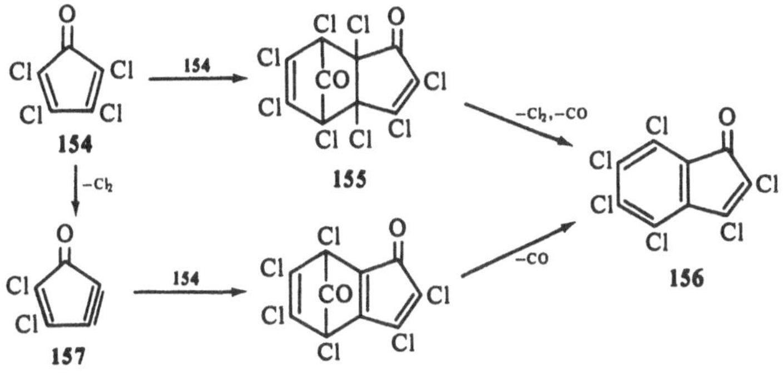

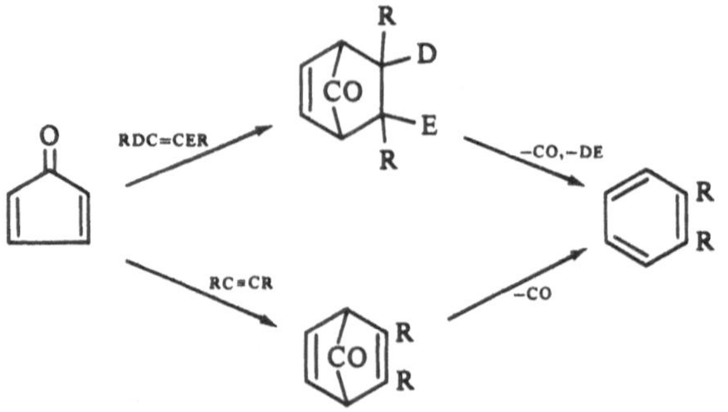

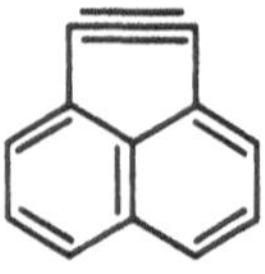

158 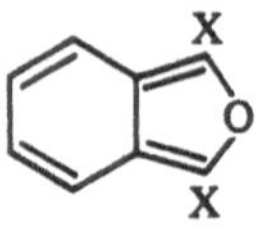 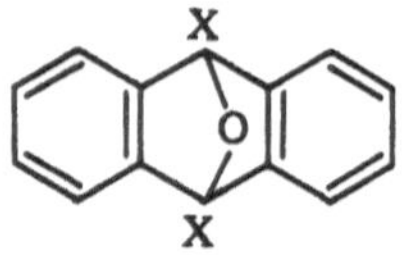   159   X = phenyl   160

expected adduct **160** forms by an elimination–addition mechanism,[311] but with the dibromoacenaphthene (**161**)[310] and the dibromocyclobutene (**162**)[312] the corresponding adducts **163 and 164** arise primarily, and probably totally, by an addition–elimination process which avoids the cycloalkynes **158** and **165**. By way of contrast the intermediacy of cyclopentyne **168** in a related process has been definitely established.[313]

Finally, 1,2,4,5-tetrazines (**169**) react with electron-rich alkenes[314] in what has been referred to[315] as the "Diels–Alder reaction with inverse electron demand" to yield, after elimination of N_2 and HY, the same pyridazine (**170**)

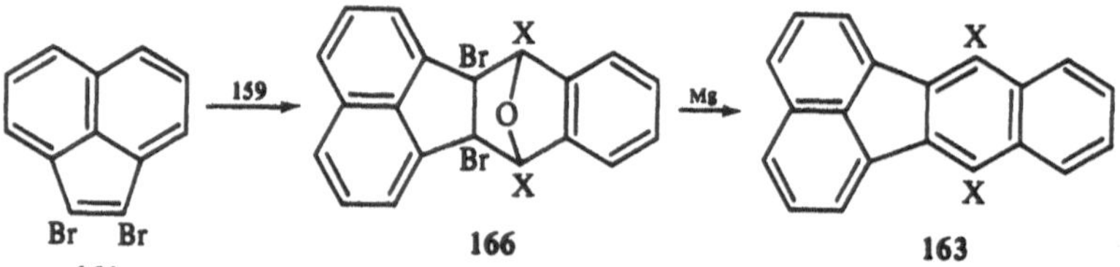

Br   Br
161   166   163

X = phenyl

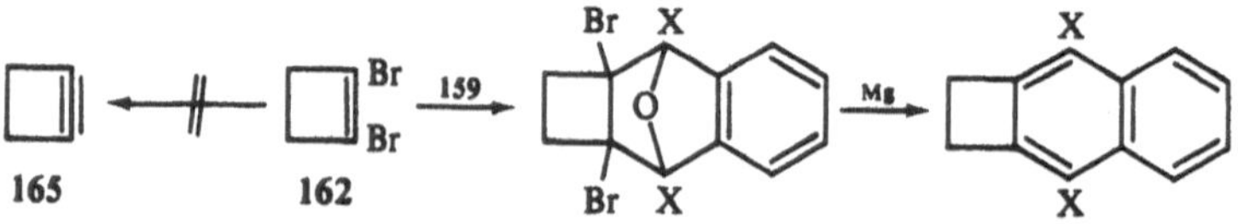

165   162   167   164

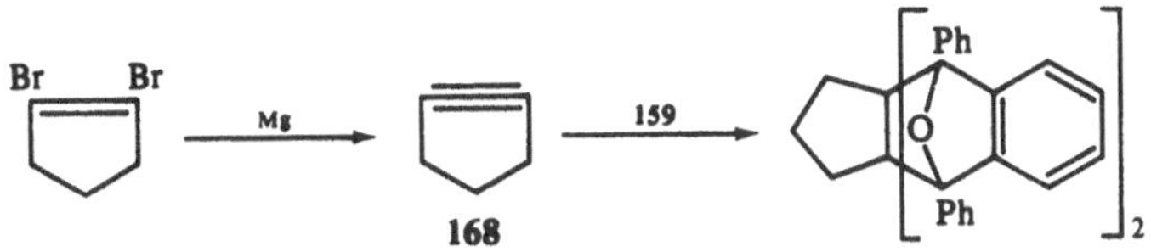

obtained directly from the corresponding alkyne.[314] Application of this reaction to electron-rich aromatics such as five-membered heterocycles[316] and their benzannelated derivatives[317,318] not surprisingly leads to similar adducts (171) identical to those expected from the corresponding five-membered hetarynes. Claims of aryne intermediates generated from electron-rich precursors which are based on the use of tetrazines as traps must, therefore, be regarded as equivocal.

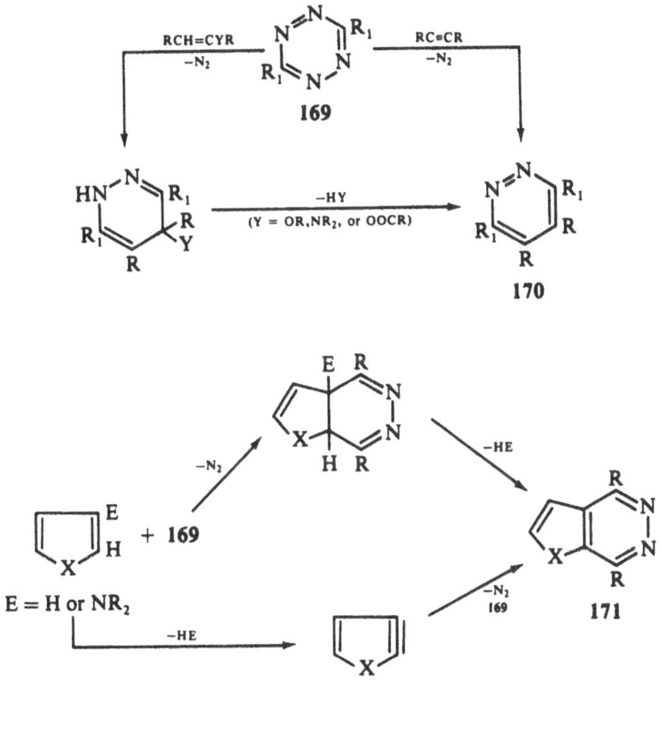

d. Suspected Aryne Traps

The well-advised caution with which the above traps should be utilized has led to other traps being suspected of leading to false aryne adducts as well. Most prominent in this category has been the first such diene trap to be discovered,[290] furan (146a). Gilman considered and presented cogent arguments

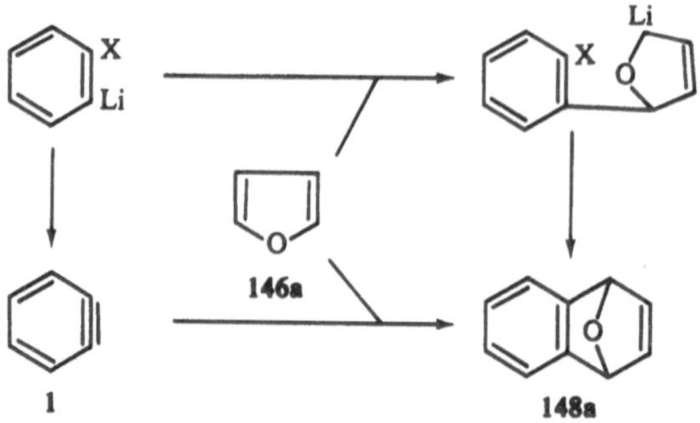

(25)

against the formation of the adduct **148a** by an addition–elimination mechanism (equation (25)).[319] Nevertheless, in some quarters[320] this possibility was considered more likely than the alternative benzyne hypothesis. The discovery of the addition–elimination reactions of isobenzofurans (**159**) with some cycloalkyne precursors,[310,312] apparently caused similar concerns about furan (**146a**).[4,303,321,322] However, the unusually high reactivity of **159** in Diels–Alder reactions[307] is due, not to the presence of the furan structure, but to the *o*-quinoidal system which upon [4+2] addition produces a benzenoid species such as **160**, **166**, or **167**. In those reactions of cycloalkyne precursors where the occurrence of an addition–elimination mechanism with furan was specifically tested for,[310,321] no supporting evidence was found. Furthermore, furan gives the aryne adduct **148a** from many different benzyne precursors,[301] a rather unlikely event if an addition–elimination mechanism such as equation (25) were involved. Finally, it has been demonstrated[57] that matrix-isolated benzyne reacts with furan to give **148a** upon warming to 50°K. At the present time, therefore, no evidence has been found which suggests that an aryne–furan adduct (**148a**) arises by any mechanism other than elimination–addition.

The failure to obtain such a furan adduct from an aryne precursor which yields other aryne–diene adducts (**148**) does not necessarily indicate the absence of arynes, however. One example of precisely this phenomenon[323] was traced to the unique ability of furan to preferentially consume the aryne precursor **41i** in a nonaryne pathway leading to 2-phenylfuran (**172**).[129a] Other examples are due to the instability of the benzyne–furan adduct **148a** to strong bases.[324]

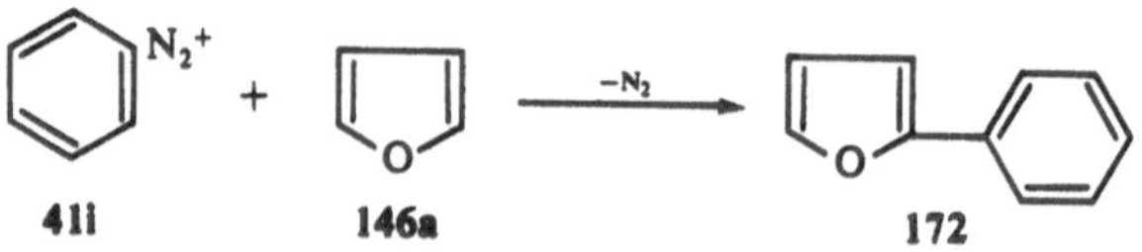

The possibility that triptycene (**149**) arose from anthracene (**147**) and some species other than benzyne (**1**) during the decomposition of *N*-nitroso-acetanilide (**41f**) was at one time considered,[323] but then rejected as the unraveling of the mechanism of this complex reaction showed that benzyne (1) was probably an intermediate.[128,129] Even the cycloalkyne precursors **161** and **162**, which reacted with tetracyclone (**151**) and isobenzofuran (**159**) by an addition–elimination mechanism, failed to react with anthracene (**147**).[310] Although the acid-catalyzed cyclization of 9-aryl-10-methylene-9,10-dihydro-anthracenes to triptycenes [equation (26)] provides a reasonable analogy for the

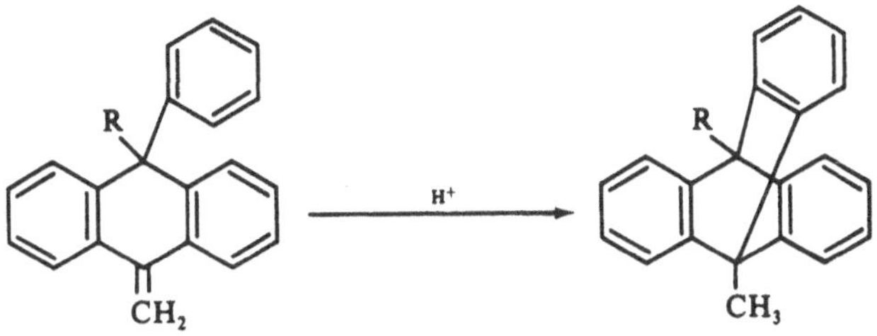

$$(26)$$

second step of an addition–elimination mechanism,[325] as with furan, the generality of triptycene (**149**) formation with different benzyne precursors[301] argues against such a mechanism. Known examples of this overall process appear to be restricted to powerful dienophiles such as tetrahalo-*p*-benzo-quinones (**173**), which lose halogen after the initial Diels–Alder reaction to give the "aryne" adducts (**174**).[326] Therefore, unless the aryne precursor has similar dienophilic properties, there is no reason to anticipate anomalous behavior for anthracene as an aryne trap.

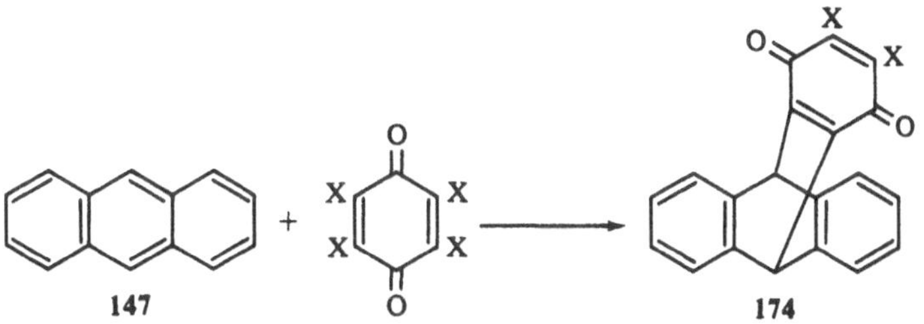

e. Precautions and Techniques

Considering the ambiguities and suspicions raised above it is worthwhile to consider some of the methods which may be used to determine if a particular "aryne" adduct (**148**) is actually formed via an aryne. Since the most common alternative involves the addition–elimination mechanism [equation (23)] this could be tested for specifically. The isolation of the initial Diels–Alder adduct (**150**)[310,312,319] and the demonstration that it goes on to the "aryne" adduct (**148**)[310,312] under the conditions of alleged aryne formation, while not eliminating the possibility of an aryne intermediate, certainly requires that other supporting evidence be presented. Conversely, the failure of the adduct (150) to go on to the "aryne" adduct (148) under the reaction conditions[313,327] may be taken as presumptive evidence that 148 must arise via an aryne. The failure to even isolate the adduct 150 from the reaction of the aryne precursor and the diene 146 is not conclusive, however, since this may simply reflect its low, but necessary, equilibrium concentration.[321] The absence of such an equilibrium can be demonstrated kinetically by the independence of the rate of decomposition of the precursor on the concentration of the diene trap 146.[321,328] It does not necessarily follow that promotion of aryne adduct 148 formation by the diene 146 indicates the absence of aryne intermediates, however. For example, in the decomposition of N-nitrosoacetanilide (41f) tetracyclone (151) serves the dual purpose of trapping the arynes formed and inhibiting a competing, non-aryne-producing, free radical chain process leading to biphenyl.[132] Nevertheless, whenever aryne formation is apparently enhanced by the presence of the trapping agent, as in the pyrolysis of nitrobenzene,[329] the possibility of alternative mechanisms should be considered.

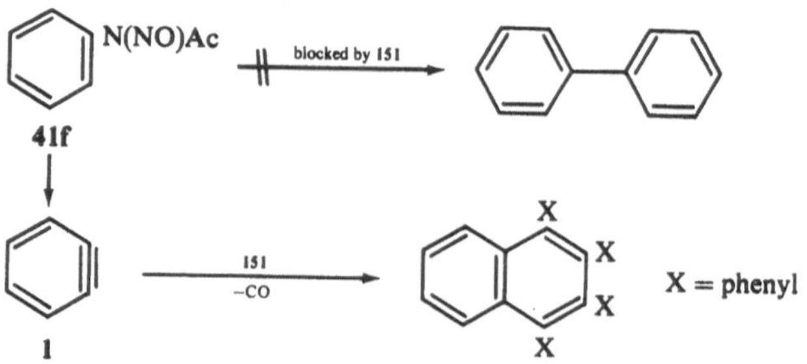

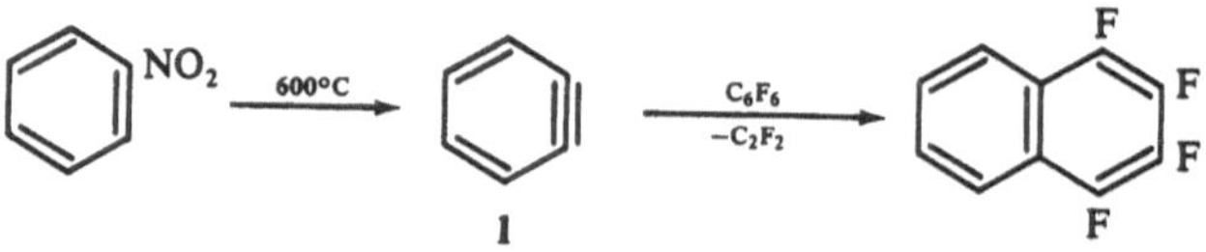

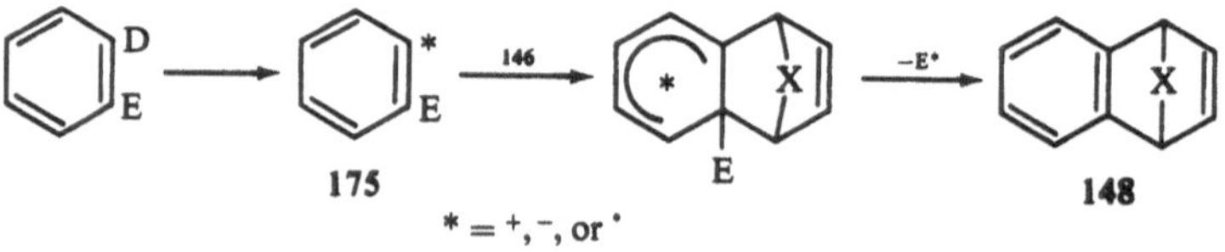

With some aryne precursors it is experimentally not possible to isolate, synthesize, or kinetically test for the Diels–Alder adduct (150). It is also conceivable that no such stable adduct exists, but that the aryne pathway is avoided by an addition–elimination process involving only reactive "arynoid" intermediates (175) [equation (27)]. In such cases a free aryne can be tested for by the competition method in which several different[55,56,330] or isomeric[331,332] precursors of the same aryne are allowed to react with a mixture of dienes[55] or a diene with several reactive sites.[55,56,330] If a common intermediate is present then the observed ratio of products should be independent of the structure of the precursors. This has been found to be the case for a sufficiently wide variety of benzyne precursors[55,56,330] that there is no question that this common intermediate is benzyne (1) and not some benzynoid species. If only a single type of aryne precursor is available then this method is naturally of little value.

With selected aryne precursors it has been possible to demonstrate the intermediacy of free arynes in cycloaddition reactions by other methods. The most convincing of these is the previously mentioned spectroscopic detection of benzyne (1) from the photolysis of phthaloyl peroxide (72) in an argon matrix at 8°K and its subsequent reaction with furan (146a) upon warming to give the adduct 148a.[57] A conceptually similar technique is pseudodilution,[333] in which

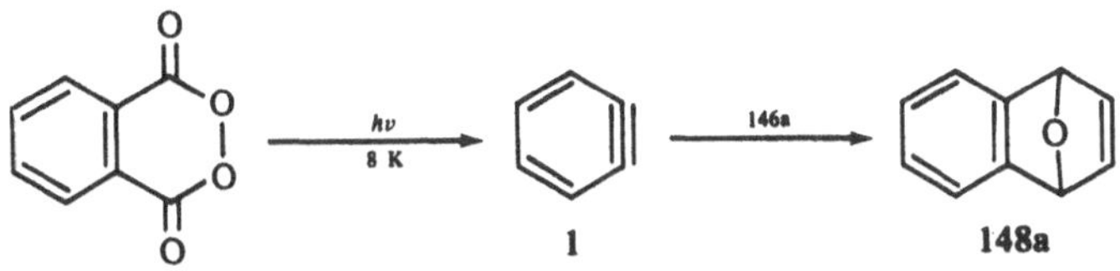

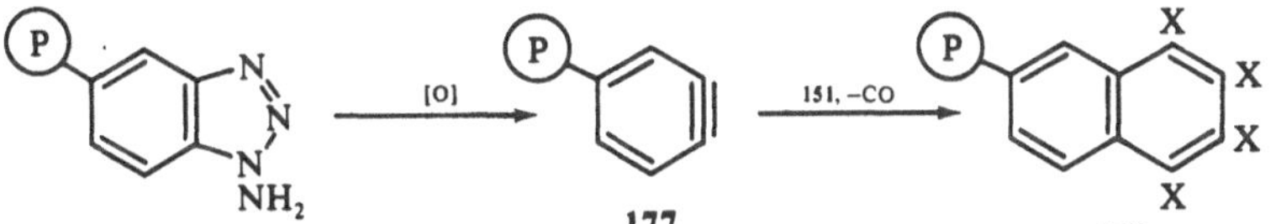

P = polymer   X = phenyl

the aryne precursor 176, and hence also the aryne 177, is immobilized on a polymer. Since the addition of the trapping agent can be delayed until after the precursor has completely reacted, the formation of the adduct (178) (which subsequently can be removed from the polymer) demonstrates that an aryne and not an arynoid species is involved. Finally, a related method for detecting free intermediates, which has not yet been applied to arynes, is the three-phase

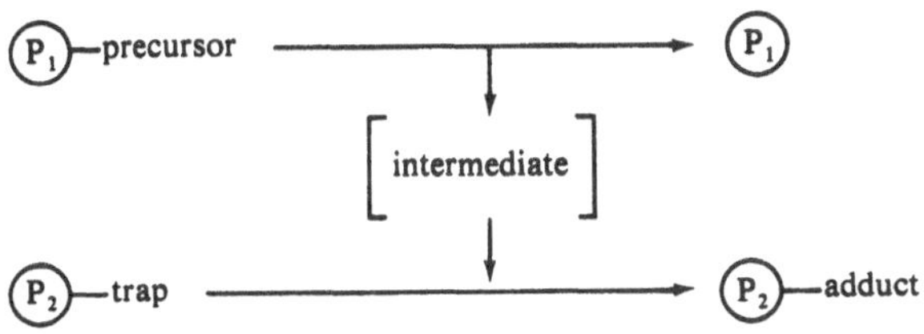

(28)

test.[334] In this technique the precursor of the intermediate and the trapping agent are each attached to different polymers, P_1 and P_2, respectively. In order for any appreciable reaction to occur a free intermediate which can diffuse from one polymer to the other must therefore be formed [equation (28)].

III. FIVE-MEMBERED HETARYNES

In this section the attempted and claimed generation of five-membered hetarynes and several closely related intermediates will be critically examined in the light of the preparative and diagnostic methods discussed in Section II. The organization, by kind and number of heteroatoms, follows that in earlier reviews on hetarynes and is based on the expectation that these factors will contribute significantly and perhaps uniquely to the chemistry of a particular five-membered hetaryne. This is certainly the case for the parent heterocycles[304] and might be even more important for hetarynes such as the 2,3-didehydro derivatives of pyrrole (179), furan (180), and thiophene (4) where interaction of the aryne orbitals with those of the adjacent heteroatom is possible.[259] For example, the extent to which such interaction stabilizes or destabilizes various polar forms of the hetaryne (Section I.2.B) ought to depend markedly on whether the heteroatom orbital involved is bonding (179) or nonbonding (180, 4), empty and "d" hybridized (4), or full and "sp^2" (180, 4), "s" (4) or "pd^2" (4) hybridized.[25,335]

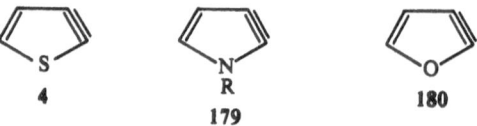

S	N	O
4	R	180
	179	

1. Oxaarynes

A. 2,3-Didehydrobenzofuran

As mentioned in Section I.3 this aryne, the first for which a structural formula was drawn, was postulated to rationalize the cine-substitution observed upon treatment of 3-bromobenzofuran (181) with ethanolic KOH or $NaOC_2H_5$ [equation (3)].[38] A more detailed examination of this oft-cited paper[1-5] reveals that the primary substitution products 182 and 183 were, in fact, never isolated except perhaps as a mixture containing 26% of the starting material (181). Their identification rests on the observation of characteristic color tests for the benzofuranones 184 and 185 formed after treatment of this mixture with gaseous HCl. The only products actually isolated in pure form from the reaction were *o*-hydroxyphenylacetic acid (186) (27% yield) and the corresponding ethyl ether (187). The acid 186 is also formed in nearly quantitative yield when 2-bromobenzofuran (188) is subjected to the above reaction with base.[38]

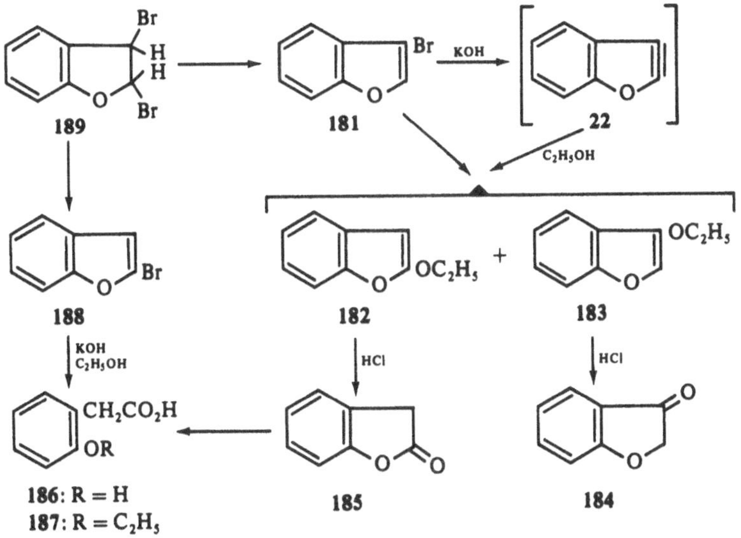

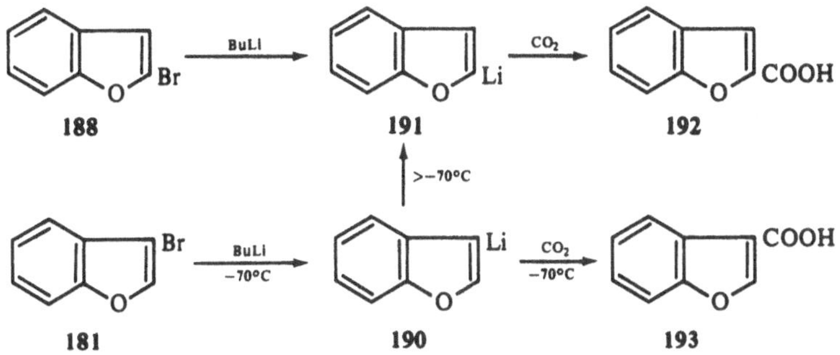

In order for the aryne interpretation of these results to be sustained it would be necessary not only to verify the formation of the primary substitution products 182 and 183 but also to demonstrate the absence of any of the other cine-substitution mechanisms discussed in Section II.2.A. Furthermore, since both bromobenzofurans 181 and 188 are prepared from the same dibromo derivative 189, the possibility that the major product 186 actually arises from contamination of 181 with 188 or 189 also must be considered. Precisely such an explanation was found to be responsible for the formation of all four products 184–187 when an assumed single chlorobenzofuran was treated with base as above.[336,337] (Interestingly, the possible formation and the structural formula of the aryne 22 were also mentioned in these earlier papers.) Because of this prior experience with such a problem, special precautions were taken by Stoermer[38] to ensure the purity of the starting materials thereby rendering the contamination hypothesis unlikely.

The possibility of rearrangement of the 3-isomer **181** to a derivative of the 2-isomer **188** under the basic reaction conditions is also possible in light of the results of some halogen–metal interconversion studies.[338] Although both bromo-benzofurans gave the expected acids **192** and **193** after carbonation at −70°C, at higher temperatures the 3-(**181**), but not the 2-isomer (**188**), gave rearranged acid **192**. The parallel with Stoermer's results[38] in this regard is striking. The mechanism of the key rearrangement step **190** to **191** was postulated[338] to proceed by what is now known[267] as the "base-catalyzed halogen dance" (BCHD) (Section II.2.A.D) involving the *o*-bromolithium compound **194**. Variations of this process which can account for the formation of the products might be operative in Stoermer's system.

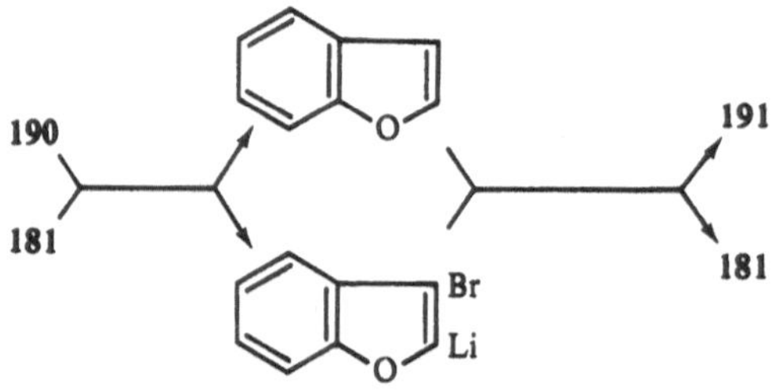

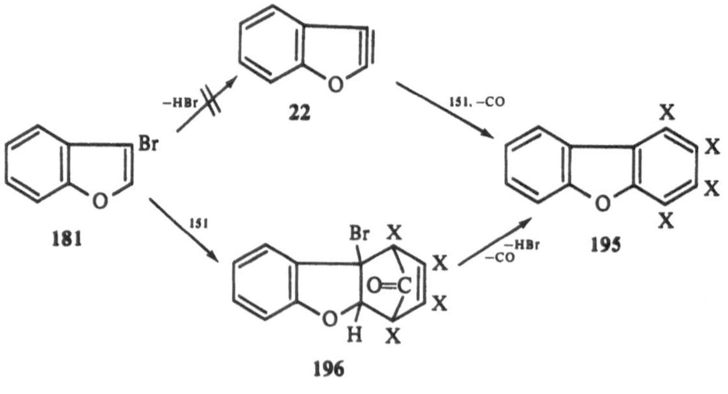

In order to probe for the aryne **22** by the cycloaddition criteria (Section II.2.B) Stoermer's reaction was carried out in the presence of tetracyclone (**151**).[339] Although the expected aryne adduct **195** was formed in 20% yield, it was subsequently shown[340] that the rate of this reaction is nearly doubled by the addition of tetracyclone (**151**). It was further demonstrated that the adduct **195** is even produced from **151** and **181** in the absence of base altogether.[341] As discussed in Section II.2.B both of these observations are indicative of an addition–elimination process via the Diels–Alder adduct **196**, not of an aryne mechanism.

The utility of o-bromoaryllithium compounds (**32b**) as aryne precursors (Section II.1.B) even at −100°C [342] suggests that the benzofuran analog **194** also might lose LiBr to give the aryne **22**. Surprisingly this compound displays the normal reactivity for an aryllithium reagent at −75°C [343] and is thermally stable at room temperature.[344] At still higher temperatures ring opening to the o-hydroxyphenylacetylene (**197**) is observed,[345] as is also the case in the reaction of **181** with n-butyllithium.[338] Presumably a BCHD intervenes to generate either **190** [346] or **198** [348] as the species which undergoes ring opening. Even if this reaction is carried out in the presence of furan (**146a**) as an aryne trap, **197** remains the major product.[346] The sodium analog of **194** displays similar thermal stability.[346]

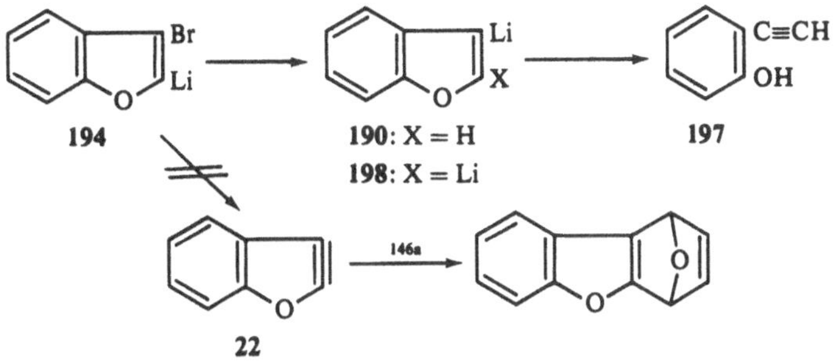

194 → 190: X = H / 198: X = Li → 197

22 / 146a

Although o-haloarylmercury derivatives (**113**) (Section II.1.F) are thermally more stable than the corresponding lithium compounds,[53] they do not undergo the BCHD and hence might survive a temperature sufficient to cause elimination of HgX$_2$ and formation of the aryne. Preliminary results indicated that the mercury compound **199** did in fact give the expected aryne adduct **195** in 70% yield when heated with tetracyclone (**151**).[347] Once again an addition–elimination mechanism was implicated by the observation that while **199** alone was stable in boiling decalin, in the presence of tetracyclone (**151**) the adduct **195** was obtained in 92% yield.[348] Clearly, **151** reacts directly with **199**

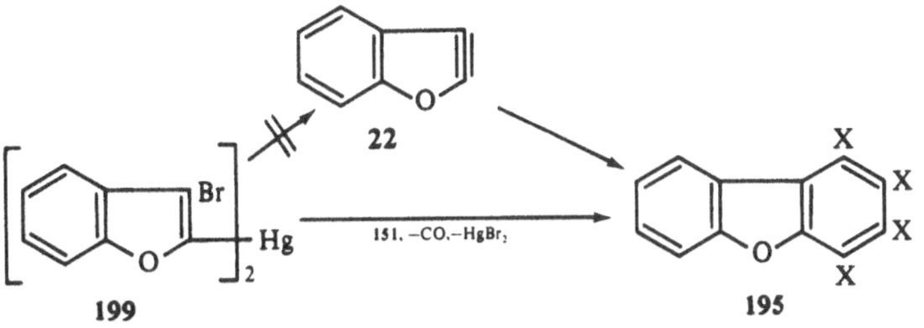

22

199   151, −CO.−HgBr$_2$ 195

and not with the hetaryne 2,3-didehydrobenzofuran (22) for whose existence no convincing evidence has yet been presented.

B. 2,3-Didehydrofuran

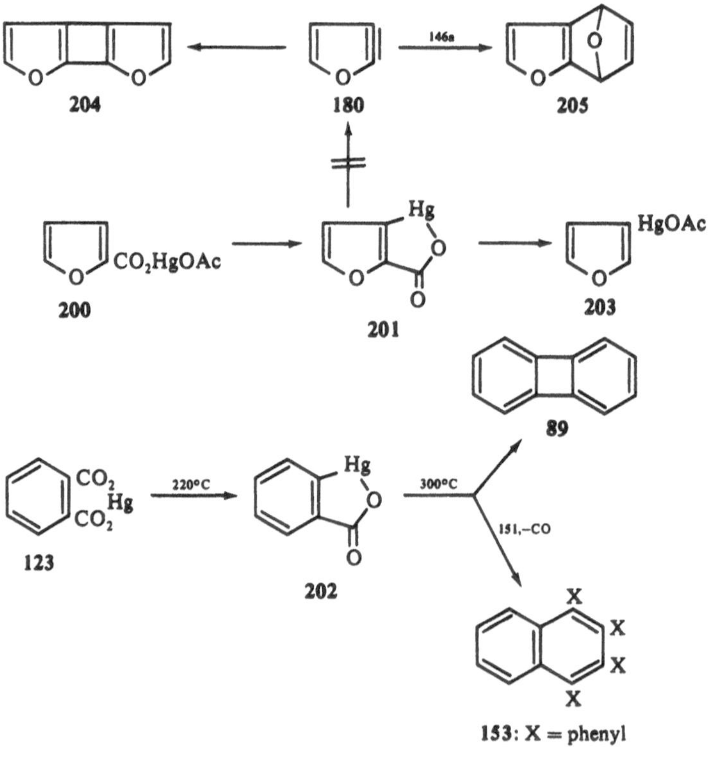

No claims for the existence of this aryne have been made, although the reactions of several potential precursors have been examined. Thermolysis of the mixed mercury salt 200 has been postulated[349] to lead to an intermediate 201, which is analogous to an apparent benzyne precursor 202. Whereas the trimer of the latter[238] loses CO_2 and mercury under more severe conditions to give biphenylenes (89)[238] or, in the presence of tetracyclone (151), the aryne adduct 153,[239] the species 201 gives primarily 3-acetoxymercurifuran (203).[348] Neither the heterocyclic biphenylene 204 nor the aryne-furan adduct 205, which might have been expected since furan (146a) is a major by-product of this reaction, were found. The conclusion that the aryne 180 is not formed in the decomposition of 200 is probably not justified by this observation, however, since modern techniques for the generation and detection of potentially unstable products such as **204** and **205** (Section II.2.B.a) have not yet been applied to this reaction.

Several 2-lithio-3-bromofurans (**206**), prepared either by halogen–metal interconversion from the 2,3-dibromofurans (**207**)[350,351] or by metallation of the 3-bromofurans (**208**),[352] show typical properties of aryllithium reagents and give no evidence for the intermediacy of 2,3-didehydrofuran (**180**).

C. Didehydromaleic Anhydride

As discussed in Section I.2.C a species such as **23** is not strictly an aryne since the parent compound, maleic anhydride, is not aromatic. Nevertheless its attempted preparations and anticipated properties are likely to be very similar to those of true five-membered hetarynes. Except for acenaphthyne (**158**),[353,354] didehydromaleic anhydride (**23**) is apparently the first species containing a triple bond in a five-membered ring which was the goal of a synthetic effort.[39] As noted in Section I.3.A attempts to dehalogenate either dibromo- (**209**) or diiodomaleic anhydrides (**210**) by unspecified methods failed to yield **23**.[39] A modern reinvestigation of this reaction revealed that either thermolysis of **210** or treatment of dichloromaleic anhydride (**211**) with Zn or Cu led only to a jet-black, conjugated polymer (**212**) whose empirical formula $(C_4O_3)_n$ corresponded to that of the aryne (**23**).[355] No trace of either the dimer (**213**) or the trimer (**214**) of the aryne (**32**) was observed. Reaction of the diiodo compound (**210**) with tetracyclone (**151**) also gave a polymer (**215**) whose empirical formula corresponded to that of the aryne adduct (**216**). Since the latter compound, tetraphenylphthalic anhydride, ought to be quite stable to polymerization under the reaction conditions, the polymer (**215**) probably arises from **212** or from the Diels–Alder adduct (**217**) formed by direct reaction of **211** with tetracyclone (**151**), as might be anticipated by the discussion in Section II.2.B.c.[309]

A closely related reaction is observed with chloromaleic anhydride (**218**), which gives the apparent aryne adduct **216** with tetracyclone (**151**) but stops at the Diels–Alder adduct **219** with other dienes such as cyclopentadiene (**146b**), butadiene (**146e**), and anthracene (**147**).[356] Not only does this result suggest the absence of an elimination–addition mechanism via the aryne (**23**) in this system,

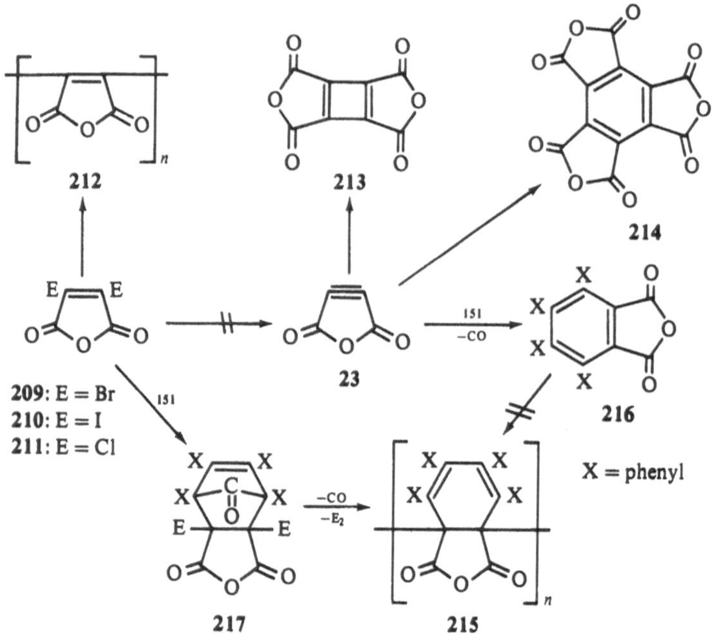

212 213 214

209: E = Br
210: E = I
211: E = Cl

23

216

X = phenyl

217 215

but it also supports the claim (Section II.2.B.d) of the relative safety of dienes such as **146b**, **146e**, and **147**, compared to tetracyclone (**151**), as a probe for arynes.

Mass spectral examination of the potential aryne precursors **210** and **211** revealed no fragments corresponding to the cation radical of the aryne (**23**).[355] Since virtually all compounds which thermolyze to arynes also display the corresponding cation radical in their mass spectrum,[177,179] the suitability of these compounds as a *thermal* source of **23** is questionable.

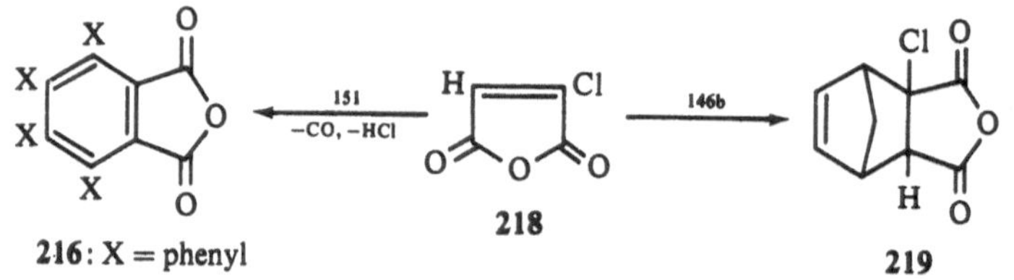

216: X = phenyl   218   219

Accordingly, the photolysis of the dihalomaleic anhydrides was examined.[357] In the solid state the dichloro compound **211** gave the same polymer **212** as was obtained from thermolysis,[355] a result which permits—but does not require—an aryne interpretation. In dilute cyclohexane solution, however, a monosubstitution product (**220**) (and no polymer) was isolated, indicating a stepwise decomposition. Irradiation of the dibromo compound **209** in the presence of N-phenylpyrrole (**221**) as a potential aryne trap[300] did not give the aryne adduct **222** but the double substitution product **223**, which once again was shown to form by a stepwise process via **224**.[358]

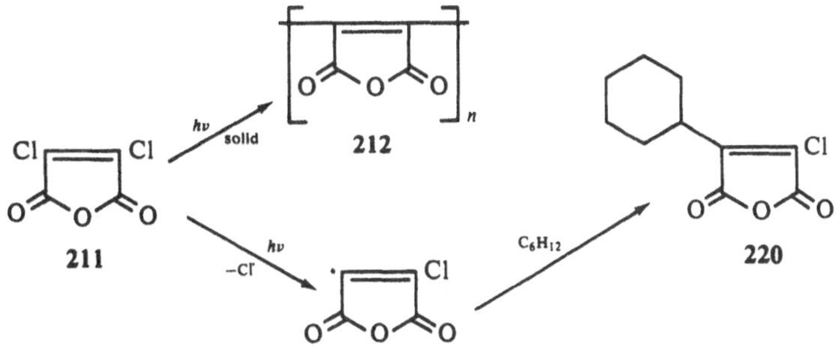

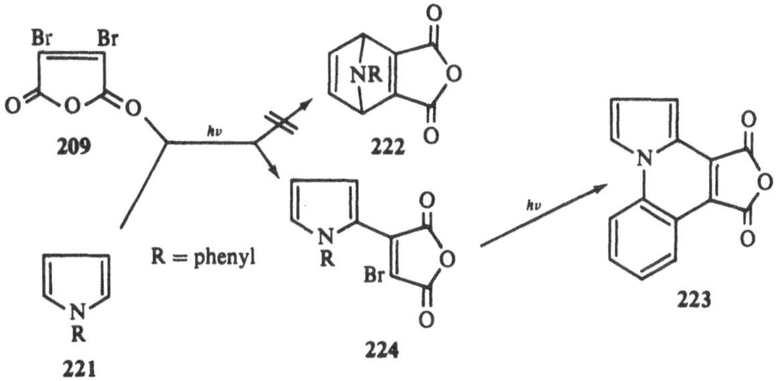

Recognition that **23** is actually the anhydride of acetylene dicarboxylic acid (**225**) led to some early attempts and claims of cyclization of this molecule with various reagents.[359] Using acetic anhydride, the product isolated was acetoxymaleic anhydride (**226**),[359] postulated[41] as possibly arising from the addition of acetic acid to **23**. Pyrolysis of either **226** or of diacetyltartaric acid

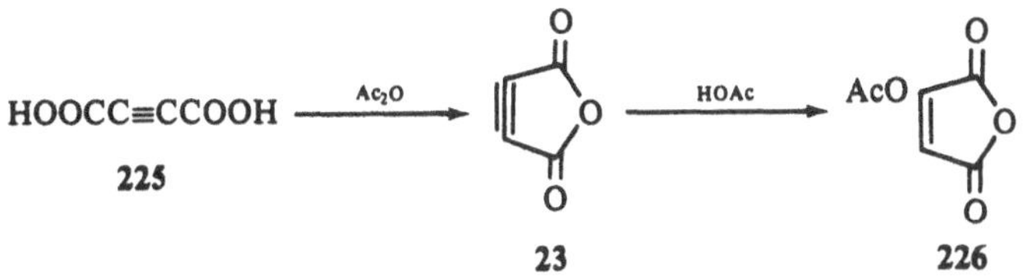

anhydride (**227**) produced carbon suboxide C_3O_2, presumably via the aryne **23** and the unlikely cyclobutyne **24**.[41] In a modern day study pyrolysis of a mixture of acetic anhydride and diethyl-4-^{14}C-oxaloacetate gave C_3O_2 containing one-half the isotopic label as would be predicted were **23** in fact an intermediate.[60] A more detailed examination of this reaction, however, revealed that the mechanism is quite complex with some of the acetic anhydride carbons ending up in the C_3O_2. Furthermore, it was shown by ^{14}C and ^{18}O labeling experiments that **226** and **227** could not lead to C_3O_2 via a symmetrical intermediate such as didehydromaleic anhydride (**23**).[360]

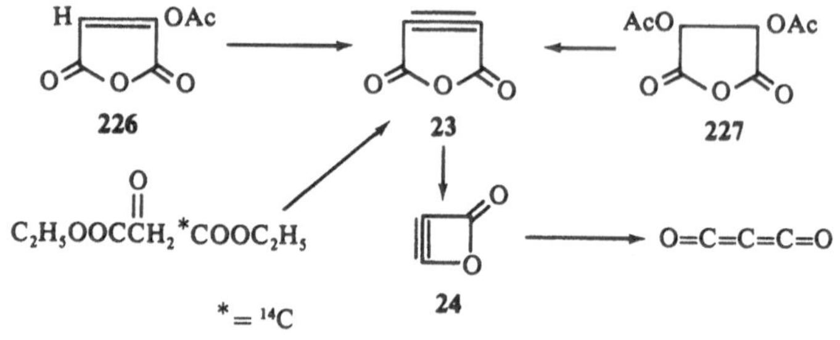

226 **23** **227**

$$C_2H_5OOCCH_2{}^*COOC_2H_5$$

$$* = {}^{14}C$$

24

$$O=C=C=C=O$$

The formal Diels–Alder adduct (**228**) of anthracene and didehydromaleic anhydride (**23**) was subjected to flash vacuum thermolysis at 600°C in the hope that a retro-Diels–Alder reaction (Section II.1.E) might generate the aryne **23**. Although the theoretical yield of anthracene was obtained, the aryne **23** could not be frozen out or indirectly detected by trapping experiments; the only other identifiable products were CO and CO_2,[361] which could have arisen directly from the anhydride moiety of the adduct **228**.[175–179]

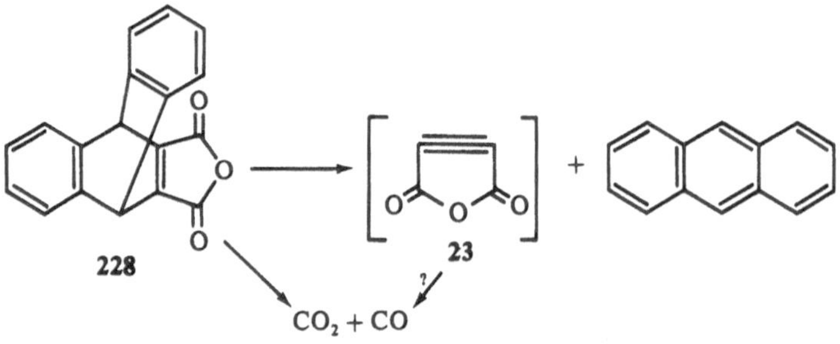

228 **23**

$$CO_2 + CO$$

One final reaction in which **23** has been considered an intermediate is the formation of phthalic anhydride (**73**) from the plasmolysis of thiophene-2,3-dicarboxylic acid anhydride (**229**) in the presence of acetylene.[362] It was suggested that a "1,3-dipolar dissociation"[363a] of **229** gives thioketene (C_2H_2S) and presumably **23**, which reacts with cyclobutadiene (known to be present during the plasmolysis of C_2H_2)[363b] to give phthalic anhydride (**73**) via **230**. Naturally, no proof for this scheme exists. For that matter, no convincing proof for the existence of any oxaaryne can be considered available at this time.

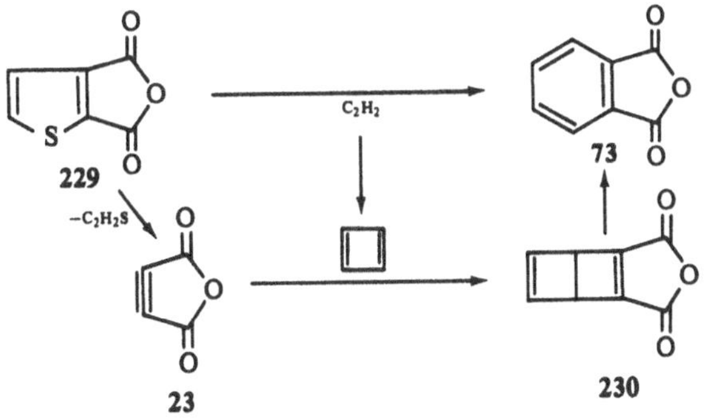

2. Azaarynes

A. 2,3-Didehydroindole

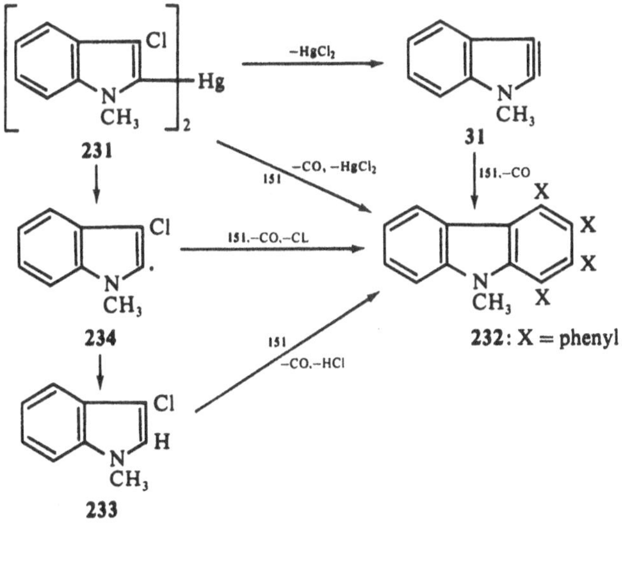

All of the attempted generations of this type of aryne to date have involved organometalic precursors or strongly basic reaction conditions, either of which is incompatible with the presence of the N–H group of the indole nucleus. Consequently only the preparation of the N-methyl derivative **31** has been investigated. Decomposition of the mercury compound **231** in the presence of tetracyclone (**151**) gave both the expected aryne adduct (**232**) and 3-chloro-N-methylindole (**233**).[364] The isolation of the latter compound indicates that the precursor **231** decomposes by a stepwise process probably via the radical **234**. Reaction of tetracyclone (**151**) with either this radical or directly with the starting material (231) was suggested as an alternative to an aryne mechanism to explain the formation of the adduct (232).[51] Although neither of these alternatives was proven it was shown that the 3-chloro compound 233 was as efficient a precursor of the adduct 232 as was the mercury compound 231.[365] Since thermal elimination of HCl from 233 at only 250°C to give the aryne (31) is highly unlikely (Section II.1.F), the most reasonable explanation for the adduct formation in this case involves an addition–elimination mechanism; and since 233 is a product of the decomposition of the mercury compound 231, the isolation of the adduct 232 can be rationalized without involving the aryne (31).

The lithium compound 235 is stable below 220°C except in the presence of the isobenzofuran 159, which causes decomposition at 180°C to give a low yield of the aryne adduct 236. Both this induced decomposition of the precursor 235 and the fact that 3-chloro-*N*-methylindole (233) itself once again reacts with the trap (159) to give the adduct 236, strongly indicate an addition–elimination process which avoids the aryne 31.[365]

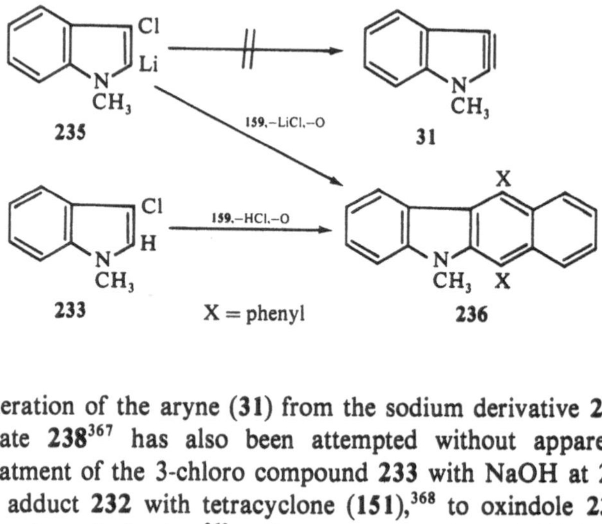

The generation of the aryne (31) from the sodium derivative 237[366] or the halocarboxylate 238[367] has also been attempted without apparent success. However, treatment of the 3-chloro compound 233 with NaOH at 200°C leads to the aryne adduct 232 with tetracyclone (151),[368] to oxindole 239 in up to 31% yield by cine-substitution,[369] and to an aryne trimer (240) in up to 16% yield.[370] As already mentioned, the fact that 233 alone reacts with tetracyclone (151) to give the adduct 232 suggests that an addition–elimination rather than an aryne process is involved in this reaction. The cine-substitution reaction of

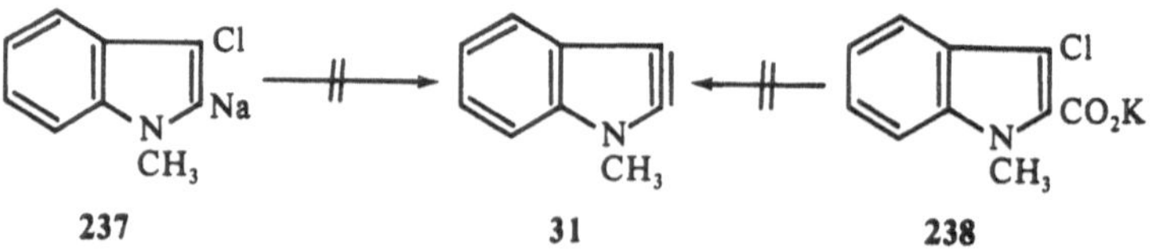

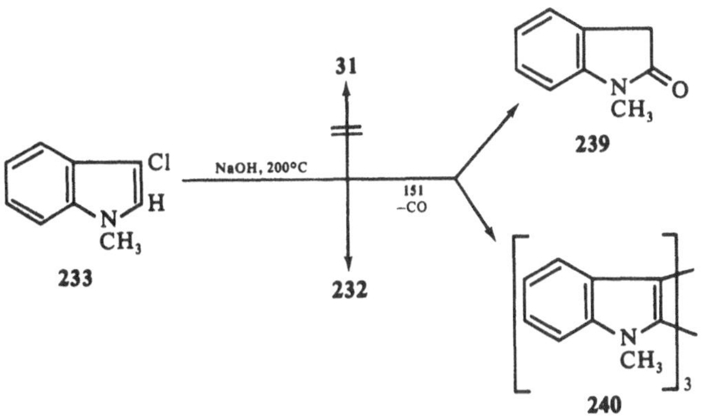

233 to 239 might proceed by any of the nonaryne mechanisms discussed in Section II.2.A and the trimerization might involve either a stepwise Wurtz coupling as discussed in Section II.2.B.a or the base-catalyzed aldol condensation of indoxyl (241) formed by normal nucleophilic aromatic substitution of 233. It has been suggested[371a] that the origin of the trimer 240 might be revealed if its exact structure were known. Whereas either of these nonaryne processes would lead to the symmetrical trimer 240a, the stepwise addition of intermediate *o*-haloanions such as 242 and 243 to the aryne 31 was predicted to give the unsymmetrical trimer 240b. This prediction was based on the assumed

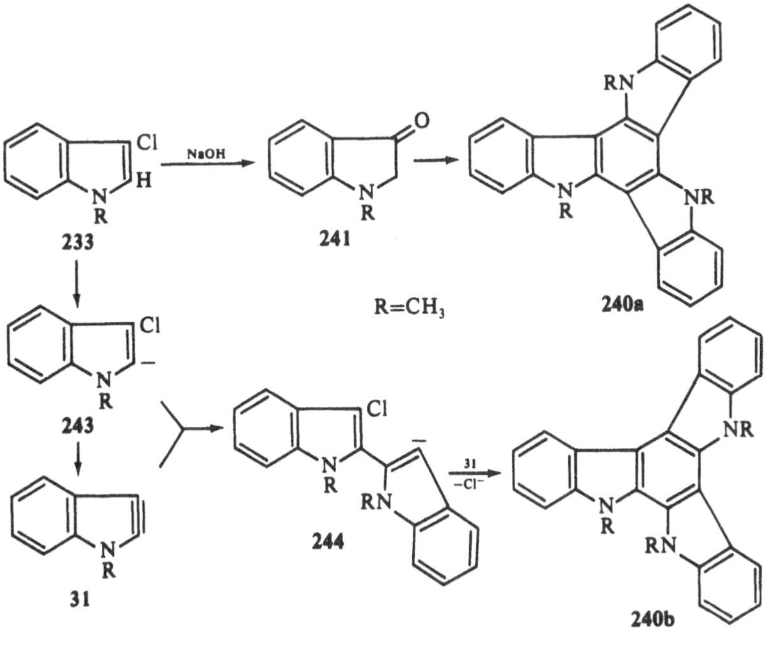

regioselective addition of nucleophiles to the 2-position of the aryne (31). Unfortunately the only basis for this assumption is the reaction of 233 to 239 for which no evidence of an aryne intermediate is as yet available. In fact, an argument can be made that if the aryne (31) were involved, then the trimer 240 would also have the symmetrical structure 240a. Nucleophiles add to arynes to give the most stable anion[371b] which, from metallation studies on *N*-methylindole,[372] should be the one (244) with the negative charge at the 2-position and the nucleophile at the 3-position and not vice-versa (245). Addition of 244 to the aryne (13) will then eventually lead to the symmetrical trimer (240a). If this analysis is correct then the structure of the trimer 240 is immaterial and other evidence to prove the existence of a 2,3-didehydroindole (31) will have to be found.

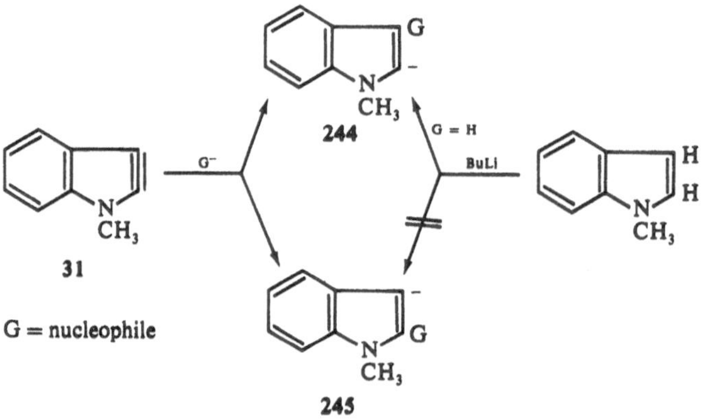

Recent independent syntheses of both the symmetrical (240a)[373a] and unsymmetrical (240b)[373b] trimers have shown that it was in fact the former which was produced in the above reactions.[370] Its preparation involves the reaction of 2-iodo-*N*-methylindole with activated copper-bronze or simply heating *N*-methyl-2-indolyl copper. Only a small amount of the unsymmetrical trimer (240b) was obtained from these reactions by a path which was concluded *not* to involve 2,2'-biindolyl species similar to 243. The mechanism of formation of the trimers (240) was proposed to proceed by the stepwise trimerization of the hetaryne–Cu complex (246).[373a] While these results are consistent with a copper-containing intermediate, it is not clear that 246 would be its best representation. An oligiomeric structure as proposed for *N*-methyl-2-indolyl copper[373a] and the *o*-phenylenemercury compound 125 [238-240] (Section II.1.F) cannot be excluded.

Since decomposition of the mercury compound 247 in the presence of activated copper–bronze also gives the symmetrical trimer (240a), trapping of the hetaryne (31) was proposed,[373a] presumably via the hetaryne–Cu complex (246) or its oligiomer. The presence of the hetaryne (31) in the reaction is suspect, however, since it is once again based on Diels–Alder adduct (232) formation[374] with the ambiguous diene, tetracyclone (151) (Section II.2.B.c). Reaction of the copper–bronze with one of the mercury-containing precursors of 31 is a more likely alternative.

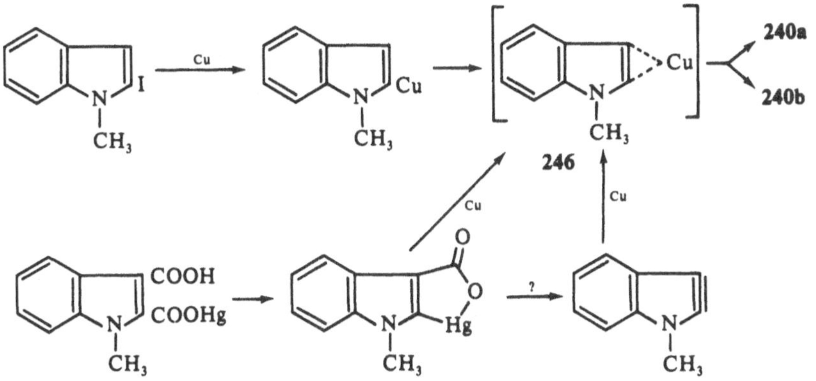

B. 2,3-Didehydropyrrole

As with other aromatic anhydrides[175–179] (Section II.1.E) the mass spectrum of the N-phenylpyrrole-2,3-dicarboxylic acid anhydride (248) shows a strong peak assigned to the cation radical of the corresponding aryne (249).[375] Upon pyrolysis of 248, however, neither aryne dimers (250), [as with phthalic anhydride (73)[175,176,179–181]] nor ring-opened aryne transformation products, (as with the anhydrides of six-membered nitrogen-containing heterocycles,[177,376]) were obtained.[375] Instead, four oxygen-containing products, 251 to 254, which were postulated to be dimers of the cyclopropenone 255, were isolated in 60% yield.[375] This result was the first example of an aromatic anhydride which apparently failed to yield an aryne upon pyrolysis. The possibility that the aryne 249 was in fact formed but reacted further to give products such as 250, which might be expected[295,296] to be unstable under the pyrolysis conditions, led to a recent reexamination of this reaction.[199] The original observations[375] were qualitatively verified, and no N-phenylindoles (256) were found when the reaction was carried out in the presence of a variety of dienes (146). Based on results in the thiophene series (Section III.3.A.f) the formation of these indoles (256) would have signaled the intermediacy of the aryne 249. Although no evidence for the generation of the aryne 249 in this reaction therefore exists, it was possible to verify the presence of the cyclopropenone species 255, or a ring-opened equivalent, by trapping with fluorinated ketones (257) to give the lactones 258.[199]

A project[377] to generate 2,3-didehydro-4,5-dimethylpyrrole (259) also failed

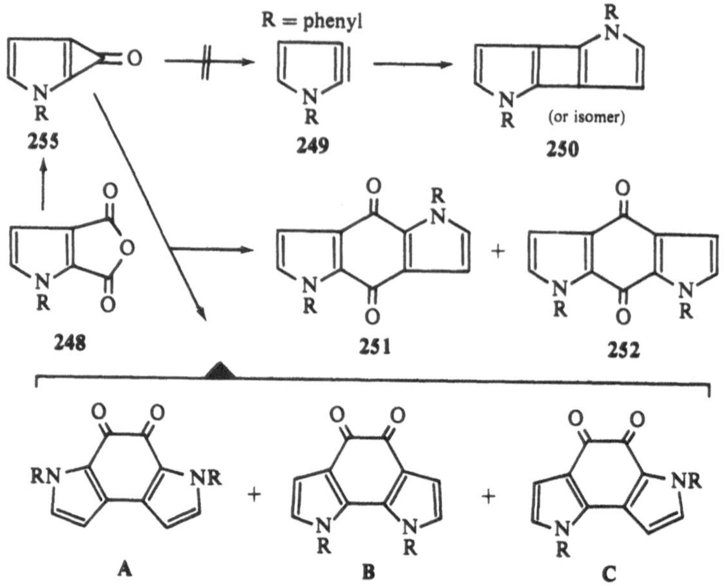

(253 and 254 are either A, B or C)

when none of the three potential precursors **260** to **262** could be prepared.[378] The one precursor (Section II.1.B) which was studied, potassium 3-iodopyrrole-2-carboxylate (**263**), was reported to give no pyrrolic products upon heating with tetracyclone (**151**).[378] This conclusion was based on infrared spectroscopy (details unspecified) and a negative Ehrlich test.[378] Since the anticipated product (**264**) of the aryne (**259**) and tetracyclone (**151**) is fully substituted in the five-membered ring, however, it would not be expected to give a positive Ehrlich test.[379] The negative result with the precursor **263** may therefore be ambiguous. For this reason and because the mass spectrometric fragmentation patterns of

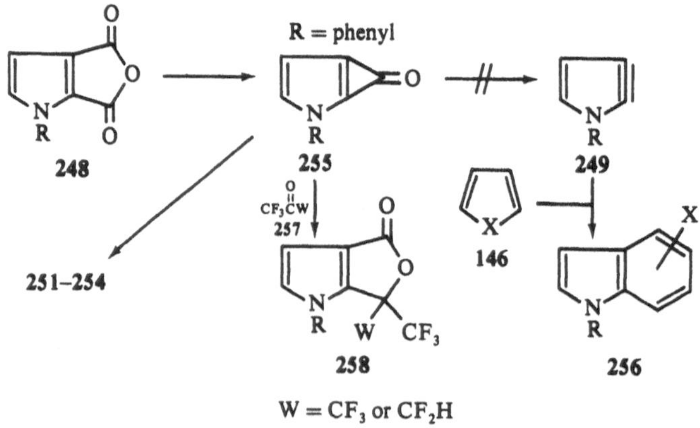

W = CF$_3$ or CF$_2$H

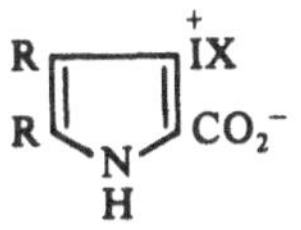

260 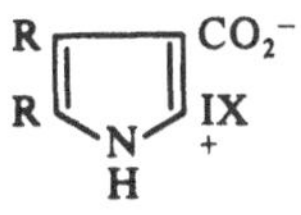 261 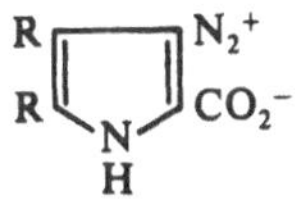 262

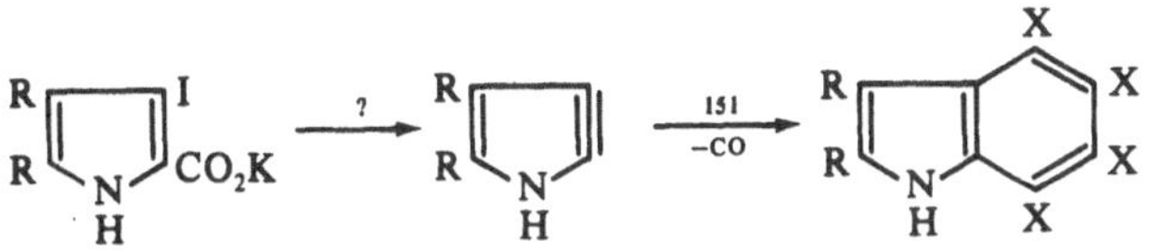

263 259 264

R = CH₃   X = phenyl

several of the iodopyrroles in this study[378] display a peak for the cation radical of the aryne **259**, further attempts to generate this intermediate may be warranted. Fragments rationalized as the ions of 2,3-didehydropyrroles (**179**) are also observed in the mass spectra of the pyrrolopyrazines (**265**).[280]

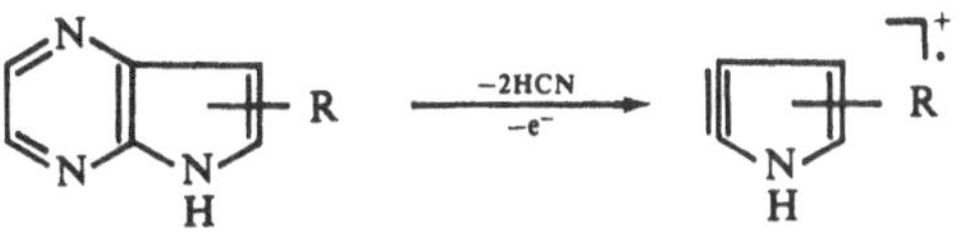

265

Recently, a cine-substitution of the 3,4-dinitropyrrole (**266**) to the 2-methoxy-4-nitropyrrole (**267**) was reported.[381] The mechanism of this transformation, which requires an acidic work-up of the original reaction mixture, was shown to proceed by an ASE mechanism (Section II.2.A.f) via the isolable pyrroline (**268**) rather than via an aryne intermediate **269**.

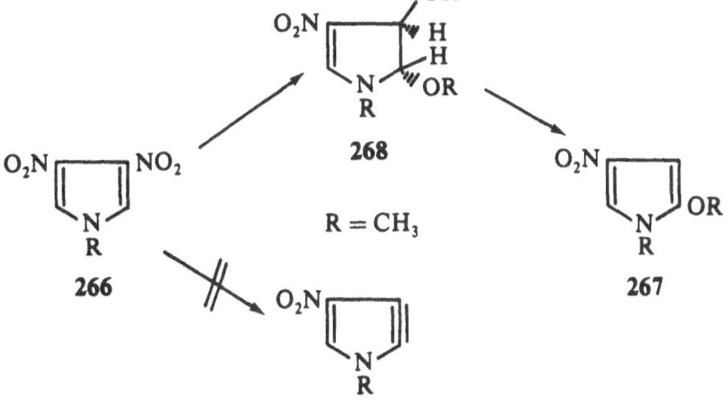

268

266

R = CH₃

267

269

C. 3,4-Didehydropyrrole

The only attempt[375] to prepare an aryne of this type has been by the pyrolysis of the 3,4-anhydride of *N*-phenylpyrrole dicarboxylic acid (270). As with the 2,3-isomer (248) the decomposition to the aryne (271) apparently was short circuited at the cyclopropenone (272) which, in this case, underwent a complex rearrangement to the furoquinoline (273) rather than lose CO.[375] Studies with a nonaromatic substituent on the nitrogen so that this reaarangement would be blocked have not yet been reported.

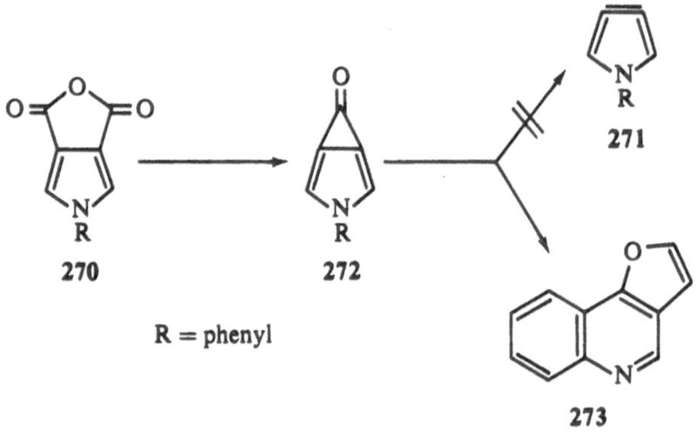

R = phenyl

D. Didehydromaleimide

As in the case of the corresponding anhydride (23) (Section III.1.C), didehydromaleimide (30) probably does not qualify as a true aryne, but is expected to be sufficiently similar in properties to be covered in this review (Section I.2.C). This species (30) was first suggested[70] as an intermediate along with the 1,2-dithiete (274) in the decomposition of the dithiin (275) in pyridine at 120°C to explain the formation of the aryne trimer (276). When this reaction was carried out in the presence of cyclopentadiene (146b), however, the failure to obtain the appropriate aryne adduct (277) was rationalized as being due to its presumed instability.[70] The only products obtained from this reaction,[70] as well as from butadiene (146e) and anthracene (147),[382] were Diels–Alder adducts such as 278 and 279 of the dithiin (275) and the dithiete (274), respectively.

In contrast to a statement in a recent review,[383] therefore, the didehydromaleimide (30) has not been trapped and only a product 276 which could arise by an alternate addition–elimination process is obtained. Both the ability of the alleged aryne precursor 275 to form Diels–Alder adducts (278) and its stability in inert solvents or upon sublimation to 300°C support this suggestion. Furthermore, several properties of 275 compared to other dithiins have led to the

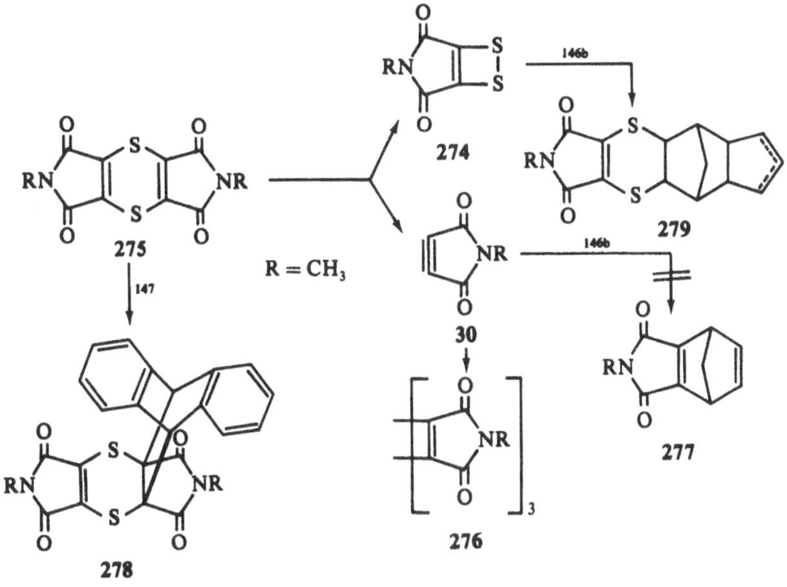

suggestion[382] that dipolar forms such as **275a** make substantial contributions to the overall structure of **274**. Perhaps the uniqueness[70] of pyridine as a solvent for the decomposition of the dithiin (**275**) is therefore related to its ability to permit such dipolar forms to undergo stepwise cyclotrimerization with attendant or subsequent loss of the dithiete (**274**) to give the aryne trimer (**276**). Regardless of the details of the mechanism this trimerization is conceptually related to the stepwise formation of triphenylene (**140**) from *o*-halophenyllithium compounds (**141**) (Section II.2.B.a) and therefore does not require, and probably does not involve, an aryne intermediate. A similar conclusion is warranted for the formation of the same aryne trimer (**276**) during the prepara-

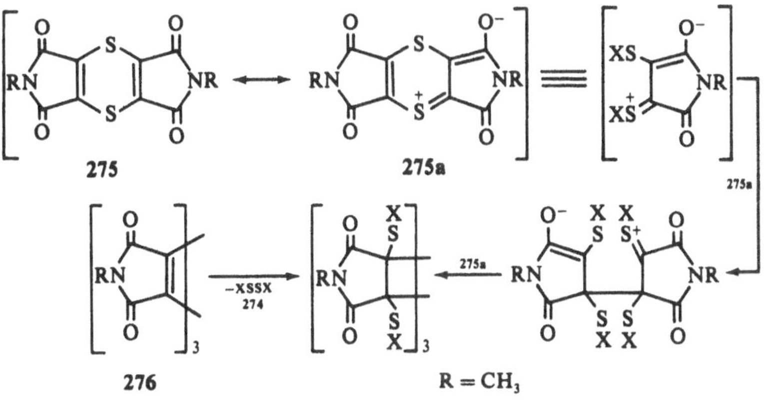

tion of the dithiin (**275**) from *N*-methyldichloromaleimide (**280a**).[70] The dibrom-omaleimide (**280b**) on photolysis in the presence of *N*-phenylpyrrole (**221**) reacts as in the maleic anhydride series (Section III.I.C) undergoing stepwise substitution and ring closure to give **281**.[383a]

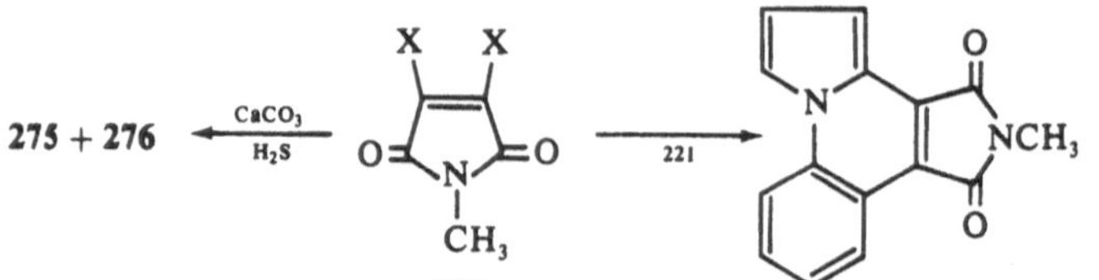

$$275 + 276 \xrightarrow[\text{H}_2\text{S}]{\text{CaCO}_3}$$

280

a: X = Cl
b: X = Br

281

Finally, thermolysis of the *N*-phenyldiiodomaleimide (**280c**) did not give any aryne trimer (**282**), but only a polymer **283** whose empirical formula was that of the aryne (**284**).[355] By analogy with the related diiodomaleic anhydride (**210**) (Section III.1.c)[355] there is, therefore, no reason to postulate the formation of the aryne (**284**) in this reaction. In fact, as was concluded for the oxaarynes (Section III.1), there is no evidence at present for the formation of any azaaryne intermediate.

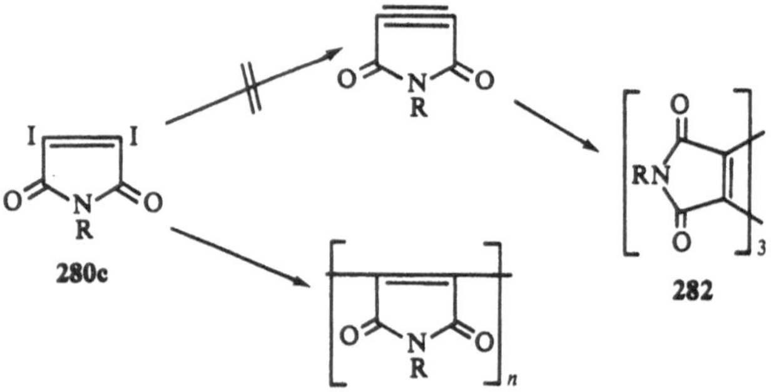

280c

282

283

3. Thiaarynes

A. 2,3-Didehydrothiophene

The preparation of 2,3-didehydrothiophene (**4**) has received more attention than that of any other five-membered hetaryne. This fact is probably due to the expectation that since thiophene is more benzene-like in its properties than any of the other five-membered heterocycles,[304] the didehydrothiophenes **4** and **25** would have the highest probability of being generated and detected. This assumption has apparently proved to be correct.

a. From *o*-Halomercury Compounds

The first claim for the generation of the aryne **4** was based on the isolation of the appropriate adduct **285** from the thermal decomposition of the mercury compound **286** in the presence of tetracyclone (**151**).[61] In light of the ambiguity of similar results[364] in the indole series (Section III.2.A), doubts as to whether (**4**) was in fact an intermediate in this reaction were raised.[1,2,51,68] Although the presence of tetracyclone (**151**) had no apparent effect on the rate of decomposition of the mercury compound (**286**),[384] it was noted[76] that the amount of 3-iodothiophene (**287**) that was isolated as a by-product decreased from 10% in the absence of the trap (**151**) to a mere trace in its presence. Significantly, the 9% yield of adduct (**285**) corresponds closely to the amount of missing 3-iodothiophene (**287**), thereby suggesting that the former originated from the reaction of the latter with tetracyclone (**151**). The possible validity of this hypothesis was demonstrated by the formation of the adduct **285** in 22% yield upon reacting 3-iodothiophene (**287**) with tetracyclone (**151**) in the presence of

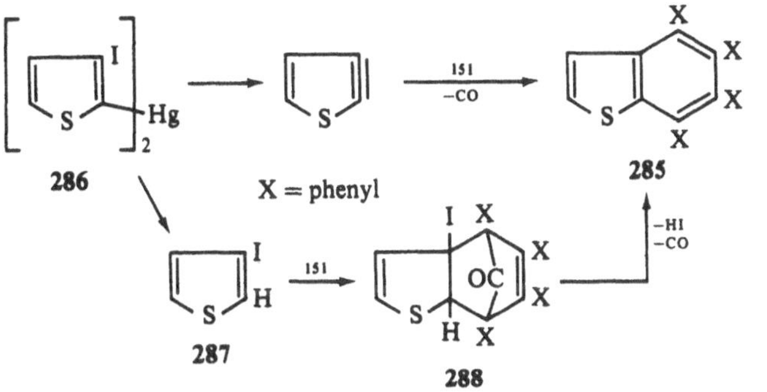

MgO. The base is unnecessary for the decomposition of **287**, which occurs in its absence, and served only to remove the HI since its strong reducing properties otherwise lead to the exclusive production of tars. Based on these observations it was concluded[76] that the formation of the adduct **285** from the mercury compound **286** does not require an elimination–addition mechanism via the aryne (**4**) but probably involves an addition–elimination reaction of the halide **287**, formed by decomposition of **286**, via the Diels–Alder adduct (**288**). A similar conclusion was reached with regard to the formation of a tetracyclone adduct **289** from the perchloro analog **290**.[385]

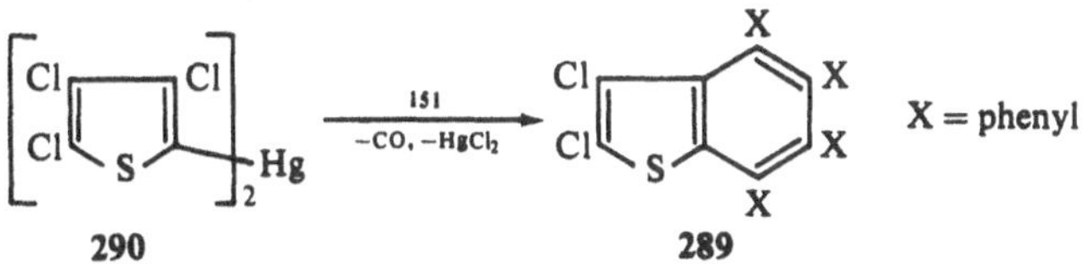

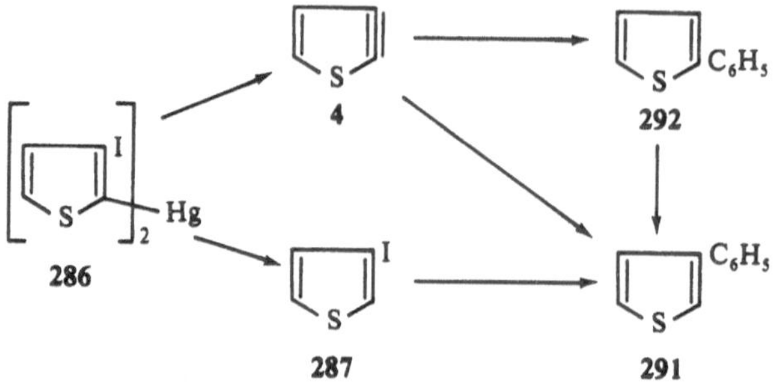

Photolysis of the mercury compound **286** in benzene leads to 3-phenyl-thiophene (**291**) and 3-iodothiophene (**287**).[386] The former could arise from an "insertion" reaction[387] of the aryne (**4**) with benzene if it is assumed[76] that the expected but missing 2-phenylthiophene **292** completely isomerizes to the 3-isomer (**291**) under the reaction conditions.[388] The argument is not convincing, however, since the 3-iodothiophene (**287**) formed as a coproduct, presumably by homolysis of the C–Hg bond, also yields 3-phenylthiophene (**291**) on photolysis in benzene, albeit in very low yield.[389] Furthermore, by analogy with the reaction of benzyne the major products should be those of cycloaddition,[390,391] which, in the case of **4**, might lead, after loss of acetylene, to thianaphthene (**293**). No such products were reported from the photolysis of the mercury compound (**286**),[76] but they have been observed[199] from other possible sources of the aryne (**4**) (Section III.3.A.f).

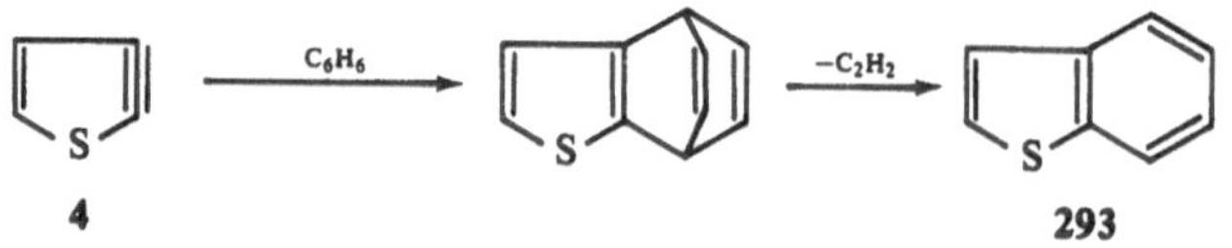

b. From Other *o*-Halometalthiophenes

By analogy with the benzene series, the elimination of alkali or alkaline earth halides from *o*-halometal thiophenes to give an aryne would be expected to proceed more easily than loss of mercuric halides.[81a] However, several early observations indicated that both 2-lithio (**294**)[392] and 2-bromomagnesium (**295**)[393] derivatives of 3-bromothiophene behaved like typical lithium and Grignard reagents with no apparent tendency toward metal halide elimination. The remarkable stability of **294** compared to the ready loss of LiBr from *o*-bromophenyllithium (**32b**) to give benzyne even at −100°C,[342] was noted,[394] and illustrated by its trapping in good yield with various electrophiles not only at −70°C[395] but also at +100°C.[386] The possibility that some 2,3-didehydro-thiophene (**4**) was nevertheless formed during the decomposition of **294** was

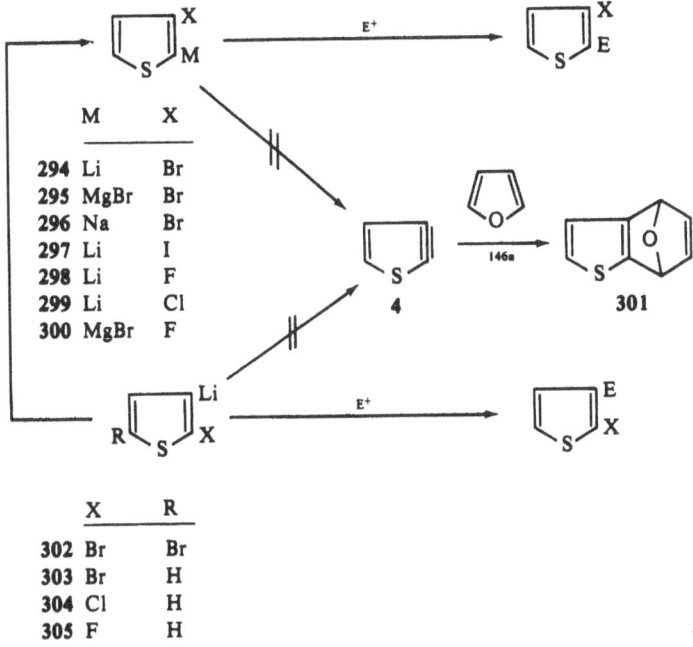

	M	X
294	Li	Br
295	MgBr	Br
296	Na	Br
297	Li	I
298	Li	F
299	Li	Cl
300	MgBr	F

	X	R
302	Br	Br
303	Br	H
304	Cl	H
305	F	H

examined by adding the diene traps furan (**146a**) and anthracene (**147**) to the reactions which were carried out for 55 hr at room temperature or for 70 hr at 120°C in the absence of solvent, respectively.[386] In neither case was an aryne adduct (**301**) or its transformation products obtained. Similar stability was observed for the sodium derivative **296**.[386] The possibility that the formation of 2,3-disubstituted thiophenes from the reaction of excess butyllithium with 3-bromothiophene might somehow involve the aryne (**4**) has also been mentioned but appropriately rejected as unlikely.[396] Although the nature of the halogen atom has a significant effect on the stability of *o*-halophenyllithium reagents (**32b**),[80] the iodothiophene derivative **297** has been characterized[62] as surprisingly unreactive to LiI loss, and the fluorine compound **298** behaves like a normal lithium reagent,[397] which gives no evidence for the formation of the aryne (**4**) upon attempted interception by furan (**146a**).[398] Similarly, heating the chloro analog **299** not only with furan (**146a**), anthracene (**147**), or benzene, but also with the much more powerful (and ambiguous; cf. Section II.2.B.c) diphenylisobenzofuran (**159**), led only to tars with no indication of an aryne adduct (**301**). Even the *o*-fluoro Grignard reagent (**300**), whose benzene analog (**32b**) leads to benzyne (**1**) more readily than any of the other *o*-halophenyl Grignard reagents,[80] failed to produce any 2,3-didehydrothiophene (**4**).[398] The apparent problem with all of these 2-metallo-3-halothiophenes **294–300** is their unusual stability, undoubtedly due to the influence of the adjacent sulfur atom on the carbanionic 2-position.[22,23]

In an attempt to circumvent this problem several of the isomeric 2-halo-3-lithiothiophenes have been examined. When the bromo compound 302 was prepared by metallation of 2,5-dibromothiophene with lithium diisopropylamide at −70°C it behaved like a normal lithium reagent.[352] If, however, the related lithium reagent 303 was generated at −70°C by the halogen–metal exchange reaction of 2-bromo-3-iodothiophene with ethyllithium,[399] then it apparently underwent a series of rapid halogen–metal exchange reactions similar to the BCHD (Section II.2.A.d),[267] and well known for other thienyllithium compounds,[395] to ultimately give the thermodynamically more stable lithium compounds such as 294. The possibility that the aryne (4) is involved in this isomerization was checked by carrying out the reaction in the presence of furan (146a) but, as before, no adduct (301) was isolated.[399] At −100°C the rearrangement of 303 can be suppressed and only the normal reactivity of a lithium reagent is observed.[399] The corresponding chloro compound 304, also prepared by halogen–metal exchange, is stable at −70°C,[400] while the fluoro analog 305 behaves normally even at −40°C.[401] At higher temperatures 304 rearranges primarily to the more stable 2-chloro-5-thienyllithium derivatives 306 and 307 by the same mechanism as is followed by the bromo analog 303.[400]

X = Br or I   304   306   307

The lone suggestion of the intermediacy of an aryne such as 4 in a reaction of an *o*-halothienyllithium compound was made to explain the formation of the bisthiophenindigo 308 in 0.17% yield from the reaction of the dilithio compound 309 with dimethylsulfate at room temperature in the presence of air.[402] Presumably the monomethylated lithium reagent 311 reacts with O$_2$ to give the peroxy anion 312, which adds to the aryne 310 with elimination of chloride ion. The resulting peroxide (313) then rearranges to the final product (308).[402] An alternate possibility which avoids the aryne (310) involves only reactions for which ample precedents are available, such as the formation[403] and oxidative coupling[404] of a 3-hydroxythiophene (314) followed by elimination of the elements of Cl$_2$ from the vicinal dichloride 315[405] to give 308. Furthermore, since the major products from 309 and its bromo analog 316 are those expected from normal lithium reagents,[402] the postulation of the aryne (310) is not required by the evidence.

c. During Cine-Substitution Reactions

Two cine-substitution reactions in the thiophene series which are known for elimination–addition mechanisms via a 2,3-didehydrothiophene (4) have

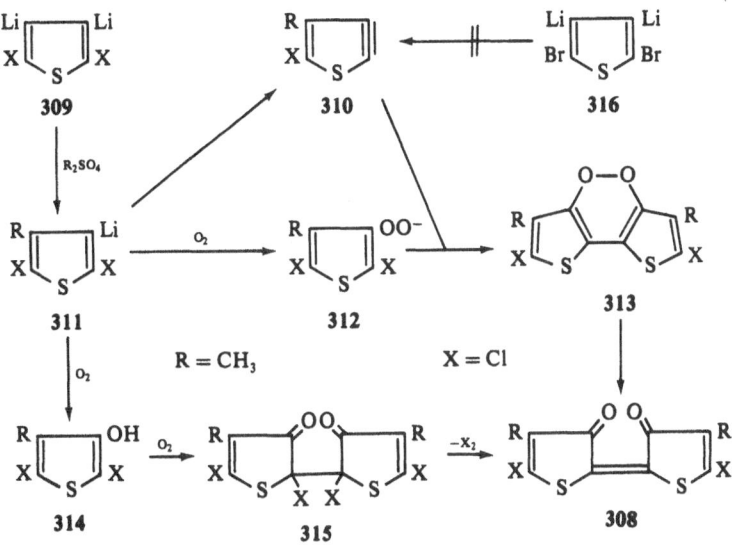

been considered. The first of these[75] involves the reaction of 2-bromothiophene (317) with KNH_2 in liquid ammonia under the precise conditions used by Roberts[12] in his classic studies establishing the intermediacy of benzyne (1) in the amination of halobenzenes. The products of this reaction, 3-amino-thiophene (318) and 3-bromothiophene (319), could be interpreted as arising from the aryne (4) by a competitive addition of NH_2^- and Br^- exclusively in the direction which gives the more stable[22,23] 2-carbanions (320) as intermediates. Although such regiospecificity of nucleophilic addition to arynes is well known[255] (Section II.2.A.b), the failure to obtain isomeric products in a constant ratio independent of the starting material does require the acquisition of additional data before an aryne intermediate can be assured (Section II.2.A.h).[288]

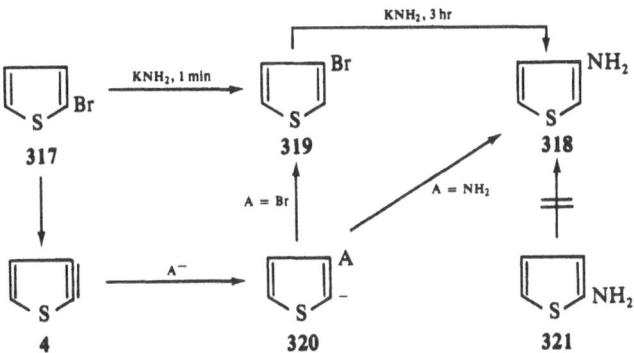

The above mechanism is apparently supported by the following observations[75]: (i) no reaction occurs in the absence of the strong base, NH_2^-, which would be required for the generation of the aryne (**4**); (ii) the cine-substitution product (**318**) does not arise by rearrangement of the normal substitution product, 2-aminothiophene (**321**) under the reaction conditions; (iii) rearrangement of the starting material (**317**) to its isomer (**319**) followed by normal substitution to give 3-aminothiophene (**318**) can be eliminated since under the reaction conditions the conversion of 3-bromothiophene (**319**) to **318** requires more than 3 hr, or at least 180 times as long as the 1-min conversion of 2-bromothiophene (**317**) to **318**; (iv) addition of a tenfold excess of NaBr to the reaction mixture increased the yield of 3-bromothiophene (**319**) from 5% to 38% at the expense of that of 3-aminothiophene (**318**), which dropped from 59% to 14% as might be expected if Br^- and NH_2^- were competing for a common intermediate such as the aryne (**4**).

The inadequacy of the above interpretation was revealed by the marked dependence of product composition on temperature, nature of the cation and base, relative and absolute concentration of the reactants, and the presence of substituents on the thiophene ring.[75,259-261,263,406] In fact, virtually any change in the reaction conditions from those used by Roberts[12] leads to the isolation of polybromo- and/or aminobromothiophenes as illustrated in Table 6.[75,259,263,406] These products suggest that the base-catalyzed halogen dance (Section II.2.A.d),[267] which is well known for thienyllithium compounds[395] (Section III.3.A.b), is occurring during the reaction of 2-bromothiophene (**317**) with potassium and sodium amides as well and therefore may be involved in the formation of both 3-amino- (**318**) and 3-bromothiophene (**319**). The latter compound would be expected to be formed, via the sodio derivative **296**, if the BCHD mechanism were operative by analogy to the isomerization of 2-bromo-3-lithiothiophene (**303**) to the more stable 3-bromo-2-lithiothiophene (**294**).[399] The validity of this hypothesis was established by a series of experiments, the most significant of which showed that many of the polybromothiophenes in Table 6 could be converted to 3-bromothiophene (**319**) under conditions which also favored its formation from **317** (low NH_2^- concentration as is obtained when the relatively insoluble $NaNH_2$ is used as a base in liquid NH_3).[75,259,263,406]

In order to explain the cine-amination product (**318**) by the BCHD mechanism a nucleophilic aromatic substitution at the β position of at least one of the polybromothiophenes in Table 6 must take place to give first an aminobromothiophene such as **328**, which then must also be able to go on to **318**,

317   318

TABLE 6. % Yield of Reaction Products from 2-Bromothiophene (317) and Metal Amides in Liquid NH₃ (X = Br; a = NH₂)

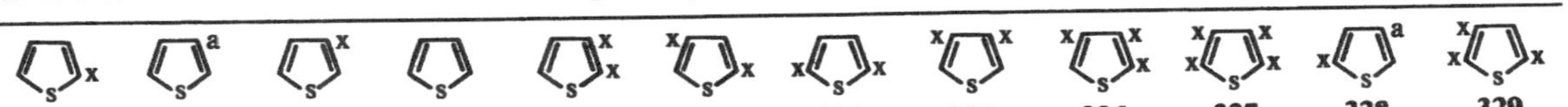

Conditions	317	318	319	321	322	323	324	325	326	327	328	329
KNH₂, 6 equiv. 15 min., −33°C		59	5									
KNH₂, 6 equiv. 1 min., −65°C	0.2		2		4	3		15	1			
NaNH₂, 1 equiv. 1 min., −33°C	0.3		25		6	4		14	3		1	Trace
NaNH₂, 0.015 equiv. 30 min., −33°C	6		3	3	6	4	1	11	21	2		3

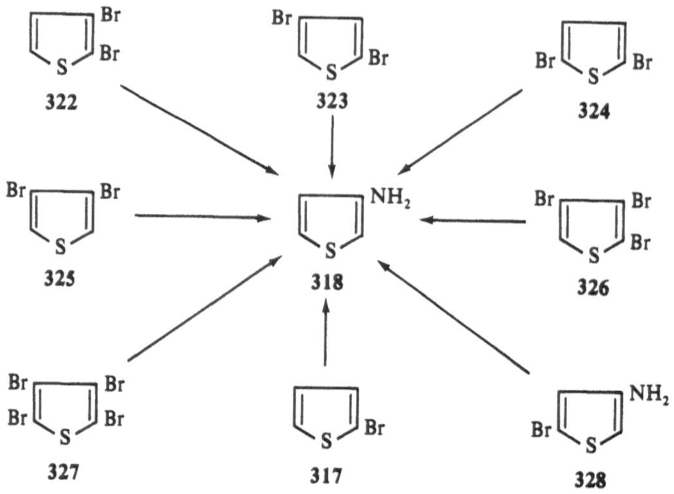

presumably by a further BCHD. The feasibility of these processes was demonstrated by converting all seven compounds **322–328** to 3-amino-thiophene (**318**) as the only isolable product under Roberts[12] reaction conditions.[75,259,263,406] This observation gives no information about the nature of the substitution step, however. Furthermore, since the polybromothiophenes in Table 6 are probably interconvertible by the BCHD under these reaction conditions, it is also not possible to specify which one or ones are directly involved in the substitution step. The fact that 4- and 5-methyl-substituted 2-bromo-thiophenes (**330**) give primarily halogen rearrangement[260] and very little

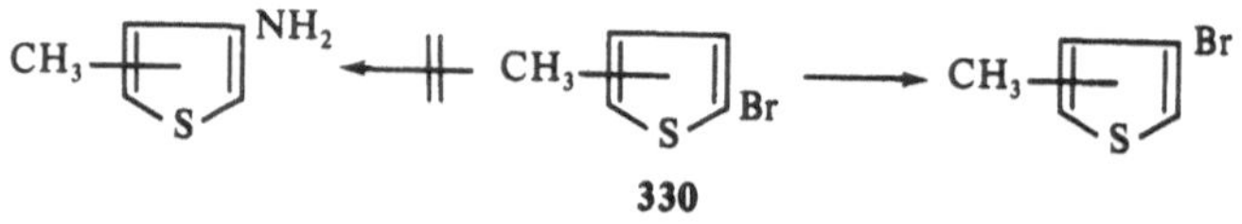

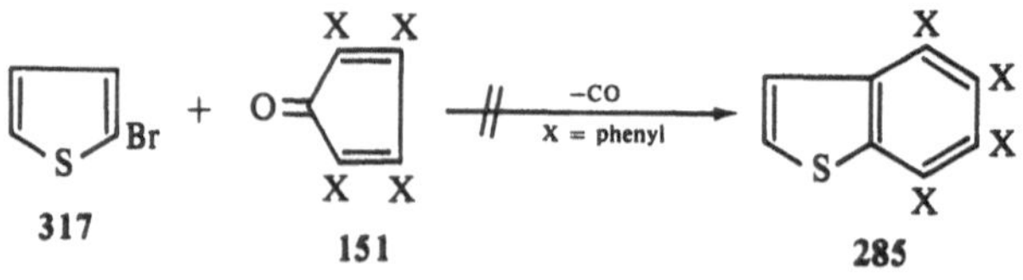

amination[261] indicates that the actual substitution step can be blocked by methyl groups on the other side of the ring. This observation is inconsistent with an elimination–addition mechanism for the substitution step[259,263] since such remote methyl groups do not prevent formation of arynes from halobenzenes[255] or halopicolines[407] under similar conditions. Furthermore, when the amination reaction was carried out in the presence of even such an ambiguous (Section II.2.B.c) aryne trap as tetracyclone (151) no evidence for the expected aryne adduct (285) was obtained.[406] Therefore, although the precise mechanism for the actual substitution step is not known, an elimination–addition process via a 2,3-didehydrothiophene such as 4 seems highly unlikely. The apparent necessity of polybromothiophenes as intermediates suggests that either an addition–elimination or an $S_{RN}{}^1$ mechanism[408] are more reasonable possibilities.

The second example of a cine-substitution of a thiophene involves the reaction of arylthiolates with 3,4-dinitrothiophene (331)[409] or 3-nitro-4-phenylsulfonylthiophene (332)[410] to give the 2,4-substituted products 333. Both an elimination–addition mechanism via the aryne (334) or an abnormal addition-elimination (AEa) mechanism (Section II.2.A.e) via the Meisenheimer complex (335) have been considered for these reactions.[409] The former is unlikely for several reasons including the lack of precedence for aryne formation from aryl nitro compounds (Section II.1) under these reaction conditions[410] and the fact that addition of the nucleophile ArS⁻ to the aryne (334) would have to proceed via the 3-thienyl anion (336) rather than via a more stable[22,23] 2-thienyl anion such as 320 as would be expected.[255] Contrariwise, cine-substitution by the AEa mechanism is favored by the ability of the complex (335) to stabilize the negative charge by delocalization to both the NO_2 group and the α position of the thiophene ring.[410] As in the pyrrole series (Section III.2.B)[381] the actual mechanism appears to be more complex, however, involving several addition and elimination steps via 337, which was recently[411] isolated from the reaction $(X = NO_2)$ and shown to go to the product 333 under the reaction conditions. It therefore appears that neither of the cine-substitutions of thiophene described in this Section proceeds via an aryne intermediate.

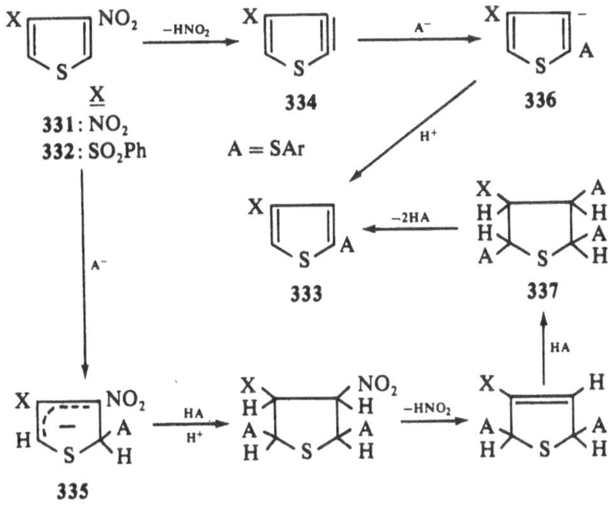

d. From Diazonium Compounds

The unsuitability of halogen compounds as precursors of 2,3-didehydro-thiophene (4) as described in the previous three sections led to investigations[78,412] of precursors which might permit generation of the aryne (4) under milder conditions. By analogy with the benzene series (Section II.1.c) the o-diazonium carboxylate 338 or one of its precursors 339 or 340 appeared to be

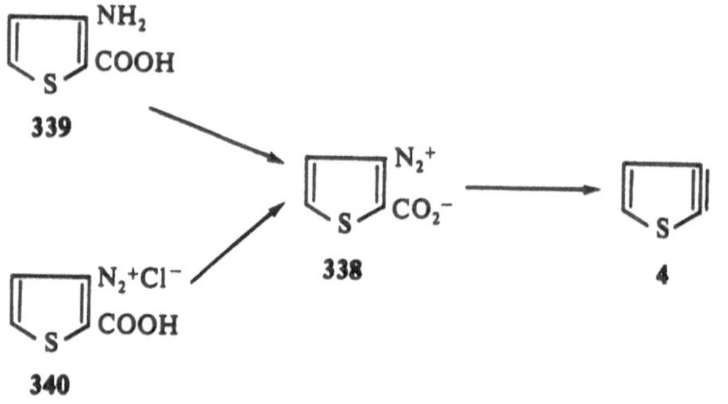

prime candidates. The known[413] unstable amino acid 339 was prepared[78,412,414] and subjected to the aprotic diazotization procedure[103] in the presence of anthracene (147) as a potential aryne trap. The only products isolated were 9-nitroanthracene (341) and the diazonium carboxylate 338 at 25°C in dioxane[78,412] or the azo-compound 342 and the three 3-thienylanthracenes 343 at 100°C in 1,2-dichloroethane as solvent.[412] No evidence for an aryne adduct (344) was obtained. A preliminary investigation with furan as a trap[412] yielded similar negative results.

In the presence of trichloroacetic acid the crude, explosive diazonium car-boxylate (338) could be isolated in up to 60% yield[78,412] from the aprotic diazotization[103] of the amino acid (339). Decomposition of 338 in the presence

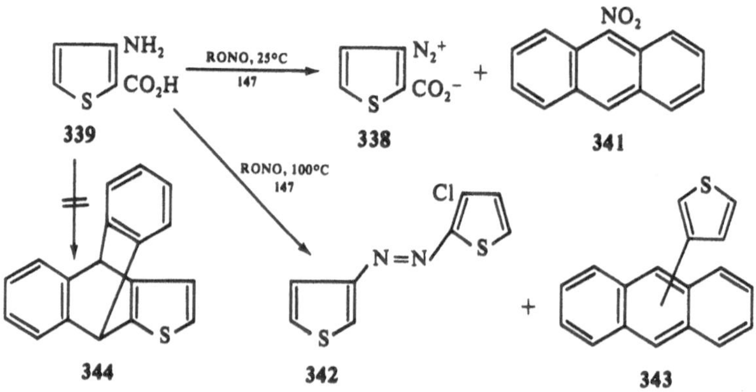

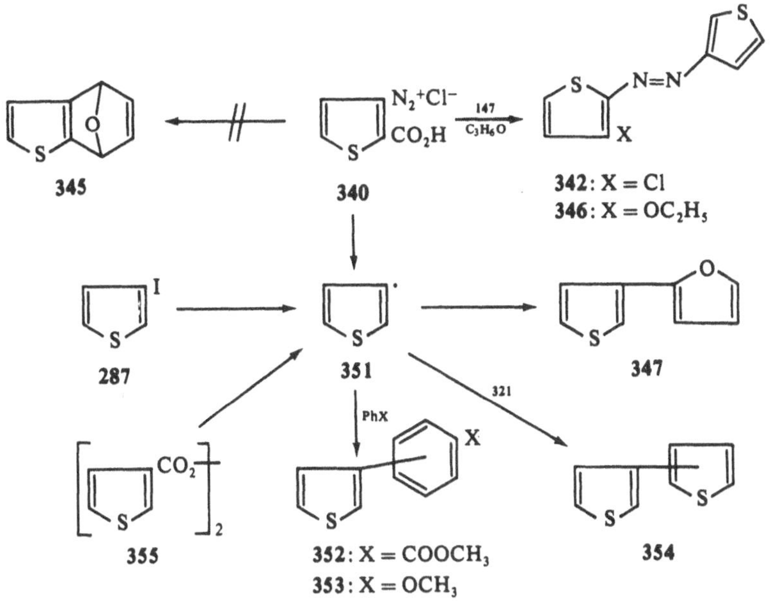

345 340 342: X = Cl
           346: X = OC$_2$H$_5$

287 351 347

355 352: X = COOCH$_3$ 354
     353: X = OCH$_3$

of anthracene and furan once again gave coupling and substitution products in low yield but no sign of aryne adducts such as **344** or **345**.[412]

Additional studies[412] were carried out with the relatively stable hydrochloride salt **340** which, in the presence of propylene oxide as an HCl scavenger, should generate the diazonium carboxylate (**338**) *in situ* and thence presumably the aryne (**4**).[105] The only products isolated from this reaction, independent of the presence of anthracene, were the azo compounds **342** and **346** (the latter when ethanol-stabilized chloroform was used as a solvent).[415] Other electron-rich aromatic acids are known to undergo similar azo-coupling accompanied by decarboxylation.[416] With furan as both the potential aryne trap and solvent the major product (64% yield) was the 2-furylthiophene (**347**). A similar reaction observed with the pyrimidine amino acid (**348**) and furan (**146a**) to give the 2-furylpyrimidine (**349**) was postulated to proceed by ionic decomposition of the diazonium carboxylate **350**.[417] The formation of **347** appears to proceed via the 3-thienyl radical **351**, however, since with anisole, methyl benzoate, or thiophene (**321**) as the solvents in place of furan (**146a**) good yields (50%) of mixtures of the isomeric 3-arylthiophenes **352**, **353**, and

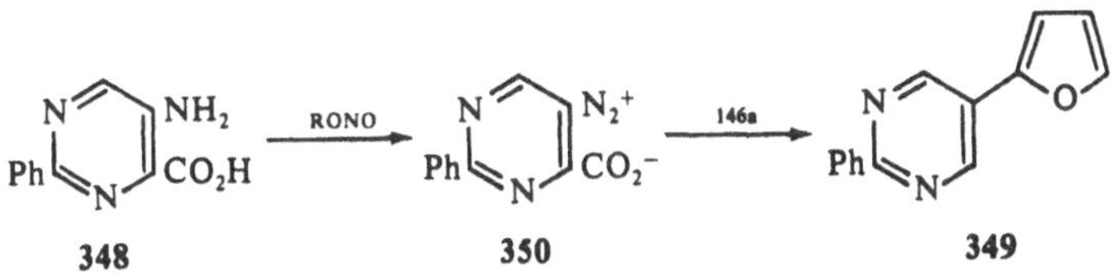

348 350 349

354 were obtained[412] in the same isomer ratio as is observed in reactions which are known to proceed via 3-thienyl radicals (**351**),[418,419] i.e., the photolysis of 3-iodothiophene (**287**)[420] and the thermolysis of 3-thenoyl peroxide (**355**).[421] Similar results were observed in the photolysis of **340** in furan–propylene oxide, except that, in addition to the biaryl **347**, the acids **356** and **357** were obtained.[412] The furan–aryne adduct **345** could not be detected in either reaction. Even in its mass spectrum the hydrochloride **340** failed to display significant peaks corresponding to the ions of the aryne **4** or its dimer as are found for the benzene analog **41c**.[422]

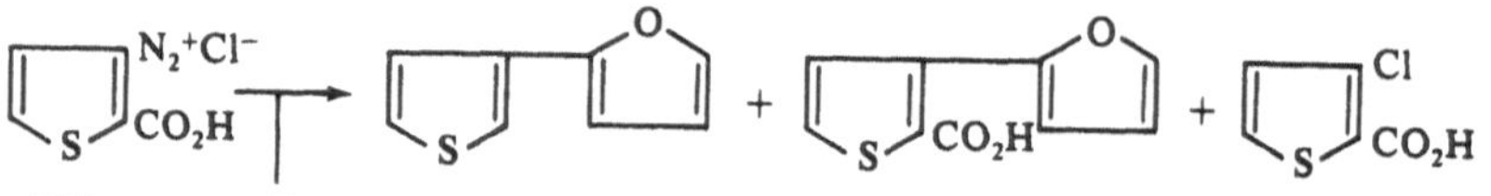

A final method involving diazonium compounds which was examined[412] as a route to the aryne (**4**) was based on the *in situ* generation and decomposition of a thiophene analog (**358**) of the *N*-nitrosoacetanilide **41f** (Section II.1.c). Reaction of 3-acetamidothiophene (**359**) with isoamyl nitrite in the presence of tetracyclone (**151**) gave no evidence of the aryne adduct (**285**) but only azo-coupling products analogous to **346** and 3-phenylthiophene (**291**), presumably formed by attack of the intermediate 3-thienyl radical **351** on solvent benzene.[412]

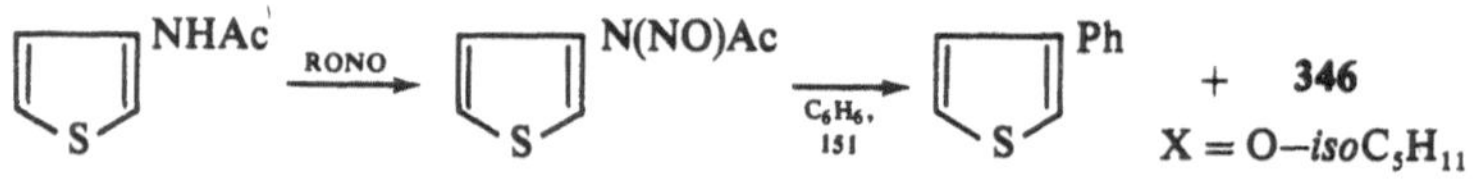

Although the experimental and structural variations of the diazonium carboxylate method as a route to 2,3-didehydrothiophens (**4**) have not yet been exhausted[78] the results cited above are not encouraging. In comparison with its benzene analog (**41a**) the diazonium carboxylate **338** appears to be relatively stable to aryne formation but much more susceptible to azo coupling and radical generation. As the reasons for the ease of these latter reactions are determined[412,423] it may be possible[78] to hinder or circumvent them entirely, thereby permitting aryne to form. Alternatively, under sufficiently severe conditions enough energy may be available to permit aryne formation to compete efficiently with reactions having lower activation energies. Studies based on this premise are described in the next two sections.

e. During Thermolysis Reactions

The generation of 2,3-didehydrothiophene (4) under thermolysis conditions has been attempted and claimed several times. Among the clearly unsuccessful attempts are the gas phase thermolyses of 2,3-diiodothiophene (360)[199,424] and 2-nitrothiophene (361)[425] which, in spite of the prominence of an ion corresponding to the aryne (4) in their mass spectra,[199,426] apparently give no products indicating an aryne intermediate even in the presence of several potential diene traps such as benzene[199,426] and furan.[199] As with the mercury compound 286 (Section III.3.A.a) the major decomposition pathway appears to involve homolysis of the bond to the 2-substituent.[199,425,426]

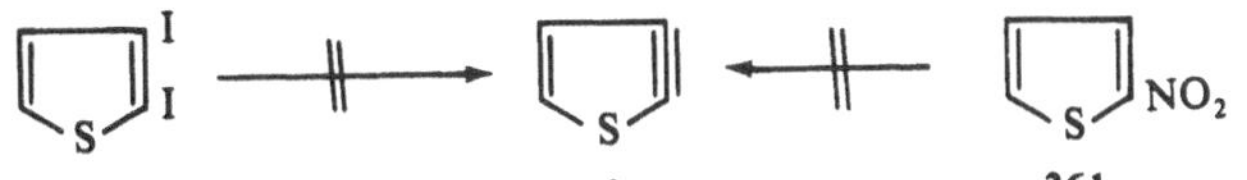

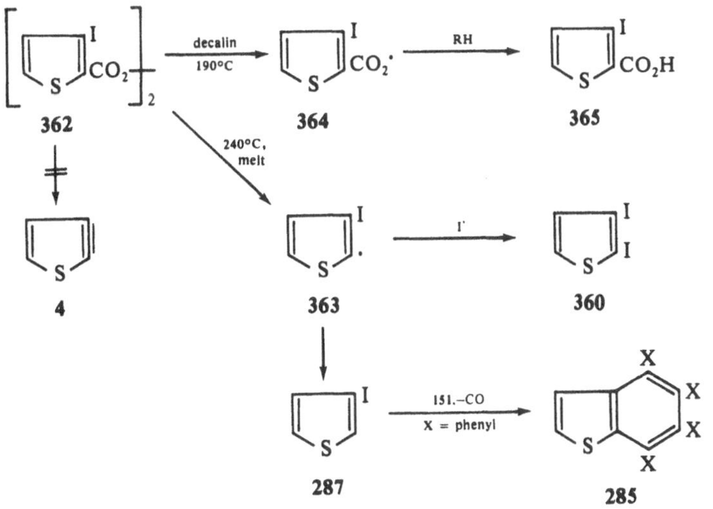

A possibly encouraging reaction is the fusion of the iodoperoxide (362) with tetracyclone (151) to give 1% of the expected aryne adduct (285) in addition to 2,3-diiodothiophene (360).[76] The formation of the latter compound, however, suggests the intermediacy of the radical 363, which upon hydrogen abstraction could give 3-iodothiophene (287), a known[76] precursor of the adduct 285 in the presence of tetracyclone (151) (Section III.3.A.a). The volatility of 287 under the reaction conditions could account for its absence as well as for the low yield of the adduct (285). Unfortunately, the presence of the radical 363 could not be verified by carrying out the decomposition of the peroxide (362) at a lower temperature in a hydrocarbon solvent, since the initially formed carboxylate radical **364** apparently abstracts hydrogen to give the iodoacid **365** faster than it decarboxylates to **363**. Nevertheless, the existence of a reasonable alternative mechanism to explain the formation of the adduct **285** from the peroxide (**362**) makes postulation of the aryne pathway unnecessary and hence unwarranted.

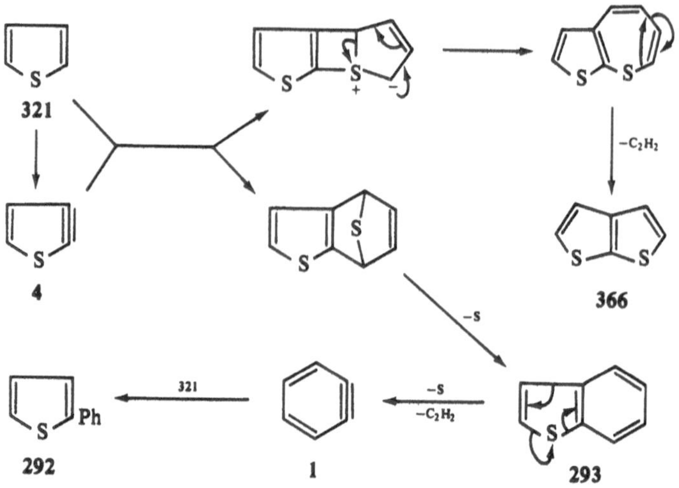

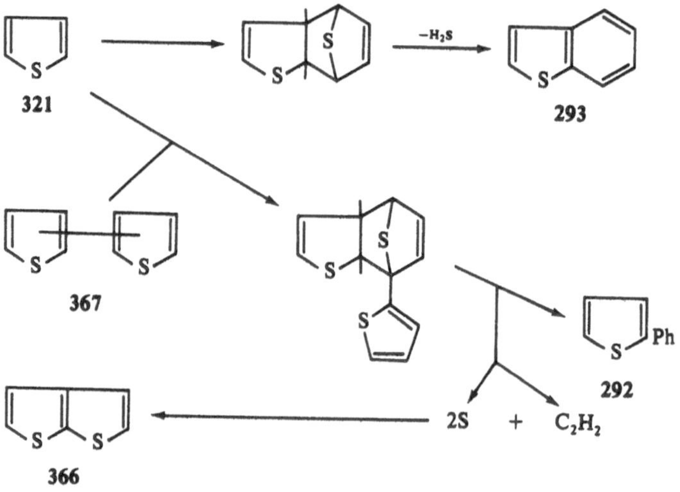

A similar reservation must be applied to the suggestion that the aryne (**4**) is involved in the formation of the minor products **292**, **293**, and **366** from the pyrolysis of thiophene (**321**).[71] The major product of this reaction is a mixture of all three possible bithienyls (**367**).[427] Although the low yields of these products and the severity of the reaction conditions make it difficult to exclude any mechanism as a possibility, the Diels–Alder self-addition hypothesis[427] appears at least as likely as the aryne suggestion.[71] Both of these mechanisms suffer from the requirement that acetylene is an expected by-product which was not observed, although it was specifically sought.[428] Therefore, in the absence of additional data, all three mechanisms proposed for the pyrolysis of thiophene[71,427,428] must be classified as highly speculative and hardly appropriate as evidence for the intermediacy of 2,3-didehydrothiophene (**4**).

Speculation concerning the formation of **4** from other arynes (Section II.1.F) is also best classified in the same way. In order to rationalize the formation of the naphthothiophenes (**368**) and the benzothiophthenes (**369**) during the copyrolysis of thiophene (**321**) and the phthalic anhydrides **73**[71] or **370**[72] an intermolecular hydrogen transfer was proposed between thiophene (**321**) and the initially formed (Section II.1.E) benzynes **1** or **371** to produce the aryne **4** and the benzenes **372** or **373**, respectively. The benzthiophthenes **369** are pictured as arising from addition of benzyne **1** to the $C_{(2)}$–S bond of thiophthene (**366**)[71] or of thiophyne (**4**) to the $C_{(2)}$–S bond of the thianaphthene (**293**).[72] A mechanism involving attack on the respective $C_{(3)}$–S bonds is also possible[428a,428b] although both reactants **293** and **366** are products of the pyrolysis of thiophene alone[71,427] as discussed in the previous paragraph. The naphthothiophenes (**368**) are rationalized as forming by Diels–Alder addition of the aryne (**4**) to the naphthalenes (**374**) followed by loss of sulfur.[71,72] The feasibility of this process has been demonstrated for both benzyne (**1**)[429] and tetrafluorobenzyne (**375**)[430] from precursors other than phthalic anhydrides. The only support, however, for

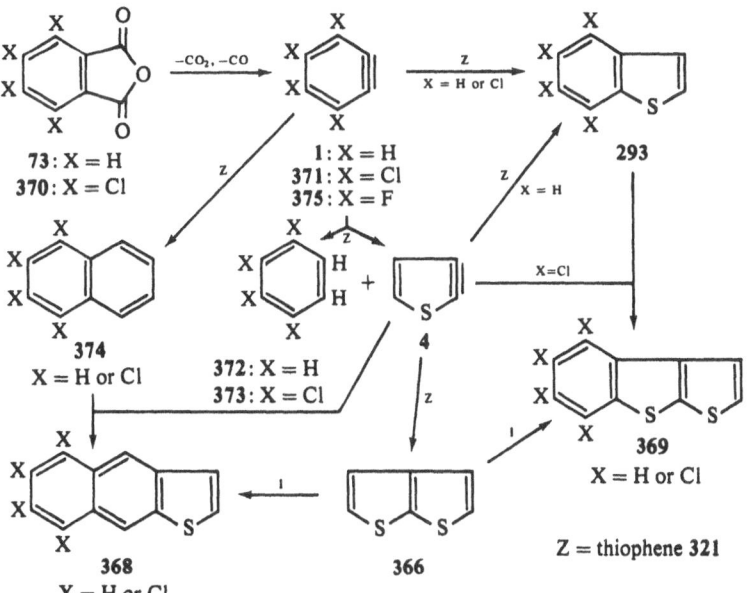

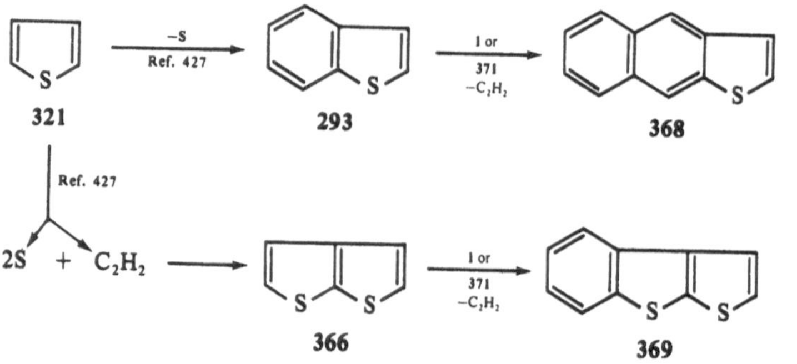

$$(29)$$

$$(30)$$

the proposed intermolecular hydrogen transfers **321** + **1** or **371** → **4** + **372** or **373** is the detection of the benzenes **372** or **373** among the pyrolysis products.[71,72] The fact that the latter is found in amounts insufficient to account for the quantity of **368** + **369** formed[72] may either indicate further decomposition of **373** under the reaction conditions[246] or that the formation of **368** and **369** are unrelated to the production of the benzenes **372** or **373** and hence of 2,3-didehydrothiophene (**4**). This latter interpretation is supported by the formation of benzene (**372**) from the pyrolysis of thiophene (**321**) alone[428] and by the ability to postulate mechanisms leading to **368** [equation (29)][246] and **369** [equation (30)][71,72] which do not involve the aryne (**4**). Although the pyrolysis studies of thiophene and phthalic anhydrides **73** and **370** therefore permit a thiophyne (**4**) interpretation,[71,72] they clearly do not require it, and therefore cannot be used to prove the intermediacy of this aryne.

The detection of minor amounts of phenylthiophene (**292**), bithienyl (**367**),

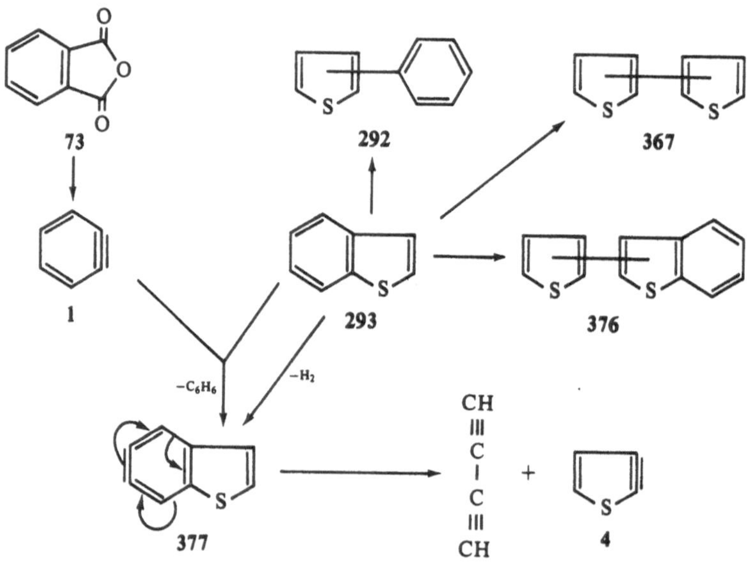

$$(31)$$

and thienylbenzothiophene (**376**) during the pyrolysis of thianaphthene (**293**) alone[246] or with phthalic anhydride (**73**)[72] seemed to require the presence of thiophene (**321**) or thiophyne (**4**) in these reactions. The latter species was postulated to arise by a retro-Diels–Alder decomposition [equation (31)] of the didehydrobenzothiophene (**377**) presumably formed from **293** by dehydrogenation or intermolecular hydrogen transfer to benzyne (**1**) as described in the previous two paragraphs. Although some evidence exists to support a process such as [equation (31)] for related species under electron impact[72,431] neither this analogy nor the detection of the minor products in the above pyrolyses of **293** provides compelling evidence for the existence of the aryne (**4**).

A final reaction which may involve the generation of the thiophyne (**4**) from another aryne was suggested to explain the formation of benzotriptycene (**378**) along with the expected thienonaphthacene (**379**) upon acid hydrolysis of the 1,4-oxide (**380**).[432] The retro-Diels–Alder reaction of **380** gives the naphthothiophene (**368**), isolated in 45% yield, and apparently the naphthalene 1,4-oxide (**148a**), which is postulated to undergo a further retro-Diels–Alder reaction (Section II.1.E) to produce benzyne (**1**). It was suggested that this benzyne may possibly displace a C_4H_2S fragment of unspecified structure from the naphthacene **379** to yield the triptycene **378**. Verification of this interesting reaction seems warranted, although the likelihood that a C_4H_2S species corresponding to 2,3-didehydrothiophene (**4**) is produced appears remote.

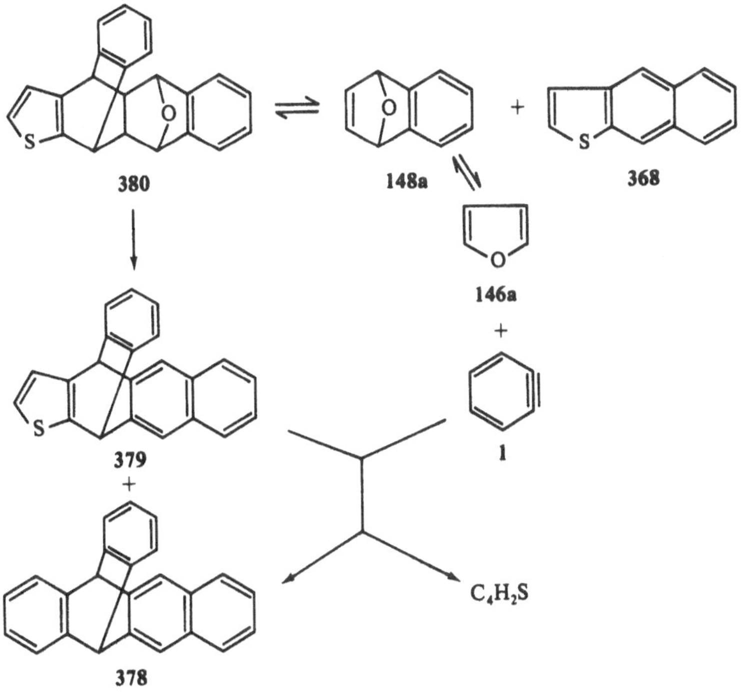

f. From Cyclic Anhydrides

The most convincing evidence to date for the generation of 2,3-didehydro-thiophene (4) comes from a study of the flash vacuum thermolysis (FVT) of thiophene-2,3-dicarboxylic acid anhydride (229) in the presence of various diene trapping agents. The FVT method[195] was chosen as one which would supply sufficient energy to the precursor in a short period of time to permit aryne formation to compete with nonaryne processes having lower activation energies (Section III.3.A.d). Trapping agents were utilized because of the expected[295 –298] instability of the aryne dimers 381 or 382 (Section II.2.B.a). The dienes used

381   382

were selected not only because of their compatibility with the FVT method (i.e., volatility and stability) but also because there was no evidence for their ambiguity as aryne probes (Sections II.2.B.c and II.2.B.d). Finally, the anhydride (229) was selected as a potential aryne precursor because of its ready availability,[433–436] the successful use of the related phthalic anhydrides (73) as a source of benzyne (1) in the gas phase (Section II.1.E),[175–181] and the observation[362,433] of a peak of significant intensity at $m/e = 82$ corresponding to the ion of aryne (4) in the mass spectrum of 229 as would be expected for a good aryne source.[177,179]

In the absence of a diene trap (N_2 as a carrier gas) the FVT of 229 led to extensive charring and the isolation of trace quantities of the fluorenone 382a and the quinone 383.[433] In the presence of any of the trapping agents considerably less carbonization occurred and several products containing a thiophene ring could be isolated.[77] With 1,3-cyclohexadiene (146c), thiophene (321), or benzene (372) thianaphthene (293) was the major product in the indicated yields. Cyclopentadiene (146b) gave a mixture of 4-(384) and 7-methylthianaphthene (385) in 7% and 6% yields, respectively, while 5,6-dimethylthianaphthene (386) was formed in 13% yield when 2,3-dimethylbuta-diene (387) was the trap. Furan (146a) gave 9% of a mixture of cyclopenteno-thiophenes (388), which probably arose by decarbonylation[437] of the initially formed thianaphthols (389) of which a small amount (1%) remained. Trace amounts (<1%) of products containing a thiophene ring were also detected from several of the other FVT reactions.

The most obvious explanation for the formation of these thianaphthenes is by aromatization of the Diels–Alder adducts (390) of the diene trap and 2,3-didehydrothiophene (4). Since the adducts 390 themselves could not be found, probably because of their instability under the reaction conditions (similar observations have been made for didehydrophenanthrene adducts),[438] support for this hypothesis must rest on the feasibility of the proposed aromatization reactions and the exclusion of any addition–elimination mechanisms (Section II.2.B.e) for the formation of the adducts (**390**). Each of these points will be taken up in turn.

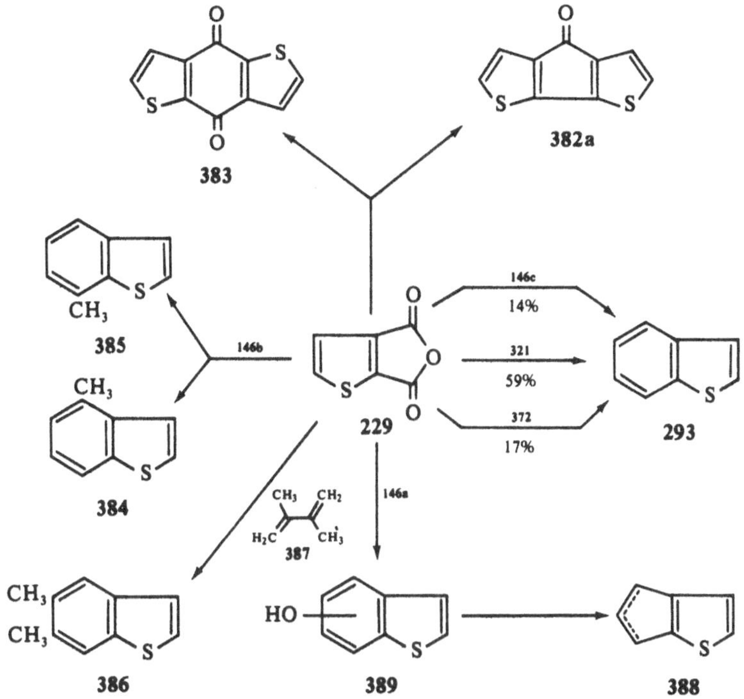

A direct demonstration of the aromatization of the aryne adducts (**390**) is

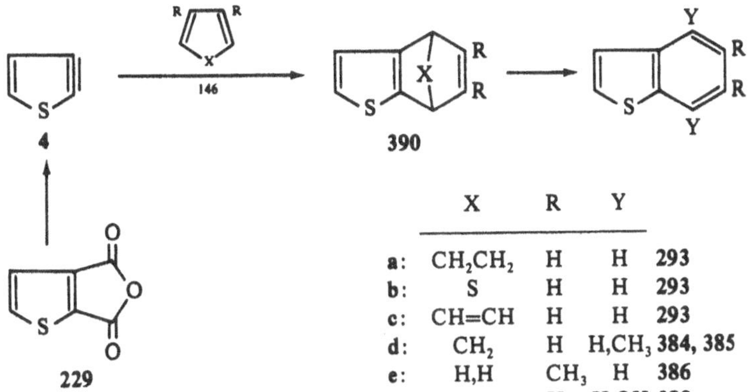

	X	R	Y	
a:	CH₂CH₂	H	H	293
b:	S	H	H	293
c:	CH=CH	H	H	293
d:	CH₂	H	H,CH₃	384, 385
e:	H,H	CH₃	H	386
f:	O	H	H,OH	389

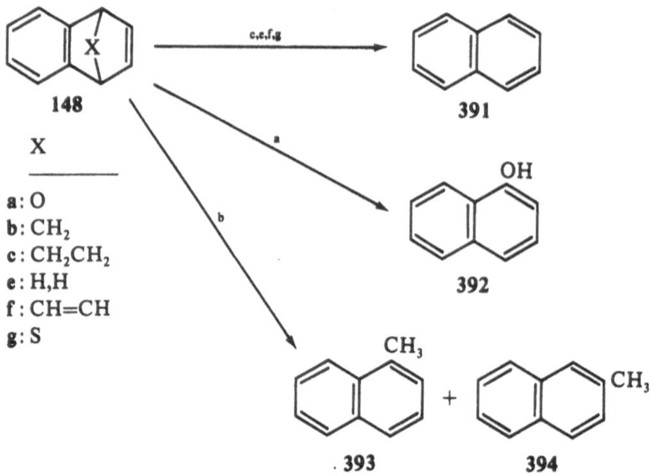
148
X
a: O
b: CH2
c: CH2CH2
e: H,H
f: CH=CH
g: S
c,e,f,g
a
b
391
OH
392
CH3
+
CH3
393
394

hindered by the probable difficulty of their synthesis via nonaryne pathways.[432,433,439] Several good analogies for the proposed aromatization reactions exist, however, particularly with the more accessible Diels–Alder adducts (**148**) of dienes and benzyne (**1**) (Section II.2.B.b). For example, both the benzyne adducts of cyclohexadienes (**148c**) and benzenes (**148f**) aromatize thermally to naphthalenes (**391**) by loss of ethylenes[440] and acetylenes,[390] respectively, via a retro Diels–Alder reaction.[194] Thermal opening of the oxygen bridge of the furan–benzyne adduct (**148a**) was postulated to account for the formation of 1-naphthol (**392**) during the gas phase pyrolysis of several benzyne precursors in the presence of furan.[53] This conversion has now been shown to proceed in 35% yield when **148a** itself was subjected to the same FVT conditions as the anhydride **229**.[77b] The thermal aromatization of the benzyne–cyclopentadiene adduct (**148b**) is considerably more complicated,[198] leading ultimately to a mixture of 1-(**393**) and 2-methylnaphthalenes (**394**), with the

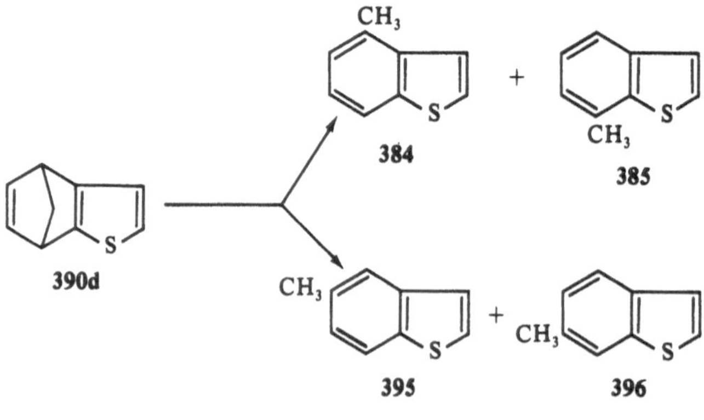
CH3
S
+
S
CH3
384
385
S
390d
CH3
S
+
CH3
S
395
396

former predominating.[77b,198] In fact, a similar product distribution is found on FVT of **229** in the presence of cyclopentadiene (**146b**) where the minor products consist of 0.5% each of 5-(**395**) and 6-methylthianaphthene (**396**).[433] This strongly supports the intermediacy of the adduct **390d** in this process. Dehydrogenation of butadiene–benzyne adducts (**148e**) to naphthalene derivatives **391** proceeds readily in the presence of noble or transition metals,[441] sulfur,[442] quinones,[443] or possibly even by thermal disproportionation.[444] Since each of these catalysts and/or hydrogen acceptors are, or may, be present during a typical FVT of the anhydride (**229**) (a bare Nichrome wire is used as the internal heat source,[77] trace quantities of the quinone **383** are often produced, and elemental sulfur is not an unlikely product from the thermolysis of thiophenes[71,427,428]), aromatization of the assumed adduct (**390e**) of thiophyne (**4**) and the butadiene **387** is a reasonable reaction. Finally, as has already been noted (Section III.3.A.e), thiophene (**321**) can serve as the diene component of Diels–Alder reactions with benzynes (**1**) generated from a variety of precursors both in the condensed[429,430] and in the gas phase,[71,72] including from indanetrione (**80**) (Section II.1.E) under the same conditions as are used for the FVT of the anhydride **229**.[199] In none of these cases has the assumed initial adduct **148g** been isolated but only the naphthalene **391** formed by desulfurization. This observation is consistent with the postulated desulfurization of the

adduct **390b** to thianaphthene (**293**) and the facile expulsion of bridged sulfur from other heterocycles.[445] Since thianaphthene (**293**) can arise from thiophene alone under certain conditions[71] (Section III.A.e), a control experiment was run and demonstrated the absence of this process under the conditions used for the FVT of the anhydride **229**.[433] In summary, therefore, the aromatization of any of the proposed adducts (**390**) must be regarded as the most reasonable and likely origin of the products obtained from the FVT of the anhydride (**229**) in the presence of diene traps.

What remains to be considered is whether the adducts **390** might arise by an addition–elimination process via **397** rather than an elimination–addition mechanism. The variety of dienes (**146**) used and their apparent lack of ambiguity as aryne probes (Sections II.2.B.c and II.2.B.d) is strong evidence (Section III.1.C) in favor of the latter mechanism unless it is argued that the anhydride **229** is an unusually powerful dienophile such as the halobenzoquinones (**173**)[326] (Section II.2.B.d). In effect **229** would have to behave like a

229 397 390 293

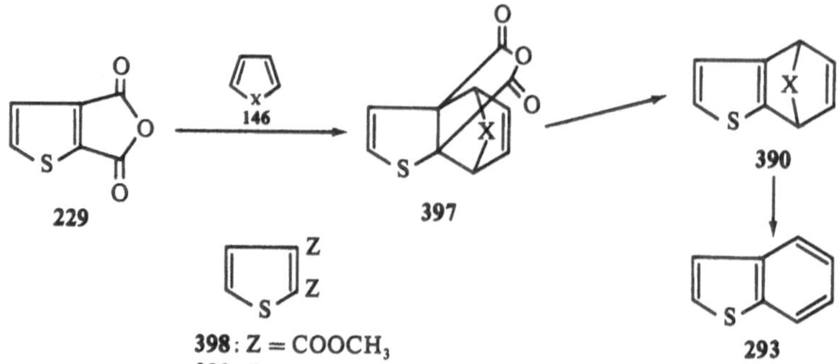

398: Z = $COOCH_3$
399: Z = CN

maleic anhydride and not an aromatic anhydride. None of the chemical or physical properties of thiophene derivatives support such a contention,[304] although [2+2] cycloaddition to the 2,3-double bond of thiophenes is known.[446] Furthermore, if it were true it might be expected that the diester **398** and dinitrile **399** would show similar behavior since dimethyl maleate and malonitrile are also effective dienophiles.[447] In fact both **398** and **399** are recovered in high yield from FVT in the presence of thiophene (**321**). Therefore, although it is negative evidence, neither experiment, precedent, nor analogy supports reaction of the anhydride **229** by an addition–elimination mechanism via **397** as a likely route to the adduct **390**.

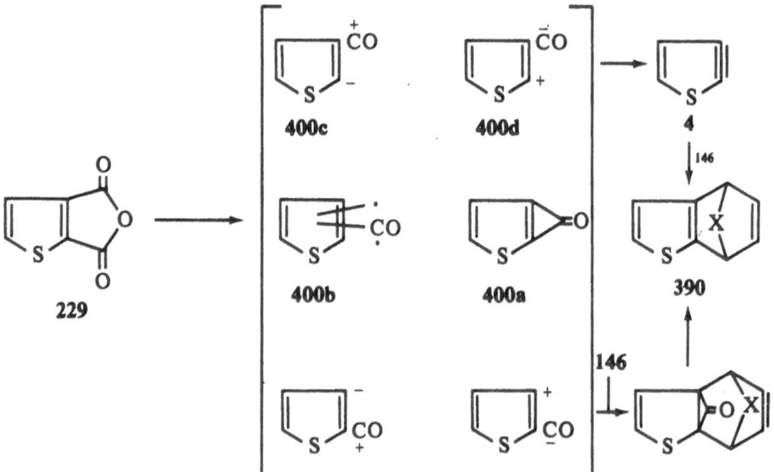

An addition–elimination mechanism involving not the anhydride (**229**) itself but an intermediate in its decomposition to the aryne (**4**) can be envisioned, however. Phthalic anhydrides (**73**) sometimes appear to decompose to arynes by a stepwise mechanism involving sequential loss of CO_2 and CO

229 400c 400d 4 400b 400a 390 400e 400f 401

and proceeding via an intermediate having the composition of a benzocyclo-propenone (**55**) (Section II.1.E).[175,179,182] By analogy, the FVT of the thiophene anhydride **229** could also involve such a carbonyl-containing species (**400**). This possibility is supported by the mass spectrum of **229**, whose base peak is at M–CO$_2$.[361,433] If this intermediate (**400**) added dienes **146** before eliminating CO, then the alleged aryne adduct (**390**) could be formed via an adduct such as **401** and hence without the intervention of the aryne (**4**). Although such dienophilic reactivity has not been detected for benzocyclopropenones (**33**), even in the presence of several traditional dienes (**146**),[155] it has been observed for cyclopropenone itself[448] (but not for diphenylcyclopropenone)[449] and for benzocyclopropene (**402**).[450] The second step of the proposed sequence **400** to

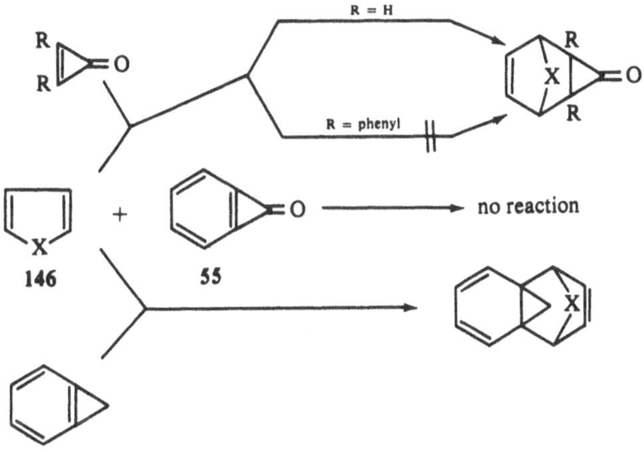

401 to 390 finds analogy in the formation of alkenes by decarbonylation of cyclopropanones,[451] although whether the more closely related formation of benzene and CO via norcaradienone (**403**) actually occurs is still an open question.[452] While these precedents are therefore somewhat strained, they are permissive of an addition–elimination mechanism via **401** and suggest that more direct evidence for the intermediacy of the thienocyclopropenone (**400**) needs to be obtained.

For benzocyclopropenones (**55**) generated from aminobenzotriazinones **49g** or **49h** [153-155] or by plasmolysis of phthalic anhydride (**73**)[182] such evidence con-

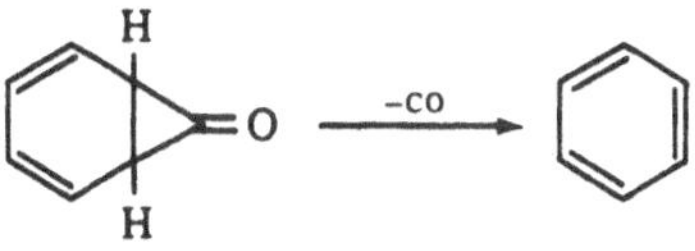

403

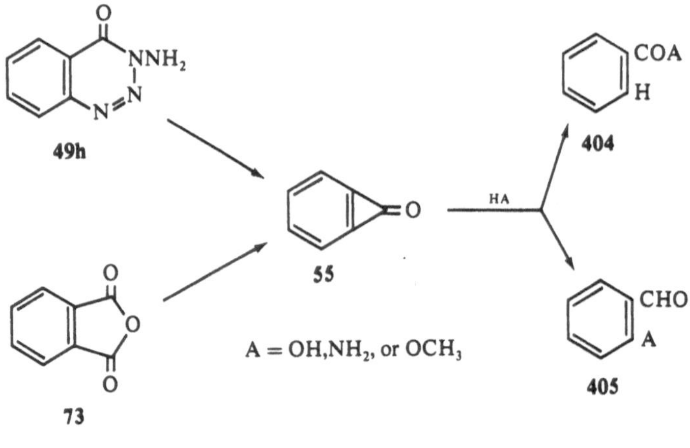

49h

73

55

A = OH,NH$_2$, or OCH$_3$

404

405

sists of trapping the intermediate with nucleophiles such as water, ammonia, or methanol to give the corresponding benzoic acid derivatives (**404**) and, in one case,[182] the *o*-substituted benzaldehyde (**405**). Although FVT of the anhydride (**229**) in the presence of such trapping agents does lead to the corresponding 3-thenoic acid derivatives (**406**),[199,433,453] these were shown to arise, at least in part, by decarboxylation of the half-acid derivatives (**407**) formed by direct reaction of the anhydride (**229**) with the nucleophilic traps. An attempt to trap **400** based on its possible diradical character (**400b**) utilized molecular hydrogen[77] or good hydrogen atom donors such as cyclohexadiene (**146c**)[433] as trapping agents during the FVT of the anhydride (**229**). No thiophene aldehydes (**408**) were found although they were demonstrated to be stable to the reaction conditions.[77] A similar trapping experiment with the *N*-phenylpyrrole anhydride **248** did give a corresponding aldehyde, presumably via the pyrrolocyclo-propenone **255** (Section III.2.B).[199] On the assumption that dipolar forms such

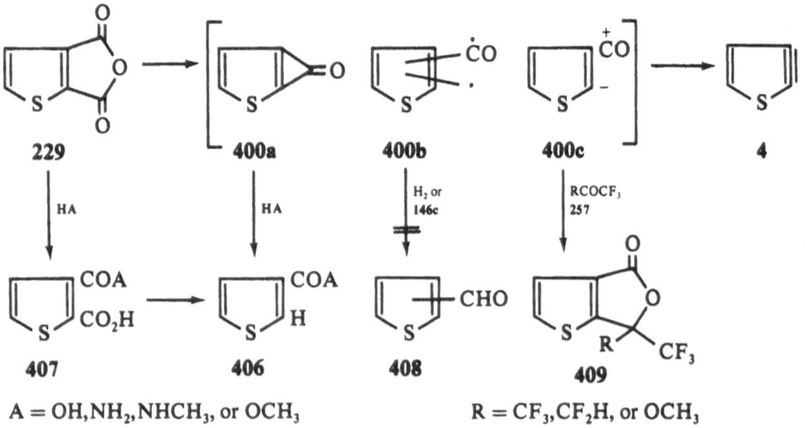

229   **400a**   **400b**   **400c**   **4**

407   **406**   **408**   **409**

A = OH,NH$_2$,NHCH$_3$, or OCH$_3$ R = CF$_3$,CF$_2$H, or OCH$_3$

as **400c** might make substantial contributions to the structure of the cyclopropenone (**400**), several trifluoromethyl carbonyl compounds (**257**) known to trap other dipolar species[454] were examined. As in the pyrrole series (Section III.2.B) evidence for the presence of the cyclopropenone intermediate **400** was obtained by the isolation of the corresponding adducts (**409**) whose structures were proved by synthesis.[453]

While this independent demonstration of the intermediacy of the cyclopropenone **400** during the FVT of the thiophene anhydride **229** is consistent with an addition–elimination mechanism via **401** as a rational for the formation of aryne adduct **390** it does not exclude an elimination–addition process via the aryne (**4**). In fact, this latter possibility is particularly supported by the observation that although both the thiophene and pyrrole anhydrides **299** and **248** decompose via the corresponding cyclopropenones **400** and **255**, only the former series leads to products (**293**) which suggest an aryne intermediate. This difference in behavior might be explained in two ways. If an addition–elimination mechanism via **401** is assumed, then the pyrrolocyclopropenone **255** must be a less reactive dienophile than the thiophene analog **400**. This conclusion would be inconsistent with both the lower aromaticity[304] and the greater strain expected[199] for **255**, since these properties should increase its dienophilic character.[307] On the other hand, if an elimination–addition mechanism is assumed then the pyrrolocyclopropenone **255** must be less reactive to decarbonylation to the aryne than is the thiophenocyclopropenone **400**, a conclusion which would be consistent with the expected greater strain of an aryne compared to a cyclopropenone[455] and of a dehydropyrrole (**299**) compared to a dehydrothiophene (**4**).[199] The validity of this analysis rests on the not unreasonable assumption[199] that the relative stabilities of these cyclopropenones and arynes will be related in some regular way to the geometry of the parent heterocycles (cf. Section IV.2.), specifically the $C_{(2)}$–$C_{(3)}$ bond dis-

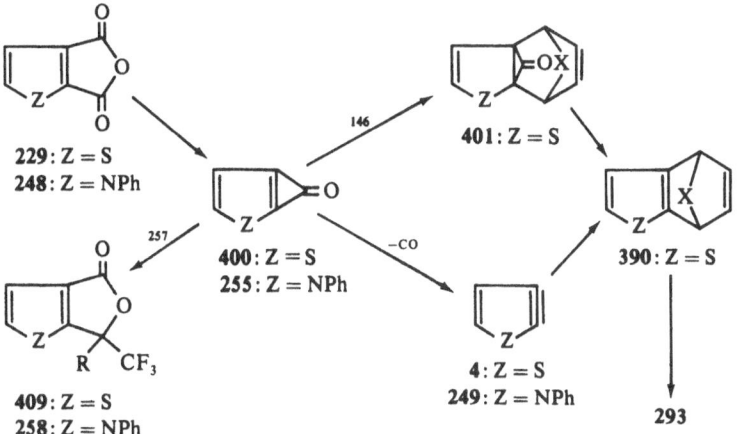

tance and the corresponding H–C–C–H angle, and hence will favor the thiophenes 4 and **400** over the pyrroles **249** and **255**. Quantitative support for this assumption, however, must await molecular orbital calculations which at present have been reported for only one of these intermediates, 2,3-didehydro-thiophene (4).[456] Experimentally, the addition–elimination mechanism via **401** can be rigorously excluded only if aryne-type products such as **390** or **293** are found from potential precursors of the aryne 4 which cannot yield the cyclopropenone **400**. While such data are not yet available in the thiophene series they have been obtained for 2,3-didehydrothianaphthene (**26**) (Section III.3.C).[78]

Although research on the FVT of the thiophene anhydride **229** has been primarily directed at obtaining evidence for the intermediacy of the aryne 4, it has also yielded some information on the chemistry of this species. The most studied reaction of thiophyne (4) thus far is with dienes **146** to give thianaphthenes **293**, presumably via the Diels–Alder adducts **390**.[77,433] The ubiquity of this reaction, as discussed above, appears to make it virtually diagnostic for the presence of the aryne (4). Thus the formation of thianaphthene (**293**) when benzene is the diene has been interpreted[77b] to involve a retro-Diels–Alder loss of acetylene from the initially formed thiophyne–benzene adduct (**390c**). Conversely the absence of thianaphthene (**293**) in the photolysis of the mercury compound **286** in benzene[76] (Section III.3.A.a) strongly suggests that the aryne 4 is not produced in this reaction.

A particularly interesting feature of the diene–aryne reaction is revealed when 2,5-dimethylthiophene (**410**) is used as the trapping agent during the FVT of the anhydride **229**.[443] In addition to the 4,7-dimethylthianaphthene (**412**)

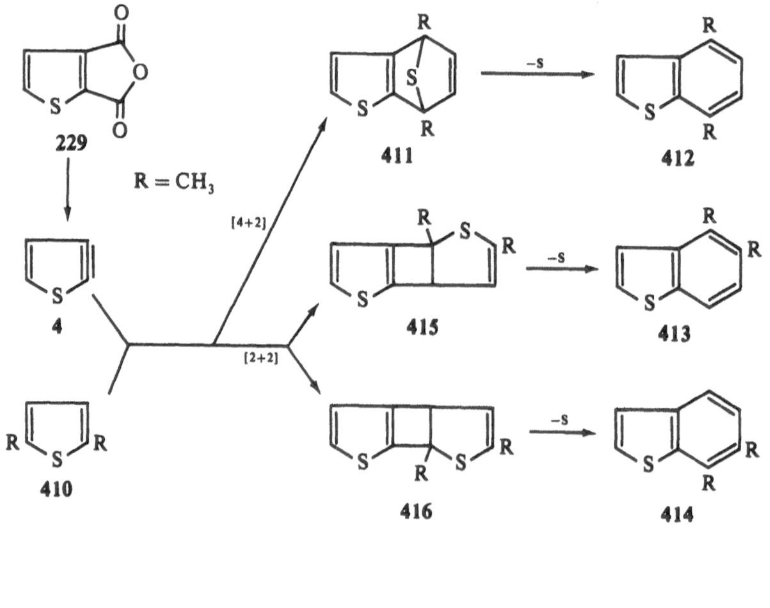

expected from aromatization of the Diels–Alder adduct **411**, the 4,5-dimethyl (**413**) and 6,7-dimethyl (**414**) isomers were also found in a ratio of 3:5:2, respectively. No other thianaphthenes could be detected and no interconversion[457a] of the three isomers **412–414** under the FVT conditions was observed. Analogous results have been obtained recently with 3,4-dimethyl-thiophene.[457b] The most consistent interpretation of these results is that the initially formed aryne **4** can react with the diene **410** by either a [4+2] cycload-dition to give the Diels–Alder adduct **411** or by two different [2+2] cycloaddi-tions to give the cyclobutene adducts **415** and **416**. Aromatization of each of the three adducts by loss of sulfur would lead to the three observed dimethyl-thianaphthenes **412–414**. Although benzynes will cycloadd to simple alkenes in a [2+2] manner[302] they usually prefer the [4+2] mode with conjugated dienes.[301] One of the most notable exceptions to this latter generalization is 2,3-dimethylbutadiene (**387**), which reacts with benzyne (**1**) to give the ben-zocyclobutene **417**.[458] The analogous products **418** or **419** from thiophyne **4** were not found from the FVT of the anhydride **229** in the presence of the diene **387**. However, in addition to the major product, 5,6-dimethylthianaphthene (**386**), small amounts of a dihydrodimethylthianaphthene **420** or **421** and its

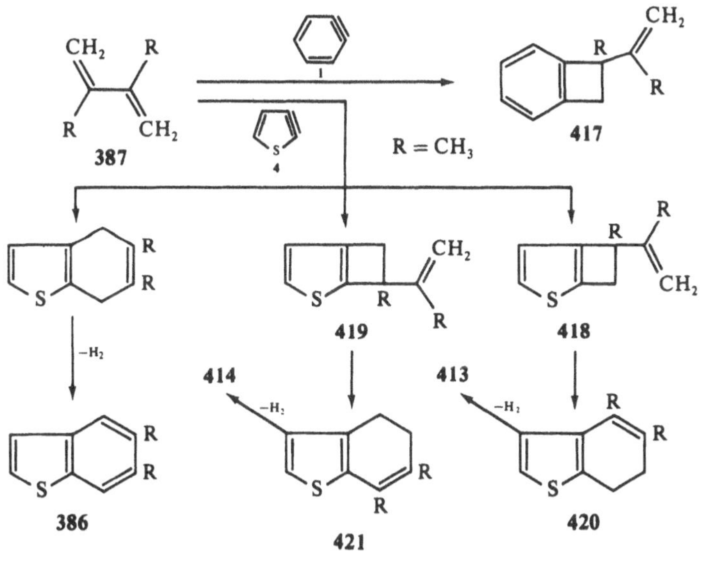

corresponding dimethylthianaphthene **413** or **414** were detected. Both of these minor products could arise under FVT conditions by the well-known[459] vinyl-cyclobutane–cyclohexene rearrangement of the proposed adduct **418** or **419** followed by dehydrogenation. If this hypothesis is correct then a second example of [2+2] cycloaddition to 2,3-didehydrothiophene (**4**) is available.

Until this reaction can be studied further with a variety of alkenes[460] and aryne generation methods, its generality and possible significance as a probe of the structure of the thiophyne (4) must remain an open question.

Another reaction with alkenes which arynes readily undergo is the ene reaction [equation (32)].[461,462] Although this process can sometimes compete

$$\text{(structure)} \longrightarrow \text{(structure)} \tag{32}$$

quite effectively with the [4+2] cycloaddition to conjugated dienes,[458-460] no ene products were observed from any of the FVT reactions of the anhydride 229 with diene traps. In the presence of methylacetylene (422), however, a mixture of thienylallenes (423) and thienylmethylacetylenes (424) was obtained.[77b] The

$$\underset{4}{\text{(structure)}} \; + \; \underset{422}{CH_3C{\equiv}CH} \; \longrightarrow \; \underset{423}{\text{(structure)}C{=}C{=}CH_2} \; + \; \underset{424}{\text{(structure)}C{\equiv}CCH_3}$$

former product is that of an ene reaction and parallels the similar behavior of benzyne [equation (33)],[172a] while the latter could arise from the former by the well-known allene–acetylene rearrangement.[463]

$$\underset{1}{\text{(structure)}} \; + \; CH_3C{\equiv}CH \; \longrightarrow \; \text{(structure)}C{=}C{=}CH_2 \tag{33}$$

A less well-known reaction of arynes is with molecular hydrogen. Benzyne (1) generated from either the gas phase pyrolysis of o-diiodobenzene (114) over zinc[220] or the plasmolysis of phthalic anhydride (73)[182] reacts with molecular hydrogen to give benzene, among other products. From the FVT of the anhydride 229 in the presence of hydrogen, thiophene (32%) and thianaphthene (15%) were the only products isolated,[77] the latter presumably arising by trapping of the thiophyne (4) by the thiophene (321) formed by its hydrogenation.

According to MO calculations the 2-position of 2,3-didehydrothiophene should have the lowest electron density and hence be the site of nucleophilic attack [equation (34)].[456] On the other hand, if the regiochemistry of the nucleophilic addition is controlled, as it is with benzyne,[255] by the stability of the initially formed anion, then attack should occur at the 3-position via the very stable 2-anion[22,23] [equation (35)]. All attempts to test these alternate hypotheses during the FVT of the anhydride 229 have thus far failed because all the nucleophiles used as traps[199,433] react preferentially with the anhydride **229** itself or perhaps with the cyclopropenone **400** to give only carbonyl-containing products **406** and **407**. Further studies to resolve this interesting question are warranted.

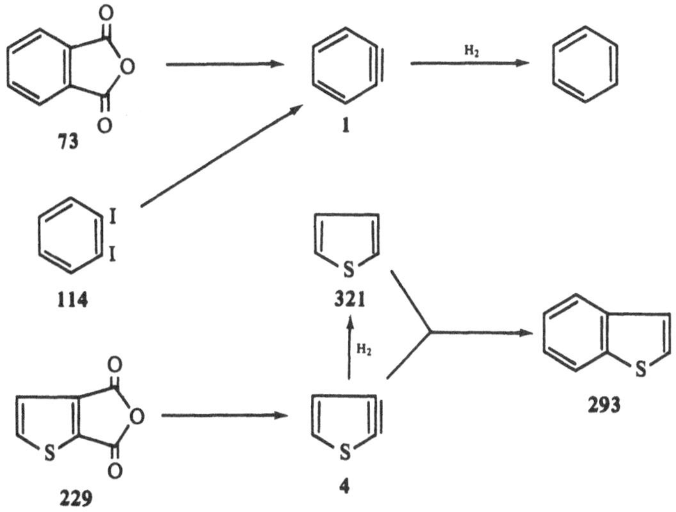

One final reaction which might involve the aryne **4** is the formation of the fluorenone analog **382** during the FVT of **229** in the absence of trapping reagents. Fluorenone itself has been observed during the thermolysis of phthalic anhydride (**73**)[175] and proposed[179] to arise by reaction of benzyne (**1**) with benzocyclopropenone (**55**). The likelihood of a similar process accounting for the presence of **382** is increased by the demonstration[433] that an alternate path via thermal decarbonylation[464] of the quinone **425**, a possible[375] but unobserved reaction product, does not take place under the FVT conditions.

Two methods other than FVT have been investigated for generating 2,3-di-dehydrothiophene (**4**) from the anhydride **229**. Heating a melt of **229** above its decomposition temperature (270°C) for several hours produces,[77b] in addition to a brittle, black polymer, 24% of only one isolable product, the same anthra-

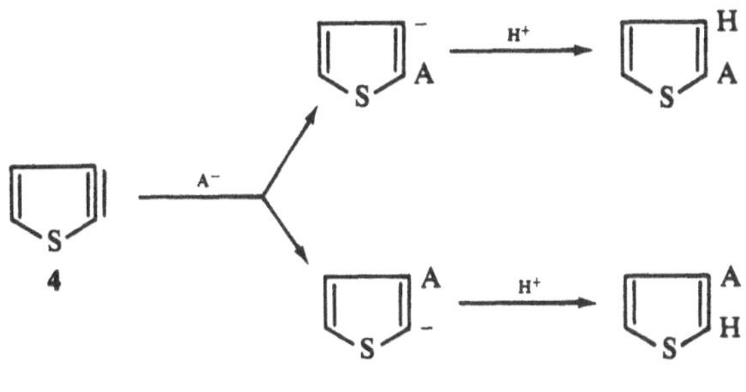

$$(34)$$

$$(35)$$

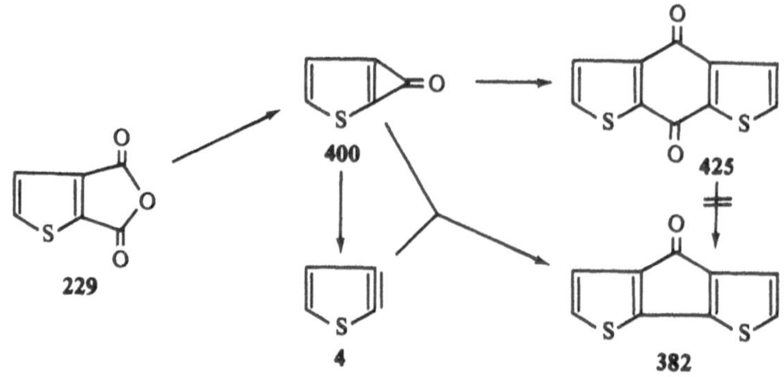

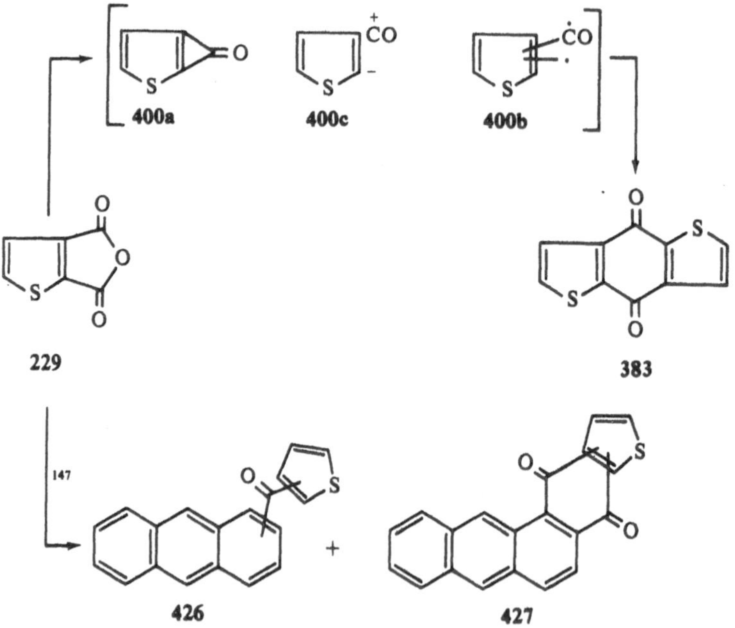

quinone analog **383** formed from the FVT of the anhydride **229** in the presence of N_2 as a carrier gas. Because of the demonstrated presence under FVT conditions of the cyclopropenone intermediate **400**,[453] and by analogy with a similar result and rationale in the pyrrole series[375] (Section III.2.B), either dimerization of **400** or its reaction with the anhydride **229** followed by loss of CO_2 seems a reasonable explanation for the formation of the quinone **383** both here and under FVT conditions. It is interesting to note that, in contrast to the pyrrole anhydride **248**,[375] only one dimer of the cyclopropenone **400** is found in this case.[77b] With anthracene (**147**) as a diluent and potential aryne trap only a thenoylanthracene (**426**) and a quinone (**427**) were found upon heating a combined melt to 340°C for a short time.[77b] Since the same two products are obtained by identical treatment of the product (**428**) of a Friedel–Crafts reaction of the anhydride (**229**) and anthracene (**147**), neither the aryne (**4**) nor the cyclopropenone (**400**) appear to be involved.[77b]

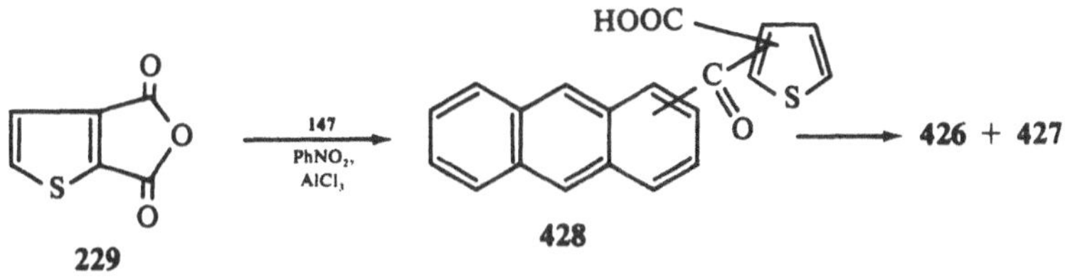

The final procedure to be discussed for generating the aryne (**4**) is the plasmolysis of the anhydride **229**.[362] In the absence of trapping agents the major product is a polymer $(C_{3.8-4.3} H_{2.9-3.5} S_{0.8-1.1} O_{0.6-1.5})^x$ and traces of two substances whose mass spectra are consistent with their being the trimers **429** and **430** of aryne **4**. Although no evidence was found for the corresponding aryne dimers **381** and **382** a coplasmolysis of the 2,3-(**229**) and 3,4-thiophene anhydrides (**431**) did give a substance whose mass spectrum was consistent with that of the mixed dimer **432**. Unfortunately, since neither the trimers nor the dimers are known, a positive structural assignment by comparison with authentic samples was not possible nor was sufficient material obtained to permit actual isolation (all analyses were carried out by GC–MS methods).

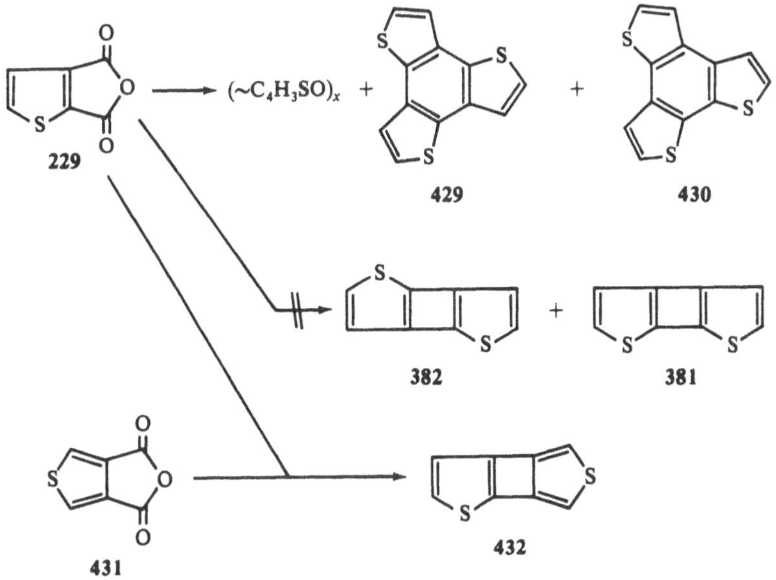

Trapping experiments were somewhat more definitive but only if the trapping agent and the anhydride **229** were introduced simultaneously into the plasma zone. If the trapping reagent was not introduced until after the plasma zone no trapping products were observed. In the presence of hydrogen both thiophene (**321**) and, as a minor product, thianaphthene (**293**) were obtained[362] analogous to the same reaction under FVT conditions.[77] These products, as well as the bithienyl (**367**) formed as the other major product, were rationalized as arising via the didehydrothiophene (**4**) as shown. An alternative exists, however, in that, in the presence of the atomic hydrogen produced in a glow discharge,[465] a diradical form (**400b**) of the cyclopropenone **400** might be more likely to be trapped as the thiophene aldehyde (**408**) than under FVT conditions,[77,433] and this aldehyde (**408**) would certainly be expected to go on to thiophene (**321**) upon plasmolysis.[466] If the thiophene (**321**) thus formed reacted upon further plasmolysis as it does on thermolysis,[71,427,428] then both the formation and relative amounts of the other two products, **293** and **367**, would be explained. In particular, the formation of bithienyls (**367**) as a major product from the plasmolysis is more consistent with such a mechanism than with an aryne pathway which, based on the FVT of the same reactants, should give no bithienyls.[77,433] An aryne mechanism cannot be ruled out, however, since plasmolytic and thermolytic behaviors of molecules do not necessarily coincide.[464] The appropriate control experiment, plasmolysis of thiophene (**321**) in the presence of hydrogen, has not been reported, although the fact that thiophene (**321**) alone gives only polymers and ring fragmentation and no bithienyls (**367**) or thianaphthene (**293**) on plasmolysis[467] does not support the above nonaryne mechanism. Further research is necessary to resolve this ambiguity.

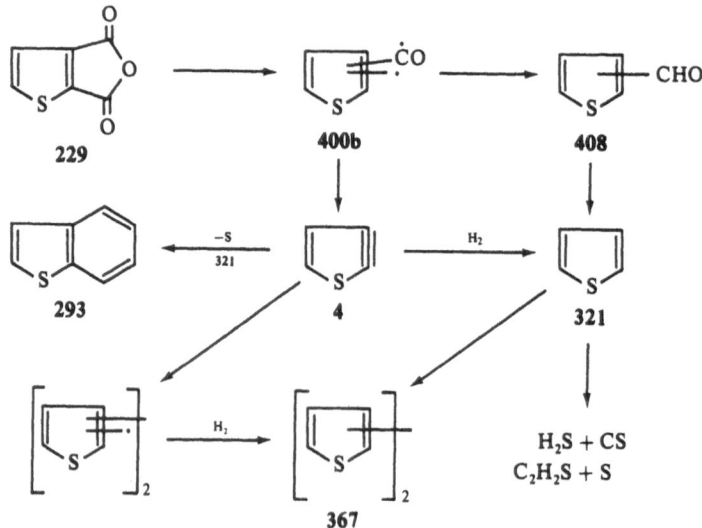

Plasmolysis of the anhydride **229** in the presence of furan (**146a**) gives a mixture of cyclopentenothiophenes (**388**), proposed[362] to arise by reaction of the aryne (**4**) with the C_3H_4 fragment known to be produced during the plasmolysis[468] of furan. Although the intermediacy of didehydrothiophene (**4**) in this reaction is highly probable, the nature of the C_3H_4 species was not specified.[362] Of the three possible C_3H_4 compounds, allene (**433**), methylacetylene (**422**), and cyclopropene (**434**), the last two have been shown to arise from furan (**146a**) by thermolysis[469] and photolysis,[470] respectively, and the first one is therefore also accessible by the allene–acetylene rearrangement.[463] By analogy with their behavior with benzyne (**1**), however, both methylacetylene (**422**) [equation (33)][172a] and allene (**433**) [equation (36)][172a,471] should undergo ene reactions, while only cyclopropene (**434**) is likely[472] to participate in the necessary cycloaddition leading to an indene [equation (37)].[473] Alternatively

$$ \tag{36} $$

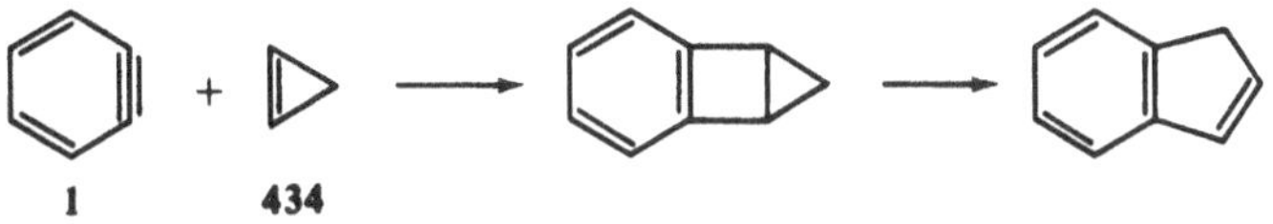

$$ \tag{37} $$

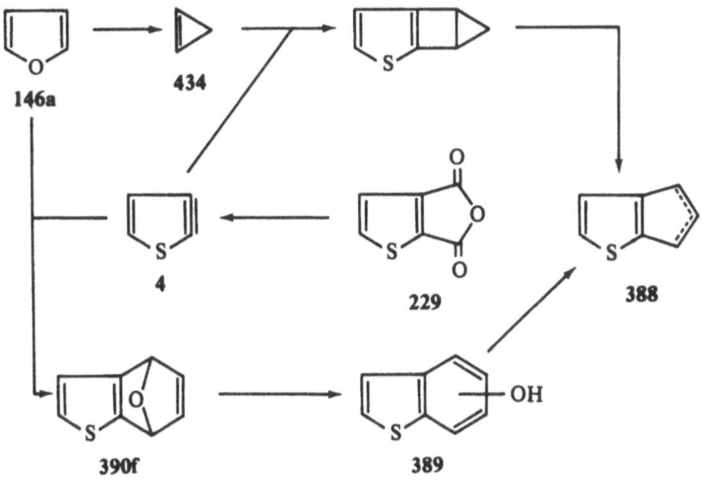

the formation of the cyclopentenothiophenes (**388**) during the plasmolysis might involve the same mechanism proposed[77] for their formation during the FVT experiments, i.e., a Diels–Alder reaction to **390f**, aromatization to **389**, and decarbonylation to **388**. This possibility is supported by the well-documented plasmolytic decarbonylation of naphthols to indenes.[474]

Acetylene is an effective trap for benzyne (**1**) generated by the plasmolysis of phthalic anhydride (**73**).[182] The major product, phenylacetylene (**435**), is

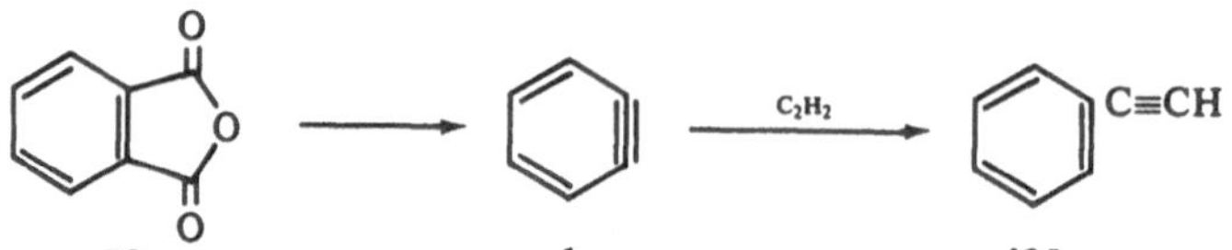

73   **1**   **435**

analogous to the thiopheneacetylene (**436**) formed during plasmolysis of the thiophene anhydride (**229**)[362] and hence is consistent with the intermediacy of 2,3-didehydrothiophene (**4**) in this reaction. On the other hand, an acetylene coupling reaction[363,475,476] with some other intermediate or by-product such as thiophene (**321**) cannot be ruled out without further work.

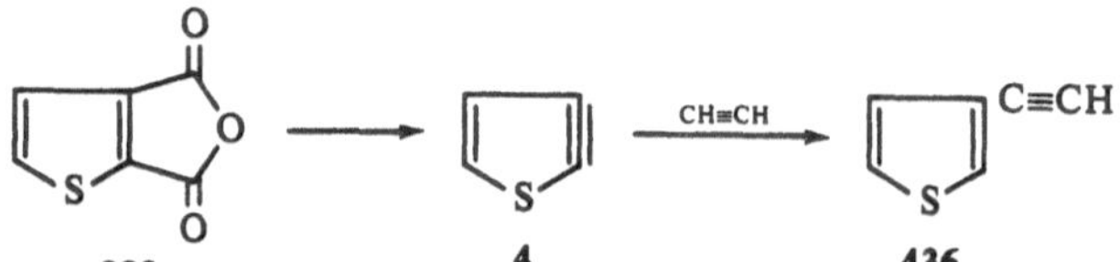

229   **4**   **436**

Although some of the above reactions might be explained easily without invoking an aryne intermediate, many of the others could be only with considerable ingenuity. It therefore seems warranted to conclude that 2,3-didehydrothiophene (**4**) probably has been generated from the FVT[77] and plasmolyses[362] of the anhydride **229** and perhaps by some of the other methods as well (Section III.3.A.e).

B. 3,4-Didehydrothiophene

By way of contrast, no evidence has been obtained to support the existence of the isomeric 3,4-didehydrothiophene (**25**). At −70°C and −40°C, respectively, 3-bromo (**437**)[394] and 3-fluoro-4-lithiothiophenes (**438**)[401] behave like normal lithium reagents with no evidence of lithium halide elimination. At room temperature the former compound rearranges[395] to the more stable 2-lithio

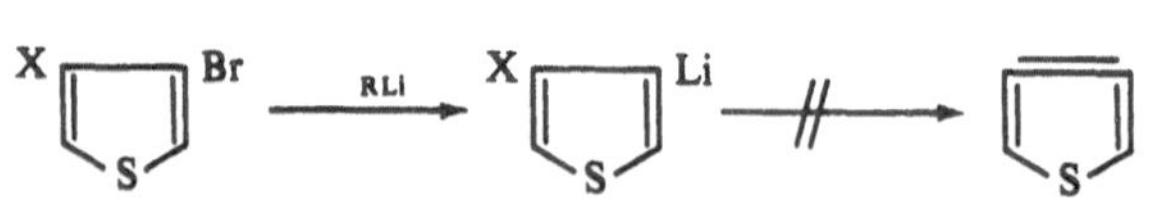

437: X = Br   **25**
438: X = F

isomers, as was also observed with 3-lithio-2-halothiophenes such as **303** (Section III.3.A.b).[399] In order to block this rearrangement the 2,5-diphenyl-3-lithio-4-iodothiophene (**439**) was prepared and its decomposition studied from −70°C to +100°C.[76] No evidence for the aryne **440** was obtained, although no potential trapping agents were apparently examined. The only reactions detected were disproportionation to the diiodo (**441**) and dilithiothiophene (**442**) and ring opening of the latter at higher temperatures to diphenylbutadiyne and lithium sulfide.[76] The generality of this ring opening of 3-lithiothiophenes is now well established[477] and does not appear to involve arynes in any way. Reaction of the related diiodothiophene **443** with *n*-butyllithium proceeds normally[402] except as discussed in Section III.3.A.b, with no indication of a 3,4-didehydro-thiophene intermediate.

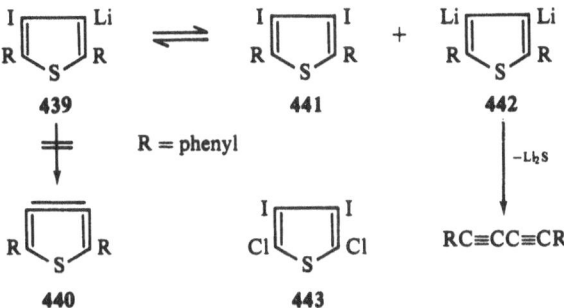

Thermolysis of the mercury compound **444** in the presence of tetracyclone (**151**) did give a tetraphenylthianaphthene which proved[62] to be the same one (**285**) as that obtained[61] from the isomeric precursor **286** and not the one (**445**) expected if the 3,4-aryne **25** was an intermediate. Since it was demonstrated that the mercury compounds **286** and **444** did not interconvert under the themolysis conditions,[386] it was postulated that an initially formed 3,4-aryne (**25**) rearranged to the apparently more stable 2,3-aryne (**4**).[62] Because of the unprecedented nature of this proposed aryne rearrangement, doubts as to the validity of this interpretation were soon raised.[1,51] A semiempirical SCF–MO calculation of the isomeric didehydrothiophenes predicted that, in contrast to the above suggestion, the 3,4-isomer **25** was more stable than the 2,3-isomer **4**.[456] The parallel behavior of the two mercury compounds **286** and **444** with respect to the formation of 3-iodothiophene (**287**) only when thermolysis is carried out in the absence of tetracyclone (**151**)[76] suggests that, as in the former case (Section III.3.A.a), the adduct **285** probably forms by reaction of tetracyclone (**151**) with 3-iodothiophene (**287**) and not the aryne (**4**). Support for this hypothesis comes from the demonstration of this last conversion, **151**+**287**→**285** (Section III.3.A.a) and the isolation of 2,5-diphenyl-3-iodo-thiophene (**446**) from thermolysis of the diphenylmercury precursor **447**.[76] In this case, none of the corresponding tetracyclone adduct **448** was obtained, presumably because of steric hindrance to addition of tetracyclone (**151**) to the iodocompound **446**.[76]. At 280°C instead of 240°C, 2,5-diphenylthiophene (**449**) becomes the major organic product (35%). Although an aryne **440** cannot be excluded as an intermediate in its formation, it appears more likely that **449** arises by further decomposition of the iodocompound **446**, whose yield drops from 16% to 1% at the higher temperature.[76]

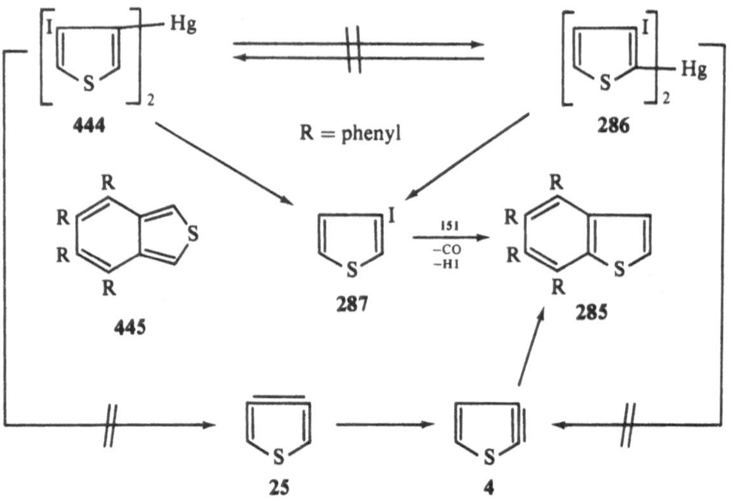

Treatment of 3-bromothiophene (**319**) with potassium amide in liquid ammonia[75] or of the iodocompound **446** with NaOH at 180°C[478a] gives no evidence for the formation of arynes **25** or **440**, respectively. In the former case a relatively slow normal substitution to 3-aminothiophene (**318**) is observed (Section III.3.A.c), and in the latter no reaction takes place. The o-dilithio-3-thienyl-tert-butylsulfone (**450**) does not eliminate lithium tert-butylsulfinate to give either of the o-lithiodidehydrothiophenes (**451**) or (**452**)[478b] although the analogous transformation does occur in the benzene series.[478c]

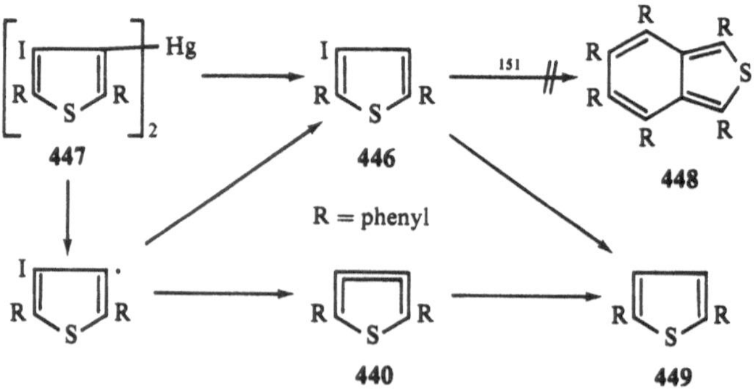

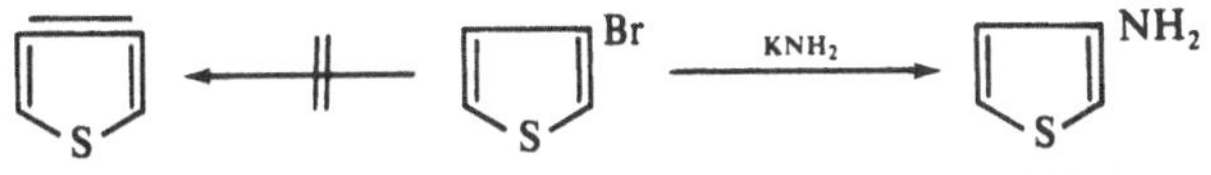

The diiodothiophenes **443** and **443a** are stable to FVT (90% recovery) at 800°C or heating with copper–bronze, except for a small amount of bithienyl formation.[402] Several other potential precursors of a 3,4-didehydrothiophene which fail to give aryne-type products on gas phase thermolysis include the iodoacid **453a**,[402] the oxazinones **454**,[192] and the anhydrides **431a**[402] and **431**.[77,402] Condensed phase thermolysis in the presence of anthracene (**147**) gives no identifiable product from the silver salt **453b**,[402] and a mixture of quinones **455** and **456** and a thenoylanthracene **426** from the anhydride **431**.[77b] The latter product appears to be identical with the one obtained from the 2,3-thiophene anhydride (**229**) from a similar reaction (Section III.3.A.f).[77b] In the absence of anthracene (**147**) a melt of the anhydride **431** was more stable than that of its 2,3-isomer **229**, requiring a temperature of 470°C instead of 260°C to undergo the decarboxylative dimerization to the quinones **457** and **383**.[77b]

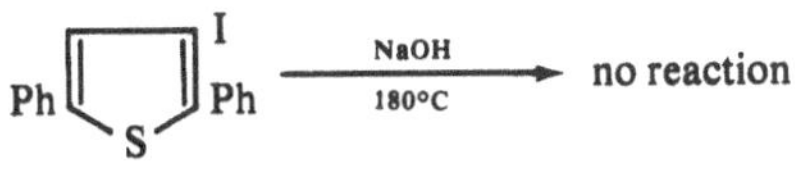

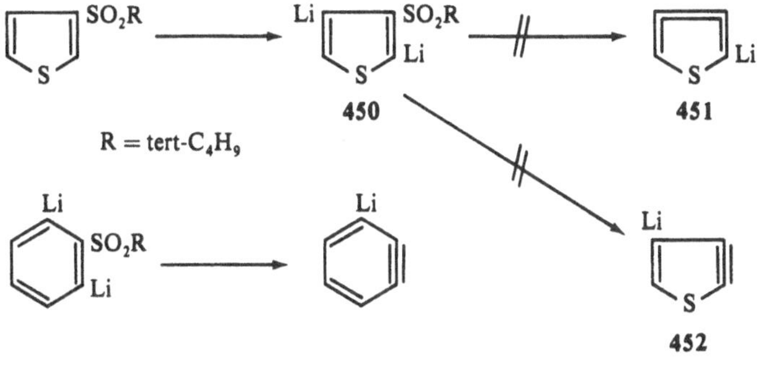

The thermolytic stability of the anhydride **431** is remarkable in view of the similarity of its mass spectrum to that of the 2,3-isomer **229** with regard to prominent M–CO_2 and aryne peaks at $m/e=110$ and 82, respectively.[77b,362] The major difference in the mass spectra of the anhydrides **229** and **431** is the presence of a significant $m/e=50$ peak in the latter case, conceivably due to loss of sulfur from the radical cation of the aryne **25**. Even if this reaction were possible thermolytically,[207] however, it should manifest itself as instability of the aryne (25) and not stability of the anhydride (431) as is observed. These facts, along with the favorable[179] M^+:aryne$^+$ intensity ratio in the mass spectrum of 431, strongly indicate that under the proper conditions this anhydride (431) should decompose to the aryne (25).

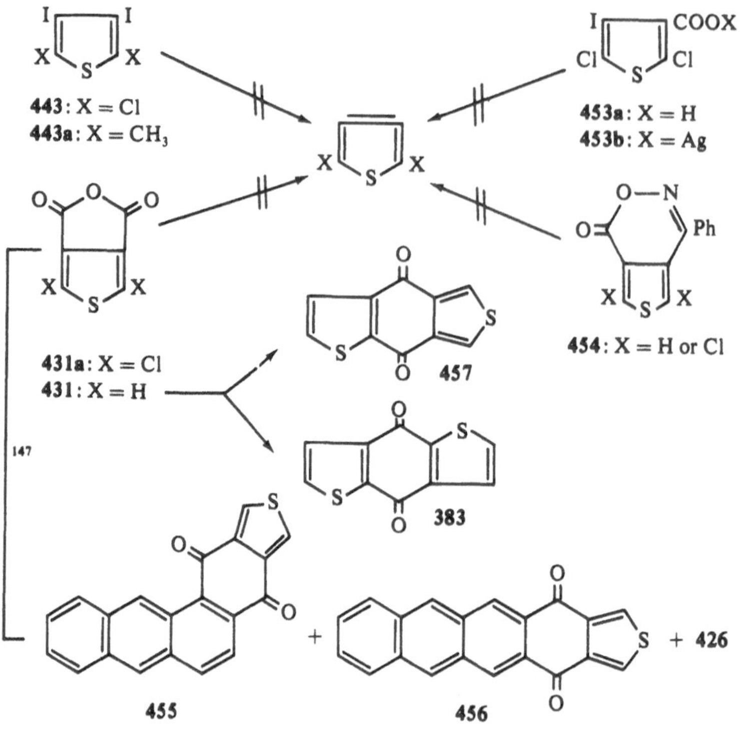

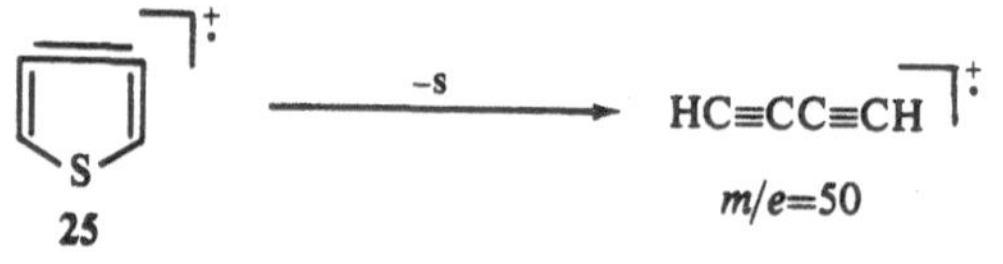

The plasmolyses[362] are unfortunately ambiguous in this regard, apparently giving results essentially parallel to those of the 2,3-anhydride 229 (Section III.3.A.f). In the absence of trapping agents, an oxygen-containing polymer is the major product, but no trimers or even the expectedly[297,298] most stable of the didehydrothiophene dimers (458) were found. Trapping with molecular hydrogen gave thiophene (321) and bithienyls (367), and with acetylene the same thienylacetylene (436) obtained from the 2,3-anhydride 229. Both of these results are permissive but do not require an aryne intermediate, as discussed in Section III.3.A.f. The best evidence for an aryne in the plasmolysis of the 2,3-

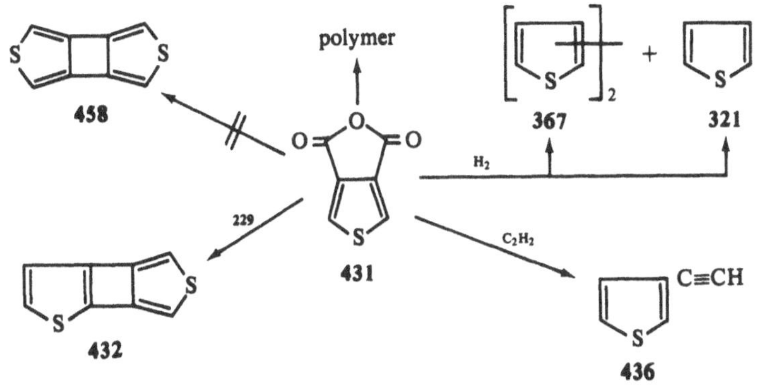

anhydride **229**, the formation of the cyclopentenothiophene (**388**) from a furan-trapping experiment, is missing for the 3,4-anhydride **431** because of the low yields of products. Although the crossed dimer **432** might indicate the inter-mediacy of the aryne (**25**) in the coplasmolysis of the anhydrides **229** and **431**, the structure of this product remains to be proved, as discussed in Section III.3.A.f.

In summary, therefore, no convincing evidence for a 3,4-didehydro-thiophene (**25**) is as yet available, although the prognosis for its preparation from precursors such as the anhydride **431** is not hopeless.

C. *2,3-Didehydrothianaphthene*

Although it reacts almost quantitatively with methoxide ion in the presence of KI and cupric oxide to give the normal substitution product **461**,[479] 3-bromothianaphthene (**459**) with alcoholic potassium hydroxide is reported to give only reduction to thianaphthene (**460**) and cine-substitution to 2,3-dihydro-2-oxothianaphthene (**462**) and its ring-opened derivatives.[63] The latter reaction was originally rationalized as involving rearrangement of the normal-substitution product **463** *in statu nascendi* but later was proposed[2,66] to proceed via 2,3-didehydrothianaphthene (**26**). In a very similar reaction with piperidine ($C_5H_{11}N$), 3-bromothianaphthene (**459**) once again underwent only reduction and cine-substitution to give thianaphthene (**460**) and 2-piperidino-thianaphthene (**464**).[65] An elimination–addition mechanism via the aryne (**26**) was once again proposed[1,3] in place of the AEa mechanism (Section II.2.A.e) tentatively suggested by the original authors[65] and favored in the comprehensive review of arynes.[480] A more convincing if somewhat mundane explanation for these apparent cine-substitution reactions came to light during a reinvestigation of the piperidine reaction[481] with the discovery that 3-bromothianaphthene (**459**) prepared by the method[482] utilized in the above studies[63,65] contains substantial quantities of thianaphthene (**460**) and 2-bromothianaphthene (**465**) even after fractional distillation. Since this latter isomer is much more reactive than is the 3-bromo compound **459** [65,481] toward piperidine, 2-piperidinothianaphthene (**464**) would be the first, and, at short reaction times,[65] the only substitution product formed. A similar explanation probably accounts for the production of the 2-oxo compound **462** from "pure" 3-bromothianaphthene (**459**).[63]

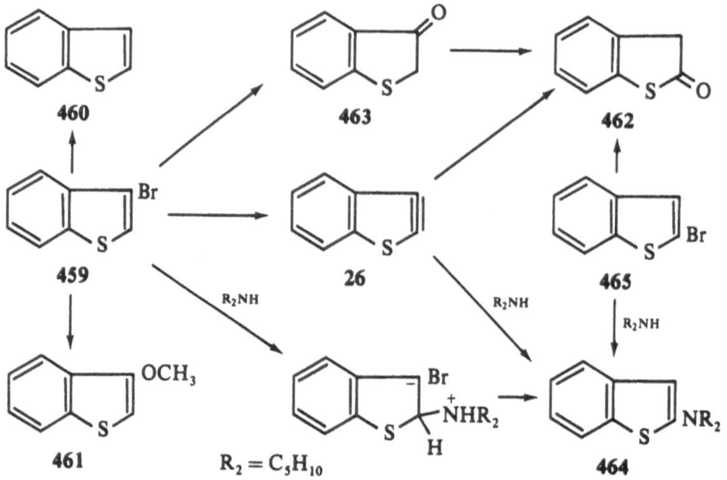

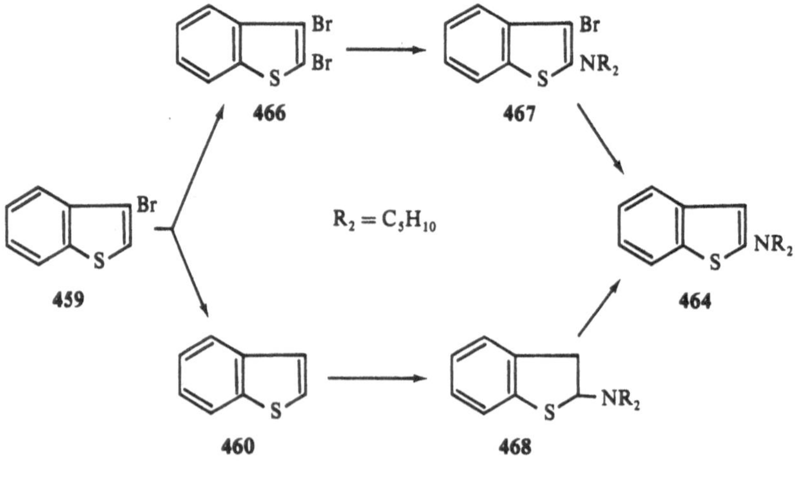

Another process must also be involved, however, since under more severe conditions with 3-bromothianaphthene (**459**) which is free of its 2-isomer (**465**), thianaphthene (**460**) and 2,3-dibromothianaphthene (**466**),[483] a small (1–5%) but significant amount of 2-piperidinothianaphthene (**464**) is still formed.[481] Although both AEa and aryne mechanisms could be responsible for this observation, a transhalogenation mechanism (Section II.2.A.d) via the dibromo- (**466**) and bromopiperidinothianaphthene (**467**) was considered more likely based on the conversion of the former to the latter compound and of both of them to the cine-substitution product **464**, but not thianaphthene (**460**), under relatively mild conditions.[481,484] The remaining step of the scheme, **459→460+466**, is analogous to the known halogen disproportionation reactions of bromothiophenes[75,263] (Section III.3.A.C) and bromothianaphthenes,[262,266] which require metal amides but proceed at much lower temperatures than the amine reactions.[65,481] This step would also account for some, but not all,[262,266,408,481] of the thianaphthene (**460**) found in these cine-substitution reactions. Thianaphthene (**460**) could in turn provide an alternative, low-yield source of the amine **464** by means of the direct addition of piperidine to give **468**, followed by dehydrogenation under the relatively severe reaction conditions.[485]Regardless of the relative contribution of these two mechanisms, however, they can explain the observed cine-substitution without invoking the intermediacy of the aryne **26** in the above reactions. There is even less reason to suggest an aryne mechanism for the reaction of halothianaphthenes with metal amides since no substitution at all, only halogen rearrangement from the 2- to the 3-position and reduction [equation (38)], is observed.[262,266]

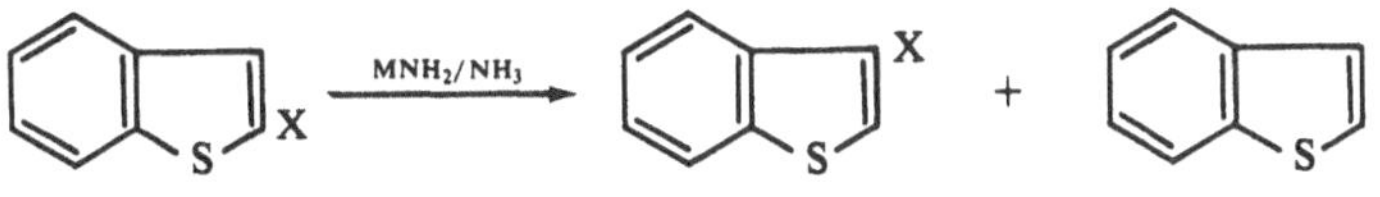

$$X = Br \text{ or } I \tag{38}$$

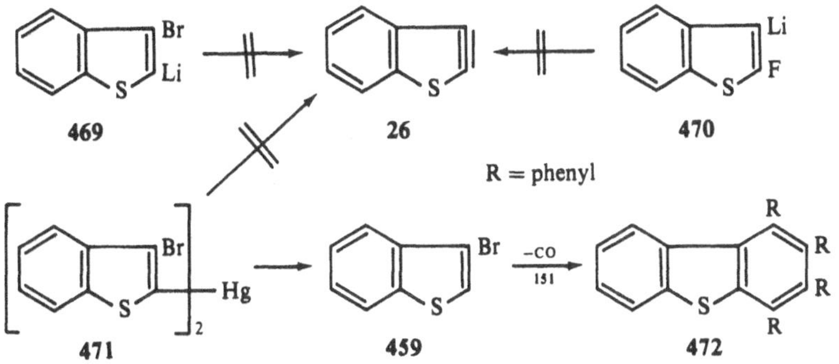

The *o*-halolithium derivatives **469** [483,486] and **470** [487] behave like normal lithium reagents and show no tendency[342] to lose lithium halide to give the aryne **26**. Neither trapping agents nor more severe conditions have been examined, although halogen rearrangement[483] and/or ring opening[488] would be the expected results. When such potential aryne-detecting conditions were used on the mercury compound **471** the anticipated tetracyclone adduct **472** was isolated in 54% yield.[76] An elimination–addition mechanism via 2,3-didehydrothianaphthene (**26**) is unlikely, however, since 3-bromothianaphthene (**459**)—the only product found when the thermolysis of **471** is carried out in the absence of tetracyclone (**151**)—also reacts with tetracyclone (**151**) to give the adduct **472** in 76% yield.[76] These observations indicate that, as in the benzofuran (Section III.1.A), indole (Section III.2.A) and thiophene (Sections III.3.A.a and III.3.B) series, the mercury compound **471** first decomposes to the halogen derivative **459**, which then reacts with the trap **151** by an addition–elimination mechanism to give the "aryne" adduct **472**. Further support for this pathway comes from the mass spectrum of the mercury compound **471** which, although it shows peaks for the aryne **26** and its dimer, loses them at 12 eV to leave only ones due to 3-bromothianaphthene **459**, thereby indicating this to be the primary thermal fragmentation product.[76,478a]

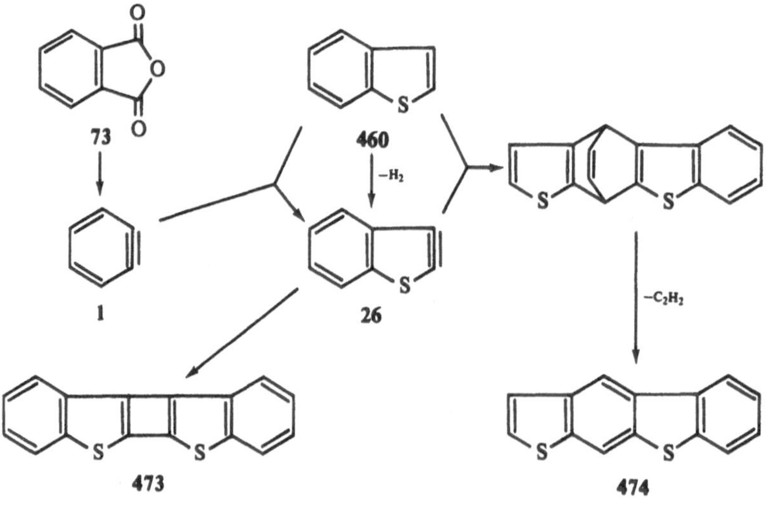

Thermolysis of a mixture of phthalic anhydride (**73**) and thianaphthene (**460**) gives several products including small amounts of two compounds which were assigned structures **473** and **474** based on their mass spectral molecular weights.[72,246] The same products are obtained in even lesser amounts from the pyrolysis of thianaphthene (**460**) alone. In both experiments 2,3-didehydro-thianaphthene (**26**) was postulated as an intermediate which arises in the first case by intermolecular hydrogen transfer to the benzyne (**1**) formed from phthalic anhydride (**73**) (Section II.1.E) and in the second case by intramolecular dehydrogenation (Section II.1.F). Dimerization of the aryne (**26**) would rationalize the presence of **473** and Diels–Alder addition to the benzene ring of thianaphthene (**460**) followed by loss of acetylene would account for (**474**). Considering the tenuous nature of the structure assignment of the products **473** and **474**,[72,246] especially in view of the expected instability[295,296] of the cyclobutadiene derivative **473**, claims for the generation of 2,3-didehydrothianaphthene (**26**) based on these thermolysis experiments appear unwarranted without additional data.

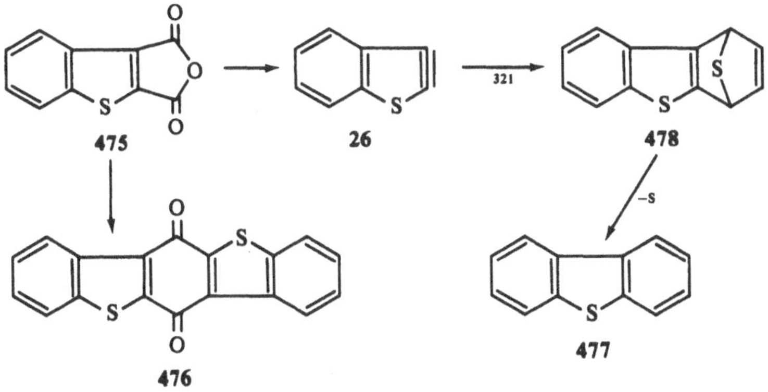

A more convincing case can be made from a study of the thermolysis of the anhydride **475**. At 270°C in a melt the quinone **476** is produced in 80% yield, but in the gas phase in the presence of thiophene (**321**) as a trap a 35% yield of dibenzothiophene (**477**) is obtained.[77b] Although no other diene traps have been examined as yet, analogy to the thiophene anhydride (**229**) experiments (Section III.3.A.f) indicates that the most reasonable explanation for this latter result is aromatization of the Diels–Alder adduct **478** of thiophene (**321**) and 2,3-didehydrothianaphthene (**26**).

Even stronger evidence for the existence of the aryne **26** has been obtained from the recent isolation of the actual Diels–Alder adducts **479** when the diazonium carboxylate (**480**) is decomposed in the presence of various anthracenes (**147**).[79] This discovery of a solution phase method for apparently generating the aryne **26** is significant beyond its use as a source of the difficultly accessible[432,439] heterotriptycenes such as **479**. In particular, it should

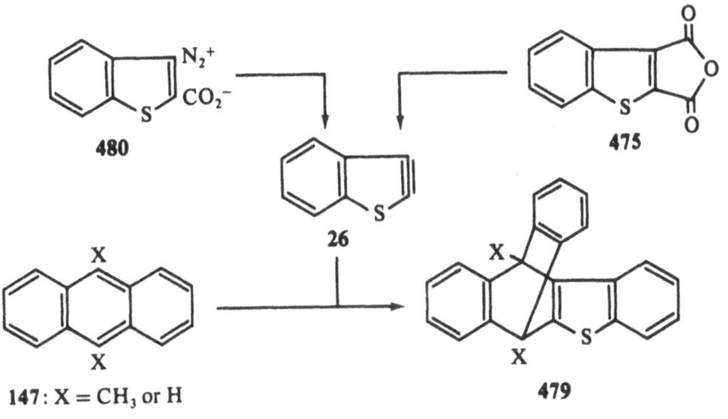

147: X = CH$_3$ or H

now be possible with the two different precursors **475** and **480** of 2,3-didehydrothianaphthene (**26**) to use the competition method[55,56,330] (Section II.2.B.e) to determine if an "arynoid" or a free aryne is being generated in these reactions.

D. 2,3-Didehydrothianaphthene-1,1-dioxide

Thianaphthene dioxides (**481**) like the corresponding thiophene dioxides would be expected to have considerably reduced aromatic character in the heterocyclic ring.[489] Not surprisingly, therefore, much of their chemistry is that

of α,β-unsaturated sulfones in that they are both good Michael acceptors and dienophiles at the 2,3 bond.[490,491] This tendency to participate in addition reactions dictates that particular care must be taken in the interpretation of any aryne-trapping experiments with either nucleophiles (Section II.2.A.h) or dienes (Section II.2.B.e). This Michael-acceptor property is probably also responsible for the fact that, in contrast to the thiophene and thianaphthene analogs, the 3-(**483**)[492,493] and not the 2-bromothianaphthene-1,1-dioxides (**482**)[64] are more reactive to nucleophilic substitution. Another difference compared to bromobenzofurans (Section III.1.A) and bromothianaphthenes (Section III.3.C) is that the 2-(**482**) not the 3-isomer (**483**) undergoes cine-substitution. This reaction has been most extensively studied with amines[64] although it has apparently been observed for various alcohols as well.[64,494] With ethanol as the solvent the Michael-addition product **484**[495] could be isolated and shown to be an intermediate in the formation of the 3-piperidino compound **485** from 2-bromothianaphthene dioxide (**482**) and piperidine,[64] thereby providing the classic prototype for the AEa mechanism of cine-substitution (Section II.2.A.e). In benzene as the solvent a considerably faster reaction occurs in which the overall reaction from **482** to **485** is more rapid than that from the addition product (**484**), thus indicating that in this solvent the latter cannot be an intermediate.[64] Although the original authors did not formulate a satisfactory mechanism for this cine-substitution of the 2-bromo compound **482**,[64] Kauffmann suggested that the aryne (**27**) might be an intermediate.[67] The role of solvent was explained[1,479] by assuming rapid reprotonation of an initially formed carbanion

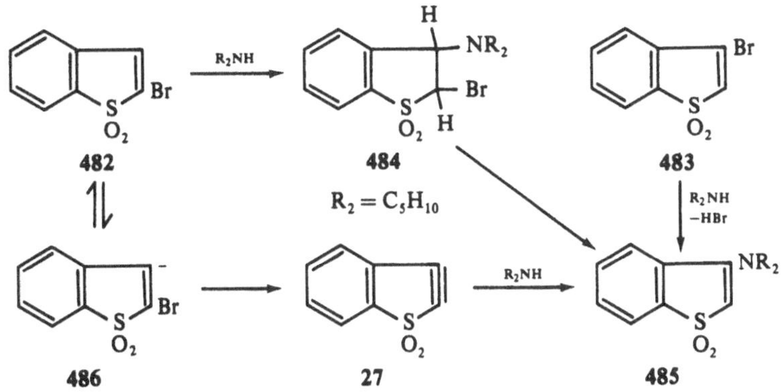

$R_2 = C_5H_{10}$

(**486**) in ethanol, but not benzene, thus forcing the reaction to proceed by the slower AEa mechanism via **484** rather than by loss of bromide ion from **486** to give 2,3-didehydrothianaphthene-1,1-dioxide (**27**). Although supporting evidence for this hypothesis is lacking, it has not been eliminated from consideration.

A transhalogenation mechanism (Section II.2.A.d) has also been proposed[497] to account for the cine-substitution of the 2-bromo compound **482** to **485** in benzene. Both the 3-bromo (**483**) and the 2,3-dibromothianaphthene-1,1-dioxides **487** are possible intermediates. The failure to detect their presence or that of the bromoamine (**488**) in the reaction[64,496] may be due to the greater reactivity of these compounds than the starting material (**482**). Furthermore, if, as in the thiophene series (Section III.3.A.c), the transhalogenation reactions proceed via intermediate carbanions such as **486**, then the retarding effect of ethanol on the conversion of **482** to **485** can be explained in the same way as proposed above for the aryne mechanism. The intermediacy of either **487** or **488**, while not rigorously excluded, was shown to be unlikely since neither compound was converted to the 3-piperidino compound **485** under the reaction conditions[496] even in the presence of thianaphthene dioxide (**481**) as a possible

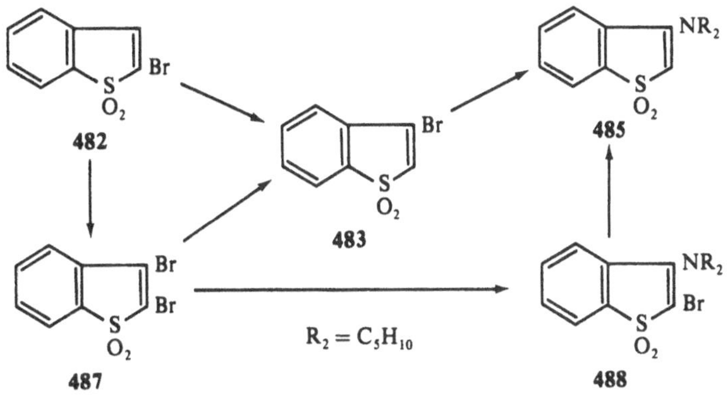

$R_2 = C_5H_{10}$

bromine-transfer agent.[266] Since the 3-bromo compound **483**, on the other hand, readily gives **485**,[492,493] its intermediacy must be considered as quite plausible. Although the facility of this substitution apparently prevents a direct demonstration of the isomerization of **482** to **483** in the presence of piperidine, this transformation does take place in 60% yield with KNH_2 in liquid ammonia at $-70°C$,[266] thereby providing further support for the transhalogenation mechanism **482→483→485**. A clear choice between this sequence and one involving the aryne **27** is not possible at this time, however.

The reaction of 2-bromothianaphthene-1,1-dioxide (**482**) with KNH_2 described above gives no substitution but only rearrangement to the 3-isomer **483** by a transhalogenation mechanism via carbanions and/or bromamine ($BrNH_2$).[266] At 110°C in liquid ammonia, however, 3-bromothianaphthene-1,1-dioxide **482** does undergo substitution to the 3-amino compound **489**, presumably by a normal addition–elimination mechanism.[492] Neither of these reactions gives any indication of an aryne intermediate, although the overall conversion **482→489** is a nucleophilic cine-substitution.

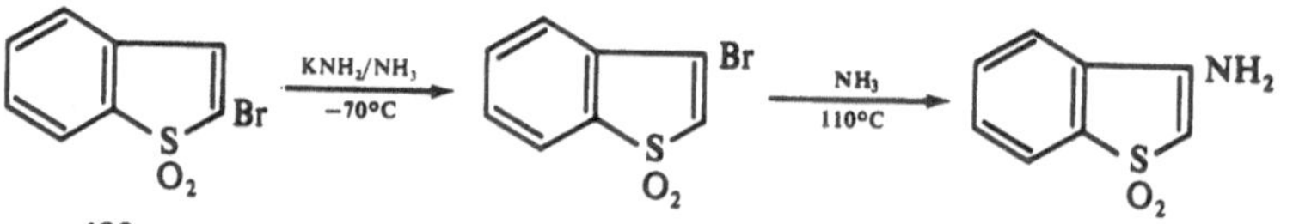

In summary, while some of the above reactions may be rationalized by invoking a 2,3-didehydrothianaphthene-1,1-dioxide intermediate (**27**), alternative, nonaryne mechanisms are of at least equal likelihood based on the available data.

4. Selenaarynes

While no concerted attempts to prepare selenaarynes have been reported, the chemistry of several potential precursors has been examined. The 2-lithio-3-bromo derivatives **490**[498] and **491**[499] react with electrophiles as expected for aryllithium reagents and show no tendency[342] at $-70°C$ to eliminate lithium bromide to form a 2,3-didehydroselenophene (**492**). Similar observations have been reported for the benzolog **493** and the possible aryne 2,3-didehydrobenzo[*b*]selenophene **494**.[500] As in the corresponding sulfur and oxygen heterocycles, this unusual[342] stability of the 2-lithio derivatives is probably related to the ability of the heteroatom to stabilize an adjacent carbanion.[22,501]

3-Lithio-2,5-dichloroselenophene (**495**) also reacts normally with electrophiles when prepared by metallation of 2,5-dichloroselenophene (**496**) with

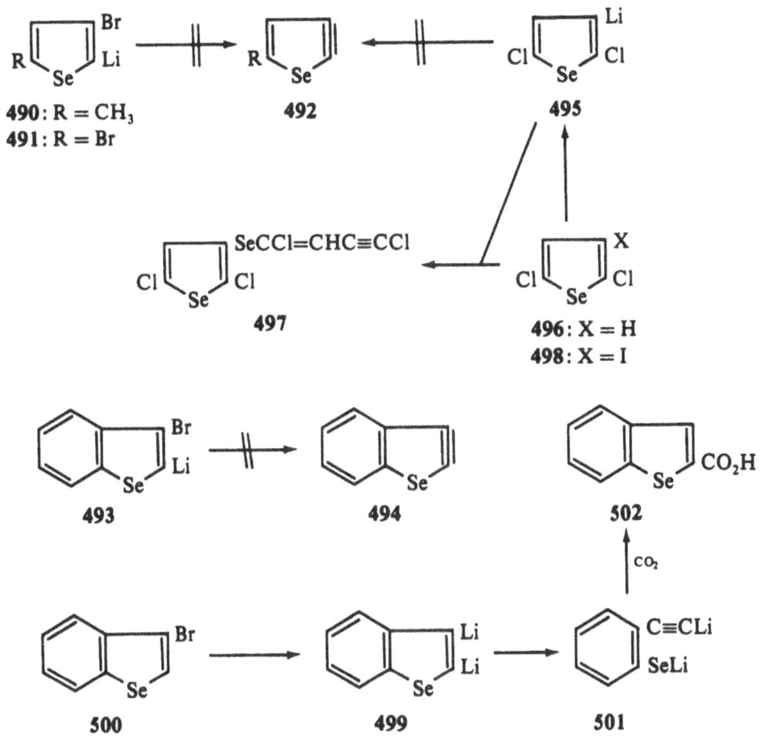

Br
Li
R
Se
490: R = CH3
491: R = Br
492
Li
Cl
495
SeCCl=CHC≡CCl
497
X
496: X = H
498: X = I
493
494
CO2H
502
CO2
500
499
C≡CLi
SeLi
501

lithium diisopropylamide, but gives a ring-opened product (**497**) when a halogen–metal exchange reaction on the iodo compound **498** is used.[498] This ring-opening reaction is different from that observed for most 3-lithio five-membered heterocycles[477,488,498,502] and appears not to involve the aryne **492** but the attack of aryllithium reagents, including **491** and phenyllithium, on the selenium atom of β-haloselenophenes.[499] The more general type of ring-opening reaction occurs with the 2,3-dilithiobenzoselenophene **499** formed by sequential halogen–metal exchange and metallation of the 3-bromo compound **500**.[503] Upon carboxylation the ring-opened species **501** recloses to give **502**, an *electrophilic* cine-substitution product hardly indicative of the intermediacy of the aryne **494**. The above results indicate that any approach to this or other selenaarynes ought best to avoid *o*-halolithium compounds as precursors.

5. *Diazaarynes*

A. Didehydroimidazole

At the time of the last review of hetarynes[4] the existence of only one five-membered hetaryne, 4,5-didehydro-*N*-methylimidazole (**28**), was still considered

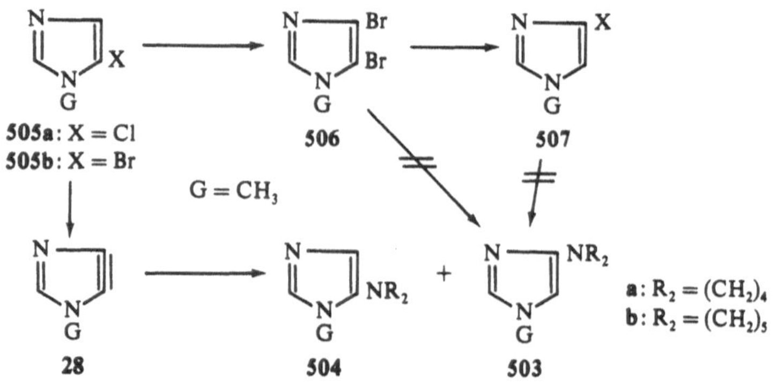

505a: X = Cl
505b: X = Br

G = CH₃

506   **507**

28   **504**   **503**

a: R₂ = (CH₂)₄
b: R₂ = (CH₂)₅

likely. This species had been proposed[68] as an intermediate to explain the apparent cine- (**503**) and normal- (**504**) substitution products observed when the 5-haloimidazoles **505** were treated with lithium amides. The possibility that the cine-substitution product **503** might arise via either the dibromoimidazole **506** or the 4-haloimidazole **507**, presumably formed by a transhalogenation mechanism,[265] was eliminated with the appropriate control experiments.[4] The original basis for the claim[68] of the aryne intermediate **28** was invalidated, however, with the discovery[18] that the alleged cine-substitution product **503b** was in fact the tele-substitution[253] product, 2-piperidino-*N*-methylimidazole (**508**).

Although an aryne (**28**) is, therefore, excluded, the competition studies[68] would seem to require that the rearranged product, regardless of whether it is **503** or **508**, be formed by a single mechanism involving a halogen-free intermediate in the product-determining step. If this conclusion is accepted then two possible mechanisms may be considered. The first of these involves a Chichibabin reaction on *N*-methylimidazole (**509**) formed by a base-catalyzed halogen disproportionation. While there is some indication that the first step of this sequence **505→509** may occur under these reaction conditions,[18,265] the

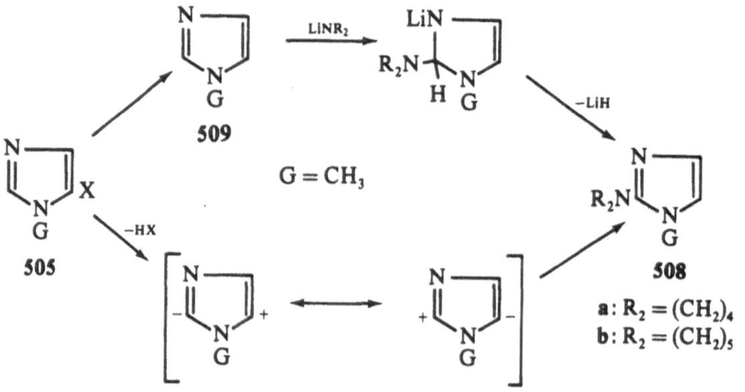

509

G = CH₃

505

508

a: R₂ = (CH₂)₄
b: R₂ = (CH₂)₅

510

second step **509→508** does not.[18] The second mechanism via a halogen-free intermediate would involve a metaaryne, 2,5-didehydro-*N*-methylimidazole (**510**). Although 1,3-didehydro species have been considered previously as intermediates in tele-substitution[253] reactions,[504,505] alternative mechanisms via AEa[4,505,506] or ANRORC processes[507] (Sections II.2.A.e and II.2.A.g) have either been demonstrated to be responsible or, at the very least, not excluded. Confirmatory evidence would therefore be desirable before the 1,3-didehydroimidazole (**510**) could be claimed as an intermediate in the conversion of **505** to **508**.

Alternative mechanisms involving a halogen-containing species in the product-determining step could be considered for this transformation if the assumption[68] that the nature of the halogen in such a species would have a significant effect on the partial competition constants[508] were invalid. The observed insensitivity of this parameter to which reactant, **505a** or **505b**, was used[68] would then not require a halogen-free intermediate. The original[68] assumption of such an element effect is reasonable, and in fact was subsequently demonstrated,[270] for an AEa mechanism leading to an actual cine-substitution product.[253] Whether a significant element effect would be noted for a tele-substitution product where the reaction centers are in a 1,3 rather than 1,2 relationship is problematical, however. Invalidating this assumption is therefore at least as justified as proposing the metaaryne **510** and permits the consideration of several nonaryne mechanisms for the tele-substitution reaction **505→508**.

A transhalogenation mechanism (Section II.2.A.d) via a 2-haloimidazole **511** is reasonable based on the high reactivity of such compounds in nucleophilic substitutions.[509] It is also supported by the exclusive formation of the 2-metalloimidazole (**512**) when 5-chloro-*N*-methylimidazole (**505a**) is treated with naphthylsodium or lithium.[510] Against this mechanism, however, is the fact that only the 4-isomer **507** and *N*-methylimidazole (**509**) are found when the 5-halo compounds **505** are treated with KNH$_2$/NH$_3$.[265] No products indicative of 2-haloimidazoles (**511**), or didehydroimidazoles (**28**) for that matter, were reported. Even more telling is the failure of the 2-bromo compound **511b** to yield a sufficient amount of the substitution product **508** under the original reaction conditions to account for the quantity formed from the 5-isomer **505b** directly ("minimal" vs. 16%, respectively).[18]

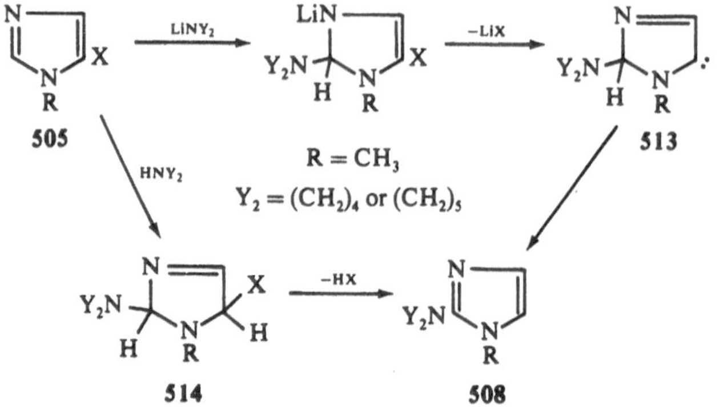

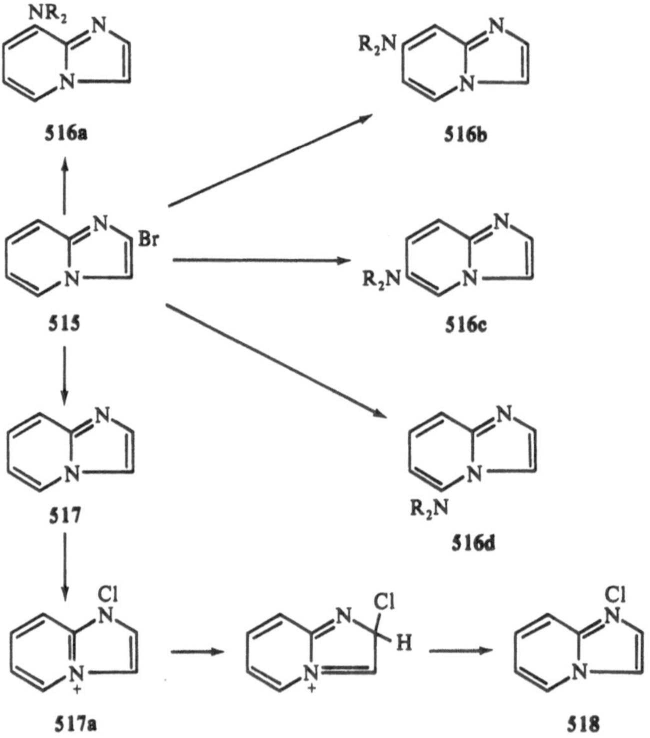

A tentative mechanism involving a carbene (**513**) has been suggested,[18] but, as already noted,[506] would be unlikely based on the lack of carbenelike side products and the expected[511] slowness of such reactions. A much more attractive possibility, with ample precedent in other heterocycles containing a C=N group,[270,506] is an AEa mechanism (Section II.2.A.e). A rate-determining 1,4 addition of amine to **505** followed by a rapid 1,4 elimination of HX from the resulting adduct **514** would lead to **508** by overall tele-substitution with a probably small element effect on the partial competition constants as observed.[68]

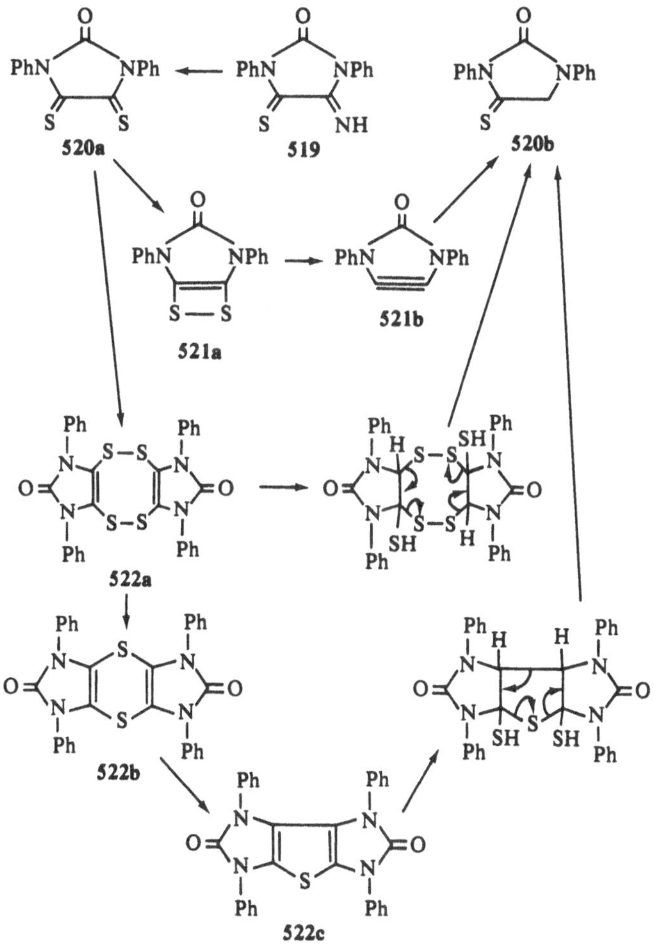

The potential complexity of nucleophilic substitution reactions of haloimidazoles is impressively illustrated by the reaction of 3-bromoimidazo-[1,2-*a*]pyridine (**515**) with metal amides.[512] No less than four tele-substitution products, **516**, and the debrominated compound **517** were formed and adequately explained without resort to didehydro intermediates. Products indicating either a normal or a cine-substitution reaction were not found, although the 2-chloro compound **518** was suggested[513] to arise by the latter process during chlorination of **517**. Once again an AEa-type mechanism on the pyridinium ion **517a** or normal electrophilic substitution are more reasonable explanations[513] for this result than the intermediacy of a didehydroimidazole species.

Such an intermediate has recently been suggested in the reaction of the iminothione (**519**) with H_2S/Et_3N, which gave, instead of the desired dithione **520a**, the reduced monothione **520b**.[513a] It was speculated that this reaction might proceed by ring closure of the dithione **520a** to the dithiete **521a**, which upon loss of S_2 gives the 4,5-didehydro-2-imidazolinone **521b**. Addition of H_2S to this five-membered hetaryne would then give the observed product **520b**. While the intermediacy of the dithione **520a** is not unreasonable, known dithione–dithiete isomerizations are observed[513b] and predicted[513c] to occur photochemically, not thermally. The products expected[513b] from **520a** are the dimer **522a** and its desulfurization products, the dithiin **522b** and the thiophene **522c**.[513d] Elimination mechanisms, such as those shown, which convert these species to the monothione **520b** are at least as reasonable as an aryne pathway and would need to be eliminated before the existence of **521b** can be considered other than speculation.

Finally, a bicyclic didehydroimidazole similar to 1,8-didehydronaphthalene (**523**)[514a] was recently proposed to explain the formation of the cyclazine **524** from the reaction of benzonitrile and the lithio compound

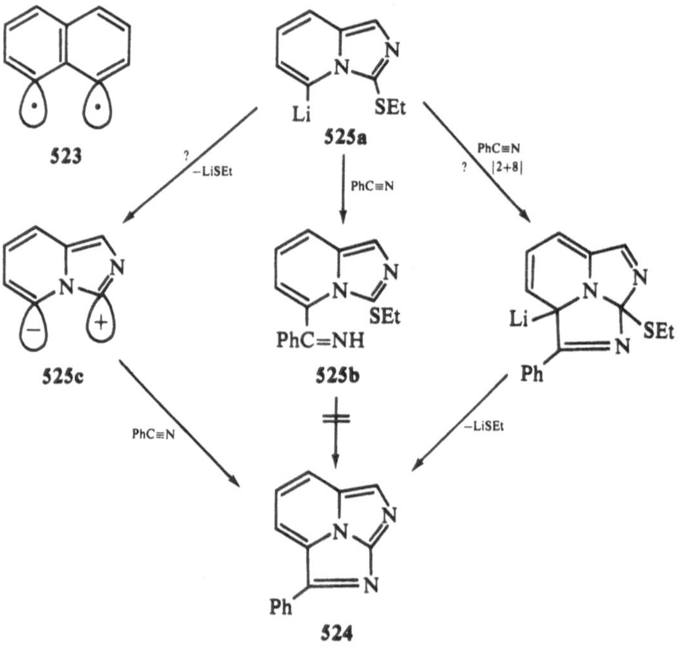

525a.[514b] Since the imine **525b** was not a precursor of the cyclazine **524**, it was concluded that C–S bond breaking must precede C–C and C–N bond formation, as would be the case if 3,5-didehydroimidazo[1,5-*a*]pyridine (**525c**) were an intermediate. An alternative explanation involving a [2+8] cycloaddition of the reactants followed by elimination of lithium thioethoxide was also considered, however, but has not yet been disproven. Precedent for such cycloaddition–elimination reactions exists in the closely related pyrrocoline system.[514c]

B. Didehydropyrazoles

Several isomers and derivatives of this aryne have been considered, but never actually claimed, as intermediates in the cine-substitution reactions of halopyrazoles and pyrazolium salts. A possible example of such a substitution was postulated in the reaction of 4-bromo-*N*-phenylpyrazole (**526a**) with lithium dimethylamide to give the cyanoamidine **527a**.[69] The intermediacy in this process of the 5-dimethylamino compound **526b**, formed by cine-substitution of **526a**, was supported by its facile ring opening to the amidine **527a** under the reaction conditions. An elimination–addition mechanism via the aryne **29** was considered as a possible route to the dimethylaminopyrazole **526b** from both the 4-bromo (**526a**) and the 5-chloropyrazole (**526c**) which also gives the amidine **527a** upon treatment with lithium dimethylamide. No firm claim for the generation of didehydropyrazole (**29**) could be made, however, not because other mechanisms for the cine-substitution (Section II.2.A) were considered, but because a route to the amidine (**527a**) not even involving a cine-substitution to (**526b**) could be envisioned. An initial normal ring opening[515] of the bromopyrazole **526a** would lead to the bromonitrile **527b**, an independently prepared sample of which was shown to be converted to the amidine **527a** with lithium dimethylamide,[69] presumably by an elimination–addition process.

An authentic example of cine-substitution has been observed in the reaction of 4-halopyrazolium salts (**528**) with hydroxide ion.[516] Several mechanisms

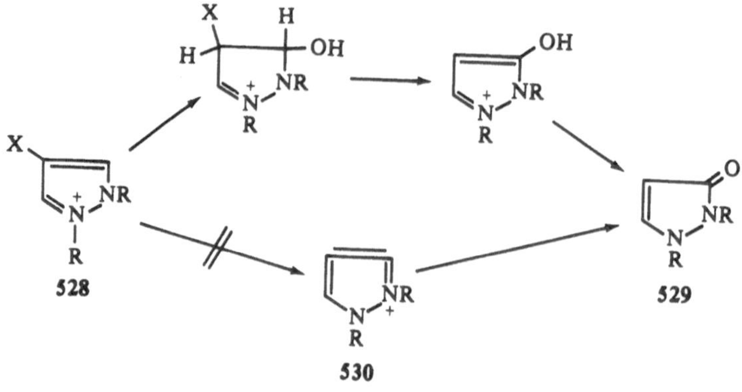

528 **530** **529**

were considered to account for the formation of the 3-pyrazolone **529**, including inter- and intramolecular transhalogenations, ANRORC, EA, and AEa process (Section II.2.A). It was concluded that the first of these was a minor[516a] and the last the major[516b] pathway for the cine-substitution reaction. The EA mechanism specifically was eliminated largely by the failure to trap the didehydro-pyrazolium ion **530** with tetracyclone (**151**) and 1,3-dipoles as well as by some specious arguments. Nevertheless, in the absence of any evidence requiring the intermediacy of the aryne (**530**) this conclusion appears justified.

No such evidence was obtained from the reaction of the bromopyrazole **531** with KNH_2/NH_3.[266] Transhalogenation and dehalogenation accounted for 85% of the reaction. It is noteworthy that the direction of halogen migration in this system, from the 5- to the 4–position, is the opposite of that which would be required to rationalize the possible cine-substitution of **526a** to **526b** by a transhalogenation mechanism,[69] thus making such a process unlikely.

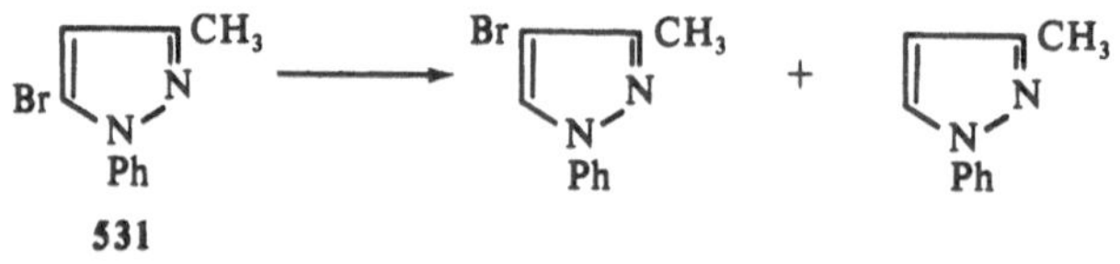

531

The difficulty of forming the didehydropyrazole **29** is illustrated by the remarkable isolation, at room temperature, of the lactone **532** from the thermal

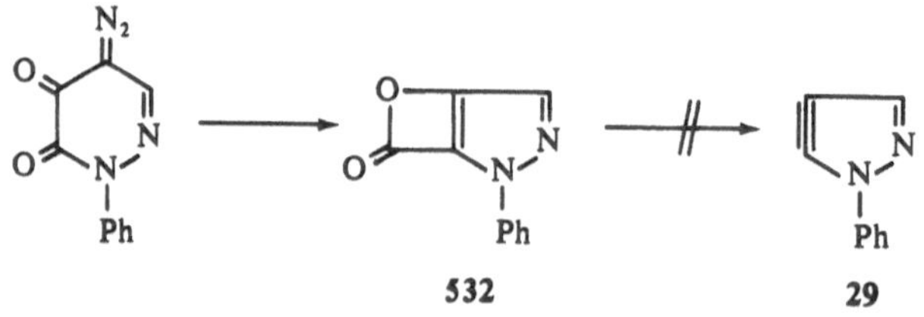

532 **29**

decomposition of the solid pyridazine diazocompound.[517a] Although reactive to nucleophilic opening of the lactone ring, 532 shows no tendency to lose CO_2 in contrast to the benzene analog 37, which has been detected in an argon matrix at $8°K$[121] and loses CO_2 to give benzyne on photolysis (Section II.1.E).[57]

6. Thiazaarynes

A. Didehydrothiazole

Only one *ortho*-aryne 533 is possible in a thiazole ring. Its intermediacy was first considered in the reaction of 4-halothiazoles (534) with methoxide ion as a possible rationale for the surprisingly similar reactivity of these compounds and the 2- and 5-halo isomers.[517b] This hypothesis was consistent with the small element effect, the rapid base-catalyzed exchange of the 5-proton,[29] and the exclusive formation of the normal substitution product 535 as would be expected if nucleophilic addition to the aryne 533 was determined[255] by the stability of the resulting anion 536 with the negative charge adjacent to the sulfur atom.[22,23] The fact that the 5-phenyl derivative 537, which cannot form an aryne 533, reacts at a comparable rate to 534 rules out the possibility of an elimination–addition mechanism, however.

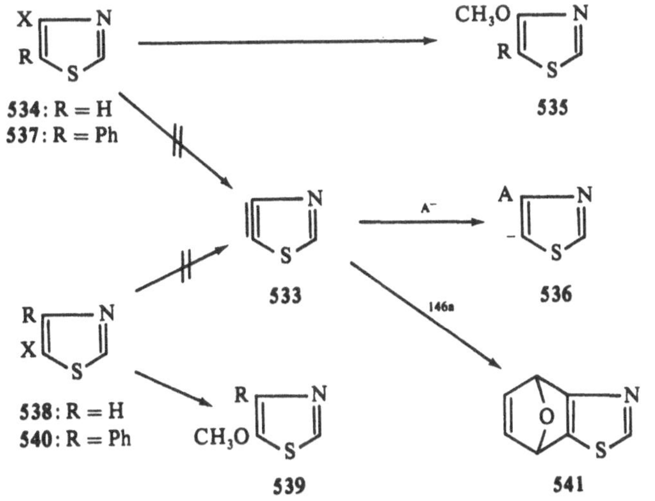

The aryne 533 was also considered and rejected as an intermediate in the nucleophilic substitution of 5-halothiazoles (538) based on the following evidence[518]: only normal substitution to 539 is obtained, inconsistent with the expected formation of 535 from the aryne 533 as described above; no exchange of the 4-proton is observed; the reactivity of the 4-phenyl derivative 540, which cannot form the aryne 533, is similar to that of 538; and finally no aryne adduct **541** was detected when the reaction was carried out in the presence of furan (**146a**).

An attempt to detect the aryne **542** from the *in situ* diazotization[103] of the amino acid **543** with isoamyl nitrite in the presence of furan (**146a**) or tetracyclone (**151**) failed to yield any of the aryne adducts **544** or **545**, respectively.[519] The only identifiable products isolated in low yield were the acids **546** and **547** and their corresponding nitro compounds **548** and **549**. Although it was considered that the latter two compounds might arise via the unlikely dipolar form (**542b**) of the aryne, it was demonstrated[519] that they were actually transformation products of the former two compounds, **546** and **547**, under the diazotization conditions. Similar decarboxylative nitrations with isoamyl nitrite[78] and nitric acid[520] have been observed in related systems. The acids **546** and **547** were proposed to arise from the dipolar species **550**,

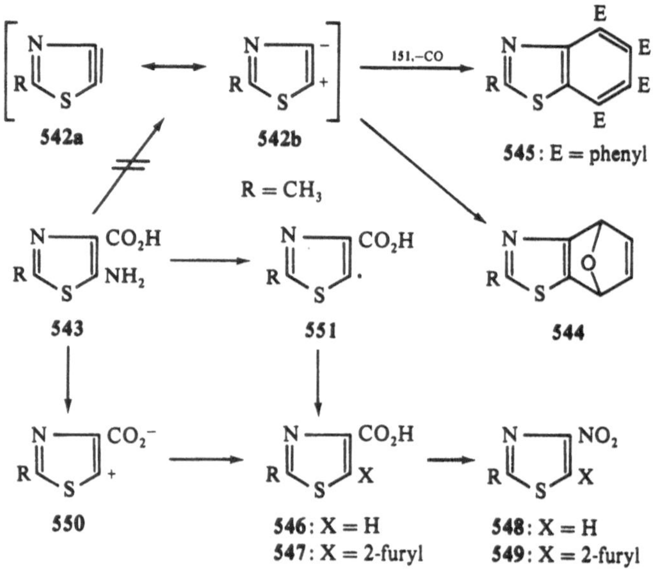

although the formation of similar products in the thiophene series (Section III.3.A.d))[412] via a radical pathway suggests that an analogous intermediate **551** might be involved here as well. In any case it is clear that no evidence supporting a didehydrothiazole (**533**) is available at present.

B. Didehydroisothiazoles

No indication for the intermediacy of either of the two possible isomers **552** and **553** of this aryne is available from the reactions of the 4-haloisothiazoles (**554**) with *n*-butyllithium.[521,522] Even if the 3-position is

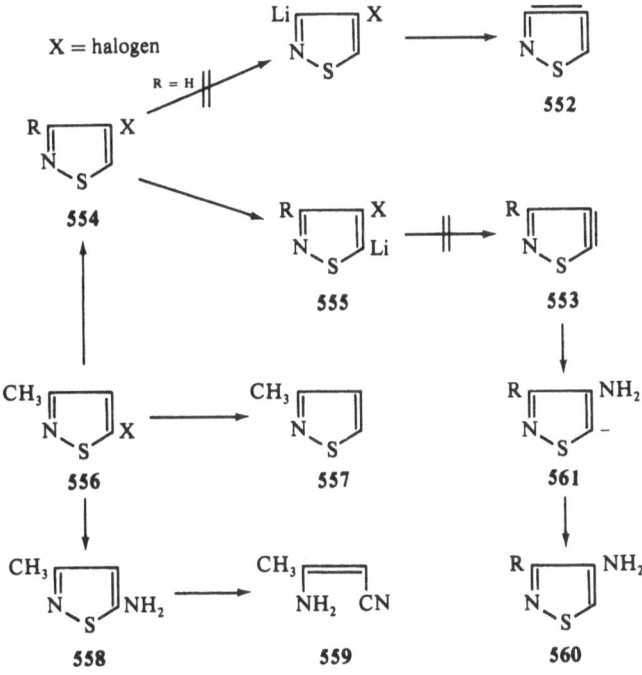

unsubstituted[521] only the more stable[29] 5-lithio derivative **555** is formed, and it behaves like a perfectly normal lithium reagent in its reaction with electrophiles. Upon treatment of the isomeric 5-haloisothiazoles (**556**) with KNH_2/NH_3 primarily transhalogenation to **554** is observed, although with the iodo compound some dehalogenation to **557** and normal substitution to **558** take place, the latter detected by the presence of the ring-opened product **559**.[264] The absence of the cine-substitution product **560**, which would be expected[255] from the addition of amide ion to aryne **553** based on the greater stability of the intermediate anion **561**[29] is inconsistent with an EA mechanism in this substitution. The preferred dehalogenation and substitution when the halogen is iodine as opposed to bromine or chlorine suggests that an $S_{RN}1$ process is likely to be involved in this case.[523]

Although no other suggestions of didehydroisothiazoles have been reported the chemistry of one possible precursor to **553** has been briefly examined.[524] Diazotization of the amino acid **562** in strong acid gives the diazonium salt **563**, which reacts normally under Sandmeyer conditions to yield the chloro acid **564** but fails to give the hydroxy acid **565** with water. The latter reaction, which may proceed via the diazonium carboxylate **566**, gives largely polymeric material and variable amounts of benzoic acid, presumably[525] arising from contamination of the amino acid **562** by its N-benzoyl precursor. The value of **566** as a source of the aryne **553** will need to be determined under more traditional aryne-producing conditions (Section II.1.C).

C. Didehydro-1,2,5-thiadiazole

Although nucleophilic substitution of 3-halo-1,2,5-thiadiazoles is known,[526] the symmetry of the heterocyclic nucleus prevents the detection of the aryne **567** by cine-substitution in the absence of isotopic labeling. An attempt to trap this species with anthracene (**147**) from the aprotic diazotization[103] of the amino acid **568** gave, instead of the heterotriptycene **569**, a mixture of 9-nitro- (**341**) and 9-thiocyanoanthracene (**570**).[527] The former compound is also observed[78,412] in a similar reaction in the thiophene series (Section III.3.A.d) and could arise by direct nitration of anthracene with nitrous acid[528] present in the alkyl nitrite.

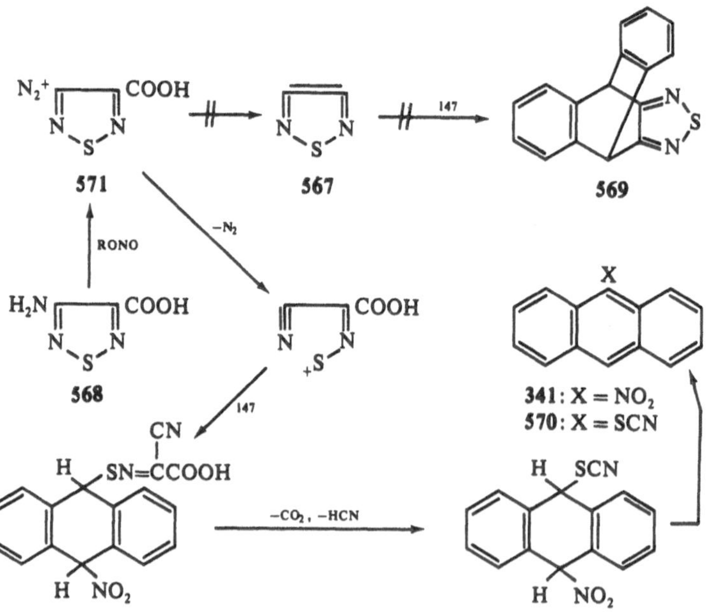

The formation of the latter product, however, must involve fragmentation of the thiadiazole ring. A mechanism accounting for both of these products which does not utilize the aryne **567**, only the diazonium salt **571** has been proposed.[527] A modification which avoids the high energy sulfenium ion intermediate involves concerted loss of nitrogen, ring opening, and attack of anthracene on the sulfur atom of **571**.

7. Benzdidehydro Five-Membered Heterocycles

The remaining categories of arynes to be considered are all on the periphery of the scope of this review. The first of these involves benzo derivatives of five-membered heterocycles where the arynic bond appears in the carbocyclic ring. As mentioned at the outset (Section I.2.B) such species are included on the assumption that the heteroatom in a neighboring ring may effect the reactivity of the aryne.

A.Benzdidehydrooxoles

The didehydrodibenzofuran **572** may have been generated almost 40 years ago during the reaction of the 3-bromo derivative **573** with butyllithium.[529]

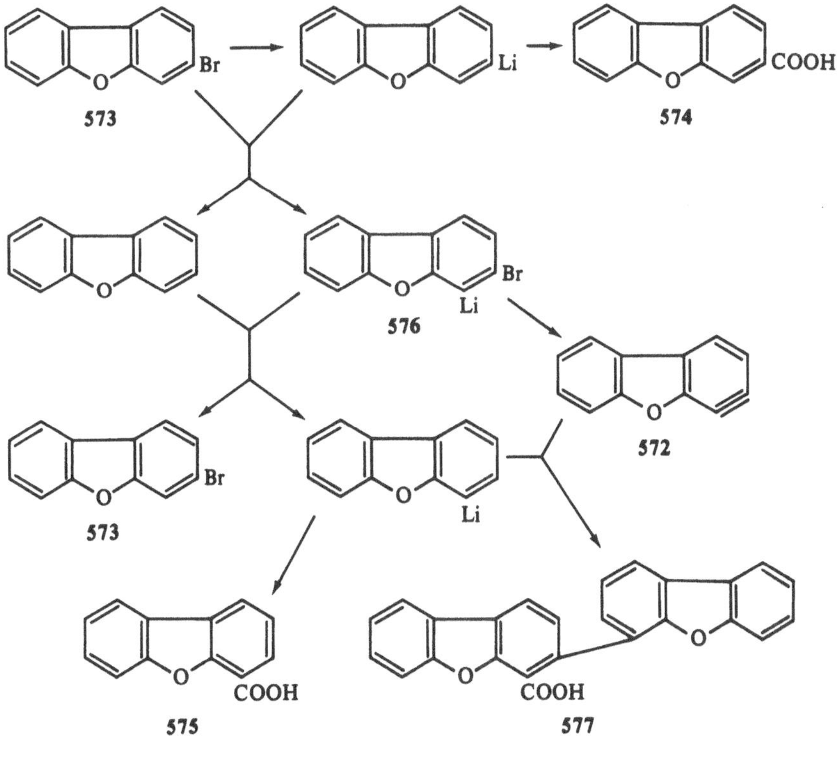

Since carbonation led to both the 3- and 4-carboxylic acids, **574** and **575**, respectively, a transhalogenation mechanism via the *o*-bromolithium compound **576** was postulated. Decomposition of this intermediate to the aryne **572** would be expected[342] and may be indicated by the isolation of an apparently dimeric acid of undetermined structure, perhaps **577**.

B. Benzdidehydroazoles

a. Benzdidehydroindole

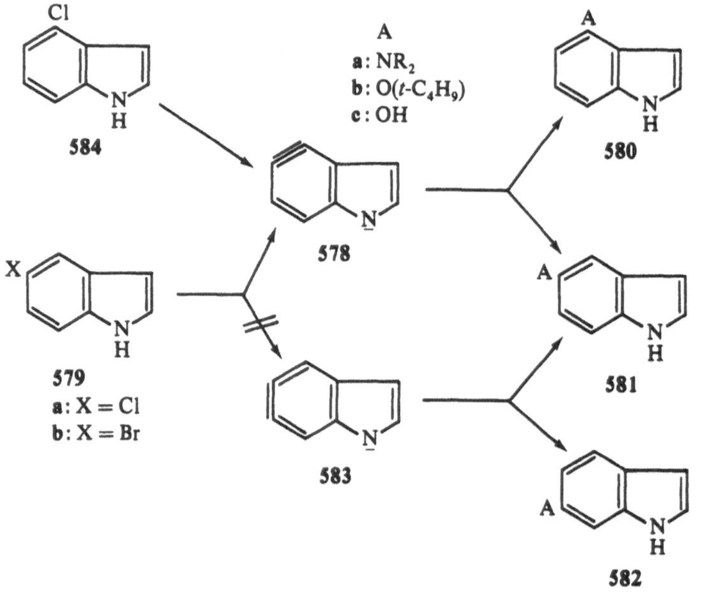

Because of its synthetic potential as a route to the lysergic acid ring system, 4,5-didehydroindole (**578**) has been extensively studied. This aryne appears to be the only one generated when 5-haloindoles (**579**) are treated with strong base, since subsequent addition of a nucleophile gives only one cine-substitution product, the 4-isomer **580**, along with the normal substitution product **581**.[530] The absence of any 6-isomer (**582**) suggests that none of the 5,6-didehydroindole (**583**) had been formed. An identical isomer ratio of **580:581**=75:25 is obtained whether the 5- (**579a**) or 4-chloroindole (**584**) is the reactant thereby clearly establishing the intermediacy of the aryne **578**. In addition to amide ions[530] the aryne **578** will react preferentially at the 4-position with stabilized carbanions,[531,532]. An apparent exception to this regioselectivity occurs with *t*-butoxide ion[531] but is probably due to partial cleavage of the initially formed aryl *t*-butyl ethers **580b** and **581b** to the phenols **580c** and **581c** under the reaction conditions.[533] With still more severe conditions complete cleavage occurs and the hydroxyindoles **580c** and **581c** are obtained in the usual 70:30 ratio, respectively.[531]

In order to generate the lysergic acid ring system **585** an intramolecular addition of a three-atom $C_{(3)}$ side chain to the 4-position of the aryne bond of **586** must be induced. Although such a cyclization was successful with a nitrogen nucleophile to form **587**[530] it failed for carbon nucleophiles.[534] By simply first reducing the indole 2,3-double bond in the precursor **588** and then generating the 4,5-didehydroindoline **589**, cyclization took place and the desired ring system **585** was obtained upon subsequent dehydrogenation of **590**. This method has been applied to a formal synthesis of *dl*-lysergic acid.[535]

b. Benzdidehydrobenzimidazole

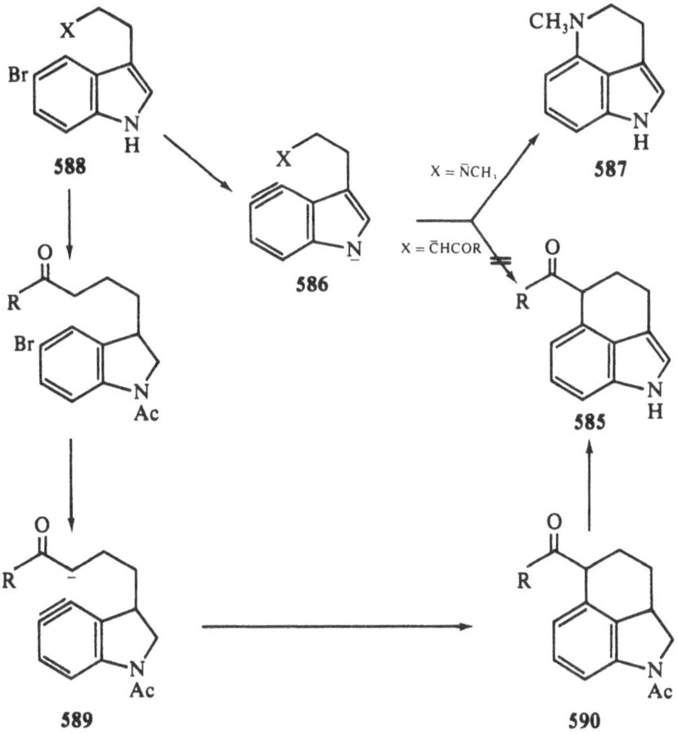

There is strong evidence that 5,6-didehydrobenzimidazoles (**591**) have been generated by both oxidation of the aminotriazoles (**592**) (Section II.1.D) and aprotic diazotization of the amino acid **593** (Section II.1.C).[536] The intermediate diazonium carboxylate **594** from the latter precursor proved to be remarkably stable but finally gave a low yield of the tetracyclone adduct **595b** upon thermolysis. Oxidation of (**592**) led to much better yields of the expected aryne a: R = CH$_3$;   b: R + R = (CH$_2$)$_4$

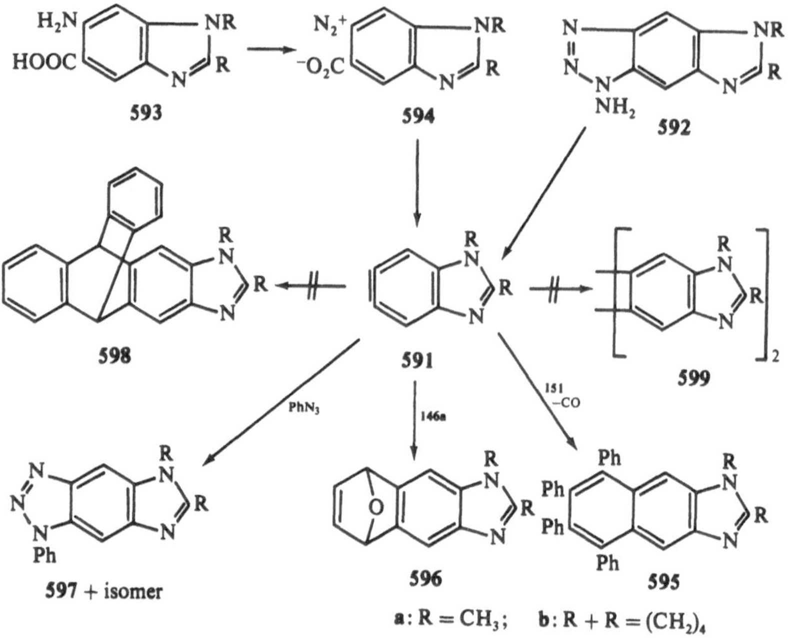

adducts **595**, **596**, and **597** with tetracyclone (**151**), furan (**146a**), and phenyl azide, respectively. The latter adduct was obtained as a 62:38 mixture of the two possible isomers (**597**). The only reported failure to trap the aryne (**591**) is with anthracene (**147**), no heterotriptycene (**598**) being found from either precursor. Another unusual feature of the aryne (**591**) as generated from **592** with lead tetraacetate is that no aryne dimers such as **599** were isolated although this method traditionally[50] gives high yields of biphenylenes (**89**) with benzyne (**1**). Nevertheless there is little question that 5,6-didehydro-benzimidazoles (**591**) have been made.

c. Benzdidehydrobenzofurazan

Either of the halobenzofurazans **600** or **601** reacts with thiomethoxide ion to give both the 4- (**602a**) and the 5-thiomethoxy (**603a**) compounds but not in identical ratios.[537] Although an elimination–addition mechanism via the aryne **604** was considered for at least part of this reaction it was rejected based on the low basicity of mercaptide ion and on the lack of deuterium incorporation from solvent in the normal-substitution products. An AEa mechanism (Section II.2.A.e) was proposed instead[537] to account for the cine-substitution products and seems to be consistent with the available data.[538]

With the more basic methoxide ion as the nucleophile, however, the isomeric iodobenzofurazans **600a** and **601a** react via the aryne **604** to give the

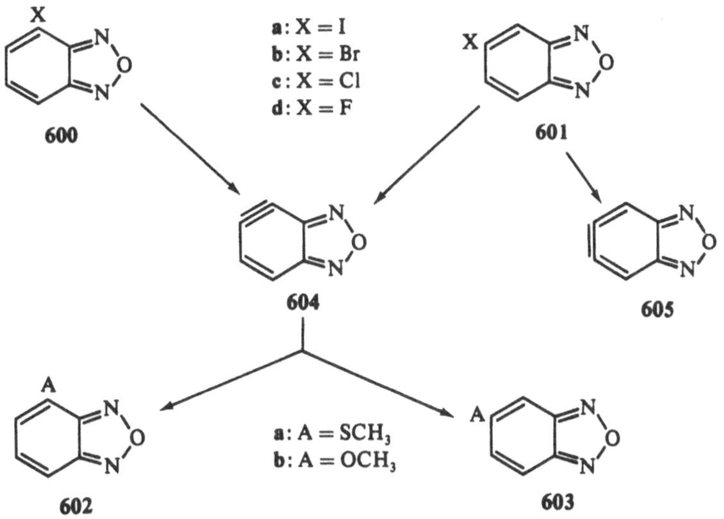

a: X = I
b: X = Br
c: X = Cl
d: X = F

600

601

604

605

a: A = SCH₃ → a: A = SCH_3
b: A = OCH₃ → b: A = OCH_3

602

603

same ratio of methoxy derivatives **602b** and **603b**.[539] In the case of the bromo derivatives **600b** and **601b** this EA mechanism competes with normal substitution, and for the other two halo compounds **600c,d** and **601c,d** this latter mechanism is followed exclusively. Because of the symmetry of the benzofurazan ring system it is not possible to determine if any products are formed via the isomeric 5,6-aryne **605**.

C. Benzdidehydrothioles

Benzene ring brominated thianaphthenes (**606**) undergo 2-lithiation followed by normal halogen–metal interchange when reacted with *n*-butyllithium.[540] Neither transhalogenation nor aryne formation via an unstable *o*-bromolithium compound[342] is indicated although certain derivatives will undergo nucleophilic substitution or reduction via an $S_{RN}1$ mechanism.[541] Similar normal behavior is observed for 5-chlorothianaphthenes.[542] Considering the success of the corresponding haloindoles as aryne sources (Section III.7.B.a) there is no reason to expect that under more appropriate conditions (Section II.1.B) the thianaphthene analogs would not also succeed.

As described previously (Section III.3.A.e) 5,6-didehydrothianaphthene (**377**) was postulated [equation (31)] as an intermediate to rationalize the

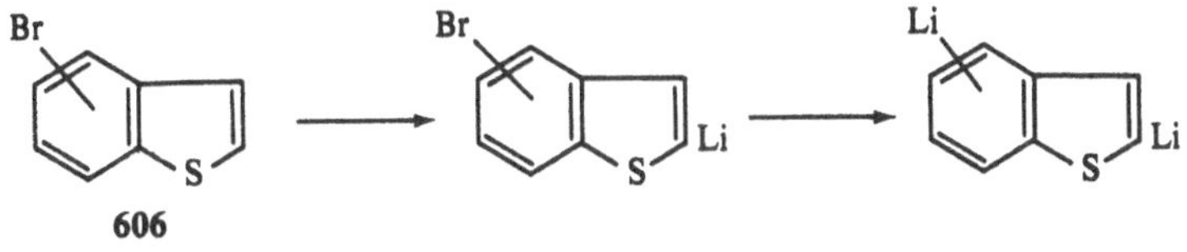

606

apparent necessity for the formation of 2,3-didehydrothiophene (4) in the thermolysis of thianaphthene (293) in the presence of phthalic anhydride (73).[72] The 4,5-didehydrothianaphthene isomer 607 was also suggested[72] with some supporting mass spectral evidence.[431] Nevertheless, neither this analogy nor the various products obtained from this pyrolysis reaction[72] provide compelling evidence for the generation of either of the arynes 377 or 607.

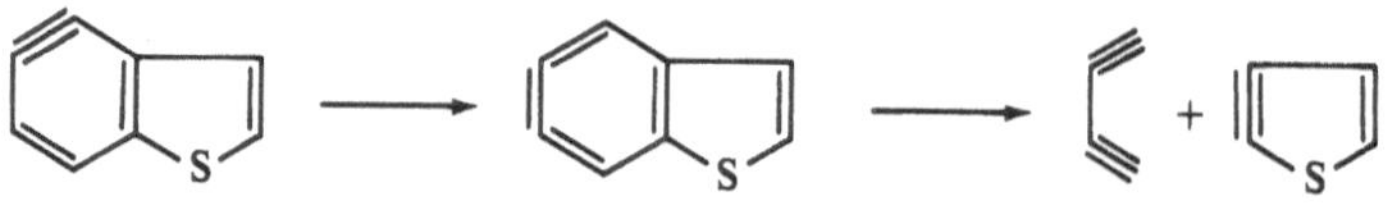

The last aryne to be discussed in this section, the didehydrocycloheptenone 608, was postulated to explain the cine-substitution observed when the bromo compound 609a reacts with secondary amines in the presence of potassium t-butoxide.[543] The possibility that a combination of an AEn and an AEa mechanism involving 610a and 610b could also account for the products 609b and 609c, respectively, was not specifically eliminated. By analogy with the similar carbocyclic aryne (611) where such precautions were taken,[321] the EA mechanism via the aryne 608 appears likely, however.

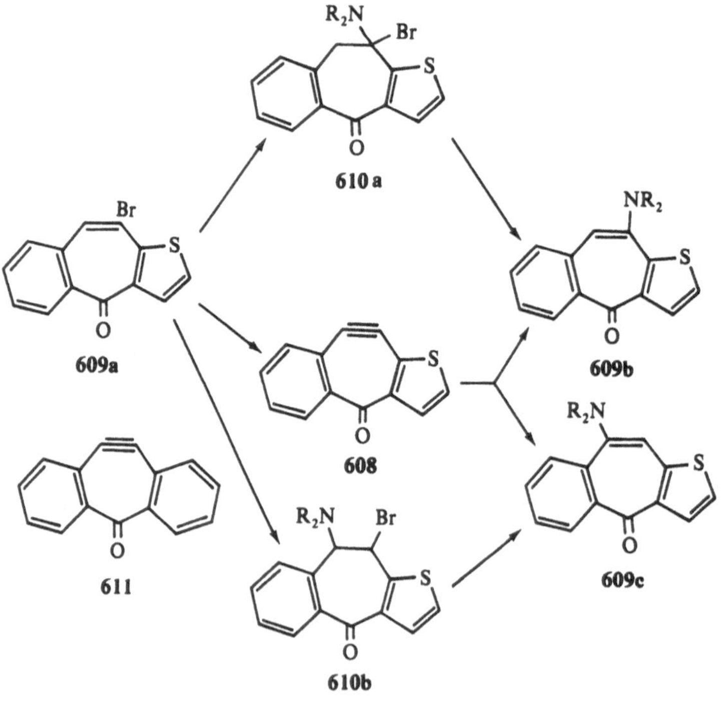

D. Didehydroazepinoazoles and Thioles

Matrix isolation techniques have shown that photolysis of aryl azides gives didehydroazepines **(612a)**[544a] and azirenes **(612b)**.[544b] With several 6- and 7-

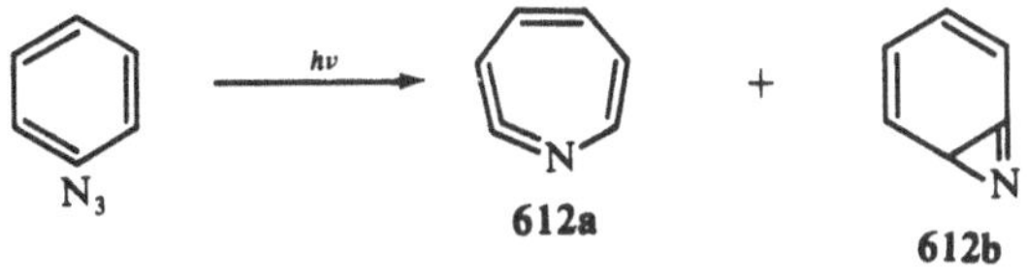

azido derivatives **(613)** of benzo[*b*]thiophene, benzothiazole, benzimidazole, and indazole, however, the photolysis products were most easily rationalized as arising via the latter type of intermediate **(614)** and not the former **(615)**.[544c]

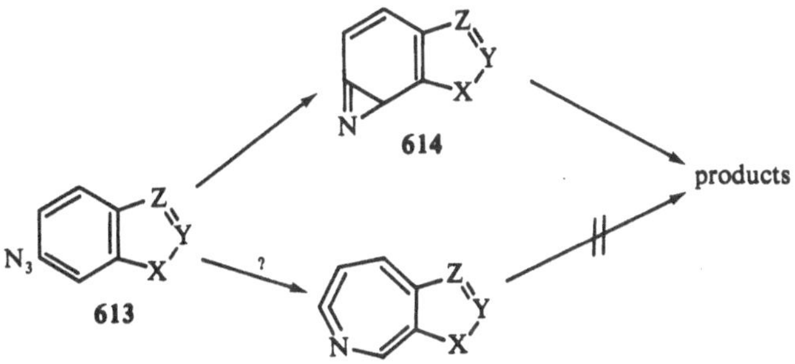

X	Y	Z
S	CH	CH
S	CH	N
NH	CH	N
NH	N	CH

8. Five-Membered Carbocyclic Arynes

A. Didehydroferrocene

Since the cyclopentadiene anion is isoelectronic with the parent five-membered heterocycles[304] it seems appropriate to consider the corresponding arynes in this review in spite of the absence of a heteroatom. Two such arynes have been suggested as intermediates, ferrocyne **(616)** and the didehydrocyclopentadiene anion **(617)** itself. The first of these was proposed as one of several possible explanations for the formation of the substitution products **618a** and **618b** from the reaction of chloroferrocene **619** with a large excess of butyllithium.[545] In the presence of added lithium piperidide some piperidylferrocene

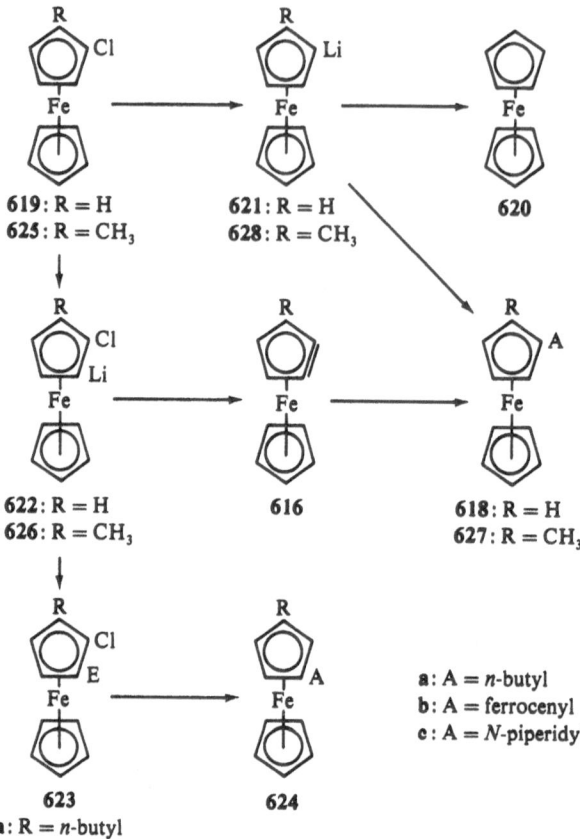

R
Cl
Fe
619: R = H
625: R = CH3
R
Li
Fe
621: R = H
628: R = CH3
Fe
620
R
Cl
Li
Fe
622: R = H
626: R = CH3
R
Fe
616
R
A
Fe
618: R = H
627: R = CH3
R
Cl
E
Fe
623
a: R = n-butyl
R
A
Fe
624
a: A = n-butyl
b: A = ferrocenyl
c: A = N-piperidyl

(**618c**) is formed as well. In both reactions considerable amounts of ferrocene (**620**) are isolated, indicating that, in spite of reports to the contrary,[546] halogen–metal interchange to ferrocenyllithium (**621**) is occurring.

Nonaryne mechanisms for the substitution of **619** which were eliminated include radical coupling of **621**, either with butyllithium to give **618a** or with itself to give **618b**, and direct displacement of chlorine by **621** or lithium piperidide to give **618b** and **618c**, respectively. A Wurtz–Fittig reaction between **621** and *n*-butyl chloride formed in the halogen–metal interchange reaction was shown to account for at least part of the butylferrocene (**618a**). A similar process involving *N*-chloropiperidine[266] (Section III.3.D) might explain the formation of **618c**, but a reaction of **619** and **621** to give biferrocenyl **618b** was specifically excluded. This latter reaction did proceed in the presence of only 1.75 equiv. (rather than 32 equiv.) of butyllithium to the exclusion of butylferrocene (**618a**) production. A termolecular coupling reaction involving butyllithium, chloroferrocene (**619**), and ferrocenyllithium (**621**) was invoked[545] to account for these results as well as those with lithium piperidide.

Such associative interactions are well known from both kinetic[547] and structural[548] studies.

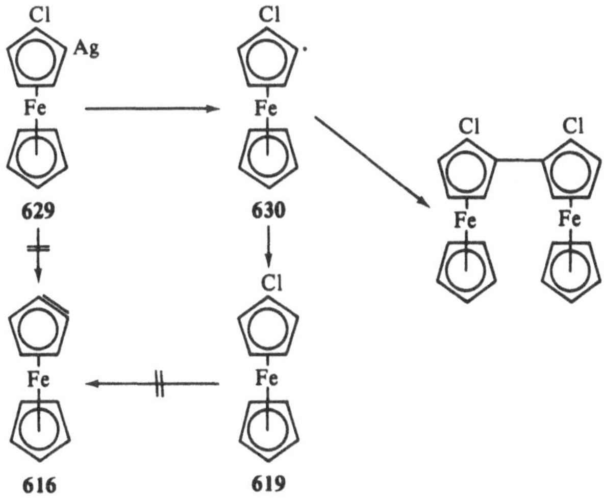

An elimination–addition mechanism via ferrocyne **616** is supported by the trapping of the α-chloro-lithium compound **622** with various electrophiles[545,549,550] to give **623** and by the observation of the cine-substitution products **624** when 2-methylchloroferrocene **625** is the reactant.[550] This latter reaction differs from that of the unsubstituted analog **619** not only in the greater stability of the initially formed lithium compound **626**, but also in the apparent absence of any biferrocenyls, **624b** and **627b**, or piperidylferrocenes, **624c** and **627c**. Presumably for this reason the termolecular mechanism proposed[545] for the reactions of **619** was not even considered in this case.[550] A Wurtz–Fittig reaction via **628** could account for the normal substitution product **627** as before, but a sequence of metallation, Wurtz–Fittig coupling, and halogen–metal interchange, **625**→**626**→**623**→**624**, would be required to account for the cine-substitution product **624**. This possibility was claimed to be highly unlikely since the required intermediate **623a** was not detected in the reaction.[550] It was further claimed that the approximately equal amounts (actually 5:3 or 2:3, depending on the reaction conditions) of normal and cine-substitution products, **624b** and **627b**, respectively, was more indicative of an aryne mechanism rather than of the competing processes outlined above. While this is clearly the most straightforward explanation of the results, the low yields (6%–8%) of substitution products and the apparent differences in reactivity of the two precursors **619** and **625** suggests that additional *positive* evidence would be desirable before the existence of the ferrocyne (**616**) can be considered likely. Attempts to generate this intermediate from chloroferrocene (**619**) with phenyllithium or alkali metal amides failed,[545] as did thermolysis of the silver compound **629**, which leads to products derived from the radical **630**.[551] No evidence for a ferrocyne intermediate (**619**) was obtained from a study of the lithioferrocene coupling reaction.[552]

B. Didehydrocyclopentadiene Anion

The claim for this aryne is based on the isolation of the appropriate adducts **631** or **632** when a crown-ether complex of the diazonium carboxylate potassium salt **633** is decomposed in the presence of tetracyclone (**151**) or the diphenyltetrazine (**169**).[553] As pointed out in Section II.2.B.c, however, both of these diene traps may be ambiguous as aryne probes because of their tendency to react with the aryne precursors by an addition–elimination process via **634** and **635** rather than by an elimination–addition mechanism involving the aryne **617**. The failure of several other dienes with unambiguous records as aryne traps (Section II.2.B.b.d), such as anthracene (**147**), to give aryne adducts with the precursor **633** tends to support the former mechanism. On the other hand,

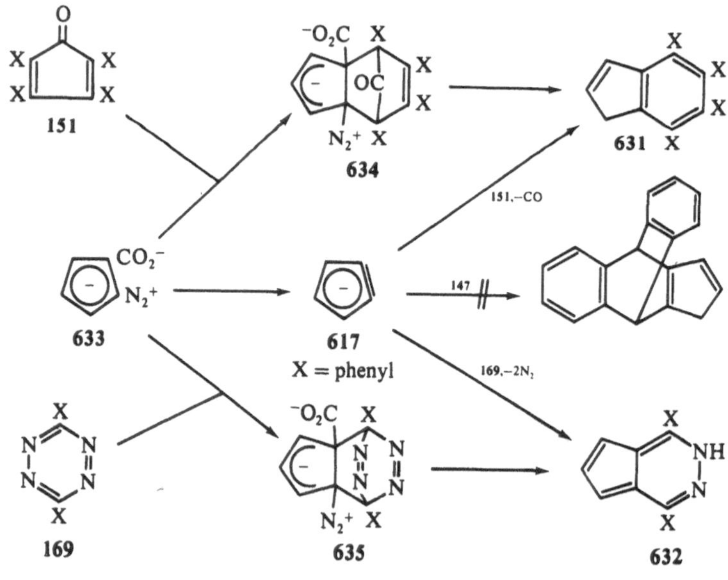

although detailed kinetic experiments (Section II.2.B.e) have not been done, the observation that the diazonium carboxylate **633** loses N_2 and CO_2 in approximately equal amounts on thermolysis *in the absence of a diene trap* constitutes at least one necessary feature of an elimination–addition mechanism. Additional studies are needed before a firm decision between the alternatives is possible.

9. Hetarynium Species

As noted in Section I.2.C, this review will cover five-membered hetarynium species such as **10** and **11** which are isoelectronic with arynes but in which the dehydro bond is exclusively between a carbon and a heteroatom and not between two carbon atoms. Such species may be cationic (**10**) or neutral (**11**) depending on the nature of the heteroatom. Hetarynium ions were first defined,[66] proposed,[554,555] and, in some cases, rejected[555] as possible reaction intermediates in the six-membered pyridine system (**9**). The observed chemistry of **9**[554] is that expected for either the α-pyridyl cation structure **9b** or, preferably,[1,556] the cyclic nitrilium ion form **9a**. No uniquely arynelike properties have been reported, which accounts for the absence of hetarynium species from Hoffmann's monograph.[27]

One example of a cine-substitution which in principle could involve a hetarynium species recently has been reported in the pyrazole series.[557] Reaction of the dinitropyrazole **636** with secondary amines leads to the cine-substitution product **637**. Although an elimination–addition process via the hetarynium species **638** was considered, an AEa mechanism (Section II.2.A.e) involving **639** and **640** was preferred and adequately accounts for all the observations.

The most likely route to five-membered hetarynium ions related to **10** is probably the dediazonization of the diazonium salts **641**, a method which represents the single successful path to phenyl cations.[100,101,558] With sulfur, oxygen, and N-substituted heterocycles, however, this reaction apparently proceeds through the related radicals **642**[559] and not cationic species such as **643**. With N-unsubstituted azoles the corresponding diazonium salts **644** are readily

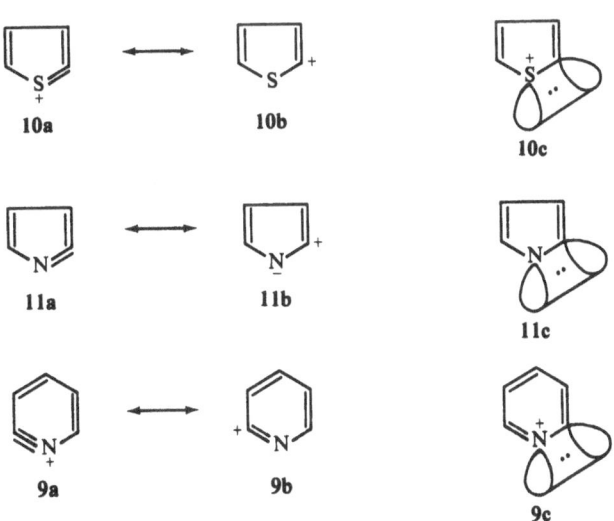

10a   10b   10c

11a   11b   11c

9a   9b   9c

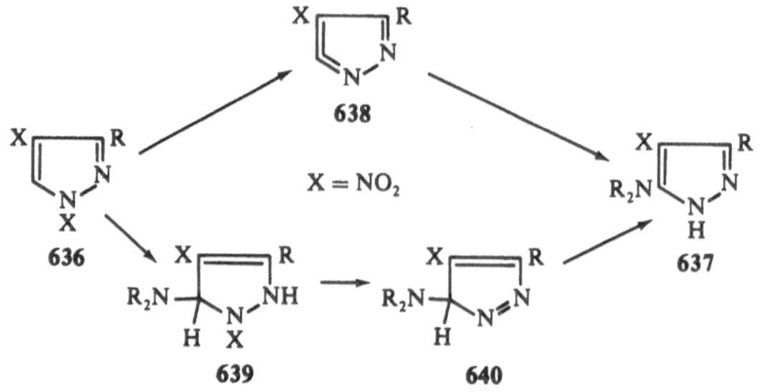

638

636

X = NO₂

637

639 **640**

deprotonated to give the neutral diazo compounds **645a** [560,561] which may be formulated as dipolar ions **645b**.[562] Loss of nitrogen from these compounds would lead to a neutral intermediate which, in principle, might be arynic (**646a**), dipolar (**646b**), or carbenic, the latter in either the singlet (**646c**) or triplet (**646d**) states.[101,563] The decomposition of several such diazoazoles (**645**) has been examined.

Dicyanodiazoimidazole (**647**) undergoes a variety of substitution reactions with loss of nitrogen.[564] Although a concerted process was not excluded, the intermediacy of a species with both dipolar (**648b**) and carbenic (**648c**) character was considered likely based on the observation of some remarkable nucleophilic substitutions and some typical insertion reactions, respectively. The latter is represented by the reaction with cyclohexane to give **649** and the former by the formation of halonium ylids **650** from the decomposition of the diazo compound **647** in halobenzenes. The fact that benzotrifluoride is attacked in the meta position to give **651** supports electrophilic rather than radical

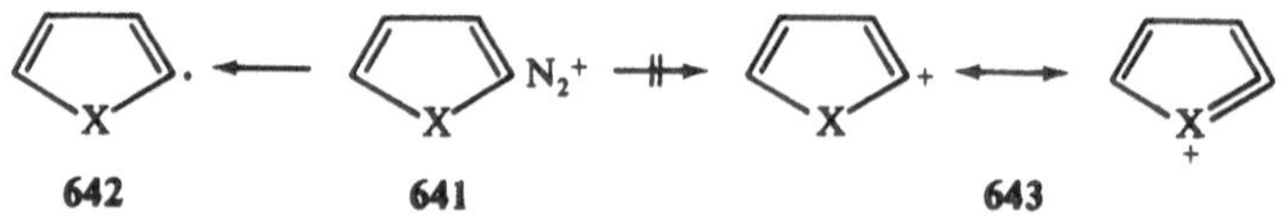

642 **641** **643**

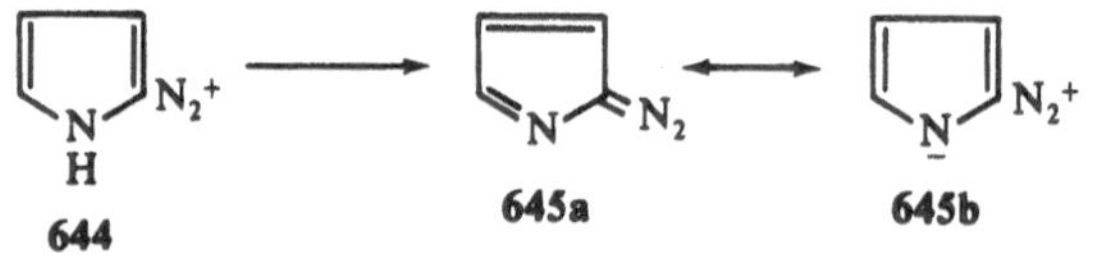

644 **645a** **645b**

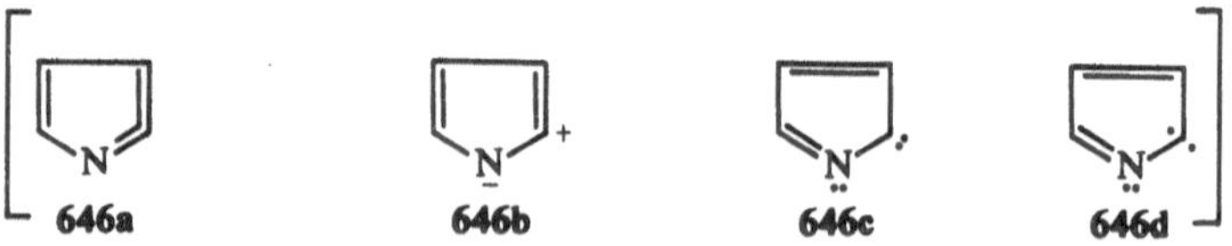

646a **646b** **646c** **646d**

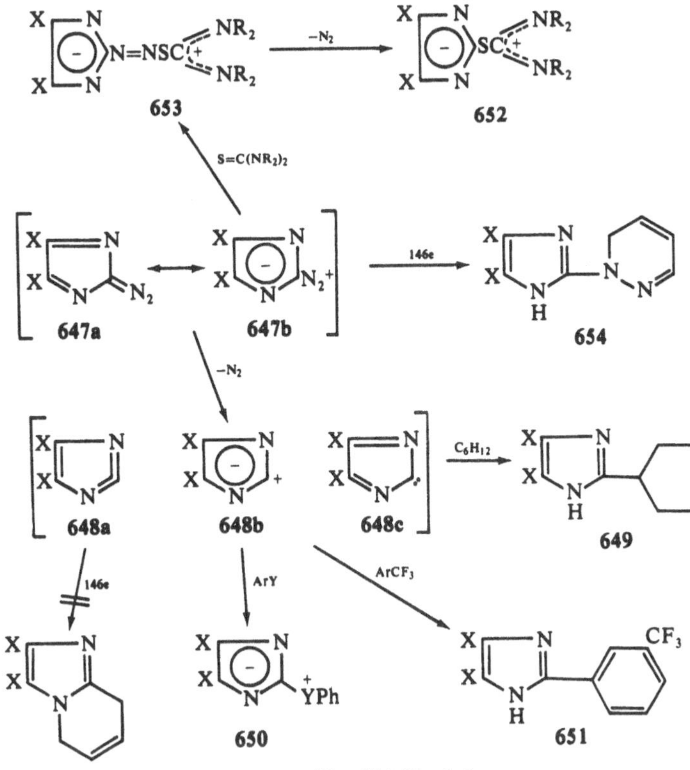

X = CN; Y = halogen

character in the intermediate **648**. The possibility that these nucleophilic substitutions do not involve initial loss of nitrogen at all, but proceed by a preliminary nucleophilic addition to the diazonium group of **647b** followed by loss of nitrogen, must also be considered in view of the recent demonstration of exactly such a process in the formation of the ylid **652** from **647** and thioureas via the adduct **653**.[565] The effect of crown ethers on the decomposition of **647** also fails to provide information on the polar character of the intermediate **648**.[566a] Evidence for arynic properties in the postulated intermediate (**648**) is also lacking since butadienes undergo Diels–Alder reactions with the diazonium group of the precursor **647** to give the dihydropyridazines **654** rather than the expected aryne adducts **655**.[564,566b]

An intermediate (**656**) with similar dual properties apparently has been generated from the decomposition of the diazopyrazole (**657**).[567] The carbenic character (**656c**) is not only suggested by a C–H insertion reaction with cyclohexane to give **658**, but also by the rearrangement to the 2*H*-azirine **659** and the reaction with benzene to give the azocine **660**. The dipolar character (**656b**) of the intermediate was proposed on the basis of its ability to cleave ethers presumably via the ylid intermediate **661**. The possibility that the ultimate product of this reaction (**662**) arose by an addition–elimination process similar to that discussed above for **652** [565] was not considered, although the first step of such a mechanism, nucleophilic addition to the diazo group of **657** to give **663**, did occur.[567] No reactions with dienes which might reveal the arynic character (**656a**) of the intermediate were reported, but activated alkenes[567,568] and alkynes[568] do undergo cycloaddition with the diazopyrazole precursors **657** themselves to give the adducts **664**.

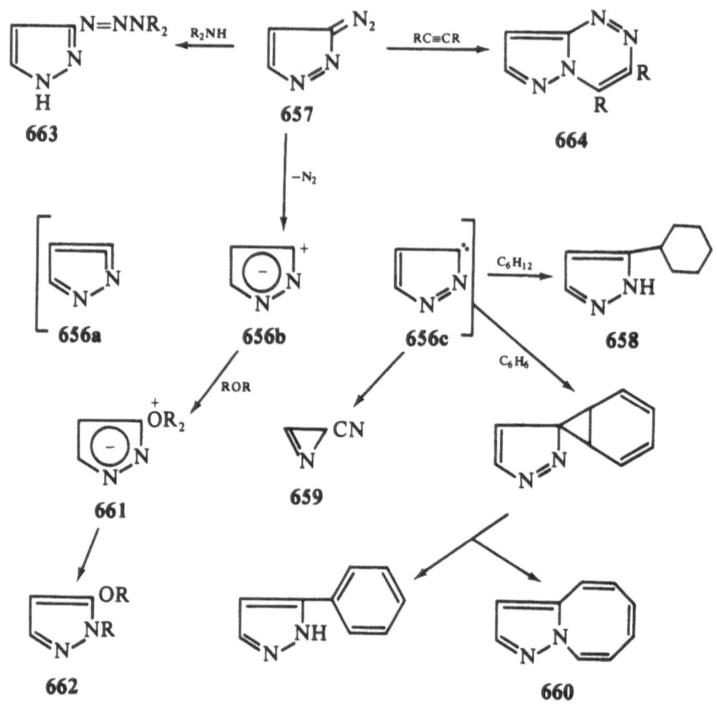

Two additional intermediates formed by loss of nitrogen from *N*-unsubstituted diazoazoles have been mentioned but no dual properties noted. Thermolysis of the diazotetrazole **665** has been suggested to proceed via the carbene **666** which disintegrates to atomic carbon and nitrogen,[569] while a radical intermediate (**667?**) is preferred, but not actually proposed, to rationalize the products of photolysis of the diazoimidazole **668**.[570]

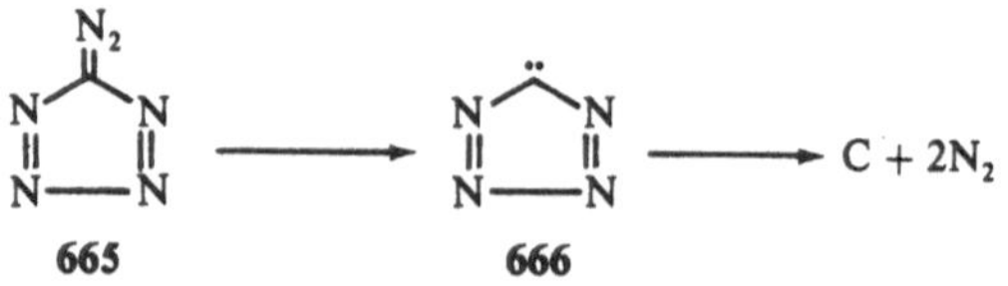

It seems clear from the above studies that intermediates isoelectronic with five-membered hetarynes can be generated in the *N*-unsubstituted azole series, but their reactivities suggest that dipolar (**646b**) or carbenic (**646c**) properties predominate rather than arynic (**646a**) ones which might arise from overlap of the peripheral orbitals as in (**11c**).

IV. SUMMARY

Although some evidence for the existence of five-membered hetarynes is now accumulating (Sections III.3.A.f and III.3.c) there is no question from the discussion in Section III that they have proved more difficult to prepare than their six-membered heterocyclic and carbocyclic analogues. The factors responsible for this situation are doubtless complex and interrelated. Nevertheless, it is probably instructive to consider each of them separately as a guide to future research in this field. For this purpose potential deleterious effects on five-membered hetaryne generation have been divided into electronic, geometric, and kinetic factors.

1. Electronic Factors

The first of these implies an inherent instability of five-membered hetarynes due to the presence of the heteroatom. Theoretical calculations on both the five[456] and six-membered hetarynes[19] do predict, in contrast to earlier conclusions,[21] that the presence of a nitrogen or sulfur atom adjacent to the aryne orbitals destabilizes the aryne compared to isomers without this arrangement. Compared to the analogous carbocyclic didehydro species, however, the presence of a nitrogen atom has been predicted to be both stabilizing[19] and destabilizing[571,572] depending on the method of calculation used. Therefore, except for the possibility that 3,4-didehydroheterocycles such as **25** might be predicted to be more accessible than the 2,3-isomers such as **4**—a conjecture in apparent contradiction to the available experimental facts (Sections III.3.A and III.3.B)—no firm theoretical basis for an adverse electronic effect in five-membered hetarynes is available. On the contrary a consideration of contributing canonical structures (Section I.2.B) suggests that interaction of the heteroatom with the aryne orbitals may even have a stabilizing influence.

2. Geometric Factors

Geometric factors might effect the stability and ease of formation of five-compared to six-membered arynes in several ways. The most obvious of these is the anticipated poorer overlap,[545] and therefore presumably poorer stabilizing interaction,[571] between the aryne orbitals in the smaller ring. In the absence of molecular orbital calculations for five-membered hetarynes with complete geometry optimization[572,573], the relative degree of aryne orbital overlap in these intermediates and their six-membered analogs can at best be estimated roughly from the geometry of the parent aromatic compounds.[19,456] One such empirical comparison[199] based on the distance between those hydrogen atoms which will be absent in the aryne (Table 7) suggests that the overlap of the aryne orbitals will decrease in the order benzyne≫3,4-thiophyne>2,3-thiophyne>3,4-pyrrolyne>2,3-pyrrolyne>2,3-furyne>3,4-furyne. Although this correlation is probably too crude to allow meaningful comparisons among the five-membered hetarynes, the rather large difference predicted between all of them and benzyne supports the suggestion of considerably poorer overlap in the former.

How energetically unfavorable this situation would be depends on whether a particular five-membered aryne is a singlet or a triplet. Available calculations[456] suggest the thiophynes 4 and 25 will have a singlet ground state but with low-lying triplets which may make appreciable contributions during a reaction. Once again calculations with complete geometry optimization would give a more accurate picture of the singlet or triplet character of these hetarynes.[572,573] The apparent tendency of 2,3-didehydrothiophene (4)[433] to undergo a preponderance of [2+2] cycloaddition (Section III.3.A.f) may indicate just such triplet character.

That a geometric factor which increases the separation of adjacent orbitals in the planar framework of smaller rings can affect the reactivity of molecules has been claimed in several pertinent cases. The considerably decreased steric inhibition of resonance[578] and ortho effects in general[579] for thiophene

TABLE 7.   H–H Distances in Aromatic Compounds

Compound	H Atoms	Distance, Å	Reference[a]
Benzene	1–2	2.46	574
Thiophene	2–3	2.63	575
	3–4	2.62	575
Pyrrole	2–3	2.69	576
	3–4	2.67	576
Furan	2–3	2.72	577
	3–4	2.75	577

[a] For experimental bond lengths and angles upon which the graphical determination[199] of the H–H distances is based.

derivatives compared to their benzene analogs has been ascribed to such geometrical considerations, as has the decreasing stability of cycloalkynes in going from an eight- to a five-membered ring.[51,580,581] On the other hand, the very fact that cyclopentyne (168) does exist[313,328,581] indicates that in and of itself this particular geometrical factor does not preclude the existence of a triple bond in a five-membered ring.

A different type of geometrical effect has been noted in bicyclic arynes. As the annelated ring is decreased in size from six- to four-membered the ease of formation of the arynes 669, 670, and 671 increases in that order.[583] Regardless of whether this observation is due to differences in the stability of the arynes themselves or of some kinetically important intermediates leading to the arynes,[584] its ultimate cause is probably related to bond fixation brought about by the geometric constraints of the annelated ring.

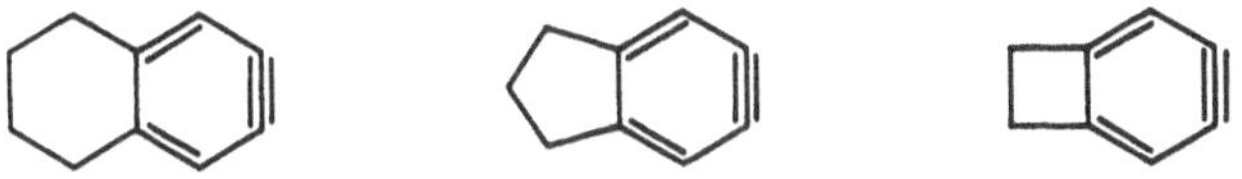

669 670 671

A fused benzene ring, however, may either retard or slightly assist the formation of a cycloalkyne depending on the particular precursor. Thus cyclopentyne (168) can be generated either from the bishydrazone 672[585] or the o-halolithium derivative 673,[313] while the indyne 674 is formed only by the former method via 675 but in slightly higher yield.[585] It therefore appears that benzannelation has no consistent geometric or conjugative effect on the ease of aryne formation, although a graphical analysis[478a] similar to that used[199] to generate the figures in Table 7 suggests that 2,3-didehydrothianaphthene (26) should have better aryne orbital overlap and hence be more stable than either of the didehydrothiophenes 4 or 25.

An additional effect of benzannelation on aryne stability can be envisioned. Although the implications of other data may be contradictory,[51]

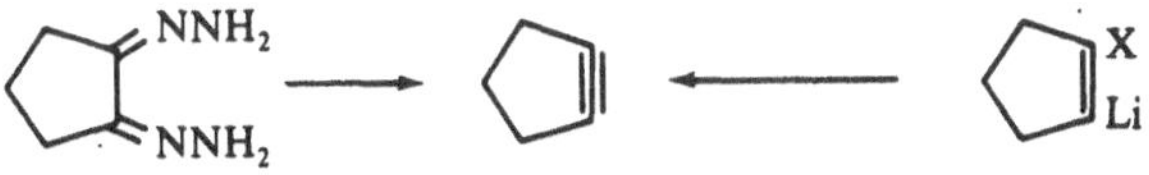

672 168 673

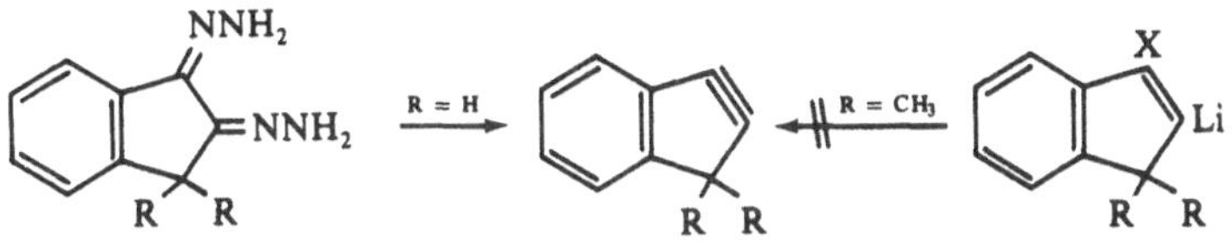

675 674 676

there is some indication that cyclohexyne (677) may be more stable than benzyne,[581] and cyclopentyne (168) has certainly been easier to generate[313,328,582] than five-membered arynes. A possible explanation for this apparent greater stability of cycloalkynes compared to the corresponding arynes may be that geometrical changes induced by forming the triple bond will increase the energy of an aryne more than of a cycloalkyne because of the loss of considerable resonance stabilization in the former due to disruption of the aromatic π system. If this hypotheis is correct it might not only be easier to prepare heterocycloalkynes such as 20 [586] and 21 but also benzofused hetarynes

677   20   21

such as 22, 26, and 31. In the former cases there would be no aromatic resonance stabilization to be lost and in the latter the fraction of the total energy lost probably would be much less than with a monocyclic hetaryne. The relative ease of partially disrupting the aromatic system of a polycyclic compared to a monocyclic aromatic compound is well known[587] and has been illustrated recently for a five-membered heterocycle and its benzo derivative.[588] The rationale proposed[588] for this observation is essentially identical to that suggested above.

3. Kinetic Factors

In a sense, the difficulty of generating five-membered hetarynes could be totally discussed in terms of kinetic factors. If these arynes are of very high energy content due to whatever electronic or geometric effect, then their rate of formation will be slow and other reactions will have the opportunity to intervene before aryne formation can occur. In this section two types of kinetic effects on potential five-membered hetaryne precursors will be discussed, those which stabilize toward aryne formation and those which labilize toward non-aryne reactions.

The most prominent example of the former situation is the unusual stability of the o-halometal derivatives where the metal is α to the heteroatom of a five-membered heterocycle. Compared to the lability of the benzene analog 32b, which loses LiBr at $-100°C$ to give benzyne,[342] heterocyclic compounds

32b   678   679

TABLE 8. Nonaryne Reactions of Five-Membered Hetaryne Precursors

1. Stable under the most severe conditions attempted: **206, 294-300, 431a, 443, 443a, 450, 453a, 469, 470, 490, 491, 493, 554, 555, 562, 563.**
2. Stepwise decomposition via nonaryne intermediates: **201, 209, 211, 226, 227, 231, 248, 270, 286, 311, 338–340, 359, 360–362, 431, 444, 447, 471, 538, 540, 543.**
3. Ring opening: **194, 439, 495, 499, 526a, 568, 571.**
4. Substitution by an addition–elimination mechanism: **266, 331, 332, 482, 505, 515, 528, 534, 537, 556.**
5. Transhalogenation: **181, 194, 302–305, 317, 437–439, 459, 482, 531, 556.**
6. Cycloaddition prior to elimination: **181, 199, 209–211, 218, 233, 235, 275, 280, 287, 290, 321, 459.**

678 and **679** discussed in Section III are stable in some cases to $+100°C$. Although, in part, this stability can be attributed[4,76,589] to the ability of the heteroatom to stabilize the adjacent carbanionic center,[590] other factors must also be involved since carbocyclic *o*-halometal compounds such as **673** and **676** show similar stability.[313,585] Presumably this is a reflection of the high energy content of the five-membered hetarynes and cycloalkynes.[589] Attempts to overcome this reluctance to form five-membered hetarynes by simply supplying more energy to the system run afoul of the second kinetic effect, the tendency of the precursors to undergo nonaryne reactions.

Several examples of such processes have been noted in Section III and include transhalogenation to the very stable α-metallated compounds **678** or **679**, ring opening, and stepwise elimination to nonaryne intermediates which can rearrange or attack the aryne traps which are present. The lower resonance energy of five-membered heterocycles compared to benzenoid compounds[489] makes them much more susceptible to addition reactions, and hence both cine-substitution and Diels–Alder products have been shown to sometimes arise by addition–elimination and not elimination–addition mechanisms. A summary of all the demonstrated nonaryne reactions of five-membered hetaryne precursors discussed in Section III is collected in Table 8.

4. Prognosis

Given the above analysis of the factors which make the generation of five-membered hetarynes difficult, what are the likely strategies for achieving this goal? One approach is to find precursors which decompose to arynes under such mild conditions as to preclude the nonaryne processes discussed in Section IV.3. Several attempts in this direction both in our laboratories[78,412] and those of others[591] have not as yet proven successful. An alternative is decomposition of the precursor with sufficient energy to permit aryne formation to compete with nonaryne reactions. This strategy appears to have some merit and has led to the strong indication of the generation of five-membered hetarynes in at least two instances (Sections III.3.A.f and III.3.C).[77,79] Further efforts along this vein are therefore warranted, particularly with those precursors which have been shown to lead to arynes under thermolysis conditions (Sections III.C, III.D, and III.E).

ACKNOWLEDGMENT

The generosity of the TCU Research Foundation and especially the Robert A. Welch Foundation in supporting the preparation of this chapter and much of the unpublished research reported herein[78,199,406,412,433,496] is gratefully acknowledged.

REFERENCES AND NOTES

1. T. Kaufmann, *Angew. Chem. Int. Ed. Engl.* **4**, 543 (1965).
2. H. J. den Hertog and H. C. van der Plas, *Adv. Heterocycl. Chem.* **4**, 121 (1965).
3. H. J. den Hertog and H. C. van der Plas, in *Chemistry of Acetylenes*, H. G. Viehe, Ed., Marcel Dekker, New York (1969), Chap. 17, p. 1149–1197.
4. T. Kauffmann and R. Wirthwein, *Angew. Chem. Int. Ed. Engl.* **10**, 20 (1971).
5. R. W. Hoffmann, *Dehydrobenzene and Cycloalkynes*, Academic Press, New York (1967).
6. R. Huisgen, in *Organometallic Chemistry*, H. Zeiss, Ed., Reinhold, New York (1960), p. 36–87.
7. J. F. Bunnett, *J. Chem. Educ.* **38**, 278 (1961); H. Heaney, *Chem. Rev.* **62**, 81 (1962); W. Czuba, *Wiad. Chem.* **19**, 157 (1965).
8. S. M. A. Hai and A. W. Qureshi, *Pak. J. Sci. Ind. Res.* **17**, 65 (1974); O. M. Nefedov, A. I. Dyachenko, and A. K. Prokofev, *Russ. Chem. Rev.* **46**, 941 (1977).
9. G. Wittig, *Angew. Chem. Int. Ed. Engl.* **4**, 731 (1965).
10. R. W. Hoffmann, in *Chemistry of Acetylenes*, H. G. Viehe, Ed., Marcel Dekker, New York (1969), Chap. 16, p. 1063–1148.
11. E. K. Fields, in *Organic Reactive Intermediates*, S. P. McManus, Ed., Academic Press, New York (1973), Chap. 7, p. 449–508.
11a. Two additional reviews on hetarynes have been written since this manuscript was completed. The first deals with all hetarynes; M. G. Reinecke, *Tetrahedron*, **38**, in press (1982). The second covers six-membered hetarynes only; H. C. van der Plas, in *The Chemistry of Triple Bonded Groups, Supplement C of the Chemistry of the Functional Groups*, S. Patai and Z. Rappoport, Eds., Wiley-Interscience, New York (1982).
12. J. D. Roberts, D. A. Semenow, H. E. Simmons, Jr., and L. A. Carlsmith, *J. Am. Chem. Soc.* **78**, 601 (1956).
13. (a) W. N. Washburn, *J. Am. Chem. Soc.* **97**, 1615 (1975); (b) R. Breslow, J. Napierski, and T. C. Clarke, *J. Am. Chem. Soc.* **97**, 6275 (1975).
14. J. D. Roberts, H. E. Simmons, Jr., L. A. Carlsmith, and C. W. Vaughn, *J. Am. Chem. Soc.* **75**, 3290 (1953).
15. G. Wittig, *Naturwissenschaften* **30**, 696 (1942).
16. T. Kauffmann, F. P. Boettcher, and J. Hansen, *Angew. Chem.* **73**, 341 (1961).
17. R. Levine and W. W. Leake, *Science* **121**, 780 (1955).
18. D. A. de Bie, H. C. van der Plas, and G. Geursten, *Rec. Trav. Chim. Pays-Bas* **90**, 594 (1971).

19. W. Adam, A. Grimison, and R. Hoffmann, *J. Am. Chem. Soc.* **91**, 2590 (1969).
20. R. D. Hamilton and E. Campaigne, in *Special Topics in Heterocyclic Chemistry*, A. Weissberger and E. C. Taylor, Eds., Wiley-Interscience, New York (1977), p. 271.
21. H. L. Jones and D. L. Beveridge, *Tetrahedron Lett.*, 1577 (1964).
22. N. N. Zatsepina and I. F. Tupitsyn, *Chem. Heterocycl. Comp. (USSR)* **10**, 1397 (1974).
23. A. Streitwieser and P. J. Scannon, *J. Am. Chem. Soc.* **95**, 6273 (1973).
24. J. A. Zoltewicz, G. Grahe, and C. L. Smith, *J. Am. Chem. Soc.* **91**, 5501 (1969); R. A. Abramovitch, I. M. Singer, and A. R. Vinutha, *Chem. Commun.*, 55 (1967).
25. W. G. Salmond, *Q. Rev. Chem. Soc.* **22**, 253 (1968).
26. R. F. Heldweg and H. Hogeveen, *Tetrahedron Lett.*, 75 (1974).
27. R. W. Hoffmann, *Dehydrobenzene and Cycloalkynes*. Academic Press, New York (1967), p. 275.
28. F. Bernardi, I. G. Csizmadia, A. Mangini, H. B. Schlegel, M. H. Whangbo, and S. Wolfe, *J. Am. Chem. Soc.* **97**, 2209 (1975).
29. R. A. Olofson, J. M. Landesberg, K. N. Houk, and J. S. Michelman, *J. Am. Chem. Soc.* **88**, 4265 (1966).
30. P. Haake, L. P. Bausher, and W. B. Miller, *J. Am. Chem. Soc.* **91**, 1113 (1969).
31. P. Haake, L. P. Bausher, and W. B. Miller, *J. Am. Chem. Soc.* **91**, 1113 (1969), footnote 39.
32. F. G. Bordwell, M. van der Puy, and N. R. Vanier, *J. Org. Chem.* **41**, 1885 (1976).
33. A. Streitwieser and S. P. Ewing, *J. Am. Chem. Soc.* **97**, 190 (1975).
34. J. A. Elvidge, J. R. Jones, C. O'Brien, E. A. Evans, and H. C. Sheppard, *Adv. Heterocycl. Chem.* **16**, 1 (1974).
35. M. Berthelot, *Justus Liebigs Ann. Chem.* **142**, 251 (1867).
36. C. deMilt, *J. Chem. Ed.* **28**, 421 (1951).
37. E. Dreher and R. Otto, *Justus Liebigs Ann. Chem.* **154**, 93 (1870).
38. R. Stoermer and B. Kahlert, *Chem. Ber.* **35**, 1633 (1902).
39. O. Diels and M. Reinbeck, *Chem. Ber.* **43**, 1271 (1910).
40. A. Michael, *Chem. Ber.* **39**, 1915 (1906).
41. E. Ott, *Chem. Ber.* **47**, 2388 (1914).
42. W. E. Bachman and H. T. Clarke, *J. Am. Chem. Soc.* **49**, 2089 (1927).
43. A. Luttringhaus and G. V. Sääf, *Justus Liebigs Ann. Chem.* **542**, 241 (1939).
44. A. A. Morton, J. B. Davidson, and B. L. Hakan, *J. Am. Chem. Soc.* **64**, 2242 (1942).
45. G. Wittig and G. Fuhrmann, *Chem. Ber.* **73**, 1197 (1940).
46. R. Huisgen and H. Rist, *Naturwissenschaften* **41**, 358 (1954); R. Huisgen and H. Rist, *Justus Liebigs Ann. Chem.* **594**, 137 (1955).
47. G. Wittig and L. Pohmer, *Chem. Ber.* **89**, 1334 (1956).
48. M. Stiles, R. G. Miller, and U. Burckhardt, *J. Am. Chem. Soc.* **85**, 1792 (1963).
49. G. Wittig and R. W. Hoffmann, *Chem. Ber.* **95**, 2718 (1962).
50. C. D. Campbell and C. W. Rees, *J. Chem. Soc. C*, 742 (1969).
51. G. Wittig, *Pure Appl. Chem.* **7**, 173 (1963).
52. R. S. Berry, G. N. Spokes, and M. Stiles, *J. Am. Chem. Soc.* **84**, 3570 (1962).
53. G. Wittig and H. F. Ebel, *Justus Liebigs Ann. Chem.* **650**, 20 (1961).
54. R. S. Berry, J. Clardy, and M. E. Schafer, *J. Am. Chem. Soc.* **86**, 2738 (1964).
55. R. Husigen and R. Knorr, *Tetrahedron Lett.*, 1017 (1963).
56. B. H. Klanderman and T. R. Criswell, *J. Am. Chem. Soc.* **91**, 510 (1969).
57. O. L. Chapman, K. Mattes, C. L. McIntosh, J. Pacansky, G. V. Calder, and G. Orr, *J. Am. Chem. Soc.* **95**, 6134 (1973).
58. R. W. Hoffmann, *Dehydrobenzene and Cycloalkynes*, Academic Press, New York (1967), Chapt. 5.
59. R. W. Hoffmann, *Dehydrobenzene and Cycloalkynes*. Academic Press, New York (1967), Chapt. 6.

60. L. B. Dashkevich, V. A. Buevich, and B. E. Kubaev, *J. Gen. Chem. USSR* **30**, 1925 (1960).
61. G. Wittig and V. Wahl, *Angew. Chem.* **73**, 492 (1961).
62. G. Wittig, *Angew. Chem. Int. Ed. Engl.* **1**, 415 (1962).
63. G. Komppa and S. Weckman, *J. Prakt. Chem.* **138**, 109 (1933).
64. F. G. Bordwell, B. B. Lampert, and W. H. McKellin, *J. Am. Chem. Soc.* **71**, 1702 (1949).
65. K. R. Brower and E. D. Amstutz, *J. Org. Chem.* **19**, 411 (1954).
66. T. Kauffmann and F. P. Boettcher, *Chem. Ber.* **95**, 949 (1962).
67. T. Kauffmann, A. Risberg, J. Schulz, and R. Weber, *Tetrahedron Lett.*, 3563 (1964).
68. T. Kauffmann, R. Nurnberg, J. Schulz, and R. Stobba, *Tetrahedron Lett.*, 4273 (1967).
69. R. Fusco and M. Bianchi, *Gazz. Chim. Ital.* **97**, 410 (1967).
70. W. Draber, *Angew. Chem. Int. Ed. Engl.* **6**, 75 (1967).
71. E. K. Fields and S. Meyerson, *Chem. Commun.*, 708 (1966).
72. E. K. Fields and S. Meyerson, in *Organosulfur Chemistry*, M. J. Janssen, Ed., Interscience Publishers, New York (1967), p. 143.
73. R. W. Hoffmann, *Dehydrobenzene and Cycloalkynes*, Academic Press, New York (1967), p. 293.
74. L. Crombie, P. A. Gilbert, and R. P. Houghton, *Chem. Commun.*, 595 (1967).
75. M. G. Reinecke and H. W. Adickes, *J. Am. Chem. Soc.* **90**, 511 (1968).
76. G. Wittig and M. Rings, *Justus Liebigs Ann. Chem.* **719**, 127 (1968).
77. (a) M. G. Reinecke and J. G. Newson, *J. Am. Chem. Soc.* **98**, 3021 (1976); (b) M. G. Reinecke, J. G. Newsom, and L. J. Chen, *J. Am. Chem. Soc.* **103**, 2760 (1981).
78. H. Ballard, unpublished results, Texas Christian University (1977).
79. M. G. Reinecke and H. Ballard, *Tetrahedron Lett.*, 4981 (1979).
80. R. W. Hoffmann, *Dehydrobenzene and Cycloalkynes*, Academic Press, New York (1967), pp. 9–31.
81. (a) R. W. Hoffmann, *Dehydrobenzene and Cycloalkynes*, Academic Press, New York (1967), pp. 43–58; (b) pp. 32–42.
82. R. F. Cunico and E. M. Dexheimer, *J. Organometal. Chem.* **59**, 153 (1973).
83. J. F. Bunnett and B. F. Hrutfiord, *J. Org. Chem.* **27**, 4152 (1962).
84. R. W. Hoffmann, *Dehydrobenzene and Cycloalkynes*, Academic Press, New York (1967), pp. 37–41.
85. (a) E. McNelis, *J. Org. Chem.* **28**, 3188 (1963); (b) G. Köbrich, *Justus Liebigs Ann. Chem.* **664**, 88 (1963).
86. (a) G. Cainelli, G. Zubiani, aand S. Morrocchi, *Chim. Ind. (Milan)* **46**, 1489 (1964); (b) F. C. Fischer and E. Havinga, *Rec. Trav. Chim. Pays-Bas* **93**, 21 (1974).
87. R. W. Hoffmann, *Dehydrobenzene and Cycloalkynes*, Academic Press, New York (1967), p. 40.
88. R. W. Hoffmann, *Dehydrobenzene and Cycloalkynes*, Academic Press, New York (1967), p. 72.
89. R. W. Hoffmann, *Dehydrobenzene and Cycloalkynes*, Academic Press, New York (1967), p. 37.
90. F. M. Beringer and S. J. Huang, *J. Org. Chem.* **29**, 445 (1964).
91. R. W. Hoffmann, *Dehydrobenzene and Cycloalkynes*, Academic Press, New York (1967), p. 238.
92. R. W. Hoffmann, *Dehydrobenzene and Cycloalkynes*, Academic Press, New York (1967), pp. 62–65.
93. F. L. Scott and R. E. Oesterling, *J. Am. Chem. Soc.* **82**, 5247 (1960).
94. T. Akiyama, Y. Imasaki, and M. Kawanisi, *Chem. Lett.*, 229 (1974).
95. J. I. G. Cadogan, A. G. Rowley, J. T. Sharp, B. Sledzinski, and N. H. Wilson, *J. Chem. Soc. Perkin Trans. 1*, 1072 (1975).
96. I. Fleming and T. Mah, *J. Chem. Soc. Perkin Trans. 1*, 1577 (1976).

97. K. Grohmann, personal communication, cited in R. W. Hoffmann, *Dehydrobenzene and Cycloalkynes*, Academic Press, New York (1967), p. 66.
98. H. Meier and K.-P. Zeller, *Angew. Chem. Int. Ed. Engl.* **16**, 835 (1977).
99. R. W. Hoffmann, *Chem. Ber.* **98**, 222 (1965); J. F. Bunnett and D. A. R. Happer, *J. Org. Chem.* **31**, 2369 (1966).
100. H. Zollinger, *Acc. Chem. Res.* **6**, 335 (1973).
101. H. Zollinger, *Angew. Chem. Int. Ed. Engl.* **17**, 141 (1978),
102. J. F. Bunnett and C. Yijima, *J. Org. Chem.* **42**, 639 (1977).
103. L. Friedman and F. M. Logullo, *J. Org. Chem.* **34**, 3089 (1969).
104. A. Hantzsch and W. B. Davidson, *Chem. Ber.* **29**, 1522 (1896).
105. (a) F. M. Logullo, dissertation, Case Institute of Technology (1965); (b) R. M. Roberts, J. C. Gilbert, L. B. Rodewald, and A. S. Wingrove, *An Introduction to Modern Experimental Organic Chemistry*, Holt, Rinehart and Winston, New York (1969), p. 196–201.
106. L. Friedman, personal communication, cited in R. W. Hoffmann, *Dehydrobenzene and Cycloalkynes*, Academic Press, New York (1967), p. 75.
107. J. Nakayama, O. Simamura, and M. Yoshida, *J. Chem. Soc. D*, 1222 (1970).
108. R. Gompper, E. Kutter, and G. Seybold, *Chem. Ber.* **101**, 2340 (1968).
109. Y. Maki, T. Furuta, M. Kuzuya, and M. Suzuki, *J. Chem. Soc. Chem. Commun.* **616** (1975).
110. R. W. Hoffmann, *Dehydrobenzene and Cycloalkynes*, Academic Press, New York (1967), pp. 74-77.
111. R. Gompper, G. Seybold, and B. Schmolke, *Angew. Chem. Int. Ed. Engl.* **7**, 389 (1968).
112. S. Yaroslavsky, *Chem. Ind. (London),* 765 (1965).
113. D. C. Dittmer and E. S. Whitman, *J. Org. Chem.* **34**, 2004 (1969).
114. H. Heaney, J. M. Japlonski, and C. T. McCarty, *J. Chem. Soc. Perkin Trans. 1*, 2903 (1972); J. M. Rao and S. Mallikarjuna, *Tetrahedron Lett.*, 283 (1979).
115. N. Dennis, A. R. Katritzky, and S. K. Parton, *J. Chem. Soc. Perkin Trans. 1,* 750 (1974).
116. The zwitterion **36** generated from **321** apparently adds to benzyne (**1**) to give benzcoumarin,[90] a product which has not been observed in reactions of **41a**.[117]
117. L. Friedman and M. Stiles, personal communications cited in R. W. Hoffmann, *Dehydrobenzene and Cycloalkynes*, Academic Press, New York (1967), p. 238.
118. T. J. Barton, A. J. Nelson, and J. Clardy, *J{φ Org. Chem.* **37**, 895 (1972).
119. S. Yaroslavsky, *Tetrahedron Lett.*, 1503 (1965).
120. R. J. Osiewicz, Dissertation, Case Western Reserve University (1972).
121. O. L. Chapman, C. L. McIntosh, J. Pecansky, G. V. Calder, and G. Orr, *J. Am. Chem. Soc.* **95**, 4061 (1973).
122. S. F. Dyke, A. J. Floyd, and S. E. Ward, *Tetrahedron* **26**, 4005 (1970).
123. R. R. Schmidt and W. Schneider, *Tetrahedron Lett.*, 5095 (1970).
124. R. A. Rossi, R. H. de Rossi, and H. E. Bertorello, *J. Org. Chem.* **36**, 2905 (1971).
125. R. Ghosh, E. B. Sheinin, C. L. Bell, and L. Bauer, *J. Heterocycl. Chem.* **12**, 203 (1975).
126. J. I. G. Cadogan, *Acc. Chem. Res.* **4**, 186 (1971).
127. P. C. Buxton and H. Heaney, *J. Chem. Soc. Chem. Commun.*, 545 (1973).
128. J. I. G. Cadogan, C. D. Murray, and J. T. Sharp, *J. Chem. Soc. Perkin Trans. 1*, 1321 (1974).
129. (a) D. L. Brydon, J. I. G. Cadogan, J. Cook, M. J. P. Harger, and J. T. Sharp, *J. Chem. Soc. B.*, 1996 (1971); (b) J. I. G. Cadogan, D. M. Smith, and J. B. Thomson, *J. Chem. Soc. Perkin Trans. 1*, 1296 (1972).
130. B. Baigrie, J. I. G. Cadogan, J. R. Mitchell, A. K. Robertson, and J. T. Sharp, *J. Chem. Soc. Perkin Trans. 1*, 2563 (1972).
131. C. Rüchardt and C. C. Tan, *Chem. Ber.* **103**, 1774 (1970).
132. J. I. G. Cadogan, C. D. Murray, and J. T. Sharp, *J. Chem. Soc. Perkin Trans. 2*, 583 (1976).

133. C. Rüchardt and C. C. Tan, *Angew. Chem. Int. Ed. Engl.* **9**, 522 (1970).
134. H. Teichmann, M. Jatkowski, and G. Hilgetag, *J. Prakt. Chem.* **314**, 129 (1972).
135. M. Kato, T. Tamano, and T. Miwa, *Bull. Chem. Soc. Jap.* **48**, 291 (1975).
136. T. E. Stevens, *J. Org. Chem.* **33**, 855 (1968).
137. E. A. Dorko and T. E. Stevens, *Chem. Commun.*, 871 (1966).
138. (a) J. A. Kampmeier and A. B. Rubin, *Tetrahedron Lett.*, 2853 (1966); D. L. Brydon and J. I. G. Cadogan, *J. Chem. Soc. C*, 819 (1968).
139. B. H. Klanderman, D. P. Maier, G. W. Clark, and J. A. Kampmeier, *J. Chem. Soc. D*, 1003 (1971).
140. G. W. Clark and J. A. Kampmeier, *J. Chem. Soc. D*, 996 (1970).
141. L. Verbit, J. S. Levy, H. Rabitz, and W. Kwalwasser, *Tetrahedron Lett.*, 1053 (1966).
142. R. A. Abramovitch and G. N. Knaus, *J. Org. Chem.* **40**, 883 (1975).
143. A. Hantzsch and R. Glogauer, *Chem. Ber.* **30**, 2548 (1897).
144. M. Stiles and R. G. Miller, *J. Am. Chem. Soc.* **82**, 3802 (1960).
145. R. W. Hoffmann, W. Sieber, and G. Guhn, *Chem. Ber.* **98**, 3470 (1965).
146. K. Ramanathan, dissertation, University of Rochester (1974); *Diss. Abstr. Int. B* **36**, 243 (1975).
147. V. Franzen and H.-I. Joschek, *Justus Liebigs Ann. Chem.* **703**, 90 (1967).
148. R. W. Hoffmann and W. Sieber, *Justus Liebigs Ann. Chem.* **703**, 96 (1967).
149. A. T. Fanning, Jr. and T. D. Roberts, *Tetrahedron Lett.*, 805 (1971).
150. A. T. Fanning, Jr., G. R. Bickford, and T. D. Roberts, *J. Am. Chem. Soc.* **94**, 8505 (1972).
151. B. M. Adger, M. Keating, C. W. Rees, and R. C. Storr, *J. Chem. Soc. Perkin Trans. 1*, 41 (1975).
152. N. Bashir and T. L. Gilchrist, *J. Chem. Soc. Perkin Trans. 1*, 868 (1973).
153. M. S. Ao, E. M. Burgess, A. Schauer, and E. A. Taylor, *Chem. Commun.*, 220 (1969).
154. J. Adamson, D. L. Forster, T. L. Gilchrist, and C. W. Rees, *Chem. Commun.*, 221 (1969).
155. J. Adamson, D. L. Forster, T. L. Gilchrist, and C. W. Rees, *J. Chem. Soc. C*, 981 (1971).
156. E. F. Ullman and E. A. Bartkus, *Chem. Ind. (London)*, 93 (1962).
157. D. L. Forster, T. L. Gilchrist, C. W. Rees, and E. Stanton, *J. Chem. Soc. D*, 695 (1971).
158. (a) C. D. Campbell and C. W. Rees, *J. Chem. Soc. C*, 752 (1969); (b) C. W. Rees and R. C. Storr, *Chem. Commun.*, 1305 (1968).
159. J. I. G. Cadogan and J. B. Thomson, *Chem. Commun.*, 770 (1969).
160. M. Keating, M. E. Peek, C. W. Rees, and R. C. Storr, *J. Chem. Soc. Perkin Trns. 1*, 1315 (1972).
161. C. W. Rees, personal communication, cited by R. W. Hoffman in *Chemistry of Acetylenes*, H. G. Viehe, Ed., Marcel Dekker, New York (1969), Chap. 16, p. 1091.
162. B. M. Adger, S. Bradbury, M. Keating, C. W. Rees, R. C. Storr, and M. T. Williams, *J. Chem. Soc. Perkin Trans. 1*, 31 (1975).
163. O. L. Chapman, C.-C. Chang, J. Kole, N. R. Rosenquist, and H. Tomioka, *J. Am. Chem. Soc.* **97**, 6586 (1975).
164. G. Ege and G. Jooss, *Chem. Ber.* **106**, 1978 (1973).
165. R. S. Atkinson and C. W. Rees, *Chem. Commun.*, 1230 (1967).
166. J. D. White and M. Kim, *Tetrahedron Lett.*, 3361 (1974).
167. J. Fout, M. Torres, H. E. Gunning, and O. P. Strausz, *J. Org. Chem.* **43**, 2487 (1978).
168. I. Lalezari, A. Shafiee, and M. Yalpani, *J. Org. Chem.* **36**, 2836 (1971).
169. M. Torres, A. Clement, J. E. Bertie, H. E. Gunning, and O. P. Strausz, *J. Org. Chem.* **43**, 2490 (1978).
170. N. P. Buu-Hoi, P. Jacquignon, and M. Mangane, *Chem. Commun.*, 624 (1965).
171. C. L. Pederson and J. Moller, *Acta Chem. Scand.* **29B**, 483 (1975).
172. (a) M. Jones, Jr. and M. R. DeCamp, *J. Org. Chem.* **36**, 1536 (1971); (b) R. T. Luibrand and R. W. Hoffmann, *J. Org. Chem.* **39**, 3887 (1974).

173. G. A. Russell, V. Malatesta, D. E. Lawson, and R. Steg, *J. Org. Chem.*, **43**, 2242 (1978).
174. F. D. Greene, *J. Am. Chem. Soc.* **78**, 2246 (1956).
175. E. K. Fields and S. Meyerson, *Chem. Commun.*, 474 (1965).
176. R. F. C. Brown, D. V. Gardner, J. F. W. McOmie, and R. K. Solly, *Chem. Commun.*, 407 (1966).
177. M. P. Cava, M. J. Mitchell, D. C. DeJongh, and R. Y. Van Fossen, *Tetrahedron Lett.*, 2947 (1966).
178. E. K. Fields and S. Meyerson, *J. Org. Chem.* **31**, 3307 (1966).
179. R. F. C. Brown, D. V. Gardner, J. F. W. McOmie, and R. K. Solly, *Aust. J. Chem.* **20**, 139 (1967).
180. (a) V. E. Platonov, T. V. Senchenko, and G. G. Yakobson, *J. Org. Chem. USSR* **12**, 816 (1976); (b) T. V. Senchenko, V. E. Platonov, and G. G. Yakobson, *J. Org. Chem. USSR* **13**, 1904 (1977).
181. A. Martineau and D. C. DeJongh, *Can. J. Chem.* **55**, 34 (1977).
182. H. Suhr and A. Szabo, *Justus Liebigs Ann. Chem.* **752**, 37 (1971).
183. G. Porter and J. I. Steinfeld, *J. Chem. Soc. A.* **877** (1968); J. Lohmann, *J. Chem. Soc. Faraday Trans. 1* **68**, 814 (1972).
184. I. R. Dunkin and J. G. MacDonald, *J. Chem. Soc. Chem. Commun.*, 772 (1979).
185. S. Meyerson, *Rec. Chem. Prog.* **26**, 257 (1965).
186. H.-F. Grützmacher and J. Lohmann, *Justus Liebigs Ann. Chem.* **733**, 88 (1970).
187. H.-F. Grützmacher and J. Lohmann, *Justus Liebigs Ann. Chem.* **726**, 47 (1969).
188. S. Meyerson and E. K. Fields, *Chem. Commun.*, 275 (1966).
189. M. P. Cava and M. J. Mitchell, *Cyclobutadine and Related Compounds*, Academic Press, New York (1967), p. 225; M. P. Cava, personal communication, cited in R. W. Hoffmann, *Dehydrobenzene and Cycloalkynes*, Academic Press, New York (1967), p. 84.
190. A. Szabo, H. Suhr, and M. Venugopalan, *Justus Liebigs Ann. Chem.* 747 (1977).
191. (a) R. F. C. Brown and R. K. Solly, *Chem. Ind. (London)*, 181 (1965); (b) R. F. C. Brown and R. K. Solly, *Chem. Ind. (London)*, 1462 (1965); (c) R. F. C. Brown and R. K. Solly, *Aust. J. Chem.* **19**, 1045 (1966).
192. M. P. David and J. F. W. McOmie, *Tetrahedron Lett.*, 1361 (1973).
193. W. Reichen, *Helv. Chim. Acta* **60**, 186 (1977).
194. J. L. Ripoll, A. Rouessac, and F. Rouessac, *Tetrahedron* **34**, 19 (1978).
195. G. Seybold, *Angew. Chem. Int. Ed. Engl.* **16**, 365 (1977).
196. G. Wittig, H. Härle, E. Knauss, and K. Niethammer, *Chem. Ber.* **93**, 951 (1960).
197. M. Ahmed and J. M. Vernon, *J. Chem. Soc. Chem. Commun.*, 462 (1976).
198. S. J. Cristol and R. Caple, *J. Org. Chem.* **31**, 585 (1966).
199. L. J. Chen, dissertation, Texas Christian University (1978).
200. T. L. Gilchrist, G. E. Gymer, and C. W. Rees, *J. Chem. Soc. Perkin Trans. 1*, 1747 (1975).
201. L. Friedman and D. F. Lindow, *J. Am. Chem. Soc.* **90**, 2324 (1969).
202. C. E. Griffin and D. C. Wysoeki, *J. Org. Chem.* **34**, 751 (1969).
203. (a) E. J. Corey, F. A. Carey, and R. A. E. Winter, *J. Am. Chem. Soc.* **87**, 934 (1965); (b) E. J. Corey and R. A. E. Winter, *Chem. Commun.*, 208 (1965).
204. E. J. Thomas, *J. Chem. Soc. Perkin Trans. 1*, 2006 (1973).
205. O. L. Chapman, C. C. Chang, and N. R. Rosenquist, *J. Am. Chem. Soc.* **98**, 261 (1976).
206. S. P. Schmidt and G. B. Schuster, *J. Org. Chem.* **43**, 1823 (1978).
207. R. C. Dougherty, *Top. Curr. Chem.* **45**, 93 (1974).
208. U. E. Wiersum and T. Nieuwenhuis, *Tetrahedron Lett.*, 2581 (1973).
209. D. C. DeJongh, R. Y. VanFossen, and C. F. Bourgeois, *Tetrahedron Lett.*, 271 (1967).
210. R. A. Abramovitch and S. Wake, *J. Chem. Soc. Chem. Commun.*, 673 (1977).
211. O. L. Chapman and C. L. McIntosh, *J. Am. Chem. Soc.* **92**, 7001 (1970).
212. O. Tsuge, M. Tashiro, S. Kanemasa, and K. Takasaki, *Chem. Lett.*, 827 (1972).

213. E. Ziegler and H. Sterk, *Monatsh. Chem.* **99**, 1958 (1968).
214. P. DeChamplain, J.-L. Luche, R. A. Marty, and P. DeMayo, *Can. J. Chem.* **54**, 3749 (1976).
215. W. Adam, *Angew. Chem. Int. Ed. Engl.* **13**, 619 (1974).
216. (a) P. R. Story, W. H. Morrison, III, T. K. Hall, J.-C. Farine, and C. E. Bishop, *Tetrahedron Lett.*, 3291 (1968); (b) P. R. Story, T. K. Hall, W. H. Morrison, III, and J.-C. Farine, *Tetrahedron Lett.*, 5397 (1968).
217. C. S. Foote, M. T. Wuesthoff, S. Wexler, I. G. Burstain, R. Denny, G. O. Schneck, and K.-H. Schulte-Elte, *Tetrahedron* **23**, 2583 (1967).
218. E. Bernatek and M. Hvatum, *Acta Chem. Scand.* **14**, 836 (1960).
219. H. F. Ebel and R. W. Hoffmann, *Justus Liebigs Ann. Chem.* **673**, 1 (1964).
220. H. Günther, *Chem. Ber.* **96**, 1801 (1963).
221. K. Iqbal and R. C. Wilson, *J. Chem. Soc. C*, 1690 (1967).
222. I. P. Fisher and F. P. Lossing, *J. Am. Chem. Soc.* **85**, 1018 (1963).
223. J. A. Kampmeier and E. Hoffmeister, *J. Am. Chem. Soc.* **84**, 3787 (1962).
224. N. Kharasch and R. K. Sharma, *Chem. Commun.*, 492 (1967).
225. R. K. Sharma and N. Kharasch, *Angew. Chem. Int. Ed. Engl.* **7**, 36 (1968).
226. J. P. N. Brewer, I. F. Eckhard, H. Heaney, M. G. Johnson, B. A. Marples, and T. J. Ward, *J. Chem. Soc. C*, 2569 (1970).
227. R. D. Youssefyeh and L. Lictenberg, *J. Chem. Soc. Perkin Trans. 1*, 2649 (1974).
228. H.-F. Grützmacher and W.-A. Lehman, *Justus Liebigs Ann. Chem.*, 2023 (1975).
229. G. Köbrich, *Chem. Ber.* **92**, 2985 (1959).
230. G. Köbrich and H. Fröhlich, *Chem. Ber.* **98**, 3637 (1965).
231. R. W. Hoffmann, *Dehydrobenzene and Cycloalkynes*, Academic Press, New York (1967), pp. 40–41.
232. H. E. Simmons, *J. Org. Chem.* **25**, 691 (1960).
233. G. Köbrich, *Chem. Ber.* **96**, 2544 (1963).
234. E. K. Fields and S. Meyerson, *J. Org. Chem.* **41**, 916 (1976).
235. B. R. Eggins, *Tetrahedron* **31**, 1191 (1975).
236. P. Sartori and A. Golloch, *Chem. Ber.* **102**, 1765 (1969).
237. E. K. Fields and S. Meyerson, *J. Org. Chem.* **42**, 1691 (1977).
238. P. Sartori and A. Golloch, *Chem. Ber.* **101**, 2004 (1968).
239. H. Müller, dissertation, University of Heidelberg (1964), cited in R. W. Hoffmann, *Dehydrobenzene and Cycloalkynes*, Academic Press, New York (1967), p. 77.
240. G. Wittig and F. Bickelhaupt, *Chem. Ber.* **91**, 883 (1958).
241. E. K. Fields and S. Meyerson, *J. Am. Chem. Soc.* **88**, 3388 (1966).
242. E. K. Fields and S. Meyerson, *J. Am. Chem. Soc.* **89**, 3224 (1967).
243. E. K. Fields and S. Meyerson, *J. Org. Chem.* **32**, 3114 (1967).
244. E. K. Fields and S. Meyerson, *J. Am. Chem. Soc.* **88**, 21 (1966).
245. H. Schüler and E. Lutz, *Z. Naturforsch. A* **16**, 57 (1961).
246. E. K. Fields and S. Meyerson, *Adv. Phys. Org. Chem.* **6**, 1 (1968).
247. R. A. Abramovitch, J. Roy, and V. Uma, *Can. J. Chem.* **43**, 3407 (1965).
248. R. A. Abramovitch and G. N. Knaus, *J. Chem. Soc. Chem. Commun.*, 238 (1974).
249. E. K. Fields and S. Meyerson, *Tetrahedron Lett.*, 571 (1967).
250. R. R. Jones and R. G. Bergman, *J. Am. Chem. Soc.* **94**, 660 (1972).
251. H.-F. Grützmacher and J. Lohmann, *Justus Liebigs Ann. Chem.* **705**, 81 (1967).
252. R. W. Hoffmann, *Dehydrobenzene and Cycloalkynes*, Academic Press, New York (1967), Chapt. 2.
253. J. F. Bunnett and R. E. Zahler, *Chem. Rev.* **49**, 273 (1951). Bunnett and Zahler originally defined a "cine substitution" as one "in which the ring position taken by the entering group is not the same as that vacated by the displaced group." By usage and convention [V. Gold, Glossary of terms used in physical organic chemistry, *Pure Appl. Chem.* **51**, 1725 (1979)],

the term *cine* has been restricted only to substitution at the position adjacent to the leaving group (previously vicinal cine-substitution) while substitution at positions more remote is referred to as tele-substitution.

254. J. Miller, *Nucleophilic Aromatic Substitution*, Elsevier Publishing, New York (1968), Chap. 3.

255. J. D. Roberts, C. W. Vaughn, L. A. Carismith, and D. A. Semenov, *J. Am. Chem. Soc.* **78**, 611 (1956).

256. M. Panar and J. D. Roberts, *J. Am. Chem. Soc.* **82**, 3629 (1960).

257. H. J. Shine, *Aromatic Rearrangements*, Elsevier Publishing, New York (1967), p. 326.

258. J. H. Wotiz and F. Huba, *J. Org. Chem.* **24**, 595 (1959).

259. M. G. Reinecke, *Am. Chem. Soc. Div. Petrol. Chem. Preps.* **14**(2), C68 (1969).

260. M. G. Reinecke, H. W. Adickes, and C. Pyun, *J. Org. Chem.* **36**, 2690 (1971).

261. M. G. Reinecke, H. W. Adickes, and C. Pyun, *J. Org. Chem.* **36**, 3820 (1971).

262. M. G. Reinecke and T. A. Hollingworth, *J. Org. Chem.* **37**, 4257 (1972).

263. H. C. van der Plas, D. A. de Bie, G. Geursten, M. G. Reinecke, and H. W. Adickes, *Rec. Trav. Chim. Pays-Bas* **93**, 33 (1974).

264. D. A. de Bie and H. C. van der Plas, *Tetrahedron Lett.*, 3905 (1968).

265. D. A. de Bie and H. C. van der Plas, *Rec. Trav. Chim. Pays-Bas* **88**, 1246 (1969).

266. D. A. de Bie, H. C. van der Plas, G. Geurtsen, and K. Nijdam, *Rec. Trav. Chim. Pays-Bas* **92**, 245 (1973).

267. J. F. Bunnett, *Acc. Chem. Res.* **5**, 139 (1972).

268. J. F. Bunnett and C. E. Moyer, Jr., *J. Am. Chem. Soc.* **92**, 1183 (1971), and following papers.

269. R. A. Benkeser and G. Schroll, *J. Am. Chem. Soc.* **75**, 3196 (1953).

270. (a) T. Kauffmann, R. Nürnberg, and K. Udluft, *Angew. Chem. Int. Ed. Engl.* **7**, 617 (1968); (b) T. Kauffmann, R. Nürnberg, and K. Udluft, *Chem. Ber.* **102**, 1177 (1969).

271. S. Senda, K. Hirota, and T. Asao, *J. Org. Chem.* **40**, 353 (1975).

272. G. Guanti, S. Thea, and C. Dell'Erba, *Tetrahedron Lett.*, 461 (1976); M. Novi, G. Guanti, S. Thea, F. Sancassan, and D. Calabro, *Tetrahedron* **35**, 1783 (1979).

273. (a) D. Bryce-Smith, A. Gilbert, and S. Krestonosich, *J. Chem. Soc. Chem. Commun.*, 405 (1976); (b) R. Nasielski-Hinkens, D. Pauwels, and J. Nasielski, *Tetrahedron Lett.*, 2125 (1978); (c) R. E. Markwell, *J. Chem. Soc. Chem. Commun.*, 428 (1979); (d) D. P. Self, D. E. West, and M. R. Stillings, *J. Chem. Soc. Chem. Commun.*, 281 (1980).

274. H. G. Viehe, personal communication quoted in R. W. Hoffmann, in *Chemistry of Acetylenes*, H. G. Viehe, Ed., Marcel Dekker, New York (1969), Chap. 16, p. 1073.

275. J. A. Zoltewicz, T. M. Oestreich, J. K. O'Halloran, and L. S. Helmick, *J. Org. Chem.* **38**, 1949 (1973).

276. H. C. van der Plas, M. Wozniak, and A. van Veldhuizen, *Rec. Trav. Chim. Pays-Bas* **96**, 151 (1977).

277. A. Albert, *Angew. Chem. Int. Ed. Engl.* **6**, 919 (1967).

278. G. S. Rork and I. H. Pitman, *J. Am. Chem. Soc.* **97**, 5559 (1975); J. W. Triplett, G. A. Digenis, W. J. Layton, and S. L. Smith, *J. Org. Chem.* **43**, 4411 (1978).

279. J. A. Zoltewicz and J. K. O'Halloran, *J. Am. Chem. Soc.* **97**, 5531 (1975).

280. G. S. Rork and I. H. Pitman, *J. Am. Chem. Soc.* **97**, 5566 (1975).

281. F. A. Sedor, D. G. Jacobson, and E. G. Sander, *J. Am. Chem. Soc.* **97**, 5572 (1975).

282. D. E. Klinge and H. C. van der Plas, *Rec. Trav. Chim. Pays-Bas* **94**, 233 (1975).

283. A. A. Morton, *Solid Organoalkali Metal Reagents*, Gordon and Breach Science Publishers, New York (1964), p. 123.

284. A. A. Morton, *J. Org. Chem.* **21**, 593 (1956).

285. J. de Valk and H. C. van der Plas, *Rec. Trav. Chim. Pays-Bas* **90**, 1239 (1971); H. C. van der Plas, *Acc. Chem. Res.* **11**, 462 (1978).

286. C. A. H. Rasmussen and H. C. van der Plas, *Rec. Trav. Chim. Pays-Bas* **96**, 101 (1977); C.

A. H. Rasmussen, H. C. van der Plas, P. Grotenhuis, and A. Koudijs, *J. Heterocycl. Chem.* **15**, 1121 (1978); C. A. H. Rasmussen and H. C. van der Plas, *Rec. Trav. Chim. Pays-Bas* **98**, 5 (1979).

287. H. C. van der Plas, P. Smit, and A. Koudijs, *Tetrahedron Lett.*, 9 (1968).

288. (a) R. W. Hoffmann, *Dehydrobenzene and Cycloalkynes*, Academic Press, New York (1967), p. 276; (b) T. Kauffmann, R. Nürnberg, and R. Wirthwein, *Chem. Ber.* **102**, 1161 (1969).

289. J. A. Zoltewicz and C. Nisi, *J. Org. Chem.* **34**, 765 (1969).

290. G. Wittig and L. Pohmer, *Angew. Chem.* **67**, 348 (1955).

291. G. Wittig and W. Herwig, *Chem. Ber.* **87**, 1511 (1954).

292. G. Wittig and H. Härle, *Justus Liebigs Ann. Chem.* **623**, 17 (1959).

293. N. Kharasch, T. G. Alston, H. B. Lewis, and W. Wolf, *Chem. Commun.* 242 (1965).

294. R. W. Hoffmann, *Dehydrobenzene and Cycloalkynes*, Academic Press, New York (1967), pp. 109–111, 203.

295. T. L. Gilchrist, E. E. Nunn, and C. W. Rees, *J. Chem. Soc. Perkin Trans. 1*, 1262 (1974); J. W. Barton and D. J. Lapham, *Tetrahedron Lett.*, 3571 (1979).

296. P. J. Garratt, *Pure Appl. Chem.* **44**, 738 (1975).

297. P. J. Garratt and K. P. C. Vollhardt, *J. Am. Chem. Soc.* **94**, 7087 (1972).

298. P. J. Garratt and S. B. Neoh, *J. Org. Chem.* **40**, 970 (1975).

299. G. Wittig and R. Ludwig, *Angew. Chem.* **68**, 40 (1956).

300. G. Wittig, *Angew. Chem.* **69**, 245 (1957).

301. R. W. Hoffmann, *Dehydrobenzene and Cycloalkynes*, Academic Press, New York (1967), Chap. 3.

302. M. Jones, Jr. and R. H. Levin, *J. Am. Chem. Soc.* **91**, 6411 (1969).

303. R. W. Hoffmann, *Dehydrobenzene and Cycloalkynes*, Academic Press, New York (1967), p. 303.

304. A. Albert, *Heterocyclic Chemistry*, 2nd ed., Oxford University Press, New York (1968).

305. E. McNelis, *J. Org. Chem.* **30**, 4324 (1965).

306. G. Wittig and E. Knauss, *Chem. Ber.* **91**, 895 (1958).

307. J. Sauer, *Angew. Chem. Int. Ed. Engl.* **5**, 211 (1966).

308. T. Zinke and H. Günther, *Justus Liebigs Ann. Chem.* **272**, 243 (1893).

309. M. A. Ogliaruso, M. G. Romanelli, and E. I. Becker, *Chem. Rev.* **65**, 261 (1965).

310. K. Rasheed, *Tetrahedron* **22**, 2957 (1966).

311. G. Wittig, E. Knauss, and K. Niethammer, *Justus Liebigs Ann. Chem.* **630**, 10 (1960).

312. G. Wittig and E. R. Wilson, *Chem. Ber.* **98**, 451 (1965).

313. G. Wittig, J. Weinlich, and E. R. Wilson, *Chem. Ber.* **98**, 458 (1965).

314. J. Sauer, A. Mielert, D. Lang, and D. Peter, *Chem. Ber.* **98**, 1435 (1965).

315. J. Sauer, *Angew. Chem. Int. Ed. Engl.* **6**, 16 (1967).

316. G. Seitz and T. Kämpchen, *Chem. Ztg.* **99**, 292 (1975).

317. D. N. Reinhoudt and C. G. Kouwenhoven, *Rec. Trav. Chim. Pays-Bas* **93**, 321 (1974).

318. G. Seitz and T. Kämpchen, *Arch. Pharm. (Weinheim)* **311**, 728 (1978).

319. H. Gilman and R. D. Gorsich, *J. Am. Chem. Soc.* **79**, 2625 (1957).

320. A. A. Morton, *Solid Organoalkali Metal Reagents*, Gordon and Breach Science Publishers, New York (1964), p. 122.

321. W. Tochterman, K. Oppenländer, and U. Walter, *Chem. Ber.* **97**, 1318 (1964).

322. J. D. Cook and B. J. Wakefield, *J. Chem. Soc. C*, 1973 (1969).

323. D. L. Brydon, J. I. G. Cadogan, D. M. Smith, and J. B. Thomson, *Chem. Commun.*, 727 (1967).

324. (a) K. L. Shepard, *Tetrahedron Lett.*, 3371 (1975); (b) D. G. Gillespie, B. J. Walker, and D. Stevens, *Tetrahedron Lett.*, 1905 (1976).

325. H. M. Walborsky and T. Bohnert, *J. Org. Chem.* **33**, 3934 (1968).

326. L. W. Butz and A. W. Rytina, *Org. React.* **5**, 136 (1949); S. L. Agarwal and S. K. Jain, *Synthesis*, 437 (1978).
327. G. Wittig and H. Heyn, *Chem. Ber.* **97**, 1609 (1964).
328. G. Wittig and J. Heyn, *Justus Liebigs Ann. Chem.* **726**, 57 (1969).
329. E. K. Fields and S. Meyerson, *Acc. Chem. Res.* **2**, 273 (1969).
330. B. H. Klanderman, *J. Am. Chem. Soc.* **87**, 4649 (1965).
331. A. Oku and A. Matsui, *Bull. Chem. Soc. Japan* **50**, 3338 (1977).
332. M. S. Newman and R. Kannan, *J. Org. Chem.* **41**, 3356 (1976).
333. P. Jayalekshmy and S. Mazur, *J. Am. Chem. Soc.* **98**, 6710 (1976); S. Mazur and P. Jayalekshmy, *J. Am. Chem. Soc.* **101**, 677 (1979).
334. J. Rebek, Jr., and F. Gavina, *J. Am. Chem. Soc.* **97**, 3453 (1975); J. Rebek, Jr., *Tetrahedron* **35**, 723 (1979).
335. N. D. Epiotis, W. R. Cherry, F. Bernardi, and W. J. Hehre, *J. Am. Chem. Soc.* **98**, 4361 (1976).
336. R. Stoermer, *Justus Liebigs Ann. Chem.* **312**, 237 (1900).
337. R. Stoermer, *Justus Liebigs Ann. Chem.* **313**, 79 (1900).
338. H. Gilman and D. S. Melstrom, *J. Am. Chem. Soc.* **70**, 1655 (1948).
339. H. Boos, unpublished results cited in G. Wittig, *Pure Appl. Chem.* **7**, 173 (1963).
340. G. Wittig and H. Boos, unpublished results cited in R. W. Hoffmann, *Dehydrobenzene and Cycloalkynes*, Academic Press, New York (1967), p. 305.
341. G. Wittig and H. Boos, unpublished results cited in G. Wittig and M. Rings, *Justus Liebigs Ann. Chem.* **719**, 127 (1968).
342. H. Gilman and R. D. Gorsich, *J. Am. Chem. Soc.* **78**, 2217 (1956).
343. M. Cugon de Sevricourt and M. Robba, *Bull. Soc. Chim. Fr.*, 142 (1977).
344. G. Wittig and G. Kolb, unpublished results cited in G. Wittig, *Angew. Chem. Int. Ed. Engl.* **1**, 415 (1962).
345. G. Wittig and G. Kolb, unpublished results cited in R. W. Hoffmann, *Dehydrobenzene and Cycloalkynes*, Academic Press, New York (1967), p. 291.
346. G. Kolb, dissertation, University of Heidelberg (1959), cited in G. Wittig and M. Rings, *Justus Liebigs Ann. Chem.* **719**, 127 (1968).
347. G. Wittig, *Angew. Chem. Int. Ed.* **1**, 415 (1962).
348. G. Wittig and H. Boos, unpublished results cited in R. W. Hoffmann, *Dehydrobenzene and Cycloalkynes*, Academic Press, New York (1967), p. 291.
349. H. Gilman and G. F. Wright, *J. Am. Chem. Soc.* **55**, 3302 (1933).
350. L. D. Tarasova and Y. L. Gol'dfarb, *Bull. Acad. Sci. USSR Div. Chem. Sci.*, 1978 (1965).
351. R. Sornay, J. Meunier, and P. Fournari, *Bull. Soc. Chim. Fr.*, 990 (1971).
352. G. M. Davies and P. S. Davies, *Tetrahedron Lett.*, 3507 (1972).
353. M. Blumenthal, *Chem. Ber.* **7**, 1092 (1874).
354. K. Dziewonski, *Chem. Ber.* **36**, 962 (1903).
355. J. I. Jones, *Chem. Commun.*, 938 (1967).
356. M. E. Synerholm, *J. Am. Chem. Soc.* **67**, 1229 (1945).
357. H.-D. Scharf, H. Leismann, and H. Lechner, *Chem. Ber.* **104**, 847 (1971).
358. T. Matsuo and S. Mihara, *Bull. Chem. Soc. Japan* **48**, 3660 (1975).
359. A. Michael and J. E. Bucher, *Chem. Ber.* **28**, 2511 (1895); R. N. Misra and S. Dutt, *J. Ind. Chem. Soc.* **13**, 98 (1936); R. N. Misra and S. Dutt, *J. Ind. Chem. Soc.* **13**, 98 (1936).
360. L. Crombie, P. A. Gilbert, and R. P. Houghton, *J. Chem. Soc. C*, 130 (1968).
361. J. L. Ripoll, P. Blickle, and H. Hopf, unpublished results cited in G. Weber, K. Menke, and H. Hopf, *Chem. Ber.* **113**, 531 (1980).
362. A. Szabo, dissertation, University of Tübingen (1975).
363. (a) J. LeFevre and J. Hamelin, *Tetrahedron Lett.*, 1757 (1979); (b) G. Rosskamp and H. Suhr, *Justus Liebigs Ann. Chem.* 1478 (1975).

364. H. Müller, dissertation, University of Heidelberg (1964) cited in G. Wittig, *Pure Appl. Chem.* **7**, 173 (1963).
365. H. Müller, dissertation, University of Heidelberg (1964), cited in R. W. Hoffmann, *Dehydrobenzene and Cycloalkynes*, Academic Press, New York (1967), p. 291.
366. H. Müller, dissertation, University of Heidelberg (1964), cited in G. Wittig and M. Rings, *Justus Liebigs Ann. Chem.* **719**, 127 (1968).
367. H. Müller, dissertation, University of Heidelberg (1964), cited in R. W. Hoffmann, *Dehydrobenzene and Cycloalkynes*, Academic Press, New York (1967), p. 289.
368. H. Müller, dissertation, University of Heidelberg (1964), cited in R. W. Hoffmann, *Dehydrobenzene and Cycloalkynes*, Academic Press, New York (1967), p. 305.
369. H. Müller, dissertation, University of Heidelberg (1964), cited in R. W. Hoffmann, *Dehydrobenzene and Cycloalkynes*, Academic Press, New York (1967), p. 299.
370. H. Müller, dissertation, University of Heidelberg (1964), cited in R. W. Hoffmann, *Dehydrobenzene and Cycloalkynes*, Academic Press, New York (1967), p. 295.
371. R. W. Hoffmann, *Dehydrobenzene and Cycloalkynes*, Academic Press, New York (1967), (a) p. 295; (b) pp. 135–145.
372. D. A. Shirley and P. A. Roussel, *J. Am. Chem. Soc.* **75**, 375 (1953).
373. (a) J. Bergman and N. Eklund, *Tetrahedron* **36**, 1439 (1980); (b) J. Bergman and N. Eklund, *Tetrahedron* **36**, 1445 (1980).
374. H. Müller, dissertation, University of Heidelberg (1964), cited in J. Bergman and N. Eklund, *Tetrahedron* **36**, 1439 (1980).
375. M. P. Cava and L. Bravo, *Tetrahedron Lett.*, 4631 (1970).
376. R. F. C. Brown, W. D. Crow, and R. K. Solly, *Chem. Ind. (London)*, 343 (1966); M. P. Cava and L. Bravo, *Chem. Commun.*, 1538 (1968).
377. M. M. King, *Ann. Rep. Petr. Res. Found.* **18**, 117 (1973).
378. R. H. Brown, dissertation, New York University (1975); *Diss. Abstr.* **36**, 5586B (1976).
379. D. Kumfer, *Anal. Biochem.* **8**, 75 (1964).
380. R. A. J. Clark and J. Parrick, *Org. Mass Spectrom.* **12**, 421 (1977).
381. P. Mencarelli and F. Stegel, *J. Chem. Soc. Chem. Commun.*, 564 (1978); L. Bonaccina, P. Mencarelli, and F. Stegel, *J. Org. Chem.* **44**, 4420 (1979); P. Mencarelli and F. Stegel, *J. Chem. Soc. Chem. Commun.*, 123 (1980).
382. W. Draber, *Chem. Ber.* **100**, 1559 (1967).
383. J. F. Liebman and A. Greenberg, *Chem. Rev.* **76**, 311 (1976).
383a. K. M. Wald, A. A. Nada, G. Szilagyi, and H. Wamhoff, *Chem. Ber.* **113**, 2884 (1980).
384. V. Wahl, dissertation, University of Heidelberg (1962), cited in R. W. Hoffmann, *Dehydrobenzene and Cycloalkynes*, Academic Press, New York (1967), p. 292.
385. M. D. Rausch, T. R. Criswell, and A. K. Ignatowicz, *J. Organometal. Chem.* **13**, 419 (1968).
386. V. Wahl, dissertation, University of Heidelberg (1962), cited in G. Wittig and M. Rings, *Justus Liebigs Ann. Chem.* **719**, 127 (1968).
387. R. W. Hoffmann, *Dehydrobenzene and Cycloalkynes*, Academic Press, New York (1967), p. 196.
388. H. Wynberg, H. van Driel, R. M. Kellog, and J. Buter, *J. Am. Chem. Soc.* **89**, 3487 (1967).
389. M. Rings, dissertation, University of Heidelberg (1966), cited in G. Wittig and M. Rings, *Justus Liebigs Ann. Chem.* **719**, 127 (1968).
390. L. Friedman and D. F. Lindow, *J. Am. Chem. Soc.* **90**, 2329 (1968).
391. I. Tabushi, H. Yamada, Z. Yoshida, and R. Oda, *Bull. Chem. Soc. Japan* **50**, 285 (1977).
392. S.-O. Lawesson, *Ark. Kemi* **11**, 373 (1957).
393. W. Steinkopf, *Justus Liebigs Ann. Chem.* **543**, 128 (1940).
394. S. Gronowitz, P. Moses, and R. Hakansson, *Ark. Kemi* **16**, 267 (1960).
395. P. Moses and S. Gronowitz, *Ark. Kemi* **18**, 119 (1961).

396. B. Ostman, *Ark. Kemi* **22**, 551 (1964).

397. S. Gronowitz and U. Rosen, *Chem. Scr.* **1**, 33 (1971).

398. S. Gronowitz and U. Rosen, unpublished results cited in S. Gronowitz and B. Holm, *Acta Chem. Scand.* **23**, 2207 (1969).

399. S. Gronowitz and B. Holm, *Acta Chem. Scand.* **23**, 2207 (1969).

400. S. Gronowitz and B. Holm, *Acta Chem. Scand. Ser. B* **30**, 505 (1976).

401. H. Christiansen, S. Gronowitz, B. Rodmar, S. Rodmar, U. Rosen, and M. K. Sharma, *Ark. Kemi* **30**, 561 (1968).

402. B. E. Ayers, S. W. Longworth, and J. F. W. McOmie, *Tetrahedron* **31**, 1755 (1975).

403. M. C. Ford and D. Mackay, *J. Chem. Soc.*, 4985 (1956).

404. A.-B. Hörnfeldt and P.-O. Sundberg, *Acta Chem. Scand.* **26**, 31 (1972).

405. L. Schmerling, *J. Am. Chem. Soc.* **67**, 1438 (1945).

406. H. W. Adickes, dissertation, Texas Christian University (1968).

407. L. van der Does and H. J. den Hertog, *Rec. Trav. Chim. Pays-Bas* **91**, 1403 (1972).

408. J. F. Bunnett and B. F. Gloor, *Heterocycles* **5**, 377 (1976).

409. C. Dell'Erba, D. Spinelli, and G. Leandri, *Gazz. Chim. Ital.* **99**, 535 (1969).

410. M. Novi, G. Guanti, C. Dell'Erba, and D. Spinelli, *J. Chem. Soc. Perkin Trans. 1*, 2264 (1976).

411. M. Novi, G. Guanti, F. Sancassan, and C. Dell'Erba, *J. Chem. Soc. Perkin Trans. 1*, 1140 (1978).

412. R. H. Walter, dissertation, Texas Christian University (1974).

413. Farbwerke Hoechst Aktiengesellschaft, Brit. Pat., 837,086 (1960); *Chem. Abstr.* **54**, 24798e (1960).

414. M. G. Reinecke and H. H. Ballard, *Abstracts of the 159th National Meeting of the American Chemical Society*, Houston, TX, American Chemical Society, ORGN 12 (1970).

415. M. G. Reinecke and R. H. Walter, *J. Chem. Soc. Chem. Commun.*, 1044 (1974).

416. H. Zollinger, *Azo and Diazo Chemistry*, Interscience, New York (1961), p. 239.

417. R. Promel, A. Cardon, M. Daniel, G. Jacques, and A. Vandersmissen, *Tetrahedron Lett.*, 3067 (1968).

418. M. Tiecco and A. Tundo, *Int. J. Sulfur Chem.* **8**, 295 (1973).

419. G. Vernin, *Bull. Soc. Chim. Fr.*, 1257 (1976).

420. G. Martelli, P. Spagnolo, and M. Tiecco, *J. Chem. Soc. B*, 901 (1968).

421. D. Mackay, *Can. J. Chem.* **44**, 2881 (1966).

422. K. Undheim, O. Thorstad, and G. Hvistendahl, *Org. Mass Spectrom.* **5**, 73 (1971).

423. M. Bartle, S. T. Gore, R. K. Mackie, and J. M. Tedder, *J. Chem. Soc. Perkin Trans. 1*, 1636 (1976); S. T. Gore, R. K. Mackie, and J. M. Tedder, *J. Chem. Soc. Perkin Trans. 1*, 1639 (1976).

424. The photolysis of 2,3-diiodothiophene (**360**) and 2,3-diiodothianaphthene (**466**, Br=I) has also been studied with a view toward generating the arynes **4** and **26**, respectively, but no results have ever been reported, apparently; H. Wynberg, H. van Driel, R. M. Kellogg, and J. Buter, *J. Am. Chem. Soc.* **89**, 3487 (1967).

425. E. K. Fields and S. Meyerson, *J. Org. Chem.* **35**, 67 (1970).

426. E. K. Fields and S. Meyerson, *Abstracts of the Fourth International Congress of Heterocyclic Chemistry*, Salt Lake City, 1973, p. 284.

427. H. Wynberg and A. Bantjes, *J. Org. Chem.* **24**, 1421 (1959).

428. C. D. Hurd, R. V. Levetan, and A. R. Macon, *J. Am. Chem. Soc.* **84**, 4515 (1962).

428a. D. Del Mazza and M. G. Reinecke, *J. Chem. Soc. Chem. Commun.* 124 (1981).

428b. M. R. Bryce and J. M. Vernon, *Adv. Heterocycl. Chem.* **28**, 183 (1981).

429. D. Del Mazza and M. G. Reinecke, *Heterocycles* **14**, 647 (1980).

430. D. D. Callander, P. L. Coe, J. C. Tatlow, and A. J. Uff, *Tetrahedron* **25**, 25 (1969).

431. S. Meyerson and E. K. Fields, *J. Org. Chem.* **33**, 847 (1968).

432. H. Wynberg, J. deWit, and H. J. M. Sinnige, *J. Org. Chem.* **35**, 711 (1970).
433. J. G. Newsom, dissertation, Texas Christian University (1977).
434. K. Tserng and L. Bauer, *J. Org. Chem.* **40**, 172 (1975).
435. D. W. H. MacDowell and F. L. Ballas, *J. Org. Chem.* **42**, 3717 (1977); D. Binder and P. Stanetty, *Synthesis*, 200 (1977).
436. M. G. Reinecke, J. G. Newsom, and K. A. Almqvist, *Synthesis*, 327 (1980).
437. G. Schaden, *Justus Liebigs Ann. Chem.* 559 (1978).
438. H.-F. Grützmacher and U. Straetmaus, *Tetrahedron* **36**, 807 (1980).
439. H. Wynberg and A. J. H. Klunder, *Rec. Trav. Chim. Pays-Bas* **88**, 328 (1969); J. deWit and H. Wynberg, *Tetrahedron* **29**, 1379 (1973).
440. W. J. Feast, W. K. R. Musgrave, and W. E. Preston, *J. Chem. Soc. Perkin Trans. 1*, 1830 (1972).
441. E. A. Braude, R. P. Linstead, and P. W. D. Mitchell, *J. Chem. Soc.*, 3578 (1954).
442. R. Weiss, *Org. Synth. Coll. Vol.* **3**, 729 (1955).
443. E. A. Braude, L. M. Jackman, and R. P. Linstead, *J. Chem. Soc.*, 3548 (1954).
444. H. von Brachel and U. Bahr, in *Houben-Weyl, Methoden der Organischen Chemie*, Vol. 5, E. Muller, Ed., George Theime Verlag, Stuttgart (1970), Part 1c, p. 967.
445. T. J. Barton, M. D. Martz, and R. G. Zika, *J. Org. Chem.* **37**, 552 (1972).
446. D. N. Reinhoudt, *Adv. Heterocycl. Chem.* **21**, 253 (1977).
447. M. C. Kloetzel, *Org. React.* **4**, 1 (1948).
448. T. Eicher and J. L. Weber, *Top. Curr. Chem.* **57**, 1 (1975).
449. K. T. Potts and J. S. Baum, *Chem. Rev.* **74**, 189 (1974).
450. B. Halton, *Chem. Rev.* **73**, 113 (1973).
451. H. H. Wasserman, G. M. Clark, and P. C. Turley, *Top. Curr. Chem.* **47**, 73 (1971).
452. A. Amano, T. Mukai, T. Nakazawa, and K. Okoyama, *Bull. Chem. Soc. Japan* **49**, 1671 (1976).
453. M. G. Reinecke, L.-J. Chen, and A. Almqvist, *J. Chem. Soc. Chem. Commun.* 585 (1980).
454. R. Wheland and P. D. Bartlett, *J. Am. Chem. Soc.* **92**, 6057 (1970); A. Padwa and P. H. J. Carlsen, *J. Am. Chem. Soc.* **99**, 1514 (1977); N. Shimizu and P. D. Bartlett, *J. Am. Chem. Soc.* **100**, 4260 (1978).
455. B. M. Trost and P. J. Whitman, *J. Am. Chem. Soc.* **96**, 7421 (1974).
456. T. Yonezawa, H. Konishi, and H. Kato, *Bull. Chem. Soc. Japan* **42**, 933 (1969).
457. (a) A. N. Korepanov, T. A. Danilova, and E. Z. Viktorova, *Chem. Heterocycl. Comp. (USSR)*, 868 (1977); (b) M. G. Reinecke, J. G. Newson, and K. A. Almqvist, *Tetrahedron* **37**, in press (1981).
458. G. Wittig and H. Dürr, *Justus Liebigs Ann. Chem.* **672**, 55 (1964).
459. J. A. Berson, P. B. Dervan, R. Malherbe, and J. A. Jenkins, *J. Am. Chem. Soc.* **98**, 5937 (1976).
460. P. Crews and J. Beard, *J. Org. Chem.* **38**, 522 (1973).
461. H. M. R. Hoffmann, *Angew. Chem. Int. Ed. Engl* **8**, 556 (1969).
462. R. W. Hoffman, *Dehydrobenzene and Cycloalkynes*, Academic Press, New York (1967), p. 197.
463. J. H. Wotiz, in *Chemistry of Acetylenes*, H. G. Viehe, Ed., Marcel Dekker, New York (1969), Chap. 7, p. 365.
464. T. Sakai and M. Hattori, *Chem. Lett.*, 617 (1974).
465. H. Suhr, *Angew. Chem. Int. Ed. Engl.* **11**, 781 (1972).
466. H. Suhr and G. Kruppa, *Justus Liebigs Ann. Chem.* **744**, 1 (1971).
467. P. E. Fjelstad and K. Undheim, *Acta Chem. Scand. Ser. B* **30**, 375 (1976).
468. H. Suhr, unpublished results cited in A. Szabo, dissertation, University of Tübingen (1975).
469. E. Sinkinson, *Ind. Eng. Chem.* **17**, 27 (1925).
470. R. Srinivasin, *J. Am. Chem. Soc.* **89**, 1758 (1967).

471. H. H. Wasserman and L. S. Keller, *Chem. Commun.*, 1483 (1970).
472. M. L. Deem, *Synthesis*, 675 (1972).
473. J. A. Berson and M. Pomerantz, *J. Am. Chem. Soc.* **86**, 3896 (1964).
474. H. Suhr and A. Szabo, *Justus Liebigs Ann. Chem.*, 342 (1975).
475. G. Martelli, P. Spagnolo, and M. Tiecco, *Chem. Commun.*, 282 (1969).
476. C. F. Cullis, D. J. Hucknall, and J. V. Shepherd, *Proc. R. Soc. Ser. A* **335**, 525 (1973).
477. S. Gronowitz and T. Frejd, *Acta Chem. Scand. Ser. B* **30**, 485 (1976).
478. (a) M. Rings, dissertation, University of Heidelberg (1966); (b) F. M. Stoyanovich, Y. L. Gol'dfarb, and G. B. Chermanova, *Bull. Acad. Sci. USSR, Div. Chem. Sci.*, 2228 (1973); (c) F. M. Stoyanovich, M. A. Marakatkina, and Y. L. Gol'dfarb, *Bull. Acad. Sci. USSR, Div. Chem. Sci.*, 2362 (1976).
479. Y. Matsuki and Y. Adachi, *Nippon Kagaku Zasshi* **89**, 192 (1968).
480. R. W. Hoffmann, *Dehydrobenzene and Cycloalkynes*, Academic Press, New York (1967), pp. 288–289.
481. M. G. Reinecke, W. B. Mohr, H. W. Adickes, D. A. deBie, H. C. van der Plas, and K. Nijdam, *J. Org. Chem.* **38**, 1365 (1973).
482. G. Komppa, *J. Prakt. Chem.* **122**, 319 (1929).
483. R. P. Dickinson and B. Iddon, *J. Chem. Soc. C*, 2733 (1968).
484. K. E. Chippendale, B. Iddon, H. Suschitzky, and D. S. Taylor, *J. Chem. Soc. Perkin Trans. 1*, 1168 (1974).
485. P. Grandclaudon and A. Lablache-Combier, *J. Org. Chem.* **43**, 4379 (1978).
486. W. Ried and H. Bender, *Chem. Ber.* **88**, 34 (1955).
487. R. D. Schuetz, D. D. Taft, J. P. O'Brien, J. L. Shea, and H. H. Mork, *J. Org. Chem.* **28**, 1420 (1963).
488. R. P. Dickinson and B. Iddon, *J. Chem. Soc. C*, 3447 (1971).
489. M. J. Cook, A. R. Katritzky, and P. Linda, *Adv. Heterocycl. Chem.* **17**, 255 (1974).
490. H. D. Hartough and S. L. Meisel, *Compounds with Condensed Thiophene Rings*, A. Weissberger, Ed., Wiley-Interscience, New York (1954), p. 159.
491. B. Iddon and R. M. Scrowston, *Adv. Heterocycl. Chem.* **11**, 177 (1970).
492. F. G. Bordwell and C. J. Albisetti, Jr., *J. Am. Chem. Soc.* **70**, 1558 (1948).
493. F. Sauter and U. Jordis, *Monatsh. Chem.* **105**, 1252 (1974).
494. A. H. Schlesinger and D. T. Mowry, *J. Am. Chem. Soc.* **73**, 2614 (1951).
495. Although the structure of this product has been questioned (footnote 29 in Reference 1) its nmr spectrum[496] is quite consistent with the assignment: $\delta = 7.7$, $m(4)$; 5.2, d, $J = 7$ Hz (1); 4.7, d, $J = 7$ Hz (1); 2.7, $m(4)$; 1.6, $m(6)$.
496. H. W. Adickes, unpublished results (1968).
497. T. A. Hollingworth, Jr., M. S. thesis, Texas Christian University (1967).
498. S. Gronowitz and T. Frejd, *Acta Chem. Scand. Ser. B* **30**, 439 (1976).
499. T. Frejd, *Chem. Scr.* **10**, 133 (1976).
500. T. Quang Minh, L. Christiaens, and M. Renson, *Bull. Soc. Chim. Fr.*, 2239 (1974).
501. A. I. Shatenshtein, A. G. Kamrad, I. O. Shapiro, Y. I. Ranneva, and E. N. Zvyagintseva, *Dokl. Chem.* **168**, 502 (1966).
502. T. L. Gilchrist and D. P. J. Pearson, *J. Chem. Soc. Perkin Trans. 1*, 989 (1976); S. Gronowitz and T. Frejd, *Chem. Heterocycl. Comp.* **14**, 353 (1978).
503. T. Quang Minh, L. Christiaens, and M. Renson, *Bull. Soc. Chim. Fr.*, 2244 (1974).
504. H. Boer and H. J. den Hertog, *Tetrahedron Lett.*, 1943 (1969).
505. H. N. M. van der Lans, H. J. den Hertog, and A. van Veldhuizen, *Tetrahedron Lett.*, 1875 (1971); G. E. Lewis, R. H. Prager, and R. H. M. Ross, *Aust. J. Chem.* **28**, 2057 (1975).
506. G. M. Sanders, M. van Dijk, and H. J. den Hertog, *Rec. Trav. Chim. Pays-Bas* **93**, 273 (1974).
507. H. W. van Meeteren and H. C. van der Plas, *Rec. Trav. Chim. Pays-Bas* **90**, 105 (1971).

508. T. Kauffmann, H. Fischer, R. Nürnberg, M. Vestweber, and R. Wirthwein, *Tetrahedron Lett.*, 2917 (1967).
509. M. R. Grimmett, *Adv. Heterocycl. Chem.* 12, 103 (1970).
510. B. A. Tertov and A. S. Morkovnik, *J. Org. Chem. USSR* 9, 2222 (1974).
511. G. M. Sanders, M. van Dijk, and H. J. den Hertog, *Tetrahedron Lett.*, 4717 (1972).
512. E. S. Hand and W. W. Paudler, *J. Org. Chem.* 43, 2900 (1978).
513. E. S. Hand and W. W. Paudler, *J. Org. Chem.* 41, 3549 (1976).
513. (a) R. Ketcham and E. Schaumann, *J. Org. Chem.* 45, 3748 (1980); (b) N. Jacobsen, P. de Mayo, and A. C. Weedon, *Nouv. J. Chim.* 2, 331 (1978); (c) J. P. Snyder, in *Organic Sulphur Chemistry*, C. J. M. Stirling, Ed., Butterworths, London (1974), p. 307; (d) H. E. Simmons, R. D. Vest, D. C. Blomstrom, J. R. Roland, and T. L. Cairns, *J. Am. Chem. Soc.* 84, 4746 (1962).
514. (a) C. W. Rees and R. C. Storr, *J. Chem. Soc. C*, 765 (1969); (b) P. Blatcher, D. Middlemiss, P. Murray-Rust, and J. Murray-Rust, *Tetrahedron Lett.*, 4193 (1980); (c) A. Galbraith, T. Small, R. A. Barnes, and V. Boekelheide, *J. Am. Chem. Soc.* 83, 453 (1961).
515. R. Fusco, V. Rosnati, and G. Pagani, *Tetrahedron Lett.*, 1739 (1966).
516. (a) M. Begtrup, *Acta Chem. Scand.* 24, 1819 (1970); (b) M. Begtrup, *Acta Chem. Scand.* 27, 2051 (1973).
517. (a) B. Stanovnik, M. Tišler, J. Bradač, B. Budič, B. Koren, and B. Mozetič-Reščič, *Heterocycles* 12, 457 (1979); (b) M. Bosco, L. Forlani, P. E. Todesco, and L. Troisi, *J. Chem. Soc. D*, 1093 (1971); L. Forlani, P. E. Todesco, and L. Troisi, *J. Chem. Soc. Perkin Trans. 2*, 1016 (1978).
518. M. Bosco, L. Forlani, P. E. Todesco, and L. Troisi, *J. Chem. Soc. Perkin Trans. 1*, 398 (1976).
519. Y. Tamura, T. Miyamoto, and M. Ikeda, *Chem. Ind. (London)*, 1439 (1971).
520. R. Motoyama, K. Sato, and E. Imoto, *Nippon Kagaku Zasshi* 78, 779 (1957); *Chem. Abstr.* 54, 22559f (1960).
521. May and Baker Ltd., Belg. Pat. 629,580; *Chem. Abstr.* 60, 15875f (1964).
522. T. Naito, S. Nakagawa, and K. Takahashi, *Chem. Pharm. Bull. (Tokyo)* 16, 148 (1968).
523. J. K. Kim and J. F. Bunnett, *J. Am. Chem. Soc.* 92, 7463 (1970).
524. R. E. Smith, dissertation, University of North Carolina (1966).
525. R. L. McKee, University of North Carolina, personal communication (1977).
526. L. M. Weinstock and P. I. Pollok, *Adv. Heterocycl. Chem.* 9, 107 (1968).
527. C. W. Bird and C. K. Wong, *Tetrahedron Lett.*, 2143 (1971).
528. S. Uemura, A. Toshimitsu, and M. Okano, *J. Chem. Soc. Perkin Trans. 1*, 1076 (1978).
529. H. Gilman, H. B. Willis, and J. Swislowsky, *J. Am. Chem. Soc.* 61, 1371 (1939).
530. M. Julia, Y. Huang, and J. Igolen, *C. R. Acad. Sci. Ser. C* 265, 110 (1967).
531. J. Igolen and A. Kolb, *C. R. Acad. Sci. Ser. C* 269, 54 (1969).
532. M. Julia, J. Igolen, and A. Kolb, *C. R. Acad. Sci. Ser. C* 273, 1776 (1971).
533. D. J. Cram and A. C. Day, *J. Org. Chem.* 31, 1227 (1966).
534. M. Julia, F. LeGoffic, J. Igolen, and M. Baillarge, *C. R. Acad. Sci. Ser. C* 264, 118 (1967); M. Julia, F. LeGoffic, J. Igolen, and M. Baillarge, *Bull. Soc. Chim. Fr.*, 1071 (1968).
535. M. Julia, F. LeGoffic, J. Igolen, and M. Baillarge, *Tetrahedron Lett.*, 1569 (1969).
536. R. C. Perera and R. K. Smalley, *J. Chem. Soc. D*, 1458 (1970).
537. L. DiNunno, S. Florio, and P. E. Todesco, *Tetrahedron* 30, 863 (1974).
538. L. DiNunno, S. Florio, and P. E. Todesco, *Tetrahedron* 32, 1037 (1976); L. DiNunno and S. Florio, *Tetrahedron* 33, 1523 (1977).
539. L. DiNunno, S. Florio, and P. E. Todesco, *J. Chem. Soc. Perkin Trans. 2*, 1171 (1974).
540. Y. Matsuki and F. Shoji, *Nippon Kagaku Zasshi* 86, 1067 (1965); *Chem. Abstr.* 65, 13638g (1966); Y. Matsuki and I. Ito, *Nippon Kagaku Zasshi* 88, 758 (1967); *Chem. Abstr.* 69, 59018 (1968).

541. F. Ciminale, G. Bruno, L. Testaferri, M. Tiecco, and G. Martelli, *J. Org. Chem.* **43**, 4509 (1978).

542. R. P. Dickinson and B. Iddon, *J. Chem. Soc. C*, 182 (1971).

543. E. Waldvogel, G. Schwarb, J. M. Bastian, and J. P. Bourquin, *Helv. Chim. Acta* **59**, 866 (1976).

544. (a) O. L. Chapman, R. S. Sheridan, and J.-P. LeRoux, *Rec. Trav. Chim. Pays-Bas* **98**, 334 (1979); (b) J. Rigaudy, C. Igier, and J. Barcelo, *Tetrahedron Lett.*, 1837 (1979); I. R. Dunkin and P. C. P. Thomson, *J. Chem. Soc., Chem. Commun.*, 499 (1980); (c) P. T. Gallagher, B. Iddon, and H. Suschitzky, *J. Chem. Soc. Perkin Trans. 1*, 2362 (1980).

545. J. W. Huffman, L. H. Keith, and R. L. Asbury, *J. Org. Chem.* **30**, 1600 (1965).

546. F. L. Hedberg and H. Rosenberg, *Tetrahedron Lett.*, 4011 (1969).

547. J. F. Eastham and G. W. Gibson, *J. Am. Chem. Soc.* **585**, 2171 (1963).

548. M. Walczak, K. Walczak, R. Mink, M. D. Rausch, and G. Stucky, *J. Am. Chem. Soc.* **100**, 6382 (1978).

549. D. W. Slocum, R. L. Marchal, and W. E. Jones, *J. Chem. Soc. Chem. Commun.*, 967 (1974).

550. J. W. Huffman and J. F. Cope, *J. Org. Chem.* **36**, 4068 (1971).

551. A. N. Nesmeyanov, B. A. Sazonova, and N. S. Sazonova, *Dokl. Chem.* **176**, 843 (1967).

552. E. W. Neuse and L. Bednarik, *Macromolecules* **12**, 187 (1979).

553. J. C. Martin and D. R. Bloch, *J. Am. Chem. Soc.* **93**, 451 (1971).

554. T. Kauffmann and H. Marhan, *Chem. Ber.* **96**, 2519 (1963).

555. R. A. Abramovitch and G. M. Singer, *J. Org. Chem.* **39**, 1795 (1974).

556. Y. Apeloig, P. v. R. Schleyer, and J. A. Pople, *J. Am. Chem. Soc.* **99**, 1291 (1977).

557. C. L. Habraken and E. K. Poels, *J. Org. Chem.* **42**, 2893 (1977); P. Cohen-Fernandes, C. Erkelens, C. G. M. van Eendenburg, J. J. Verhoeven, and C. L. Habraken, *J. Org. Chem.* **44**, 4156 (1979).

558. C. G. Swain, J. E. Sheats, and K. G. Harbison, *J. Am. Chem. Soc.* **97**, 783 (1975); L. R. Subramanian, M. Hanack, L. W. K. Chang, M. A. Imhoff, P. v. R. Schleyer, F. Effenberger, W. Kurtz, P. J. Stang, and T. E. Dueber, *J. Org. Chem.* **41**, 4099 (1976).

559. R. N. Butler, *Chem. Rev.* **75**, 241 (1975).

560. J. M. Tedder, *Adv. Heterocycl. Chem.* **8**, 1 (1967).

561. M. Tišler and B. Stanovnik, *Heterocycles* **4**, 1115 (1976).

562. R. O. Duthaler, H. G. Forster, and J. D. Roberts, *J. Am. Chem. Soc.* **100**, 4974 (1978).

563. G. F. Koser, *J. Org. Chem.* **42**, 1474 (1977); R. A. Abramovitch and J. G. Saha, *Can. J. Chem.* **43**, 3269 (1965); R. W. Taft, *J. Am. Chem. Soc.* **83**, 3350 (1961).

564. W. A. Sheppard and O. W. Webster, *J. Am. Chem. Soc.* **95**, 2695 (1973).

565. P. Gronski and K. Hartke, *Chem. Ber.* **111**, 272 (1978).

566. M. Kocevar, B. Stanovnik, and M. Tisler, *Heterocycles* **6**, 681 (1977).

566a. W. A. Sheppard, G. W. Gokel, O. W. Webster, K. Betterton, and J. W. Timberake, *J. Org. Chem.* **44**, 1717 (1979).

566b. M. Kocevar, B. Stanovnik, and M. Tisler, *Heterocycles* **6**, 681 (1977).

567. W. L. Magee and H. Shechter, *J. Am. Chem. Soc.* **99**, 633 (1977).

568. H. Durr and H. Schmitz, *Chem. Ber.* **111**, 2258 (1978).

569. P. B. Shevlin, *J. Am. Chem. Soc.* **94**, 1379 (1972); *Reactive Intermediates*, R. A. Abramovitch, Ed., Vol. I, Plenum Press, New York (1980), Chap. 1.

570. K. L. Kirk and L. Cohen, *J. Am. Chem. Soc.* **95**, 4619 (1973).

571. R. Huisgen, W. Mack, and L. Mobius, *Tetrahedron* **9**, 29 (1960).

572. M. J. S. Dewar and W.-K. Li, *J. Am. Chem. Soc.* **96**, 5569 (1974).

573. M. J. S. Dewar and G. P. Ford, *J. Chem. Soc. Chem. Commun.*, 539 (1977).

574. V. Schomaker and L. Pauling, *J. Am. Chem. Soc.* **61**, 1769 (1939).

575. B. Bak, D. Christensen, L. Hansen-Nygaard, and J. Rastrup-Andersen, *J. Mol. Spectrosc.* **7**, 58 (1961).

576. B. Bak, D. Christensen, L. Hansen, and J. Rastrup-Andersen, *J. Chem. Phys.* **24**, 720 (1956).
577. B. Bak, D. Christensen, W. B. Dixon, L. Hansen-Nygaard, J. Rastrup-Andersen, and M. Schottlander, *J. Mol. Spectrosc.* **9**, 124 (1962).
578. D. Spinelli, G. Guanti, and C. Dell'Erba, *J. Heterocycl. Chem.* **5**, 323 (1968).
579. D. Spinelli, R. Noto, G. Consiglio, and A. Storace, *J. Chem. Soc. Perkin Trans. 2*, 1805 (1976); G. Consiglio, S. Gronowitz, A. B. Hörnfeldt, B. Maltesson, R. Noto, and D. Spinelli, *Chem. Scr.* **11**, 175 (1977); G. Consiglio, S. Gronowitz, A. B. Hörnfeldt, R. Noto, K. Pettersson, and D. Spinelli, *Chem. Scr.* **13**, 20 (1978).
580. R. W. Hoffmann, *Dehydrobenzene and Cycloalkynes*, Academic Press, New York (1967), pp. 356–357.
581. L. K. Montgomery and L. E. Applegate, *J. Am. Chem. Soc.* **89**, 5305 (1967).
582. L. K. Montgomery, F. Scardiglia, and J. D. Roberts, *J. Am. Chem. Soc.* **87**, 1917 (1965).
583. R. L. Hillard III and K. P. C. Vollhardt, *J. Am. Chem. Soc.* **98**, 3579 (1976).
584. G. Eigenmann and H. Zollinger, *Helv. Chim. Acta* **48**, 1795 (1965).
585. G. Wittig and H. Heyn, *Chem. Ber.* **97**, 1609 (1964).
586. J. Bolster and R. M. Kellogg, to be published, cited in J. Bolster and R. M. Kellogg, *J. Org. Chem.* **45**, 4804 (1980). *J. Am. Chem. Soc.* **103**, 2868 (1981) have recently reported the generation of a derivative of 20 (X=S).
587. E. H. Rodd, Ed., *Chemistry of Carbon Compounds*, Vol. IIIB, Elsevier Publishing, New York (1956), Chaps. XX, XXI, and XXII.
588. A. Banerji, J. C. Cass, and A. R. Katritzky, *J. Chem. Soc. Perkin Trans. 1*, 1162 (1977).
589. R. W. Hoffmann, *Dehydrobenzene and Cycloalkynes*, Academic Press, New York (1967), p. 290.
590. H. E. Zieger and G. Wittig, *J. Org. Chem.* **27**, 3270 (1962).
591. G. Wittig and R. W. Bashe, in preparation, cited in G. Wittig and M. Rings, *Justus Liebigs Ann. Chem.* **719**, 127 (1968), footnote 27.

6

A Survey of Favorskii Rearrangement Mechanisms:
Influence of the Nature and Strain of the Skeleton

André Baretta and Bernard Waegell

I. INTRODUCTION

The rearrangement of α-haloketones under basic conditions was discovered by Favorskii in 1892.[1] This reaction, which can be fairly useful in synthetic processes, has been reviewed in several articles.[2]

When cyclic α-haloketones are submitted to the Favorskii reaction conditions, a ring contraction can be observed. This efficient procedure has been applied to constrained substrates. The most spectacular example is probably the cubane synthesis achieved in 1964 by Eaton and Cole[3]:

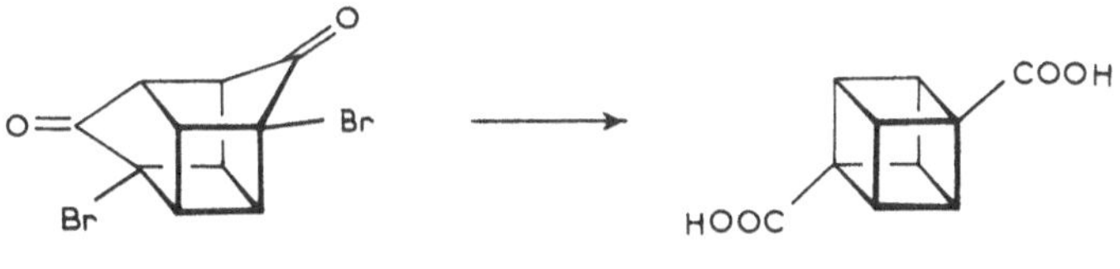

Although the Favorskii reaction is pretty well known, studies concerning its mechanism are still in progress, especially in view of the current interest in reaction intermediates such as zwitterion, carbanion, and cyclopropanones.

André Baretta and *Bernard Waegell* ● Université d'Aix-Marseille, Centre de Saint-Jérôme, Laboratoire de Stéréochimie, rue H. Poincaré, 13397 Marseille Cédex 4, France.

The purpose of the present chapter is not to present an extensive report concerning all the studies related to the Favorskii rearrangement. It is limited to cyclohexane and constrained polycyclic molecules. However, as an introduction, we will survey rapidly the various processes which have been considered for this reaction. We will then present a general discussion of these mechanisms. The behavior of linear aliphatic α-haloketones will be compared with those of cyclic ones when submitted to basic conditions. We will emphasize the influence of the structure upon the nature of the reaction mechanism involved, as well as on the product distribution. It will be shown that the strain in polycyclic α-haloketones has a decisive influence on the rearrangement mechanism involved.

Chemists have generally been concerned with the Favorskii rearrangement because of its synthetic potential, which is particularly useful in three main directions:

(i) Using aliphatic and juxtacyclic† α-haloketones it is possible to prepare carboxylic acids or esters having a tertiary carbon next to the carbonyl:

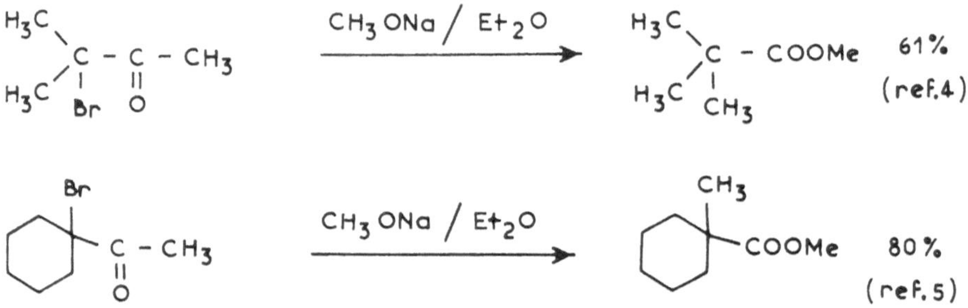

(ii) Ring contraction can be efficiently achieved using cyclic α-haloketones:

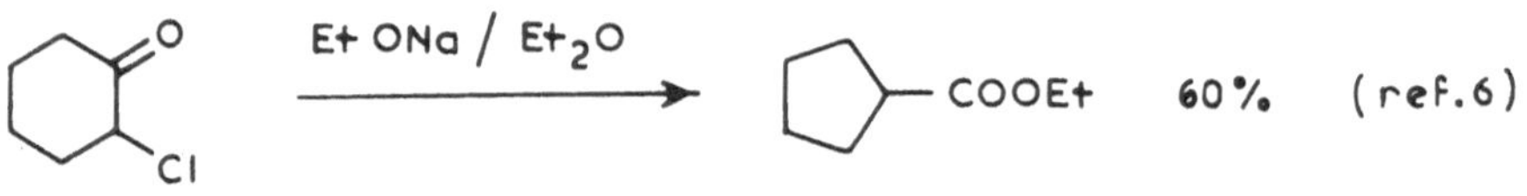

(iii) With α,α- or α,α'-dihaloketones it is possible to have an entry into the preparation of *cis-α,β-unsaturated acids:*

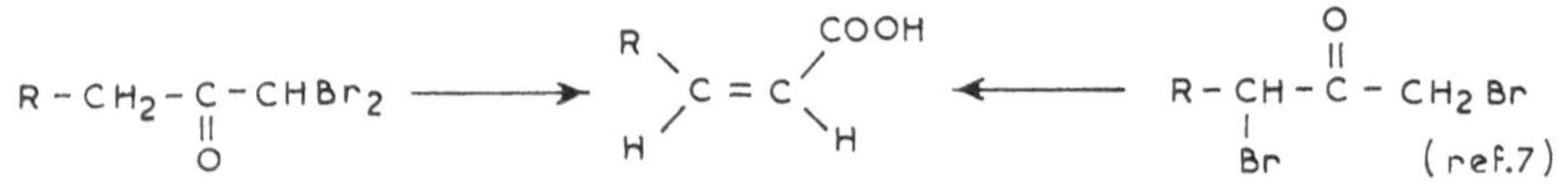

† Juxtacyclic α-haloketone: this is an aliphatic haloketone with a ring α to the carbonyl group.

This latter synthesis generally gives fairly good yields of rearranged products. However, this is not always the case. In fact, a general consideration of the large amount of experimental results concerning the Favorskii rearrangement clearly shows that the outcome of the reaction depends on (i) the α-haloketone structure; (ii) the nature of the base used; and (iii) the solvent in which the reaction is carried out. Consequently, the acid or its transformation product is seldom isolated as the only product of the Favorskii rearrangement. The following side products are often present in the reaction mixture:

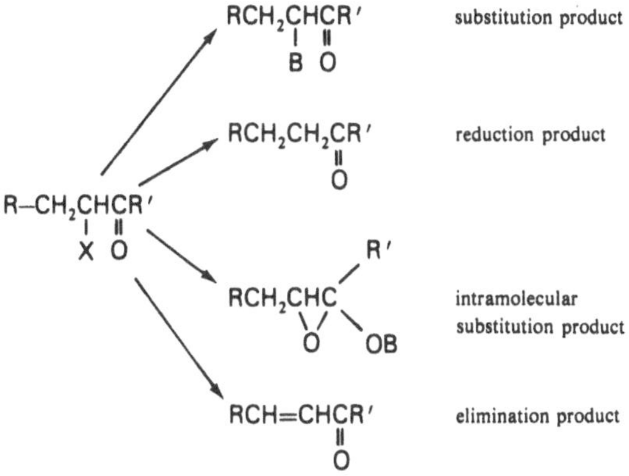

In order to carry out the Favorskii rearrangement, the α-haloketone is generally added to a solution or a suspension of the base. Many bases have been used: alkali and alkali earth hydroxides, alcoholates, neutral and acidic alkali carbonates, ammonia and amines. The reaction can be carried out either in protic solvents (water, alcohols) or aprotic solvents (ether, dioxane, dimethoxyethane, ...).

It is fairly difficult to recommend a definite set of ideal experimental conditions, as the outcome of the reaction, as well as the mechanism involved, strongly depend on the α-haloketone structure. It is actually the latter which decides the reaction conditions.

1. Favorskii Reaction Mechanisms

Several mechanisms have been suggested for this rearrangement.[2e] We shall only discuss here the two main reaction pathways which are generally considered. These are summarized in Table 1. The type of mechanism involved results naturally from the experimental conditions chosen (type of solvent, type of base). But the overall structural features of the α-haloketone certainly have a decisive influence on the process involved. The following are of particular importance: (i) presence or absence of a hydrogen α' to the carbonyl;

TABLE 1. Mechanisms of the Favorskii Rearrangement

Cyclopropanone pathway

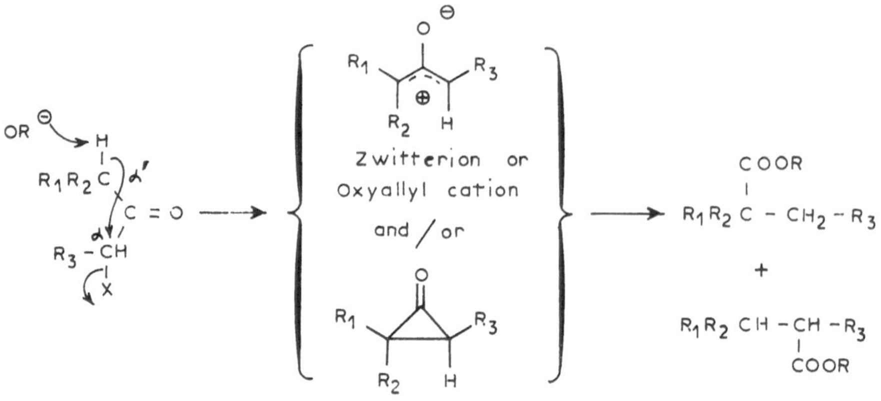

Zwitterion or Oxyallyl cation

and / or

$R_1 R_2 C = O$

$COOR$

$R_1 R_2 C - CH_2 - R_3$

+

$R_1 R_2 CH - CH - R_3$
$COOR$

Semibenzylic pathway

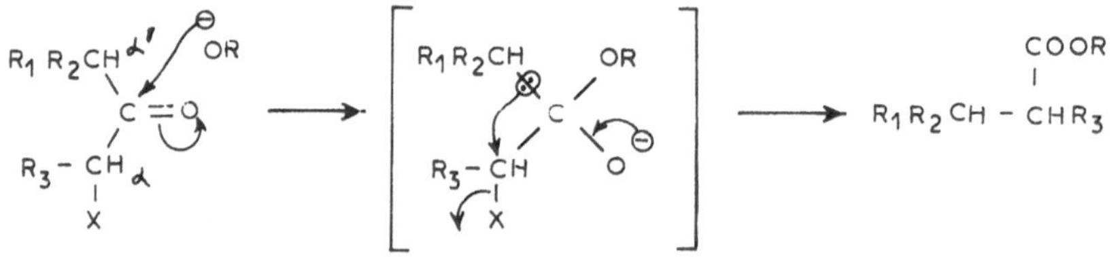

$COOR$

$R_1 R_2 CH - CHR_3$

(ii) halogen stereochemistry, and (iii) strain in the substrate which might result from substitution at the reactive sites or (and) from ring strain itself.

The presence or the absence of these features will actually decide which one of the two processes mentioned in Table 1 will be operative.

The *symmetrical* reaction pathway requires an acidic hydrogen α' to the carbonyl group. This mechanism, also known as the cyclopropanone mechanism, will involve a cyclopropanone or a zwitterionic intermediate. We will discuss the reactivity of acyclic substrates as a function of solvation, where the factors playing a decisive role are not the same as those involved in the cyclohexane series. In the latter case, it is the halogen stereochemistry and the ring strain which are essential. The reactivity of cyclopentane derivatives can be explained on the basis of the cyclopropanone–zwitterion equilibrium, which is itself ring strain dependent. A continuous variation of reactivity can be described going from acyclic to cyclohexane and cyclopentane α-haloketones. We will suggest a mechanism essentially based on structural features and solvent effects.

The *unsymmetrical* mechanism is also called the semibenzilic mechanism. Generally, a weakly acidic hydrogen is present α' to the carbonyl. When there is no such α'-hydrogen, the reaction is called a quasi-Favorskii, and involves the same type of mechanism. The geometry which is necessary for this process to occur is the same in both the acyclic and cyclic series. Obviously in the latter case the reactivity will be directly related to the halogen stereochemistry. Ring strain will also play an important role and will often be the decisive factor in determining which of the semibenzilic and the cyclopropanone mechanisms obtain.

II. SYMMETRICAL MECHANISM

1. Cyclopropanone Mechanism and Experimental Evidence for the Symmetrical Intermediate

McPhee and Klingsberg[8] were the first to suggest that the reaction proceeds via symmetrical intermediate, as the same rearranged ester was formed starting from two isomeric α-haloketones differing in the halogen's position α or α' to the carbonyl:

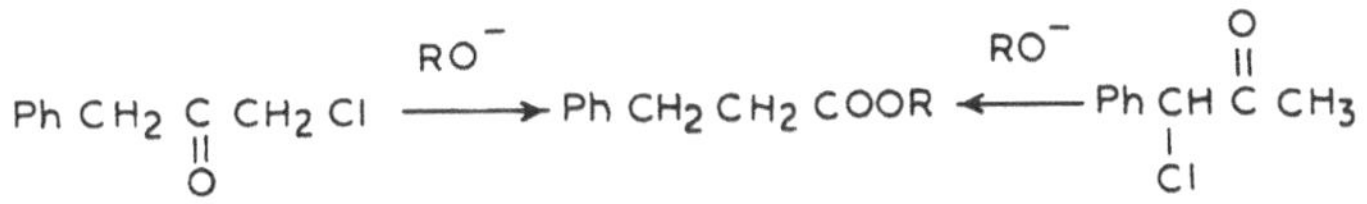

$$Ph\ CH_2\ \underset{\underset{O}{\|}}{C}\ CH_2\ Cl \xrightarrow{RO^-} Ph\ CH_2\ CH_2\ COOR \xleftarrow{RO^-} Ph\ \underset{\underset{Cl}{|}}{CH}\ \overset{\overset{O}{\|}}{C}\ CH_3$$

A very elegant demonstration of the existence of this symmetrical intermediate was given by Loftfield.[9] He used as starting material 2-chlorocyclohexanone labeled with ^{14}C at both functionalized carbons. Treatment with sodium amylate in amylic alcohol gave the following cyclopentane esters:

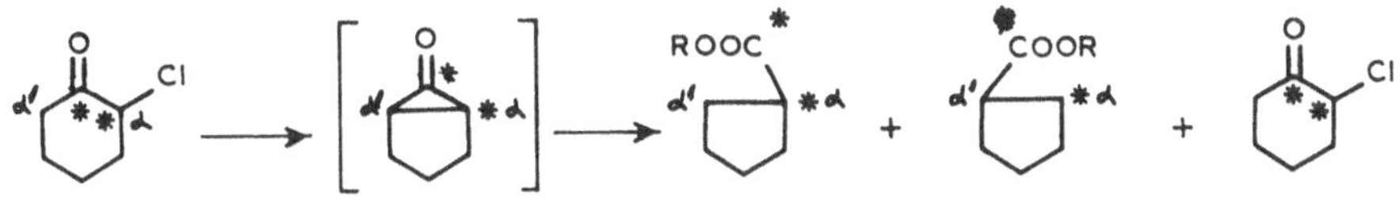

Analysis of the 2-chlorocyclohexanone recovered at half-reaction time showed that there was no isomerization of the halogen from the α to the α' carbon. The same isotopic distribution was found in the two ring-contracted esters. Consequently, Loftfield assumed that a cyclopropanone was the common intermediate. It should be noticed that a 2-oxyallyl cation or a zwitterion (Table 1) would lead to the same experimental results.

For quite a long time, the essential argument against the Loftfield mechanism was the fact that there was no experimental evidence for such a cyclopropanone. However, development of the chemistry of cyclopropane derivatives[10] recently allowed the isolation of a cyclopropanone under Favorskii rearrangement conditions[11]:

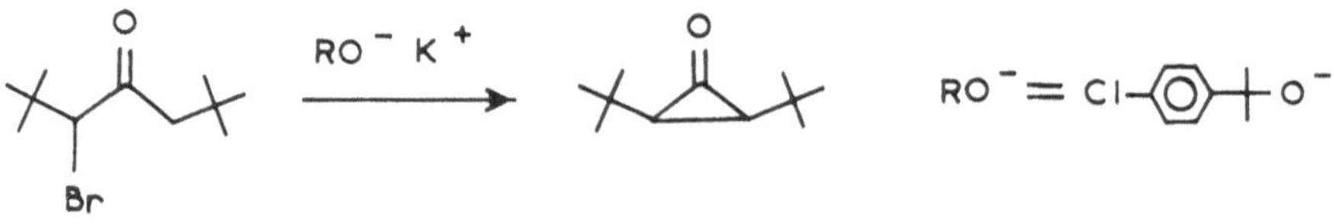

$$RO^- = Cl\!-\!\langle O \rangle\!-\!\!\!\top\!\!-O^-$$

Moreover, this type of compound, when placed in basic media, yields exactly the same kind of rearranged products as are observed from Favorskii rearrangement. This is the case for the tetramethylcyclopropanone studied by Turro and Hammond[12]:

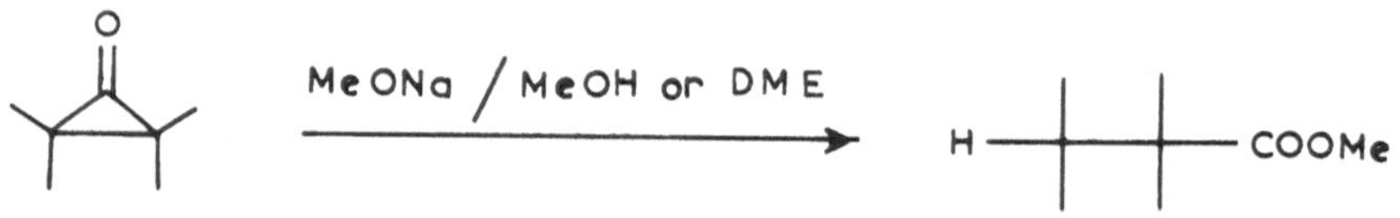

MeONa / MeOH or DME

These different experiments clearly show that the cyclopropanone is indeed an intermediate in the symmetrical mechanism, as assumed by Loftfield. Consequently, the kinetic and stereochemical studies of the Favorskii rearrangement are generally understood in terms of a cyclopropanone intermediate, even in some of the cases where the three-membered intermediate cannot be isolated as such. The overall process is generally considered as involving four distinct steps, as shown in Scheme 1. In the first step the α-haloketone undergoes an enolization in the α′ position. The intermediate cyclopropanone is formed during the second step by loss of halide ion. The cyclopropanone is ring opened by a base in the third step to give a carbanion. The latter reacts with the conjugate acid of the base to give the final product (fourth step). The two last steps are thought to be so rapid that they do not have an influence on the overall rate of the reaction. However, this mechanism cannot explain all the kinetic and stereochemical results observed. We will now discuss the two first steps of the rearrangement.

2. First Step of the Mechanism

Since the occurrence of a cyclopropanone mechanism was now considered to be beyond doubt, the studies carried out on the Favorskii rearrangement were directed toward a better understanding of the process leading to this cyclopropanone. The first step of the symmetrical mechanism leads to the formation of a carbanion enolate (Scheme 2) by an α′ hydrogen abstraction with

SCHEME 1

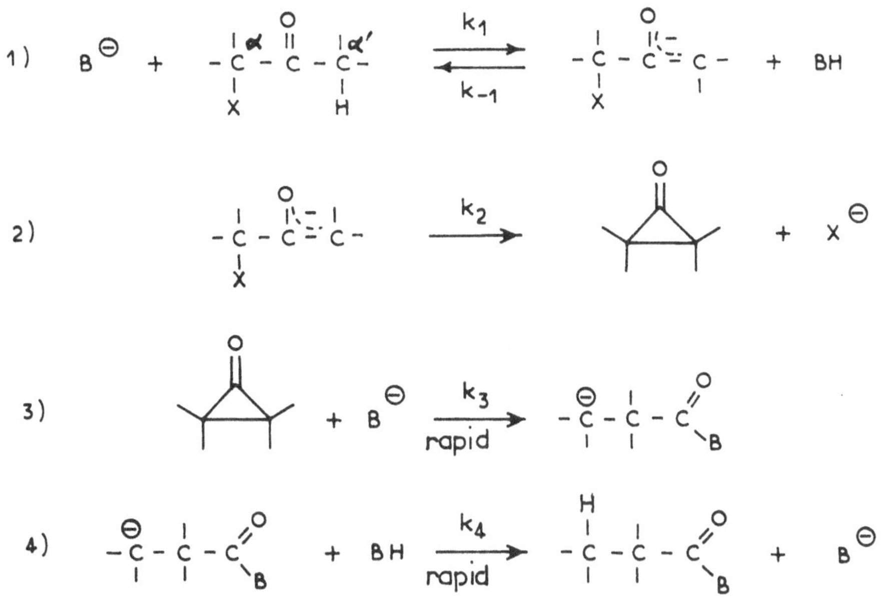

1) $B^{\ominus}$ + (α-halo ketone with X on α carbon and H on α′ carbon) $\underset{k_{-1}}{\overset{k_1}{\rightleftharpoons}}$ (enolate) + BH

2) (enolate with X) $\overset{k_2}{\longrightarrow}$ (cyclopropanone) + $X^{\ominus}$

3) (cyclopropanone) + $B^{\ominus}$ $\underset{\text{rapid}}{\overset{k_3}{\longrightarrow}}$ $-\overset{\ominus}{\underset{|}{C}}-\overset{|}{\underset{|}{C}}-C\overset{\diagup O}{\diagdown_{B}}$

4) $-\overset{\ominus}{\underset{|}{C}}-\overset{|}{\underset{|}{C}}-C\overset{\diagup O}{\diagdown_{B}}$ + BH $\underset{\text{rapid}}{\overset{k_4}{\longrightarrow}}$ $-\overset{|}{\underset{|}{C}}-\overset{\overset{H}{|}}{\underset{|}{C}}-C\overset{\diagup O}{\diagdown_{B}}$ + $B^{\ominus}$

the base. However, it is also possible to consider the attack of the base on the proton geminate to the α halogen. As a consequence, it is possible to observe a modification of the halogen stereochemistry which plays a decisive role in the outcome of the reaction in cyclohexane derivatives (see below). An epimerization of the halogen during the first step of rearrangement has to be seriously considered.

A. Cyclohexane Substrates

The reactivity of the α and α' hydrogens during the first step can be analyzed by deuterium exchange with the reaction medium. There are two ways of considering the problem: the reaction medium can be deuterated, or the starting haloketone can be labeled with deuterium at the α and α' positions and submitted to rearrangement in a protic solvent. One would expect the exchange to be more likely at the position α to the halogen rather than at the α' position.[13]

a. Axial Halogen

According to the above discussion, one should observe the epimerization of substrates with an axial halogen (Table 2, step C). Furthermore, under iden-

TABLE 2. Epimerization of α-Bromocyclohexanone under Basic Conditions

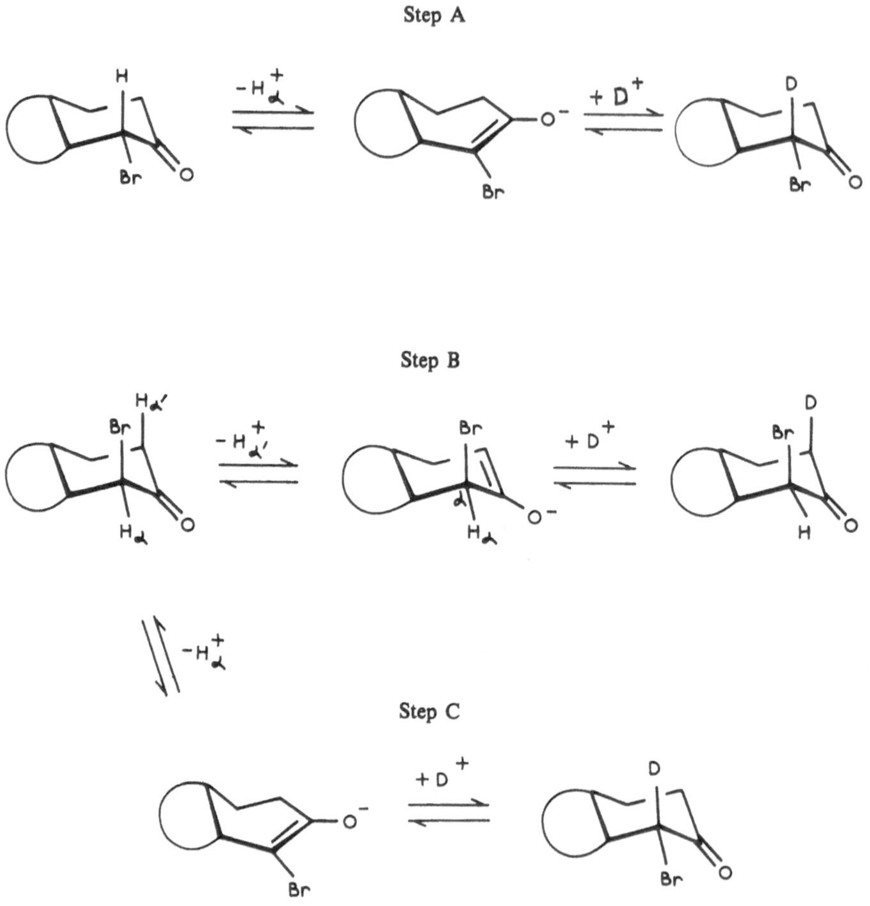

Step A

Step B

Step C

tical experimental conditions, derivatives with axial halogens should give exactly the same reaction mixture as their epimers having an equatorial halogen. We shall see later on that this is not the case. In protic solvents, rearrangement products are obtained from substrates with an equatorial halogen whereas, under exactly the same conditions, substitution products are obtained from substrates with an axial halogen. As the starting materials have the same cyclohexane skeleton, the difference in reactivity can only be a consequence of the different conformations of the halogen. It seems reasonable to assume that the axial halogen derivatives do not epimerize before they rearrange, or, in other words, that the axial α′ proton is removed before the equatorial α proton

SCHEME 2

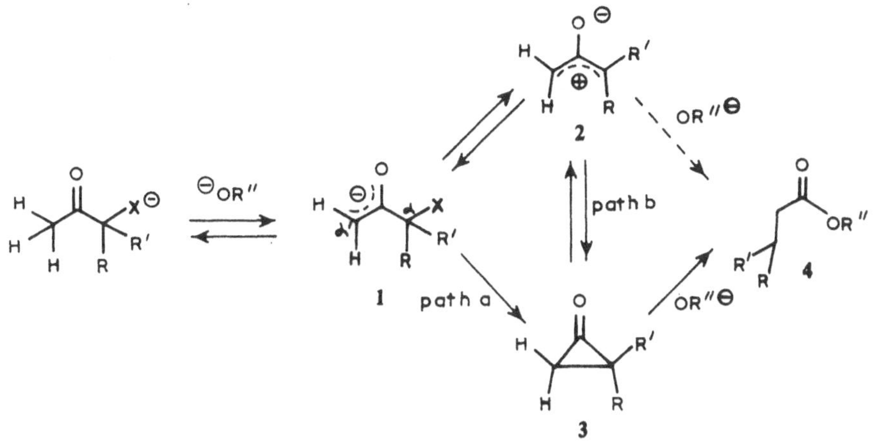

(Table 2, step B). It is known that the attack of an electrophile on an enolate occurs axially in deuteration, halogenation, or alkylation reaction[†]. According to the principle of microscopic reversibility, removal of the axial proton should be favored. Recent work by Trimitsis and Van Dam[§] is in agreement with such a process. It was shown experimentally that in 4-tert-butylcyclohexanone, axial protons α and α' to the carbonyl are exchanged more rapidly than the equatorial ones in the presence of base and D_2O. This process is actually subject to stereoelectronic control.[17] Calculations following the least movement principle[18] also show that it is the loss of an axial proton which costs the least energy during the enolization of a ketone. It appears that the exchange of axial α' protons in ketones is definitely favored by electronic factors.

b. Equatorial Halogen

When the cyclic substrate has an equatorial halogen (Table 2, step A), the experiment shows that the enolization at the α position does not lead to epimerization.[19] This again is in perfect agreement with the stereoelectronic control which favors fixation of the deuterium by the enolate in the axial direction. This enolization does not seem to have any further influence on the Favorskii rearrangement.[2e]

---

† See References 14-15. For a broader discussion of stereoelectronic control and "perpendicular attack," see L. Velluz, J. Valls, and G. Nominee, *Angew. Chem. Int. Ed. Eng.* 4, 181 (1965).

§ See Reference 16. In neutral media, only the exchange of axial protons is observed.

3. Second Step of the Mechanism

Intermediate **1** is transformed into the cyclopropanone **3**, following two different pathways (Scheme 2): (i) a direct intramolecular nucleophilic substitution (path *a*),[9,20] and (ii) a zwitterionic intermediate **2**[21] (path *b*).

It has even been suggested that the ester **4** is formed directly from the zwitterion **2**.[22] In our opinion this does not seem very likely. We will discuss successively the second step of this mechanism in juxtacyclic[†] linear series, as well as in cyclic series. In linear substrates the transformation of **1** via the cyclopropanone intermediate **3**, or via the zwitterion **2**, will depend on the base strength and the solvation. In cyclic substrates, structural features will have to be considered in addition.

A. Open-Chain Juxtacyclic Substrates

a. Role of the Solvent

Aprotic Medium. The cyclopropanone formed—whether by one or the other mechanism—is extremely sensitive to the nature of the solvent. In this respect the most significant results are those obtained with 1-acetyl-1-chloro-2-methylcyclohexanes **5** and **6**[23] (Scheme 3) in an aprotic nonpolar solvent such as ether. Under these conditions, the two epimers **5** and **6** rearrange stereospecifically. In both acids **7** and **8**, the new bond C–C=O has a stereochemistry which is the reverse of that of the corresponding bond in the initial α-haloketone. These results clearly show that the cyclopropanones **9** and **10** are the only possible intermediates. They are formed by an intramolecular

SCHEME 3

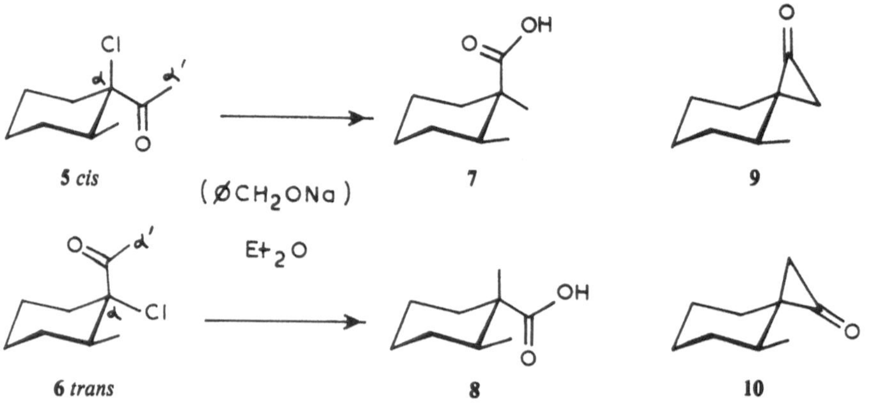

$$(\emptyset CH_2 ONa)$$

$$Et_2O$$

5 *cis*   **7**   **9**

6 *trans*   **8**   **10**

[†] See footnote, p. 528.

SCHEME 4

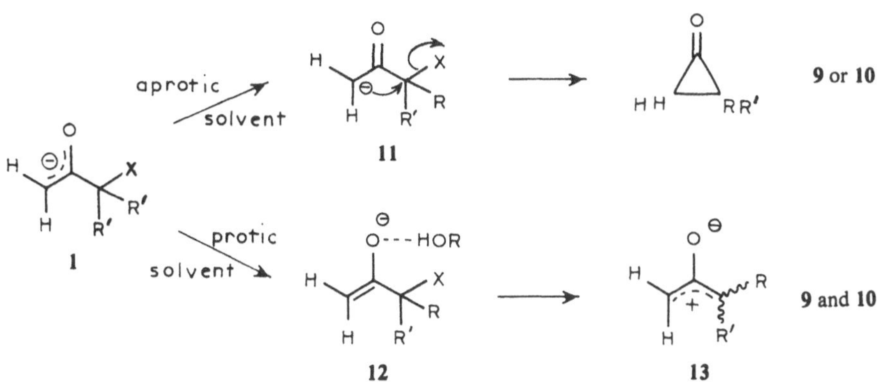

S_N2 mechanism. The carbanion formed at α' attacks the α carbon resulting in an inversion of configuration at the carbon bearing the halogen.

Protic Medium. The stereospecificity which is characteristic of an aprotic medium is no longer observed when the rearrangement is carried out in a protic solvent such as methanol.[24] The *trans* derivative **6** gives a 1:1 mixture of the acids **7** and **8**. We will discuss two interpretations which were given in order to explain this behavior, namely, House's interpretation that suggests that the cyclopropanone is solvated, and Tchoubar's hypothesis that it is the enolate **1** which is influenced by the solvent.

(i) House's interpretation.[24] In this case, the cyclopropanone is solvated. The protic solvent (MeOH) will favor the cleavage of the $C_{(2)}$–$C_{(3)}$ bond opposite the carbonyl. This cleavage will yield the zwitterion **13** (Scheme 4) and then, by ring closure, both cyclopropanones **9** and **10**. However, recent results relative to cyclopropanone chemistry show that this hypothesis is not reasonable. Indeed, in the presence of hydroxylated derivatives, cyclopropanones do not ring open. In fact, the corresponding hemiketals are found as unique products. They are perfectly stable and can be isolated. This is the case with 2,2-dimethylcyclopropanone which has a structure very close to that of intermediates **9** and **10**[25]:

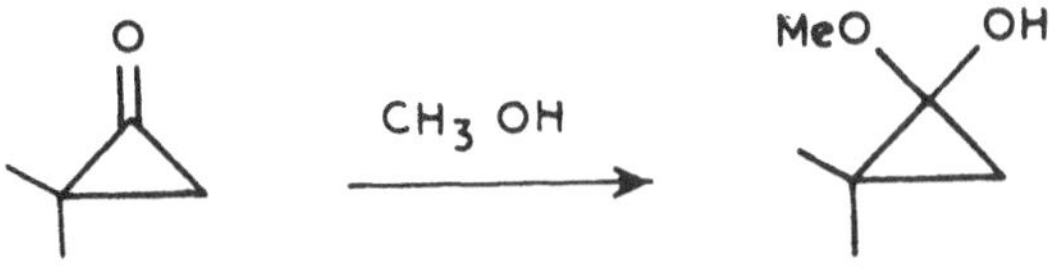

On the other hand, cleavage of the cyclopropanone bond opposite to the carbonyl generally requires acidic conditions.[26] The same is true for the corresponding hemiketals. However, when the cyclopropanones are sterically crowded, ring cleavage can be achieved with methanol under reflux; substituted derivatives are obtained as the main products as in tetramethylcyclopropanone:

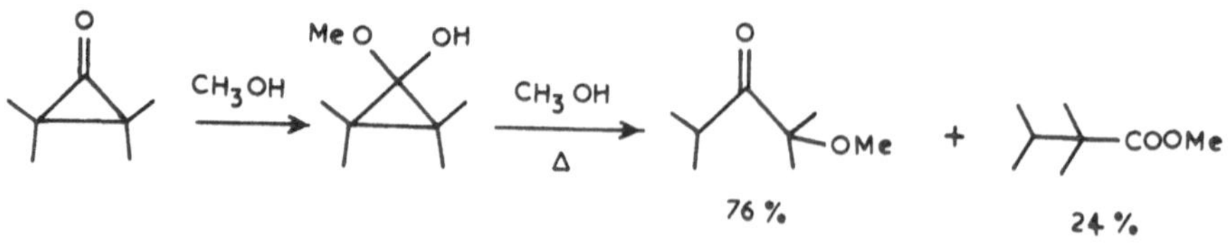

All these arguments do not favor House's interpretation.[24]
(ii) Tchoubar's interpretation.[27] According to the work of Tchoubar and coworkers using compound 6 (Scheme 3), the solvent exerts its influence on the enolate 1 [27] (Scheme 4).

In an aprotic solvent, the enolate 1 is only slightly solvated, if at all, and is thus free. Under these conditions the reactive species is similar to carbanion 11, which by ejection of the halogen, leads to 10 stereospecifically via an intramolecular S_N2 mechanism. On the other hand, in a protic solvent, the enolate 1 is solvated by hydrogen bonding. Now the reactive species looks like intermediate 12 (Scheme 4), which favors the formation of the zwitterion 13.[28,29] In the case of compound 6, the ionization of the halogen leads to the zwitterions 13a and 13b (Scheme 5). Thus, formation of acids 7 and 8 in equal amounts in a protic medium via cyclopropanones 9 and 10 can be accounted for in terms of these intermediates.

Solvation of enolate 1 is a hypothesis which is perfectly coherent with the

SCHEME 5

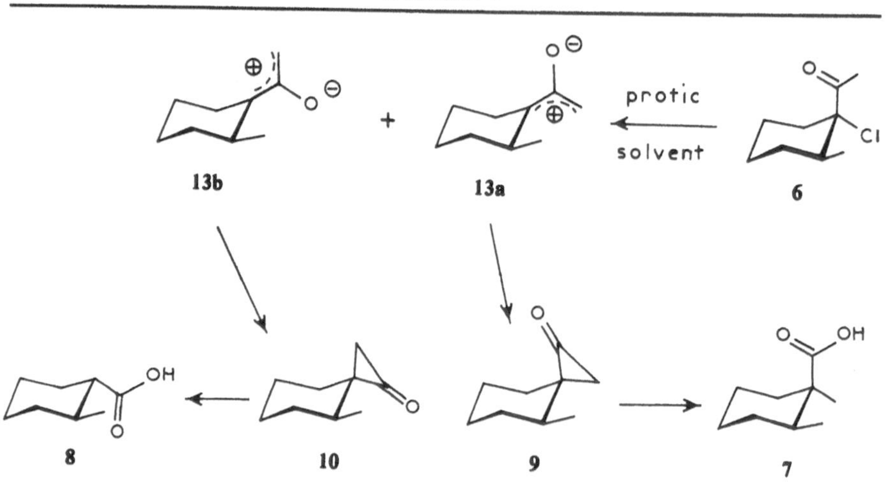

observed reactivity of **5** and **6** in aprotic media and of **6** in protic media. As shown in the following experiment, the solvation of **1** is also a function of base concentration.

b. Influence of the Base Concentration in a Protic Medium

Another result described by Tchoubar is related to the concentration of base in methanolic medium. The latter has a definite influence on the stereoselectivity of the process observed during the rearrangement of **6**.[30] The stereoselectivity increases with base concentration; the major product observed is the acid **8**, which is also obtained in aprotic medium with inversion of configuration.[23] The same influence is observed when increasing quantities of inert salts are added to a medium having a constant base concentration. The anions are strongly solvated by these salts.[30,31]

These results can be understood in the following way: the high concentration of small anions (CH_3O^- or X^-) induces solvation with methanol by hydrogen bonding. Consequently the activity coefficient of methanol is decreased; in other words, the amount of available free OH groups of methanol is reduced.[32] Even though the reaction is proceeding in a protic medium, the enolate **1** cannot be solvated further by hydrogen bonding. Under these conditions its reactivity is comparable to that observed in an aprotic solvent. These results are in perfect agreement with Scheme 4, where it is shown that with linear haloketones **5** and **6**, solvation exerts its influence at the level of the enolate **1**. In aprotic solvents the cyclopropanone is preferentially formed by an intramolecular S_N2 process initiated by the free carbanionic charge. In a protic solvent, the cyclopropanone is much more likely to form via the zwitterion.

B. Cyclohexane Substrates

It was of interest to check whether the results just discussed in the open-chain series would apply to cyclohexane derivatives. However, in such compounds, the stereochemistry of the halogen has to be considered since we have

SCHEME 6

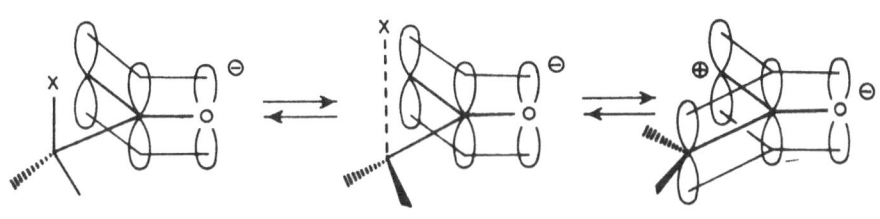

just discussed (see above) how its epimerization can occur. Cyclopropanone formation by an intramolecular S_N2 mechanism requires an equatorial halogen, whereas an axial halogen will favor zwitterion formation. Such an intermediate has been proposed by Aston[21] and by Dewar.[22] More recently, more details about zwitterion formation in open-chain compounds were discussed by Bordwell.[28] As shown in Scheme 6, the carbon halogen bond ionizes with participation of the enolate π system, provided that both are parallel. In other words, in open-chain compounds zwitterion formation requires a halogen carbon bond perpendicular to the plane formed by the three carbons C_α–$\underset{\underset{O}{\|}}{C}$–$C_{\alpha'}$.

In cyclohexane derivatives this can only be achieved with an axial orientation of the halogen. Consequently a derivative with an equatorial halogen has a semi-W geometry[33] adequate for an intramolecular S_N2 reaction. The reactive species is a developing carbanion, where the negative charge is developing in an initially axial sp^3 orbital, properly oriented to attack the antibonding lobe of the carbon orbital describing the C–Cl bond. Such an attack results in an inversion of configuration with the formation of an *anti* cyclopropanone **14**. Thus, an equatorial α-chlorocyclohexanone, when treated with base in an aprotic solvent, should yield a ring contraction product such as **14a**.

On the other hand, a substrate with an axial halogen has a semi-U geometry.[33] In the presence of a base, the developing carbanion (in the transition state), which can initially be described by an sp^3 orbital, does not have the appropriate stereochemistry to react with the axial C–Cl bond (see Scheme 7). In an aprotic medium the carbanion carbon has rehybridized into a p orbital stabilized by overlap with the carbonyl p orbitals, and the formation of a zwitterion is observed. Cyclization of the latter will lead to the two isomeric cyclopropanones **14** and **15**, which will yield the two rearranged derivatives **14a** and **15a** by ring opening.

The above discussion shows that cyclopropanone formation is a consequence of the halogen orientation and the nature of the solvent. Therefore the reaction mechanisms can be predicted and verified by conformational analysis of the stereochemistry of the rearranged products experimentally obtained. These ideas were at the origin of a great deal of research using cyclic α-haloketones where the halogen stereochemistry was known. The results are discussed in the following sections.

a. Results

In Tables 3 and 4 we have collected the results obtained when Favorskii rearrangement conditions were applied to cyclohexane α-haloketones of known stereochemistry.

The behavior of derivatives with an equatorial halogen is described in

SCHEME 7

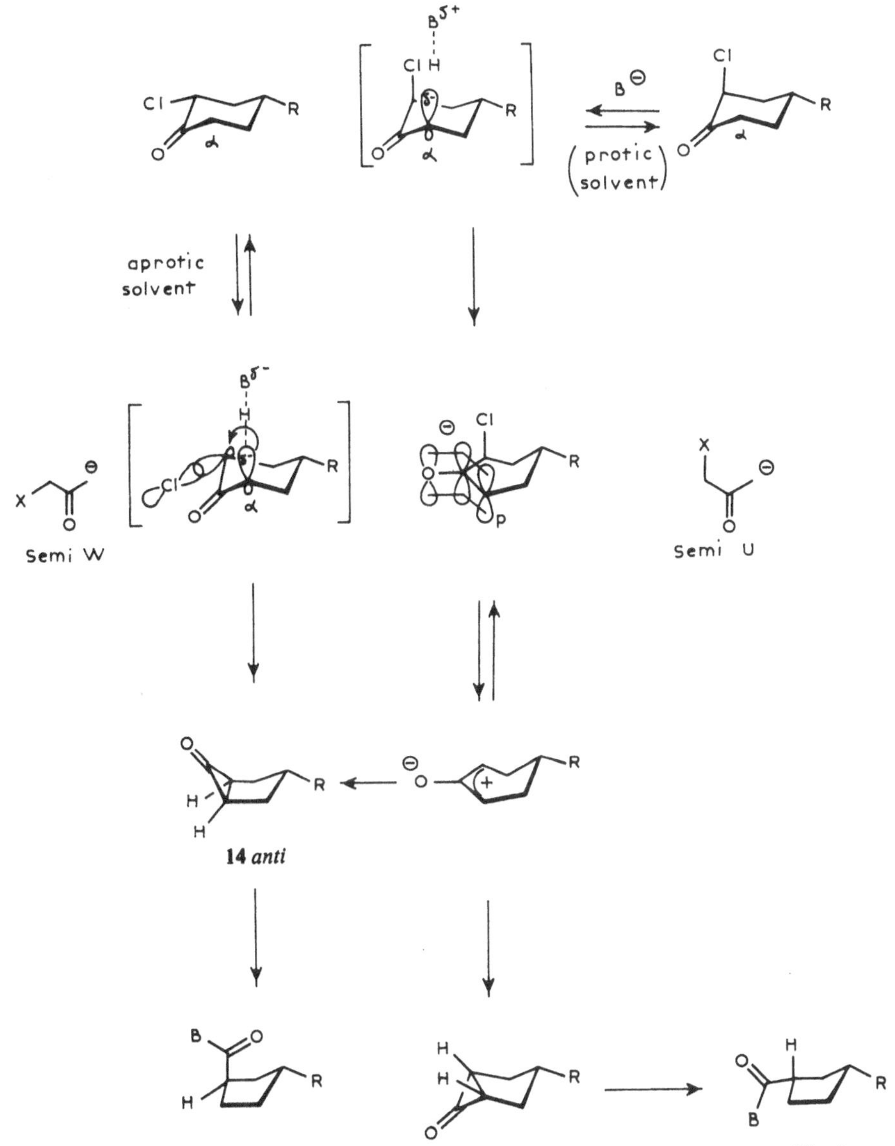

Semi W

Semi U

14 *anti*

14a *trans*

15 *syn*

15a *cis*

TABLE 3. Favorskii Rearrangement of α-Halocyclohexanone Derivatives Having an Equatorial Halogen

α-Haloketone	Base, solvent	Rearranged compounds	Substitution products	Reference
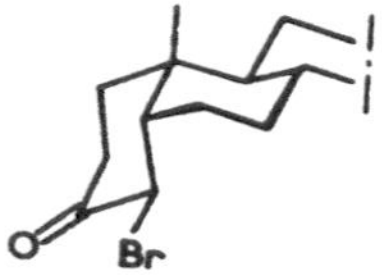 16	MeONa Et₂O/MeOH	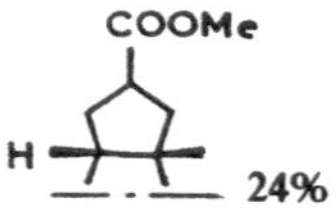 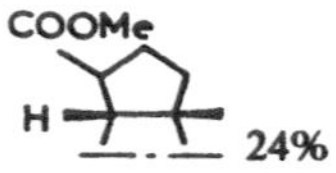 COOMe 24% ret. / COOMe 24% ret.	No substitution	34
	MeOH	60%		
	MeOH 98%	18%		
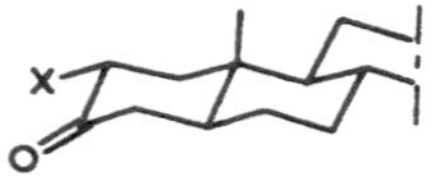 17	MeONa	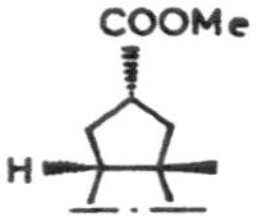 COOMe ret. 25% / 50%	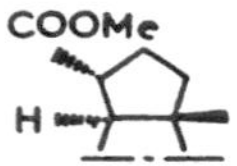 HO⋯ 25%	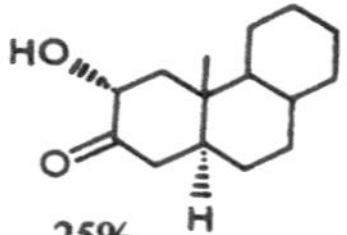 34
	Et₂O/MeOH EtOH	COOMe ret. 2% < / 50%		
	EtONa/EtOH	X = Cl 66%		35
		X = Br 66%		
		X = I 78%		
	MeONa/MeOH	X = Br 35–40%		36
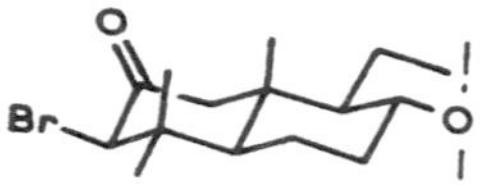 18e	MeONa DME/MeOH	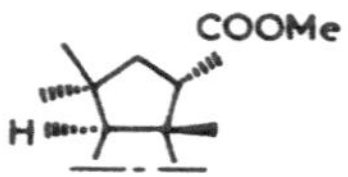 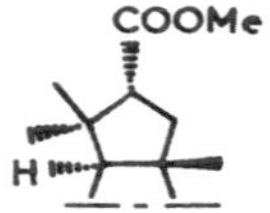 COOMe inv. 19% / COOMe inv. 9%	No substitution	37

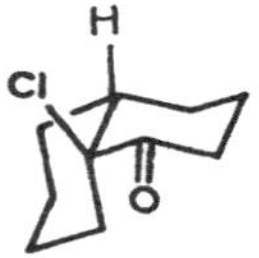

MeONa
MeOH

32% ret.   57% ret.

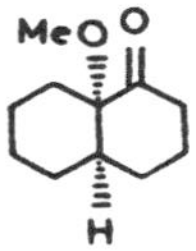

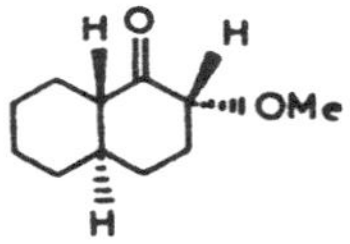

39%   58%

38

DME

Substitution

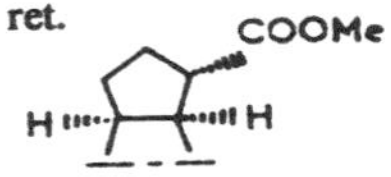

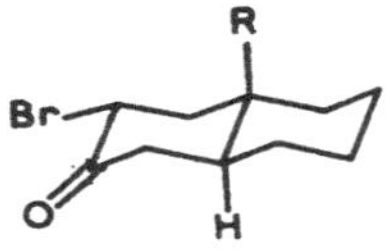

R = H 50%   R = CH₃ 43%
R = H 45%   R = CH₃ 35%

39

19e

EtONa
EtOH
DME

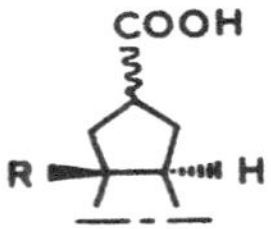

R = H 11%   R = CH₃ 38%
R = H 13%   R = CH₃ 44%

20e

φCH₂ONa
φCH₂OH

66%

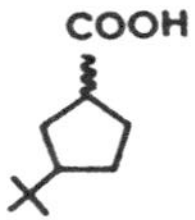

40

21

(continued)

TABLE 3.—cont.

α-Haloketone	Base, solvent	Rearranged compounds	Substitution products	References
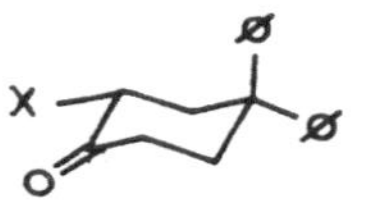 **22:** X = Br X = Cl	MeONa MeOH	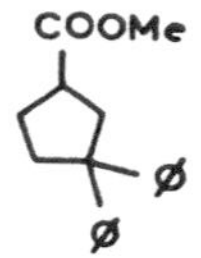	No substitution	28a
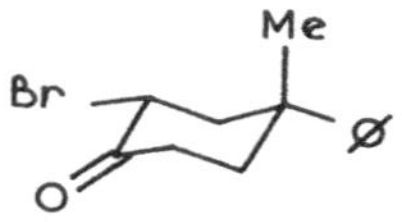 **23**	MeONa MeOH	inv. 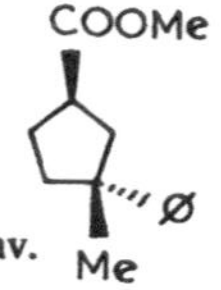 ret. 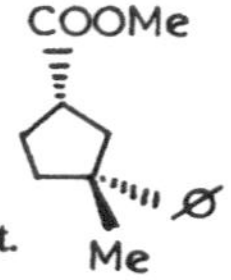	No substitution	41a
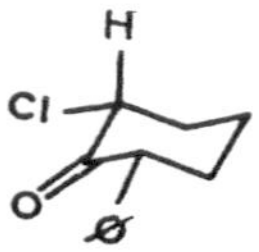 **24**	MeONa MeOH or DME		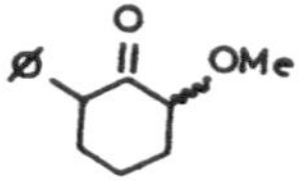 *cis* and *trans*	41b

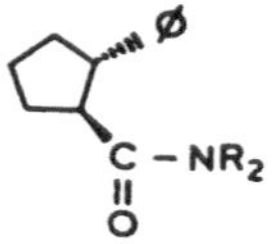

TABLE 4. Favorskii Rearrangement of α-Halocyclohexanone Derivatives Having an Axial Halogen

α-Haloketone	Base, solvent	Rearranged compounds	Substitution products	References
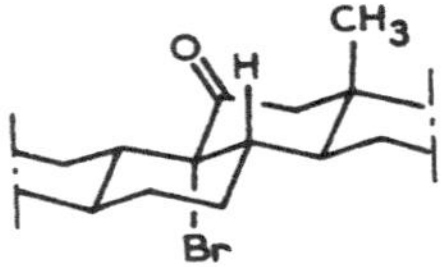 **25**	MeONa MeOH	No rearrangement	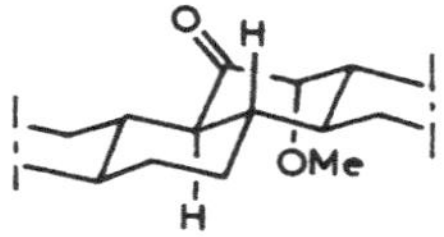	42
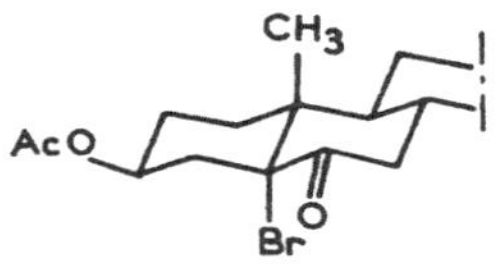 **26**	RO⁻ ROH	No rearrangement	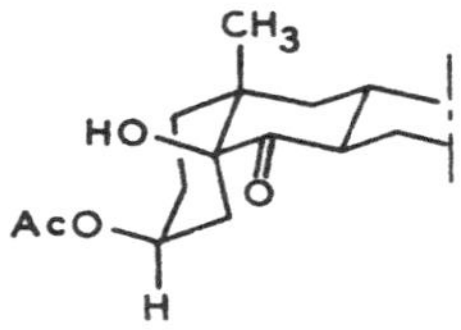	43
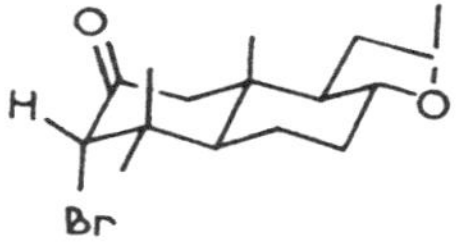 **18a**	MeONa MeOH/DME	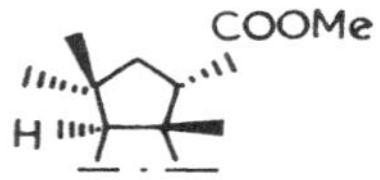 Ret.  Ret.	No substitution	37

(continued)

TABLE 4.—cont.

α-Haloketone	Base, solvent	Rearranged compounds	Substitution products	References
27	t-AmONa t-AmOH	COOR R = t-Am 65%		20
	MeONa/MeOH φCH₂ONa φCH₂OH	R = Me 3%		41a
		R = CH₂φ 75%	61% 28%	23
	RO⁻ ROH	R = Me		23
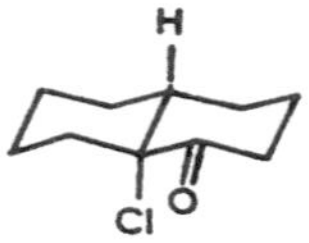 **19a**	MeONa MeOH		 30% 39% 17%	
	MeONa DME	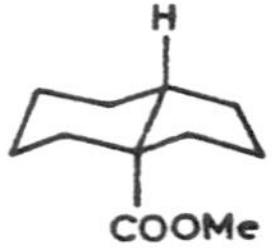 88%	No substitution	38

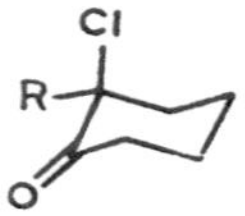

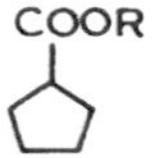

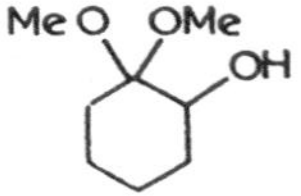

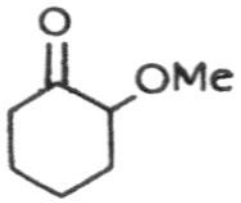

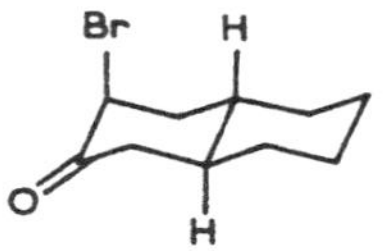

39

44

major

No rearrangement

21%

7%

29%–33%

46%

36%–47%

36%

49%

59%–67%

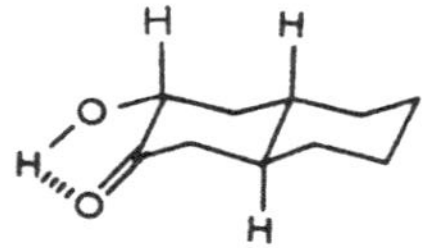

EtONa
EtOH

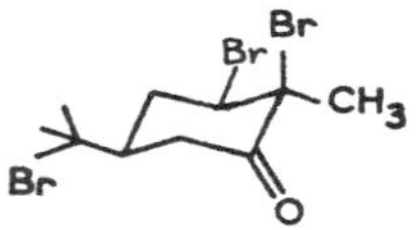

EtONa
DME

MeONa
MeOH

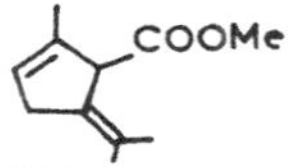

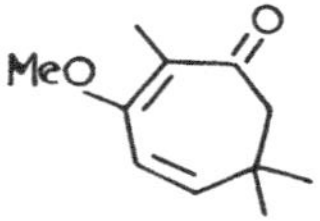

MeONa
Et₂O

MeONa
MeOH

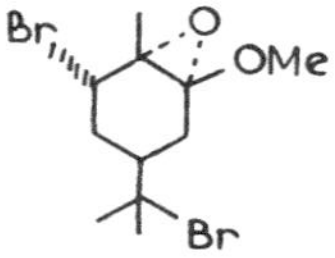

MeONa
Et₂O

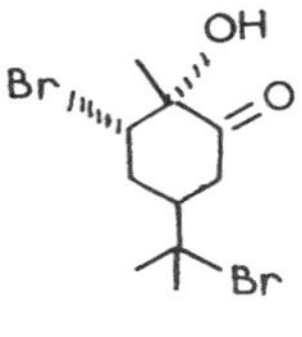

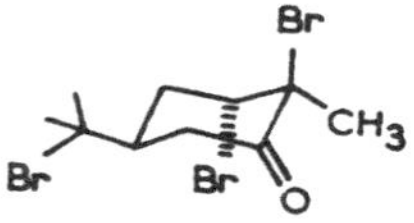

20a

28e

28a

OH

Br

OMe

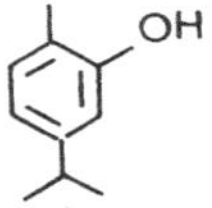

MeO

COOMe

COOMe

CH₃

H

Table 3. The halogen is secondary in all cases except in **19e** where it is tertiary; all the substrates have protons in α'. These are apt to be removed by the base to give the enolate in α' according to the cyclopropanone mechanism. From an inspection of Table 3, it appears that the overall behavior of this kind of substrate is the same. They practically all give ring contraction products whatever the nature of the solvent, with the exception of compound **19e** which has a tertiary halogen. In this case the reactivity is solvent dependant: in protic media, substitution is observed, whereas ring contraction occurs under aprotic conditions.

·The reactivity of axial halogen derivatives is reported in Table 4. From the structural point of view all derivatives mentioned have a tertiary halogen except in the cases of **20a** and **18a**; enolization is also possible at the α' position. The behavior of these derivatives is quite different from that observed with the compounds having an equatorial halogen: in protic media only substitution is observed except for **28e** and **28a**. However, in aprotic solvents, it is difficult to have a general idea of their behavior, since all compounds listed in Table 4 were not treated under these conditions. Compounds **18a**, **19a**, **28a**, and **28e** give ring contraction products in aprotic solvents. Tribrominated derivatives **28e** and **28a** give rearranged products in protic and aprotic media. It is very likely that the presence of a β-bromine relative to the carbonyl has an influence on the transition state. Generally speaking, in the case of cyclic derivatives with axial halogen, the solvent has an influence on the reactivity.

b. Discussion

Analysis of the results reported in Table 3 and 4 shows that the reactivity of cyclohexanyl α haloketones differs from that observed with open-chain juxtacyclic series; in particular, the outcome of the reaction is not solely solvent dependent.

The solvent's nature does not have a decisive influence on the chemical behavior of equatorial halogen derivatives (Table 3) which always undergo some ring contraction. On the other hand, the axial halogen derivatives give substitution in protic media, except in special cases which we will discuss later, and ring contraction under aprotic conditions. From this first analysis, it appears that this difference in reactivity is related to the halogen stereochemistry. In other words, under the basic conditions used, the halogens do not epimerize fast enough, otherwise they would have behaved similarly.

The results obtained with cyclohexane derivatives with an equatorial halogen reported in Table 3 show that the rearranged esters are obtained whatever the nature of the solvent. Compounds **16**, **17**, and **19e** do not react by an S_N2 mechanism, as they yield esters with retention of configuration at the carbon initially bearing the halogen. It is only 3β-bromo-8,13-epoxy-2-labdanone (**18e**) which gives products with inversion of configuration.[37] It should be noticed that its axial epimer **18a** (Table 4) gives the same esters. This would be in agreement with Loftfield's S_N2 process,[9] according to which an equatorial halogen would lead to inversion of configuration and an axial one to retention. However, the esters formed from **18e** and **18a** as a result of the opening of the cyclopropanone have the α configuration: this corresponds to the less hindered side of the molecule. A similar observation can be made for the haloketones **16, 17,** and **19e,** which give the rearranged esters via the less hindered cyclopropanones. This means that the regioselective formation of the cyclopropanone is essentially determined by overall steric factors, rather than by the stereochemistry of the halogen.

Nevertheless, according to Bordwell,[28a] cyclohexane derivatives with an equatorial halogen would react through a zwitterionic intermediate if the molecule can adopt a boat conformation in which the halogen is axial; to reach this geometry McGrath[37] proposed a conformational equilibrium of the α' enolate. Dreiding models of α' enolates of compounds **16, 17, 19e, 18e,** and **20e** show that the molecules can adopt a boat conformation; the intermediate cyclopropanone could, therefore, result from a zwitterion precursor. However, in the case of **18e,** it is not possible to exclude an S_N2 mechanim. In the case of **16, 17,** and **19e,** the possibility of disrotatory "in" and "out" cyclization of the zwitterion[45] could explain the regioselective formation of the less hindered cyclopropanone. In fact, the formation of α-hydroxyketone from **17** and α-methoxyketone from **19e** shows that the axial stereochemistry required of the halogen in order to form these secondary products (see below) can indeed be reached in these substrates.

In cyclohexanyl haloketones **21, 22, 23,** and **24** the very bulky equatorial substituents 1,3 to the halogen prohibit the existence of any chair–chair equilibration. However, these derivatives can adopt a boat conformation in which the axial halogen is not sterically hindered, except in the case of derivative **24,** which does not ring contract in the presence of sodium methoxide in methanol.[41b,†] When zwitterion formation is favored under such conditions, its disrotatory "in" or "out" cyclization can give rise to both α- and β-cyclopropanones. Ring opening of the latter will lead to esters with configurations[45] opposite to those of the initial carbon–bromine bond. Such a hypothesis is in accord with the reactivity observed in the case of **23.**[41a]

Kinetic studies do not allow one to differentiate between the two mechanisms of cyclopropanone formation depending on the halogen stereochemistry. A difference in behavior can be observed, however, between equatorial brominated and chlorinated derivatives. Loftfield's process (Scheme 1), according to which the first step is reversible, is verified in the case of α-bromocyclohexanones. So in **17**[35] and **22**[28a] (X = Br) hydrogen–deuterium

---

† Compound **24** undergoes ring contraction with piperidine via a process different from the Favorskii mechanism.[41b]

exchange is small or nonexistent before rearrangement: the slow step of the reaction is, therefore, enolate formation. However, this mechanism is not verified for α-chlorocyclohexanones such as **22** (X = Cl) in which deuterium exchange is important, or total, in α' before occurrence of the rearrangement.[28a,46] In the latter case, the slow step of the reaction is halogen ionization.

On the other hand, it is clear that the axial stereochemistry favors zwitterion formation (Schemes 7 and 8). This zwitterion is the precursor[1a,22,41a] of the cyclopropanone, or is in equilibrium with it.[41a†] The "in" or "out" cyclization of the zwitterion would then be controlled by steric factors in such a way that only the less hindered cyclopropane would be formed.[45] This is the case of the esters formed with retention of configuration from 3α-bromo-8,13-epoxy-2-labdanone (**18a**) [37] and from trans-9-chlorodecalone **19a**.[38]

It should be noted that the α-haloketones reported in Table 4 are characterized by their aptitude to yield secondary substitution products. This peculiar reactivity is related to the fact that the halogen is axial, which favors formation of α-methoxyketone and α-hydroxyketone[2] simultaneously.

In the presence of a protic solvent such as methanol, Bordwell suggests that the α-methoxyketones are formed by the methanolysis of the enolized α-bromoketone via a 2-hydroxy allylic intermediate.[28d,47] The latter is actually the protonated form of the dipolar ion (Scheme 8).

Normally, the α-methoxyketone is obtained with retention of configuration. In the presence of a symmetrical intermediate such as 2-hydroxyallyl, derivatives methoxylated in α and α' can be formed: this is the case for the derivatives of **19a**.[38] However, steric factors can orient the substitution selec-

---

† It is quite a critical question to know whether the cyclopropanone and the zwitterion are mesomeric forms, or two real chemical species in equilibrium with each other. According to Burr and Dewar[22] the unsubstituted zwitterion is not a mesomeric representation of the unsubstituted cyclopropanone; in fact, the zwitterion is an individual chemical species which is the precursor of the cyclopropanone, Unfortunately, there is no experimental proof of this hypothesis as cycloaddition reactions are not observed when nonsubstituted cyclopropanones are interacted with dienophiles. It should be noted that this kind of cycloaddition reaction probably proceeds via a dipolar mechanism.[68]

However, evidence for the existence of a zwitterionic intermediate has been obtained by Bordwell[41a] when the zwitterion is substituted. In this case, Bordwell assumed an equilibrium between the zwitterion and the corresponding cyclopropanone. This was confirmed by Fort,[56b] who was able to trap the zwitterion by a cycloaddition reaction.

Consequently, in absence of substitution, the zwitterion is either a mesomeric form of the cyclopropanone or a species in equilibrium with the cyclopropanone; but in the latter case, this equilibrium would be totally displaced toward the cyclopropanone, a view in agreement with the theoretical results of Greenberg and Liebman.[48]

When the zwitterion or the cyclopropanone is included in a ring system, we think that there is indeed an equilibrium between the two individual species. The latter will depend on the ring strain, which is directly related to the ring size. For example, in the case of five-membered rings, the equilibrium is totally displaced toward the zwitterion.[77-80]

SCHEME 8

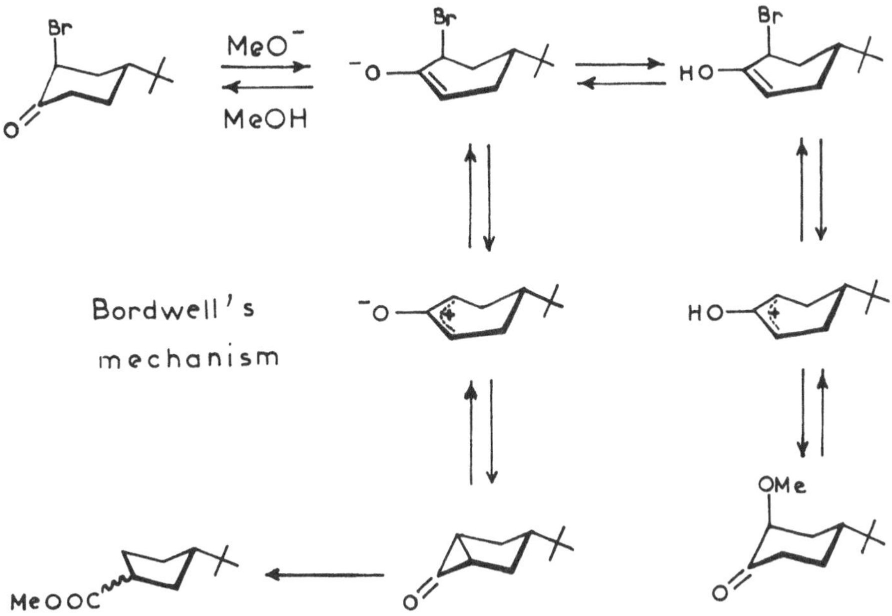

Bordwell's
mechanism

tively to α or α', as is the case for the 12-α-methoxyketone formed from the 9-α-bromo steroid **25**.[42]

The mechanism which allows one to understand how the α-hydroxyketone is formed is known, and involves the opening of an intermediate epoxyether. This reaction, which is solvent independent, is observed for the haloketones **26**, **27**, **20a**, and **28e** (Table 4)[2,28a]:

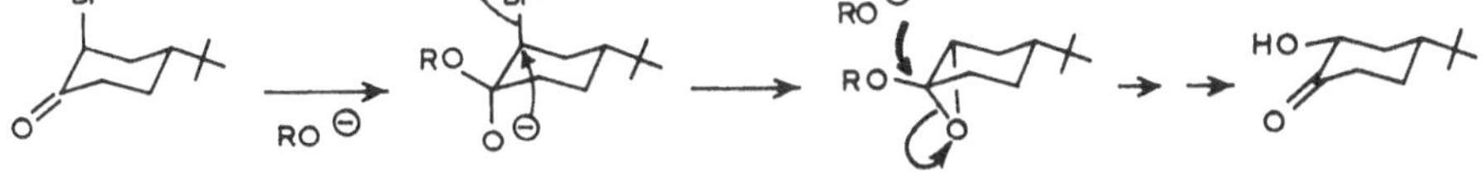

The following conclusions arise from a comparison of the mechanisms explaining the cyclopropanone formation in the cyclohexane and open-chain series.

Although an S_N2 mechanism cannot be excluded in the case of **18e**, the zwitterionic mechanism is the only one which gives a coherent explanation of the reactivity of axial and equatorial halogen derivatives. However, the halogen stereochemistry plays an important role. An axial halogen favors the secondary substitution reactions and the formation of delocalized species; these can lead to ring contraction in aprotic media and to substitution reaction in protic solvents. On the other hand, an equatorial halogen always gives rise to rearrangement products.

The general mechanism proposed by Bordwell (Scheme 8) is in agreement with the reactivity of derivatives in which the halogen can become axial.[28d,47] However, several examples discussed below will show that this mechanism explains neither the substitutions observed in aprotic media—in strongly basic media generally favoring rearrangement processes[2]—nor the difference in reactivity between cyclohexane derivatives and juxtacyclic open-chain derivatives (Schemes 3 and 5) studied by Tchoubar and coworkers.[27,30] In the latter case, there is no doubt that other structural features play a decisive role in this difference of reactivity. Recent results concerning zwitterion stability relative to the cyclopropanone as a function of substitution allow a better understanding of such a structural influence. We will see later that substitution in open-chain compounds and (or) strain (in cyclic compounds) stabilize the zwitterion and destabilize the cyclopropanone.[48]

Taking into account the results discussed up to now, we shall consider the influence of steric factors on the development of the Favorskii rearrangement.

C. Influence of Structural Features on the Reactivity of Acyclic and Cyclic Substrates

The formation of α-alkoxyketones according to Bordwell's mechanism (Scheme 8) requires a protic medium; in fact 2-chloro-6-phenylcyclohexanone (24) (Table 3) gives α-methoxyketones in both protic and aprotic media.[41b] Similarly, 2-chlorocyclopentanones yield solely α-alkoxyketones whatever the experimental conditions,[49] whereas the α-halocycloalkanones from C_7 to C_{12} as also $C_{(16)}$[50] give essentially ring contraction. Finally, in the cyclic series, the absence of ring contraction is observed (independently of the nature of the solvent) with either cyclopentane or substituted cyclohexane substrates.

In open-chain compounds, where rearrangement products are more easily obtained, 2-bromo-2,4-dimethyl-3-pentanone gives the α-alkoxyketone as a major product even when the reaction is carried out under aprotic conditions.[12,51] Consequently two factors seem to be decisive in inhibiting cyclopropanone formation and yielding the α-alkoxyketone even in an aprotic solvent. The first is the degree of substitution, which can play a role in open-chain or cyclic compounds. The second operates only in cyclic systems and is ring strain which can be calculated by force field techniques.[52] This kind of strain can be due to valence angle deformations or to transannular nonbonded interactions or bonded interactions.[53]

According to what has been discussed until now, structural features should be capable of stabilizing a zwitterionic intermediate. Evidence for such an intermediate has been obtained by Smith and Gonzalez in the case of substitution

SCHEME 9

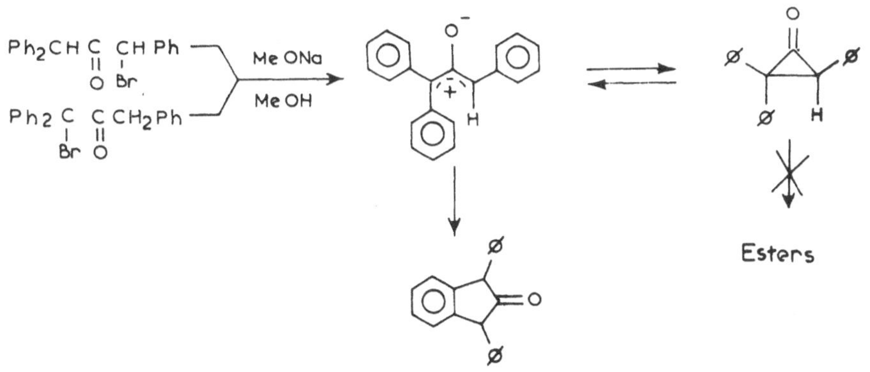

reactions carried out in basic medium on the 2-chlorocyclohexanone ^{14}C labeled at $C_{(1)}$ and $C_{(2)}$.[54]

Another example shows that under typical Favorskii rearrangement conditions, substituents stabilize the zwitterion and inhibit cyclopropanone formation: 1-bromo- and 3-bromo-1,1,3-triphenylpropanones[55] do not yield esters. In these cases the equilibrium between the cyclopropanone and the zwitterion is totally displaced in favor of the latter. (See Scheme 9.)

α-Alkoxyketone formation by substitution of the zwitterion has been proposed by Fort[56] and by Turro[53] in aprotic media. From all preceding arguments, it is reasonable to assume that in substrates where the degree of substitution and/or the ring strain are important, the contribution of the zwitterion is more significant than that of the corresponding cyclopropanone; in other words, the zwitterion–cyclopropanone hybrid is more weighted toward the zwitterion.[†] Under such conditions, the reaction will evolve toward the α-alkoxyketone even in aprotic media.

Accordingly it appears that the results concerning the relative stabilities of the zwitterion and the cyclopropanone as a function of substitution can be extended to the Favorskii rearrangement in open-chain or cyclic compounds. In the latter case, both substitution and ring strain have to be considered.

It is reasonable to predict that in compounds not bearing many substituents or in unstrained cyclic compounds, the contribution of the cyclopropanone will be more important than that of the zwitterion and

---

[†] Cyclopropanone ring opening to give the ester is generally considered [2e] to be rapid and not to have any influence on the overall rate of the reaction. Similarly, direct substitution of the zwitterion is not considered to be the slow step in the reaction.[28] As the reaction rate of both cyclopropanone and zwitterion are comparable, it is reasonable to assume that the ratio of the products formed reflects the ratio of the intermediates from which they are issued.

rearrangement will be the only process occurring, whatever the solvent. The reactivity of juxtacyclic open-chain substrates (Schemes 3 and 5) and of medium and large rings from C_7 to C_{12} as well as C_{16},[50] is in good agreement with such an interpretation.

Between these two extreme situations, however, it is possible to consider substrates where the reactivity is a function of the base concentration and the nature of solvent. For instance, among open-chain compounds when 29a, 29b, and 29c were subjected to a low concentration of MeONa in MeOH they gave the ester, a mixture of the ester and the α-methoxyketone, and only the α methoxyketone, respectively.[27,28d] It is clear that, compared to compound 29a, the introduction of a methyl and a phenyl substituent progressively increases the proportion of α-methoxyketone until it is the main and unique reaction product. The kinetic data show that there is a rate acceleration for compounds 29b and 29c relative to 29a. These rate increases are sufficiently significant for us to reasonably consider that the methyl and phenyl group have a stabilizing influence on the positive charge which progressively develops on the α carbon in the transition state.[28d] Nevertheless, when submitted to strongly basic conditions, the three α-haloketones give essentially only the corresponding esters.

$$PhCH_2CCH_2X \qquad PhCH_2CCHCH_3 \qquad PhCH_2CCHPh$$

(structures: 29a with C=O; 29b with C=O and CHCH$_3$ bearing X; 29c with C=O and CHPh bearing X)

29a   29b   29c

In cyclic compounds, particularly with cyclohexane derivatives, the formation of α-alkoxyketones is also sensitive to the base strength and to solvation. Under weakly basic (MeONa–MeOH) conditions, the 2-chlorocyclohexanone 27 (Table 4) gives 28% of methoxyketone besides the esters. When the base concentration is increased,[41a] the esters are the only products. When the base strength is modified similar results can be observed. With strong bases such as t-AmONa and PhCH$_2$ONa, respectively, in the corresponding alcohols, 27 gives the ring-contracted ester in yields of 65%[2a] and 75%.[23] On the other hand, in the presence of a weak base such as PhONa in phenol, the α-phenoxyketone is obtained as the only product.[54] Compounds 29a, 29b, 29c, and 2-chlorocyclohexanone 27 react via a zwitterion intermediate.[28d,37,54] The stabilization of the latter by a methyl (29b) or phenyl group (29c), or by a ring (27), is certainly less important than that observed with (i) two phenyl groups on the zwitterion, as is formed from 1- and 3-bromo-1,1,3-triphenylpropanones, or (ii) one phenyl and a ring, as in 2-chloro-6-phenylcyclohexanone (24) (Table 3). These three haloketones yield the substitution product via the zwitterion, even in an aprotic solvent. Actually in such a medium, or in a protic medium and a high base concentration, compounds 29a, 29b, and 29c and 2-chlorocyclohexanone 27 give mainly rearrangement products. In such cases, it can be concluded that stabilization due to different features does not modify the relative stability of the zwitterion and the cyclopropanone, the latter always being the more stable. However, it can be reasonably assumed that the energy difference between the two species is small, and that the additional influence of the medium can modify the relative stabilities. Thus, in a protic medium, substitution and rearrangement products are obtained simultaneously from **29b** and **27**, whereas the α-alkoxyketones are the only products formed from **29c**. The fact that the proportion of substitution side products decreases with the protic character of the reaction medium shows that a solvation effect on the zwitterion is likely to be involved. This effect would reverse the relative stability of the intermediates in favor of the zwitterion. It is not possible to state precisely whether the zwitterion is stabilized by hydrogen-bonding solvation or by 2-hydroxyallyl cation formation. We favor this latter hypothesis without rejecting the solvation concept.

Accordingly, the evolution of the Favorskii rearrangement can explained as being a function of the relative stability of the zwitterion relative to the cyclopropanone. This stability is itself dependent on the nature of solvent (protic or aprotic) and on structural features (substitution and/or ring strain).

In order to explain all the experimental results described up to now, we suggest a general mechanism (Scheme 10) which simultaneously explains the formation of the rearrangement esters and of the α-alkoxyketones.

When the cyclopropanone is more stable than the zwitterion[†] the reaction will give the rearrangement products (step b) (examples: substrates bearing few substituents, juxtacyclic open-chain compounds, or those of the type $ArCH_2CCH_2X$ and $ArCHCCH_3$).

$$ArCH_2\underset{\underset{O}{\|}}{C}CH_2X \quad \text{and} \quad Ar\underset{\underset{X}{|}}{C}H\underset{\underset{O}{\|}}{C}CH_3$$

If the zwitterion is more stable than the cyclopropanone only substitution products will be obtained in protic (step *a*) and aprotic (step *c*) media (examples: highly substituted or constrained substrate: 1-bromo- or 3-bromo-1,1,3-triphenylpropanones, α-halocyclopentanones, derivatives such as **24** (Table 3). When the energy difference between the cyclopropanone and the zwitterion is small, the rearrangement will be the only process observed when a high base concentration is used in either a protic or aprotic medium (step *b*).

In a protic medium at low base concentration substitution will be the only process observed (step *a*) (examples: compounds **19e** or **19a**) unless the substitution (step *a*) is occurring simultaneously with the rearrangement (step *b*) (examples: **29b**; $PhCHCCH_2CH_3$, 2-chlorocyclohexanone, and all derivatives in

$$Ph\underset{\underset{X}{|}}{C}H\underset{\underset{O}{\|}}{C}CH_2CH_3$$

Table 3 except **19e** and **24**).

[†] See footnote on p. 550.

SCHEME 10

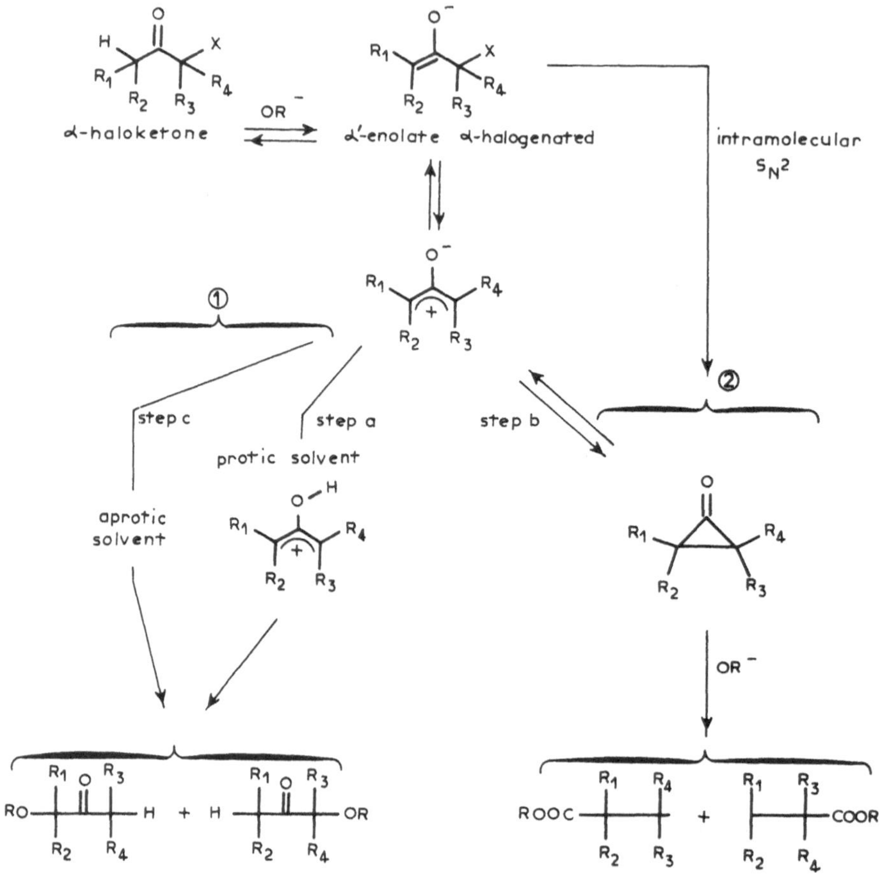

(1) if R_1, R_2, R_3 and $R_4 \neq H$ or bulky substituents, or strained rings.

(2) if R_1, R_2, R_3 and $R_4 = H$ or small substituents or unstrained rings, no
    participation of the medium (protic or aprotic).

This mechanism is also in agreement with the reactivity observed for cyclohexane derivatives with an axial halogen. Compounds **25, 26, 27** (R = Me), **19a**, **28e**, and **28a** have a tertiary halogen, and as **19e** and **19a** they are sensitive to solvation; in protic media it is only substitution that is observed. In aprotic media, **19a** yields the ring contraction product, whereas the yield of the esters formed from **28e** and **28a** increases. In **18a**, the halogen is secondary and reacts as in 2-chlorocyclohexanone, whereas the structure of **20** favors the side reaction giving rise to an epoxyether.

We will now discuss the stability of the zwitterion relative to the cyclopropanone.

4. Relative Stability of Zwitterion and Cyclopropanone

A. Acyclic Substrates

Fort was the first to obtain evidence for the zwitterion by studying the behavior of chlorodibenzylketone **29c** under Favorskii rearrangement conditions. In the presence of a weak base such as 2,6-lutidine in MeOH **29c** yields the methoxydibenzylketone **30**.[56a] With a stronger base such as EtONa in EtOH, the rearranged derivative **31** is obtained.[57] On the other hand, in the presence of lutidine and furan in DMF, Fort isolated the adduct **33**.[56b] The reactivity of compound **29c** is similar to that of 2-chlorocyclohexanone (**27**) (Table 4), which forms a symmetrical intermediate.[9,54] The similarity in reactivity is confirmed by the isolation of adduct **33**, which provides evidence for the intervention of a symmetrical delocalized intermediate which is most likely to be the zwitterion **32**, stabilized by two phenyl groups.

As in lutidine-methanol, 1-chloro-2-propanone undergoes methanolysis much more slowly than does **29c**; this difference in reactivity is due to the effect of the two phenyl groups in zwitterion **32**.[56a]

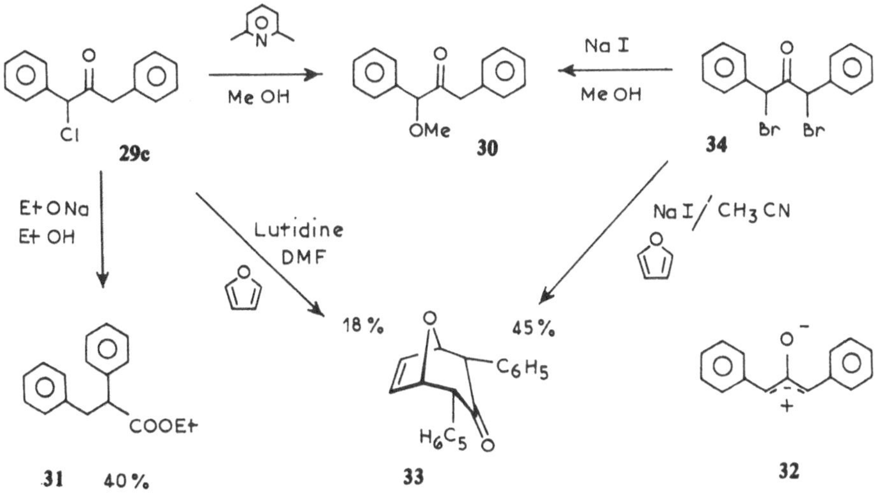

It is of interest to note that the reduction of $\alpha\alpha'$-disubstituted and -dihalogenated ketones by sodium iodide,[56a,58] zinc–copper,[59] or iron nonacarbonyl[60] involves a zwitterionic intermediate. The capture of the latter by a diene is possible only when the zwitterion is stabilized by phenyl or alkyl substituents.[60] A typical example of the formation of this kind of zwitterion is shown in the case of the reaction of the $\alpha\alpha$-dibrominated ketone **34** with sodium iodide to give **32**.

Besides these experimental results, many research groups have tried to analyze the problem of the zwitterion–cyclopropanone duality in open-chain compounds using different theoretical approaches. A few years ago, Burr and Dewar[22] made the assumption that in the case of the Favorskii rearrangement, a nonclassical delocalized intermediate—less stable than the corresponding cyclopropanone—was involved. These results, based on a simple approach, seem to have been confirmed experimentally by Fort.[56] Hoffmann[61] found a reverse order of stability which has been disproved experimentally since cyclopropanone could be observed in the vapor phase.[62] Since then, numerous calculations (Table 5) have tended to confirm the initial results of Burr and Dewar[22] as giving a better estimate of the difference in energy between the zwitterion and the cyclopropanone, which according to Table 5 seems rather high. Experimental results are in agreement with these differences in stability, as unsubstituted cyclopropanones[68] do not give cycloaddition reactions.[10,68a] As a consequence it is reasonable to assume that the contribution of zwitterion is small.[69b]

When the cyclopropanone is substituted by methyl groups, however, the corresponding derivatives undergo cycloaddition or ring-opening reactions[10,68] much more easily. Indeed, by using thermodynamic data Greenberg and Liebman[48] showed that the stability of the zwitterion is higher as its degree of

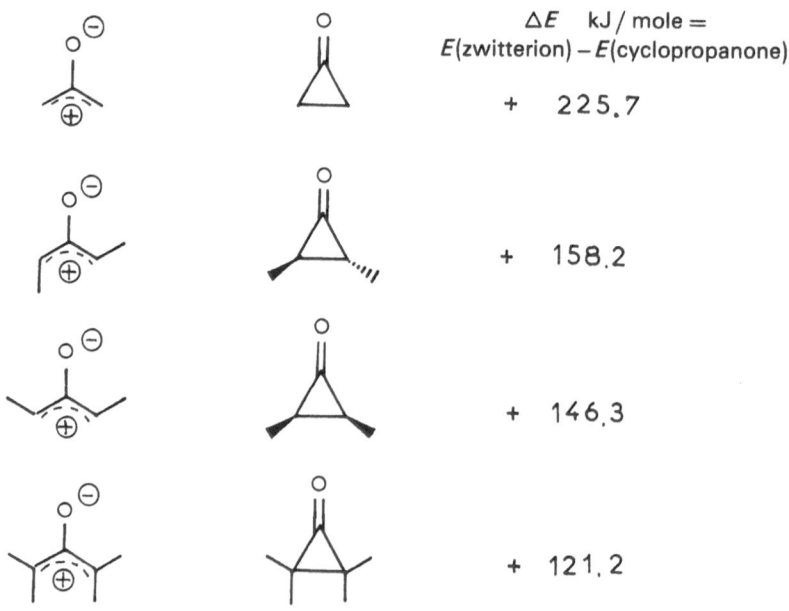

$$\Delta E \quad \text{kJ/mole} =$$
$$E(\text{zwitterion}) - E(\text{cyclopropanone})$$

+   225.7

+   158.2

+   146.3

+   121.2

TABLE 5. Theoretical Calculations of the Difference in Energy Between
Zwitterion and Cyclopropanone

References	Type of calculation	$\alpha_A{}^a$	$\alpha_B{}^b$	$E_A{}^c$	$E_B{}^c$	$\Delta E = E_A - E_B{}^c$
63	MINDO/2 [d]		64°24′ [e]	+184.3	−142.5	436.8
64	INDO	90	62°30′			919.6
65	INDO	90	65°			969.9
	Ab initio	105	65°			346.9
66	MINDO	70	64°36′			150.4
48	Thermodynamic [f] calculations					225.7
67	MINDO/3 [g]		61°52′ [e]	+195.2	−81.5	276.7

[a] Zwitterion angle in degrees corresponding to energy minimum.
[b] Angle opposite to the CH_2–CH_2 bond calculated or adopted for the cyclopropanone, expressed in degrees and minutes. The experimental angle determined by microwave study is 64°36′ ± 50′.[62]
[c] Energies expressed in kJ/mol.
[d] Calculations of heats of formation at 25°C.
[e] Calculated from the bond length given by the authors.
[f] Estimated heats of formation by bond-additivity schemes.
[g] Calculations of heats of formation.

substitution increases. Other reactions, such as the racemization[69] of the optically active *trans*-2,3-di-*tert*-butylcyclopropanone are explained in terms of the lower-energy oxyallyl intermediate (free energy of activation ≃112.9 kJ/mol).

The numerical values should be considered as having only relative value; however, they clearly show the trend toward stabilization of the delocalized intermediate when it is tetramethylated.

In conclusion, calculations based on a theoretical approach and experiments show that in open-chain compounds the zwitterionic contribution increases with substitution.

B. Cyclic Substrates

The relative stability of the zwitterion–cyclopropanone couple is also sensitive to ring strain. For instance, it is well known that an α-halocyclopentanone does not give a ring contraction product under Favorskii rearrangement conditions, but gives instead mainly substitution products.[49] Dipole moment measurements[70] and infrared data[71] obtained on these compounds give an angle of 77° between the carbon-halogen and C=O bond; this value is in excellent agreement with that of 78° calculated for a half-chair conformation.[52,72] Consequently, the carbon–halogen bond certainly has some axial character, which allows one to compare reasonably the α-halocyclopentanones to the derivatives in Table 4. An axial halogen favors zwitterion formation. As we will show, this zwitterion is actually more stable than the corresponding cyclopropanone (that is, in fact, bicyclo[2.1.0]pentan-5-one (35), which has never been isolated).[73] The zwitterionic structure 36 exhibits

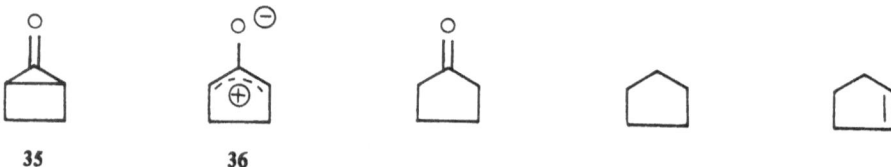

35   36

three sp^2 carbons, a fact which implies two more hybridization changes from sp^3 to sp^2 relative to the cyclopentanone. According to the internal strain (I-strain) concept, these hybridization changes are accompanied by new strains—angular strain, transannular strain, strain due to eclipsed bonds—which can have a decisive influence on structural changes.[74]

In the case of five-membered rings, the introduction of one or two sp^2 carbon(s) is generally easy, simply because the product thus obtained is more stable[72] than the original. This can be confirmed by strain energy calculations, using the force field technique[75]: cyclopentanone (strain energy: 27.7 kJ/mol) and cyclopentene (strain energy: 28.4 kJ/mol) are more stable than cyclopentane (strain energy: 31.4 kJ/mol).[52,75] The introduction of a third sp^2 carbon in the cyclopentene renders the ring planar, and introduces strain due to the eclipsing of the methylene carbon–hydrogen bonds. However, the strain increase is balanced by the resonance energy of the zwitterion. If, nevertheless, there is any strain increase, it is certainly negligible compared to the strain developed in bicyclo[2.1.0]pentan-5-one (35), which involves a cyclobutane. The strain in the latter is estimated at 109.8 kJ/mol by the same technique[76]; furthermore a cyclopropanone would also be included in the same structure. Although there is no accurate strain estimation for cyclopropanone, it would expected to be similar to that in methylenecyclopropane, estimated to be 174.3 kJ/mol.[75] Consequently it is not unreasonable to assign bicyclo[2.1.0]pentan-5-one (35) a higher strain energy than that of zwitterion **36**. This is confirmed by experiment. Indeed, Paquette and coworkers have obtained experimental evidence for a zwitterionic intermediate which is part of a five-membered ring by trapping it in an intramolecular process.[77] This is the way they explain the formation of adduct **38** from the α-halogenated cyclopentanone **37** under Favorskii rearrangement conditions.

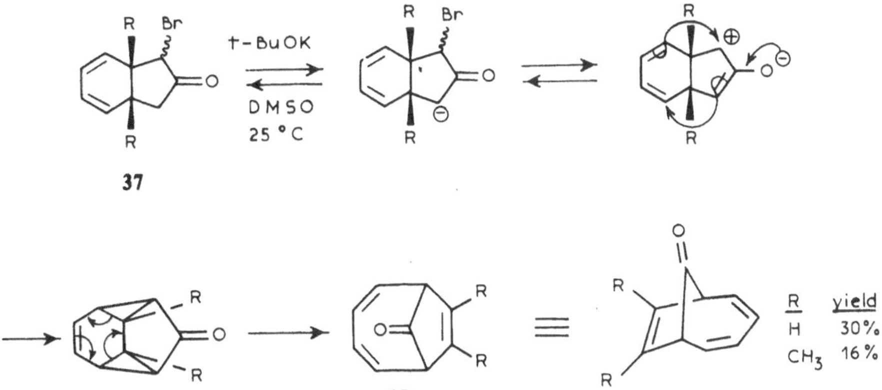

Mention should also be made of the formation of a similar zwitterion by irradiation of the dienone **39**.[78] This intermediate yields the adduct **40** by addition to furan.

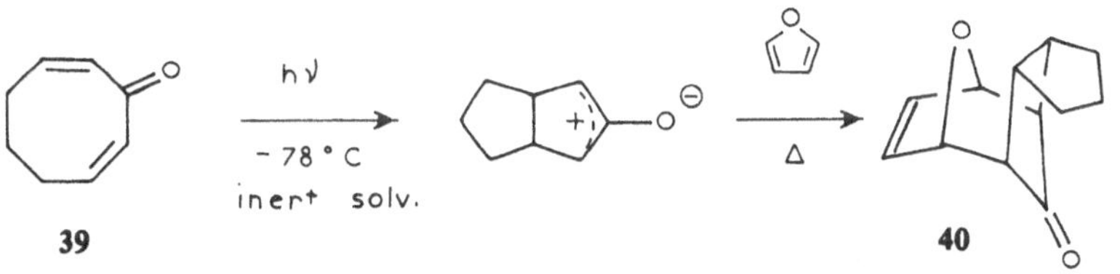

This relative stability of cyclopropanone–zwitterion in five-membered rings can be generalized, as a similar behavior is found, for instance, in the Ramberg–Bäcklund reaction of α-halogenated sulfones.[79] (The mechanism of this reaction is known, and involves the formation of an episulfone heterocyclic analog of the cyclopropanone.) When this intermediate is formed in a strained system such as propellane **41**, Paquette and coworkers[80] suggest that it opens to a zwitterion in order to explain the formation of the substitution product **42**.

Contrary to what is observed in the Favorskii rearrangement, such a substitution product is seldom observed in the Ramberg–Bäcklund reaction, as cyclic α-halogenated sulfones generally give only the ring-contracted alkene under basic conditions. The sulfone which has a skeleton similar to the diene **37** yields, as does the latter, the intramolecular cycloaddition product formed by the interaction of the zwitterionic episulfone with the diene.[77]

It can be concluded that the zwitterion is more stable than the cyclopropanone when it is generated in five-membered rings.

One could actually wonder why the α-halogenated cyclopentanone does not undergo ring contraction by a semibenzilic mechanism. As discussed below, the conditions required for such a mechanism to be operative are very narrowly related to the carbon halogen stereochemistry which must be equatorial. We have just seen that in α-halocyclopentanones, the carbon halogen bond is pseudoaxial, a stereochemistry which is more suitable to zwitterion formation.

At this point, it is interesting to compare the absence of reactivity of α-halocyclopentanones under Favorskii rearrangement conditions, with the ring contraction observed when such α-halocyclopentanones are included in cage structures used as precursors of cubane derivatives. We will show in what follows that this marked difference in reactivity can be explained by the equatorial stereochemistry of the halogen in the cage α-cyclopentanone derivatives.

5. Conclusion

By comparing the various parameters which are operative in the symmetrical mechanism of the Favorskii rearrangement in open-chain and cyclic α-haloketones, we have been able to stress the decisive influence of the structural features on the outcome of the reaction.

The symmetrical mechanism implies the occurrence of a zwitterion and/or a cyclopropanone. Experimental facts and theory tend to show that there is not always a strict equivalence between the cyclopropanone and the corresponding zwitterion.[†] Instead, two different species can actually coexist, separated by a potential barrier depending on steric and electronic factors.[48] These species will have a decisive effect on the outcome of the symmetrical mechanisms which will result from a compromise between two major influences:

[†] The formation of the protonated zwitterion (the 2-hydroxyallyl cation) has been proved by nmr spectroscopy of the cyclopropanone in super acid medium;[116] see footnote p. 550.

(i) Structural effects, i.e., substitution and ring strain, which modify the potential barrier between the zwitterion and the cyclopropanone. When the overall influence is important, the zwitterion can become more stable than the cyclopropanone. In some extreme cases, the cyclopropanone does not even have a real physical or chemical existence.[117] (ii) Effects resulting from the nature of medium. In a protic solvent, the zwitterion is stabilized by solvation relative to the cyclopropanone. In aprotic media there is no solvation of the negative charge, and the relative stabilities of the zwitterion and the cyclopropanone depend only on structural effects.

The mechanism proposed in Scheme 10 explains the formation of both the rearranged product and the α-alkoxyketones. Solvation exerts its influence on the zwitterion. This scheme includes also an intramolecular S_N2 process and Tchoubar's results[27] (see Schemes 3 and 5) and the reactivity under various conditions of the derivatives reported in Table 3 and 4, which has already been discussed. In the cyclohexane series and with an axial halogen, the zwitterion contribution is expected to be favored and to depend upon the ring strain.

In this respect, the behavior of the following three α-bromoketones[117] is of particular interest as increasing ring strain is observed going from *b* and *c* to *a*:

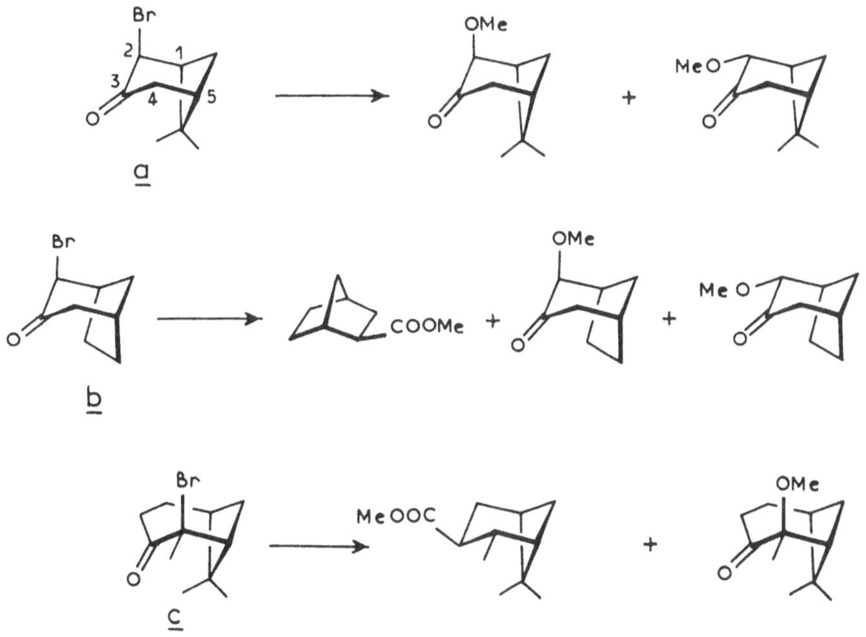

Thus the α-bromoisonopinone derivative *a*, which is definitely more constrained than *b* or *c*, yields only substitution products under Favorskii rearrangement conditions, whatever the solvent or the base. Obviously only the zwitterion is formed; its transformation into the corresponding cyclopropanone is totally prohibited by the molecular structure of the substrate and particularly by the large distance between $C_{(2)}$ and $C_{(4)}$.

On the other hand, derivatives *b* and *c*, which are less strained than *a*, only give ring contraction in aprotic media, whereas ring contraction and substitution are observed in a protic solvent. Both the cyclopropanone and zwitterion intermediates are formed. Furthermore, the proportion of rearranged ester increases strikingly with the increase in base concentration, a fact which is in agreement with the mechanism proposed in Scheme 10.

There is a continuous variation in the stability of the cyclopropanone compared to the corresponding zwitterion. In juxtacyclic compounds the cyclopropanone is more stable than the zwitterion. In somewhat strained bridged bicyclic series (compounds *b* and *c*) both the cyclopropanone and the zwitterion can occur. In more highly strained derivatives such as α-bromocyclopentanone, or α-bromoisonopinone (*a*), the zwitterion is more stable, and the cyclopropanone cannot be formed. There is a point at which the order of stability of the two species is reversed. In Scheme 11 we attempt to show these variations of stability as a function of the structural features of the substrates involved, i.e., substitution and ring strain. This diagram has been established by taking into account the energy differences calculated by Greenberg and Liebman[48] for various substituted zwitterion-cyclopropanone couples which correspond to open-chain α-haloketones. These data have then been extrapolated to cyclopropanone systems submitted to external ring strain.

The different areas shown in Scheme 11 cannot be defined very accurately owing to lack in numerical data. However, the general trend is fairly precise, as shown by the agreement observed for numerous cyclohexane and constrained derivatives taken from the literature.

Three different areas can be defined on this diagram. In area I, the cyclopropanone is more stable than the zwitterion. Theoretically—and this is generally experimentally observed—one should only obtain rearrangement products. However this area corresponds to open-chain α-haloketones, and it is not possible to exclude cyclopropanone formation by a direct intramolecular S_N2 reaction, or to rule out substitution processes involving S_N2 mechanisms.

Moving to the right (Scheme 11), one reaches area II, where the difference in energy between the cyclopropanone and the zwitterion is small. In many cases, the two species will not have real independant existence.[†] α-Bromobicyclo[3.2.1]octan-3-one (*b*) and 2-bromo-2-methylbicyclo[4.2.1]octan-3-one (*c*) belong in this area, as do the usual cyclohexane derivatives and some juxtacyclic compounds. In this area, either the zwitterion or the cyclopropanone may be favored according to the nature of the solvent and the base strength.

[†] See footnote p. 550.

SCHEME 11

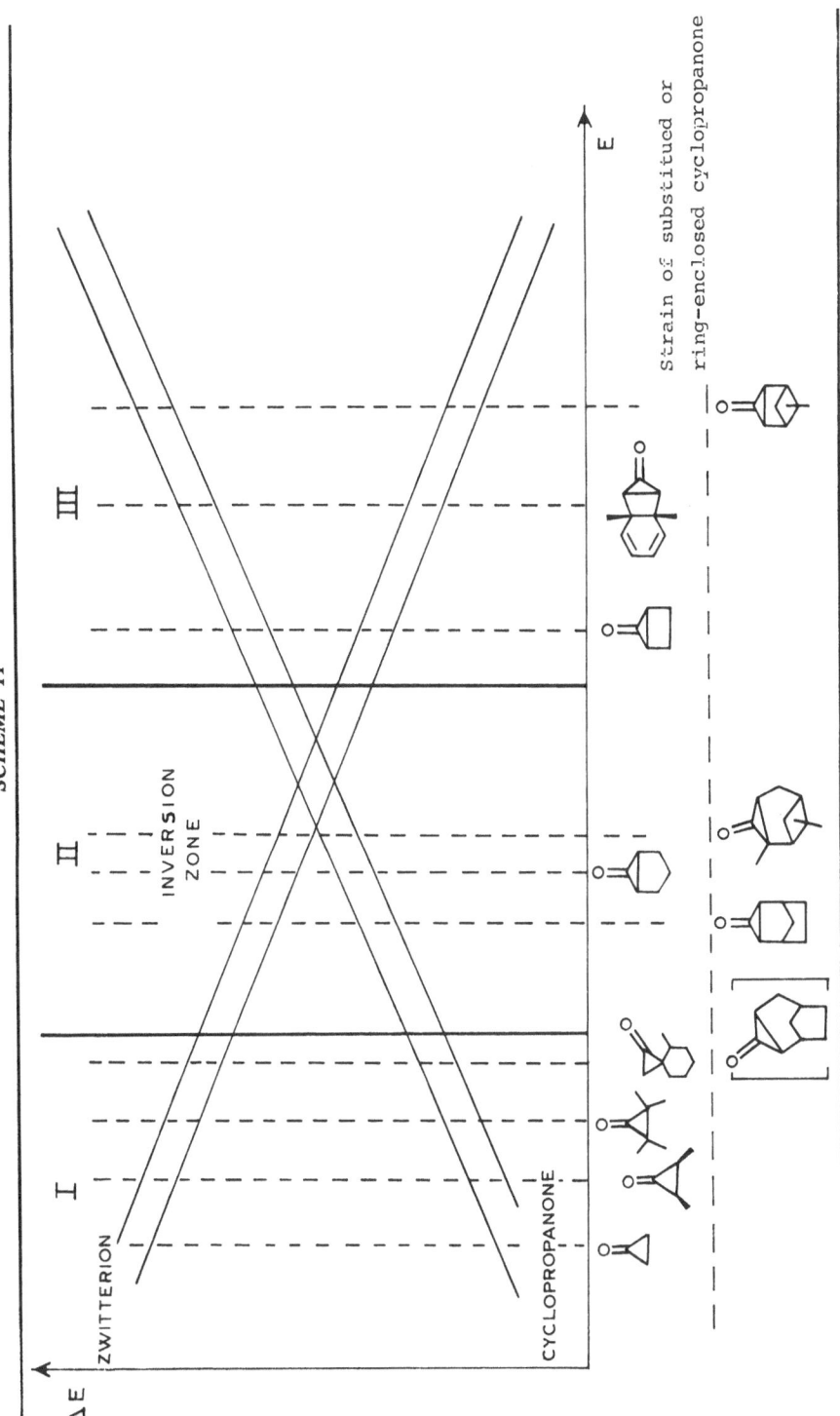

In area III on the extreme right, one finds substrates which would lead to highly strained intermediate cyclopropanones, which are unlikely to exist. It is actually the zwitterion which is formed exclusively, so that only substitution is observed. As there is no possibility of zwitterion–cyclopropanone interconversion, modification of experimental conditions will be without influence. Both α-bromocyclopentanone and α-bromoisonopinone (*a*) are illustrative examples of substrates found in this area.

III. UNSYMMETRICAL MECHANISM

1. Occurrence

The semibenzilic mechanism[81] requires (Scheme 12) a nucleophilic addition at the carbonyl carbon. Then, in a concerted process, the halogen is ejected by the 1,2-anionotropic migration of the $C_\alpha'-\underset{\underset{O}{\|}}{C}$ bond.

Several structural requirements appear to be necessary in order to be able to observe the semibenzilic mechanism. Thus the acidity of the α' hydrogen plays an important role. Indeed, it should be weakly acid in order to favor nucleophilic attack at the carbonyl carbon. Of course, this kind of mechanism will be facilitated by the absence of an α' hydrogen; the rearrangement is then known under the name of "quasi-Favorskii." Since both substrates just discussed react via the semibenzilic mechanism, we shall discuss them simultaneously.

The carbon–halogen bond must have a given geometry. Indeed, it is clear from Scheme 12 that, in the second step, the carbon–carbon bond which migrates and the carbon–halogen bond must be antiparallel so that the rearrangement can occur.[82] In order to achieve this kind of geometry in cyclic compounds, it is imperative to have an equatorial halogen.[83] Also the α-

SCHEME 12

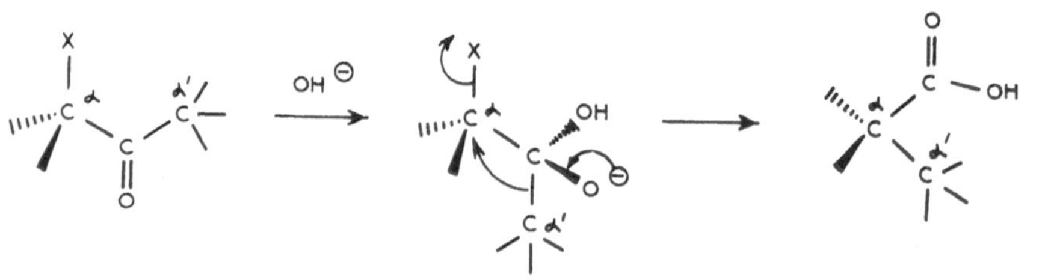

haloketones must be such that the formation of an intermediate cyclopropanone is inhibited. This is the case in substrates having an α' hydrogen which is sufficiently sterically hindered that the symmetrical mechanism is not possible. In cyclic compounds, an equatorial halogen and important ring strain will favor the intervention of the semibenzilic mechanism.

Close examination of the structure of the rearranged product (Scheme 12) shows that the semibenzilic mechanism is regioselective; the carboxyl or ester group is selectively bound to the α carbon initially bearing the halogen in the starting material. However this criterion is not sufficient in itself to differentiate this mechanism from the one involving a cyclopropanone. The latter, under the influence of appropriate electronic factors, can ring-open selectively to give an ester fixed on the α carbon.

Finally, let us recall the experimental conditions used for these reactions: sodium or potassium hydroxide are used as bases in the presence of protic (water,alcohol) or aprotic (xylene, toluene) solvents. Silver or mercuric salts can also be used; they facilitate the ionization of the carbon halogen bond and introduce a modification of the semibenzilic mechanism, called push–pull mechanism.[84] As these latter reagents are no longer the bases usually employed for the Favorskii rearrangement, we will not discuss their use. Sodium or potassium hydroxide in an inert solvent are the best reagents to use in the rearrangement of substrates not bearing an α' hydrogen (quasi-Favorskii).[85]

2. Evidence for the Semibenzilic Mechanism

A. Open-chain Juxtacyclic Substrates

The semibenzilic mechanism implying an inversion of configuration at the α carbon initially bearing the halogen (Scheme 12) has been discussed by Tchoubar and Sackur.[81] However Smissman and coworkers[86] ruled out a semibenzilic mechanism, and proposed an α-ketocarbenium ion intermediate on the basis of the obtention of racemic rearranged products from optically active juxtacyclic α-haloketones. In fact, the α-haloketones studied by Smissman and coworkers[86] contained a nitrogen atom which facilitated ketocarbenium ion formation. The racemization of the product can be explained by this peculiar feature and this is a special case of the quasi-Favorskii rearrangement. However, as recently shown unambiguously by Charpentier-Morize and coworkers,[87] an inversion of configuration occurs at the α carbon when compounds 43 and 44 are treated under basic conditions. These results provide final proof for the occurrence of the semibenzilic mechanism in open-chain juxtacyclic compounds.

The inversion of configuration observed at the α-carbon is an important fact as it presupposes (i) that the carbon–bromine and the migrating carbon–carbon bonds are antiparallel, and (ii) that the carbon–bromine and the

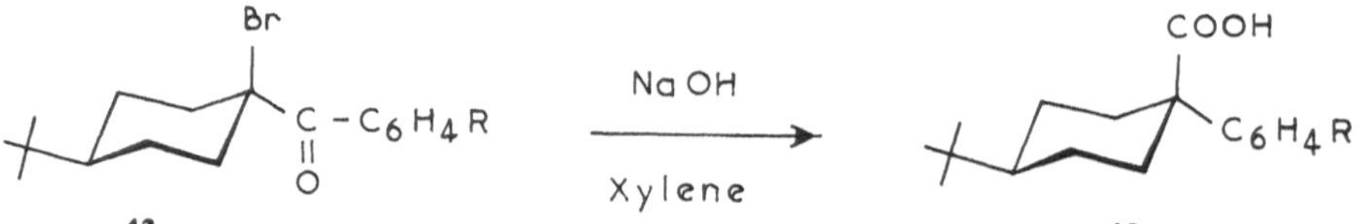

43

45

44

46

carbon–O⁻ bonds are *syn*, as shown in conformation **47**. If the *anti* conformation **48** is assumed by the substrate, epoxy-alcohols – which subsequently yield
secondary substitution products[87]—are formed.

That the conformation **47** suitable for rearrangement is frozen is attributed
to a surface effect on sodium hydroxide.[81,87]

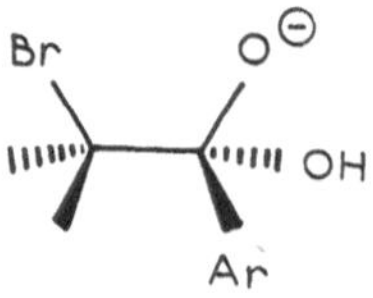

47

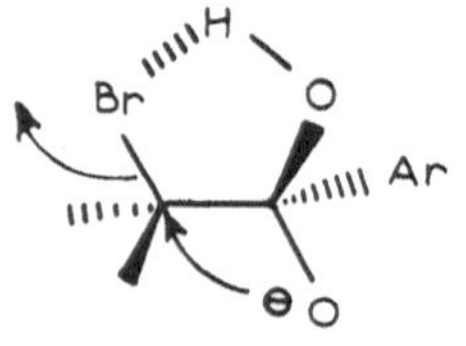

48

B. Cyclic Substrates

In order to achieve antiparallelism between the carbon–halogen and the
carbon–carbon bonds in cyclic substrates, it is necessary to have an equatorial
halogen. But we have seen earlier, that this kind of substrate is able to react
according to an S$_N$2 mechanism to yield the cyclopropanone intermediate. In
order to avoid the formation of such an intermediate, the cyclic derivatives with
an equatorial halogen must have important ring strain; in such a case, they can
undergo a semibenzilic process, as in the typical example of 2-
bromocyclobutanone **49**. Conia and coworkers,[88] have elegantly shown that, in
heavy water, this bromoketone reacts according to the semibenzilic mechanism
to yield the cyclopropane carboxylic acid **50**, as there is no deuterium incorporated in the rearranged cyclopropane ring.

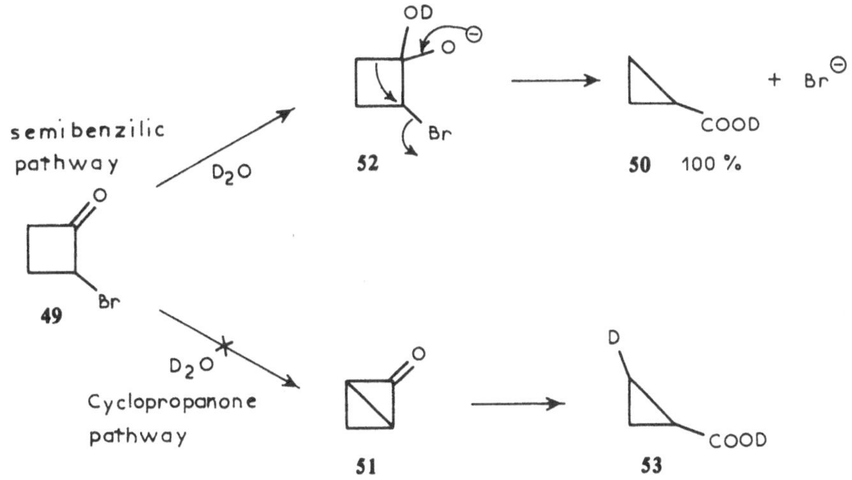

As a matter of fact, in a symmetrical mechanism, the intermediate cyclopropanone **51** would be a bicyclo[1.1.0]butan-4-one, quite highly strained, so that its formation would be very difficult, or almost impossible from the thermodynamic point of view. The strain in bicyclo[1.1.0]butane has been estimated to be 278 kJ/mol.[76] The strain in the corresponding ketone is at least equivalent or even higher, as the internal strain will be increased by the introduction of an sp^2 hybridized carbon. The fact that **49** ring contracts under basic conditions according to a semibenzilic mechanism is in good agreement with the known aptitude of the cyclobutanone carbonyl to undergo addition reactions easily.[89] This behavior is likely to be a consequence of the important internal ring strain of this ring, which is estimated at 121.3 kJ/mol, and is lowered to 113.5 kJ/mol by changing the hybridization of one carbon from sp^2 to sp^3.[52] Thanks to the folded geometry of the cyclobutane ring in **49**, the carbon–bromide bond is able to adopt an equatorial conformation.[90] But cyclobutanones **54**, **55**, and **56**, in which the halogen is axial or pseudoaxial, do not react in basic medium,[91] as a result of the conformation of the halogen, although steric hindrance might also play an important role. However, in the spirocyclobutanones **57** and **58**, the axial stereochemistry of the halogen can explain the formation of substitution

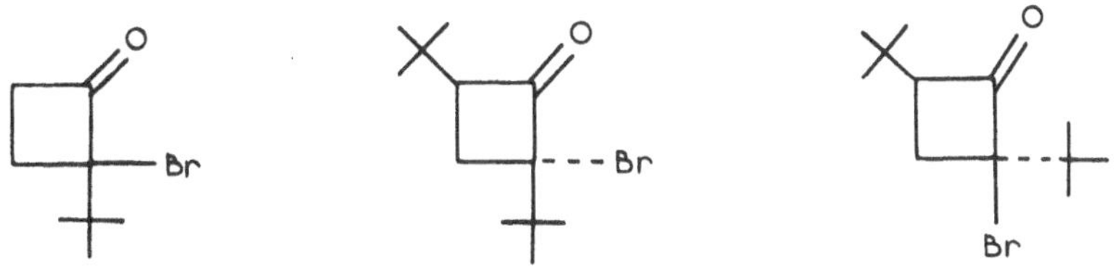

54   **55**   **56**

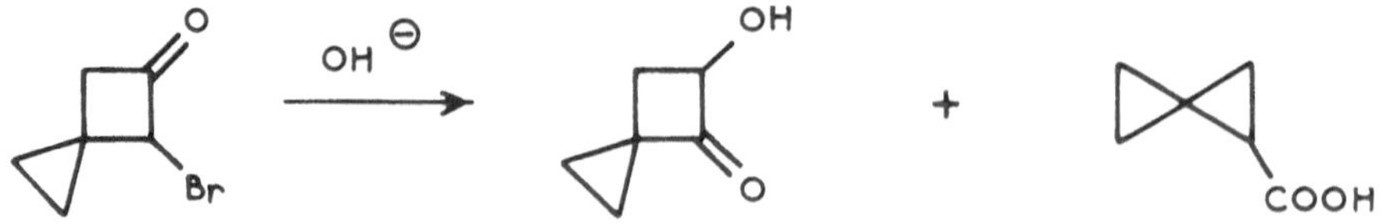

products.[91] In the latter case, it is difficult to know whether these derivatives are formed by direct halogen substitution, or through the zwitterionic intermediate.

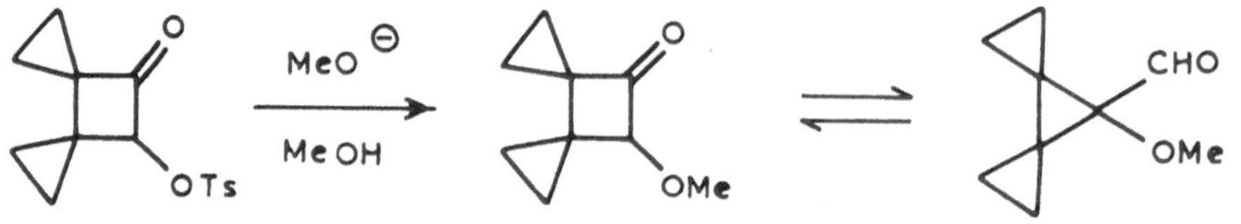

In short, if the halogen is equatorial, ring strain controls the choice between two processes: (i) a semibenzylic mechanism occurring when ring strain is important, (ii) a symmetrical mechanism occurring when the strain is sufficiently low so that the intermediate cyclopropanone can form.

3. Competition of the Cyclopropanone and the Semibenzilic Mechanisms as a Function of the Ring Strain and of the Base Strength

The Favorskii rearrangement of α-haloketones arising from bicyclo[3.3.1]nonanone (**59**), bicyclo[4.3.1]decanone (**60**), and bicyclo[5.3.1]undecanone (**61**), has been studied by Warnhoff, Wong, and Tai.[92] It should be noted that all these derivatives have two common structural features, one which does not vary, that is, the equatorial or pseudoequatorial orientation of the halogen due to their bridgehead position, and one which is variable, namely the ring strain which decreases when the ring size increases,[76] and accounts for the choice of mechanisms during the rearrangement.

These mechanisms can be identified by analysis of deuterium incorporation (Scheme 13) as in the experiments of Conia and coworkers.[88] According to Bredt's rule the deuterium exchange of α' hydrogen before rearrangement (see also experiment 5 in Scheme 13) is difficult. Consequently, deuterium incorporation is a direct reflection of the reaction process involved.

In the presence of sodium hydroxide and deuterated solvents (experiments 1 and 3), derivatives **59** and **60** undergo the Favorskii rearrangement via the semibenzilic mechanism, whereas it is the symmetrical mechanism which is involved with **61** in experiment 7. At first glance, it is only the difference in ring strain (the only variable factor) which is at the origin of this different

SCHEME 13

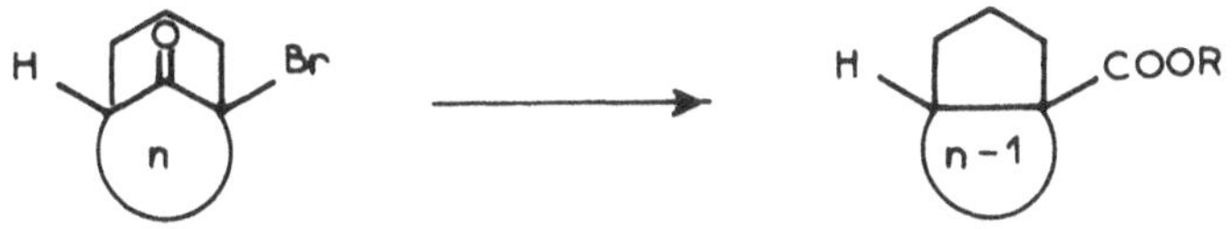

59: $n = 6$
60: $n = 7$
61: $n = 8$

62: $n = 6$
63: $n = 7$
64: $n = 8$

Reactant	Experiment	Base	Solvent	Products	Yield %	D/molecule	Mechanisms
59	1	NaOD	EtOD–D$_2$O	62, R=H	95	0.00	Semibenzylic
	2	t-BuOK	t-BuOD	62, R = t-Bu	95	0.026	Semibenzylic
60	3	NaOD	EtOD–D$_2$O	63, R = H	80	0.01	Semibenzylic
	4	NaOMe	MeOD	63, R = Me	96	0.00	Semibenzylic
	5	t-BuOK	t-BuOD	63, R = t-Bu	67	0.83	Cyclopropanone
				60	—	0.00	
	6	AgNO$_3$	EtOD–D$_2$O	63, R = H	85	0.00	Semibenzylic
61	7	NaOD	EtOD–D$_2$O	64, R = H	88	0.90	Cyclopropanone
	8	AgNO$_3$	EtOD–D$_2$O	64, R = H	25	0.005	Semibenzylic

mechanistic behavior. In the case of the less constrained compound 61, it is possible to form the cyclopropanone, a process rendered impossible in the case of the more strained derivatives 59 and 60. A closer inspection of the results reported in Scheme 13 shows, however, that the base strength is also playing an important role.

Indeed, in the presence of a stronger base than sodium hydroxide, namely, sodium t-butoxide[93] in deuterated t-butanol (experiments 2 and 5), 59 and 60 follow a different reaction pathway. Compound 59 still reacts according to a semibenzilic mechanism, whereas 60 rearranges via the cyclopropanone process. In the latter case, the increase in the base strength allows the removal of the α' hydrogen, whereas in presence of sodium hydroxide (experiment 3) it is the semibenzilic process which is favored. These experiments clearly show the relative importance of base strength and ring strain. With a compound such as 60, where the relatively low ring strain allows cyclopropanone formation, it is the base strength which decides the fate of the reaction. For a more strained substrate such as 59, it is no longer the base strength but the ring strain which is decisive in the selection of the reaction mechanism. Consequently, experiments 1, 3, and 7 allow the prediction that 61 is less constrained than 59 and 60; nevertheless, as 59 and 60 are also differently strained, the mechanism will be dependent of the base strength which must, therefore, be specified.

It is known that the acidic character of a hydrogen, and particularly that of a bridgehead hydrogen, is more marked when bound to a carbon having more s character.[94]

One could then wonder if it is possible to draw a relationship between this carbon hybridization and the overall strain of the substrate. In fact, hybridization is certainly related to the angular deformation strain around the α' carbon; but the latter is only a component of the overall constraint of the substrates, which also involves transannular and eclipsing interactions.[5]

4. Conclusion

The analysis of the Favorskii rearrangement semibenzilic mechanism sheds light on some experimental and structural factors which allow a differentiation from the symmetrical mechanism.

The semibenzilic mechanism will be preferentially involved (i) in open-chain substrates which do not have an α' hydrogen (ii) or in cyclic substrates exhibiting simultaneously an equatorial halogen and ring strain prohibiting cyclopropanone formation. In order to observe this mechanism, it is also necessary to have an antiparallelism of the rearranging groups. This has a bearing on the experimental conditions in open-chain compounds where surface effects in the base play a decisive role in this arrangement. Consequently open-chain substrates should be treated under heterogeneous conditions such as those obtained with sodium or potassium hydroxide in aprotic medium. On the other hand, in cyclic compounds, the migrating groups concerned in the rearrangement process, are already in a suitable position for the semibenzilic mechanism, when the halogen is equatorial, and it is then possible to work in a homogeneous medium as shown by the experiments of Warnhoff and coworkers.[92] Ring strain will play a decisive role in the selection of the reaction mechanism and can be estimated at 56 kJ/mol(Reference 52) for the haloketone **59**. One can ask whether or not this value is representative of the minimal strain required for semibenzilic mechanism to be followed. A value of 41.8 kJ/mol has been calculated by Allinger[52] for the ring strain in cyclopentanone in the envelope conformation. It is interesting to note that such α-halocyclopentanones undergo ring contraction via a semibenzilic mechanism when they are involved in cage compounds such as the dibromocubanone studied by Eaton and Cole[3]. But in these compounds the halogens are equatorial, whereas in isolated α-halocyclopentanones the halogen is axial. No ring contraction is observed in the latter case as neither the semibenzilic mechanism nor the cyclopropanone mechanism can occur; the first is prohibited by the wrong orientation of the halogen and the second by the drastic strain increase. So it seems that it is really the equatorial orientation of the halogen which is decisive, and it seems difficult to set a precise barrier for the minimal ring strain to permit the observation of the semibenzilic mechanism.

IV. THE FAVORSKII REARRANGEMENT IN BRIDGED POLYCYCLIC AND CAGE COMPOUNDS

1. Results and Discussion

We report in Table 6 the results observed when various bridged polycyclic and cage halo- or dihaloketones were submitted to Favorskii rearrangement conditions. Three classes of compounds can be distinguished: (i) cubane derivatives, (ii) bicyclo[2.2.1]heptane derivatives, and (iii) adamantane derivatives. These kind of derivatives are more strained than the cyclohexane substrates reported in Tables 3 and 4, as shown by force field calculations carried out by Schleyer and Allinger.[52,75] These authors set up a strain scale of various constrained hydrocarbons and ketones, according to which homocubane (Table 7), which has a skeleton similar to that of α-haloketone **65**, has a strain energy estimated at 493–505 kJ/mol, whereas the strain of cubane corresponding to the skeleton of the rearranged product **79** is estimated at 693–698 kJ/mol. Similarly, going from bicyclo[2.2.1]heptane derivatives (same skeleton as **73**, **74**, and **75**) to bicyclo[2.2.0]hexane (same skeleton as **87**, **88**, and **89**) involves a strain increase of 73–221 kJ/mol. On the other hand, hydrocarbons of the adamantane series are less constrained. Thus the rearrangement of adamantane (skeleton of **77**) into noradamantane (skeleton of **92**) is related with a strain increase going from 29–31 to 79–83 kJ/mol. The transformation of homoadamantane (skeleton of **78**) into the twist-brendane (skeleton of **94**) is actually the result of two ring contractions. The strain related with this process increases from 58–71 to 125–142 kJ/mol. For comparative purposes, let us recall that the strain energy of cyclohexanone is 26.7 kJ/mol.

Comparatively, the strain differences between the starting hydrocarbons and the rearranged skeleton are fairly important. In view of these quite large differences, it does not seems unreasonable to consider that the strains of the hydrocarbons and the corresponding α-haloketones are of the same order of magnitude. Thus the interest of the Favorskii rearrangement is particularly stressed as a ring contraction technique allowing an entry into derivatives which are more constrained than the original starting α-haloketones (Table 7).

It should, however, be noted that the various classes of compounds reported in Table 6 exhibit reactivity differences. It is indeed possible to differentiate the compounds which are able to undergo a double ring contraction in the cubane and adamantane series. The 1,4-dihalogenated diketones **67** and **68** of the cubane series give products of double ring contraction, but there is no evidence for the intermediates which would result from a single ring contraction. It is interesting to note that it is the cyclopentanone which ring contracts, whereas it is known that isolated α-halocyclopentanones not included in a cage compound do not undergo ring contraction under Favorskii rearrangement conditions. On the other hand, the less strained 1,5-dihalogenated adamantane **77**

TABLE 6. Favorskii Rearrangement in Constrained and Cage-Type α-Haloketones

α-Haloketone	Base/solvent	Yield %	Product	Ref.
65 X = H		55		95
	25% aq.KOH, Δ	79		
X = H, Br, CH$_3$				96
66 X = CH$_2$, Y = H	30% aq.KOH, Δ	98		97,98
Y = H	10% aq KOH, Δ	62 **80**		99
X =		95		95
Y = Br	10% aq.KOH, Δ			
67	50% aq.KOH, Δ	**81**		3,100
68	aq.KOH, Δ	80 **82**		101
69	NaOH, toluene, Δ	65 **83**		102
70	NaOH, benzene, Δ	55 **84**		99
71	10% aq.KOH, Δ	85 **85**		103
72	KOH, xylene, Δ or 10% aq.KOH, Δ	85 **86**		103

(continued)

TABLE 6.—cont.

α-Haloketone		Base/solvent	Yield %		Product	Ref.
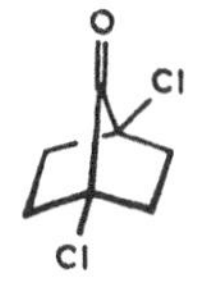	73	NaOH–THF, 0°C	93	87	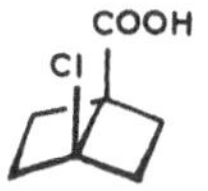	104-105
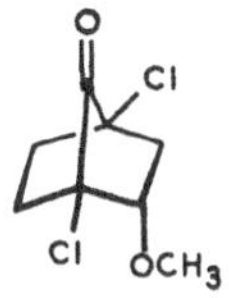	74	NaOH–THF, 40°C 40% aq. NaOH THF 40°C		88	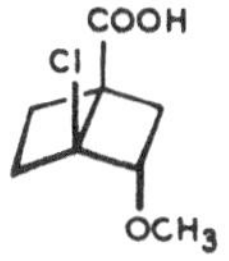	106
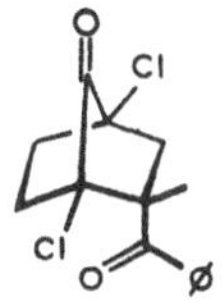	75	NaOH, THF, 0°C	70	89	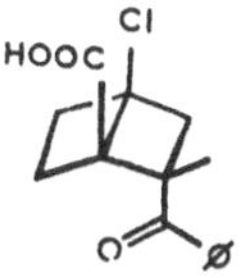	107
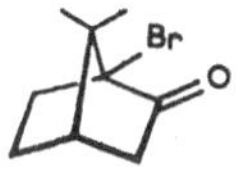	76	KOH, 2 M, 155°C	79	90		108
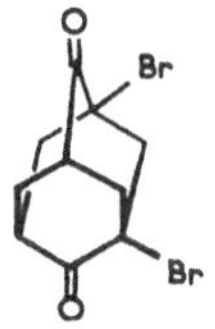	77	KOH–EtOH/ H₂O, Δ 4 hr	86	91	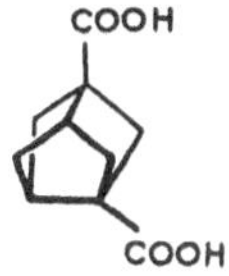	109
		KOH–EtOH/ H₂O, Δ 2 hr	85	92	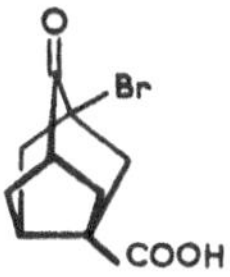	110
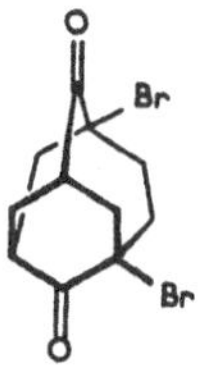	78	aq. NaHCO₃ Δ		93	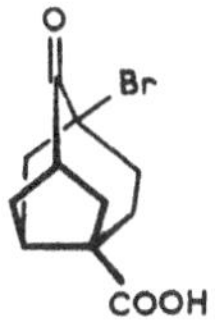	111
		50% aq. NaOH EtOH, Δ	98	94	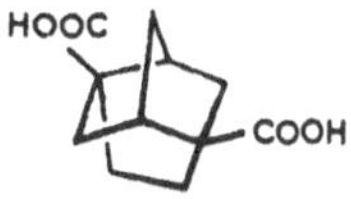	112

TABLE 7. Different Constrained and Cage-Type Substrates

Substrate	S.E.$_1$*^{a,b}	Substrate	S.E.$_2$*^{a,b}	$\Delta E = E_2 - E_1$
 Homocubane	493–505	 Cubane	693–698	200–193
 Bicyclo[2.2.1]heptane	73.1	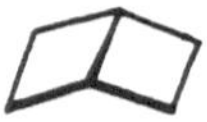 Bicyclo[2.2.0]hexane	221.5	148.4
 Adamantane	29–37	 Noradamantane	79–83	50–46
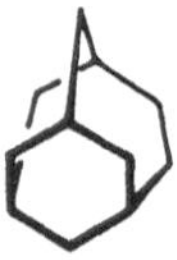 Homoadamantane	58–71	 Twist-brendane	125–142	67–71 For two rearrangement

^{a} S.E., strain energy. Values are given in kJ/mole.
^{b} References 75 and 52.

and homoadamantane 78 yield the single contraction derivatives 92 and 93, or those of double contraction 91 and 94, depending upon the experimental conditions. In this series, it is the cyclohexanone or the cycloheptanone which ring contracts. It is likely that the differences in strain between these two classes of α-haloketo derivatives are at the origin of this difference.

From the structural point of view, the substrates reported in Table 6 have halogens which are bonded to bridgehead carbons and are equatorial or pseudoequatorial. This stereochemistry allows us to compare the results obtained with these derivatives with those reported in Table 3 for cyclohexanes which also have equatorial halogens. Contrary to the cyclohexane derivatives, the strained α-haloketones give exclusively ring-contracted products regioselectively. This regioselectively, as well as the absence of secondary substitution products, differentiates clearly the behavior of the particularly strained derivatives of Table 6 from the cyclohexane substrates reported in Table 3. The experimental conditions also are distinctive. The compounds in Table 6 react in the presence of sodium or potassium hydroxide at room temperature. This differentiation raises the question as to whether the processes involved in the two series of compounds are identical or not.

The three types of compounds reported in Table 6 are more strained than those in Table 3. This factor, as well as others, such as the differences in reactivity and in experimental conditions, the steric and electronic factors, are all much more favorable for a semibenzilic rather than for a cyclopropanone mechanism. For derivative 67 of the cubane series, Eaton and Cole[3] proposed such a mechanism. For the bicyclo[2.2.1]heptane derivatives, with the exception of 76, no α or α' hydrogens are available, so that the semibenzilic mechanism is the only one possible (quasi-Favorskii). However, in the latter compounds, the $\alpha\alpha'$-dihalogenated or the polyhalogenated derivatives can undergo ring opening. We will now discuss the influence of the steric and electronic factors on this secondary reaction.

2. Secondary Reactions in Bridged Bicyclic and Cage Compounds

In these constrained molecules, the Favorskii rearrangement leads essentially to the ring-contracted acid or ester in good yields. However, the formation of ring-opened products has been reported when polyhalogenated strained ketones react according to the quasi-Favorskii mechanism. Results using this type of substrate are reported in Table 8.

This secondary reaction cannot be compared satisfactorily with the Haller–Bauer reaction,[113] as has sometimes been done in the literature.[99,103] In the presence of a strong base, rupture of the C–C=O bond is observed in nonenolizable ketones in the Haller–Bauer reaction. If the latter would occur, all the nonenolizable ketones of Table 6 would lead to ring-opened products and not to rearranged derivatives, as are experimentally observed. As shown by Dauben and coworkers,[114] this behavior is the result of the combined influence of steric and inductive effects of several chlorine atoms. The derivatives in Table 8 are tetra- or octachlorinated, so that many C–Cl and C–H bonds are, respectively, or mutually eclipsed. Torsional angle strain and nonbonded interactions will consequently add to the carbon skeleton ring strain. This is what is observed in the case of the octachlorohomocubanone 96, which undergoes ring opening, whereas, the monobrominated homocubanone 65 yields only the ring-contracted cubanoic acid 79. Compound 96 is more strained because of the presence of eight chlorine atoms. The ring opening which yields the intermediate carbanion 103 is not only sterically but also electronically favored; indeed the negative charge on the intermediate is stabilized by the inductive electron withdrawal by the chlorine atoms.[114]

TABLE 8. Secondary Reactions Observed in Favorskii Rearrangement of Constrained and Cage-Type Systems

α-Haloketone	Base/solvent	Yield, %	Products	Reference
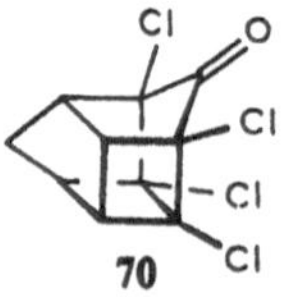 **70**	KOH, H$_2$O, Δ or KOH, φH, Δ	94	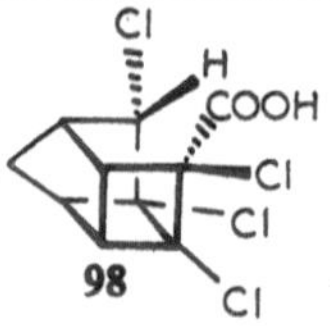 **98** (55%)	99 114
	NaOH, φH, Δ		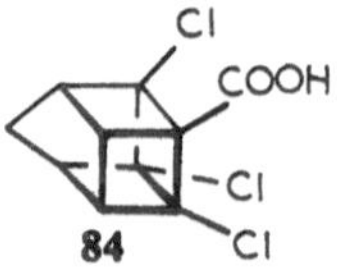 **84** (40%)	99
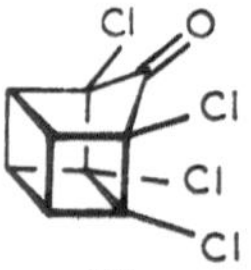 **95**	KOH, H$_2$O 6 hr reflux 3 days, 75°C	92	**99** 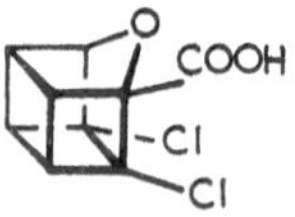	114
	Na OH, φH, Δ		**100** 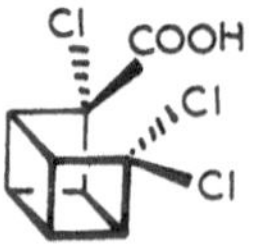	
 **96**	10% aq. KOH	>50	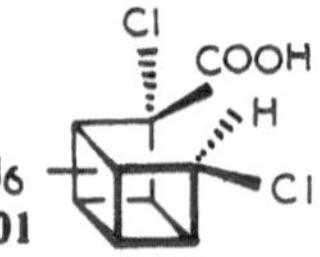 **101**	103
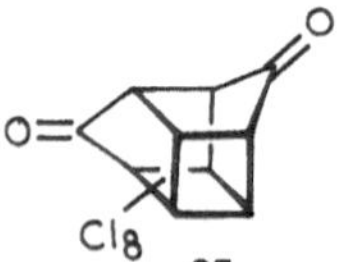 **97**	KOH, xylene, Δ or 10% aq. KOH, Δ	50	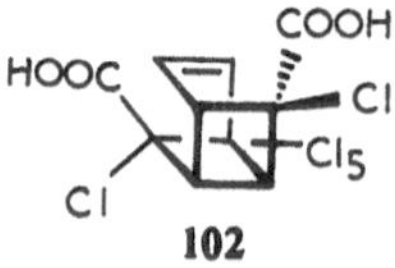 **102**	103
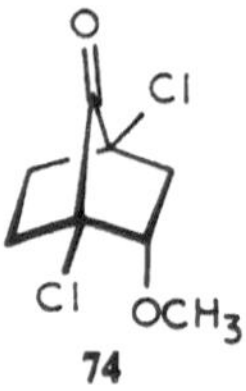 **74**	NaOH, THF, 40°C		**104** 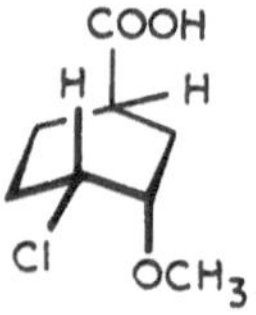 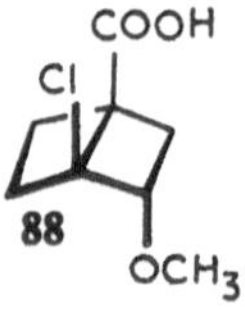 **88**	106

Let us also note that the ring-opening products reported in Table 8 are formed with retention of configuration as the hydrogen atom and the carboxyl group are located inside the cage compound.

The same type of arguments can explain how compound **102** is formed from the octachloroketone **97** by a double ring opening followed by an elimination reaction. Comparison of the reactivity of the tetra- (**95**) and octachlorinated derivatives (**96**) of homocubanone confirms the influence of the accumulation of chlorine atoms on the overall ring strain. On one hand, potassium hydroxide treatment of **95** gives the homooxycubane **99** and the ring-opened derivative **100**, while on the other, no derivative similar to **99** is obtained from the octachloroketone; this is most likely due to the fact that such a compound would be more strained than the ring-opened derivatives **101**. In fact, it is not unlikely that **101** is more strained than **100** because of the presence of additional chlorine atoms.

96 **103** **101**

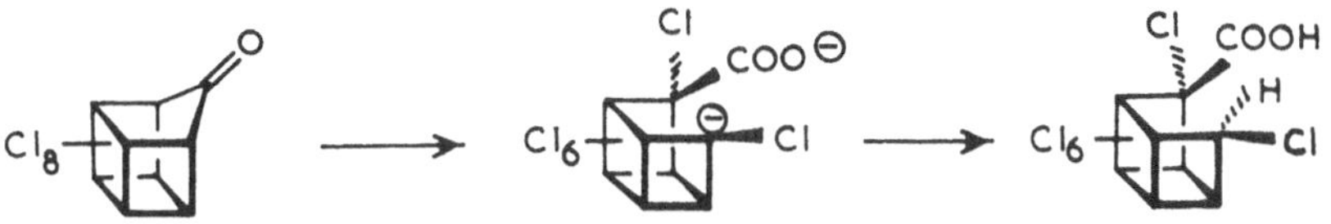

This ring-opening reaction has also been observed in the less strained bicyclo[2.2.1]heptane series.[106,107] When **74** bearing a methoxyl group at $C_{(2)}$ is subjected to basic conditions stereospecific formation of the ring-opened derivative **104** and of the Favorskii rearrangement product is observed.[106] While the nonconcerted semibenzilic mechanism is involved, carbanion **105** is formed because it is somewhat stabilized by the chlorine atom. When this is not the case, exclusive ring contraction leading to **87** is observed (see Table 6).[104] The proportion of ring-opened product **104** increases when a protic medium is used,[106] which is in agreement with the formation of the intermediate carbanion **105**.

74

	105	104	88
NaOH THF		11 %	89 %
95 % aq. NaOH THF		34 %	64 %

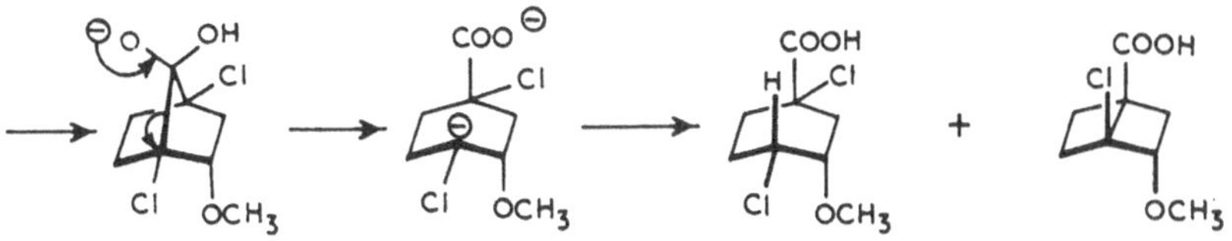

However, as shown in Table 8, the outcome of the reaction in the cubane series is not solvent dependent.

3. Conclusion

The role of steric and electronic factors which we have just discussed for the case of cage substrates can be further illustrated by the chemical behavior of 1-chlorobicyclo[2.1.1]hexan-2-one (106). The rearrangement product 108 is not observed and the ring strain is obviously relieved by the formation of the ring-opened product 107.[115]

This case is likely to represent the limit at which the Favorskii reaction can be used as an access to a very constrained product. At the other extreme of the other compounds mentioned in Table 8, 106 is enolizable and not polyhalogenated.

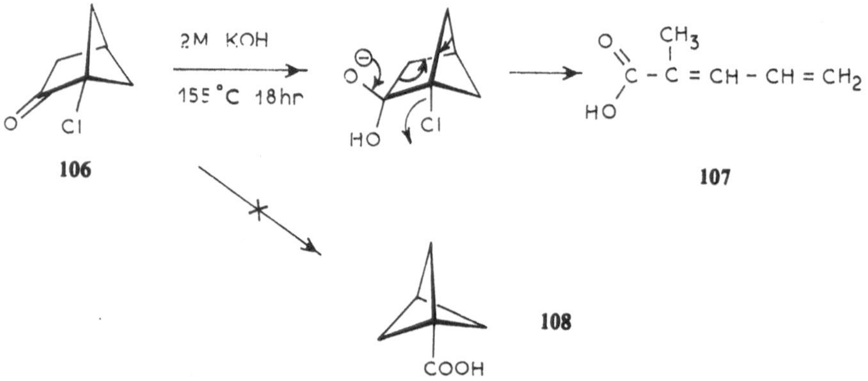

It is possible to explain this peculiar reactivity as being the consequence of halogen stereochemistry in 106 and/or the important strain energy which would develop in the ring contraction product 108. Molecular models show that the migrating C—C=O bond and the equatorial halogen are coplanar and antiparallel, an arrangement which is known to be the most favorable for the semibenzilic mechanism. The strain in the bicyclo[1.1.1]pentane is estimated at 388 kJ/mol,[75] which is 217 kJ/mol more than the 171 kJ/mol assigned as strain energy to the bicyclo[2.1.1]hexane. This will represent an important barrier which is actually higher than the one involved in the ring contractions occurring in the cubane series. Accordingly, it is most likely that the important ring strain of 108 relatively to 106 is at the origin of the ring opening observed in the bicyclo[2.1.1]hexane derivative 106.

V. GENERAL CONCLUSION

The comparative analysis of the development of the Favorskii rearrangement mechanism in cyclohexane and constrained polycyclic substrates allows one to stress the decisive influences of the halogen stereochemistry, of ring constraint, and of substitution on the final outcome of this reaction.

When a symmetrical process—which requires α'-enolizable α haloketones—is involved, the difference in reactivity between the juxtacyclic open chain series, the open chain phenyl-substituted series, and the cyclohexane series is best explained by the relative stability of the zwitterion–cyclopropanone couple. The latter depends on structural factors and on solvation. In order to rationalize all the results concerning these three categories of α-haloketones, we have proposed a general mechanism (Scheme 10) which is essentially based on the aforementioned interpretation. As shown by the results obtained with cyclohexanes, the stereochemistry of the halogen plays a decisive role. Indeed, the amount of secondary products is small with equatorial halogen derivatives, and large with axial halogen substrates.

The equatorial α-haloketones undergo ring contraction by an S_N2 or by a zwitterionic mechanism; in the latter case, it is likely that the molecule reacts in a boat conformation where the halogen has become axial. However, one can wonder about the reasons why the equatorial halogen derivatives— which react via a zwitterion similarly to their axial halogen epimers—do not give mainly secondary products.

When the halogen is axial in α-halocyclohexanones or α-cyclopentanones the formation of the zwitterion is favored. This process is apparently very sensitive to solvation and to ring strain. This interpretation is in agreement with the behavior of several α-haloketone derivatives in the bicyclo[3.1.1]heptane, bicyclo[3.2.1]octane, and bicyclo[4.1.1]octane series under basic conditions where the ring strain can be continuously varied.[117]

The semibenzilic mechanism is observed when a certain number of structural features are present in the α-haloketone. For open-chain derivatives it is necessary not to have an enolizable α' hydrogen. In cyclic compounds the α' hydrogen should be weakly acidic or absent, whereas the halogen ought to be equatorial. In fact, such an equatorial halogen also favors the symmetrical mechanism. In order to differentiate between these two mechanisms, two factors are determining: (i) ring strain and (ii) the strength of the base. When cyclopropanone formation is sterically possible, it is the strength of the base which will determine the reaction mechanism: a strong base will favor the symmetrical mechanism, whereas a weak base will orient the reaction toward a semibenzilic mechanism. When ring strain prohibits cyclopropanone formation, it is the semibenzilic mechanism which is observed whatever the base strength. Here also the halogen orientation plays an important role. A strained substrate with an equatorial halogen will undergo rearrangement, whereas substitution will be observed with an axial halogen.

Finally, in polycyclic strained derivatives of the cage type, the semibenzilic mechanism will be preferred because of the equatorial halogen and the fairly high ring strain, which actually prohibits cyclopropanone formation. In most of these derivatives, it is the cyclopentanone ring with an equatorial halogen which undergoes contraction. When these substrates are polyhalogenated the semibenzilic mechanism observed is no longer fully concerted. Steric and electronic factors will then also contribute to the ring opening in order to release strain.

REFERENCES

1. A. E. Favorskii, *Zh. Russ. Khim. Obshch.* **24**, 254 (1892); A. E. Favorskii, *Zh. Russ. Khim. Obshch.* **26**, 556 (1894).
2. (a) A. E. Favorski, *J. Soc. Chem. Phys. R* **26**, 590 (1894); (b) R. Jacquier, *Bull. Soc. Chim. Fr.*, 35 (1950); (c) B. Tchoubar, *Bull. Soc. Chim. Fr.*, 1363 (1955); (d) A. S. Kende, *Org. React.* **11**, 261 (1960); (e) A. A. Akhrem, T. K. Ustynyuk, and Yu. A. Titiv, *Russ. Chem. Rev.* **39**, 732 (1970); (f) H. Hart and G. J. Karabatsos, in *Advances in Alyciclic Chemistry*, vol. 3, Academic Press, London (1971), p. 46; (g) C. Rappe, in *The Chemistry of the Carbon–halogen Bond*, S. Patai, Ed., John Wiley and Sons, London (1973), p. 1084; (h) E. Buncel, in *Carbanions: Mechanistic and Isotopic Aspects*, Elsevier, Amsterdam (1975), p. 143; (i) P. J. Chenier, *J. Chem. Educ.* **55**, 286 (1978).
3. P. E. Eaton and T. W. Cole, *J. Am. Chem. Soc.* **86**, 962 (1964).
4. J. C. Aston and R. B. Greenburg, *J. Am. Chem. Soc.* **62**, 2590 (1940).
5. R. B. Wagner and J. A. Moore, *J. Am. Chem. Soc.* **72**, 2884 (1950).
6. D. W. Goheen and W. R. Vaughan, *Org. Synth. Collect.* **4**, 594 (1963).
7. C. Rappe, *Acta Chem. Scand.* **17**, 2766 (1963); C. Rappe and R. Adestrom, *Acta Chem. Scand.* **19**, 383 (1965).
8. W. D. McPhee and E. Klingsberg, *J. Am. Chem. Soc.* **66**, 1132 (1944).
9. R. B. Loftfield, *J. Am. Chem. Soc.* **73**, 4707 (1951).
10. H. H. Wasserman, G. M. Clark, and P. C. Turley, *Top. Curr. Chem.* **47**, 73 (1974).
11. J. F. Pazos and F. D. Greene, *J. Am. Chem. Soc.* **89**, 1030 (1967).
12. N. J. Turro and W. B. Hammond, *J. Am. Chem. Soc.* **87**, 3258 (1965).
13. W. H. Sach, *Acta Chem. Scand.* **25**, 2643 (1971); J. Julien and Nguyen Thoi-Lai, *Bull. Soc. Chim. Fr.*, 3948 (1970); R. P. Bell and O. M. Lidwell, *Proc. R. Soc. London Ser. A* **176**, 88 (1940); E. Gould, in *Mechanism and Structure in Organic Chemistry*, Holt, Rinehart and Winston, New York (1959), p. 383.
14. H. O. House, *Modern Synthetic Reaction*, 2nd ed., W. A. Benjamin, San Francisco (1972), p. 586.
15. H. E. Ferran Jr., R. D. Roberts, J. N. Jacob, and T. A. Spencer, *J. Chem. Soc. Chem. Commun.*, 49 (1978); R. R. Fraser and P. J. Champagne, *J. Am. Chem. Soc.* **100**, 657 (1978).
16. G. B. Trimitsis and E. M. Van Dam, *J. Chem. Soc. Chem. Commun.*, 610 (1974); A. Rassat and J. Ronzaud, *Tetrahedron* **32**, 239 (1976).
17. E. J. Corey, *J. Am. Chem. Soc.* **76**, 175 (1954); E. J. Corey and R. A. Sneen, *J. Am. Chem. Soc.* **78**, 6269 (1956).
18. O. S. Tee, *J. Am. Chem. Soc.* **91**, 7144 (1969); O. S. Tee and K. Yates, *J. Am. Chem. Soc.* **94**, 3074 (1972); O. S. Tee, J. A. Altmann, and K. Yates, *J. Am. Chem. Soc.* **96**, 3141 (1974).
19. H. R. Nace and B. A. Olsen, *J. Org. Chem.* **32**, 3438 (1967).
20. R. B. Loftfield, *J. Am. Chem. Soc.* **72**, 632 (1950).
21. J. G. Aston and J. D. Newkirk, *J. Am. Chem. Soc.* **73**, 3900 (1951).
22. J. G. Burr and M. J. S. Dewar, *J. Chem. Soc.*, 1201 (1954).

23. G. Stork and I. J. Borowitz, *J. Am. Chem. Soc.* **82**, 4307 (1960).
24. H. O. House and W. F. Gilmore, *J. Am. Chem. Soc.* **83**, 3980 (1961).
25. N. J. Turro and W. B. Hammond, *J. Am. Chem. Soc.* **88**, 2880 (1966).
26. N. J. Turro and W. B. Hammond, *Tetrahedron* **24**, 6029 (1968).
27. A. Gaudemer, J. Parello, A. Skrobek, and B. Tchoubar, *Bull. Soc. Chim. Fr.*, 2405 (1963).
28. (a) F. G. Bordwell, R. R. Frame, R. G. Scamehorn, J. G. Strong, and S. Meyerson, *J. Am. Chem. Soc.* **89**, 6704 (1967); (b) F. G. Bordwell, R. G. Scamehorn, and W. A. Springer, *J. Am. Chem. Soc.* **90**, 6751 (1968); (c) F. G. Bordwell, R. G. Scamehorn, and W. A. Springer, *J. Am. Chem. Soc.* **91**, 2087 (1969); (d) F. G. Bordwell and M. W. Carlson, *J. Am. Chem. Soc.* **92**, 3370 (1970); (e) F. G. Bordwell, M. Carlson, and A. C. Knipe, *J. Am. Chem. Soc.* **91**, 3949, 3951 (1969).
29. E. W. Garbish Jr. and J. Wohllebe, *Chem. Commun.*, 306 (1968).
30. A. Skrobek and B. Tchoubar, *C. R. Acad. Sci. Paris Ser. C* **263**, 80 (1966).
31. J. Baliarda and B. Tchoubar, *C. R. Acad. Sci. Paris Ser. C* **267**, 582 (1968).
32. G. Nee and B. Tchoubar, *C. R. Acad. Sci. Paris Ser. C* **283**, 223 (1976).
33. A. Nickon and N. H. Werstiuk, *J. Am. Chem. Soc.* **89**, 3914 (1967).
34. D. E. Evans, A. C. De Paulet, C. W. Shoppee, and F. Winternitz, *J. Chem. Soc.*, 1451 (1957).
35. H. R. Nace and B. A. Olsen, *J. Org. Chem.* **32**, 3438 (1967).
36. N. Pappas and H. R. Nace, *J. Am. Chem. Soc.* **81**, 4556 (1959).
37. M. J. A. McGrath, *Tetrahedron* **32**, 377 (1976).
38. H. O. House and H. W. Thompson, *J. Org. Chem.* **28**, 164 (1963).
39. E. E. Smissman, T. L. Lemke, and O. Kristiansen, *J. Am. Chem. Soc.* **88**, 334 (1966).
40. B. A. Olsen, *Diss. Abstr.* **25**, 4413 (1965).
41. (a) F. G. Bordwell and J. G. Strong, *J. Org. Chem.* **38**, 579 (1973); (b) F. G. Bordwell and J. Almy, *J. Org. Chem.* **38**, 571 (1973).
42. J. S. Cox, *J. Chem. Soc.*, 4508 (1960).
43. A. T. Rowland, *J. Org. Chem.* **27**, 1135 (1962).
44. J. Wolinsky and R. O. Hutchins, *J. Org. Chem.* **37**, 21 (1972).
45. R. B. Woodward and R. Hoffmann, *J. Am. Chem. Soc.* **87**, 395 (1965).
46. H. Ginsburg, *Bull. Soc. Chim. Fr.*, 3645 (1965).
47. F. G. Bordwell and M. W. Carlson, *J. Am. Chem. Soc.* **92**, 3377 (1970).
48. J. F. Liebman and A. Greenberg, *J. Org. Chem.* **39**, 123 (1974).
49. A. Favorskii and A. Bozhovski, *J. Russ. Phys. Chem. Soc.* **50**, 582 (1920); A. Mousseron and R. Jacquier, *Bull. Soc. Chim. Fr.*, 202 (1949); A. Mousseron, R. Jacquier, and A. Fontaine, *C. R. Acad. Sci. Paris Ser. C* **231**, 864 (1950); C. Rappe, *Acta Chem. Scand.* **19**, 270 (1965); K. Sato, S. Inoue, S. Kuranami, and M. Ohaschi, *J. Chem. Soc. Perkin Trans. 1*, 1666 (1977); (b) A. Barco, G. De Guili, and G. P. Pollini, *Synthesis*, 626 (1972).
50. G. D. Gutsche, *J. Am. Chem. Soc.* **71**, 3513 (1949); C. Hesse and F. Urbanek, *Chem. Ber.* **91**, 2733 (1958); Rhône-Poulenc S. A., *Neth. Appl.* **6–605**, 908 (1966); *Chem. Abstr.* **66**, 85538s (1967); G. Buchi, U. Hochstrasser, and W. Pawlak, *J. Org. Chem.* **38**, 4348 (1973); W. Ziegenbein, *Chem. Ber.* **94**, 2989 (1961).
51. H. O. House and G. A. Frank, *J. Org. Chem.* **30**, 2948 (1965).
52. N. L. Allinger, M. T. Tribble, and M. A. Miller, *Tetrahedron* **28**, 1173 (1972); and N. L. Allinger, M. T. Tribble, M. A. Miller, and D. W. Wertz, *J. Am. Chem. Soc.* **93**, 1637 (1971); E. M. Engler, J. D. Andose, and P. von R. Schleyer, *J. Am. Chem. Soc.* **95**, 8005 (1973).
53. J. March, in *Advanced Organic Chemistry: Reactions, Mechanisms and Structure*, 2nd ed., McGraw–Hill, London (1977), p. 140.
54. W. B. Smith and C. Gonzales, *Tetrahedron Lett.*, 5751 (1966).
55. F. G. Bordwell and R. G. Scamehorn, *J. Am. Chem. Soc.* **93**, 3410 (1971).

56. (a) A. W. Fort, *J. Am. Chem. Soc.* **84**, 2620, 2625 (1962); (b) A. W. Fort, *J. Am. Chem. Soc.* **84**, 4979 (1962).
57. J. Julien and P. Fauche, *Bull. Soc. Chim. Fr.*, 374 (1953).
58. R. C. Cookson, M. J. Nye, and G. Subrahmanyan, *J. Chem. Soc.*, 473 (1967).
59. H. M. R. Hoffmann, *Angew. Chem. Int. Ed. Engl.* **12**, 819 (1973).
60. R. Noyori, S. Makino, and H. Takaya, *J. Am. Chem. Soc.* **93**, 1272 (1971). R. G. Pearson, M. Sobel, and J. Songstad, *J. Am. Chem. Soc.* **90**, 319 (1968). R. Noyori, Y. Hayakawa, H. Takayan, S. Murai, and R. Kobayashi, *J. Am. Chem. Soc.* **100**, 1759 (1978).
61. R. Hoffmann, *J. Am. Chem. Soc.* **90**, 1475 (1968).
62. J. M. Pochan, J. E. Baldwin, and W. H. Flygare, *J. Am. Chem. Soc.* **90**, 1072 (1968); *J. Am. Chem. Soc.* **91**, 1896 (1969).
63. N. Bodor, M. J. S. Dewar, A. Harget, and E. Haselbach, *J. Am. Chem. Soc.* **92**, 3854 (1970).
64. J. F. Olsen, S. Kang, and L. Burnelle, *J. Mol. Struct.* **9**, 305 (1971).
65. A. Liberles, A. Greenberg, and A. Lesk, *J. Am. Chem. Soc.* **94**, 8685 (1972).
66. A. Liberles, A. Greenberg, and S. Kang, *J. Org. Chem.* **38**, 1922 (1973).
67. R. C. Bingham, M. J. S. Dewar, and D. M. Lo, *J. Am. Chem. Soc.* **97**, 1302 (1975).
68. (a) S. S. Edelson and N. J. Turro, *J. Am. Chem. Soc.* **92**, 2770 (1970); (b) N. J. Turro, *Acc. Chem. Res.* **2**, 25 (1969).
69. D. S. Sclove, J. F. Pazos, R. L. Camp, and F. D. Greene, *J. Am. Chem. Soc.* **92**, 7488 (1970).
70. McClellan, in *Tables of Experimental Dipole Moments*, W. H. Freeman, San Francisco (1963).
71. L. J. Bellamy, in *The Infrared Spectra of Complex Molecules*, John Wiley and Sons, New York (1958), p. 149.
72. E. L. Eliel, N. L. Allinger, S. J. Angyal, and G. A. Morrison, *Conformational Analysis*, John Wiley and Sons, New York (1967), p. 203.
73. A. S. Kende and E. E. Riechke, *J. Chem. Soc. Chem. Commun.*, 383 (1974).
74. M. C. Brown, R. S. Fletcher, and R. B. Johannesen, *J. Am. Chem. Soc.* **73**, 212 (1951); E. L. Eliel, in *Steric Effects in Organic Chemistry*, M. S. Newman, Ed., John Wiley and Sons, New York (1956), p. 541; see also Reference 52.
75. S. J. Chang, D. McNally, S. Shary-Tehrany, M. J. Mickey, and R. M. Boyd, *J. Am. Chem. Soc.* **92**, 3109 (1970); see also References 52 and 74.
76. P. von R. Schleyer, J. E. Williams, and K. R. Blanchard, *J. Am. Chem. Soc.* **92**, 2377 (1970).
77. L. A. Paquette, R. H. Meiseinger, and R. E. Wingard Jr., *J. Am. Chem. Soc.* **95**, 2230 (1973).
78. J. K. Crandall and R. P. Haseltine, *J. Am. Chem. Soc.* **90**, 6251 (1968).
79. L. A. Paquette, *Acc. Chem. Res.* **1**, 209 (1968); F. G. Bordwell, in *Organosulfur Chemistry*, M. J. Janssen, Ed., Wiley–Interscience, New York (1967), p. 271.
80. L. A. Paquette and R. W. Houser, *J. Am. Chem. Soc.* **93**, 4522 (1971).
81. B. Tchoubar and O. Sackur, *C. R. Acad. Sci. Paris Ser. C* **208**, 1020 (1939).
82. K. B. Wiberg and V. A. Hess, *J. Am. Chem. Soc.* **89**, 3015 (1967).
83. A. Nickon and R. C. Weglein, *J. Am. Chem. Soc.* **97**, 1271 (1975).
84. A. C. Cope and E. S. Graham, *J. Am. Chem. Soc.* **73**, 4702 (1951).
85. C. L. Stevens and E. Farkas, *J. Am. Chem. Soc.* **74**, 5352 (1952).
86. E. E. Smissman and J. L. Diebold, *J. Org. Chem.* **30**, 4005 (1965); E. E. Smissman and G. Hite, *J. Am. Chem. Soc.* **82**, 3375 (1960).
87. D. Baudry, J. P. Begue, and M. Charpentier-Morize, *Bull. Soc. Chim. Fr.*, 1416 (1971); D. Baudry and M. Charpentier-Morize, *C. R. Acad. Sci. Paris Ser. C* **269**, 561 (1969).
88. J. M. Conia and J. L. Ripoll, *Bull. Soc. Chim. Fr.*, 755, 763 (1963). J. M. Conia and J.

Salaun, *Bull. Soc. Chim. Fr.*, 1957 (1964); C. Rappe and L. Knutsson, *Acta Chem. Scand.* **21**, 163 (1967).

89. A. Burger and H. H. Ong, *J. Org. Chem.* **29**, 2588 (1964).
90. J. M. Conia and J. Salaun, *Acc. Chem. Res.* **5**, 33 (1972).
91. J. M. Conia and J. M. Denis, *Tetrahedron Lett.*, 3545 (1969); 2845 (1971).
92. E. W. Warnhoff, C. M. Wong, and W. T. Tai, *J. Am. Chem. Soc.* **90**, 514 (1968).
93. D. E. Pearson and C. A. Buehler, *Chem. Rev.* **74**, 47 (1974).
94. G. L. Closs and L. E. Closs, *J. Am. Chem. Soc.* **85**, 2022 (1963).
95. P. E. Eaton and T. W. Cole, *J. Am. Chem. Soc.* **86**, 3157 (1964).
96. J. T. Edward, P. G. Farell, and G. E. Langford, *J. Am. Chem. Soc.* **98**, 3075 (1976).
97. G. L. Dunn, V. J. Dipasquo, and J. R. E. Hoover, *Tetrahedron Lett.*, 3737 (1966).
98. P. von R. Schleyer, J. J. Harper, G. L. Dunn, V. J. Dipasquo, and J. R. E. Hoover, *J. Am. Chem. Soc.* **89**, 698 (1967).
99. G. L. Dunn, V. J. Dispaquo, and J. R. E. Hoover, *J. Org. Chem.* **33**, 1454 (1968).
100. Tien-Yau Luh and L. M. Stock, *J. Org. Chem.* **37**, 338 (1972).
101. J. C. Barborak, L. Watts, and R. Pettit, *J. Am. Chem. Soc.* **88**, 1328 (1966).
102. R. J. Stedman, L. S. Miller, and J. R. E. Hoover, *Tetrahedron Lett.*, 2721 (1966).
103. K. V. Scherer, R. S. Lunt III, and G. A. Ungefug, *Tetrahedron Lett.*, 1199 (1965); K. V. Scherer, G. A. Ungefug, and M. G. Ly, Abstracts of the National Meeting of the American Chemical Society, Miami Beach, April 1967, p. 9.
104. K. V. Scherer, *Tetrahedron Lett.*, 5685 (1966).
105. D. G. Dauben, J. L. Chitwood, and K. V. Scherer, *J. Am. Chem. Soc.* **90**, 1014 (1968).
106. R. N. McDonald and C. A. Curi, *Tetrahedron Lett.*, 1423 (1976).
107. C. L. Perrin and Mong-Tseng Hsia, *Tetrahedron Lett.*, 751 (1975).
108. W. C. Fong, R. Thomas, and K. V. Scherer, *Tetrahedron Lett.*, 3789 (1971).
109. O. W. Webster and L. H. Sommer, *J. Org. Chem.* **29**, 3103 (1964).
110. B. R. Vogt and J. R. E. Hoover, *Tetrahedron Lett.*, 2841 (1967).
111. B. R. Vogt, *Tetrahedron Lett.*, 1575 (1968).
112. B. R. Vogt, *Tetrahedron Lett.*, 1579 (1968).
113. K. E. Hamlin and A. W. Weston, *Org. React.* **9**, 1 (1957).
114. W. G. Dauben and L. N. Reitman, *J. Org. Chem.* **40**, 841 (1975).
115. F. T. Bond and C. Y. Ho, *J. Org. Chem.* **41**, 1421 (1976).
116. G. A. Olah and M. Calin, *J. Am. Chem. Soc.* **90**, 938 (1968).
117. A. Baretta and B. Waegell, *Tetrahedron Lett.*, 753 (1976); A. Baretta, Ph. D. thesis, Marseille (1978).

Index

587